Smart Biosensor Technology

Edited by
George K. Knopf
Amarjeet S. Bassi

CRC Press
Taylor & Francis Group
Boca Raton London New York

CRC Press is an imprint of the
Taylor & Francis Group, an informa business

OPTICAL SCIENCE AND ENGINEERING

1. Electron and Ion Microscopy and Microanalysis: Principles and Applications, *Lawrence E. Murr*
2. Acousto-Optic Signal Processing: Theory and Implementation, *edited by Norman J. Berg and John N. Lee*
3. Electro-Optic and Acousto-Optic Scanning and Deflection, *Milton Gottlieb, Clive L. M. Ireland, and John Martin Ley*
4. Single-Mode Fiber Optics: Principles and Applications, *Luc B. Jeunhomme*
5. Pulse Code Formats for Fiber Optical Data Communication: Basic Principles and Applications, *David J. Morris*
6. Optical Materials: An Introduction to Selection and Application, *Solomon Musikant*
7. Infrared Methods for Gaseous Measurements: Theory and Practice, *edited by Joda Wormhoudt*
8. Laser Beam Scanning: Opto-Mechanical Devices, Systems, and Data Storage Optics, *edited by Gerald F. Marshall*
9. Opto-Mechanical Systems Design, *Paul R. Yoder, Jr.*
10. Optical Fiber Splices and Connectors: Theory and Methods, *Calvin M. Miller with Stephen C. Mettler and Ian A. White*
11. Laser Spectroscopy and Its Applications, *edited by Leon J. Radziemski, Richard W. Solarz, and Jeffrey A. Paisner*
12. Infrared Optoelectronics: Devices and Applications, *William Nunley and J. Scott Bechtel*
13. Integrated Optical Circuits and Components: Design and Applications, *edited by Lynn D. Hutcheson*
14. Handbook of Molecular Lasers, *edited by Peter K. Cheo*
15. Handbook of Optical Fibers and Cables, *Hiroshi Murata*
16. Acousto-Optics, *Adrian Korpel*
17. Procedures in Applied Optics, *John Strong*
18. Handbook of Solid-State Lasers, *edited by Peter K. Cheo*
19. Optical Computing: Digital and Symbolic, *edited by Raymond Arrathoon*
20. Laser Applications in Physical Chemistry, *edited by D. K. Evans*
21. Laser-Induced Plasmas and Applications, *edited by Leon J. Radziemski and David A. Cremers*
22. Infrared Technology Fundamentals, *Irving J. Spiro and Monroe Schlessinger*
23. Single-Mode Fiber Optics: Principles and Applications, Second Edition, Revised and Expanded, *Luc B. Jeunhomme*
24. Image Analysis Applications, *edited by Rangachar Kasturi and Mohan M. Trivedi*
25. Photoconductivity: Art, Science, and Technology, *N. V. Joshi*
26. Principles of Optical Circuit Engineering, *Mark A. Mentzer*
27. Lens Design, *Milton Laikin*

28. Optical Components, Systems, and Measurement Techniques, *Rajpal S. Sirohi and M. P. Kothiyal*
29. Electron and Ion Microscopy and Microanalysis: Principles and Applications, Second Edition, Revised and Expanded, *Lawrence E. Murr*
30. Handbook of Infrared Optical Materials, *edited by Paul Klocek*
31. Optical Scanning, *edited by Gerald F. Marshall*
32. Polymers for Lightwave and Integrated Optics: Technology and Applications, *edited by Lawrence A. Hornak*
33. Electro-Optical Displays, *edited by Mohammad A. Karim*
34. Mathematical Morphology in Image Processing, *edited by Edward R. Dougherty*
35. Opto-Mechanical Systems Design: Second Edition, Revised and Expanded, *Paul R. Yoder, Jr.*
36. Polarized Light: Fundamentals and Applications, *Edward Collett*
37. Rare Earth Doped Fiber Lasers and Amplifiers, *edited by Michel J. F. Digonnet*
38. Speckle Metrology, *edited by Rajpal S. Sirohi*
39. Organic Photoreceptors for Imaging Systems, *Paul M. Borsenberger and David S. Weiss*
40. Photonic Switching and Interconnects, *edited by Abdellatif Marrakchi*
41. Design and Fabrication of Acousto-Optic Devices, *edited by Akis P. Goutzoulis and Dennis R. Pape*
42. Digital Image Processing Methods, *edited by Edward R. Dougherty*
43. Visual Science and Engineering: Models and Applications, *edited by D. H. Kelly*
44. Handbook of Lens Design, *Daniel Malacara and Zacarias Malacara*
45. Photonic Devices and Systems, *edited by Robert G. Hunsberger*
46. Infrared Technology Fundamentals: Second Edition, Revised and Expanded, *edited by Monroe Schlessinger*
47. Spatial Light Modulator Technology: Materials, Devices, and Applications, *edited by Uzi Efron*
48. Lens Design: Second Edition, Revised and Expanded, *Milton Laikin*
49. Thin Films for Optical Systems, *edited by Francoise R. Flory*
50. Tunable Laser Applications, *edited by F. J. Duarte*
51. Acousto-Optic Signal Processing: Theory and Implementation, Second Edition, *edited by Norman J. Berg and John M. Pellegrino*
52. Handbook of Nonlinear Optics, *Richard L. Sutherland*
53. Handbook of Optical Fibers and Cables: Second Edition, *Hiroshi Murata*
54. Optical Storage and Retrieval: Memory, Neural Networks, and Fractals, *edited by Francis T. S. Yu and Suganda Jutamulia*
55. Devices for Optoelectronics, *Wallace B. Leigh*
56. Practical Design and Production of Optical Thin Films, *Ronald R. Willey*
57. Acousto-Optics: Second Edition, *Adrian Korpel*
58. Diffraction Gratings and Applications, *Erwin G. Loewen and Evgeny Popov*
59. Organic Photoreceptors for Xerography, *Paul M. Borsenberger and David S. Weiss*
60. Characterization Techniques and Tabulations for Organic Nonlinear Optical Materials, *edited by Mark G. Kuzyk and Carl W. Dirk*
61. Interferogram Analysis for Optical Testing, *Daniel Malacara, Manuel Servin, and Zacarias Malacara*
62. Computational Modeling of Vision: The Role of Combination, *William R. Uttal, Ramakrishna Kakarala, Spiram Dayanand, Thomas Shepherd, Jagadeesh Kalki, Charles F. Lunskis, Jr., and Ning Liu*
63. Microoptics Technology: Fabrication and Applications of Lens Arrays and Devices, *Nicholas Borrelli*
64. Visual Information Representation, Communication, and Image Processing, *edited by Chang Wen Chen and Ya-Qin Zhang*
65. Optical Methods of Measurement, *Rajpal S. Sirohi and F. S. Chau*

66. Integrated Optical Circuits and Components: Design and Applications, *edited by Edmond J. Murphy*
67. Adaptive Optics Engineering Handbook, *edited by Robert K. Tyson*
68. Entropy and Information Optics, *Francis T. S. Yu*
69. Computational Methods for Electromagnetic and Optical Systems, *John M. Jarem and Partha P. Banerjee*
70. Laser Beam Shaping, *Fred M. Dickey and Scott C. Holswade*
71. Rare-Earth-Doped Fiber Lasers and Amplifiers: Second Edition, Revised and Expanded, *edited by Michel J. F. Digonnet*
72. Lens Design: Third Edition, Revised and Expanded, *Milton Laikin*
73. Handbook of Optical Engineering, *edited by Daniel Malacara and Brian J. Thompson*
74. Handbook of Imaging Materials: Second Edition, Revised and Expanded, *edited by Arthur S. Diamond and David S. Weiss*
75. Handbook of Image Quality: Characterization and Prediction, *Brian W. Keelan*
76. Fiber Optic Sensors, *edited by Francis T. S. Yu and Shizhuo Yin*
77. Optical Switching/Networking and Computing for Multimedia Systems, *edited by Mohsen Guizani and Abdella Battou*
78. Image Recognition and Classification: Algorithms, Systems, and Applications, *edited by Bahram Javidi*
79. Practical Design and Production of Optical Thin Films: Second Edition, Revised and Expanded, *Ronald R. Willey*
80. Ultrafast Lasers: Technology and Applications, *edited by Martin E. Fermann, Almantas Galvanauskas, and Gregg Sucha*
81. Light Propagation in Periodic Media: Differential Theory and Design, *Michel Nevière and Evgeny Popov*
82. Handbook of Nonlinear Optics, Second Edition, Revised and Expanded, *Richard L. Sutherland*
83. Polarized Light: Second Edition, Revised and Expanded, *Dennis Goldstein*
84. Optical Remote Sensing: Science and Technology, *Walter Egan*
85. Handbook of Optical Design: Second Edition, *Daniel Malacara and Zacarias Malacara*
86. Nonlinear Optics: Theory, Numerical Modeling, and Applications, *Partha P. Banerjee*
87. Semiconductor and Metal Nanocrystals: Synthesis and Electronic and Optical Properties, edited by *Victor I. Klimov*
88. High-Performance Backbone Network Technology, *edited by Naoaki Yamanaka*
89. Semiconductor Laser Fundamentals, *Toshiaki Suhara*
90. Handbook of Optical and Laser Scanning, *edited by Gerald F. Marshall*
91. Organic Light-Emitting Diodes: Principles, Characteristics, and Processes, *Jan Kalinowski*
92. Micro-Optomechatronics, *Hiroshi Hosaka, Yoshitada Katagiri, Terunao Hirota, and Kiyoshi Itao*
93. Microoptics Technology: Second Edition, *Nicholas F. Borrelli*
94. Organic Electroluminescence, *edited by Zakya Kafafi*
95. Engineering Thin Films and Nanostructures with Ion Beams, *Emile Knystautas*
96. Interferogram Analysis for Optical Testing, Second Edition, *Daniel Malacara, Manuel Sercin, and Zacarias Malacara*
97. Laser Remote Sensing, *edited by Takashi Fujii and Tetsuo Fukuchi*
98. Passive Micro-Optical Alignment Methods, *edited by Robert A. Boudreau and Sharon M. Boudreau*
99. Organic Photovoltaics: Mechanism, Materials, and Devices, *edited by Sam-Shajing Sun and Niyazi Serdar Saracftci*
100. Handbook of Optical Interconnects, *edited by Shigeru Kawai*
101. GMPLS Technologies: Broadband Backbone Networks and Systems, *Naoaki Yamanaka, Kohei Shiomoto, and Eiji Oki*

102. Laser Beam Shaping Applications, *edited by Fred M. Dickey, Scott C. Holswade and David L. Shealy*
103. Electromagnetic Theory and Applications for Photonic Crystals, *Kiyotoshi Yasumoto*
104. Physics of Optoelectronics, *Michael A. Parker*
105. Opto-Mechanical Systems Design: Third Edition, *Paul R. Yoder, Jr.*
106. Color Desktop Printer Technology, *edited by Mitchell Rosen and Noboru Ohta*
107. Laser Safety Management, *Ken Barat*
108. Optics in Magnetic Multilayers and Nanostructures, *Štefan Višňovský*
109. Optical Inspection of Microsystems, *edited by Wolfgang Osten*
110. Applied Microphotonics, *edited by Wes R. Jamroz, Roman Kruzelecky, and Emile I. Haddad*
111. Organic Light-Emitting Materials and Devices, *edited by Zhigang Li and Hong Meng*
112. Silicon Nanoelectronics, *edited by Shunri Oda and David Ferry*
113. Image Sensors and Signal Processor for Digital Still Cameras, *Junichi Nakamura*
114. Encyclopedic Handbook of Integrated Circuits, *edited by Kenichi Iga and Yasuo Kokubun*
115. Quantum Communications and Cryptography, *edited by Alexander V. Sergienko*
116. Optical Code Division Multiple Access: Fundamentals and Applications, *edited by Paul R. Prucnal*
117. Polymer Fiber Optics: Materials, Physics, and Applications, *Mark G. Kuzyk*
118. Smart Biosensor Technology, *edited by George K. Knopf and Amarjeet S. Bassi*

CRC Press
Taylor & Francis Group
6000 Broken Sound Parkway NW, Suite 300
Boca Raton, FL 33487-2742

Printed in the United States of America on acid-free paper
10 9 8 7 6 5 4 3 2 1

International Standard Book Number-10: 0-8493-3759-3 (Hardcover)
International Standard Book Number-13: 978-0-8493-3759-8 (Hardcover)

Library of Congress Cataloging-in-Publication Data

Smart biosensor technology / edited by George K. Knopf and Amarjeet S. Bassi.
p. cm. -- (Optical science and engineering)
Includes bibliographical references and index.
ISBN-13: 978-0-8493-3759-8 (alk. paper)
ISBN-10: 0-8493-3759-3 (alk. paper)
1. Biosensors--Materials. 2. Optical transducers--Design and construction. 3. Molecular electronics. 4. Smart materials. 5. Biomedical engineering--Instruments. 6. Bioelectronics. I. Knopf, George K. II. Bassi, Amarjeet S. III. Title.

R857.B54S6345 2007
610.28--dc22 2006031905

Visit the Taylor & Francis Web site at
http://www.taylorandfrancis.com

and the CRC Press Web site at
http://www.crcpress.com

Foreword

The topic of this book is smart biosensors, analytical devices that combine a biological or biomimetic sensing element with a separate signal transducer system. These biosensors have ultimate applications in biological research as well as medical and environmental monitoring. The range of applications is so extensive that it is hard to find one involving the monitoring of an external chemical or biological agent that would not benefit from the availability of a smart biosensor. However, such sensors are rarely available as commercial devices except in the field of medical monitoring (e.g., blood oxygen and glucose sensors). Current efforts in this field are limited primarily to academic, government, and a few commercial research laboratories. Much remains to be done before smart biosensors advance to the point where they have more global comparative advantage. Thus, the present volume is both timely and important as it will further research and interaction among the groups working in this important field.

The classical meaning of the term "biosensor" references the sensory apparatus of a living organism. It is useful to examine the evolution of such biosensors as it provides some insight into the current effort of scientists and engineers to make artificial or biomimetic biosensors. Evolutionary biologists have concluded that archaea and bacteria evolved from a common ancestor. While there is no fossil record of this ancestor, a comparison of the genetic codes of archaea and bacteria provides little doubt. Such studies indicate that many if not most of the "biosensors" that are expressed in modern prokaryotes and eukaryotes were present in a common form in this early ancestor. An important exception is lens-based image-resolving eyes, which require a more complex skeleton than is possible within the single-cellular domain of the early prokaryotes. Nevertheless, more recent studies have identified similarities in the genetic information associated with complex eye development that suggests a monophyletic origin. This remarkable continuity of developmental genes suggests that even the most complex of the biosensors in living organisms have a common ancestor that dates back to the archaeal era, 3.5 billion years ago. The inescapable conclusion is that life requires a complex and highly evolved biosensor network to succeed, and that much of the early work of evolution was dedicated to the exploration and optimization of these biosensors. It was only after the key biosensors had evolved that life was able to expand into the complex domains that are present today.

The development of biosensors by scientists and engineers is a recent endeavor that has progressed for less than one hundred millionth the time period that nature has devoted to the optimization process. It would therefore be logical to assume that nature is better at creating biosensors than we are. And indeed this is true in many areas. The most impressive sensors of visual or chemical information are biological, not human engineered. In contrast, human engineering has created more accurate sensors of temperature, position, and time. But the presumption is that if biology needed more accurate measurement of these three variables, it could provide the necessary resolution. Natural selection only seeks comparative advantage, not perfection. Nevertheless, natural sensors of the environment are exquisitely sensitive and approach perfection in terms of efficiency, sensitivity, and longevity.

This book explores the world of "human-engineered" biosensors. In most cases, success is due in part to taking advantage of what nature has provided, either in the form of materials or through biomimetics. Biomimetic sensors are devices that are modeled after biological sensory architectures or involve the use of natural materials that have been modified via chemical or genetic manipulation. It is a blurry line that separates biomimetic from biological sensors, and it is best not to waste energy trying to make a formal distinction. Improved methods in genetic engineering now allow more efficient and directed modification of biological molecules to enhance their performance in sensor applications. These capabilities increase the importance of biomimetics in the creation of biosensors.

Nanotechnology is also having an increasingly important role in current biosensor development. It is now possible to make insulating or semiconducting nanoscale structures with spherical or cylindrical geometry. These structures can be inserted into cells to monitor localized ionic strength or electrostatic charge. Alternatively, reporter molecules can be attached to the tip of cylindrical nanostructures and inserted inside a cell to provide photovoltaic, fluorescent, or electrochemical reporting of local conditions. The coupling of nanotechnology and bioelectronics has generated a new field of study called nano-bioelectronics. We anticipate that biosensors will benefit significantly from the exploration of nano-bioelectronics.

The chapters included in this volume provide an excellent introduction to and perspective on current research and developments in biosensors. A question that people might want to ponder as they read these chapters is the extent to which we have been successful in making "smart" biosensors. As noted above, a smart biosensor is defined as an analytical device that combines a biological, or biologically derived, sensing element with an electrical or optical signal transducer system. But when these sensors achieve their design goals, are they really "smart"? I suggest that in many cases of biosensor design, the "smart" derives from the use of biological molecules that have an elegance of structure and function that would only be possible after billions of years of evolutionary optimization.

Robert R. Birge
Coventry, Connecticut

Preface

A smart biosensor is a compact analytical device that combines a biological, or biologically derived, sensing element with an electrical, optical, or chemical transducer. Early biosensors had simple designs that affixed soluble enzymes or enzyme membranes to the surface of electrochemical transducers. Many of these systems exploited the direct covalent bonding of organized biomolecular monolayers on the surface of semiconductor devices and optical waveguides. In recent years, however, the functionality and design of the biosensor have greatly changed with the introduction of new intelligent materials and rapid advances in computing technology and microfabrication techniques.

Biosensors are now integrated devices that closely connect the biomolecular recognition element with a physical transducer to create an electrical or optical signal that is proportional to the concentration of a specific chemical or biological agent being measured. These microsystems include numerous sensors, transducers, and microfluidic actuators on a single integrated chip. The technology has evolved to sophisticated biochips where a collection of microarrays, or miniature test sites, can be fabricated on a single solid substrate to allow a large number of tests to be performed simultaneously at multiple sites. This design principle has greatly increased the speed of biochemical analysis. These biochips have been able to perform and gather data on thousands of biological reactions in only a few seconds permitting scientists to undertake previously impossible tasks such as decoding genes.

Modern biosensor design stresses technology integration and smart function for enhanced system performance. In essence, it is not a single technology or specific material that makes a biosensor smart or intelligent, but the synergistic interaction between the constituent components that comprise the sensory system. Seamless functional integration requires the development team to understand fundamental scientific and design principles derived from chemistry, physics, biology, material science, electronics and optics, and informatics. These seemingly diverse branches of science, engineering, and information science all provide unique perspectives on the meaning of "intelligent biosensors," their design, and potential applications.

This volume of *Smart Biosensor Technology* provides a multidisciplinary, multiple-author perspective of this rapidly evolving field. The scope of this book is to provide both the essential background to understanding biosensors and introduce new ideas of intelligent biosensor design and the leading-edge technologies used for realizing solutions in a wide variety of applications. This book is a collection of 23 chapters organized in five parts covering the fundamentals of smart biosensor technology, issues related to material design and selection, bioelectronics, and applications. The chapters are written by an international group of leading experts from academia and industry representing North America, Europe, and Asia.

Part I: Overview and Fundamentals

Part I is composed of six chapters that provide the reader with key definitions, and an overview of the fundamental functions and characteristics of smart biosensors. The discussion focuses on different perspectives of intelligence and how smart functions can be incorporated into sensors or sensor systems. The contributing authors represent a variety of scientific and engineering disciplines with backgrounds as diverse as engineering design, informatics, theoretical chemistry, and biological sciences. Topics include an introduction to the fundamental characteristics of smart biosensors, understanding and exploiting the intelligent properties of biological macromolecules, nanoscale optical methods for detecting single molecules, the design of BioFET sensors, the role that machine learning can play in analyzing biosensor data, and an introduction to innovative neural network biosensor designs.

Part II: Material Design and Selection I

Future advances in smart biosensors are dependent upon rapidly evolving materials that broaden bioreceptor selectivity, enhance sensitivity to diverse chemical compounds, enable biosensors to be seamlessly interfaced with the environment being monitored, and permit sensor system designs that have never been previously envisioned. This part contains four chapters that introduce and explore material-dependent technologies that will lead the development of functionally intelligent biosensors. Specific issues related to smart material design, molecular film assembly, nanostructured organic matrices, and biotic/abiotic interfaces are discussed in this part of the book.

Part III: Material Design and Selection II

Advances in microfabrication techniques have also led to the development of new biosensor designs and previously unheard-of technologies. This part presents three chapters that describe new nanotechnologies that will greatly change the functionality of biosensors, including porous silicon, carbon nanotubes and nanowires, and electrochemical biosensor designs that exploit innovative nanotubes.

Part IV: Bioelectronics

Bioelectronics is an emerging discipline that exploits the signal detection and processing capabilities of biological materials. In general, biomolecular electronics is defined as a technology that uses chromophore and protein molecules to encode, manipulate, and retrieve information at the molecular or macromolecular level. Biomolecular electronics provides an opportunity for product designers to create new hybrid technologies and computing architectures that can perform sophisticated information processing tasks faster and more energy efficiently. This part presents four chapters that explore the scientific basis of photoelectric biosensors, discuss the role of bacteriorhodopsin as a biophotonic material and its application to chemical sensing, how this bioelectric material can be used for color-sensitive imaging, and the design and fabrication of new bioelectronic sensing arrays on flexible substrates.

Part V: Applications in Detection and Monitoring

Although biosensor technology has been utilized over the past several decades, it was not until the introduction of new biomaterials and nanofabrication technologies that "smarter," more effective biosensors were developed. This part consists of six chapters that describe the application of biosensors to solving real-world problems in analyte detection and environment monitoring. The chapters include a discussion on how optical biosensors can be used for detecting foodborne pathogens, a multiarray biosensor is utilized for monitoring toxicity, aptasensors are used to detect allergens, and biosensors can be used for detecting and identifying viruses. The book concludes with a chapter that gives the contributing authors an opportunity to provide their viewpoints as to the future prospects of biosensor technology and innovative applications. It is hoped that this more speculative chapter will spur discussion and provide young researchers with insight on future directions. This presentation may also be of interest to individuals who wish to see a snapshot of views of a rapidly changing technology in the early part of the twenty-first century.

Editors

George K. Knopf is the associate dean (academic) in the faculty of engineering at The University of Western Ontario, and professor in the Department of Mechanical and Materials Engineering. He received his B.A. degree in humanities and B.E. degree in mechanical engineering in 1984, and then obtained his M.Sc. and Ph.D. degrees in 1987 and 1991, respectively, from the University of Saskatchewan.

Dr. Knopf's teaching and research interests lie in the general area of product design and manufacturing. His main fields of expertise are 3D shape reconstruction, adaptive geometric modeling, interactive data visualization, optical sensors and actuators, and bioelectronics. He has authored and coauthored numerous journal and conference papers, and coedited an IEEE Press volume entitled *Neuro-Vision Systems: Principles and Applications* and several SPIE *Proceedings on Optomechatronic Systems*. His recent work on an innovative three-dimensional scanning system has led to a US patent (No. 6,542,249).

He has acted as a technical reviewer for a variety of refereed journals, conferences, and granting agencies and has cochaired several international conferences. Over the past several years, he has been on the Canadian Advisory Committee (CAC) on Robots for Manufacturing Environment (TC 184/SC2) for the Standards Council of Canada (SCC).

Amarjeet S. Bassi is a professor of chemical and biochemical engineering in the faculty of engineering at The University of Western Ontario. He received his B.Tech degree in chemical engineering from the Institute of Technology, Banaras Hindu University, Varanasi, India in 1980, an M.E.Sc. degree in chemical engineering from the University of Saskatchewan in 1989, and a Ph.D. from the University of Western Ontario in 1993. He was a post-doctoral fellow at the University of California, Riverside, 1993 to 1994.

Dr. Bassi's teaching and research interests are environmental biosensor design and biocatalysis. His main fields of expertise are biosensors and nanosensors, enzymatic biocatalysis, and bioseparations. He has authored and coauthored over 75 refereed journal and conference papers, and several book chapters, and holds two patents on his recent work. Over the past several years, he has graduated a large number of M.E.Sc. and Ph.D. students and currently supervises a large group.

He has acted as a technical reviewer for numerous refereed journals, conferences, and granting agencies and has been involved in organizing several conferences, including the Canadian Conference in Chemical Engineering.

Contributors

Manuela Adami Fondazione Elba, Rome, Italy

Silvana Andreescu Department of Chemistry, State University of New York, Binghamton

Amarjeet S. Bassi Department of Chemical and Biochemical Engineering, The University of Western Ontario, London, Ontario, Canada

Shekhar Bhansali Department of Electrical Engineering, University of South Florida, Tampa

Pallab Bhattacharia Department of Electrical Engineering and Computer Science, University of Michigan, Ann Arbor

Arun K. Bhunia Department of Food Science, Purdue University, West Lafayette, Indiana

Robert R. Birge Department of Chemistry, University of Connecticut, Storrs

Emile Brys Cranfield Health, Cranfield University, Silsoe, Bedfordshire, UK

Wilfred Chen Department of Chemical and Environmental Engineering, University of California, Riverside

Brian M. Cullum Department of Chemistry and Biochemistry, University of Maryland–Baltimore County, Baltimore

Michael Frydrych Laboratory of Computational Engineering, Helsinki University of Technology, Helsinki, Finland

Tao Geng Department of Food Science, Purdue University, West Lafayette, Indiana

Jeffrey Girard Department of Chemistry, Syracuse University, Syracuse, New York

Guenter W. Gross Center for Network Neuroscience, Department of Biological Sciences, University of North Texas, Denton

Jason R. Hillebrecht Department of Molecular and Cell Biology, University of Connecticut, Storrs

Steven Hoath Division of Neonatology, Skin Sciences Institute, Cincinnati Children's Hospital Medical Center, Cincinnati, Ohio

Felix T. Hong Department of Physiology, Wayne State University – Medicine, Detroit, Michigan

Timo Jaaskelainen Department of Physics, University of Joensuu, Finland

Jason Karasinski Department of Chemistry, State University of New York, Binghamton

Puneet Khanna Department of Electrical Engineering, University of South Florida, Tampa

George K. Knopf Department of Mechanical and Materials Engineering, University of Western Ontario, London, Ontario, Canada

Jeremy F. Koscielecki Department of Chemistry, University of Connecticut, Storrs

Ulrich J. Krull Department of Chemical and Physical Sciences, University of Toronto, Mississauga, Ontario, Canada

Won-Yong Lee Department of Chemistry, Yonsei University, Seoul, South Korea

Lasse Lensu Department of Information Technology, Lappeenranta University of Technology, Lappeenranta, Finland

Chen-Zhong Li Biomedical Engineering Department, Florida International University

Duane L. Marcy Department of Electrical Engineering and Computer Science, Syracuse University, Syracuse, New York

Kenneth A. Marx Center for Intelligent Biomaterials, Department of Chemistry, University of Massachusetts, Lowell

Marco Mascini Dipartimento di Chimica, Università di Firenze, Italy

Melissa Massey Department of Chemical and Physical Sciences, University of Toronto, Mississauga, Ontario, Canada

Benjamin L. Miller Department of Dermatology and the Center for Future Health, University of Rochester, New York

Maria Minunni Dipartimento di Chimica, Università di Firenze, Italy

Ashok Mulchandani Department of Chemical and Environmental Engineering, University of California, Riverside

Nosang V. Myung Department of Chemical and Environmental Engineering, University of California, Riverside

Claudio Nicolini Nanoworld Institute and Biophysics Division, University of Genoa, Italy

Joseph P. Pancrazio National Institute of Neurological Disorders and Stroke, Neuroscience Center, Rockville, Maryland

Jussi Parkkinen Department of Computer Science, University of Joensuu, Finland

Sinikka Parkkinen Department of Biology, University of Joensuu, Finland

Cristina Paternolli Nanoworld Institute and Biophysics Division, University of Genoa, Italy

Paul A.E. Piunno Department of Chemical and Physical Sciences, University of Toronto, Mississauga, Ontario, Canada

Omowunmi A. Sadik Department of Chemistry, State University of New York, Binghamton

Ravil A. Sitdikov Department of Chemical and Nuclear Engineering, University of New Mexico, Albuquerque

Rod Smallwood Medical Physics and Engineering, University of Sheffield, UK

Jeffrey A. Stuart Department of Chemistry, University of Connecticut, Storrs

Xueliang Sun Department of Mechanical and Materials Engineering, The University of Western Ontario, London, Ontario, Canada

Shin-ichiro Suye Graduate School of Engineering, University of Fukui, Japan

William Tetley Department of Electrical Engineering and Computer Science, Syracuse University, Syracuse, New York

Sara Tombelli Dipartimento di Chimica, Università di Firenze, Italy

Anthony P.F. Turner Cranfield Health, Cranfield University, Silsoe, Bedfordshire, UK

Adam K. Wanekaya Department of Chemistry, Missouri State University, Springfield

Wei Wei Wang Department of Mechanical and Materials Engineering, The University of Western Ontario, London, Ontario, Canada

Xiaomeng Wang Department of Chemistry, University of Toronto, Ontario, Canada

Ebtisam S. Wilkins Department of Chemical and Nuclear Engineering, University of New Mexico, Albuquerque

Brian J. Yordy Department of Chemistry, Syracuse University, Syracuse, New York

Haitao Zheng School of Material Science and Chemical Engineering, Tianjin Polytechnic University, Tianjin, China

Acknowledgments

We express our sincere and heartfelt thanks to all the contributors of this volume for their time and effort in preparing the chapters. Their excellent work is very much appreciated. We would also like to express our sincere gratitude to Ms. Jessica Vakili, production coordinator, Taylor & Francis, for her assistance, advice, and patience.

Indeed we are very much indebted to our families and wives, Eirin and Simer, who have generously supported this project at each step by letting us use family time during evenings, weekends, and holidays.

George K. Knopf
Amarjeet S. Bassi

Contents

Part III Material Design and Selection II

Part IV Bioelectronics

Part V Applications in Detection and Monitoring

Part I

Overview and Fundamentals

This section provides some of the latest directions in smart biosensor development which include single molecule detection, nanowires, and intelligence and pattern recognition. The first chapter by Professor Kenneth A. Marx from the Centre for Intelligent Biomaterials (CIB), University of Massachusetts Lowell, represents a broad overview of different approaches in biosensor design but with a focus on examples of actual projects being carried out or completed at CIB. The fundamental principles of biosensors are first considered. This is followed by a discussion on immobilization methods and properties of "intelligent" materials. Specific biosensor designs including optical biosensors, electrochemical, piezoelectric, and quartz crystal microbalance (QCM) biosensors are presented. In particular, photodynamic proteins such as phycobiliproteins, bacteriorhodopsin, and chemi-luminescent probes are discussed. The author then presents applications related to organo-pesticide detection, metal ion detection, and DNA interactions. Chapter 1 also presents some work on "Smart bandages" and use of DNA in smart structures. Finally, concepts of machine learning and nanobiosensors are also presented. Biosensors for single molecule detection represent the latest generation of such platforms being developed. Professor Ulli Krull and others discuss the development of such platforms in Chapter 2. These biosensors can provide information not previously accessible and as well can lead to new understanding in biological functions such as DNA interactions and physiological processes. Both the instrumentation and procedures for single molecule detection by optical methods, and applications of such methods in bioanalytical chemistry and biosensor development are presented.

In Chapter 3, Professor Brian M. Cullum provides a critical review of optical based nanosensors and biochips, and discusses significant advancements in nanoscale biosensor design and development. The chapter reviews the evolution of biochip technology and nanosensors from their beginning to the present. The latter includes biosensors capable of probing subcellular compartments of individual cells. The application of such systems to biological measurements is presented and as well as future directions in nanosensing are considered. In Chapter 4, Professor Ashok Mulchandani and others present biosensors based on conducting polymers nanowires (CP NWs). The authors discuss and present approaches to synthesize and fabricate CP NWs. They also compare the limitations of various methods from different laboratories and attempts which have been made to address these limitations. The functionalization and assembly of these 1-D nanostructures is presented and their application for labelfree detection of various biological molecules at high sensitivity and low detection limits is demonstrated. Finally, some of the challenges in the 1-D nanostructured material design are highlighted. Multiple signal generation using array technologies can lead to large amounts of data. In particular, real-time monitoring involves complex systems and environmental conditions. Pattern recognition algorithms and data exploration techniques provide a method of data analysis which can be embedded in the sensors using programmable microcontrollers or implemented off-line on graphical workstations. In Chapter 5, George K. Knopf provides an overview of intelligence and pattern recognition and how they can be linked to biosensor design. These concepts include adaptive signal processing algorithms, adaptive control, intelligent decision making using an integrated and functional design approach. In Chapter 6, Professor Guenter W. Gross from the Center for Network Neuroscience (CNNS), University of North Texas and Joseph P. Pancrazio focus on neuronal network biosensors (NNBS) formed using functional, spontaneously active neural cell based networks on substrate-integrated microelectrode arrays. These NNBS are biosensors which respond to a diverse array of compounds that attack neural cells. These compounds include bacterial toxins, metabolic poisons, toxic metals, neuropharmacological compounds, hallucinatory drugs, and epileptogenic agents. These sensors are being applied for investigations of neurotoxicity and cytotoxicity, and for exploiting their properties for environmental toxicology, drug development, and even in defense as broadband biosensors. This approach leads to multisite electrophysiological data, for example action potential patterns and waveshapes, as well as simultaneous cytological information through high power microscopy and fluorescence. This chapter provides a description of prototype systems developed or under development at CNNS.

1

Toward Understanding the Intelligent Properties of Biological Macromolecules—Implications for Their Design Into Biosensors

Kenneth A. Marx

CONTENTS

1.1 Introduction to Smart Biosensors

The focus for this review chapter on smart biosensor technology centers around the broad scope of research projects carried out in the Center for Intelligent Biomaterials at the University of Massachusetts Lowell. After a brief section introducing smart biosensors, we proceed to describe biosensor systems that have been created and studied in our Center. They include optical, electrochemical, and piezoelectric-based systems designed to detect specific analyte molecules in solution. These biosensor systems are discussed only within the broad context of biosensor research, since our intent in this chapter is not to write a comprehensive review of all different areas of solution-based biosensor research. Looking to the future, we then devote the latter part of this review to our more recent use of specific computational approaches to understand the properties of biological elements. In this discussion, we have particularly focused on DNA and DNA–protein complexes. In the future, the ability to predict the intelligent properties of biomolecules would greatly facilitate the design and integration of these biological elements into functioning smart biosensors, without the necessity for performing exhaustive characterization experiments prior to their use.

1.1.1 Components of Biosensors

In terms of the simplest definition of its components, biosensors are typically thought to be comprised of a biological element(s) that is usually attached permanently to an underlying substrate. These are integrated with an appropriate signal transduction platform that provides a mechanism whereby the presence and usually the concentration of analyte being sensed by the biological element is converted into some type of quantitative electronic signal or output.

1.1.1.1 Biological Elements

The biological elements that have been used in the creation of biosensors vary widely in type and include proteins (enzymes), nucleic acids—primarily DNA, lipids and membranes, carbohydrates, complexes between these individual components, and living cells. One may ask what the advantage is of integrating biological elements as opposed to purely chemical recognition elements into the design of a sensor? A compelling answer is that in most instances biological macromolecules, the current endpoints of evolution, provide overall superior properties compared with chemical systems developed to carry out equivalent functions. In general, nature has designed far better systems for tasks such as recognition specificity, catalytic efficiency, electron transfer, and other complex integrated functions, than the talented bench scientist is capable of creating with current design and synthesis approaches. This is the case even where biomimetic studies form a part of the design process. As we describe in more detail in a later section, biological macromolecules and their complexes, as well as living cells, possess subsets of or all of the intelligent properties that we wish to exploit in the design of smart biosensors. However, there are recognized drawbacks to the use of biological macromolecules in smart biosensors. For one, they tend to be functionally less robust than chemical systems to specific factors in their environment. These include extremes of pH, temperature, the presence of oxidizing agents, as well as enzymatic degradation, to name just a few. Another drawback is that one has been limited traditionally by the function(s) nature evolved into the available biological systems. However, these stability and functional limitations are currently being overcome through modern approaches that seek to modify biological macromolecules through a targeted design approach. Methods such as Directed Evolution (1,2), which involve the repeated sequence evolution of existing native proteins coupled to a criterion-based selection protocol, have provided new approaches for overcoming nature's design limitations. In fact, not only can stability be enhanced greatly, but also new function can be evolved through the use of these techniques. This is not a topic we deal with explicitly in any more detail in this review. But it does represent a type of technical approach that will result in novel and improved biosensor components in the future.

1.1.1.2 Immobilization Methods

Once the appropriate biological element has been identified and the platform chosen, a suitable surface attachment strategy must be devised. An effective attachment strategy should preserve the function and create a stable environment for the biological element, as well as facilitate and enhance coupling of the signal from the biological element to the platform to maximize the biosensor sensitivity to the analyte. There have been quite a large number and variety of physical and chemical attachment methods used in developing biosensors and we will not review them here. However, we will focus on one of our major thematic approaches in the Center for Intelligent Biomaterials, which has been to utilize polymeric thin films bound to the biosensor platform surface to immobilize biological elements. We have utilized both conducting and nonconducting polymers in various thin-film formats to carry out immobilizations of different biological elements. As an example, some of these strategies have involved chemical synthesis of conducting polymers derivatized with biotin, which functions as a pendant biological attachment site. In Figure 1.1a, a general single-step reaction used widely to create derivatized conducting polythiophenes is presented. In Figure 1.1b, we outline the two-step synthesis of a representative biotinylated (B) polythiophene copolymer, B-PUMT, where the biotin is connected by an ester linkage to the pendant CH_2OH group on the thiophene polymer backbone (3). These biotin ligands recognize and bind each of the four subunits of the tetrameric proteins streptavidin and avidin with affinity constants approaching those of covalent bonds, around 10^{15} per M (4). These two proteins have been widely used to covalently derivatize many biological macromolecules, leading to an attachment system where multiple derivatized biological elements can be immobilized upon the biotinylated polymer chain, following prior immobilization of the polymer upon the surface of the chosen platform.

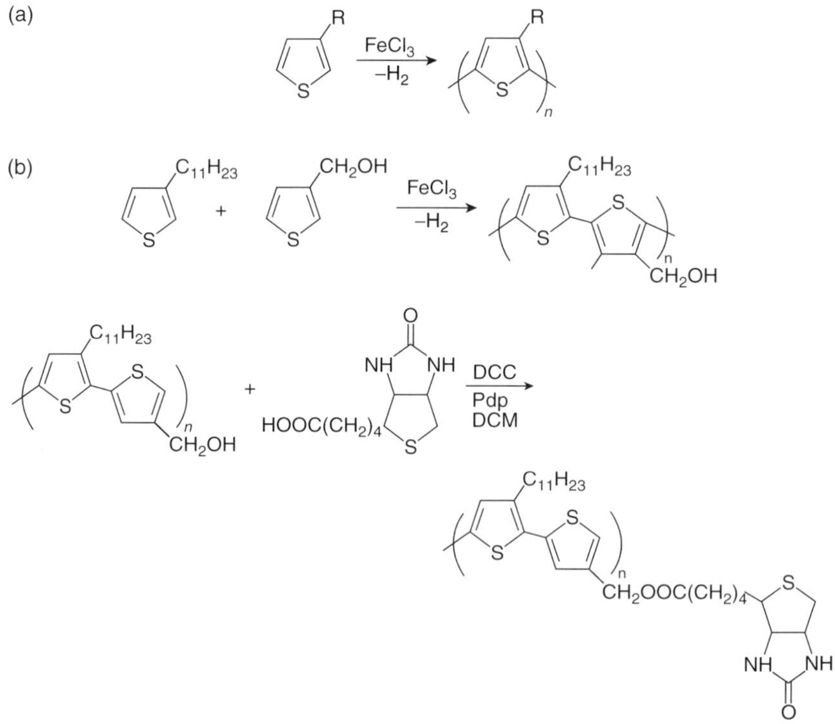

FIGURE 1.1
Synthetic schemes for creating thiophene polymers: (a) general synthetic scheme for poly(3-alkylthiophenes) for any alkyl (R)-derivatized monomer; (b) two-step synthesis of B-PUMT, biotinylated poly(3-undecylthiophene-co-3-methanolthiophene). The biotin ligand (B) is added in the second esterification step. Reprinted with permission from Marx, K.A., Samuelson, L.A., Kamath, M., Lim, J.O., Sengupta, S., Kaplan, D., Kumar, J., Tripathy, S.K. (1994). Intelligent Biomaterials Based on Langmuir–Blodgett Monolayer Films. In: Birge, B.R., ed. *Molecular and Biomolecular Electronics.* Advances in Chemical Series, Vol. 240, American Chemical Society Books, Washington, D.C., 395–412. Copyright (1994) American Chemical Society.

1.1.1.3 Signal Transduction Mechanisms and Biosensor Output

A number of general physicochemical classes of signal transduction mechanisms have been used to create biosensors. The most commonly used classes are optical, electrochemical, and piezoelectric. Thermometric and magnetic mechanisms have also been used. Microelectromechanical mechanisms and integrated systems for liquid handling leading to preprocessing of samples as well as their subsequent analysis represent a burgeoning area of investigation for microscale analytical chemistry as well as biosensor development. Each of these signal transduction mechanism classes have different advantages, disadvantages, and limiting sensitivities, some aspects of which we describe in the sections that follow. At the Center for Intelligent Biomaterials, we have utilized the three most common signal transduction mechanisms in the development of biosensors and we select some of them for discussion in this review. In the generic biosensor, once an analyte at a specific concentration has been detected by an array of the biological elements, the physicochemical signal is transduced by the platform's mechanism to create an output for the end user. However, in principle, more complex biosensors can be designed involving nonlinear systems where inputs can be obtained from multiple channels. In these cases, there is the need for further 'intelligent' analysis involving statistical or algorithmic processing of the biosensor input to make accurate interpretations and provide clear quantitative output to the end user of what is being sensed. Therefore, we present a discussion of the importance of informatics and data mining approaches later in this review. Our purpose is twofold. First, we highlight the importance of this capability for smart biosensor data processing in some cases prior to signal output. Second, we indicate that these approaches can aid in the original biosensor design process. This can be carried out through the analysis of large complex data sets to understand and possibly predict the intelligent properties of biological elements for potential integration into the biosensor.

1.1.2 Intelligent Properties of Biological Macromolecules and Systems

Systems comprising biological macromolecules, assemblies of these molecules, living cells, and certainly entire organisms possess some or all of what can be termed intelligent properties. In fact, individual cells and certainly whole organisms represent paradigms of systems possessing intelligent properties. These properties include *template-based self-assembly, self-multiplication, self-repair, self-degradation (selective), redundancy, self-diagnosis, learning, and prediction/notification.* While single biological macromolecules or small assemblies of them might contain limited subsets of these properties, living cells, and entire organisms possess various manifestations of all of them. Intelligent properties incorporated into smart biosensors ideally would enable the biosensors to be responsive in real time to their environment and be capable of integrating multiple functions such as recognition/discrimination, feedback, standby, appropriate response, to name a few.

That we focus on living cells and their constituent macromolecules in the Center for Intelligent Biomaterials is due to the following important fact. Evolution, acting over nearly 1.5 billion years time, has achieved highly sophisticated levels of hierarchical organization and complex integrated function in the biochemical subassemblies found in cells, comprising complexes of DNA, RNA, proteins, carbohydrates, and lipid membranes (5). Despite decades of design attempts by synthetic chemists and materials scientists to duplicate these systems' properties using biomimetic approaches, the level of organization and complex integrated functions found in cells remains unsurpassed. These facts have helped define our research direction, which has been to identify important and appropriate biological macromolecular systems for study, modification, and design into biosensors and biomaterials. In the following section, we describe the experiments our Center has carried out incorporating specific proteins, DNA, and cellular systems into biosensors for small molecule detection. Then in Section 1.3, we describe our most recent efforts to understand the properties of some specific DNA systems using computational approaches, with the ultimate aim of intelligent property prediction that can be used in the design of both intelligent biosensors and biomaterials. In Section 1.4, we describe informatics and data mining approaches and how these techniques can be applied to understand complex nonlinear data systems. The idea here is to apply these techniques, where appropriate, to help decipher the behavior of complex biosensors to create an optimal signal output and also to improve the design of biosensors by better understanding and selecting optimal biological components. Finally, in Section 1.5, we describe future prospects for the creation of newer and smaller biosensors with superior properties.

1.2 Creating Biosensors That Detect Small and Large Molecules Using Different Signal Transduction Mechanisms

1.2.1 Optical-Based Biosensors

Some of the Center's initial research centered upon the creation of systems involving optical signal transduction. The optical elements were of two basic types. In one, the chromophores were integral moieties of naturally occurring proteins. In the second, the chromophore was an enzymatically activated chemiluminescent molecule that emitted a visible photon. In the first type, the phycobiliproteins had the advantage of being proteins that evolved to possess efficient optical absorption at low light levels and had high-fluorescence quantum yields (6). These were studied in a number of immobilization formats. We also studied the very stable membrane-bound protein bacteriorhodopsin (bR) (7) that possesses a complex photocycle involving well-defined protein conformation-chromophore states. In the case of the second type of optical element, we studied a particular molecule that was developed to be capable of undergoing enzymatic cleavage to the chemiluminescent product species by phosphatase enzymes such as alkaline phosphatase. The alkaline phosphatase we used was primarily immobilized via conducting polymers to sensitively detect low solution concentrations of organophosphate pesticides and certain metal ions using a competition strategy. Examples of these biosensor systems are presented in the following subsections.

1.2.1.1 Chromophore-Containing Proteins in Biosensor Applications

1.2.1.1.1 The Phycobiliproteins

The phycobiliproteins represent an interesting class of photodynamic proteins that have evolved to function with extremely high light collection efficiency in low-light-level environments, such as deep underwater, where their host algae are often found in highly competitive ecological niches. A closely related family of proteins, the phycobiliproteins—phycoerythrin, phycocyanin, and allophycocyanin—in that order, are found in vivo in supramolecular assemblies in an antennalike structure called the phycobilisome. Each protein absorbs in their respective region in the visible spectrum and progressively transfers the absorbed light energy with high efficiency ($>90\%$ quantum yield) via a Forster-type transfer mechanism down the phycobilisome and into Photosystem II to drive photosynthesis (8). The chromophores found in the individual subunits of the three different phycobiliproteins are variants of open-chain tetrapyrroles coupled to specific protein residues through thioether linkages, as is shown in Figure 1.2. As optical elements, these phycobiliprotein chromophores possess distinct advantages. These include intense fluorescence, nearly 20-fold greater than that of a fluorescein molecule, high quantum yields, and large Stokes shifts—some 2.7-fold greater than that found in fluorescein (9). Because of these desirable properties, some of the phycobiliproteins have found commercial uses as biochemical and biotechnological probes (10). Here we describe experiments that demonstrated they retained their desirable optical properties after being surface immobilized using a number of different biosensor compatible strategies. These include binding to Langmuir–Blodgett (LB) monolayer films, entrapment within optically accessible sol-gel glasses, and binding to conducting polymers immobilized upon optical fiber surfaces. Therefore, the phycobiliproteins have potential for use in biomaterials and smart biosensor applications.

One of the earliest phycobiliprotein immobilization strategies we investigated was to LB monolayer films. The LB technique creates a monolayer of amphiphilic molecules at the air–water interface above an aqueous subphase within an LB trough device. In the first system we investigated, the monolayer was created using a biotinylated phospholipid, *N*-biotinoyl-*O*-dipalmitoyl-L-alpha-phosphatidyl ethanolamine, triethylammonium salt (B-DPPE). In Figure 1.3, we show representative pressure–area isotherms for LB trough compression experiments where monolayers have been created by close packing the B-DPPE molecules (3,11–14). In all cases, the isotherms displayed a relatively steep slope above a pressure of 15 mN/m, which corresponds to an area/molecule of just over 100 $Å^2$. This is approximately the two-dimensional (2-D) cross-section area of the B-DPPE molecule oriented vertically at the air–water interface. Phycoerythrin (PE) was chosen for this study, since its absorption spectra matched the 496.5-nm output of the argon ion laser used for excitation in the detection scheme. We immobilized the PE by derivatizing it with either streptavidin (Str) or avidin (Av) proteins. Both of these proteins are tetrameric and each of the four subunits binds biotin. The biotin-derivatized ends of the closely packed B-DPPE molecules in the LB monomayer film are hydrophilic and are oriented downward into the subphase of the LB trough. The biotin of each B-DPPE molecule can bind each subunit of

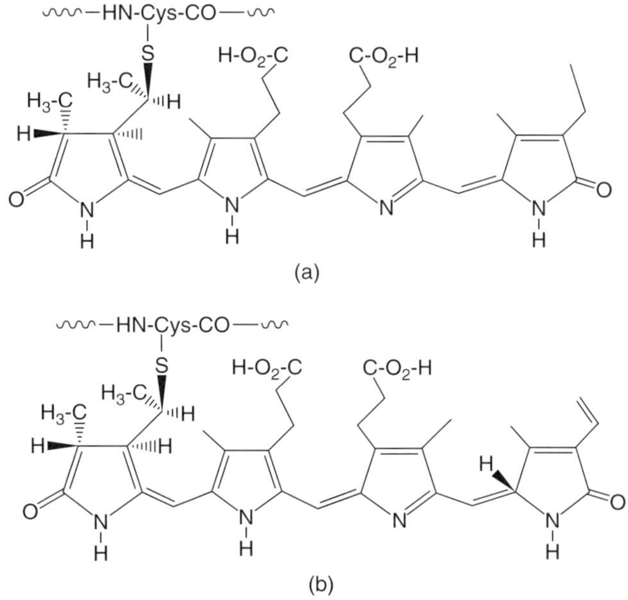

FIGURE 1.2
Chromophores of two of the phycobiliproteins: (a) phycocyanin; (b) phycoerythrin. The highly conjugated tetrapyrrole ring systems of these two phycobiliproteins are shown as well as their covalent thioether linkage to a cysteine residue of the protein amino acid side chain. Reprinted from Beladakere, N.N., Ravindran, T., Bihari, B., Sengupta, S., Marx, K.A., Kumar, J., Tripathy, S.K. (1993). Photovoltaic Effects and Charge Transport Studies in Phycobiliproteins. In: Viney, C., Case, S.T., Waite, J.H., eds. Biomolecular Materials, *Proc. Mat. Res. Soc.*, 292:193–198. With permission from the Materials Research Society.

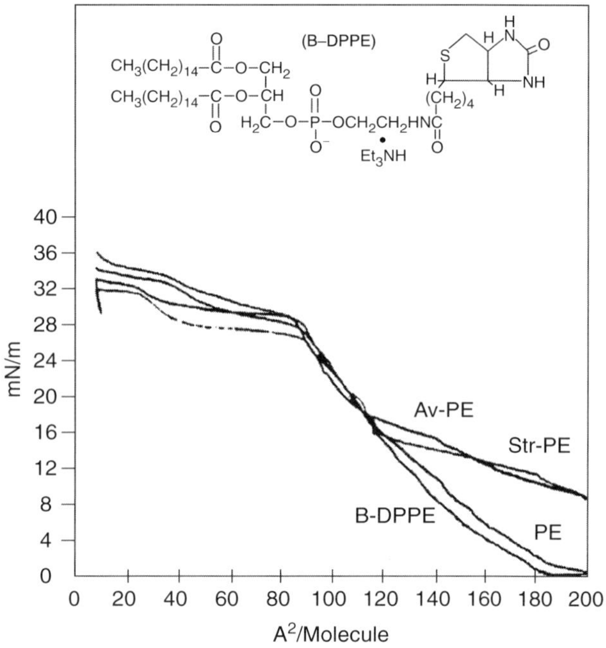

FIGURE 1.3
Isotherms of pressure vs. surface area/B-DPPE molecule from Langmuir–Blodgett (LB)-trough measurements of the B-DPPE phospholipid monolayer system. In the region of expanded area before contraction to an organized monolayer occurred, a series of four different subphase injections were carried out in different experiments. These were a control subphase injection labeled B-DPPE and subphase injections of either PE protein, Av-PE or Str-PE protein conjugates. The subphase contained 0.1 mM sodium phosphate buffer pH 6.8, 0.1 M NaCl. The inset shows the structure of the B-DPPE molecule. Reprinted with permission from Samuelson, L.A., Miller, P., Galotti, D., Marx, K.A., Kumar, J., Tripathy, S.K., Kaplan, D.L. (1992). The Monomolecular Organization of a Photodynamic Protein System through Specific Surface Recognition of Streptavidin by Biotinylated Langmuir–Blodgett Films. *Langmuir* 8:604–608. Copyright (1992) American Chemical Society.

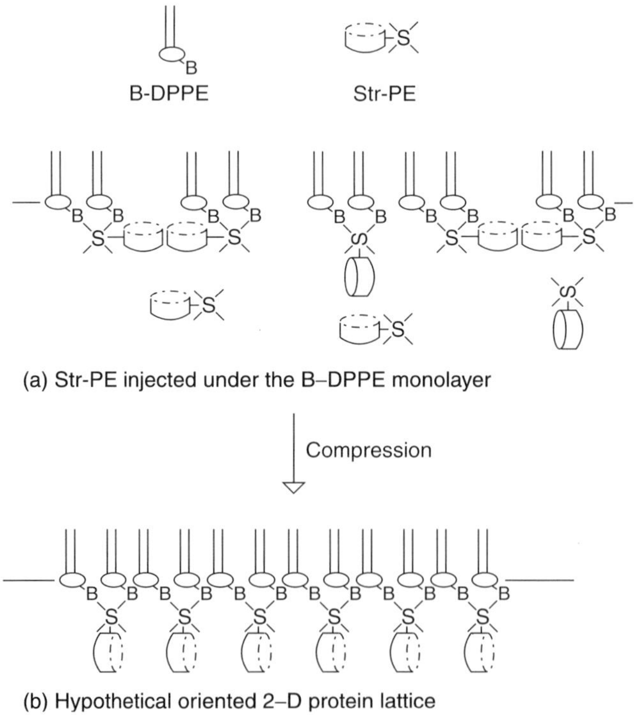

FIGURE 1.4
Idealized schematic of the two-dimensional (2-D) ordering of subphase injected Str-PE protein conjugate: (a) as it binds to the biotin (B) ligand of the phospholipid B-DPPE in the expanded film and (b) as it becomes ordered above 15 mN/m in the compressed LB-monolayer film. Reprinted with permission from Samuelson, L.A., Miller, P., Galotti, D., Marx, K.A., Kumar, J., Tripathy, S.K., Kaplan, D.L. (1992). The Monomolecular Organization of a Photodynamic Protein System through Specific Surface Recognition of Streptavidin by Biotinylated Langmuir–Blodgett Films. *Langmuir* 8:604–608. Copyright (1992) American Chemical Society.

the Str-PE or Av-PE injected into the subphase solution with high affinity ($K_a = 10^{15}$), producing a stability level nearly that of a covalent bond. Upon carefully injecting the Str-PE or Av-PE into the subphase below the film, an altered low-pressure response to film formation was observed for the LB film in the expanded state. This produced an increased pressure compared with either the pure B-DPPE film or injecting the underivatized PE into the subphase. This behavior suggests that the bulky derivatized PE, having formed a complex with biotin, is somehow inserting itself into the expanded state of the LB film during its formation. The PE protein is large (240,000 Da) and is disk shaped with dimensions of 6.0 by 12.0 nm. In Figure 1.4 we present a schematic representation of how this might happen (12). As the pressure increased above 15 mN/m, then all LB films exhibited similar behavior. We believe this was due to the Str-PE, bound to the pendant biotin, being pushed down into the subphase away from the B-DPPE molecules beginning to close pack in the LB film. After forming the Str-PE immobilized LB monolayer film, it was transferred to the surface of a hydrophilic glass slide and allowed to dry, and the fluorescence spectrum determined using a system that comprised an argon ion laser for excitation, monochromator, and photocounter as shown schematically in Figure 1.5 (12). As expected, the control films formed with either pure B-DPPE or PE injected under the B-DPPE while forming the film exhibited no fluorescence, as shown in Figure 1.5. A normal native fluorescence spectrum with emission maximum around 576 nm was observed for both the LB films where either Str-PE or Av-PE (not shown) was injected into the subphase to allow binding to biotin during film formation. Therefore, these phycobiliproteins retained their optical properties through all of the binding and drying steps.

We have also taken an alternate film-based approach to the immobilization of fluorescent phycobiliproteins and other proteins. In this instance, we also utilized the biotin–streptavidin interaction, but here the biotin was a pendant ligand covalently

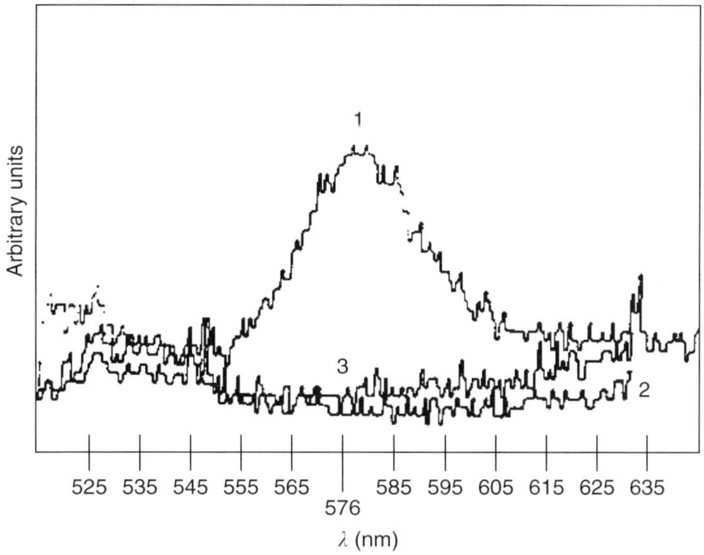

FIGURE 1.5
Fluorescence spectra of LB-monolayer films picked up onto glass substrates: (1) B-DPPE plus subphase injected Str-PE protein conjugate; (2) B-DPPE plus subphase injected PE protein; (3) DPPE (phospholipid lacking biotin) plus subphase injected Str-PE protein conjugate. Reprinted with permission from Samuelson, L.A., Miller, P., Galotti, D., Marx, K.A., Kumar, J., Tripathy, S.K., Kaplan, D.L. (1992). The Monomolecular Organization of a Photodynamic Protein System through Specific Surface Recognition of Streptavidin by Biotinylated Langmuir–Blodgett Films. *Langmuir* 8:604–608. Copyright (1992) American Chemical Society.

attached to a conducting polymer chain. We synthesized a number of different copolymers for phycobiliprotein surface immobilization using the biotin–streptavidin interaction (3,15). In the case of immobilization onto an LB polymer film, the film was first formed by compressing an amphiphilic polymer, B-PUMT, synthesized as presented in Figure 1.1b by the copolymerization of 3-undecylthiophene with 3-thiophenemethanol followed by biotin derivatization. As shown in Figure 1.6, during its compression to a monolayer, the B-PUMT copolymer exhibited a series of different complex micron-scale phase patterns in the presence of different proteins at varying LB trough pressures and incubation times (16). Upon injection of Str-PE or various protein controls into the subphase, then picking up the film onto a glass substrate for optical measurement, phycoerythrin immobilization was demonstrated to depend upon the presence of both the biotin derivatization of the copolymer and the streptavidin derivatization of the phycoerythrin, as was the case in the biotinylated B-DPPE LB monolayer film discussed previously.

Using the same copolymer, B-PUMT, we demonstrated that the Str-PE immobilization could be carried out on the surface of optical fibers by a self-assembly technique much simpler than the LB film experiment described above (17). As described in Figure 1.7, the fibers were first declad to expose the optical surface, which was then silanized with chlorodimethyloctadecylsilane. Then, the B-PUMT was bound to the optical fiber surface through hydrophobic interactions between its pendant undecyl chains and the octadecyl chains of the silanizing agent. Biotin ligands on the copolymer, pendant in aqueous solution, bound added streptavidin in the Str-PE complexes. The native fluorescence

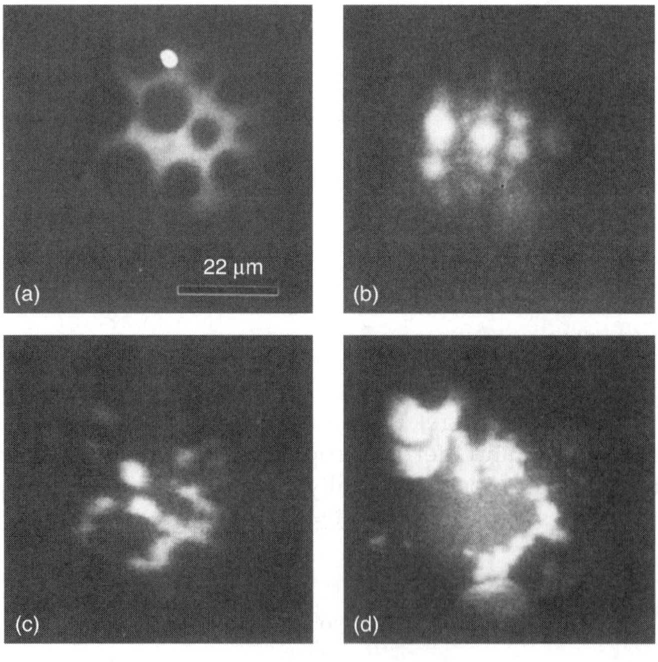

FIGURE 1.6
Fluorescence microscopic images of LB-monolayer films obtained in situ on the LB trough surface during film assembly from: (a) the thiophene polymer B-PUMT (see Figure 1.1) expanded monolayer (<2 mN/m); (b) B-PUMT compressed monolayer (15 mN/m); (c) expanded B-PUMT monolayer within 2 h after subphase injection of Str-PE; (d) expanded B-PUMT monolayer 24 hr after subphase injection of Str-PE. Reprinted from Samuelson, L.A., Kaplan, D.L., Lim, J.O., Kamath, M., Marx, K.A., Tripathy, S.K. (1994). Molecular Recognition Between a Biotinylated Polythiophene Copolymer and Phycoerythrin Utilizing the Biotin–Streptavidin Interaction. *Thin Solid Films* 242:50–55. With permission from Elsevier Publishing.

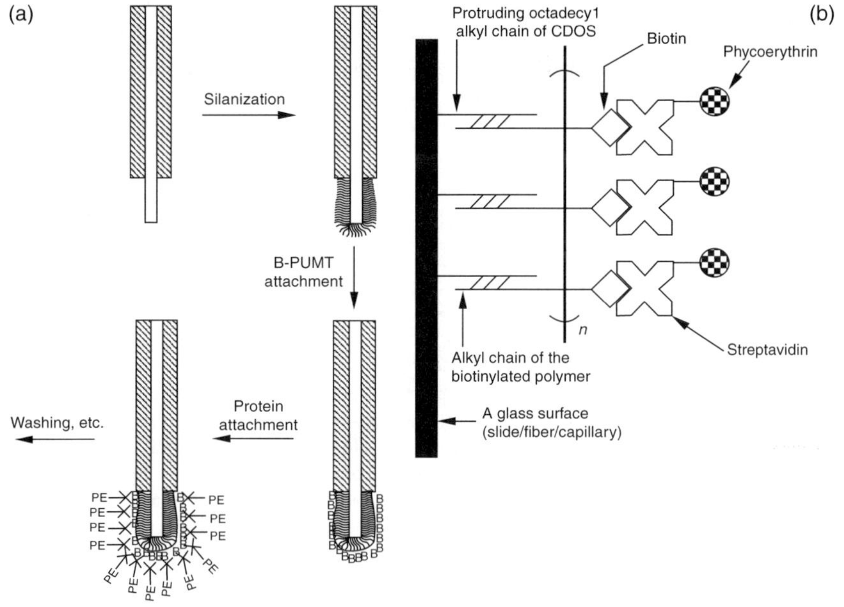

FIGURE 1.7
(a) Schematic of the steps involved in B-PUMT self-assembly on the surface of declad optical fibers, where X-PE is Str-PE protein conjugate. On the right is a diagram (b) of the structure built up on the optical fiber surface as a result of the assembly process. Reprinted from Ayyagari, M.S., Pande, R., Kamtekar, S., Gao, H., Marx, K.A., Kumar, J., Tripathy, S.K., Akkara, J., Kaplan, D.L. (1995). Molecular Assembly of Proteins and Conjugated Polymers: Toward Development of Biosensors. *Biotech. Bioeng.* 45:116–121. With permission of Wiley-Liss, Inc., a subsidiary of John Wiley & Sons, Inc.

spectrum of phycoerythrin was then measured via the optical coupling between evanescent waves extending from the optical fiber and the immobilized phycoerythrin chomophores (17,18). One of the advantages of the attachment system involving the biotin–streptavidin interaction is the fact that these derivatizations can be carried out for all biological molecules and polymers. It results in a very useful and versatile cassette methodology for surface immobilization of any photodynamic protein, as well as other proteins, and nucleic acids.

Another surface immobilization procedure we successfully investigated was the use of sol-gels for the direct entrapment of phycobiliproteins. The sol-gel formation technique we used involved the reactions shown in Figure 1.8: the hydrolysis and polycondensation of silicon alkoxides (19,20). The sol-gel technique produced transparent glass possessing excellent optical, mechanical, and thermal properties, without any necessity for high-temperature processing. PE, entrapped physically within this sol-gel, possessed absorption and fluorescence properties of the native protein. However, in a more extensive study of all three phycobiliproteins entrapped in the sol-gel matrices, the optical properties of the phycocyanin and allophycocyanin species were found to undergo minor changes upon sol-gel entrapment that we associated with changes in aggregation state (21). In the same study, we also showed that phycoerythrin could be entrapped in a thin film of the sol-gel matrix on the surface of an optical fiber, allowing retention of its native optical properties. The normal fluorescence spectrum was measured through the optical coupling between the protein and the evanescent wave of the optical fiber.

We have also presented evidence for two-photon absorption-induced fluorescence from phycoerythrin both in solution and entrapped in sol-gels (19,22). In this environment, the protein's fluorescence, following two-photon absorption, is nearly identical to the fluorescence resulting from single photon absorption, as shown in Figure 1.9. Moreover, it is

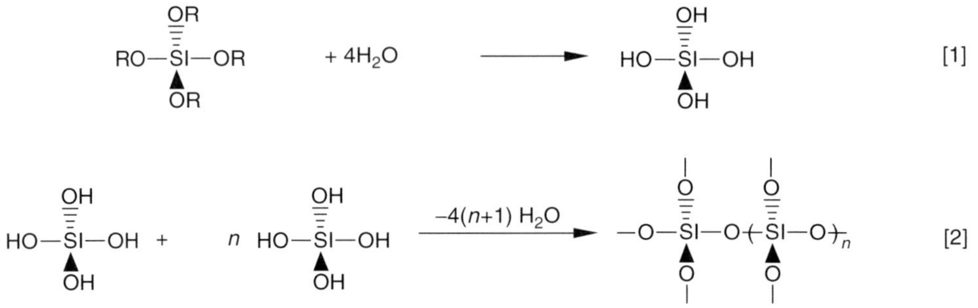

FIGURE 1.8
Reaction scheme leading to the formation of transparent sol-gel glasses at room temperature. In step [1], an alkoxide is hydrolyzed to form silanol groups (Si-OH) and alcohol (not shown). Subsequent condensation reactions between silanol groups take place in step [2] to produce siloxane bonds (Si–O–Si). Polycondensation events continue during aging and the sol-gel is then dried to produce a transparent glass. Reprinted from Chen, Z.P., Chittibabu, K.G., Marx, K.A., Kumar, J., Tripathy, S.K., Samuelson, L.A., Akarra, J., Kaplan, (1994). Photodynamic Protein Incorporated in Conducting Polymer and Sol-Gel Matrices: Toward Smart Materials for Information Storage and Processing. In: Varadan, V.K., ed., Smart Materials and Structures, *Proc. SPIE*, 2189:105–115. With permission of the International Society for Optical Engineering.

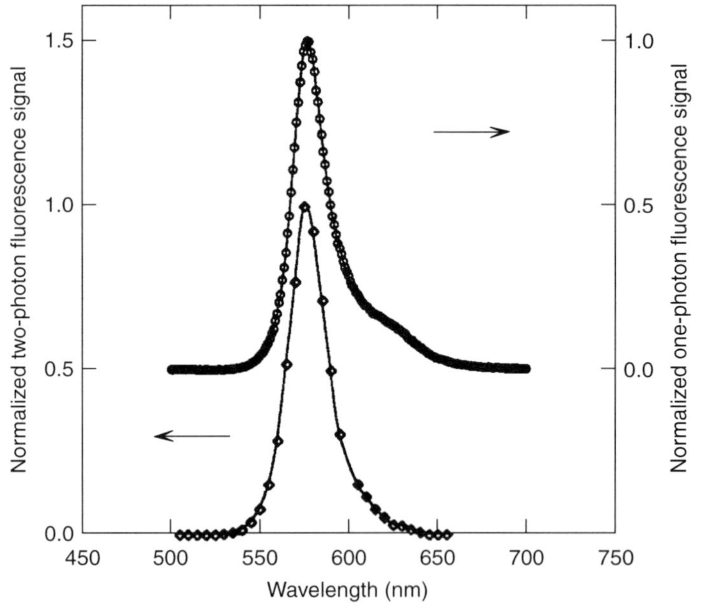

FIGURE 1.9
Fluorescence spectra of PE protein induced by one or two photons in buffer solution. Open circles (upper curve) are data for two-photon-induced fluorescence, with excitation at 1.06 μm. Open diamonds (lower curve) are data for one-photon-induced fluorescence excited at 490 nm. Reprinted from Chen, Z.P., Kaplan, D.L., Yang, K., Kumar, J., Marx, K.A., Tripathy, S.K. (1997). Two-Photon Induced Fluorescence From the Phycoerythrin Protein. *Appl. Optics* 36:1655–1659. With permission from the Optical Society of America.

photostable to 3×10^5 repetitions of the two-photon optical experiment, not exhibiting any loss of intensity of the type that is observed in single photon fluorescence. The measured two-photon cross-section of phycoerythrin was observed to be 20-fold larger in magnitude than that of the dye rhodamine 6G. Therefore, there appears to be potential applications of phycobiliproteins for two-photon fluorescence microscopy of three-dimensional (3-D) biological samples as well as potentially in 3-D optical memories.

Photovoltaic properties of phycobiliproteins have also been studied. We investigated charge transport phenomena by analyzing the dark current–voltage and photocurrent characteristics obtained across gold–phycobiliprotein–gold samples (23). A photovoltaic effect was observed for the gold–phycoerythrin–gold sample. At low intensity levels, the photocurrent closely followed Onsager's law of geminate recombination in three dimensions.

1.2.1.1.2 Bacteriorhodopsin

Bacteriorhodopsin (bR) is the light-transducing integral membrane-bound protein found in the purple membrane of *Halobacterium halobium*. The single subunit protein is 248 amino acids in length and contains a retinylidene chromophore covalently linked via Schiff's base linkage to the lysine-216 amino acid residue (24). In the bacterium, bR acts as a light-driven proton pump. Upon absorption of a photon, bR undergoes a photocycle characterized by a series of well-defined intermediates with different absorption spectra (25). In Figure 1.10, we present a schematic of the photocycle. The defined optical states are characterized by the changes in absorption maximum indicated by the states' subscript wavelengths. The solid arrows show the thermal decay pathways and decay times. Dashed lines correspond to photon-driven processes with some quantum yields indicated. When oriented, bR films possess very interesting optoelectronic properties that have the potential to be used in a number of different application areas including real-time holography, artificial retinas, optical neural networks, and image processing (26–32). Like all proteins, bR has the advantage that genetic variants with potentially improved properties for a particular application can be screened to select for a desired alteration. Since the study of this protein was not a major focus of our Center, we do not discuss its interesting properties comprehensively in any further detail, but focus only on our use of it in specific applications.

Aside from its interesting photocycle, one of the very attractive features of this integral membrane protein is its exceptional stability. By the standards of protein chemistry, the bR protein and its proton pumping photocycle are extremely robust to denaturation or normal oxidative damage. A second attractive feature is its ability to be ordered. The bR protein occurs in a regular hexagonal lattice within the bacterial membrane and is uniquely oriented to the two membranes faces. The membrane fragments found in a typical bR preparation can be easily oriented using electric fields (33), a property that results from the membrane possessing differential charge characteristics upon its intracellular and extracellular faces. Taking advantage of these features, we investigated the bR

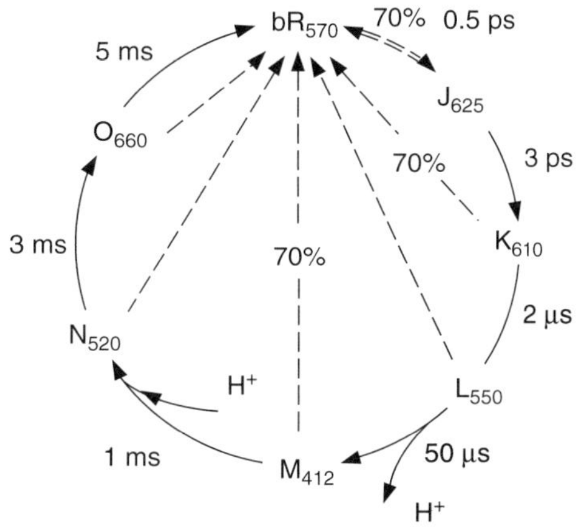

FIGURE 1.10
Photochemical cycle of the bacteriorhodopsin (bR) molecule. In this schematic summary, dashed arrows indicate photon-driven processes. The solid arrows show the thermal decay path. Also indicated are the lifetimes of the intermediate states, as well as the quantum yields for the forward and backward photoreactions of the bR, J, K, and M states. Reprinted from Chen, Z., Lewis, A., Kumar, J., Tripathy, S.K., Marx, K.A., Akkara, J., Kaplan, D.L. (1994). Second Harmonic Generation of Bacteriorhodopsin and Its Application for Three-Dimensional Optical Memory. In: M. Alper, H. Bayley, D. Kaplan, M. Navia, eds. Biomolecular Materials by Design, *Proc. Mat. Res. Soc.*, 330:263–268. With permission from the Materials Research Society.

protein in a number of different formats. We studied its optical properties incorporated within a sol-gel matrix, interacting with a conducting polymer and oriented between parallel plate electrodes. In the study of bR incorporation into sol-gel matrices, the absorption and fluorescence properties of bR were unaltered following incorporation (20). Interaction of bR with a soluble anionic conducting polymer, poly(3-thienylacetic acid)—a thiophene monomer-based system, was carried out and the complex was cast on either quartz windows or interdigitated electrodes, dried, and then studied. Within the polymer matrix, the bR was photoactive and the difference absorption spectra indicated that the M state of the bR photocycle was preferentially stabilized. In fact, bR could be switched back and forth between the photochromic M and bR states using photons of appropriate wavelength. Through photocurrent studies upon interdigitated electrodes, bR was shown to increase its charge transport in the presence of the conducting polymer, compared with bR alone (20). This is in agreement with the expectation that the positive-charged defect structures existing within the π electron conjugation system of the conducting polymer backbone should facilitate charge transport from bR upon its photoexcitation. It also confirms the potential for using bR-conducting polymer conjugates in biosensor designs where charge transport is required.

We also investigated the potential of oriented bR films for use in two-photon 3-D erasable optical memory systems (20,34). The bR protein exhibits an unusually large optical nonlinearity that is due to a large change in the light-induced dipole moment. We studied a bR-based 3-D erasable optical memory storage system based upon this nonlinear optical property to carry out nondestructive reading. Specifically, a two-photon-induced photochromic transition is used for writing and the second harmonic generation is used for reading. As indicated in Figure 1.11, the writing is carried out by transitions between the bR and M states of the photocycle, representing 0 and 1 binary information states possible in the optical memory. The bR protein system is a good material for this type of 3-D memory for a number of reasons. It possesses long-term thermal stability and has a high threshold to photochemical degradation. Also, it exhibits large quantum efficiencies, ϕ, for the photochromic transitions and has a large

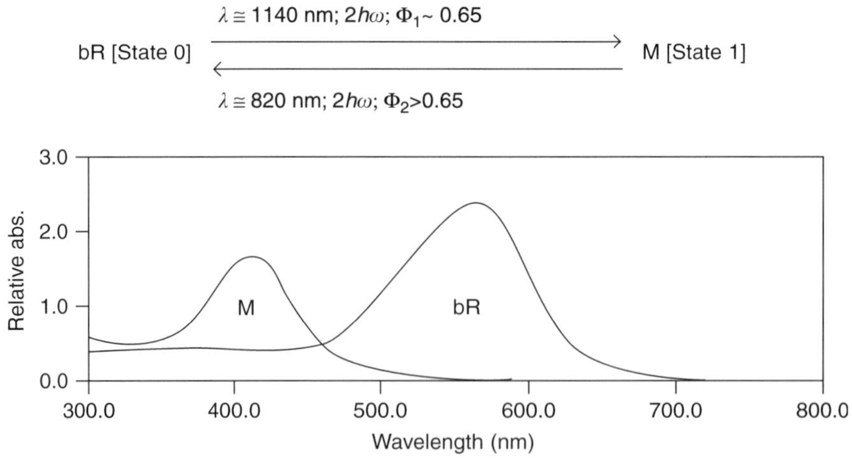

FIGURE 1.11
Two-photon photochromic states bR [state 0] and M [state 1] of bR, the two-photon transition characteristics (wavelength and quantum yield) between states, and the absorption spectra corresponding to those states. Reprinted from Chen, Z., Lewis, A., Kumar, J., Tripathy, S.K., Marx, K.A., Akkara, J., Kaplan, D.L. (1994). Second Harmonic Generation of Bacteriorhodopsin and Its Application for Three-Dimensional Optical Memory. In: M. Alper, H. Bayley, D. Kaplan, M. Navia, eds. Biomolecular Materials by Design, *Proc. Mat. Res. Soc.*, 330:263–268. With permission from the Materials Research Society.

two-photon absorption cross-section and second-order optical nonlinearity. Figure 1.12 presents a schematic diagram of a two-photon 3-D optical memory that could be used for reading and writing of information. The 3-D voxel of information would be located as shown (magnified) within a matrix of stacked 2-D planes of oriented bR membranes located within a polymer matrix. A 3-D disk could be adapted to a rotating disc format and with a total thickness of 1 mm could hold 100 layers of data, assuming that the separation of each data stack is less than 10 μm. A 3-D storage format has a tremendous advantage over 2-D storage for equivalent surface area storage devices, since for visible or IR lasers the storage density limit is ~10^8 bits/cm^2 for the 2-D format, but rises to ~10^{12} bits/cm^3 for the 3-D case. Because both two-photon excitation and second harmonic generation depend quadratically on intensity, writing and reading with 3-D resolution can be accomplished by this nonlinear process either with a focused beam or with the overlap of two beams, as shown in Figure 1.12. The writing process would access a voxel of information to write 0 or 1 by the two-photon process at the depth location in the disc determined by where the two photons from the separate writing beams were brought to focus. Everywhere except inside the voxel, the two isolated unfocused beams would produce no optical state changes and no writing information changes within the disc.

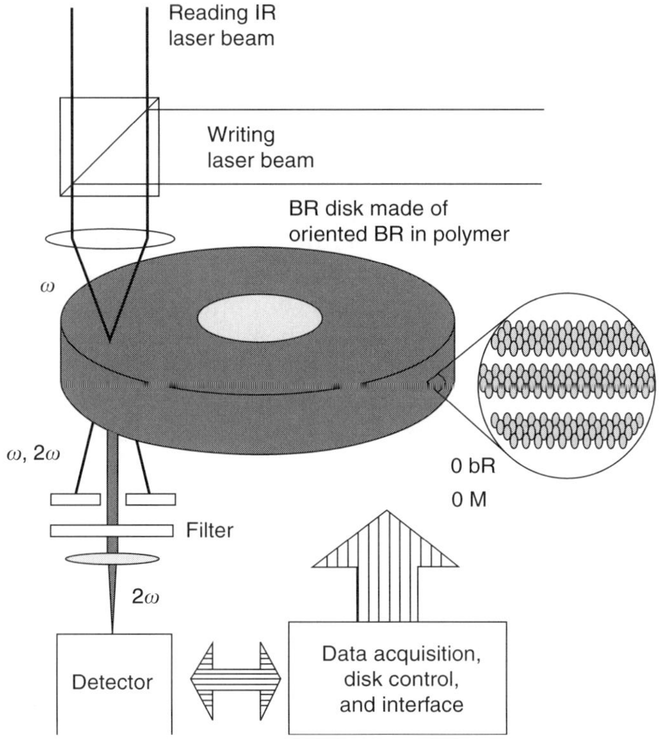

FIGURE 1.12
Schematic diagram of a potential two-photon three-dimensional (3-D) optical memory device based upon bR membranes. The storage medium is a disk made from oriented bR membranes in a polymer matrix. A build-up of bR membrane stacks creates the 3-D disk, as the right-hand expanded view indicates. The data storage capacity equals the data density of each layer multiplied by the number of layers in the 3-D disk. The writing operation in this memory is performed by two-photon absorption and reading is accomplished by second harmonic generation. Reprinted from Chen, Z., Lewis, A., Kumar, J., Tripathy, S.K., Marx, K.A., Akkara, J., Kaplan, D.L. (1994). Second Harmonic Generation of Bacteriorhodopsin and Its Application for Three-Dimensional Optical Memory. In: M. Alper, H. Bayley, D. Kaplan, M. Navia, eds. Biomolecular Materials by Design, *Proc. Mat. Res. Soc.*, 330:263–268. With permission from the Materials Research Society.

Reading the information within a voxel would be carried out by second harmonic generation using a 1.54-μm laser. This has a wavelength of the fundamental and second harmonic photon outside the absorption band of the molecule. Figure 1.13 shows the second harmonic generated from bR in the M and bR states. L1 is the light to switch from the bR to the M state. L2 is the light to convert from the M back to the bR state. The first peak intensity corresponds to the second harmonic signal from molecules in the bR state. When L1 is turned on, molecules in the bR state are photochemically pumped to the M state. The second harmonic intensity decreases to that of the M state. This change is due to a large difference in the second-order nonlinearity between the bR and M states. The contrast ratio between the bR and M states is about 6, which is large enough to reliably assign each voxel its information state during the reading operation. In addition to the nondestructive reading operation using second harmonic generation, this method has the advantage of fast speed and potential for parallel access. Because no molecular transition is involved in the reading operation, the second harmonic photon is generated instantaneously and the reading cycle is likely limited only by the second harmonic photon detection device. Other methods for two-photon optical memories have been proposed using organic, semiconductor, and biomolecular materials (35–40). However, some of these suffer from optical cross talk, or in the case of a bR system based upon its two-photon-induced photoelectric signal to perform the reading operation, the need for reset operations to prevent the bR from switching states. Clearly, bR offers interesting optical properties as well as distinct advantages over other protein systems that make it especially attractive for integration into smart biosensors.

1.2.1.2 Chemiluminescent Probe-Based Enzyme Biosensors

Chemiluminescent molecules emit visible light, typically accompanying the breakage of an unstable high-energy bond(s). Because of this property, chemiluminescent molecules have a wide range of applications, including those in the field of biotechnology. In our studies, we have utilized a particularly interesting example: chloro 3-(4-methoxy spiro [1,2-dioxetane-3-2′-tricyclo-[3.3.1.1.]-decan]-4-yl) phenyl phosphate (CSPD). This phenyl phosphate species is capable of being dephosphorylated by members of a class of phosphatase enzymes. We have used alkaline phosphatase in our biosensor studies and it carries out the initial dephosphorylation step enzymatically, as shown in Figure 1.14. The dephosphorylated reaction product contains an unstable highly strained four-atom

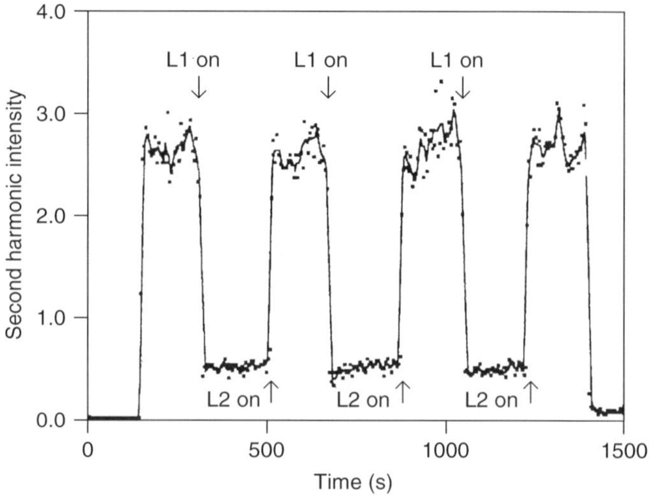

FIGURE 1.13
Second harmonic signal from the photochromic states of bR as a function of time. The times indicated by L1 and L2 are the times when the light beams are turned on to switch, respectively, from the bR to the M state and from the M to the bR state. Reprinted from Chen, Z., Lewis, A., Kumar, J., Tripathy, S.K., Marx, K.A., Akkara, J., Kaplan, D.L. (1994). Second Harmonic Generation of Bacteriorhodopsin and Its Application for Three-Dimensional Optical Memory. In: M. Alper, H. Bayley, D. Kaplan, M. Navia, eds. Biomolecular Materials by Design, *Proc. Mat. Res. Soc.*, 330:263–268. With permission from the Materials Research Society.

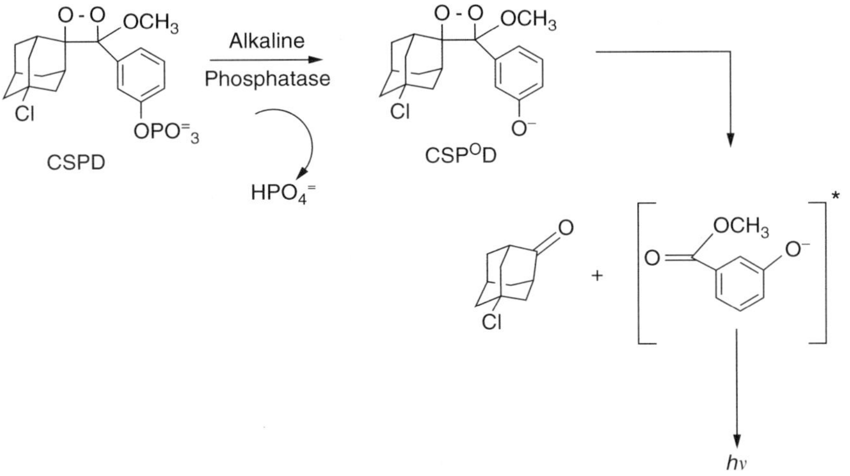

FIGURE 1.14
Reaction sequence starting with chloro 3-(4-methoxy spiro [1,2-dioxetane-3-2′-tricyclo-[3.3.1.1.]-decan]-4-yl) phenyl phosphate (CSPD) that results in emission of a chemiluminescent photon from the unstable product species. The reaction is initiated by dephosphorylation of the CSPD substrate by the enzyme alkaline phosphatase. Reprinted from Ayyagari, M.S., Kamtekar, S., Pande, R., Marx, K.A., Kumar, J., Tripathy, S.K., Kaplan, D.L. (1994). Chemileuminescence-Based Pesticide Biosensor Utilizing the Intelligent Evolved Properties of the Enzyme Alkaline Phosphatase. In: Rogers, C., ed. *Proceedings Second International Conference on Intelligent Materials.* Technomic Press, pp. 85–96.

dioxetane ring that undergoes subsequent ring cleavage at the weak -O-O- bond in a second step to form two separate product molecules. The unstable phenolate species then emits a chemiluminescent photon over a broad wavelength range. Chemileuminescence, as an optical process for biosensor use, has distinct advantages over absorption or fluorescence. To begin with, no light source is needed because the chemiluminescent molecule already contains a photon energy equivalent to the free energy stored in its internal electronic structure. There is a concomitant advantage that no optical background source noise can interfere with the chemiluminescent signal being detected, as is often the case with widely used fluorescence detection schemes.

1.2.1.2.1 Organophosphorus Pesticide Detection

The organophosphorus pesticides were developed in the 1950s as an alternative to the chlorinated pesticides (DDT and lindane) that were available first, but suffered from their ecological persistence in soil. As a class, organophosporus molecules are considerably more water soluble than the chlorinated compounds (41). Consequently, their continued use poses a current threat to aquatic life. Given this situation, it is essential to be able to monitor the concentrations of these compounds in various environments to determine compliance with EPA regulations as well as the efficacy of remediation treatments (42). With this in mind, we developed an enzyme-based biosensor that utilizes chemileuminescence output for the quantitation of organophosphorus-based pesticides. Detection was carried out in our biosensor design by the pesticide's inhibition of alkaline phosphatase's (AP) enzymatic catalysis of the CSPD substrate to its chemiluminescent reaction product. Our initial studies characterized the solution kinetics of the chemileuminescence output of CSPD catalysis by AP and then its inhibition by varying concentrations of two representative organophosphorus pesticides—paraoxon and methyl parathion (18,41,43). Using a Michaelis–Menten analysis of the kinetics, we showed that a mixed competitive and noncompetitive inhibition was exhibited by the AP in the presence of varying concentrations of either inhibitor—paraoxon or methyl parathion. Without attempting serious optimization of the system, the lower limits of detection in solution via this inhibition approach was determined to be 50 ppb for paraoxon and 80 ppb for methyl parathion.

Next, we studied this system after carrying out immobilization of the AP onto the interior walls of small-bore and small-volume glass capillaries using a biotinylated conducting polymer attachment mechanism. This method had the advantage of being technically simple, involving a few dipping steps of the optical surface into a succession of solutions (17,44,45). We show a schematic of the self-assembly process system in Figure 1.15. The interior wall glass surfaces were first silanized with chlorodimethyloctadecyl-silane and then stable hydrophobic interactions were formed in a second step with the surface binding of the conducting copolymer poly(3-undecylthiophene-co-3-thiophenecarboxaldehyde)6-biotinamidohexanohydrazone (B-PUHT). Within this immobilized polymer, the biotin ligand is pendant in solution because of a long hydrophilic spacer group attaching it to the polymer backbone, making the ligand readily available for efficient streptavidin binding. Following the streptavidin derivatization of AP to form Str-AP, this complex was then bound to the immobilized copolymer in a third step, completing the core of the biosensor. In a variation of the polythiophene polymer attachment method just discussed, we first silanized the glass surface with 3-aminopropyl trimethoxysilane, which created a pendant amino group. Polythiophene was then grown in situ by chemical polymerization upon the pendant amino group, before a final biotinylation step (46). In all of these experiments, the small-bore capillary was used for convenience since it provided a high surface to volume ratio. The long length presented a large surface area for immobilizing a significant number of Str-AP molecules and the small bore and interior volume (100 μL) brought the CSPD everywhere in solution into close proximity to the surface-immobilized AP for enzymatic catalysis to form the product that emitted chemiluminescent photons. Using this simple set-up, we detected the activity of as little as 0.1 fmol of AP. In Figure 1.16, we present data

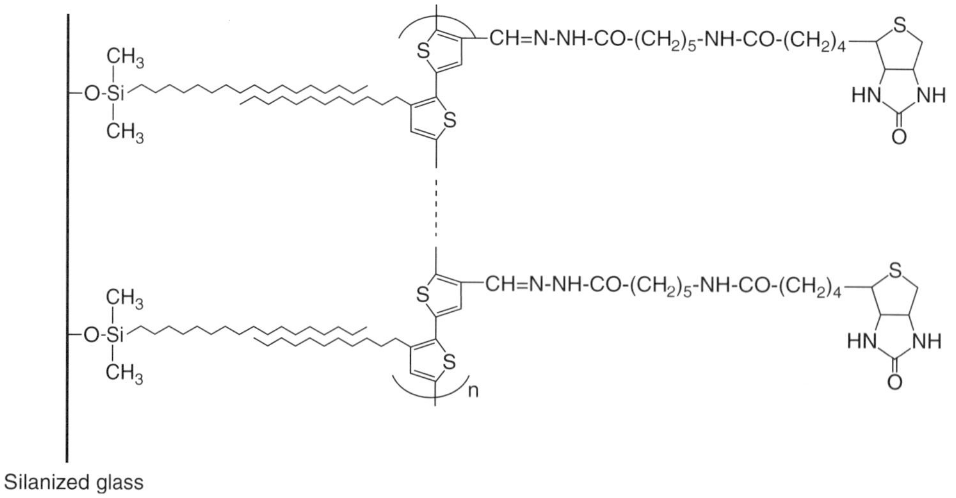

FIGURE 1.15
Schematic representation of the hydrophobic interaction-based surface modification of alkyl-chain-silanized glass by the undecyl chain of the biotinylated 3-thiophene copolymer, B-PUHT. Reprinted with permission from Pande, R., Kamtekar, S., Ayyagari, M.S., Kamath, M., Marx, K.A., Kumar, J., Tripathy, S.K., Kaplan, D.L. (1996). A Biotinylated Undecylthiophene Copolymer Bioconjugate for Surface Immobilization: Creating an Alkaline Phosphatase Chemileuminescence-Based Biosensor. *Bioconjugate Chem.* 7:159–164. Copyright (1996) American Chemical Society.

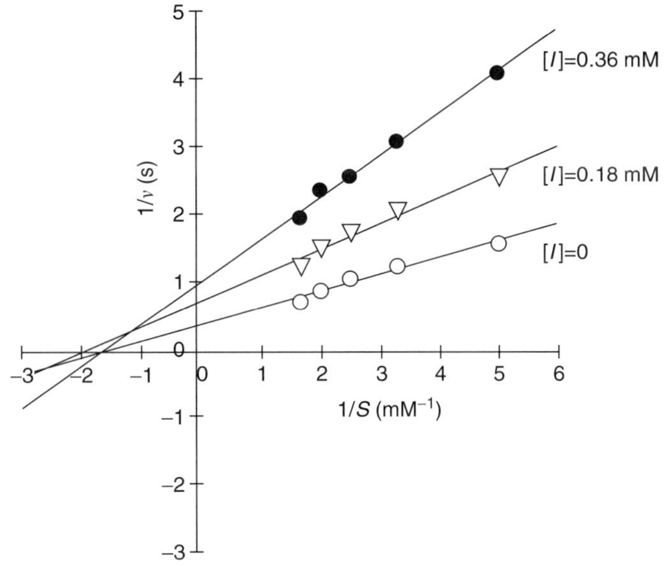

FIGURE 1.16
Lineweaver–Burk plot for paraoxon (*I*)-mediated inhibition of alkaline phosphatase-catalyzed CSPD (*S*) reaction rates (*v*). The reactions for different (*S*) concentrations were studied at the three different paraoxon (*I*) concentrations indicated. Reprinted with permission from Pande, R., Kamtekar, S., Ayyagari, M.S., Kamath, M., Marx, K.A., Kumar, J., Tripathy, S.K., Kaplan, D.L. (1996). A Biotinylated Undecylthiophene Copolymer Bioconjugate for Surface Immobilization: Creating an Alkaline Phosphatase Chemileuminescence-Based Biosensor. *Bioconjugate Chem.* 7:159–164. Copyright (1996) American Chemical Society.

indicating that this immobilized AP system exhibited the same type of mixed competitive and noncompetitive inhibition with paraoxon as we demonstrated previously for AP in solution studies (43). Furthermore, the immobilized AP system exhibited a robust character, retaining 75% of the initial AP activity for at least 30 days when stored in buffer at 4°C (44). As shown in Figure 1.17, this simple biosensor is sensitive to paraoxon inhibition, especially at concentrations lower than 2 ppm, as the inset shows. We were able to achieve lower detection limits of 500 ppb for paraoxon and 700 ppb for methyl parathion.

1.2.1.2.2 *Detecting Zn^{2+}, Be^{2+}, and Bi^{3+} Ions Competitively*

In solution studies, we further demonstrated that the AP-catalyzed CSPD chemileuminescence biosensor system was capable of detecting certain metal ions (47,48). The AP enzyme is a dimeric metalloenzyme containing four Zn^{2+} cations, two coordinated within each of the AP's active sites. Therefore, we could determine Zn^{2+} by reactivation of the alkaline phosphatase activity within the biosensor following prior Zn^{2+} removal (48). The cations Zn^{2+}, Be^{2+}, and Bi^{3+} are all known to be inhibitors of native AP activity (49–51). Therefore, their concentrations could be determined by their inhibition of the native alkaline phosphatase within the biosensor. As an example, in Figure 1.18 we show the inhibitory effects measured for the three cations on the AP system. Using the Zn^{2+} reactivation approach, sub-ppb Zn^{2+} sensitivities were achieved. Using the inhibition of native AP, the detection limits achieved were 170 ppb for Zn^{2+}, 1 ppb for Be^{2+}, and 1.8 ppm for Bi^{3+} (47,48). Also, we showed that selective determination of Zn^{2+} in the presence of Be^{2+} could be achieved by masking the interfering ions with acetylacetone and sodium fluoride. Although we did not pursue an immobilization strategy, we have already demonstrated that successful copolymer-based immobilization of AP can be achieved to form a chemileuminescence-based biosensor. This biosensor would likely be sensitive for detecting these metal ions.

1.2.2 Electrochemical-Based Biosensors

There are distinct advantages that electrochemical methods provide over other signal transduction methodologies for use in biosensors. One advantage is electrochemical control over

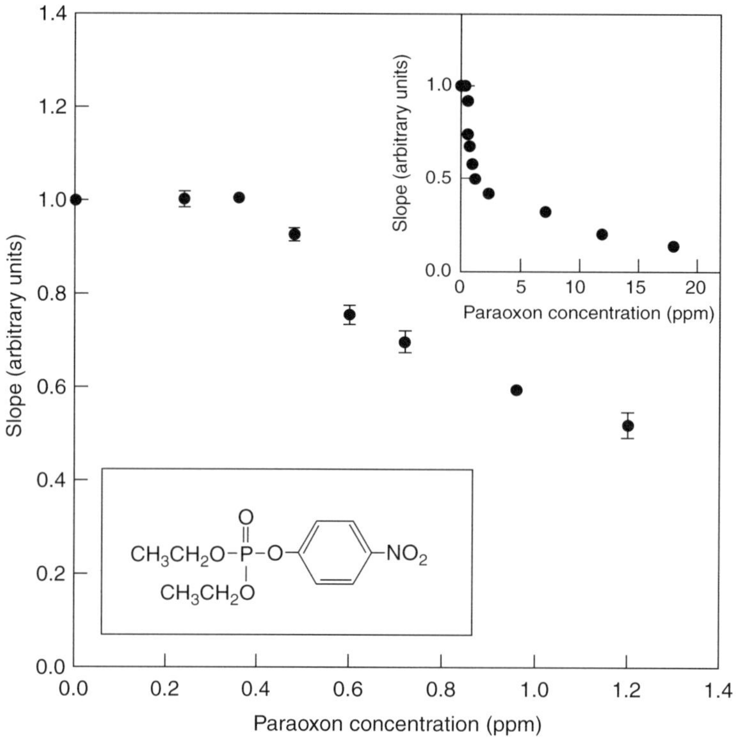

FIGURE 1.17
Calibration curve for paraoxon inhibitor using the alkaline phosphatase biosensor created by assembling Str-Alk phosphatase onto the biotinylated polymer as presented in Figure 1.15. The conditions of 0.4 mM CSPD and 10% enhancer were used as described in Ref. (44). Insets display the higher concentration range of paraoxon inhibition effects as well as the chemical structure of paraoxon. Reprinted with permission from Pande, R., Kamtekar, S., Ayyagari, M.S., Kamath, M., Marx, K.A., Kumar, J., Tripathy, S.K., Kaplan, D.L. (1996). A Biotinylated Undecylthiophene Copolymer Bioconjugate for Surface Immobilization: Creating an Alkaline Phosphatase Chemileuminescence-Based Biosensor. *Bioconjugate Chem.* 7:159–164. Copyright (1996) American Chemical Society.

the electrode potential and therefore selectivity over the analyte species to be quantitated, typically by the current detected at the electrode. Another is that the electrochemical sensing via potential control is confined to the electrode surface. And the electrode itself may be whatever geometry and size is required, including microscale to nanoscale dimensions. As a result, electrochemical methods are an attractive methodology for creating biosensors. They provide an additional advantage in that electropolymerization is possible upon an electrode, allowing the creation of thin polymeric films from monomers in solution through controlled electron transfer. Such films can be utilized for both immobilization of biological elements as well as the underlying signal transduction process of the biosensor, where electron transfer can be employed to generate a signal dependent upon analyte concentration. Electrochemical-based biosensors were some of the earliest types of biosensors commercialized (52,53). A good example is the glucose sensor developed for monitoring blood glucose levels in diabetic and prediabetic patients. In the Center for Intelligent Biomaterials, we have carried out basic studies of the electropolymerization of thin films, as well as employed strategies utilizing these films and electrochemical signal transduction in a number of studies of different biosensor systems. We describe some of these systems in the following sections.

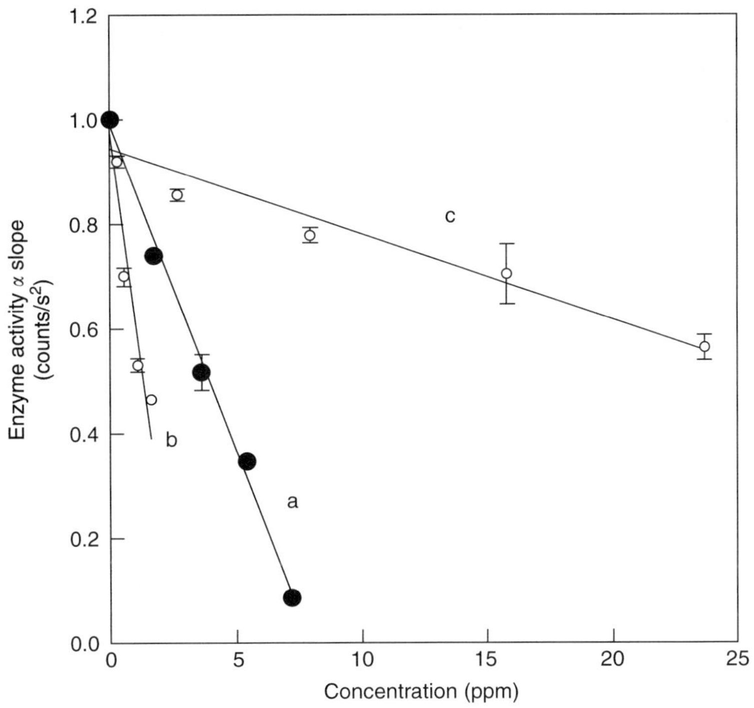

FIGURE 1.18
Inhibitory effect of (a) Zn(II), (b) Be(II), and (c) Bi(III) ions on the chemileuminescence signal from alkaline phosphatase. Reprinted with permission from Kamtekar, S., Pande, R., Ayyagari, M.S., Kaplan, D.L., Marx, K.A., Kumar, J., Tripathy, S.K.(1996). Trace Analysis of Zn(II), Be(II), and Bi(III) by Enzyme-Catalyzed Chemileuminescence. *Anal. Chem.* 68:216–220. Copyright (1996) American Chemical Society.

1.2.2.1 DNA Interaction With Electropolymerized Conducting Polymers—Immobilization by Electrostatic Interactions

Conducting polymers are an interesting class of polymeric materials that have been investigated extensively. Most often they have been synthesized by chemical means. However, the monomers possess oxidation potentials that allow them to be electropolymerized to form conducting polymers at an electrode surface (54,55). Electropolymerization is preferred over synthetic chemical approaches since electropolymerization can be thought of as a more Green Chemistry polymerization method. It eliminates the use of many harsh chemicals involved in the chemical polymerization methods, although some solvents such as acetonitrile are still needed to solubilize certain monomers. Once polymerized, conducting polymers exhibit extensive electron delocalization between multiple monomer units along the polymer backbone. In all cases, the extent of electron delocalization and the degree of planarity depend upon the 3-D chain conformation of the particular polymer. Take as an example polypyrrole. Scanning tunneling microscopy studies of polypyrrole have demonstrated the presence of semicrystalline domains and the presence of helical chain structures within these domains (56). When oxidized to form a cationic species, individual positive-charged 'defects' are created in the electronic structure, as shown in Figure 1.19. Evidence for these structures has been obtained by ESR spectroscopy (57). These positive-charged defect structures are capable of migration along the highly conjugated backbone, giving rise to substantial electrical conductivity. This can range from 10^{-3} to 10^{3} Ω/cm for electropolymerized polypyrrole, depending upon synthesis conditions (58).

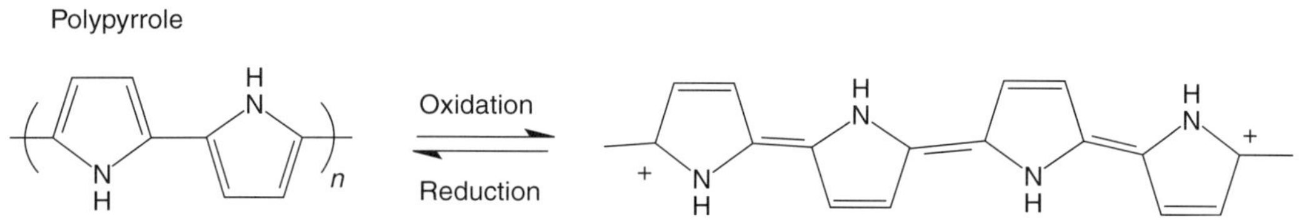

FIGURE 1.19
A representation of the oxidized form of Polypyrrole is shown, where the local oxidized species depicted is a positive-charged 'defect' structure. These structures can be eliminated upon reduction back to the neutral species using either chemical or electrochemical methods. Reprinted from Marx, K.A., Lim, J.O., Minehan, D., Pande, R., Kamath, M, Tripathy, S., Kaplan, D. (1994). Intelligent Materials Properties of DNA and Strategies for It's Incorporation into Electroactive Polymeric Thin Film Systems. *J. Intelligent Mater. Systems Struct.* 5:447–454.

Electrical conductivity is a property that can be taken advantage of in the design of biosensors. Polycationic polymers, such as polypyrrole, should be capable of rearranging their mobile charge distribution in the presence of a polyanion, to form energetically favorable electrostatic complexes. With this as an immobilization strategy, in a number of studies we have examined two conducting polymers, polypyrrole and polythiophene—both polycations, for their ability to interact with and immobilize DNA—a polyanion.

Using low-resolution SEM, we studied electropolymerized polypyrrole films (~100 μm thick) removed from their synthesis electrodes (59). Two strikingly different surface morphologies were observed for these films. A rough surface was exhibited on the solution exposed polymerization growth face and a smooth morphology on the Pt electrode proximal face. The interaction of DNA with these electropolymerized polypyrrole films was then studied. In Figure 1.20 we show a series of images of linear pBR 322 plasmid DNA bound in an extended conformation to the surface of an electropolymerized and cationic polypyrrole film (60). Images such as these were obtained from samples where DNA binding kinetics were measured. These studies revealed that the mass levels of DNA bound/cm^2 surface area to polypyrrole and polythiophene films depended directly upon both the film electropolymerization conditions and resulting measured conductivity level, reflecting the positive-charged defect density (61–64). In the case of polypyrrole binding to DNA, the surface binding level kinetics were observed to exhibit a $t^{0.5}$ dependence and to depend directly on DNA concentration, as Figure 1.21 shows. This concentration-dependent behavior provided evidence to support surface binding by a simple diffusion-limited DNA binding model with very low activation energy (62). That this interaction was largely electrostatic in nature was demonstrated in our study by competition experiments using other anion and polyanion competitors during DNA binding. Once bound, DNA could be competitively removed from the polypyrrole surface, albeit at a very slow rate. This slow rate undoubtedly resulted from the multiple electrostatic bonds formed between both polypyrrole and the DNA polymers.

In a further study, using radiolabeled DNA uptake by electropolymerized polypyrrole films, we showed that the rough and smooth faces of polypyrrole-immobilized DNA from solution at very different rates (60). The rate differences were correlated with differences in the films' smooth and rough face surface areas and the number of voids and channels penetrating throughout the films' interior. Furthermore, we provided evidence for the internal migration of DNA through channels in the polypyrrole films. Therefore, the electrostatic interactions between DNA and the mobile cations of conducting polymer matrices serves as a simple and convenient immobilization strategy, placing DNA in close contact with the conducting polymer chain for use in signal transduction schemes. A number of studies have investigated the properties of DNA in contact with polypyrrole and other conducting polymers used in biosensor applications (65).

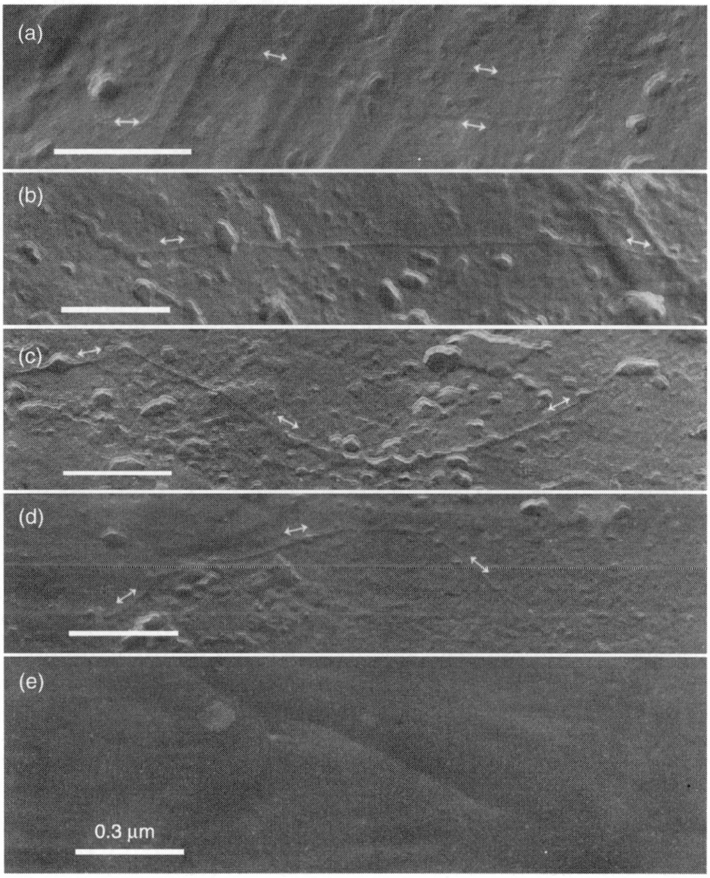

FIGURE 1.20
Transmission electron micrographs from freeze fracture preparations of linear pBR 322 plasmid DNAs bound to electropolymerized polypyrrole films. Panels (a)–(d) show examples where the entire length of individual pBR 322 plasmid DNAs on the polypyrrole surface is visible, highlighted by the arrows shown over portions of their path. Panel (e) is a control sample of the polypyrrole surface without DNA bound. The magnification bar in all panels indicates 0.3 μm length. Reprinted from Pande, R., Ruben, G.C., Lim, J.O., Tripathy, S., Marx, K.A. (1998). DNA Bound to Polypyrrole Films: High Resolution Imaging, DNA Binding Kinetics and Internal Migration. *Biomaterials* 19:1657–1667. With permission of Elsevier Publishing.

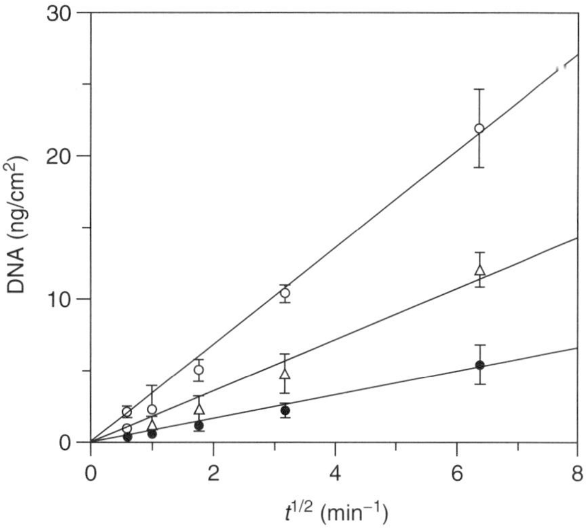

FIGURE 1.21
DNA adsorption plotted as a function of $t^{0.5}$ onto electropolymerized polypyrrole films performed in 1 mM EDTA, 10 mM Tris, pH 8.0. Three DNA concentrations were studied: (diamonds) 0.4 μg/mL, (triangles) 0.2 μg/mL, and (circles) 0.1 μg/mL. Reprinted with permission from Minehan, D.S., Marx, K.A., Tripathy, S.K. (1994). Kinetics of DNA Binding to Electrically Conducting Polypyrrole Films. *Macromolecules* 27:777–783. Copyright (1994) American Chemical Society.

1.2.2.2 Enzyme Electrode Biosensor—Enzyme Entrapment During Electropolymerization of Thin Phenolic Films for Hydrogen Peroxide Biosensing

The electropolymerization of thin films using phenolic-based monomers has a number of attractive features. Phenolic monomers tend to form self-limiting polymer thin films on the electrode surface that don't exceed thicknesses of about 100 nm (66). This is due to the inability of the monomers to penetrate the growing film matrix beyond a certain thickness to be oxidized at the electrode surface. Other significant advantages previously mentioned are that electropolymerizations can be controlled via the applied potential as well as spatially, occurring solely at the electrode surface. Finally, these monomers have the advantage of being soluble in purely aqueous solution, allowing electropolymerization without the need for additional harsh polymerization initiators or oxidizing agents. This latter feature makes it a Green Chemistry method preferred over chemical synthesis for making polymeric films. It also has the advantage that many mainly hydrophilic biological molecules can be added into the solution in the electrochemical cell during the electropolymerization step, leading to their stable incorporation by physical entrapment within the growing film matrix. As an immobilization method, this approach is simple, convenient, and somewhat unique. It has real value since many biological molecules, especially proteins, often have their fragile native structures and biological activities stabilized upon immobilization. Once entrapped in thin films or immobilized by other methods upon an electrode, a biological macromolecule can then be accessed by electron transfer methods at a specific potential. Biosensor output, typically the current, then can reveal the particular electroactive analyte concentration being detected.

A requirement for this type of biosensor is that the biological molecule involved in electron transfer be efficiently coupled electronically to the electrode surface by a suitable transfer mechanism. It is known that electron transfer rates between electron donator and receptor species decrease exponentially as their separation distance increases, as described by Marcus theory. This requirement for close approach of the electroactive enzyme active site to the electrode surface has spurred the development of a number of innovative methods to enhance electronic coupling via transfer agents, thereby eliminating the need for physical proximity of the active site to the electrode surface. One specific example involves the use of hydrophilic polymers containing repeating spaced osmium redox centers to shuttle electrons between the enzyme active site and the electrode surface via transfers between successive osmium redox groups along the polymer chain (67).

We have created and studied an enzyme-based biosensor of the small molecule analyte H_2O_2. The enzyme horseradish peroxidase [HRP: (68) review] was physically entrapped during electropolymerization of various thin films formed from the phenolic-based monomers shown in Figure 1.22. All of the monomers shown—phenol, substituted phenols, and tyrosine-based amino acid derivatives—possessed the phenol ring moiety, with the tyrosine monomers having this moiety as a pendant portion of their amino acid side chains (69). The films were all formed via cyclic potential sweeping across the potential range for oxidation of the monomers, which had broad peak maxima between +0.6 and +0.8 V at neutral pH. In this electrochemical process, films were formed by a free radical-based coupling of monomers. The insoluble polymerization products that formed were either direct phenol ring–ring C–C linkages, ether ring C to ring -OH linkages or a mixture of both linkages and the film formation rates were found to be monomer concentration dependent (70,71). The films electropolymerized on the electrode surface were to varying extents self-limiting in thickness and pinhole free. We studied the enzyme electrode sensitivities to H_2O_2 at two different potentials. The first corresponds to enzymatic catalysis sensing at −0.05 V. Following the direct two-electron electroreduction of H_2O_2 by entrapped HRP, the HRP active oxidation state is regenerated at –0.05 V and this current was shown to be proportional to H_2O_2 concentration. In Figure 1.23, we show that the current detection sensitivities for HRP-entrapped tyrosine derivative films are in the low millimolar range for H_2O_2 and could result in micromolar determinations at nanoampere current detection levels. Clearly, there is background detection of H_2O_2 by direct electroreduction at the bare Pt electrode or at the electrode

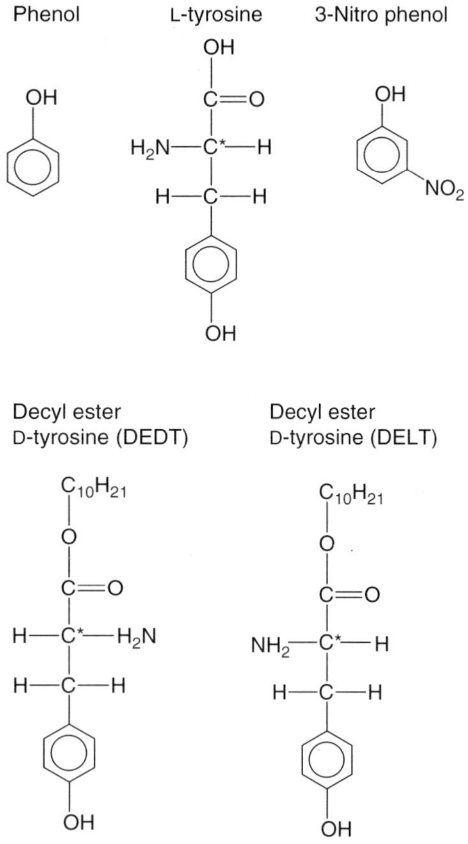

FIGURE 1.22
Structures of phenolic monomers used to create thin films by electropolymerization. The films could be used to entrap enzymes for producing enzyme electrode biosensors. Reprinted from Long, D.D., Marx, K.A., Zhou, T. (2001). Amperometric Hydrogen Peroxide Sensor Electrodes Coated With Electropolymerized Tyrosine Derivative and Phenolic films. *J. Electroanal. Chem.* 501:107–113. With permission of Elsevier Publishing.

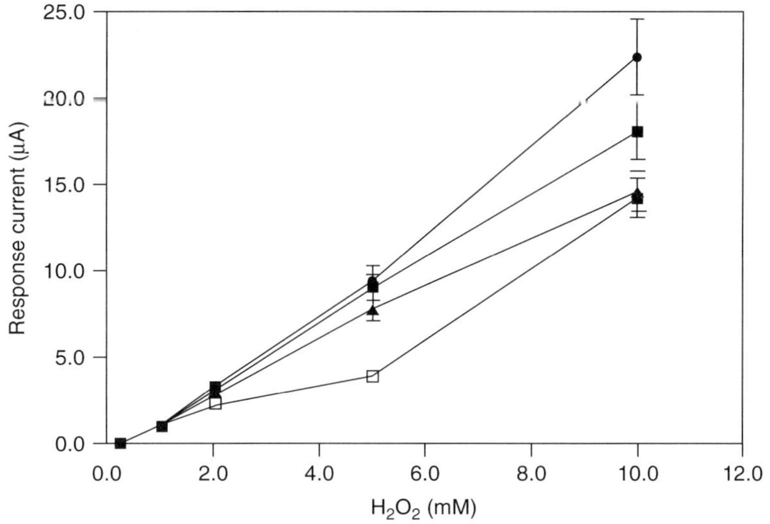

FIGURE 1.23
Current responses at –0.05 V on a Pt electrode from enzyme electrode biosensors (±)horseradish peroxidase (HRP) entrapped in different electropolymerized films: (filled diamonds) HRP-poly-decyl ester derivatives of D-tyrosine (DEDT) film; (filled squares) HRP-poly-tyrosine film; (open squares) no HRP or film; (open triangles) poly-DEDT film. Reprinted from Long, D.D., Marx, K.A., Zhou, T. (2001). Amperometric Hydrogen Peroxide Sensor Electrodes Coated With Electropolymerized Tyrosine Derivative and Phenolic films. *J. Electroanal. Chem.* 501:107–113. With permission of Elsevier Publishing.

covered with films containing no HRP. However, the HRP-containing films are overall the most sensitive for H_2O_2 detection. The second H_2O_2 detection method we studied is not formally a biosensor since it relies upon a process independent of the HRP, that is direct electrode sensing of H_2O_2 by a two-electron oxidation process at +0.85 V. Both methods detect H_2O_2 in the low millimolar range, although the enzymatic biosensor detection method is more sensitive and is less prone to potential interferences from biological species such as uric acid and ascorbic acid.

It is worth noting that there is a rather large range of potential monomers that could be electropolymerized to form biosensor films containing entrapped enzymes. As an example, we studied a series of these including phenol, phenylenediamine, *o*-dianisidine, benzene dialdehyde, and glutaraldehyde. These were electropolymerized to form films that entrapped a range of protein and enzyme systems (72). These included: phycoerythrin, which retained its native fluorescence spectra; alkaline phosphatase, which catalyzed the organophosphorus pesticides paraoxon and methyl parathion to *p*-nitrophenol products that were detected electrochemically at +0.8 V; and HRP and glucose oxidase enzymes that were used to detect H_2O_2 and glucose using a colorimetric approach.

It is interesting to consider that in the system just described, HRP is itself capable of forming polymers by catalysis of phenolic monomers in the presence of H_2O_2. This is a solution enzymatic process stoichiometrically dependent upon H_2O_2 and no electron transfer at the electrode surface is involved. However, it is of interest since it is an aqueous-based Green Chemistry method for forming polymers and can be used as an alternate method to electropolymerization of films. We have carried out a number of studies of HRP-based polymerizations. In the case of LB monolayer films formed from amphiphilic anilinic and phenolic monomers, such as 4-*o*-tetradecyloxyphenol, HRP added to the subphase was shown to catalyze the conversion of the LB monomer film to an LB polymer film (73–76). This process formed films with a high degree of order. FT-IR experiments, using s and p polarized radiation as orientation probes, were performed upon the monolayers picked up from the LB trough with a quartz substrate and demonstrated a significant degree of order within the LB polymer film monolayers. In fact, the ordered highly conjugated polymeric products produced by the LB trough method possessed enhancements of their third-order nonlinear optical properties compared with bulk polymerized monomers (77,78). Following assembly, monolayers comprising these polymer films could be picked up from the LB trough surface and used to create biosensors. In the case of the monomer discussed above, this pickup process would rely upon the adhesion of the alkyl chains to a hydrophobic substrate surface. As an example, we have shown using AFM and x-ray photoelectron spectroscopy that related amphiphilic phenol-based monomers quantitatively coat hydrophobic gold surfaces and have significantly different morphologies before and after HRP polymerization (79).

HRP is the most widely used of the peroxidase enzymes. Because of its broad substrate specificity, it forms the basis for many commercial biotechnology applications (80). Using genetic engineering techniques, its substrate specificity can be manipulated. For example, the enantioselectivity of HRP for oxidation of alkyl aryl sulfides has been increased through genetic engineering approaches (81). Many of the widespread HRP applications occur in immunohistochemical assays for research and clinical use and involve colorimetric- or fluorimetric-based product assays. In the colorimetric systems, an optical signal is the output where the polymerization products have much increased extinction coefficients at particular wavelengths in the visible. We have studied such an optical system, involving hydroxyquinoline-5-sulfonic acid as a substrate for HRP that could form the basis of a biosensor for Fe(III) in the 10^{-5} M concentration range (82).

1.2.2.3 Electropolymerization of Comonomer Mixtures of Biomimetic Tyrosine and Tyrosine-Containing Peptides Forms Thin Films Possessing Specific Cell Attachment/Localization Properties

One of the potential uses of the thin-film electropolymerization strategy we consider is to combine monomers to form copolymerized product films that increase the resulting film specificity for biosensor applications. At the Center for Intelligent Biomaterials, we have pursued this approach. We copolymerized cell recognition peptides containing an added terminal tyrosine amino acid with tyrosine and phenolic-based monomers to form films upon the electrode surface. The films formed have the peptide pendant in solution and are thus available for recognition by a given cell type's specific surface receptors in a subsequent step involving cell addition and binding. One example of the specific cell recognition films formed was the following. Electropolymerizing of the two monomers, tyrosineamide and the terminal tyrosine found in the tetrapeptide arginine–glycine–aspartic acid–tyrosine (Arg-Gly-Asp-Tyr or RGDY), was carried out as shown in the upper panel of Figure 1.24 (83). The net decrease in the oscillating Δf evident with each synthesis cycle of voltage sweep along the time axis corresponds to the addition of film mass to the electrode surface of the quartz crystal microbalance (QCM) device used to make these measurements. The Arg-Gly-Asp tripeptide portion of the tetrapeptide monomer is the biological recognition sequence found within fibronectin and vitronectin. These proteins are located within the extracellular matrix (ECM) that binds to surface integrin protein sites on normal endothelial cells (ECs) found in vivo located on the interior surface of blood vessels (84). In the lower panel of Figure 1.24, there is a significant Δf decrease observed following the addition of ECs to the QCM device that is due to cells binding to the electropolymerized film. As Figure 1.25 also clearly illustrates, ECs do bind to these biomimetic electropolymerized films in substantial numbers, exhibiting normal growth properties and possessing a normal attached and spread light microscopic phenotype. However, this occurs only when the RGDY recognition sequence is electropolymerized into these copolymer films at sufficiently high concentration (the 1:3 monomer ratio case for example) to provide a sufficient number of stable attachment sites. We believe that this electropolymerization film forming strategy could be utilized to build into those films different binding specificity for particular cell types. Films formed in this way could be tailored to specific biosensor designs for different cell types since they possess a cassettelike quality where only the recognition sequence portion of the peptide, not the electropolymerized tyrosine, would need to be varied to achieve each different desired cell binding result. The only requirement for this cassette system, for any particular application, would be prior knowledge of the peptide recognition sequence for that specific cell type.

1.2.3 Piezoelectric-Based Biosensors

In this signal transduction mechanism, any surface mass change is sensitively detected by an alteration in the resonant frequency of an oscillating crystal. The piezoelectric effect has served as the basis for many devices and biosensors. One of the most frequently used devices for biosensor creation has been the QCM. In this device, an AT-cut quartz crystal of varying thickness operates at its resonant frequency, ranging between about 5 and 20 MHz, driven by an AC oscillator circuit operating through electrodes placed on either face of the crystal. In Figure 1.26, we present a schematic of an electrochemical quartz crystal microbalance (EQCM). This is a variant of the QCM technique that allows the mass changes measured by the standard QCM technique at the upper electrode, to be combined with electrochemical processes at that electrode, which also serves as a working electrode

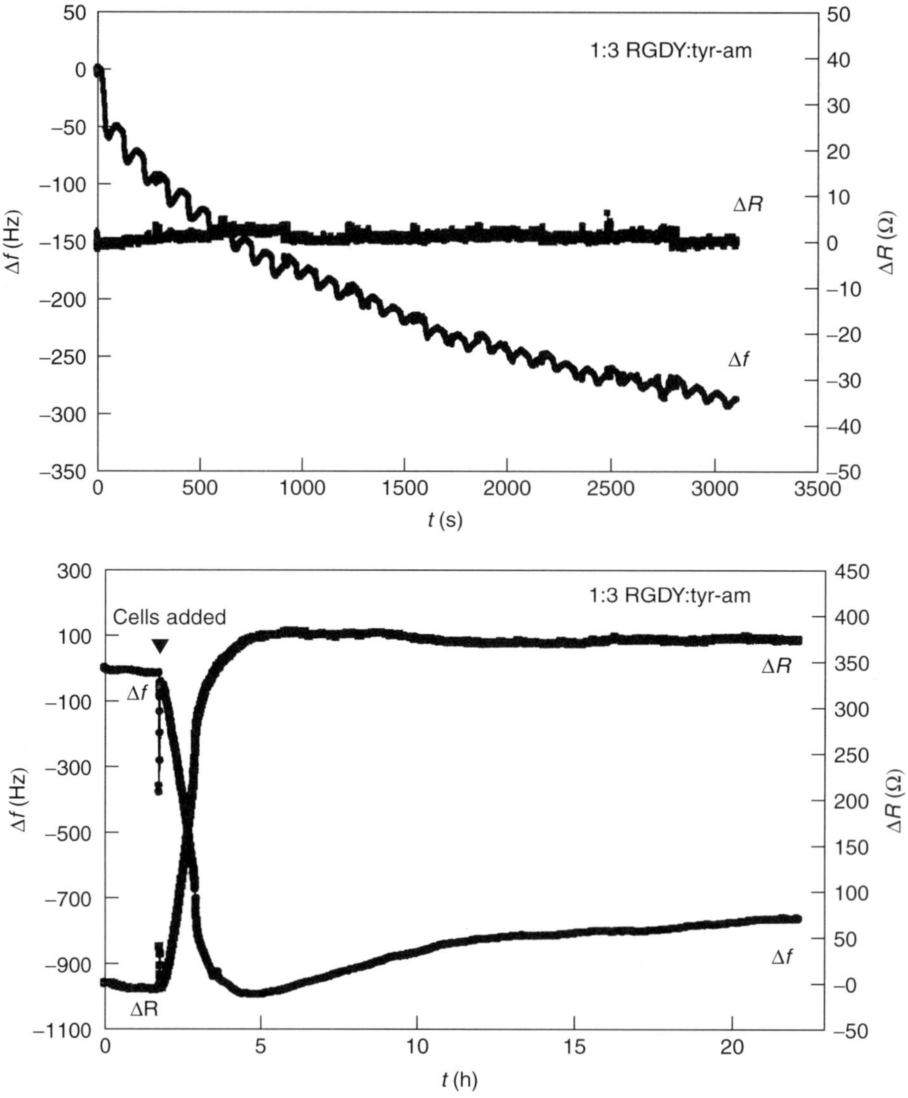

FIGURE 1.24
Time course of film formation is presented in the upper panel measured via the Δf and ΔR shifts of an electrochemical quartz crystal microbalance (EQCM) device during electropolymerization by a series of cyclic voltammetry (0–1.0 V) sweeps of the comonomer mixture, tetrapeptide Arg-Gly-Asp-Tyr (RGDY) and tyrosineamide in a 1:3 ratio. In the lower panel, the time course of endothelial cell (EC) binding to the electropolymerized film is presented for 20,000 cells added at the time indicated by arrowhead following a 2 h baseline signal with only media and serum.

in a three-electrode electrochemical cell. This set-up can sensitively detect changes in mass deposition down into the nanogram range via changes in the resonant frequency, f, and motional resistance, R, of the oscillating quartz substrate. Originally developed for gas-phase use, in the early 1980s solution-phase QCM was developed and has been applied to a number of fundamental studies of biomolecular systems, polymeric systems and whole cells. For a detailed discussion, we refer the reader to an extensive recent review from the Center for Intelligent Biomaterials on the use of the QCM and EQCM in solution-based biosensor and biomaterials characterization applications (65).

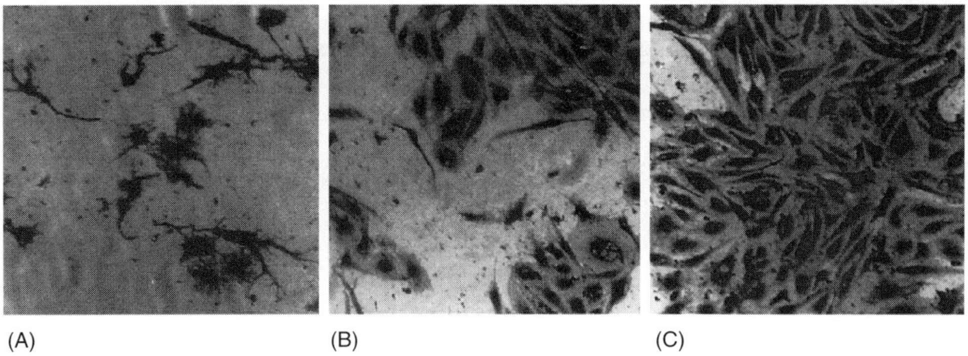

FIGURE 1.25
(See color insert following page 330) Light micrographs of Coomassie blue-stained ECs attached to the surface of three different electropolymerized films: (A) poly-tyrosineamide film, (B) poly-RGDY:tyrosineamide film (1:20 molar ratio), (C) poly-RGDY:tyrosineamide film (1:3 molar ratio). Reprinted from Marx, K.A., Zhou, T., McIntosh, D., Braunhut. S.J. (2004). Electropolymerization of Biomimetic Peptide-Tyrosineamide Polymer Films for Specific Cell Attachment. *Arch. Appl. Biomater. Biomol. Mater., Mater. Res. Soc.* EXS-1:169–171. With permission of the Materials Research Society.

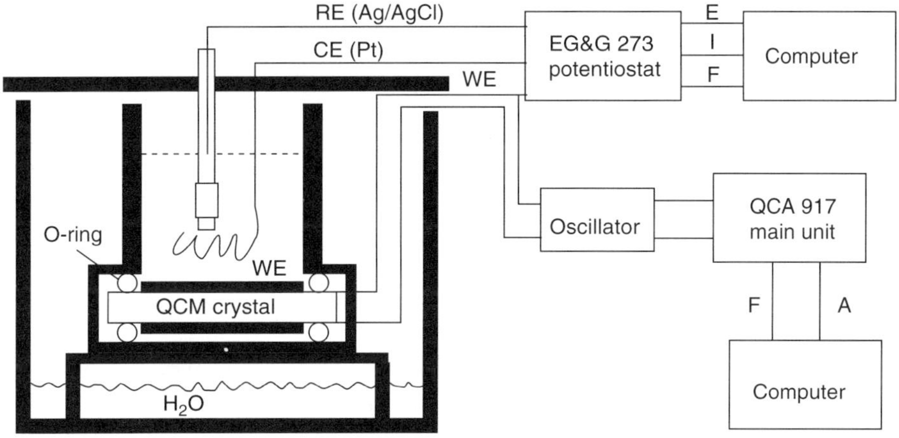

FIGURE 1.26
Electrochemical quartz crystal microbalance device schematic is presented. This electrochemical set-up comprised the three-electrode configuration and associated potentiostat for computer-controlled electrosynthesis used to electropolymerize a number of films. The oscillator was used for measuring the Δf and ΔR of the oscillating quartz crystal during and following film synthesis and in studies of biosensor applications, including growing cells upon films following their synthesis as in Figures 1.24 and 1.25. For the cell growth application, we have shown the EQCM device placed inside a cell culture incubator. Reprinted from Marx, K.A., Zhou, T., McIntosh, D., Braunhut. S.J. (2004). Electropolymerization of Biomimetic Peptide-Tyrosineamide Polymer Films for Specific Cell Attachment. *Arch. Appl. Biomater. Biomol. Mater., Mater. Res. Soc.* EXS-1:169–171. With permission of the Materials Research Society.

1.2.3.1 Quartz Crystal Microbalance Biosensor Detects Enzymatic Polymerization

We have studied both electrochemical and enzymatic polymerization and the thin-film properties of a number of biopolymer systems. Some examples of these we discussed in the prior section on electrochemical signal transduction methods. For one of these systems, the decyl ester derivatives of D-tyrosine (DEDT) shown in Figure 1.22 (85), we have also shown that the QCM can act in a novel way as a sensitive biosensor of enzymatic polymerization taking place in solution and on the gold electrode surface above the oscillating quartz crystal. We had previously demonstrated that these amphiphilic monomers self-assemble into long fibrillike 2 μm average width aggregate structures in aqueous buffered solution (82,86,87). In these studies, following enzymatic polymerization by HRP, we measured the continued presence in solution of long fibrillike monomer aggregates of DEDT.

The DEDT monomer self-assembly process that forms fibers was studied at the QCM gold electrode surface as pH values were increased from 3 to 7, followed by HRP polymerization of the fibrils. In this biosensor, as a result of HRP catalysis the self-assembled monomers' gold surface binding properties (mass) as well as their viscoelastic properties above the crystal surface have changed upon forming self-assembled polymers. This change was detected through shifts in the frequency, Δf, and motional resistance, ΔR, parameters of the QCM device, providing a sensitive measure of the time course of the polymerization reaction. As shown in Figure 1.27, in this system prior to polymerization (follow arrows up to point labeled 2), the frequency decreases as monomer fibril mass is increasingly bound to the gold QCM electrode, but the motional resistance at the QCM surface has only reached 200 Ω. Consequently, only a low level of energy was dissipated by the self-assembled monomer DEDT at these pH values. Once fibrillar aggregates were bound to the QCM surface in a quasiequilibrium state, we initiated polymerization at the points labeled 2–4 by adding a series of small aliquots of HRP and hydrogen peroxide, as indicated in Figure 1.27. Enzymatic polymerization resulted in significant changes in the frequency and motional resistance, especially the latter parameter, indicating that a significantly greater energy was being dissipated. There was a shift from 200 Ω prior to polymerization to nearly 500 Ω at the end of polymerization. These changed values reflected changes in the viscoelastic properties of the bound and surface proximal fibrillar aggregates as polymerization of monomers proceeded. A final polymerized state was achieved in which the altered physical properties of the polymerized fibrillike aggregates made the solution immediately above the gold QCM surface behave as a Newtonian fluid, producing a nearly pure density–viscosity energy dissipative effect, $(\rho\eta)^{1/2}$, on the measured crystal frequency and motional resistance values.

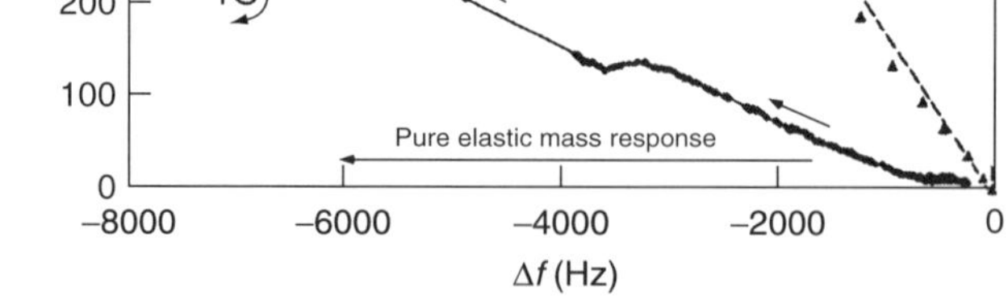

FIGURE 1.27
A ΔR–Δf diagram of the time course of binding and HRP polymerization of DEDT monomer (see Figure 1.22) aggregates at pH 7.0. Arrows indicate the time course of the data points starting at the lower right with the $t = 0$ point being the most rightward. At the point labeled 1, HRP was added. At circled points 2, 3, and 4, H_2O_2 was added. The final circled time point at 143 h after the initiation of the experiments is also indicated. The ΔR–Δf behavior of a pure elastic mass is indicated by the horizontal line. The ΔR–Δf points (filled triangles) determined experimentally for a series of increasing concentration sucrose solutions are depicted with the best fit dashed line. This is the behavior of a Newtonian fluid producing a pure $(\rho\eta)^{1/2}$ density–viscosity effect on the QCM ΔR–Δf values. Reprinted with permission from Marx, K.A., Zhou, T., Sarma, R. (1999). Quartz Crystal Microbalance Measurement of Self-Assembled Micellar Tubules of the Amphiphilic Decyl Ester of D-Tyrosine and Their Enzymatic Polymerization. *Biotechnol. Progress* 15:522–528.. Copyright (1999) American Chemical Society.

Viewed in this way, it can be thought that on going from self-assembled monomers to self-assembled polymers, there would necessarily be an increase in the effective $(\rho\eta)^{1/2}$ since η certainly would increase on going from a monomer to a polymer and so might ρ.

Amphiphilic phenolic and tyrosine-based monomers, possessing aliphatic decyl chains, are capable of self-assembling into aggregates, beginning at their critical micelle concentrations (cmc). We studied these monomers both electrochemically and using AFM. The DEDT monomer (Figure 1.22), at concentrations below its cmc, was clearly shown to form films in the EQCM device via electropolymerization, but it also adsorbed at significant levels onto Pt electrodes in the absence of an applied potential (71). This did not occur with the tyrosine monomer, even at 100-fold higher concentrations. Using the EQCM, we further characterized the concentration dependence of the DELT and DEDT isomer derivatives of tyrosine (70). As monomer concentration was increased the films were found to grow thicker until the cmc was reached, whereupon a saturation type of effect was observed. Also, in the absence of monomer in solution the films formed above the cmc were found to be unstable, undergoing a slow time-dependent desorption of adsorbed monomers from the film. We also have studied copolymers formed from combining the monomers DELT and L-tyrosineamide (79). As the Figure 1.28 schematic of the coupled self-assembly equilibria illustrates, aggregation occurs above the cmc of DELT, but it may not involve the other monomer L-tyrosineamide. Above the cmc and before polymerization with HRP, the DELT aggregate structures we observed in the comonomer mixtures possessed a smooth amorphous morphology of varying shape and dimension, following their self-assembly and adsorption to a gold surface. After HRP copolymerization, these complexes formed structured interacting spherical aggregates upon the gold surface, as Figure 1.29 indicates, that had diameters tightly centered around 1.4 μm. Moreover, these structures now possessed obvious nanoscale surface structure. The self-assembled monomers and copolymerized structures also completely covered gold surfaces, even below their cmc, as measured with x-ray photoelectron spectroscopy (88).

What is novel and important about the use of this QCM biosensor device for kinetic measurements of a polymerization process is that these measurements can be performed upon optically opaque solutions, where more widely used optical sensing techniques may not be possible. Optical opacity was certainly the case for the self-assembling DELT system we studied, since these measurements were performed well above the known cmc of this monomer (87). Detection of the kinetics of the enzymatic polymerization process relied upon the QCM device detecting an alteration in the underlying physical properties

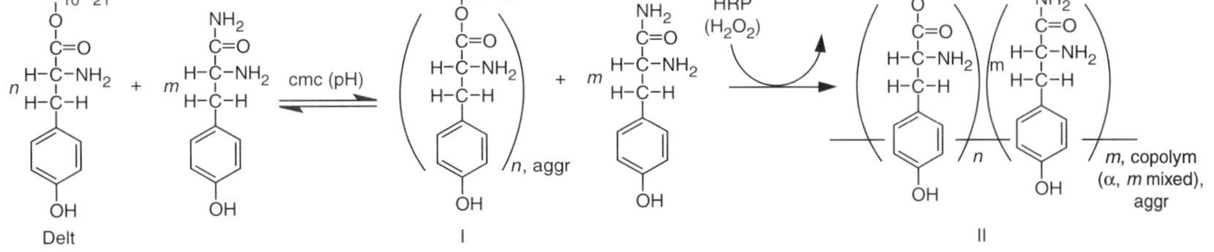

FIGURE 1.28
Schematic view of equilibria involved in the comonomer mixture of DELT and L-tyrosineamide including cmc-dependent aggregation to form structure I and subsequent HRP copolymerization to form structure II. Structure II is a projected copolymer state where the final copolymer linkages (*o* vs. *m* ring–ring linkages) have not been specified, since a mixture of the two is likely to occur. Reprinted with permission from Marx, K.A., Lee, J.S., Sung, C. (2004). Enzymatic Copolymerization Alters the Structure of Unpolymerized Mixtures of the Biomimetic Monomers: The Amphiphilic Decyl Ester of L-Tyrosine and L-Tyrosineamide—An AFM Investigation of Nano- to Micrometer Scale Structure Differences. *Biomacromolecules* 5:1869–1876. Copyright (2004) American Chemical Society.

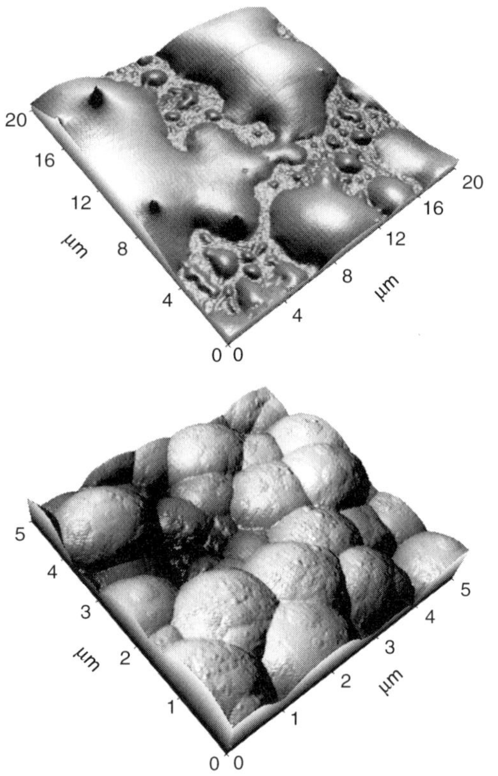

FIGURE 1.29
(See color insert) Noncontact AFM three-dimensional views of a 1:1 comonomer mixture of DELT (above cmc) and L-tyrosineamide prior to (upper image) and following (lower image) enzymatic polymerization with HRP and subsequent 24-h immersion of a gold-coated mica substrate to immobilize the complexes upon. Reprinted with permission from Marx, K.A., Lee, J.S., Sung, C. (2004). Enzymatic Copolymerization Alters the Structure of Unpolymerized Mixtures of the Biomimetic Monomers: The Amphiphilic Decyl Ester of L-Tyrosine and L-Tyrosineamide—An AFM Investigation of Nano- to Micrometer Scale Structure Differences. *Biomacromolecules* 5:1869–1876. Copyright (2004) American Chemical Society.

of the long-fibrillar aggregates as the reaction product was being formed. This QCM biosensor approach has the potential for application to many other experimental situations where optical techniques cannot be employed for sensing.

1.2.3.2 Quartz Crystal Microbalance Cell Biosensor for Cell Characterization and Drug Discovery Applications

With the advent of solution-based QCM devices in the 1980s, investigators began to apply the technology sporadically to studies of living cells. Two types of studies were performed. The first was aimed at understanding fundamental cellular processes, and the second was the use of the QCM as a cellular biosensor of some analyte. A number of recent comprehensive reviews, including one from this Center, have described this cell-oriented research area (65,89–91). At the Center for Intelligent Biomaterials, we have carried out studies of both fundamental cellular processes as well as the integration of cells into QCM biosensors. Our interest in the creation and characterization of cell QCM biosensors was based upon our desire to exploit the wide range of intelligent properties possessed by living cells. The current end products of evolution, cells, are exquisitely complex biological elements that await more widespread integration into biosensors of varying design.

1.2.3.2.1 Measuring the Fundamental Process of Cell Attachment During Biosensor Formation

The sensitive mass detection and surface motional resistance measurement capability of the QCM signal transduction platform provides a unique capability for illuminating the process of cellular attachment to surfaces. During the development of our cell QCM biosensor, we studied the behavior of normal ECs during their attachment to the gold upper electrode surface of the QCM platform (92,93). As Figure 1.30 illustrates, we first demonstrated that the QCM device could be used as a sensitive and continuous

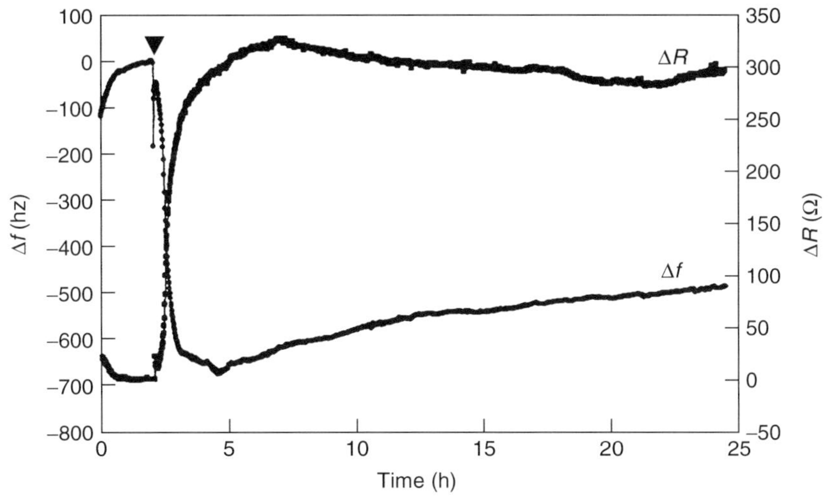

FIGURE 1.30
Kinetics of Δf and ΔR shifts produced in a QCM biosensor during its formation by the addition of 50,000 ECs at the time indicated by the arrowhead position. Prior to cell addition, media and serum-lacking cells were incubated for 2 h to establish baseline f and R values for calculation of the Δf and ΔR shift values from the differences at all times following cell addition. Reprinted with permission from Marx, K.A., Zhou, T., Warren, M., Braunhut, S. J. (2003). Quartz Crystal Microbalance Study of Endothelial Cell Number Dependent Differences in Initial Adhesion and Steady-State Behavior: Evidence for Cell–Cell Cooperativity in Initial Adhesion and Spreading. *Biotechnol. Prog*. 19:987–999. Copyright (2003) American Chemical Society.

monitoring tool to study the time-dependent surface attachment of ECs on the gold QCM surface. This involved first determining the requirements for cell attachment and maintenance in growth media and serum in a sterile environment. Then we were able to determine the time course required to achieve steady-state attachment values of the measure Δf and ΔR shift parameters. Also, we determined that with increasing cell number added, the magnitude of these QCM shift parameters were well correlated with the number of electronically counted cells released following trypsinization from the QCM surface (93,94). In this way, the biosensor accurately reflects the population of ECs, up to about 20,000/0.196 cm^2, stably attached to the QCM surface. We demonstrated that during their transition from initial contact and adhesion to stable attachment, cells exhibited increasing amounts of energy dissipation as expressed through increasing levels of motional resistance measured for the QCM crystal. Viewed as a progression of time points in Figure 1.31 (following the arrows), the frequency and resistance shift data for cell attachment showed that the cells at the surface initially exhibited pure liquid behavior [$(\rho\eta)^{1/2}$ dependence] following the linear fit to the pure sucrose measured data. With increasing time, the data points evolve to indicate that motional resistance and energy dissipation properties at the crystal surface have increased significantly, to values characteristic of the steady state of the cells. These properties are consistent with the following known qualitative picture of anchorage-dependent normal cell attachment behavior. In the cell attachment process, an ECM of specific proteins is secreted by the cell upon a surface. Integral membrane protein receptors on the cell surface termed integrins then recognize specific peptide sites within ECM and a stable attachment is formed. The internal protein cytoskeleton of the attached cell, using these external integrin anchor points to ECM, then forms a networklike structure that maintains the spread shape of the attached cell upon the matrix surface. The cell's internal protein structure responsible for the spread shape has been termed the tensegrity structure (95). Tensegrity structures formed within the cells are consistent with the significantly increased motional resistance

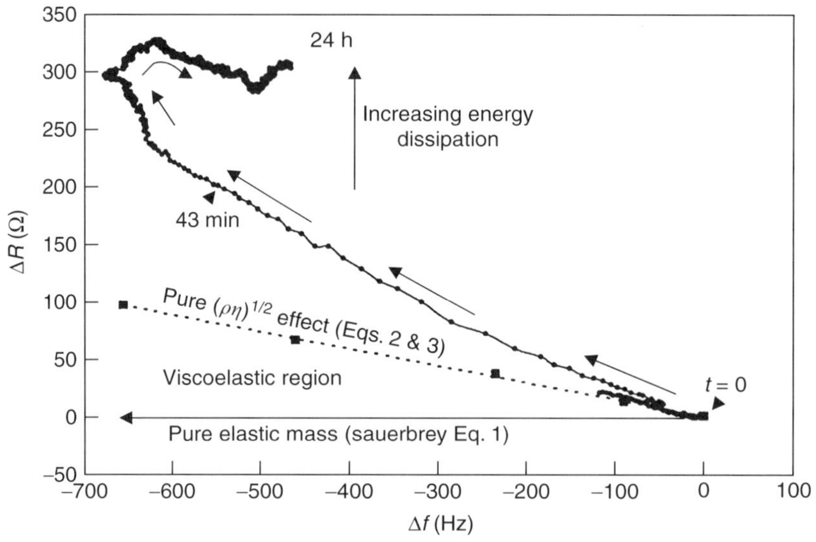

FIGURE 1.31
A ΔR–Δf diagram of the time course following addition of 30,000 ECs to the QCM surface. Arrows indicate the time course of the data points starting at the lower right with the $t = 0$ point at the arrowhead. The final time point occurs for the cells at 24 h at their steady state where maximum energy dissipation is exhibited. The ΔR–Δf behavior of a pure elastic mass is indicated by the horizontal line. The ΔR–Δf points (filled squares) determined experimentally for a series of increasing concentration sucrose solutions are depicted with the best fit dashed line. This is the behavior of a Newtonian fluid producing a pure $(\rho\eta)^{1/2}$ density–viscosity effect on the QCM ΔR–Δf values. Reprinted with permission from Zhou, T., Marx, K.A., Warren, M., Schulze, H., Braunhut, S. J. (2000). The Quartz Crystal Microbalance as a Continuous Monitoring Tool for the Study of Endothelial Cell Surface Attachment and Growth. *Biotechnol. Prog.* 16:268-277. Copyright (2000) American Chemical Society.

and energy dissipation properties exhibited by the steady-state cells that we have observed on the oscillating quartz crystal. It is noteworthy that the QCM biosensor is detecting energy dissipation levels for cells reaching steady-state attachment in Figure 1.31 that are above the pure density–viscosity effect, $(\rho\eta)^{1/2}$, line. This is significantly greater viscoelastic behavior than the energy dissipated by unpolymerized or even HRP-polymerized fibrillike DEDT structures binding to the gold surface (Figure 1.27) that produced measured values lying below this $(\rho\eta)^{1/2}$ line.

In a more detailed QCM study, we examined the normal EC attachment process as a function of cell number added (93). We observed a short 10-min time lag in the attachment process before cells were able to produce measurable QCM frequency and motional resistance shifts. Also, the QCM provided evidence for the existence of novel cell–cell cooperativity behavior in the initial stages of cell adhesion and spreading (93). This was manifested in the Figure 1.32 Δf and ΔR shift data plotted against cell number trypsinized from the QCM surface in any given experiment. This trypsinized cell number reflected an accurate indication of the number of firmly attached cells. The curve shape best fit to the data at 1 h postcell addition was sigmoid shaped. However, the 24 h postcell addition, or steady-state condition, had a best-fit hyperbolic curve shape. The sigmoid curve shape behavior at 1 h following cell addition is characteristic of a cooperative process. It likely is caused by the dependence of cell attachment on a mechanism involving some type of cell–cell communication that is number or density dependent. Once established at their steady state upon the surface, we also studied EC growth stimulation by the effect of adding fibroblast growth factor protein and observing cell growth continuously for up to 4 days (92).

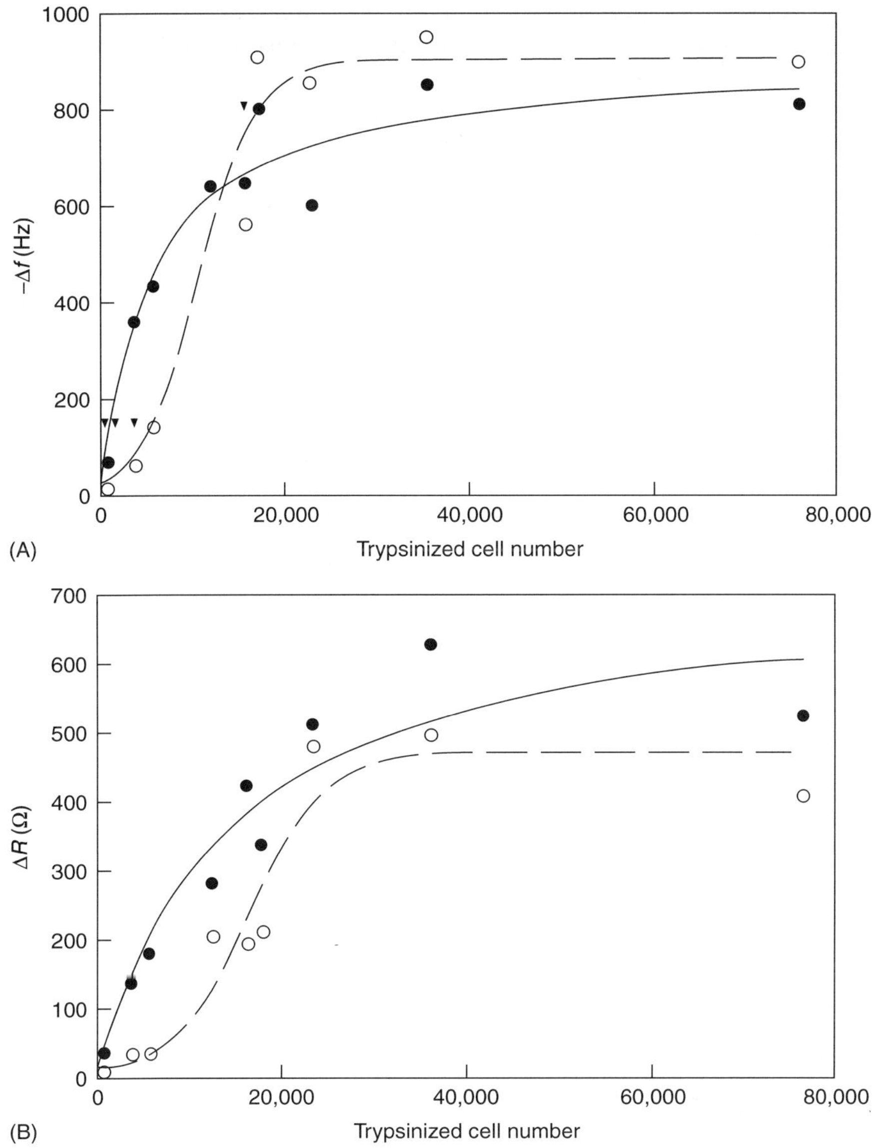

FIGURE 1.32
Plots of the Δf and ΔR decreases for ECs attaching to hydrophilically treated gold QCM surfaces. In (A) Δf and in (B) ΔR are plotted vs. the number of electronically counted ECs requiring trypsinization to be removed from the QCM gold surface. In both panels, the open circles represent 1-h data values (fitted with a three component sigmoid curve) and the filled circles represent 24-h data values (fitted with a hyperbolic curve) following cell addition to the QCM surface. In each case these fits represent the best fits to the data. Reprinted with permission from Marx, K.A., Zhou, T., Warren, M., Braunhut, S. J. (2003). Quartz Crystal Microbalance Study of Endothelial Cell Number Dependent Differences in Initial Adhesion and Steady-State Behavior: Evidence for Cell–Cell Cooperativity in Initial Adhesion and Spreading. *Biotechnol. Prog.* 19:987–999. Copyright (2003) American Chemical Society.

Therefore, during the creation of the cell QCM biosensor, by studying the complex behavior involved in normal cell adhesion and spreading we reported the novel observations just discussed. Having formed the cell QCM biosensor, we applied it as a drug biosensor, as we now describe in the following section. In summary, the QCM provides a unique signal transduction platform for studying the behavior of cells during their attachment to surfaces as well as their behavior once fully established on the surface. However, the complexity of living cells makes stringent demands upon the investigator creating the biosensor. As biological elements in a biosensor, living cells are far more subtle in their sensitivity to complex environment effects than are individual biological macromolecules as biosensor components. At a minimum, cells typically must be maintained at constant temperature within defined growth media inside a sterile environment, to provide accurate output responses and to prevent their contamination with microrganisms. On the other hand, the reward is that cell-containing biosensors can be far more valuable for specific applications because they exploit the full range of intelligent properties possessed by living cells.

1.2.3.2.2 *Cell Quartz Crystal Microbalance Biosensor Detection of Cytoskeleton Binding Drugs*

The above studies examining novel details of the cell attachment process during formation of the QCM biosensor were preparatory to our main goal of creating sensitive cell QCM biosensors for drug discovery applications. To implement this, we used as the biological element ECs attached to the QCM device surface at the steady-state condition corresponding to ~20 h postcell addition, where constant frequency and motional resistance values were exhibited. We show a schematic diagram of this biosensor set-up in Figure 1.33A (96). An integral feature of this biosensor is the requirement for endothelial cells to maintain

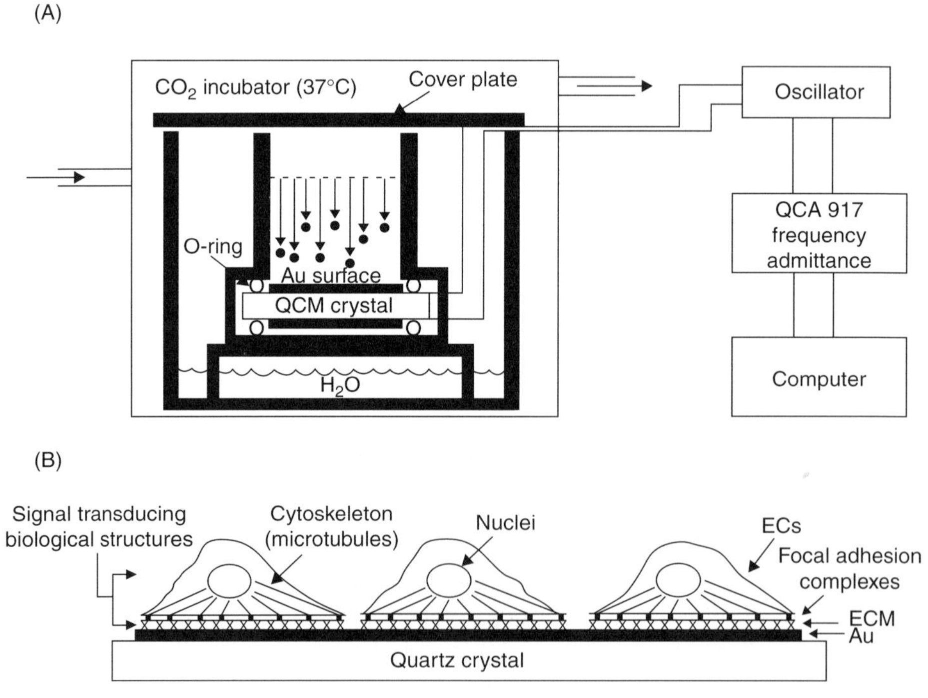

FIGURE 1.33
Schematic diagrams of (A) the entire cell QCM biosensor measurement system and (B) cell QCM biosensor surface with labeled signal transduction elements. Reprinted from Marx, K.A., Zhou, T., Montrone, A., Schulze, H., Braunhut, S. J. (2001). A Quartz Crystal Microbalance Cell Biosensor: Detection of Microtubule Alterations in Living Cells at nM Nocodazole Concentrations. *Biosen. Bioelectron.* 16:773–782. With permission from Elsevier Publishing.

their normal attachment phenotype on the QCM surface. To do this, the cell relies upon its internal cytoskeleton attached via integrin cell surface receptors to its underlying ECM. The multiple important components of the complex signal transduction system in the biosensor are shown labeled in Figure 33B, which represents a schematic of individual ECs attached to their ECM on the gold surface of the QCM biosensor. Small molecule drugs that disrupt different components of this cellular structure, the cytoskeleton, the ECM, or their integrin-mediated connectivity would be expected to have major effects upon the cell QCM biosensor frequency and motional resistance values. To test the validity of this hypothesis, we examined the small molecule drug nocodazole. The cellular targets of nocodazole are the tubulin dimer components that comprise the majority of the mass of the intracellular microtubules. Microtubules are long protein polymer structural components of the cellular cytoskeleton that are required to maintain the spread state of the cell following attachment to a surface. Upon binding, nocodazole is known from other studies to cause dissociation of the microtubules (97).

Given the significant effect of nocodazole on the cellular cytoskeleton, we used the EC QCM biosensor to examine the effect of adding varying nocodazole concentrations upon the biosensor's frequency and motional resistance values (96). In Figure 1.34, we show the significant detection capability of the biosensor when 2 μM nocodazole was added (second arrowhead) to a steady-state population of 20,000 cells on the biosensor surface. In forming the biosensor, there was a rapid decrease in the frequency magnitude and increase in motional resistance, from the values the attached cells exhibited immediately after addition to the constant values after about 20 h. The 20 h constant values represented the steady state of cell attachment characteristic of the biosensor that we used in all the nocodazole experiments. The response to 2 μM nocodazole addition at 22 h (second arrowhead) is shown to be significant in magnitude. In different experiments between 0.11 and 15 μM nocodazole, we observed the frequency values of this biosensor to decrease significantly in magnitude,

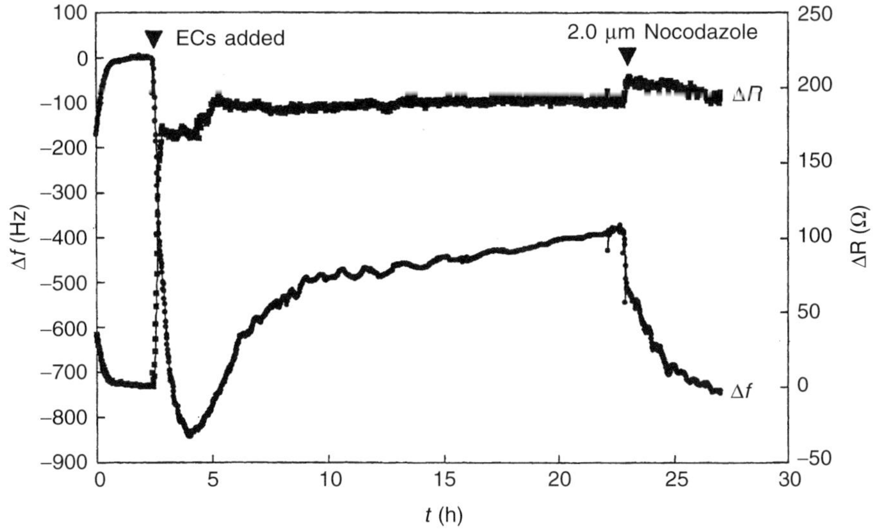

FIGURE 1.34
Time-dependent Δf and ΔR behavior of the cell QCM biosensor created by the addition of 20,000 ECs at the first arrowhead. After establishing stable steady-state properties, nocodazole was added to the biosensor at the second arrowhead to a final concentration of 2 μM. Reprinted from Marx, K.A., Zhou, T., Montrone, A., Schulze, H., Braunhut, S. J. (2001). A Quartz Crystal Microbalance Cell Biosensor: Detection of Microtubule Alterations in Living Cells at nM Nocodazole Concentrations. *Biosen. Bioelectron.* 16:773–782. With permission from Elsevier Publishing.

to a limiting value that changed nearly 100% of the original magnitude at the highest doses used. The response occurred in a dose-dependent fashion, over a 5–6-h period following drug addition. This effect is consistent with nocodazole's disruption of microtubules. By contrast, using this biosensor with a 10 μM taxol addition, an FDA-approved and widely used cancer chemotherapeutic drug, caused little initial time-dependent alteration in frequency, consistent with it's known microtubule hyperstabilization effect (98,99). When we plotted the total biosensor frequency decreases vs. the increasing nocodazole concentration added, as shown in Figure 1.35, we observed a dose–response curve. This sigmoid curve shape fit to the data is typical for small-molecule drugs interacting with their biological targets. The midpoint of this transition is around 900 nM. We corroborated that our cell QCM biosensor was yielding accurate biological effects by examining the nocodazole range of this dose curve with fluorescence light microscopy to visualize cellular actin using a rhodamine-labeled phalloidin chromophore acting as an actin stain. Actin is a known cell stress marker for surface detachment in these ECs. The range of moderate to progressively more severe rhodamine–phalloidin staining alterations in cellular actin fibers that we observed mimicked the range of nocodazole concentrations spanning the biosensor dose curve. Also in good agreement with the dose curve are nocodazole's cytological effects observed by light microscopy in various cell types in a number of other studies (97,100). When the cell QCM biosensor frequency shifts were presented normalized by the number of cells found firmly attached to the QCM surface following trypsin removal and electronic counting, then the dose curve did not fall off at the highest concentrations (not shown in Fig. 1.35). In this case, the curve midpoint was shifted to lower nocodazole concentrations, indicating that this additional trypsin/cell count measurement creates a more sensitive drug biosensor.

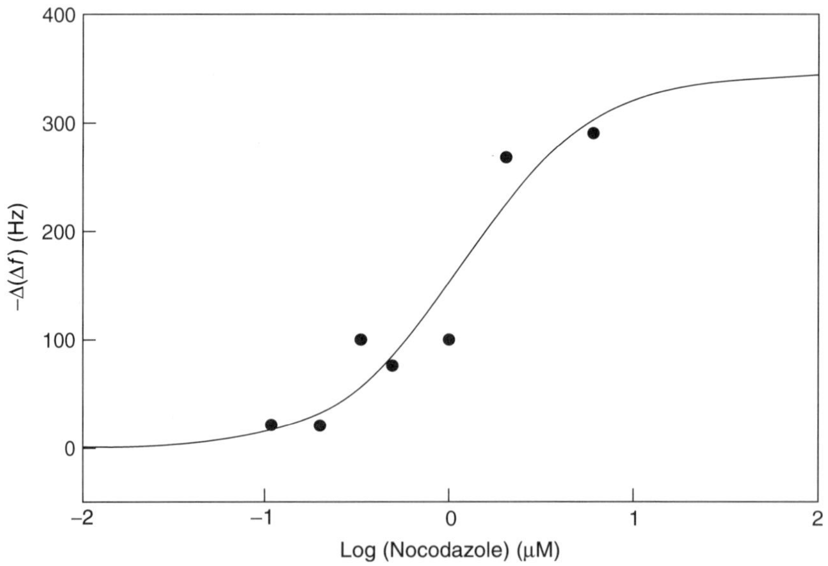

FIGURE 1.35
Dose curve of $-\Delta(\Delta f)$ vs. log[Nocodazole] concentration where the maximum Δf shift values produced have been plotted for different experiments carried out at the indicated nocodazole concentrations. A three-component sigmoid curve has been fit to these data where $R^2 = 0.91$. Reprinted from Marx, K.A., Zhou, T., Montrone, A., Schulze, H., Braunhut, S. J. (2001). A Quartz Crystal Microbalance Cell Biosensor: Detection of Microtubule Alterations in Living Cells at nM Nocodazole Concentrations. *Biosen. Bioelectron.* 16:773–782. With permission from Elsevier Publishing.

Next, we investigated the dynamics of the EC QCM biosensor response by studying the kinetics of the frequency decrease in individual experiments at increasing nocodazole concentrations. Over the entire drug range, for each experiment the frequency decrease remaining as a function of time measured by the EC QCM biosensor was found to possess similar kinetics and was well fit by a single exponential decay equation. We present an example of these single exponential fit data for 2 μM nocodazole in Figure 1.36 (101). For all the nocodazole doses studied, $t_{0.5}$, the time to reach half the total signal change, was invariant, averaging $t_{0.5} = 0.83 \pm 0.14$ h. The exponential decay was 99% complete by $t_{0.99} = 5.52 \pm 0.93$ h, on average. The functional dependence of these data demonstrated that a single dynamic sensing system within the cell, the microtubules, responded to the addition of nocodazole. This fact was confirmed by a fluorescence light microscopy investigation of the cells with increasing nocodazole dose, using a fluorescent antibody to tubulin. Finally, reversibility of the cell QCM biosensor was demonstrated after a number of hours by the return of the frequency and motional resistance shifts to their original values before drug addition. And in another example of the cell QCM biosensor's versatility, this involving cell growth, we demonstrated that the effect of adding fibroblast growth factor protein to stimulate ECs to grow could be followed continuously for up to 4 days (92).

These results indicated that the EC QCM biosensor could be used to detect and study cytoskeletal alterations and dynamics. More broadly, we have previously suggested the novel use of this class of cell QCM biosensor for the real-time identification or screening of all classes of biologically active small-molecule drugs or biological macromolecules that affect cellular attachment, regardless of their molecular mechanism of action (96,98). For example, different cell types could be used that would make the biosensor more sensitive to particular cytoskeleton-affecting drugs or classes of drugs. In a drug screening type of format, cell QCM biosensors could be studied comparing the response of a wild-type cell with

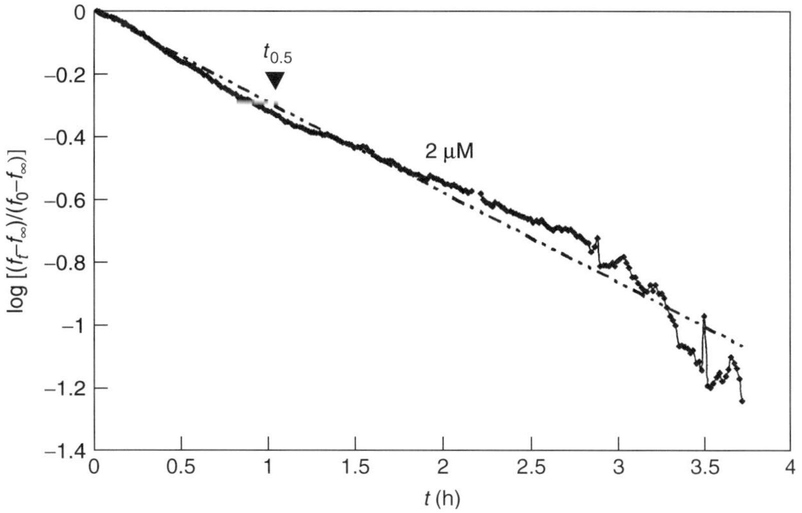

FIGURE 1.36
The logarithm of the fraction of remaining Δf change, $\log[(f_t - f_{\text{infinity}})/(f_0 - f_{\text{infinity}})]$, is plotted for 1-min interval measurements as a function of time for the EC QCM biosensor following addition of 2 μM nocodazole. A single exponential decay fit to the data is presented, as is the position of the calculated half-time for the single exponential decay, $t_{0.5}$. Reprinted from Marx, K.A., Zhou, T., Montrone, A., Braunhut, S. J. (2002). A Quartz Crystal Microbalance Cell Biosensor: Detecting Nocodazole Dependent Microtubule Disruption Dynamics in Living Cells. *Adv. Biomater. Character. Tissue Eng. Complex., Mater. Res. Soc.* 711:125–132. With permission from the Materials Research Society.

mutants of that cell specific for a particular cytoskeletal target. In this way, drugs could be screened for their differential effect upon the particular target within the biological context of these two cell types as the biological components of the biosensor. Finally, clinical cell biopsies might be capable of being screened or diagnosed where these biopsy-derived cells would represent the biological element in cell QCM biosensors.

1.2.3.3 Quartz Crystal Microbalance Cell Biosensor—Removing Cells Yields Intact Extracellular Matrix: A Natural Intelligent Biomaterial With Potential for Creating a Smart Bandage

Many normal cells, including ECs, synthesize an ECM underneath themselves, to which they remain stably attached in vivo. Not only is the ECM a defined structural protein matrix for cell attachment, but it also serves as a repository for many important biological-signaling factors (102). These protein and peptide signaling factors are stored in the ECM in vivo in a stable form until they are needed to be quickly mobilized in response to particular biological demands, such as *repair* needed for injuries or surface wounds. These properties of the ECM impart to them some of the aspects of intelligent materials or systems. It appears that in the case of an injury such as a wound, this abnormal local state is sensed by the ECM at the wound site. In response, the stored biological factors are released to initiate a cascade of cellular and biochemical processes that result in repair of the wound (103). However, the multiple mechanism(s) involved in this process and how the ECM carries out its particular evolved functions is not well understood.

1.2.3.3.1 Isolating and Studying the Extracellular Matrix—A Natural Intelligent Biomaterial

In an early study of ours, normal ECs attached to their underlying ECM on a conducting ITO electrode surface were stimulated by small positive and negative electrochemical potentials (104). We demonstrated that the cells' growth rate and appearance were significantly affected by these different treatments. The negative potential was stimulatory, and the positive potential was inhibitory to cell growth. One possible explanation for these effects was the differential release of biologically active peptide or protein factors from the ECM during the different electrochemical stimulations. Given the importance of the ECM in vivo as a natural biomaterial possessing intelligent properties, we therefore decided to study it devoid of attached cells after first isolating it carefully and intact upon the cell QCM biosensor surface. In these studies, the cell QCM biosensor comprised either ECs or human breast cancer cells (MCF-7), but primarily the former. We studied the relative contributions of the cells and their underlying ECM to the measured QCM Δf and ΔR shift values. The detachment of the ECs from their ECM was carried out using a simple nonproteolytic method involving EGTA complexation of Ca^{2+}. This resulted in disruption of the Ca^{2+}-dependent bonds between the cells' integrin receptors and their binding sites in the underlying ECM. With the cell ECM biosensor, we were able to follow the disruption of the cells' integrin–ECM linkages and thereby monitor the process of cellular detachment (105,106). Representative cell QCM biosensor data for ECs undergoing this process are presented in Figure 1.37. We could apportion the total frequency decrease of the biosensor into contributions from cell attachment to the ECM as well as from the mass of intact underlying ECM following cell removal. That the frequency shift remaining after EC removal does correspond to the native ECM remaining on the QCM surface was confirmed in fluorescence antibody staining experiments by the presence of fibronectin protein, one of the principle structural components of intact ECM.

During the process of cell detachment, we observed an interesting and novel transient increase in visoelastic behavior expressed in the cell QCM biosensor output as a transient

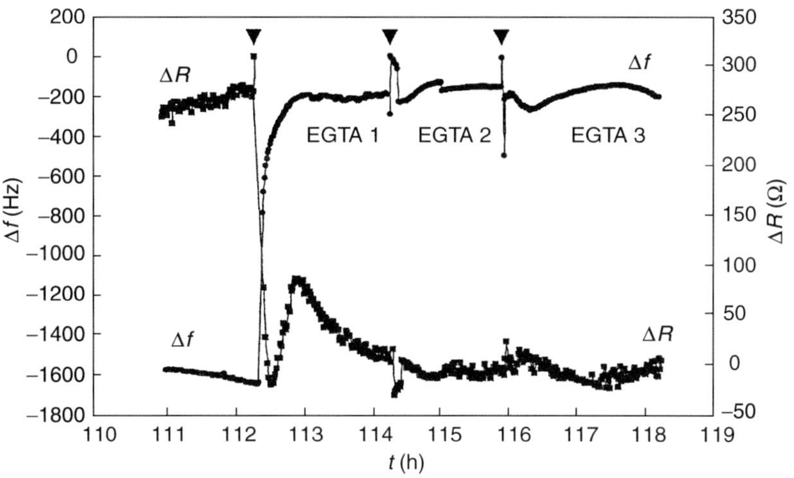

FIGURE 1.37
Time-dependent Δf and ΔR shifts of the EC QCM biosensor formed using 50,000 cells after 112 h of attachment are shown as the attached ECs were removed with three successive additions of EGTA in PBS at three arrowhead positions. Reprinted from Marx, K.A., Zhou, T., Montrone, A., McIntosh, D., Braunhut, S.J. (2005). Extracellular Matrix Studied In Situ With the Quartz Crystal Microbalance: Documenting Cellular Stress During Detachment and the Elastic Behavior of the Matrix. *Anal. Biochem.* 343:23–34. With permission of Elsevier Publishing.

increase in the motional resistance, R, parameter. This occurred during the first EGTA release cycle, as is clear in Figure 1.37. To better understand this phenomenon, we showed that the transient viscoelastic ΔR increase correlated with cellular stress exhibited by the cells in response to their removal from the ECM surface with EGTA. The subsequent decrease in ΔR following its transient rise corresponds to cells finally losing their attachment and lifting from the ECM, which results in lower energy dissipation at the crystal surface. Using the ECM intact on the QCM biosensor surface, we demonstrated as shown in Figure 1.38 that this protein matrix, from which all cells had been removed, did not exhibit any significant level of energy dissipation, exhibiting constant R or elastic mass behavior, in contrast to the ECM complexed with attached cells (Figure 1.31).

1.2.3.3.2 *Releasing Wound-Healing Factors by Potential Stimulation of the Acellular Extracellular Matrix—The Smart Bandage Concept*

Having described in the previous subsection the method for producing intact ECM on the QCM biosensor surface, we carried out the following experiment utilizing this intelligent biomaterial's properties that evolved for use in response to injury or the need for repair. Since the ECM is thought to store biological factors that participate in surface wound healing at the site of injury, we decided to determine whether these protein factors could be released by applying an electrochemical potential to the upper gold electrode upon which the ECM, without any cells, was adhering. To carry out this experiment, we took advantage of the EQCM feature of the biosensor device, creating a three-electrode electrochemical cell, as we have previously shown in Figure 1.26. In this set-up the working electrode was the upper gold electrode upon which we grew cells, which were removed using the EGTA procedure, and to which the acellular ECM remained intact. A Pt counterelectrode and Ag/AgCl reference electrode completed the three-electrode set-up. By applying a −0.3 V potential to the ECM on the working electrode in this cell in the presence of phosphate buffered saline solution, we were able to release protein, peptide, or other biologically active factors that promoted accelerated wound healing in a mouse surface

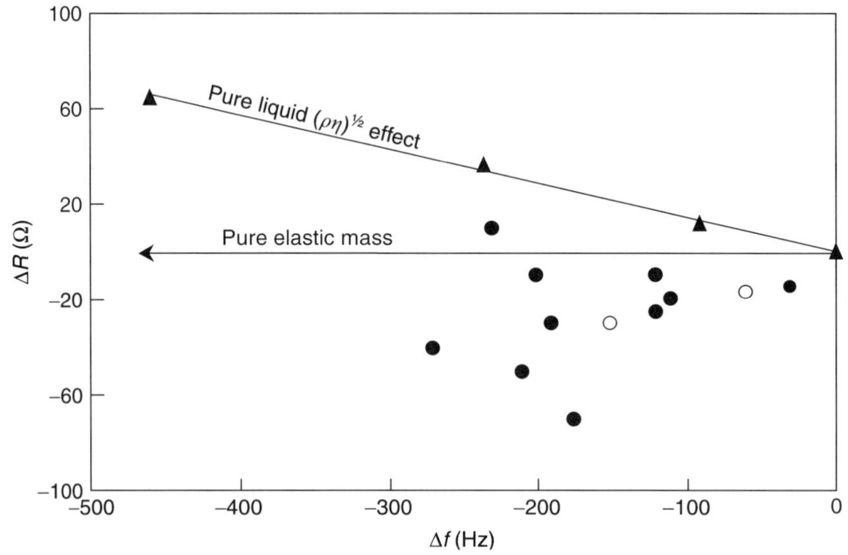

FIGURE 1.38
A Δf–ΔR representation of the shift values for a series of different biosensors due to ECM remaining on the QCM surface following complete cell detachment from each of the QCM biosensors. The ECM from ECs are filled circles; the ECM from MCF-7 cells are unfilled circles. The Δf–ΔR behavior of a pure elastic mass bound at the QCM surface is indicated by the zero-slope horizontal line. The Δf–ΔR triangle points determined experimentally for a series of increasing sucrose concentration solutions is depicted with a best linear fit line. This represents the behavior of a Newtonian fluid producing a pure $(\rho\eta)^{1/2}$ density–viscosity effect on the QCM Δf–ΔR values. Reprinted from Marx, K.A., Zhou, T., Montrone, A., McIntosh, D., Braunhut, S.J. (2005). Extracellular Matrix Studied In Situ With the Quartz Crystal Microbalance: Documenting Cellular Stress During Detachment and the Elastic Behavior of the Matrix. *Anal. Biochem.* 343:23–34. With permission of Elsevier Publishing.

wound-healing model system. These data are presented in Figure 1.39. Nearly a twofold acceleration of the wound healing rate was evident by day 9, after only a single administration of the solution containing factors released from the ECM (107,108). Since they are present at a very low concentration, the identity of the active species involved in the accelerated wound healing we observed is not currently known. However, experiments are underway to discover their identity. These experiments involving the intelligent biomaterial, ECM, suggest that electrochemical potential stimulation of an ECM-containing system could be created to accelerate wound healing for use in human injuries. Experiments are underway to develop this concept, which we term the 'smart bandage', into a practical device.

1.3 Simulating the Properties of DNA—Potential for Predicting Intelligent Properties and Evaluating Their Role in Biosensors

1.3.1 Intelligent Properties of DNA

Biological macromolecules, their cellular assemblies, as well as living cells and whole organisms possess intelligent properties. In this section, we present a discussion focused on the intelligent properties of the DNA molecule. DNA plays the central role in the

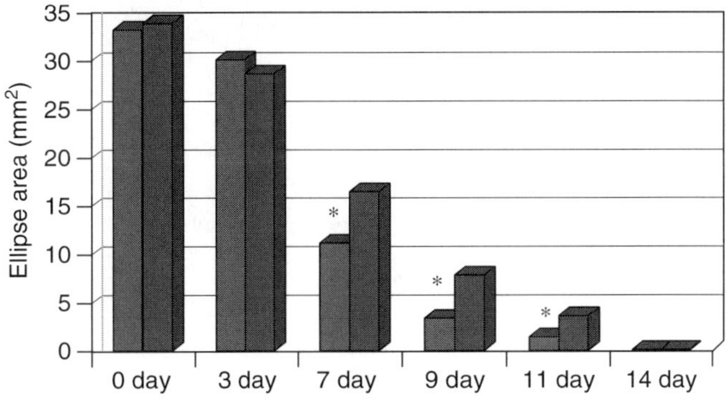

FIGURE 1.39
Histogram of the measured surface areas of mouse wounds measured at various times during healing. Wounds on a number of mice at day 0 were treated with a single application of either PBS releasate from 1 h –0.3 V stimulated ECM (1st bar) or unstimulated control PBS (2nd bar). The '*' indicates healing times where the two conditions have achieved wound-healing areas that are significantly different at the $p < 0.05$ confidence level. Reprinted from Braunhut, S.J., McIntosh, D., Vorotnikova, K., Zhou, T., Marx, K.A. (2004). Development of a Smart Bandage Applying Electrical Potential to Selectively Release Wound Healing Growth Factors from Cell Free Extracellular Matrix. In: Architecture and Applications of Biomaterials and Biomolecular Materials, *Proc. Mater. Res. Soc.*, EXS-1:403–405. With permission from Elsevier Publishing.

information storage, flow, and control functions within all living cells. Our aim in the following discussion is to illuminate the advantages of being able to simulate the intelligent properties of molecules such as DNA, prior to their incorporation and testing in biosensors. In the Center for Intelligent Biomaterials at the University of Massachusetts Lowell, we have focused efforts on characterizing and simulating certain sequence-specific properties of DNA as well as attempting to understand DNA–protein complex stability by investigating computed physical property parameters of the underlying DNA sequence. This is a far easier task than modeling DNA structure and dynamics using all atom molecular simulation methods on specific sequences. Such methods provide far greater detail about the system, but they are greatly limited by computational constraints on the system size and the time scale that can be modeled. With the physical property parameter approach, we hope to eventually understand at a meta-level the basis for the intelligent properties of DNA. Such an understanding will provide insight as well as predictive capabilities and allow the rational design of specific DNA and DNA–protein complexes into smart biosensor systems.

The DNA double helix is a regular repeating 3-D structure comprising two right-handed interwound polynucleotide single strands. Each single strand contains a linear string of nitrogenous bases that form complementary hydrogen bonds to the corresponding bases on the opposite single strand. Together, these bases create the A-T and C-G base pairs comprising the central core of the double helix. At the time of the original publication by Watson & Crick of the fiber diffraction-based 3-D structure of DNA, and for the following two decades, little structural variability was imagined for the DNA double helix. However, we know today primarily from the numerous determinations of 3-D structures for short DNAs by NMR and x-ray crystallography that DNA forms a variety of repeating secondary structure types, some of which are interconvertable. We also know that DNA, while still considered an overall rigid structure on the length scale of tens of base pairs, and with a persistence length in solution on the order of <200 bp, is known to be capable of considerable generic as well as sequence-based variation in structure and dynamic motion (109,110). The sequence-based properties and dynamics of the DNA double helix underlie

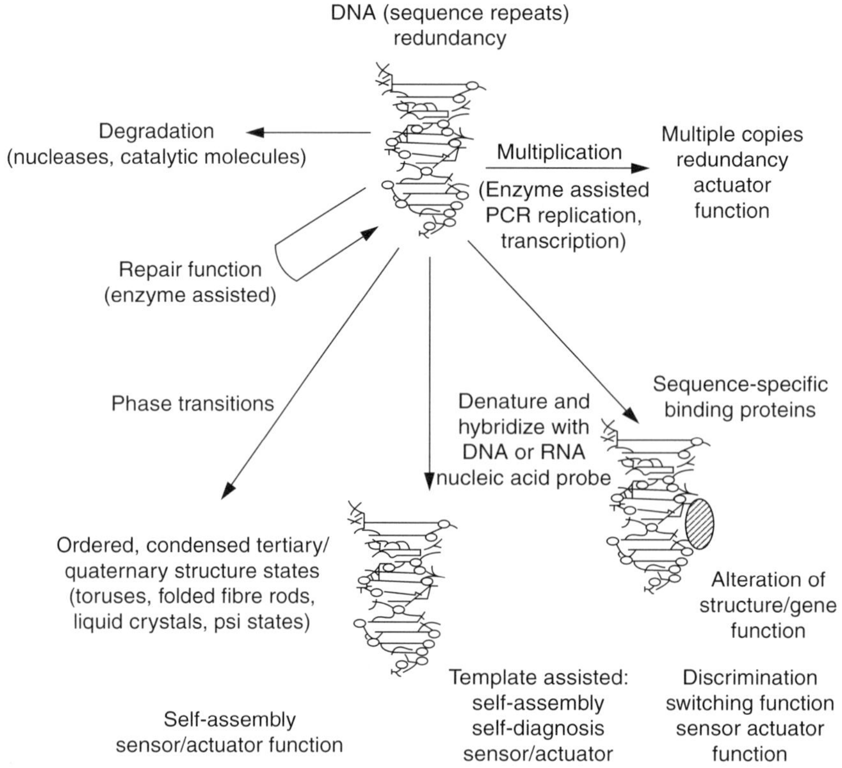

FIGURE 1.40
Schematic representation of the intelligent materials properties of DNA and its complexes with other molecules is presented. Reprinted from Marx, K.A., Lim, J.O., Minehan, D., Pande, R., Kamath, M, Tripathy, S., Kaplan, D. (1994). Intelligent Materials Properties of DNA and Strategies for It's Incorporation into Electroactive Polymeric Thin Film Systems. *J. Intelligent Mater. Systems Struct.* 5:447–454. With permission of Sage Publications.

its intelligent properties, a number of which we present schematically in Figure 1.40. For emphasis in the discussion that follows, we have italicized the individual intelligent properties of DNA as we discuss them.

The evolved DNA sequences of any organism (its genome) represent the genetic program that orchestrates the intelligent properties of that cell or whole organism. In humans, the genome is comprised of around 3×10^9 base pairs organized into 23 long linear DNA pieces called chromosomes. In normal cells, two copies of each of the 23 chromosomes are found within the nucleus of every cell. With the success of the international Human Genome Project the genomes of more than 100 organisms, including humans, have been successfully sequenced (111). In simple terms, this means that we know the linear order of bases, the information content, along each DNA strand in the chromosomes of those organisms. What is more difficult to know with certainty is the answer to the first question of interest once this sequence is in hand. What is the exact location in the DNA sequence of all of the genes? This is an important question to ask since genes possess the protein coding information necessary for making all the proteins that carry out their myriad of functions in the cell. Fortunately, this is a tractable problem that is currently being solved. However, a far more difficult task is to understand how the DNA comprising the genome is regulated. That is, when and in which cell types are specific genes copied to make proteins? These gene regulation processes occur via highly organized control mechanisms involving regulatory proteins and are the means by which intelligent properties are exhibited by living cells and whole organisms. With this genomic regulatory information in hand, the modern techniques of Molecular Biology make possible the rational modification and augmentation of the intelligent properties of DNA and, as a result, its host organism.

In the cell, DNA is acted upon by numerous sequence-specific binding proteins as well as enzymes that have evolved to carry out specific functions. These functions form part of the array of intelligent properties that DNA possesses, as Figure 1.40 depicts in a schematic fashion. In vivo, DNA is capable of copy number *multiplication* by multiprotein enzymatic processes. In the laboratory setting, investigator-determined selective DNA sequence geometric amplification is routinely carried out by the polymerase chain reaction (PCR) technology. Specific *redundancy* features can be designed into DNA regions as repeating sequence motifs by taking advantage of modern cloning and PCR amplification techniques. Through the action of specific repair protein systems, DNA is capable of *self-diagnosis* and *repair* functions that restore its native structure and sequence following mutational errors. This occurs via the repair protein systems that have evolved to discriminate correct from incorrect (mutations) base complementarity features of the double helical structure. In contrast, there exist specific degradative enzymes that are capable of completely degrading DNA, or other enzymes that can be used to make desired covalent modifications to native DNA. One example of a naturally occurring modification is enzymatic methylation at the 5-position of the cytosine base located in the DNA major groove. This DNA modification occurs naturally in bacteria and helps to protect those genomes from bacteriophage (viruses that infect bacteria) infection. In higher organisms, methylation is used to regulate mRNA transcription of specific genes (112).

DNA forms the molecular basis of stored evolutionary control information at the cellular and organism levels. Therefore, in the broadest sense, DNA is part of a macromolecular system capable of *learning or adaptation* as a consequence of evolutionary molecular mechanisms. That DNA is the organism's repository of stored evolutionary information has been known and accepted for decades. However, only within the past decade have large numbers of DNA sequences in any given organism been characterized directly and at the highest resolution by high-throughput sequencing. This detailed sequence knowledge has allowed DNA to be characterized from the perspective of Information Theory. As measured by entropylike metrics such as the Mutual Information Function, DNA in the gene-containing regions from many genomes has been shown to have a significantly higher *information* content than DNA in nongene-containing regions (113). This general property is exactly the behavior one would expect since genes represent the specific information repositories for protein sequences. The Mutual Information Function is a logarithmic-based measure of the length-dependent persistence of base correlations (information) at all pairwise neighbor positions along a DNA sequence. It can be calculated in different DNA base reading frames and averaged. As shown in Figure 1.41, this information correlation is high and persists over long sequence lengths in primate exon- or gene-containing regions, but is low in magnitude at all sequence lengths in primate nongene or intron regions. There is nearly a threefold higher average mutual information value for exons compared with introns. Another striking feature is the marked three-base periodicity of the exon mutual information. This results from the nonuniform frequency of base usage in the three-base frame positions of the nonoverlapping codons that comprise the exon's protein coding information content. This mutual information periodicity is completely absent in the intron sequences that lack any codon structure. Given this threefold difference in mutual information, approaches like the Mutual Information Function metric can serve as the basis for classifying raw DNA sequences by functional DNA sequence category—coding vs. noncoding. In fact, we have carried out such classifications of human coding and noncoding sequences using Information Theory-related approaches with a relatively high accuracy of

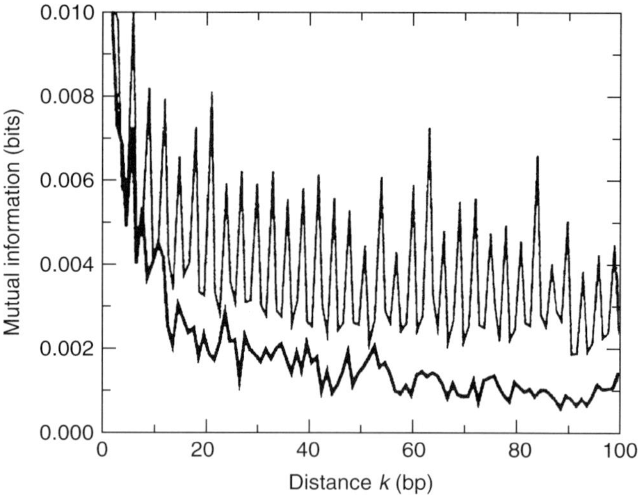

FIGURE 1.41
Mutual Information calculated for primate exons (upper line) and for primate introns (lower line) longer than 500 base pairs (bp) as a function of separation length k bp. Using a finite length correction, each DNA sequence was cut into a 500-bp piece and the mutual information function value (DNA codon frame) calculated and position averaged to present the arithmetic mean of the functions. Reprinted from Grosse, I., Herzel, H., Buldyrev, S.V., Stanley, H.E. (2000). Species Independence of Mutual Information in Coding and Noncoding DNA. *Phys. Rev. E* 61:5624–5629. With permission of the American Physical Society.

correct classification (114). However, as of yet, there is no algorithmic method that can, a priori, predict all of the gene sequence locations within a string of raw nonannotated DNA from any organism. Therefore, the identification of all genes in an organism must still be made ultimately by experimental-based approaches. Once genes are defined, a considerably more difficult problem remains of how they are regulated within cells to carry out their specific functions. Then, once all of the important regulatory control elements have been identified for a given gene, the intelligent properties of *prediction* and *notification* can be designed into specific DNA systems containing that gene, usually via combination of the DNA with other small ligands or sequence-specific recognition and control proteins that include *reporter* functions, allowing the gene system to function as part of a biosensor.

1.3.2 *Redundancy* of Single Base Repeating Tracts—The Simplest Repeating Sequences

For decades preceding the Human Genome Project, human DNA and many other eukaryotic DNAs (higher organisms that possess a distinct nucleus) had been known to possess repetitive sequences of varying unit lengths and repetition frequencies. Therefore, to a first approximation, naturally occurring DNA possesses the property of *redundancy*. Historically, repetitive sequences in eukaryotic organisms were first discovered through analysis of the organism's DNA reassociation kinetics following DNA denaturation (115). These studies showed that human DNA repetitive sequences were not remarkable, in that they resembled sequence repetition frequency distributions from many other organisms (116,117). More recently, higher resolution views of repetitive sequences have been obtained through restriction enzyme analysis and direct DNA sequencing. By both methods, a range of repetitive sequences of varying repeat lengths (unit repeat sizes from as few as one to hundreds of thousands of base pairs) and repetition frequencies (from a few to millions of copies/genome) have been characterized in a wide range of eukaryotic organisms, including humans.

The simplest possible repeating sequences in any DNA genome are the homopolymer repeats (tracts)—where a single base is repeated on one single strand and its base complement repeats on the opposite strand. In addition to the obvious minimal information content within homopolymer tracts, these sequences have been found to possess unusual secondary structures and physical properties (118). These and other short repeats, 2–6 bp in unit length, are thought to arise within organisms to their equilibrium frequency distributions by a mechanism involving DNA single-strand slippage during replication combined with point mutations (119). We have studied the occurrence of these simple repeats in the DNA sequences found in the Genbank public databases (111) for a large population of eukaryotes—27 organisms of (G + C)% varying across the range from about 20 to 60% (120). In all the eukaryotes examined, we found that all four of the homopolymer tract types beginning at lengths of 3 bp and greater are found at significantly higher normalized frequencies/base pair (bp) of total DNA than would be expected based upon random occurrence frequencies in a randomized sequence of the equivalent base composition. This trend terminates at around lengths of 5–6 bp for G and C tracts where longer tracts are simply not observed. But the trend increases in magnitude for A and T tracts of length longer than 3 bp. Specific examples of these long tracts are shown underlined in Figure 1.42 where we present an intron sequence occurring within the genome of the very (A + T)% base-rich organism *Dictyostelium discoideum*. This sequence contains

```
tcgattttaa agattggttt ttttttttta attattttt ttttttttt tttttattt
tttgaaaaaa tttaattacg ccgttttaat cgttcggcaa attttaaaaa atttattttt
tttttatttt tatttttttt ttactttttt tttttaacg gaaaaaggga aaaaaaaaaa
aaaaaaaaaa aaaaaaaaaa ttaaaactcg ttgtgtgtgt gtttgaaaat aataataata
ataataataa
```

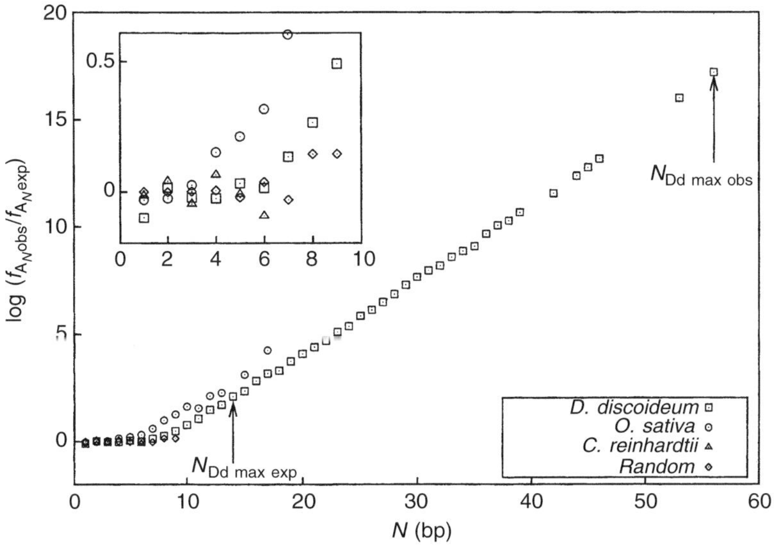

FIGURE 1.42

Homopolymer A_N tract repetitive sequences present at significantly greater than expected frequencies in eukaryotes. At the top we present an example of an intron sequence found in the extremely (A + T)% base-rich organism *Dictyostelium discoideum*, the slime mold, that contains two long homopolymer tracts (underlined). The graph presents homopolymer A tract results calculated for three different eukaryotes (symbol box) compared with random generated sequences. The *Y*-axis presents the $\log(fA_{Nobs}/fA_{Nexp})$, the logarithm of the calculated observed frequency/base pair of A tracts of size N found in the real sequence population divided by the frequency/base pair of A tracts of size *N* expected for a random sequence of base composition equivalent to the real sequences. By this metric, an observed A tract frequency identical to the expected frequency would be at 0 on the *Y*-axis, exactly the case observed for low *N* values for the three organisms. The inset shows the low-*N* region in greater detail. The $N_{Dd\ max\ exp}$ point indicates the maximum A tract size expected for a random sequence genome of the base composition of *D. discoideum*. The $N_{Dd\ max\ obs}$ point indicates the actual maximum A tract size we observed in the real *D. discoideum* sequence population. Reprinted from Zhou, Y., Bizzaro, J. W., Marx, K.A. (2004). Homopolymer Tract Length Dependent Enrichments in Functional Regions of 27 Eukaryotes and their Novel Dependence on the Organism DNA (G+C)% Composition. *BMC Genom.* 5:95–106. With permission of National Academy of Science and authors.

both a long T (22 mer) and a long A (30 mer) tract. As the Figure 1.42 graph inset clearly shows, by around a length of 4–8 bp the frequency/bp for A tracts from three different eukaryotes has significantly increased relative to frequencies expected for random tract occurrence in an organism comprised of random sequences of the same overall base composition. However, for *D. discoideum* sequences the relative frequency increase is remarkable. Tracts of A above 40 bp length were observed, with two A tracts even greater than 50 bp, representing a frequency/bp that is 10^{15}-fold above that expected to occur within random sequence DNA of equivalent base composition (120,121). In this study, nearly as long a tract distribution was observed within the genome of the (A + T)% base-rich organism *Plasmodium falciparum* (the malaria parasite). The *redundancy* built into the occurrence of these homopolymer tracts has specific functions in the DNA of these organisms that is just beginning to be elucidated. Some of the functions may result from the unusual structures of these DNA tracts. Short A and T run homopolymer tracts (3–5 bp) are known to exhibit helix axis bending of around 15–20° (122). In some biological situations, such as in the kinteoplast DNA of primitive eukaryotes, these tract bends are found in phase with the turn of the helix at lengths of 10–11 bp, resulting in the large-scale macroscopic bending of long stretches of DNA hundreds of base pairs in length (123). More commonly observed functions for A and T homopolymer tracts at lengths of 10 bp or greater are the transcriptional regulation of genes (124), where tract-specific recognition proteins bind the tracts and act as promoters of specific gene transcription.

We have also observed that the longer A and T tracts (>8–10 bp) are not located randomly in genomic DNA but are found on average to be spaced at preferred distances apart. In the case of *D. discoideum* (125) and a few other eukaryotes (126), this spacing corresponds to the measured average spacing of the organism's nucleosomes, the repeating nucleoprotein structural complex found in the chromosomes of all eukaryotic organisms. This fact suggests that these homopolymer tracts may exist within specific cellular chromosomal packaging environments in the cell as well as possess functions that are distinct from the average DNA sequence found in every organism.

1.3.3 DNA Tertiary Structure *Self-Assembly*—Counterion Condensation Drives Intramolecular DNA Collapse and Helps Determine Electrophoretic Mobilities

Self-assembly is a property of DNA that is exhibited in a number of different ways. Fundamentally, it underlies the double helical structure of DNA, via the molecular recognition involved in complementary base pairing between single strands. This is taken advantage of in countless ways in the molecular biology research laboratory and in the biotechnology industry, through the use of DNA hybridization assays in various research protocols. The DNA hybridization process has also been incorporated into the design of biosensors. A common strategy is to hybridize a single-stranded sequence probe, often immobilized onto a solid support, with an analyte single-stranded sequence from a sample where knowledge of the presence and sometimes concentration of the nucleic acid is desired. This *self-assembly* hybridization strategy is *discriminating in its recognition* and sensor function. The effect of increasing base mismatch level in imperfectly matched sequence hybrids produces thermodynamically unstable hybrids that are not sensed. This feature is typically taken advantage of in biosensor systems and provides them with a *discrimination* capability for detection of only the perfectly matched correct analyte sequence of interest. As an example, biosensor systems incorporating a hybridization strategy can result in detection of human pathogens possessing DNA sequences unique from our own. Therefore, hybridization-based biosensors can be used to identify infections in patient samples (127).

Self-assembly is also involved in certain tertiary-structure DNA phase transitions. At low DNA concentrations, these phase transitions involve intramolecular collapse from the extended wormlike coil structure of DNA in solution to condensed insoluble states characterized by maximized close packing of DNA double helical segments. While normally quite soluble in aqueous solution, the DNA polyanion can be driven to condense in the presence of a critical solution concentration of trivalent cations (128). These *self-assembled* charge-neutralized DNAs possess a high degree of order. They are similar in dimension and are therefore thought to resemble the collapsed state of DNA within native bacteriophage heads and certain viral capsids (129). Self-assembly of DNA to condensed states is also of more recent interest as potential packaging states for gene therapy delivery systems (130), where polymeric and lipid-based cations have been shown to induce collapse of DNA into aggregates that cells readily take up from solution. At high concentrations and under appropriate solute conditions, DNA can also be driven to *self-assemble* via an intermolecular process into liquid crystalline ordered phases (131).

In a number of published experimental studies, we have reported on the *self-assembly* process involving trivalent counterion-induced intramolecular DNA collapse. This is a general physical change of state induced by trivalent cations and is largely independent of the size and specific DNA sequence used. Condensed DNA tertiary structures are produced when a critical level of DNA charge neutralization, around 88–90%, is achieved by any combination of solution cations (128). This charge neutralization level is consistent with Manning's counterion condensation theory of polyelectrolytes (132). In a series of studies (133–136), we have shown by high-resolution transmission electron microscopy (TEM) that condensed DNA forms toroidal-shaped structures about the size of a bacteriophage head

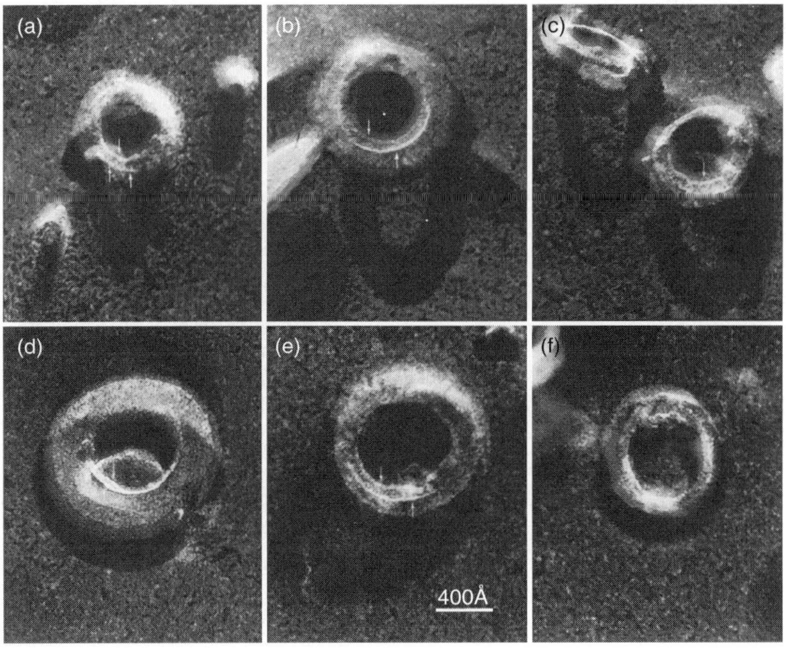

FIGURE 1.43
Toroidal DNA condensates formed by condensing $\phi\chi$-174 RFII:*Xho*I-digested linear plasmid DNA with the organic trivalent cation spermidine^{3+}. Images are from transmission electron microscopy (TEM) carried out on freeze fracture-deep etched specimens with low metal deposition level (Pt/C 0.9 nm thick). Reprinted from Marx, K.A., Ruben, G.C. (1986). A Study of PhiX-174 DNA Torus and Lambda DNA Torus Tertiary Structure and the Implications for DNA Self-Assembly. *J. Biomol. Struct. Dynamics* 4:23–39. With permission from Adenine Press.

some 100 nm in diameter, which are comprised circumferentially wrapped DNA. In Figure 1.43, we present a number of examples of toroidal-shaped condensed linear $\phi\chi$-174 RFII:*Xho*I-digested plasmid DNAs (5386 bp) that were condensed by the biological trivalent cation spermidine^{3+}. The structures were visualized by TEM after preparation by the freeze-fracture, deep-etch technique to yield Pt-shadowed replicas of the hydrated frozen toroidal DNA specimens. In these images, individual DNA segments, highlighted with arrows, can be observed wrapping around the toroidal condensates. Using stereoscopic techniques, we were also able to map the *z*-axis topology of individual DNA strands wrapping circumferentially in different toroidal DNA condensates (137). As Figure 1.44 shows, with this technique we could map the relative *z*-axis heights of interesting DNA strand features on the torus surface as well as features on the ice substrate surface.

Different toroidal DNA systems of varying length and topological state (calf thymus DNA, bacteriophage λ DNA, $\phi\chi$-174 bacteriophage linear and circular DNAs) were also studied using nonspecific deoxyribonuclease enzyme cleavage of the DNAs (138–141). These studies invariably showed an arithmetic length series of fragments generated by digestion of spermidine^{3+}–DNA condensates from different organisms. Figure 1.45 presents data on the DNA fragment sizes resulting from different micrococcal nuclease digestion times for spermidine^{3+} condensed linear ϕX-174 RFII:*Xho*I-digested plasmid DNAs. Following digestion, the fragment sizes were obtained relative to molecular weight standards by electrophoresis on agarose gels. Extrapolation of each fragment's digestion time to 0 provides a biochemical estimate, immediately prior to digestion, of the arithmetic fragment sizes corresponding to integer multiples of DNA length wrapping around the toroidal DNA condensates (~800 bp: monomer; 1600 bp: dimer; and 2400 bp: trimer). The extrapolated monomer DNA length around 800 bp is exactly the length needed to wrap once around the average toroidal circumference measured from the TEM images we presented for these DNA toroidal condensates in Figure 1.43. To account for these digestion data, we proposed a general model for how an arithmetic series of DNA fragments can be generated in these experiments (134,138). By applying a topological argument to the micrococcal nuclease cleavage study results of the linear and circular DNA–spermidine^{3+}

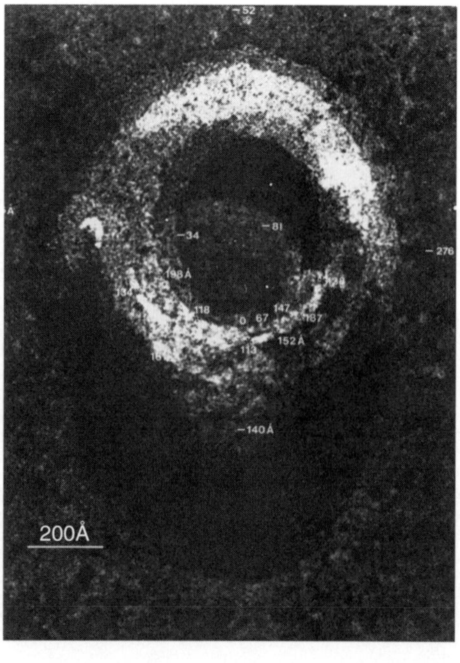

FIGURE 1.44
Stereoscopic measurements performed on a toroidal spermidine^{3+}-calf thymus DNA condensate using two 10° tilted TEM images for the stereoscopic measurements as described in and reprinted from Ruben, G.C., Marx, K.A. (1984). Parallax Measurements on Stereomicrographs of Hydrated Single Molecules: Their Accuracy and Precision at High Magnification. *J. Microsc. Res. Tech.* 1:373–385. With permission of Wiley-Liss, Inc., a subsidiary of John Wiley & Sons, Inc.

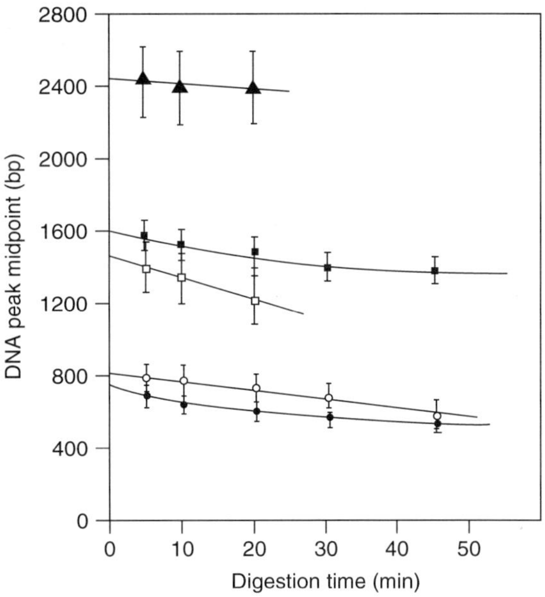

FIGURE 1.45
DNA fragment peak sizes as a function of digestion time with micrococcal nuclease. The agarose gel fragment midpoints are plotted vs. digestion time for the monomers (circles), dimers (squares), and trimers (triangle) resulting from spermidine^{3+}-condensed $\phi\chi$-174 DNA, the RFII circular (filled symbols) and the linear *Xho*I-digested (open symbols) complexes. Reprinted from Marx, K.A., Reynolds, T.C. (1982) Spermidine-Condensed PhiX-174 DNA Cleavage by Micrococcal Nuclease: Torus Cleavage Model and Evidence for Unidirectional Circumferential DNA Wrapping. *Proc. Natl. Acad. Sci. USA* 79:6484–6488. With permission of National Academy of Science, copyright 1982, and authors.

condensates, combined with the TEM results, these data presented strong evidence for a model of toroidal-shaped DNA structure formation by a continuous circumferential wrapping of DNA (138).

We have also used Manning counterion condensation theory (132) to explain the ion concentration dependence of other experimental DNA physical properties, such as DNA end-to-end distance changes and the electrophoretic mobilities of noncollapsed DNA in solution. This was done by a modification of the Manning Theory to apply to two different cations, monovalent and divalent, that occur in the electrophoresis experiments below the critical counterion condensation levels needed for DNA collapse to toroidal structures. We discovered that the measured length-dependent electrophoretic properties could be accounted for by the magnitude of the DNA phosphate charge fraction not neutralized as determined from the calculations (142–145).

Recently, the Seeman lab has pioneered the development of new kinds of self-assembling regular geometric DNA nanoscale structures. They have been formed from short synthetic DNA sequence constructs by a combination of the concepts of DNA self-assembly along with rationally designed and positioned sequence complementarity. The structures created include DNA cubes, octahedrons, and dimorphic structures capable of being switched between states by changes in counterion type and concentration. They also include tiling structures that can serve as self-assembling DNA computing systems (146,147). These novel structures with their variable self-assembly properties have the potential to serve as new types of biological elements in smart biosensors.

1.3.4 Computational Simulation of DNA Melting—Reversing the *Self-Assembly* Process of the Double Helix

Melting is an order–disorder phase transition of the DNA double helix that can be driven to the disordered single-stranded state by increasing the temperature. The melting transition can be characterized most simply by its T_m, the melting temperature at the transition midpoint. Measured primarily by optical methods, T_m values have been obtained for innumerable whole genomic DNAs, genomic DNA fractions, and short synthetic DNA duplexes. DNA melting carried out experimentally was one of the early biophysical tools used to characterize genomic DNA (148), before the development of the modern techniques of molecular biology such as restriction enzyme analysis, cloning, and fast sequence analysis. With the advent of solid-phase DNA synthesis and PCR technology in the 1980s, the use of short DNA single-stranded sequences, ranging from around 10–300 nt in size, has rapidly increased. These short DNA sequences are in widespread use in both research and diagnostic applications as biosensor probes. They can be used to detect the occurrence and often the concentration of DNA from an infectious agent or abnormal tissue state found in cells or patient samples in clinical settings (149). The thermodynamic stability, represented by the T_m of the hybrids formed in these assays as well as that of the short DNA duplexes synthesized in the ubiquitous PCR reaction, are important quantities for investigators to estimate or to calculate for use in experimental protocols. Typically, these T_m values are determined experimentally or are estimated using nonrigorous rules employing the simple base composition of the sequence as input in the estimation.

It has been known for some time that experimental DNA melting could be simulated using a statistical mechanical approach. The Poland–Scheraga algorithm is a methodology that was developed to describe the energetically cooperative process of equilibrium DNA melting. This approach uses a two-state model where a series of fundamental equilibrium energetic states, shown in Figure 1.46, represent the complex process of long DNA sequences of different local base composition (melting blocks), melting as thermodynamic blocks at different T_m values (150). In the program MELTSIM, this algorithm has been implemented and parameterized for long DNA sequences with the experimentally determined thermodynamic quantities for all 16 [(4 possible bases)^2] of the possible DNA dinucleotide pairs (151).

At the Center for Intelligent Biomaterials, we have utilized MELTSIM to study a number of features of the melting of long DNA sequences. In the broadest genomic survey conducted to date, we used MELTSIM to study the melting behavior of whole chromosomes and populations of long single gene-containing sequences from 28 eukaryotic organisms covering the entire (G + C)% composition range from about 20 to 60% (152). For over 8000 sequences in these organisms, we demonstrated a high correlation between each sequence's calculated T_m value and its fraction (G + C) base composition. Through a series of melting simulations of control sequences created as a result of

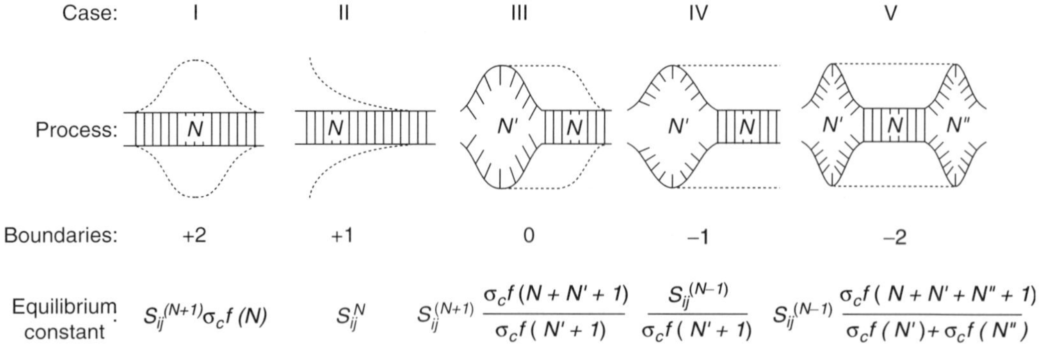

FIGURE 1.46
Schematic representation of five of the six cases of thermal melting processes of double helical DNA, with changes in the numbers of boundaries and equilibrium constants indicated. Reprinted from Blake, R.D., Bizzaro, J.W., Blake, J.D., Day, G.R., Delcourt, S.G., Knowles, J., Marx, K.A., SantaLucia, Jr., J. (1999). Statistical Mechanical Simulation of Polymeric DNA Melting With MELTSIM. *Bioinformatics* 15:370–375. With permission of Oxford University Press.

various shufflings of the original sequence's bases to new positions, we showed that nearest-neighbor dinucleotide pair frequency conservation alone was sufficient to account for nearly all of the average T_m melting behavior of the real sequences. This is in agreement with the accepted dependence observed in experimental DNA melting. However, rather than the single classic Marmur–Doty linear relationship often cited in the literature (148,151) to exist for all DNAs, we demonstrated that different relationships existed, one for exon sequences (coding regions) and another for intron sequences (noncoding regions). These two sequence classes were statistically different by 6.6% in their slopes. These two dependencies could be explained by their slightly different biases in nearest-neighbor dinucleotide pair frequency distributions. This study helped establish the widespread utility of MELTSIM to accurately simulate the sequence dependence of DNA melting behavior.

In 1996, *Saccharomyces cerevisiae* (baker's yeast) was the first eukaryotic genome (12,067,277 bp) to be fully sequenced (153). We have shown that MELTSIM accurately simulates the experimentally determined T_m value of this entire genome (154). An important feature of MELTSIM allows the simulation and display of the positional melting behavior along any DNA sequence. We have used this positional mapping feature to demonstrate melting behavior along the 16 individual yeast chromosomes (154,155). As Figure 1.47 demonstrates, the positional mapping feature of MELTSIM can also reveal the positions of block DNA melting exhibited, for example, within a single gene: in this case a diphosphatase enzyme from *D. discoideum*. There is a close correspondence between the positions of introns and UTR regions of the diphosphatase gene and low-melting temperature block behavior. Exon positions correspond to the high-melting block regions of the diphosphatase gene. These block melting data are consistent with the known base composition and dinucleotide composition of exons and introns and their correlations with DNA melting temperature (121,151,152). This exon vs. intron discrimination within the diphosphatase gene is just one of a number of such cases we identified within sequences of the (A + T)% base -rich *D. discoideum* (156,157) and *Schistosoma mansoni* (158) genomic DNAs based upon their differential positional melting behaviors. Such behavior is likely to be a general property of (A + T)% base-rich eukaryotic genomes.

We believe that the ability to simulate the melting behavior of long DNA sequences using MELTSIM has important potential uses in the understanding of DNA stability for sequences of considerable length incorporated into biosensors. More complex biosensor systems will eventually incorporate longer DNA sequences, lengths for which T_m calculation programs designed for short oligonucleotide lengths are not appropriately parameterized (159). Given that the thermodynamic equilibrium block melting size for DNA can be as little as 100 bp, we believe that in the future the understanding of complex melting will be necessary to successfully deploy the intelligent properties of long DNA sequences within biosensors. As a practical example of this importance, we describe the current situation with microarrays. Nucleic acid-based microarrays are recognized as increasingly important measurement tools in the basic research, biotechnology, and pharmaceutical industries for characterization of the complex mRNA transcriptional status of tens of thousands of genes within cells and tissues (160). It is perhaps not widely appreciated that these microarrays are fundamentally biosensors. They rely typically upon single-stranded nucleic acid probe sequences being bound to small spotted areas [$(\sim 70\ \mu m)^2$ area], one sequence/spot, upon a suitable substrate. The spots are grouped into large-array formats (tens of thousands of spots) that allow for simultaneous detection and quantitation of huge numbers of gene product mRNA sequences. Following total mRNA isolation from a biological sample of interest, fluorescent chromophore-labeled cDNA copies of these gene products are typically placed upon the

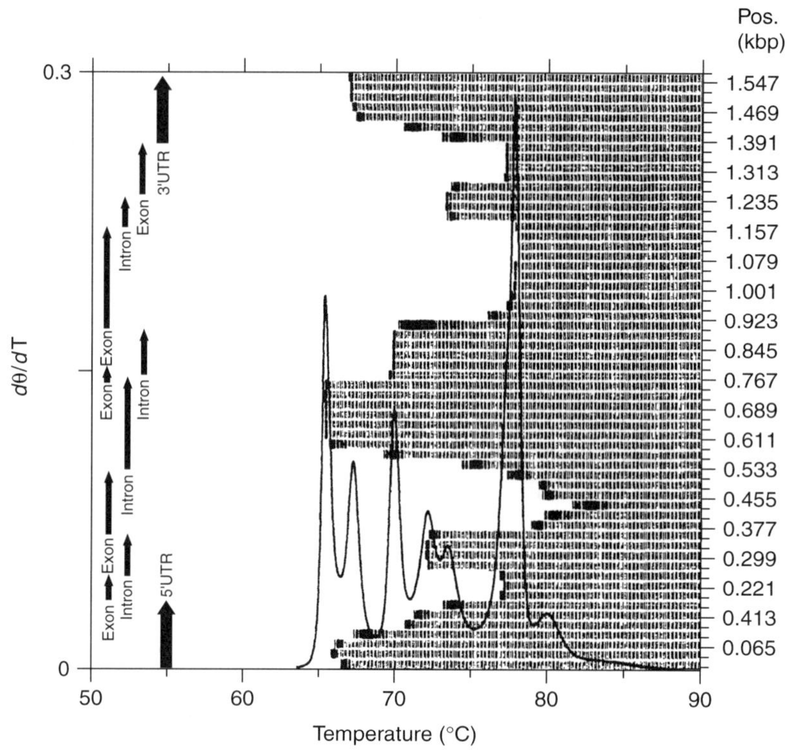

FIGURE 1.47
MELTSIM simulated first-derivative melting plot of the fraction theta of the *D. discoideum* diphosphatase gene melted (left-hand *Y*-axis) as a function of temperature presented as the solid line. The right-hand *Y*-axis shows the positions along the diphosphatase gene, as does the exon and intron locations described by the labeled vertical arrows on the left-hand portion of the figure. The positional melting simulation results along the diphosphatase gene are indicated by the temperatures of the leftward bolded edges of the shaded areas that correspond to the melting at that gene position segment. The bolded melting regions can be seen to correspond to the positions of peaks in the total gene melting description depicted by the solid line. Reprinted from Marx, K.A., Assil, I.Q., Bizzaro, J.W., Blake, R.D. (1998). Comparison of Experimental to MELTSIM Calculated DNA Melting of the (A+T) Rich *Dictyostelium discoideum* Genome: Denaturation Maps Distinguish Exons from Introns. *J. Biomol. Struct. Dynamics* 16:329–340. With permission from Adenine Press.

array and allowed to hybridize to their immobilized complementary nucleic acid sequence, followed by washing to remove unhybridized sequences. The fluorescence intensities of all hybridized spots, reflecting the in vivo mRNA levels, are then determined. This type of large-scale gene product analysis is critically important for understanding the molecular basis of serious and complex biomedical conditions, such as cancer, where interconnected gene networks can produce nonlinear molecular effects characteristic of the biomedical state being analyzed. Increasingly, the nucleic acids being incorporated into microarray spots can be quite long, hundreds of base pairs, considerably longer than the thermodynamic block melting size found in typical DNAs. Therefore, complex multiblock melting behavior similar to that being simulated by MELTSIM will increasingly occur within hybrids formed on the microarray chips. For reasons such as this, we believe that our demonstration of the utility of MELTSIM to accurately simulate long DNA sequence melting in a position-dependent fashion along the sequence will facilitate the rational incorporation of long DNAs into smart biosensor designs.

1.3.5 Proteins Interacting With DNA act as *Switches of DNA Effector or Actuator Function*—Understanding Stability and Functional Outcomes of the Complexes by Examining the DNA Sequence Physical Properties

Through interactions with other molecules, DNA possesses mechanisms that can act as *switches of effector or actuator function*. For example, DNA has the ability to bind small molecules. When small planar molecules possess appropriate functional group arrays and geometries, including multiple fused aromatic rings, they can intercalate into the DNA helix. In the intercalation process, two planar DNA base pairs stacked on top of one another in the interior of the double helix move apart from each other as the double helix unwinds. When separation is sufficient, this allows the planar intercalating molecule to insert itself and 'stack' with the adjacent planar base pairs at the thermodynamically most favorable 0.34 nm Van der Waals interaction distance observed for stacked base pairs in DNA. Investigators have demonstrated that the intercalation process can involve a limited sequence specificity in certain systems involving more complex ligands. Following intercalation, the optical properties of the intercalating ligand are often very different and the physical properties and intelligent properties of DNA can be changed (161). The same consequences can result from sequence-specific binding for some nonintercalating ligands that bind within the major and minor grooves of DNA.

Another more biological mechanism leading to a potential *switch of effector or actuator function* in DNA is the binding of sequence-specific proteins to specific conserved DNA binding sites. This class of interactions, found in all cells, has evolved as a critical part of the underlying biological mechanism involved in the transcriptional regulation of specific genes being copied into mRNA for subsequent protein synthesis. In many instances of complex genomic regulation in cells, the initial DNA sequence-specific protein(s) binding their DNA recognition sites are followed by subsequent protein–protein recognition and binding events. The assembly of regulatory multiprotein complexes upon the initial DNA–protein complexes brings about a change in the transcriptional status of the regulated gene(s). Given these facts, our understanding of the basis for the energetic stability of sequence-specific DNA–protein complexes is an important first step in being able to manipulate the intelligent properties of DNA and its *actuation* or *switching* by proteins. However, understanding why a specific protein sequence, correctly folded in solution, recognizes and interacts favorably with a particular DNA sequence is well beyond the ability of any current theory to provide adequate insights or answers (162). An important activity on the pathway to this understanding has been the cataloging into databases of members of the different classes of DNA–protein complexes whose 3-D structures have been obtained from experimental NMR and x-ray crystallography studies (163). The ultimate goal envisioned for the database is that the study of DNA–protein interaction classes will facilitate an understanding of the 'universe' of DNA–protein interaction types. In turn, this understanding will enable the discovery of important structural and energetic rules underlying these interactions.

One of the insights that has resulted from analysis of these high-resolution structure databases has been a general recognition of how proteins in sequence-specific DNA–protein complexes distort their DNA recognition sites. An example of this type of DNA distortion is shown in Figure 1.48 where an x-ray structure of the bacteriophage cro protein is presented on the right in the form of a ribbon diagram of its secondary structure. It is depicted binding to its 14-bp DNA recognition sequence (164). Note that at either end of the DNA region contacted by the protein dimer, the DNA helix axis bends to the right. Distortion of DNA binding sites by sequence-specific recognition proteins is a commonly observed phenomenon (165,166). It has led to the calculation of a DNA physical property parameter called the protein deformability (PD). The magnitude of any local PD region in

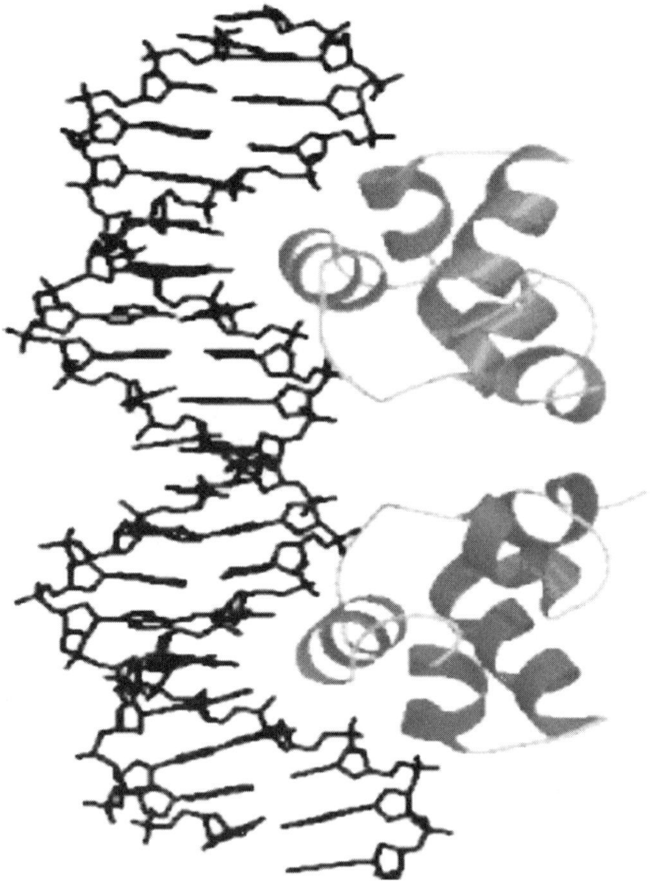

FIGURE 1.48
Cro protein–DNA structure from x-ray crystallography. On the left is the DNA double helix segment containing the 14-bp binding site with two pseudosymmetric conserved 4-bp segments. On the right bound to the DNA is the cro protein dimer with only the alpha helical segments shown connected by nonalpha helical polypeptide backbone regions. Note that the DNA helix axis is bent to the right at both of the end regions contacting the cro dimer. Reprinted from Supplementary website of Luscombe, N.M., Austin, S.E., Berman, H.M., Thornton, J.M. (2000). An Overview of the Structure of Protein–DNA Complexes. *Genome Biol.* 1:1–37.

a DNA binding site has been shown to be very much a local property that is dependent upon the dinucleotide neighbor pair sequence. That is, on average, sequence-specific proteins deform DNA not based upon which base *j* occurs at a specific position for the protein to interact with, but based upon the two dinucleotides *ij* and *jk* at that specific position *j*. The PD values for all 16 possible dinucleotide pairs in DNA have been determined. They are based upon averaged representations of how each protein specifically deformed its DNA recognition sequence in a dataset of nearly 100 high-resolution DNA–protein structures (167).

At the Center for Intelligent Biomaterials, we have taken the approach that DNA sequence-based parameters, such as PD, provide a simple and potentially useful way to eventually understand the intelligent properties underlying DNA–protein energetics and stability. Therefore, we have initially carried out studies to establish whether the PD scale provides insight into the energetic stability of single base pair variants of a specific DNA sequence interacting with their recognition protein. The DNA–protein system we studied is that shown inset in Figure 1.49, the relatively simple yeast kinetochore, a centromere DNA–multiprotein complex responsible for binding to the microtubules of the yeast mitotic spindle (168). The result of this binding is correct segregation into daughter cells of each of the 16 yeast chromosomes during cell division. In this system, we have focused on two particular conserved sequence DNA–protein interactions. The first is between the short symmetric DNA sequence labeled CDEI and the dimeric protein CBF1. The second is between the longer symmetric CDEIII DNA and a complex of binding proteins labeled

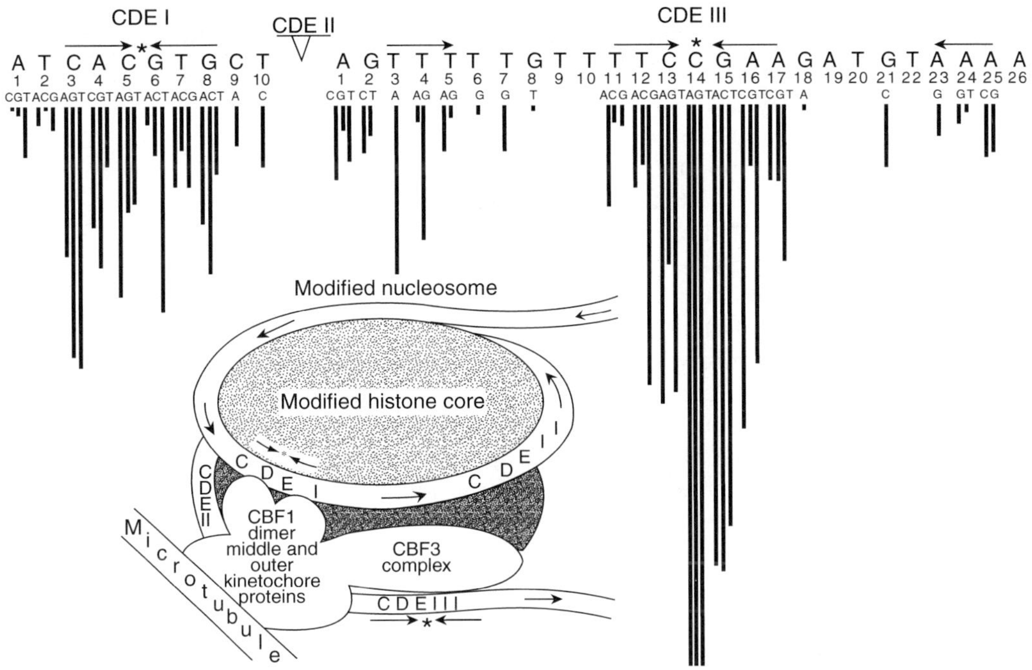

FIGURE 1.49
Schematic representation in the inset of the simple centromere structure of the yeast kinetochore emphasizing the positions of the two DNA–protein binding sites CDEI and CDEIII wrapped around the nucleosome core comprised of the histone octamer. Indicated in a general way are the site binding proteins and the multiprotein complex forming the kinetochore that binds to the microtubule responsible for separating the chromosome to the appropriate daughter cell during cell division. Above this inset schematic are indicated the DNA sequences of CDEI and CDEIII sites, with their centers of symmetry (asterisks) and symmetry elements (arrows) highlighted. Below the sequence is indicated the binding site position number and below that the three possible mutants for that position. All of the single base pair mutants of the CDEI and CDEIII sites were studied and the chromosome loss rates resulting from each of the indicated mutants were determined (169). The logarithm of each chromosomal loss rate is depicted by the length of each vertical line below each mutant position. Adapted from Hegeman, J.H., Flieg, U.N. (1993). The Centromere of Budding Yeast. *BioEssays* 15:451–460. With permission of Wiley-Liss, Inc., a subsidiary of John Wiley & Sons, Inc.

CBF 3. As shown in the inset schematic, these complexes are in close proximity, since they occur wrapped to form a modified centromere nucleosome found at the CEN sequence location on each yeast chromosome. Both complexes subsequently participate in the self-assembly of a complex multiprotein, multi-DNA sequence kinetochore on each chromosome. This large complex binds a long spindle microtubule in the cell, which then correctly segregates the chromosome during cell division.

In a published report, the yeast chromosome 6 centromere DNA and dozens of CDEI and CDEIII single base pair mutants were studied (169). Their relative chromosome loss rates were measured during yeast cell division and the logarithm of those rates are presented in Figure 1.49 as the vertical line lengths beneath each base mutant site position for both the CDEI and CDEIII binding sites. In general, the more significantly the DNA mutant affects interaction stability with the binding protein, the greater is the resulting chromosomal loss. We reasoned that the mutants resulting in greater deformation of the original DNA sequence by the protein would likely be those producing greater chromosome loss rates through larger defects in kinetochore structure and energetics. Therefore, from the sequences we calculated the change in absolute value of PD for all of the single base CDEI and CDEIII mutants at all positions compared with the original wild-type sequence. We demonstrated that these values correlated linearly with the logarithm of the measured chromosomal loss rates (170). This type of correlation between a simple calculable DNA physical property for any altered DNA sequence in a DNA–protein complex and the measured functional property of the altered sequence complex was a novel observation. Also, for each CDEI and CDEIII binding site position, plots of the change in PD values vs. log(chromosomal loss rates) for the three possible mutants and the wild-type sequence were made and their slopes determined. As Figure 1.50 illustrates, these slope magnitudes for each CDEI and CDEIII site position were found to be symmetric about the center of symmetry of these binding sites. That is, position pairs 3 and 8, 4 and 7, and 5 and 6 possess similar slopes in CDEI; and position pairs 11 and 17, 12 and 16, and 13 and 15 possess similar slopes in CDEIII (170). This symmetric site behavior agrees with the known rotational symmetry of the conserved CDEI and CDEIII binding sites and provides stronger validation that our simple PD-scale-based computational approach is revealing important quantitative information about DNA–protein complexes.

These results support the validity of our approach using the PD scale parameters to begin to understand the energetic basis of DNA sequences interacting with their recognition proteins. One advantage this approach provides is that DNA sequence-based physical properties are far easier to calculate than protein physical properties. This results from the fact that all DNA as it occurs inside cells has a regular repeating secondary structure close to the B secondary structure proposed originally by Watson and Crick (171). As a consequence, a number of different dinucleotide- and trinucleotide-based DNA physical property data sets (PD, bendability, and Position Preference) have been calculated from experimental data (172). Using these data sets, the local physical properties for any DNA sequence can be calculated. No such equivalent simple computational rules exist for proteins, which lack the overall energetic and structural uniformity that DNA possesses. The computational approach we have described has important implications for the use of DNA–protein complexes in biosensors. Stability and property differences for DNA

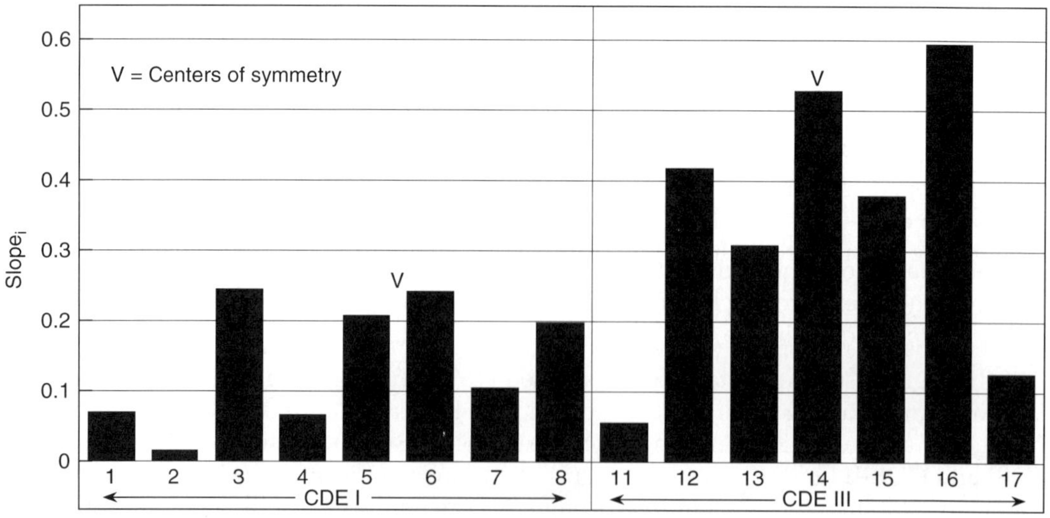

FIGURE 1.50
Histogram plot of the slopes determined from the experimental $\log_e$ (chromosomal loss rates) vs. ΔPD plots calculated for each binding site position of CDEI and CDEIII. The site position numbers are indicated along with the center of symmetry (V symbol) for each binding site. From Henemuth, B., Marx, K.A. (2006). Protein Deformability Properties of Single Base Mutants of Yeast DNA Kinetochore Protein Binding Sites Correlate with Measured Chromosomal Loss Rates for the Mutants in Yeast Cells. *BMC Mol. Biol.* 7:12–20. With permission.

sequence variants of DNA–protein complexes may eventually be predictable through simple calculations. This would be a valuable capability, since the laborious experimental generation of altered DNA sequences and testing of their DNA–protein stabilities and properties would not need to be performed prior to their use in biosensors.

1.4 The Importance of Informatics and Data Mining Approaches in Understanding Biological Macromolecules and in Biosensor Design and Operation

The discipline of Computer Science provides ever more sophisticated and important software applications to all areas of science. Increasingly important are informatics and data mining approaches applied to the understanding of large complex data sets related to biological macromolecules' structures, functions, and intelligent properties. This understanding is important both for the macromolecules' potential integration into biosensors as well as in the analysis of a complex smart biosensors' input data, prior to providing a simple output report to the user. While the analysis of complex biosensor input is important in some instances, it represents a situation we did not encounter in our biosensor development to date. Therefore, we do not discuss this in great detail. Rather, we illustrate the importance of these data mining techniques by discussing two representative examples where complex high-dimensionality biosensor data can only be properly addressed using an informatics and data mining approach involving machine learning techniques.

1.4.1 Machine Learning Approaches

A number of the machine learning techniques that are sometimes used in the analysis of biosensor output are also important in the analysis of large multidimensional, nonlinear biological, biochemical, and chemical datasets. Such analyses can be used to understand the intelligent properties of specific biological macromolecules as well as to incorporate them into the design of biosensors. The use of informatics and data mining activities applied to data sets in these domain areas have increasingly been termed, respectively, bioinformatics and cheminformatics. The latter term has overlap with and has somewhat superseded an older term, chemometrics, in the chemical literature.

Machine learning is the derivation of general knowledge from specific data sets using statistical techniques and analytical algorithms. The data sets are searched using potential hypotheses with the goal of building descriptive and predictive models general enough that they may allow one to draw similar conclusions from other related data sets. Machine learning can either involve classification techniques using supervised learning methods or clustering techniques in the case of unsupervised learning methods. Supervised machine learning methods can be utilized in the analyses of complex biosensor data where classes of analytes are already known and are used to train the entire system of biosensor inputs. Thus, machine learning methods could be used to recognize the biosensor input signal attributes that most accurately identify the correct analyte from among many incorrect possibilities. The resulting learned combination of input signal attributes that accurately determine the analyte identity and concentration would then be applied to unknown samples, to provide the routine biosensor output to the end user. Classification techniques vary from simple testing of sample features for statistical significance to sophisticated probabalistic modeling techniques. Some algorithmic methods widely used in machine learning approaches include: Naive Bayes, neural networks, support vector machines, instance-based learning (*K*-nearest neighbor), and logistic regression, to name a few (173).

Since the basis for the learning approaches inherent in these different algorithms is beyond the scope of the present review, we have left out the details of these methods.

Once machine learning has been carried out, there remains the problem of validation. Validation involves the task of determining the expected error rate from applying the trained classifier to a new data set. For large data sets, where generating new copies is not possible, the common machine learning validation technique called cross-validation is often used. For example, in tenfold cross-validation, the entire data set is divided into ten groups. The classifier model is created using nine of the groups and the classifier is then tested on the 10th group. This process is repeated iteratively, until each of the ten groups has served as the test group. The ten estimates are then averaged, providing an overall accuracy distribution and error estimate of the predictive power of the classifier. Another variant of this technique is the leave-one-out validation method. In this technique, one data point is left out of each of the iterations used to create the classifier model. This data point is then used to test the classifier. The process is repeated until every data point has been left out once for testing the classifier and the accuracy estimates are then averaged.

In machine learning approaches, classifiers do best when the number of dimensions or variables within the data set is small (less than 100) and the number of data points is large (greater than 1000)—a minimum ratio of 1:10. However, in many data sets this is not the case. In fact, the very opposite may be true and we illustrate this below with the case of nucleic acid microarray biosensor measurements.

1.4.2 Application of Machine Learning to the Analysis of High-Dimensionality Data From Microarray Biosensors

As we discussed previously, nucleic acid microarrays are a type of optical-based biosensor employing hybridization of analyte nucleic acids from an unknown sample to complementary surface-immobilized nucleic acid sequence probes from known genes (160). Within a single microarray area, the number of immobilized probes, or dimensions of data obtained from them, may be as high as 30,000. In the case of human gene probes forming the microarray, this would represent nearly the entire human gene complement. In contrast, the number of data points determined for each data dimension, representing different patient's tissue samples as one example, may be as little as 30. For this example the data dimension:data point ratio is 1000:1, very much the reverse of the desired ratio (1:10) in data mining situations that we discussed above. Even though each of the 30,000 probes are immobilized on a small surface area, around $(70\ \mu m)^2$/probe, these nucleic acid 'biochips' totalling a few cm^2, are costly and collecting many data points in this type of biosensor experiment may entail unreasonable expense. In Figure 1.51 we show a representative example of the hybridization intensities from a microarray spotted surface (174). Each spot signal intensity is measured using an automated reading system. Initially, simple plotting is often performed of the typical quantitative fluorescence output of the biosensor's 30,000 individual hybridization signals, to ascertain the quantitative level of each of the complementary mRNA sequences in the biological sample. However, to use machine learning effectively on these data sets, one is often faced with reducing the dimensionality of the data set. A number of statistical techniques, such as pairwise t and F statistics, can be used to reduce the dimensionality. The resulting aim is to select the minimum number of dimensions that best discriminate among classes and eliminate all those dimensions that provide no class discrimination or in fact prevent it (175).

The Center for Intelligent Biomaterials has been involved in informatics and data mining analyses of large data sets' from nucleic acid array biosensors as well as whole cell assay biosensors. This was prior to and subsequent to the formation of AnVil, Inc., a venture-backed

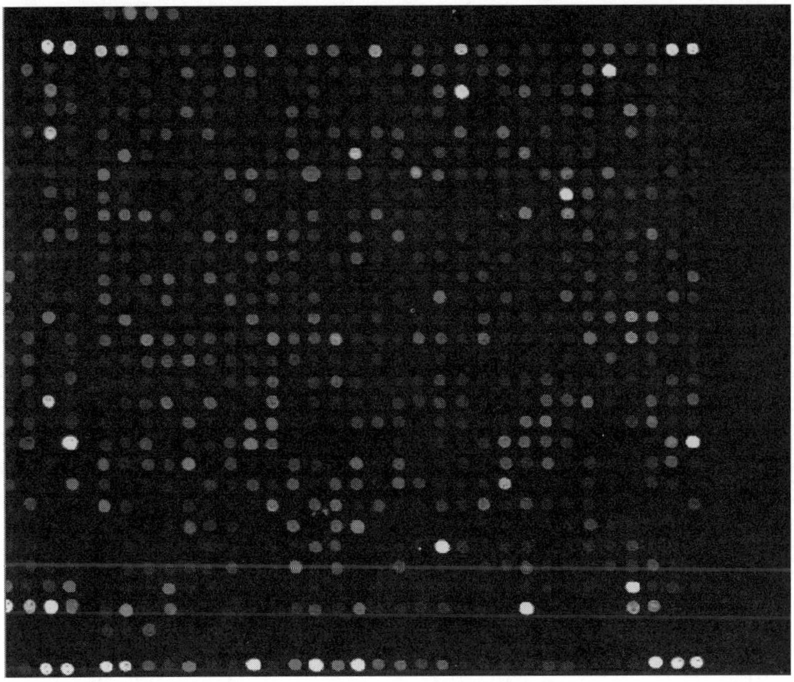

FIGURE 1.51
(See color insert) A low-resolution fluorescence light microscope image of the spotted array surface of a typical microarray experiment. Reprinted from website http://www.microarray.org/sfgf/jsp/home.jsp (accessed May 27, 2005). with permission of website.

data analysis company cofounded by the author, that specialized in the application of integrated high-dimensional machine learning and visualization techniques to partners' data sets in the biotechnology and pharmaceutical industry. Working collaboratively with AnVil, we applied these data mining techniques to the following two biosensor systems as examples. The first application was to the microarray biosensor analysis of mRNAs from known cancer tissue samples to create a cancer tissue classifier for undiagnosed clinical samples. The second application was to the analysis of whole cell biosensor output from the cellular testing of tens of thousands of compounds for cancer efficacy in the National Cancer Institute's (NCI) repository, to create compound subsets that target-specific cancer cell types.

In the first biosensor case, the object was to create a diagnostic three-class classifier using a supervised machine learning approach that could distinguish between normal lung tissue and two subclasses of nonsmall cell lung carcinoma (non-SCLC)—adenocarcinoma and squamous cell carcinoma. The classifier would be based upon the results of nucleic acid microarray biosensor experiments testing the mRNA populations from a number of each of these three clinical tissue types diagnosed via traditional pathological criteria (176,177). A total of 75 patient samples were studied, consisting of 17 normals, 30 adenocarcinomas, and 28 squamous cell carcinomas. A second data set used was in the public domain and consisted of 157 samples, including 17 normals, 139 adenocarcinomas, and 21 squamous cell carcinomas (178). The data were obtained using Chip A from the Affymetrix Human Genome U95 Chip Set (179). Chip A is a microarray containing spotted sequences for 12,000 full-length genes as well as controls. In a preliminary two-class (normal lung vs. non-SCLC tissue) examination of the data, the measured microarray biosensor levels from ~2000 genes were found to be statistically significant ($p < 0.05$) in differentiating clinical samples between these two tissue classes. For the more interesting and informative three-class classification of normal vs. adenocarcinoma vs. squamous cell carcinoma tumor types, it is more difficult to achieve accurate classification than for the two-class system.

In Figure 1.52, we present a representative RadViz display of the classification output for the three-class problem. Here, only 15 genes most effective at tissue classification were selected from the entire starting population by the statistical algorithm underlying the RadViz tool (177). This RadViz display illuminates the overall classification results nicely and in a visually intuitive manner. In this display, the specific genes located on the particular labeled circumference regions are those most effective at discriminating that tissue type from the other two types. How the RadViz display achieves this class discrimination for each clinical sample or point in the display is as follows. The genes are selected from the large starting population by an algorithm that uses varying statistical criteria to rank the genes best able to discriminate all the tissues (points) into their known class. The location of each tissue sample point shown in the display is determined using a Hooke's spring paradigm to vectorially sum its measured mRNA expression magnitudes from the microarray biosensor experiment for all 15 genes arrayed at their specific positions around the circle. The data points for all members of the three tissue type samples from the larger data set we analyzed are shown in the RadViz display. By and large, for the result shown

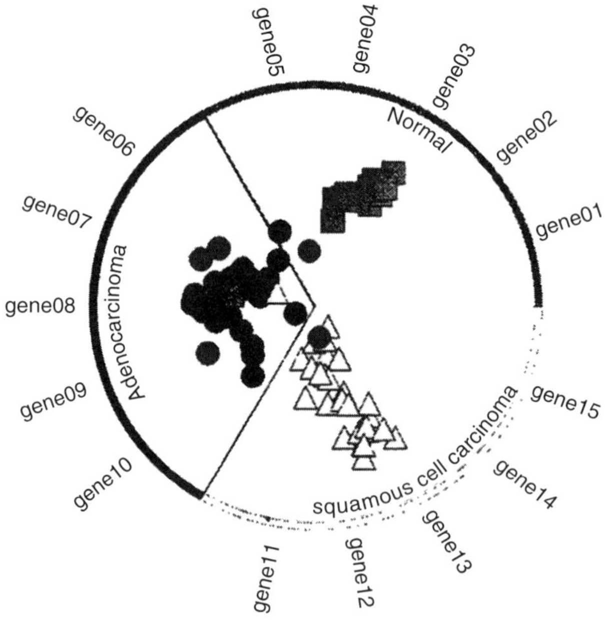

FIGURE 1.52
Supervised learning using RadViz for the three-class classification of cancer tissue biopsy samples into normal (squares), adenocarcinoma (circles), and squamous cell carcinoma (triangles) tissue classes. The RadViz algorithm classified the known cancer cell tissue types based upon pathologists's clinical determinations using microarray mRNA transcript level determinations on each tissue sample to carry out the classification by selecting those genes whose mRNA transcripts are best at classification. The genes whose mRNA transcript levels were most effective and met the statistical criterion for classification are shown in each tissue class sector. In the RadViz algorithm, the positions of these genes are what pulls each of the tissues (points) into their respective positions, forming a visual output of the supervised clustering algorithm. Reprinted from McCarthy, J.F., Marx, K.A., Hoffman, P.E., Gee, A.G., O'Neil, P., Ujwal, M.L., Hotchkiss, J. (2004). Applications of Machine Learning and High-Dimensional Visualization in Cancer Detection, Diagnosis and Mangement. In: Umar, A., Kapetanovic, I., Khan, J., eds. *The Applications of Bioinformatics in Cancer Detection, Ann. N.Y. Acad. Sci.* 1020:239–262.

the symbols are confined to their appropriate tissue type circle segment. This will be the case when a particular clinical sample or point is correctly classified. Only a few individual tissue samples do not fall into the correct circle segment (filled circles mostly). That only a few of these samples are misclassified reflects the high degree of accuracy—a 96% correct assignment to known clinical cancer class—we achieved using this three-class classifier on this microarray biosensor-derived clinical cancer data set. Similar high accuracies were obtained by applying the same techniques to a two-class prediction of lymphoma cell subtypes based upon analysis of microarray biosensor experiments (180). This example clearly demonstrates the power of supervised machine learning approaches to be trained to correctly classify complex biosensor input from known class examples. The trained algorithm can then be applied to yield accurate output predictions for data from new samples.

1.4.3 Applying Supervised Machine Learning to the NCI Compounds' Effects on Cancer Cells

In the second example applying a supervised machine learning approach, we examined the output from the National Cancer Institute's internal testing program involving tens of thousands of compounds in their cancer compound library being tested against 60 cancer cell lines grown in tissue culture. For each of the 60 cancer cell types, the NCI determined each compound's GI50 value—the compound's concentration needed to achieve 50% growth inhibition of that cell type. These 60 cancer cell lines were selected as representatives from nine clinically different cancer tissue types. In this data mining example, we examined the 60 tests from each of the 1400 different compounds (181). This restricted compound subset of the total NCI library resulted from the fact that significant numbers of missing values were present for most of the compounds in the larger compound library. An example of our approach is exemplified in Figure 1.53, where we present the RadViz statistical classification output from a three-class classifier. This output shows that we identified small subsets of the 1400 compounds whose GI50 inhibition test data were most highly effective at discriminating between or separating the three cancer cell line classes—the melanoma cell line class (14 compounds), the leukemia cell line class (30 compounds), and the nonmelanoma, nonleukemia cell line class (8 compounds). The NSC identification numbers of these effective compounds are shown arrayed around the circumference at their specific positions within the RadViz sections corresponding to these three tissue class types. It was these NSC compounds' data, vectorially summed, that were used to position the tested cell lines (points) within the RadViz display. Interestingly, in this study we discovered a pattern to the chemical character of the two compound subsets best at mutually discriminating melanoma cells from leukemia cells from nonmelanoma, nonleukemia cells. For the 14 compounds most effective against melanoma, 11 were identified as substituted *p*-quinones and all 11 had an internal quinone ring structure of the type shown in Figure 1.54. In contrast, of the 30 compounds most effective against leukemia, eight were identified as substituted *p*-quinones and six of the eight had an external *p*-quinone ring structure as we illustrate in Figure 1.54. The discovery of the quinone compounds' cancer type specificity and its implications are thoroughly discussed elsewhere (177,181).

This example analyzing data on chemical compounds, coupled with the previous example analyzing microarray biosensor data, illustrates the power and broad applicability of machine learning approaches to discover interesting patterns in any kind of data, including complex biosensor inputs. In fact, the discoveries we described would not have been possible by traditional analysis approaches. Although these particular examples contained

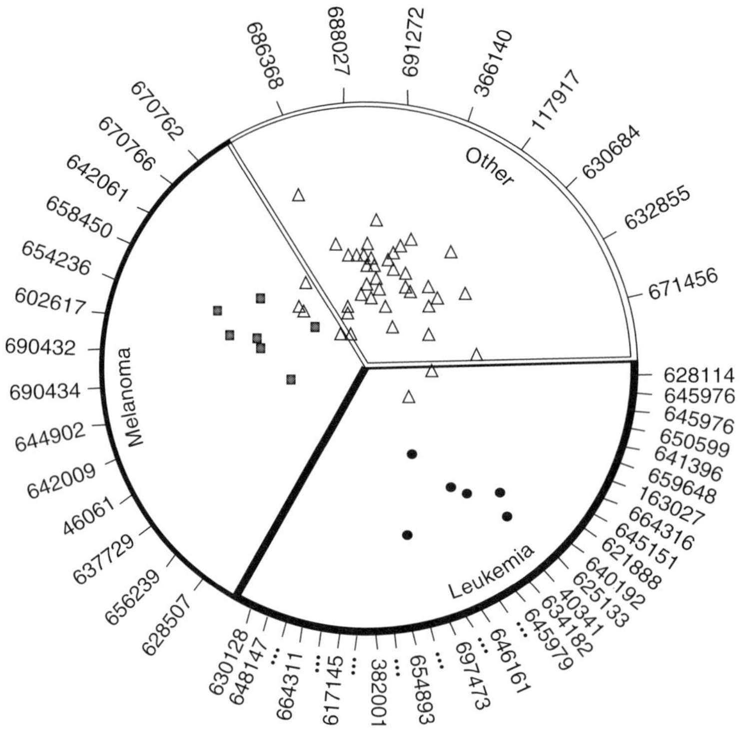

FIGURE 1.53
Supervised learning using RadViz for the three-class classification of cancer cell lines into melanoma (squares), leukemia (circles), and other or nonmelanoma, nonleukemia (triangles) classes. The RadViz algorithm classified the known cancer cell lines using a 1400-compound subset of the NCI database. The compounds' GI50 values were used to carry out the classification by selecting those compounds best at classification at the $p < 0.01$ statistical criterion with Bonnferroni correction for multiple tests. The compounds whose GI50 values were most effective and met the statistical criterion for classification have their NSC numbers shown in the class sector. In the RadViz algorithm, the positions of these compounds and the GI50 values are what pulls each of the 60 cells (points) into their respective positions, forming a visual output of the supervised clustering algorithm. Reprinted from McCarthy, J.F., Marx, K.A., Hoffman, P.E., Gee, A.G., O'Neil, P., Ujwal, M.L., Hotchkiss, J. (2004). Applications of Machine Learning and High-Dimensional Visualization in Cancer Detection, Diagnosis and Mangement. In: Umar, A., Kapetanovic, I., Khan, J., eds. *The Applications of Bioinformatics in Cancer Detection, Ann. N.Y. Acad. Sci.* 1020:239–262.

biosensor data relevant to drug discovery, the process is general and potentially applies to all biosensors possessing complex multidimensional data inputs.

1.5 Future Prospects for Biosensors

Looking to the future, there are many ways in which improvements can be made in creating smart biosensors. First, progress is continually being made in improving biosensor platforms. For example, significant efforts are underway to miniaturize individual biosensor components using microfabrication strategies and integrating them into MEMS devices (182). As a result, in a single MEMS device samples can be preprocessed and, in principle, be subsequently analyzed by multiple biosensor tests. In many cases, the miniaturization

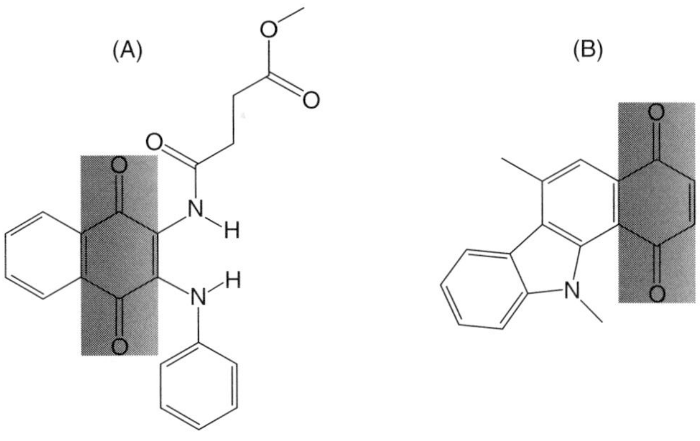

FIGURE 1.54
Most potent representatives of the quinone subtypes found effective against the melanoma and leukemia cancer cell lines in Figure 1.53 are shown. (A) Internal quinone NSC 670762, effective against melanoma; (B) external quinone NSC648147, effective against leukemia. The quinone ring portion is highlighted by the shaded region in each compound. Reprinted from McCarthy, J.F., Marx, K.A., Hoffman, P.E., Gee, A.G., O'Neil, P., Ujwal, M.L., Hotchkiss, J. (2004). Applications of Machine Learning and High-Dimensional Visualization in Cancer Detection, Diagnosis and Mangement. In: Umar, A., Kapetanovic, I., Khan, J., eds. *The Applications of Bioinformatics in Cancer Detection, Ann. N.Y. Acad. Sci.* 1020:239–262.

process will also result in an increased precision of processing and component separations, increased device sensitivity to the analyte, increased overall speed, and lowered cost of the analysis. Going beyond MEMS to the nanoscale, so-called NEMS devices are being proposed based upon current research characterizing various nanoscale biosensor systems. For nanoscale structures, exploiting the novel properties exhibited by materials components on this size scale promises to provide unique sensing mechanisms as well as enhanced sensitivity and robustness of the components. As an example, biosensors using Si nanowire signal transduction elements have been shown to selectively detect individual virus particles in solution (183). Figure 1.55a shows schematically how this biosensor works based upon virus-specific antibodies bound to the Si nanowire surface. Significant and measurable conductance changes occur when a single virus is bound to the nanowire-immobilized antibodies. In panel (b) are shown examples of the biosensor output that demonstrate viral detection specificity for two antibodies, one specific to influenza type A and the other to adenovirus, placed on two different Si nanowires. Detection events of average duration 16 s before release of the virus from a given Si nanowire can be seen in the output examples. Only the correct virus type is bound by its recognition antibody on the appropriate Si nanowire and is recorded by the biosensor. Single virus detection sensitivity has also been reported using an antibody biosensor that measures the acoustic energy associated with the rupture of the virus–antibody complex from the surface of the biosensor platform. This technique has been termed rupture event scanning and is based upon a novel energy scanning variant of the QCM technique discussed previously (184). The Si nanowire biosensor system discussed above was also shown to be capable of sensitive DNA detection. By immobilizing a 12-mer single-stranded recognition probe on the Si nanowire, hybridization of the complementary strand could be detected from a 25 pM DNA solution and single base mismatched hybridizations with the 12-mer sequence could also be discriminated (185).

Another example of nanoscale enhancement of biosensor detection sensitivity is in glucose sensing using carbon nanotubes. Single semiconducting carbon nanotubes exhibit strong luminescence in the near infrared. Electroactive mediators, such as $Fe(CN)_6^{3-}$, irreversibly bind to the nanotubes and quench the light emission after photoexcitation. When $Fe(CN)_6^{3-}$ is

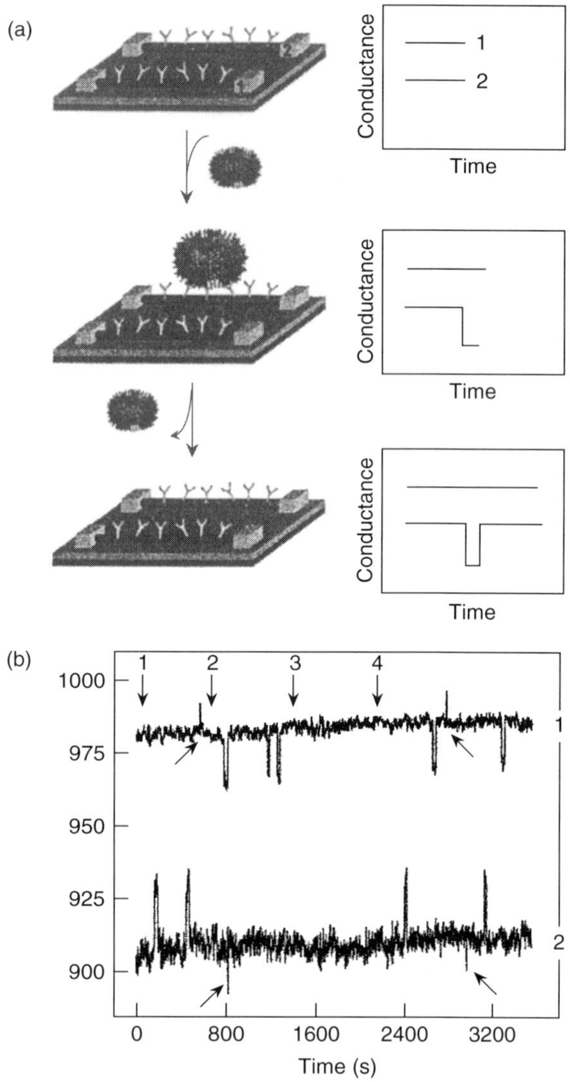

FIGURE 1.55
(a) Schematic for a biosensor based upon Si nanowire detection of single viruses via immobilized antibodies. Two nanowires are shown each with a different immobilized antibody. Nanowire 1 contains anti-influenza type A antibody, and nanowire 2 contains antiadenovirus group III antibody. The temporal signal response to the binding of an adenovirus and its subsequent release is shown to the right. (b) Si nanowires 1 and 2 outputs are shown. The vertical black arrows correspond to the times of addition of: 1, adenovirus; 2, influenza A virus; 3, pure buffer; and 4, 1:1 mixture of adenovirus and influenza A virus. Smaller arrows indicate lower intensity short-duration conductance changes due to virus particles passing by the Si nanowire but not binding to antibody. Reprinted from Patolsky, F, Zheng, G., Hayden, O., Lakadamyali, M., Zhuang, X., Lieber, C.M. (2004). Electrical Detection of Single Viruses. *Proc. Natl. Acad. Sci. USA* 101:14017–14022. With permission of Elsevier Publishing and National Academy of Sciences (2004) copyright.

reduced by H_2O_2, the nanotube luminescence is restored. Therefore, glucose oxidase catalysis of increasing glucose concentrations to form increasing H_2O_2 product results in an increasing luminescence signal (186). This nanoscale biosensor allows sensitive detection of glucose down to 35 pM. An additional advantage is that the near IR optical signal (994 nm emission max.) may be detectable in vivo up to centimeters below the skin. In this wavelength region, human tissue and biological fluids such as blood are particularly transparent. Aside from the progress being made in nanoscale biosensors and in increasing the sensitivity of different signal transduction mechanisms, new immobilization strategies are being developed to couple ever more specific and sensitive biological components to the biosensor platforms. Some involve creative innovations in microfabrication through surface-based self-assembly methods following specific chemical or adsorptive treatments of the surface (187,188).

To achieve miniaturization of feature sizes to the nanoscale on a biosensor platform for commercial applications the current methods employed in the research laboratory must be evolved to incorporate new and rapid manufacturing processes allowing nanomanufacturing of platforms. At the University of Massachusetts Lowell, we and two partner institutions (University of New Hampshire and Northeastern University) have recently initiated an NSF funded collaboration to advance nanoscale manufacturing. In general, the goals of the collaboration are to create novel high-rate/high-volume nanomanufacturing processes. Some of the nanoscale technical issues that must be solved include robustness, reliability/testing, tooling and interconnects, registration and nanosite identification, and nanoscale reading of the signal. One proof-of-principle outcome of this project will be to create a nanoscale patterned polymeric surface, suitable for attachment of biological elements such as antibodies to form a nanoscale biosensor array. The nanoscale patterned surface will be created in an injection molding device. During cooling of the polymer melt, patterning will result from phase separation and solidification of each of two component polymers possessing different surface energies at energetically compatible regions upon a solid nanotemplate surface. The follow-on biosensor proof-of-principle will evolve upon this platform into an antibody-based biosensor diagnostic for different clinical indications. Part of the development of this biosensor will involve creating more sensitive and nanoscale resolution detection methodologies that would be capable of subsequently quantitating indication-specific biomarker analytes binding to nanoscale spot sizes.

Many investigators are taking highly creative approaches to incorporating novel synthetic molecules and engineered biological macromolecules into structures that could be incorporated into the future design of biosensors. A few notable examples are worth mentioning. As we briefly mentioned earlier, a prime mover in the field of DNA nanoscale topological structures has been the Seeman lab (146,147). They developed an approach involving the design of appropriately positioned base complementary regions within multiple short DNA sequences that could be transformed by hybridization-coupled self-assembly to form regular 3-D geometric structures. Their efforts have yielded examples such as the formation of a DNA cube, truncated octahedron, and dimorphic switchable linear structures. In Figure 1.56, an example of the self-assembled DNA structure forming a truncated octahedron is presented (189). Structures like these, appropriately modified,

FIGURE 1.56

(See color insert) A schematic diagram of the top of a truncated DNA octahedron. The molecule contains 14 cyclic strands of DNA with each face of the octahedron comprising squares and hexagons, corresponding to a different cyclic strand. This view is looking down the fourfold axis of one of the squares comprising the octahedron. The approximate molecular weight of the DNA octahedron is about 790,000 Da. Reprinted with permission from Zhang, Y., Seeman, N.C. (1994). Construction of a DNA-Truncated Octahedron. *J. Am. Chem. Soc.* 116:1661–1669 and Dr. Ned Seeman. Copyright (1994) American Chemical Society.

could find unique applications such as providing novel structural scaffolds to create unique reporter systems in biosensors. Synthetic DNAs have also been shown to possess self-assembly properties that result in micrometer-scale arrays with nanoscale features (190). A DNA self-assembly system proceeding by cooperative binding at multiple weak domains has been proposed as a Turing universal biomolecular computing system (191). In this device, computations are carried out using the nanoscale DNA 2- or 3-D self-assembly or 'tiling' properties of short DNAs, representing the algorithmic basis for the device. Progress in this area will require, among other advances, methods to reduce errors in self-assembly, a way to extend 2- to 3-D self-assembly, and methods to connect nanoelectronics to nanometer-scale DNA assemblies. Other types of nanoscale DNA and nucleic acid self-assembly are being investigated for biosensors. For example, DNA hybridization can mediate the assembly of nanosuperstructures formed from gold nanoparticles (192) or inorganic nanocrystals (193). Via analyte nucleic acid cohybridization to different short single-stranded DNAs immobilized on different particles, the transition to assembled nanoparticle structures producing optical changes could sensitively indicate the presence of a specific analyte nucleic acid of interest. Lastly, in an interesting new system called hybridization chain reaction, a DNA hybridization system self-assembles to form double helical segments when an initiator nucleic acid strand is sensed—the analyte (194). This is capable of functioning in a biosensor as an amplifying transducer when coupled to aptamer nucleic acid triggers.

Other creative approaches with potential applications to biosensors involve the design and formation of supramolecular structures via the self-assembly of carefully designed molecular units possessing complementary, favorably interacting functional groups. For example, from the Stoddart lab, large organic molecules such as polyrotaxanes and catenanes have been demonstrated to form a variety of interlocked and threaded structures. A specific example of one complex topologically linked system they have synthesized is the five-ring interlocked structure aptly named Olympiadone (195). These supramolecular structures often do not themselves involve biological macromolecules. However, they could be used as scaffolding for incorporating biological macromolecules or as actuator components of biosensors. Another interesting example is a chiral molecular motor that can be caused to rotate about an internal double bond in a mechanism driven by light. When the molecule is incorporated into a liquid crystal, the light-driven motor perturbs the liquid crystal phase. By controlling the motion of the molecular motor, the resulting color of the liquid crystal can be controlled over the entire visible spectrum (196). By coupling some analyte detection scheme to this molecular motor, the system could serve as the optical output for a biosensor.

We believe that the complete set of evolved intelligent properties that living cells possess holds great promise in the future for the increasing incorporation of specific cell types into biosensors. This fact has resulted in our most recent biosensor focus being on the use of living cells attached to the piezoelectric QCM platform. The cell QCM biosensor systems have allowed us to propose novel ways in which they could be used to screen small-molecule drugs that differentially affect the cellular cytoskeleton, as well as to be potentially able to discriminate normal from cancerous cells contained in human biopsy samples (96,98). Also, we described the application of specific applied electrochemical potentials to the isolated ECM from normal ECs, to release specific biological factors. When applied to surface wounds, these factors enhanced nearly twofold the rate of wound healing in a mouse model system (107,108).

In the last few years, our most recent research direction at the Center for Intelligent Biomaterials at the University of Massachusetts has been to employ novel computational strategies to understand the intelligent properties of specific biological molecules. We have particularly focused on analyzing DNA and DNA-sequence-specific protein interactions through the use of DNA neighbor base pair sequence property metrics. A related approach we have taken in the Center for Intelligent Biomaterials and at AnVil, Inc. is the application of high-dimensional visualization techniques integrated with machine learning approaches to important problems possessing large multidimensional data sets. This methodology is particularly well suited to the discovery of patterns and trends not easily observed unaided within large complex data sets, where highly nonlinear effects may exist.

Important discoveries have already been made using approaches where machine learning and informatics applied to large data sets has played a prominent role. One example is a biosensor called TIGER from Isis Pharmaceuticals, Inc., which is reaching commercialization. The development of this biosensor first involved discovering bacterial RNA sequence motifs that are conserved to varying extents and that can reveal the presence of different bacteria and different strains within a given species. In patient fluid samples, the sequences to be detected are first amplified using the PCR reaction. Then, in a high-throughput mode, the amplified samples are examined using high-resolution mass spectrometry. Subsequent informatics treatment of the data reveals characteristic sequence motifs for different bacterial species and strains. Using this biosensor, these investigators were able to detect three different species of pathogenic bacteria within the respiratory fluid samples from a disease outbreak at a US military base (197). They identified a particularly virulent strain of *Streptococcus pyrogenes* as the primary cause of the outbreak in all the disease samples analyzed. The TIGER biosensor has unique capabilities. Without using bacterial culturing techniques, it is able to rapidly detect different bacterial species and strains within species, as well as provide rapid identification of the source of infections, even in cases where the bacteria has not yet been characterized.

Another example where machine learning and informatics applied to large data sets has played a prominent discovery role is in the problem of identifying important protein biomarkers in complex biological fluids. In this instance, proteomic data from mass spectrometry of serum proteins has been used to accurately diagnose ovarian cancer (198). This need for pattern discovery in complex data is the case in many examples found in biological systems. For example, large numbers of proteins interact in complex interconnected regulatory networks within all cells. This is the case for both small-molecule metabolic pathways and complex biological gene control networks comprising DNA sequence sites interacting with control proteins. These two critical cellular systems are just two examples where informatics and data mining approaches are necessary, even critical to comprehending how they function. An understanding of how such systems operate will enable their future practical use, where subsets of these biological systems may in fact be incorporated into ever more sophisticated biosensors. The need for machine learning techniques coupled with novel ways of displaying the data may also be critical for the future operation of complex biosensors. Analyses of a biosensor's measured complex nonlinear input signals and the algorithmic learning of a biosensor's analyte class behavior may be necessary to produce an accurate easily interpretable output for the end user.

Acknowledgments

Over the years, the author has worked with many creative faculty colleagues, postdoctoral fellows, and graduate students on the projects described in this chapter. In particular, he would like to single out for acknowledgement Prof. Susan Braunhut, Prof. Georges Grinstein, Prof. Sukant Tripathy, and Prof. George Ruben for their stimulating and productive collaborations. The author gratefully acknowledges major funding support from a number of federal agencies including: Army Research Office Grants: DAAL03-91-G-0064, DAAH04-94-0003 and the Joint Army/USDA Advance Materials Program Grant DAAK60-93-K0012; NIH Grants CA 23108, GM25886, AI17586, AI24110 and GM58583; and a DARPA Grant. Smaller funding support is acknowledged from the UML Chancellor's Fund and Research Foundation Seed grants, NSF NSEC Grant EEC-0425826 and collaborative effort from other UML Centers & Institutes. For some aspects of the research described herein, financial support is acknowledged from the following companies: AnVil, Pfizer, Aventis, Genetics Institute, and Millennium Pharmaceuticals.

References

1. Arnold, F.H. (1998). Design by Directed Evolution. *Acct. Chem. Res.* 31:125–131.
2. Waldo, G.S. (2003). Genetic Screens and Directed evolution for Protein Stability. *Curr. Opinion Chem. Biol.* 7:33–38.
3. Marx, K.A., Samuelson, L.A., Kamath, M., Lim, J.O., Sengupta, S., Kaplan, D., Kumar, J., Tripathy, S.K. (1994). Intelligent Biomaterials Based on Langmuir–Blodgett Monolayer Films. In: Birge, B.R., ed. *Molecular and Biomolecular Electronics.* Advances in Chemical Series, Vol. 240. Washington, DC: American Chemical Society Books, pp. 395–412.
4. Wilchek, M., Bayer, E.A. (1989). Avidin-Biotin Technology Ten Years on: Has it Lived up to its Expectations? *Trends Biochem. Sci.* 14:408–412.
5. Alberts, B., Bray, D., Lewis, J., Raff, M., Roberts, K., Watson, J.D. (1994). *Molecular Biology of the Cell.* New York, NY: Garland Publishing, p. 16.
6. Gantt, E. (1981). Phycobilisomes. *Ann. Rev. Plant Physiol.* 32:327–347.
7. Stoeckenius, W., Bogomolni, R. (1982). Bacteriorhodopsin and Related Pigments of Halobacteria. *Ann. Rev. Biochem.* 52:587–617.
8. Glazer, A.N. (1985). Light Harvesting by Phycobilisomes. *Ann. Rev. Biophys. Biophys. Chem.* 14:47–77.
9. Suter, G.W., Holzwarth, A.R. (1987). A Kinetic Model for the Energy Transfer in Phycobilisomes. *Biophys. J.* 52:673–683.
10. Haugland, R. (1994). *Handbook of Fluorescent Probes and Research Chemicals,* 5th edn. Eugene, OR: Molecular Probes, Inc.
11. Samuelson, L.A., Miller, P., Galotti, D., Marx, K.A., Kumar, J., Tripathy, S.K., Kaplan, D.L. (1991). Oriented Fluorescent Streptavidin Conjugated Phycoerythrin Protein on Biotinylated Lipid L-B Monolayer Films. In: Renugopalakrishnan, V., Carey, P.R. Smith, I.C.P., Huang, S.G., Storer, A.C., eds. *Proteins-Structure, Dynamics and Design.* Leiden, The Netherlands: ESCOM, pp. 160–164.
12. Samuelson, L.A., Miller, P., Galotti, D., Marx, K.A., Kumar, J., Tripathy, S.K., Kaplan, D.L. (1992). The Monomolecular Organization of a Photodynamic Protein System through Specific Surface Recognition of Streptavidin by Biotinylated Langmuir–Blodgett Films. *Langmuir* 8:604–608.
13. Samuelson, L.A., Yang, Y., Marx, K.A., Kumar, J., Tripathy, S.K., Kaplan, D.L. (1992). Surface-Specific Monolayer Attachment of a Photodynamic Protein Through Biotin-Streptavidin Complexation: Demonstration of a Cassette Approach. *Biomimetics* 1:51–61.
14. Samuelson, L.A., Miller, P., Galotti, D., Marx, K.A., Kumar, J., Tripathy, S.K., Kaplan, D.L. (1992). Langmuir–Blodgett Films of Streptavidin-Conjugated Phycoerythrin Bound to Biotinylated Monolayers. *Thin Solid Films* 210/211:796–798.
15. Kamath, M., Lim, J.O., Chittibabu, K.G., Sarma, R., Samuelson, L.A., Akkara, J.A., Kaplan, D.L., Kumar, J., Marx, K.A., Tripathy, S.K. (1993). Biotinylated Poly(3-hexylthiophene-co-3-methanolthiophene): A Langmuir Monolayer-Forming Copolymer. *J. Macromol. Sci., Pure Appl. Chem. A* 30(8):493–502.

16. Samuelson, L.A., Kaplan, D.L., Lim, J.O., Kamath, M., Marx, K.A., Tripathy, S.K. (1994). Molecular Recognition Between a Biotinylated Polythiophene Copolymer and Phycoerythrin Utilizing the Biotin–Streptavidin Interaction. *Thin Solid Films* 242:50–55.
17. Ayyagari, M.S., Pande, R., Kamtekar, S., Gao, H., Marx, K.A., Kumar, J., Tripathy, S.K., Akkara, J., Kaplan, D.L. (1995). Molecular Assembly of Proteins and Conjugated Polymers: Toward Development of Biosensors. *Biotech. Bioeng.* 45:116–121.
18. Ayyagari, M.S., Kamtekar, S., Pande, R., Marx, K.A., Kumar, J., Tripathy, S.K., Kaplan, D.L. (1994). Chemileuminescence-Based Pesticide Biosensor Utilizing the Intelligent Evolved Properties of the Enzyme Alkaline Phosphatase. In: Rogers, C., ed. *Proceedings Second International Conference on Intelligent Materials.* Lancaster, PA, Technomic Press, pp. 85–96.
19. Chen, Z.P., Samuelson, L.A., Akarra, J., Kaplan, D.L., Gao, H., Kumar, J., Marx, K.A., Tripathy, S.K. (1995). Sol-Gel Encapsulated Light-Transducing Protein Phycoerythrin: A New Biomaterial. *Chem. Mater.* 7:1779–1783.
20. Chen, Z.P., Chittibabu, K.G., Marx, K.A., Kumar, J., Tripathy, S.K., Samuelson, L.A., Akarra, J., Kaplan, D.L. (1994). Photodynamic Protein Incorporated in Conducting Polymer and Sol-Gel Matrices: Toward mart Materials for Information Storage and Processing. In: Varadan, V.K., ed., Smart Materials and Structures, *Proc. SPIE*, 2189:105–115.
21. Chen, Z.P., Kaplan, D.L., Yang, K., Kumar, J., Marx, K.A., Tripathy, S.K. (1996). Phycobiliproteins Encapsulated in Sol-Gel Glass. *J. Sol-Gel Sci. Tech.* 7:99–108.
22. Chen, Z.P., Kaplan, D.L., Yang, K., Kumar, J., Marx, K.A., Tripathy, S.K. (1997). Two-Photon Induced Fluorescence From the Phycoerythrin Protein. *Appl. Optics* 36:1655–1659.
23. Beladakere, N.N., Ravindran, T., Bihari, B., Sengupta, S., Marx, K.A., Kumar, J., Tripathy, S.K. (1993). Photovoltaic Effects and Charge Transport Studies in Phycobiliproteins. In: Viney, C., Case, S.T., Waite, J.H., eds. Biomolecular Materials, *Proc. Mat. Res. Soc.*, 292:193–198.
24. Spudich, J.L.,. Bogomolni, R.A. (1998). Sensory Rhodopsins of Halobacteria. *Ann. Rev. Biophys. Biophys. Chem.* 17:183–215.
25. Lanyi, J.K. (1992). From Time Resolved Difference Spectra to Kinetics, Mechanism and Thermodynamics in the Bacteriorhodopsin Photocycle. In: Pullman, A., Jortner, J., Pullman, B., eds. *Membrane Proteins, Structures Interactions and Models.* Dordrecht, Holland: Kluwer Academic Publishers, pp. 85–96.
26. Hampp, N. (2000). Bacteriorhodopsin as a Photochromic Retinal Protein for Optical Memories. *Chem. Rev.* 100:1755–1776.
27. Chen, Z., Lewis, A., Takei, H., Nabenzahl, I. (1991). Bacteriorhodopsin Oriented in Polyvinyl Alcohol Films as an Erasable Optical Storage Medium. *Appl. Opt.* 30:5188–5197.
28. Chen, Z., Birge, B. (1993). Protein Based Artificial Retina. *Trends in Biotechnol.* 11:292–296.
29. Birge, R.R. (1990). Photophysics and Molecular Electronic Applications of the Rhodopsins. *Ann. Rev. Phys. Chem.* 41:683–695.
30. Oesterhelt, D., Brauchle, C., Hampp, N. (1991). Bacteriorhodopsin: A Biological Material for Information Processing. *Quart. Rev. Biophys.* 24:425–478.
31. Takei, H., Lewis, A., Chen, Z., Nebenzahl, I. (1991). Implementing Receptive Fields with Excitatory and Inhibitory Optoelectronic Responses of Bacteriorhodopsin Films. *Appl. Opt.* 30:500–507.
32. Chen, Z., Takei, H., Lewis, A. (1990). Optical Implementation of Neural Networks with Wavelength-Encoded Bipolar Weight using Bacteriorhodopsin. *Proceedings, Inter. Joint Conf. Neural Networks* 11:803–809.
33. Nicolini, C., Erokhin, V., Padden, S., Sartore, M. (1998). Towards a Light-Addressable Transducer Bacteriorhodopsin Based. *Nanotechnology* 9:223–227.
34. Chen, Z., Lewis, A., Kumar, J., Tripathy, S.K., Marx, K.A., Akkara, J., Kaplan, D.L. (1994). Second Harmonic Generation of Bacteriorhodopsin and Its Application for Three-Dimensional Optical Memory. In: M. Alper, H. Bayley, D. Kaplan, M. Navia, eds. Biomolecular Materials by Design, *Proc. Mat. Res. Soc.*, 330:263–268.
35. Parthenopoulos, D.A., Rentzepis, P.M. (1989). Three-Dimensional Optical Storage Memory. *Science* 245: 843–845.

36. Strickler, J.H., Webb, W.W. (1991). Three-Dimensional Optical Storage in Refractive Media by Two-Photon Point Excitation. *Opt. Lett.* 16:1780–1782.
37. Birge, R.R. (1992). Protein-Based Optical Computing and Memories. *IEEE Computer* 35:56–67.
38. Birge, R.R., Gillespie, N.B., Izaguirre, E.W., Kuznetsov, A., Lawrence, A.F., Singh, D., Song, Q.W., Schmidt, E., Sturat, J.A., Seetharaman, S., Wise, K.J. (1999). Biomolecular Electronics: Protein-Based Associative Processors and Volumetric Memories. *J. Phys. Chem. B* 103:10746–10766.
39. McCormick, F.B., Cokgor, I., Esener, S.C., Dvornikov, A.S., Rentzepis, P.M. (1996) Two-Photon Absorption Based 3-D Optical Memories. In: Schwartz, T.A., Francis, M., eds. High Density Data Recording and Retrieval Technologies, *Proc. SPIE*, 2604:2332.
40. Lee, B.K., Chih-Jen, C. R., Liang-Chian, C.D., Cheng, J., Yeong-Ning, C.I., Beyette, F.R., Stecki, A.J. (2001). High-Density Er-Implanted GaN Optical Memory Devices. *Appl. Opt.* 40:3552–3558.
41. Ayyagari, M.S., Kamtekar, S., Pande, R., Marx, K.A., Kumar, J., Tripathy, S.K., Kaplan, D.L. (1995). Biosensors for Pesticide Detection Based on Alkaline Phosphatase-Catalyzed Chemileuminescence. *Mater. Sci. Eng. C* 2:191–196.
42. EPA Report EP1.29/2, (1980).
43. Ayyagari, M.S., Kamtekar, S., Pande, R., Marx, K.A., Kumar, J., Tripathy, S.K., Akkara, J., Kaplan, D.L. (1995). Chemileuminescence-Based Inhibition Kinetics of Alkaline Phosphatase in the Development of a Pesticide Biosensor. *Biotech. Prog.* 11:699–703.
44. Pande, R., Kamtekar, S., Ayyagari, M.S., Kamath, M., Marx, K.A., Kumar, J., Tripathy, S.K., Kaplan, D.L. (1996). A Biotinylated Undecylthiophene Copolymer Bioconjugate for Surface Immobilization: Creating an Alkaline Phosphatase Chemileuminescence-Based Biosensor. *Bioconjugate Chem.* 7:159–164.
45. Pande, R., Kamtekar, S., Ayyagari., Marx, K.A., Kumar, J., Tripathy, S.K., Kaplan, D.L. (1995) Integrating Biotinylated Polyalkylthiophene Thin Films with Biological Macromolecules: Biosensing Organophosphorus Pesticides and Metal Ions with Surface Immobilized Alkaline Phosphatase Utilizing Chemileuminescence Measurements. *Smart Mater. SPIE* 2441:12–22.
46. Ayyagari, M.S., Gao, H., Bihari, B., Chittibabu, K.G., Kumar, J., Marx, K.A., Kaplan, D.L., Tripathy, S.K. (1994). Molecular Self-Assembly on Optical Fibre-based Fluorescence Sensors. In: Lieberman, R.A., ed. Chemical, Biochemical and Environmental Fibre Sensors, *Proc. SPIE*, 2068:168–178.
47. Kamtekar, S., Pande, R., Ayyagari., M.S., Marx, K.A., Kaplan, D.L., Kumar, J., Tripathy, S.K.(1995). A Chemileuminescence-Based Biosensor for Metal Ion Detection. *Mater. Sci. Eng. C* 3:79–83.
48. Kamtekar, S., Pande, R., Ayyagari, M.S., Kaplan, D.L., Marx, K.A., Kumar, J., Tripathy, S.K.(1996). Trace Analysis of Zn(II), Be(II), and Bi(III) by Enzyme-Catalyzed Chemileuminescence. *Anal. Chem.* 68:216–220.
49. Townshend, A., Vaughan, A. (1970). Detection of Traces of Zinc by Activation of Apoalkaline Phosphatase. *Anal. Chim. Acta* 49:366–367.
50. Guilbault, G.G., Sadar, M.H., Zimmer, M. (1969). Analytical Applications of the Phosphatase Enzyme System. Determination of Bismuth, Beryllium and Pesticides. *Anal. Chim. Acta* 44:361–367.
51. Townshend, A., Vaughn, A. (1969). Applications of Enzyme-Catalyzed Reactions in Trace Analysis-IV-Determination of Beryllium and Zinc by Their Inhibition of Calf-Intestinal Alkaline Phosphatase. *Talanta* 16:929–937.
52. Magner, E. (1998). Trends in Electrochemical Biosensors. *Analyst* 123:1967–1970.
53. Kuhn, L.S. (1998). Biosensors: Blockbuster or Bomb? Electrochemical Biosensors for Diabetes Monitoring. *Electrochem.l Soc. Interface*, Winter:26–31.
54. Diaz, A.F., Hall. B. (1983). Mechanical Properties of Electrochemically Prepared Polypyrrole Films. *IBM J. Res. Dev.* 27:342–345.
55. Warren, L.F., Walker, J.A., Anderson, D.P., Rhodes, C.G., Buckley, L. (1989). A Study of Conducting Polymer Morphology. *J. Electrochem. Soc.* 136:2286–2295.

56. Yang, R., Evans, D.F., Christensen, L., Hendricksen, W.A. (1990). Scanning Tunneling Microscopy (STM) Evidence of Semicrystalline and Helical Conducting Polymer Structures. *J. Phys. Chem.* 94:6117–6122.
57. DeVreux, F., Genoud, F., Nechtschein, M., Vilberet, B. (1987). ESR Investigation of Polarons and Bipolarons in Conducting Polymers: the Case of Polypyrrole. *Synth. Metals* 18:89–93.
58. Diaz, A.F., Bargon, J. (1986). Electrochemical Synthesis of Conducting Polymers. In: Skotheim, T.A., ed. *Handbook of Conducting Polymers.* New York: Marcel Dekker, pp. 1–81.
59. Pande, R., Lim, J.O., Marx, K.A., Tripathy, S.K., Kaplan, D.L. (1993). Comparison of Single and Double Stranded DNA Binding to Polypyrrole. In: Viney, C., Case, S.T., Waite, J.H., eds. *Biomolecular Materials, Mater. Res. Soc.* 292:135–140.
60. Pande, R., Ruben, G.C., Lim, J.O., Tripathy, S., Marx, K.A. (1998). DNA Bound to Polypyrrole Films: High Resolution Imaging, DNA Binding Kinetics and Internal Migration. *Biomaterials* 19:1657–1667.
61. Marx, K.A., Lim, J.O., Minehan, D., Pande, R., Kamath, M, Tripathy, S., Kaplan, D. (1994). Intelligent Materials Properties of DNA and Strategies for It's Incorporation into Electroactive Polymeric Thin Film Systems. *J. Intelligent Mater. Systems Struct.* 5:447–454.
62. Minehan, D.S., Marx, K.A., Tripathy, S.K. (1994). Kinetics of DNA Binding to Electrically Conducting Polypyrrole Films. *Macromolecules* 27:777–783.
63. Minehan, D.S., Marx, K.A., Tripathy, S.K. (2001). DNA Binding to Electropolymerized Polypyrrole: The Dependence on Film Characteristics. *J. Macromol. Sci. Pure Appl. Chem. A* 38:1245–1258.
64. Lim, J.O., Minehan, D., Kamath, M., Marx, K.A., Tripathy, S.K. (1992). DNA Binding to Condcuting Polymer Films. In: Aksay, I., Baer, E., Sarikaya, M., Tirrel, D.A., eds. Hierarchically Structured Materials, *Proc. Mater. Res. Soc.*, 225:195–200.
65. Marx, K.A. (2003). The Quartz Crystal Microbalance: A Valuable Tool for Characterizing Thin Polymeric Films and Biomolecular Systems at the Solution–Surface Interface. *Biomacromolecules* 4:1099–1120.
66. Bartlett, P.N., Cooper, J.M. (1992). A Review of the Immobilization of Enzymes in Electropolymerized Films. *J. Electroanal. Chem.* 362:1–12.
67. Yang, L., Janie, E., Huang, T., Gitzen, J., Kissinger, P.T., Vreeke, M., Heller, A. (1995). Applications of Wired Peroxidase Electrodes for Peroxide Determination in Liquid Chromatography Coupled to Oxidase Immobilized Enzyme Reactors. *Anal. Chem.* 67:1326–1331.
68. Dunford, B.H. (1990). Horseradish Peroxidase: Structure and Kinetic Properties. In: Everse, J., Everse, K.E., Grishham, M.B., eds. *Peroxidases in Chemistry and Biology II.* Boca Raton, FL: CRC Press, p. 1.
69. Long, D.D., Marx, K.A., Zhou, T. (2001). Amperometric Hydrogen Peroxide Sensor Electrodes Coated With Electropolymerized Tyrosine Derivative and Phenolic films. *J. Electroanal. Chem.* 501:107–113.
70. Marx, K.A., Zhou, T., Long, D.D. (2005). Electrochemical Quartz Crystal Microbalance Study of the Films Formed from the Decyl Esters of D- and L-Tyrosine: A Comparison of Monomer Concentration Dependent Formation by Electropolymerization vs. Adsorption and the Film Energy Dissipation and Surface Hydrophilicity. *Biomacromolecules* 6:1698–1706.
71. Marx, K.A., Zhou, T. (2002). Comparative Study of Electropolymerization Versus Adsorption of Tyrosine and the Decyl Ester of Tyrosine on Platinum Electrodes. *J. Electroanal. Chem.* 521:53–60.
72. Alva, K.S., Marx, K.A., Samuelson, L.A., Kumar, J., Tripathy, S.K., Kaplan, D.L. (1996). Novel Immobilization Techniques in the Fabrication of Efficient Electrochemical Biosensors. In: Crowson, A. ed. Smart Structures and Materials, *Proc. SPIE*, 2716:152–163.
73. Bruno, F.F., Akkara, J.A., Samuelson, L.A., Kaplan, D.L., Marx, K.A., Tripathy, S.K. (1993). The Enzymatic Mediated Polymerization of Phenol and Aniline Derivatives on a Langmuir Trough. In: Viney, C., Case, S.T., Waite, J.H., eds. Biomolecular Materials, *Proc. Mater. Res. Soc.*, 292:147–153.
74. Bruno, F.F., Marx, K.A., Tripathy, S.K., Akkara, J.A., Samuelson, L.A., Kaplan, D.L. (1994). Enzymatic Catalyzed Polymerization of Phenol and Aniline Derivatives on a Langmuir Trough to Form Ordered 2-D Polymer Films. *J. Intel. Mater. Sys. Str.* 5:631–634.

75. Bruno, F.F., Akkara, J.A., Samuelson, L.A., Kaplan, D.L., Mandal, B.K., Marx, K.A., Kumar, J., Tripathy, S.K. (1995). Enzymatic Synthesis of Conjugated Polymers at the Langmuir Trough Air–Water Interface. *Langmuir* 11:889–892.
76. Bruno, F.F., Akkara, J.A., Kaplan, D.L., Sekher, P., Marx, K.A., Tripathy, S.K. (1995). Enzyme-Mediated Two-Dimensional Polymerization of Aromatic Derivatives on a Langmuir Trough. *Ind. Eng. Chem. Res.* 34:4009–4015.
77. Akkara, J.A., Bruno, F.F., Samuelson, L.A., Ayyagari, M., Tripathy, S.K., Marx, K.A., John, V.T., Kaplan, D.L. (1993). Synthesis of Two-Dimensional Electrooptic Polymer Networks Through Biocatalysis. *Polymer Preprints, ACS Award Symp.* 34:775–776.
78. Akkara, J.A., Ayyagari, M., Bruno, F.F., Samuelson, L.A., John, V.T., Karayigitoglu, C., Tripathy, S.K., Marx, K.A., Rao, D.V.G.L.N., Kaplan, D.L. (1994). Biomimetic Membrane and Interface Templates for Enzyme-Based Polymerization Reactions. *Biomimetics* 2:331–339.
79. Marx, K.A., Lee, J.S., Sung, C. (2004). Enzymatic Copolymerization Alters the Structure of Unpolymerized Mixtures of the Biomimetic Monomers: The Amphiphilic Decyl Ester of L-Tyrosine and L-Tyrosineamide—An AFM Investigation of Nano- to Micrometer Scale Structure Differences. *Biomacromolecules* 5:1869–1876.
80. Colonna, S., Gaggero, N., Richelmi, C., Pasta, P. (1999). Recent Biotechnological Developments in the Use of Peroxidases. *TIBTECH* 17:163–168.
81. Ozaki, S., Ortiz de Montellano, P.R. (1995). Molecular Engineering of Horseradish Peroxidase: Thioether Sulfoxidation and Styrene Epoxidation by Phe-41 Leucine and Threonine Mutants. *J. Am. Chem. Soc.* 117:7056–7064.
82. Alva, K.S., Sarma, R., Marx, K.A., Kumar, J., Tripathy, S.K., Akkara, J.A., Kaplan, D.L. (1997). Biochemically Designed Polymers as Self Organized Materials. *Smart Materials Technol., SPIE* 3040:200–210.
83. Marx, K.A., Zhou, T., McIntosh, D., Braunhut. S.J. (2004). Electropolymerization of Biomimetic Peptide-Tyrosineamide Polymer Films for Specific Cell Attachment. *Arch. Appl. Biomater. Biomol. Mater., Mater. Res. Soc.* EXS-1:169–171.
84. Rosso, F., Giordano, A., Barbarisi, N., Barbarisi, A. (2004). From Cell-ECM Interactions to Tissue Engineering. *J. Cell Physiol.* 199:174–180.
85. Marx, K.A., Zhou, T., Sarma, R. (1999). Quartz Crystal Microbalance Measurement of Self-Assembled Micellar Tubules of the Amphiphilic Decyl Ester of D-Tyrosine and Their Enzymatic Polymerization. *Biotechnol. Progress* 15:522–528.
86. Sarma, R., Alva, K.S., Marx, K.A., Tripathy, S.K., Akkara, J.A., Kaplan, D.L., (1996). Enzymatic Polymerization of Amphiphilic Alkyl Tyrosine Derivatives From Emulsions. *Mater. Sci. Eng. C* 4:189–192.
87. Marx, K.A., Alva, K.S., Sarma, R. (2000). Self-Assembled Micron Scale Tubule Structures Formed by Amphiphilic Alkyl Tyrosine Derivatives Prior to and Following Enzymatic Polymerization. *Mater. Sci. Eng. C* 11:155–163.
88. Angelopoulos, A., Marx, K.A., Oh, S. (2002). The Use of Aqueous Enzymatic Polymerization of Amphiphilic Alkyl Tyrosine Derivatives as Environmentally Benign Coatings in the Microelectronics Industry. Polym. Interf. Thin Films *Mater. Res. Soc.* 710:207–212.
89. Janshoff, A., Galla, H.J., Steinem, C. (2000). Piezoelectric Mass Sensing Devices as Biosensors-an Alternative to Optical Biosensors? *Angew. Chem. Int. Ed.* 39:4004–4032.
90. Wegener, J., Janshoff, A., Steinem, C. (2001). The Quartz Crystal Microbalance as a Novel Neans to Study Cell-Substrate Alterations in situ. *Cell Biochem. Biophys.* 34:121–151.
91. Cavic, B.A., Hayward, G.I., Thompson, M. (1999). Acoustic Waves and the Study of Biochemical Macromolecules and Cells at the Sensor-Liquid Interface. *Analyst* 124:1405–1420.
92. Zhou, T., Marx, K.A., Warren, M., Schulze, H., Braunhut, S. J. (2000). The Quartz Crystal Microbalance as a Continuous Monitoring Tool for the Study of Endothelial Cell Surface Attachment and Growth. *Biotechnol. Prog.* 16:268-277.
93. Marx, K.A., Zhou, T., Warren, M., Braunhut, S. J. (2003). Quartz Crystal Microbalance Study of Endothelial Cell Number Dependent Differences in Initial Adhesion and Steady-State Behavior: Evidence for Cell–Cell Cooperativity in Initial Adhesion and Spreading. *Biotechnol. Prog.* 19:987–999.

94. Zhou, T., Braunhut, S. J., Medeiros, D., Marx, K.A. (1999). Cellular Adhesion and Spreading of Endothelial Cells Monitored in Real Time Using the Quartz Crystal Microbalance. In: Neenan, T., Marcolongo, M., Valentini, R.F., eds. Biomedical Materials—Drug Delivery, Implants and Tissue Engineering, *Proc. Mater. Res. Soc.*, 550:177–182.
95. Ingber, D.E. (1997). Tensegrity: The Architectural Basis of Cellular Mechanotransduction. *Ann. Rev. Physiol.* 59:575–599.
96. Marx, K.A., Zhou, T., Montrone, A., Schulze, H., Braunhut, S. J. (2001). A Quartz Crystal Microbalance Cell Biosensor: Detection of Microtubule Alterations in Living Cells at nM Nocodazole Concentrations. *Biosen. Bioelectron.* 16:773–782.
97. DeBrabander, M.J., Van de Viere, R.M.L., Aerts, F.E.M., Borgers, M., Jannsen, P.A. (1976). The Effects of Nocodazole: a New Synthetic Antitumoral Drug Interfering With Microtubules, on Mammalian Cells Cultured in vitro. *Cancer Res.* 36:905–916.
98. Braunhut, S.J., McIntosh, D., Vorotnikova, K., Zhou, T., Marx, K.A. (2005). Detection of Apoptosis and Drug Resistance of Human Breast Cancer Cells to Taxane Treatments Using Quartz Crystal Microbalance Biosensor Technology. *Assay Drug Dev.* 3:77–88.
99. Drukman, S., Kayallaris, M. (2002). Microtubule Alterations and Resistance to Tubulin-Binding Agents (Review). *Int. J. Oncol.* 21:621–628.
100. Vasques, R.J., Howell, B., Yvon, A.M.C., Wadsworth. P., Cassimeris, L. (1997). Nanomolar Concentrations of Nocodazole Alter Microtubule Dynamic Instability in vivo and in vitro. *Mol. Biol. Cell* 8:973–985.
101. Marx, K.A., Zhou, T., Montrone, A., Braunhut, S. J. (2002). A Quartz Crystal Microbalance Cell Biosensor: Detecting Nocodazole Dependent Microtubule Disruption Dynamics in Living Cells. *Adv. Biomater. Character. Tissue Eng. Complex., Mater. Res. Soc.* 711:125–132.
102. Raghow, R. (1994). The Role of Extracellular Natrix in Postinflammatory Wound Healing and Fibrosis. *FASEB J.* 8:823–831.
103. Bullard, K.M., Longaker, M.T., Lorenz, H.P. (2003). Fetal Wound Healing: Current Biology. *World J. Surg.* 27:54–61.
104. Zhou, T., Braunhut, J., Medeiros, D., Marx, K.A. (1998). Potential Dependent Endothelial Cell Adhesion, Growth and Cytoskeletal Rearrangements. In: *Materials Science of the Cell*, Pittsburgh, PA: Material Research Society, 489:211–216.
105. Marx, K.A., Zhou, T., McIntosh, D., Braunhut, S.J. (2004). Extracellular Matrix Studied In Situ with the Quartz Crystal Microbalance: Documenting Cellular Stress During Removal and the Elastic Behavior of the Matrix. In: Architecture and Applications of Biomaterials and Biomolecular Materials, *Proc. Mater. Res. Soc.*, EXS-1:111–113.
106. Marx, K.A., Zhou, T., Montrone, A., McIntosh, D., Braunhut, S.J. (2005). Extracellular Matrix Studied In Situ With the Quartz Crystal Microbalance: Documenting Cellular Stress During Detachment and the Elastic Behavior of the Matrix. *Anal. Biochem.* 343:23–34.
107. Braunhut, S.J., McIntosh, D., Vorotnikova, K., Zhou, T., Marx, K.A. (2004). Development of a Smart Bandage Applying Electrical Potential to Selectively Release Wound Healing Growth Factors from Cell Free Extracellular Matrix. In: Architecture and Applications of Biomaterials and Biomolecular Materials, *Proc. Mater. Res. Soc.*, EXS-1:403–405.
108. Braunhut, S.J., McIntosh, D., Vorotnikova, K., Zhou, T., Marx, K.A. (2005). Development of a Smart Bandage: Accelerated Wound Healing by Growth Factors Selectively Released from Cell Free Extracellular Matrix by Applied Electrical Potential. University of Massachusetts Lowell, Lowell, MA.
109. Wahl, M.C., Sundaralingam, M. (1995). New Crystal Structures of Nucleic Acids and Their Complexes. *Curr. Opin. Struc. Biol.* 5:282–295.
110. Kopchoyan, M., Leroy, J.L. (1995). Hydration and Solution Structure of Nucleic Acids. *Curr. Opin. Struc. Biol.* 5:329–333.
111. http://www.ncbi.nlm.nih.gov/genbank/
112. Attwood, J.T., Yung, R.L., Richardson, B.C. (2002). DNA Methylation and the Regulation of Gene Transcription. *Cell Mol. Life Sci.* 59:241–257.
113. Grosse, I., Herzel, H., Buldyrev, S.V., Stanley, H.E. (2000). Species Independence of Mutual Information in Coding and Noncoding DNA. *Phys. Rev. E* 61:5624–5629.
114. Hoffman, P., Grinstein, G., Marx, K.A., Grosse, I., Stanley, E. (1997). DNA Visual and Analytic Data Mining. *Proc. IEEE Visualization* 97:437–441.

115. Britten, R.J., Kohne, D.E. (1968). Repeated Sequences in DNA. *Science* 161:529–540.
116. Hearst, J.E., Cech, T.C., Marx, K.A., Rosenfeld, A., Allen, J.R. (1973). Characterization of the Rapidly Renaturing Sequences in the Main CsCl Density Bands of *Drosophila*, Mouse and the Human DNA. *Cold Spring Harbor Symp. Quant. Biol.* 38:329–339.
117. Marx, K.A., Allen, J.R., Hearst, J.E. (1976). Characterization of the Repetitious Human DNA Families and Evidence of Their Satellite DNA Equivalents. *Biochim. Biophys. Acta.* 425:139–147.
118. Hagerman, P.J. (1990). Sequence Directed Curvature of DNA. *Ann. Rev. Biochem.* 59:755–781.
119. Kruglak, S., Durrett, R.T., Schug, M.D., Aquadro, C.F. (1998). Equilibrium Distributions of Microsatellite Repeat Lengths Resulting from a Balance Between Slippage Events and Point Mutations. *Proc. Nat'l. Acad. Sci. USA* 95:10774–10778.
120. Zhou, Y., Bizzaro, J.W., Marx, K.A. (2004). Homopolymer Tract Length Dependent Enrichments in Functional Regions of 27 Eukaryotes and their Novel Dependence on the Organism DNA (G+C)% Composition. *BMC Genom.* 5:95–106.
121. Marx, K.A., Hess, S.T., Blake, R.D. (1993). Characteristics of the Large (dA).(dT) Homopolymer Tracts in *D. discoideum* Gene Flanking and Intron Sequences. *J. Biomol. Struc. Dyn.* 11:576–566.
122. Koo, H.S., Wu, H.M., Crothers, D.M. (1986). DNA Bending at Adenine×Thymine Tracts. *Nature* 320:501–506.
123. Marini, J.C., Levene, S.A., Crothers, D.M., Englund, P.T. (1983). A Bent Helix in Kinetoplast DNA. *Cold Spring Harbor Symp. Quant. Biol.* 47:279–283.
124. Iyer, V., Struhl, K. (1995). Poly(dA:dT). A Ubiquitous Promoter Element that Stimulates Transcription Via its Intrinsic DNA Structure. *EMBO J.* 14:2570–2579.
125. Marx, K.A., Hess, S., Blake, R.D. (1994). Alignment of (dA).(dT) Homopolymer Tracts in Gene Flanking Sequences Suggests Nucleosomal Periodicity in *D. discoideum* DNA. *J. Biomol. Str. Dyn.* 12:235–246.
126. Marx, K.A., Zhou, Y., Kishawi, I.Q. (2005). Evidence for Long Poly(dA).Poly(dT) Tracts in *D. discoideum* DNA at High Frequencies and Their Preferential Avoidance of Nucleosomal DNA Core Regions. *J. Biomol. Str. Dyn.* 23:429–446.
127. Bernstam, V.A. (1992). *Handbook of Gene Level Diagnostics in Clinical Practice.* Boca Raton, FL: CRC Press.
128. Wilson, R.W., Bloomfield, V.A. (1979). Counterion Induced Condensation of Deoxyribonucleic Acid-A Light Scattering Study. *Biochemistry* 18:2192–2196.
129. Earnshaw, W.C., Harrison, S.C. (1977). DNA Arrangement in Isometric Phage Heads. *Nature* 268:598–602.
130. Kai, E., Slack Nelle, L., Ahmad, A., Evan, H.M., Liu, A.J. Samuel, C.E., Safinya, C.A. (2004). Cationic Lipid–DNA Complexes for Gene Therapy: Understanding the Relationship Between Complex Structure and Gene Delivery Pathways at the Molecular Level. *Curr. Med. Chem.* 11:133–149.
131. Strzeleca, T.E., Davidson, M.W., Rill, R.L. (1988). Multiple Liquid Crystal Phases of DNA at High Concentrations. *Nature* 331:457–460.
132. Manning, G.S. (1978). The Molecular Theory of Polyelectrolyte Solutions With Applications to the Electrostatic Properties of Polynucleotides. *Quart. Rev. Biophys.* 11:179–246.
133. Marx, K.A., Ruben, G.C. (1983). Evidence for Hydrated Spermidine-Calf Thymus DNA Toruses Organized by Circumferential DNA Wrapping. *Nucl. Acids Res.* 11:1839–1853.
134. Marx, K.A., Ruben, G.C. (1984). Studies of DNA Organization in Hydrated Spermidine-condensed DNA Toruses and Spermidine-DNA Fibres. *J. Biomo. Struct. Dynamics* 1:1109–1132.
135. Marx, K.A., Ruben, G.C. (1986). A Study of PhiX-174 DNA Torus and Lambda DNA Torus Tertiary Structure and the Implications for DNA Self-Assembly. *J. Biomol. Struct. Dynamics* 4:23–39.
136. Marx, K.A., Ruben, G.C. (1986). Hydrated Macromolecular Assembly Structure Revealed by Freeze-Etch Stereo-Electron Microscopy. *Biophys. J.* 49:53–55.
137. Ruben, G.C., Marx, K.A. (1984). Parallax Measurements on Stereomicrographs of Hydrated Single Molecules: Their Accuracy and Precision at High Magnification. *J. Microsc. Res. Tech.* 1:373–385.

138. Marx, K.A., Reynolds, T.C. (1982) Spermidine-Condensed PhiX-174 DNA Cleavage by Micrococcal Nuclease: Torus Cleavage Model and Evidence for Unidirectional Circumferential DNA Wrapping. *Proc. Natl. Acad. Sci. USA* 79:6484–6488.
139. Marx, K.A., Reynolds, T.C. (1983). Ion Competition and Micrococcal Nuclease Digestion Studies of Spermidine-Condensed DNA: Evidence for Torus Organization by Circumferential DNA Wrapping. *Biochim. Biophys. Acta* 741:279–287.
140. Marx, K.A., Reynolds, T.C. (1989). A Micrococcal Nuclease Digestion Study of Spermidine-Condensed DNA. *Intern. J. Biol. Macromol.* 11:241–248.
141. Marx, K.A. (1987). Toroidal DNA Condensates; Aspects of Formation, Structure, Dynamics and Biological Implications. In: Nicolini, C., ed. *Structure and Dynamics of Biopolymers, NATO ASI Series E 133,* Martinus Nijhoff Publishers, Dordrecht, The Netherlands, pp 137–168.
142. Li, A., Huang, H., Marx, K.A. (1997). DNA End-to-End Distance Change Due to Divalent Counterion Condensation Studied by Pulse Gel Electrophoresis. Statistical Mechanics in Physics and Biology, *Material Research Society*, Pittsburgh, PA, 463:67–72.
143. Li, A., Marx, K.A., Qi, L., Zhang, H. (1996). Trivalent Counterion Condensation on DNA Measured by Pulse Gel Electrophoresis. *Biopolymers* 38:367–376.
144. Li, A., Huang, H., Re, X., Qi, L.J., Marx, K.A. (1998). A Gel Electrophoresis Study of the Competitive Effects of Monovalent Counterion on the Extent of Divalent Counterions Binding to DNA. *Biophys. J.* 74:964–973.
145. Li, A.Z., Marx, K.A. (1999). The Iso-Competition Point for Counterion Competition Binding to DNA: Calculated Multivalent vs. Monovalent Cation Binding Equivalence. *Biophys. J.* 77:114–122.
146. Seeman, N.C. (2003). DNA in a Material World? *Nature* 421:33–37.
147. Seeman, N.C. (2003). Structural DNA Nanotechnology: A New Organizing Principle for Advanced Nanomaterials. *Mater. Today* 6:24–29.
148. Marmur, J., Doty, P. (1959). Heterogeneity in Deoxyribonucleic Acids. *Nature* 183:1427–1431.
149. Mantripragach, K.K., Buckley, P.G., de Stahl, T.D., Dumanski, J.P. (2004). Genomic Arrays in the Spotlight. *Trends Genet.* 20:87–94.
150. Poland, D., Scheraga, H.A. (1970). *Theory of Helix-Coil Transitions in Biopolymers.* New York: Academic Press.
151. Blake, R.D., Bizzaro, J.W., Blake, J.D., Day, G.R., Delcourt, S.G., Knowles, J., Marx, K.A., SantaLucia, Jr., J. (1999). Statistical Mechanical Simulation of Polymeric DNA Melting With MELTSIM. *Bioinformatics* 15:370–375.
152. Long, D.D., Grosse, I., Marx, K.A. (2004). Coding and Non Coding DNA Thermal Stability Differences in Eukaryotes Studied By Melting Simulation, Base Shuffling and DNA Nearest Neighbor Frequency Analysis. *Biophys. Chem.* 110:25–38.
153. Multiple Authors, Multiple Articles. (1997). The Yeast Genome Directory. *Nature* 387:1–105.
154. Bizzaro, J.W., Marx, K.A., Blake, R.D. (1998). Comparison of Experimental With Theoretical Melting of the Yeast Genome and Individual Yeast Chromosome Denaturation Mapping Using the Program MELTSIM. In: Materials Science of the Cell, *Proc. Mater. Res. Soc,.* 489:73–78.
155. Bizzaro, J.W., Marx, K.A., Blake, R.D. (2005). Comparison of Experimental With Theoretical Melting of the 16 Yeast Chromosomes: Denaturation Mapping Using the Program MELTSIM. University of Massachusetts Lowell, Lowell, MA.
156. Marx, K.A., Bizzaro, J.W., Blake, R.D. (1997). Comparison of Experimental and Theoretical Melting Behavior of DNA. In: Statistical Mechanics in Physics and Biology, *Proc. Mater. Res. Soc.*, 463:147–152.
157. Marx, K.A., Assil, I.Q., Bizzaro, J.W., Blake, R.D. (1998). Comparison of Experimental to MELTSIM Calculated DNA Melting of the (A+T) Rich *Dictyostelium discoideum* Genome: Denaturation Maps Distinguish Exons from Introns. *J. Biomol. Struct. Dynamics* 16:329–340.
158. Marx, K.A., Bizzaro, J.W., Blake, R.D., Tsai, M.H., Tao, L.F. (2000). Experimental Melting Behavior of the Three Major Schistosoma Species. *Mol. Biochem. Parasitol.* 107:303–307.
159. Blake, R.D., Delcourt, SG. (1998). Thermal Stability of DNA. *Nucl. Acids Res.* 14:3323–3332.
160. Howbrook, D.N., van der Valk, A.M., O'Shaughnessy, M.C., Sarkar, D.K., Baker, S.C., Lloyd, A.W. (2003). Developments in Microarray Technologies. *Drug Disc. Today* 8:642–651.

161. Wang, A.H.J. (1992). Intercalative Drug Binding to DNA. *Curr. Opin. Struc. Biol.* 2:361–368.
162. Benos, P.V., Lapedes, A.S., Stormo, G.D. (2002). Is there a Code for Protein-DNA Recognition? Probab(ilistical)ly. *BioEssays* 24:466–475.
163. Berman, H.M., Olson, W.K., Beveridge, D.M., Westbrook, J., Gelbin, A., Dememy, T., Hsieh, S.H., Srinivasan, A.R., Schneider, B. (1992). The Nucleic Acid Database: A Comprehensive Relational Database of Three-Dimensional Structures of Nucleic Acids. *Biophys. J.* 63:751–759. URL: http://ndbserver.rutgers.edu
164. Luscombe, N.M., Austin, S.E., Berman, H.M., Thornton, J.M. (2000). An Overview of the Structure of Protein–DNA Complexes. *Genome Biol.* 1:1–37.
165. Olson, W.K., Zhurkin, V.B. (2000). Modeling DNA Deformations. *Curr. Opin. Struc. Biol.* 10:286–297.
166. Jones, S., Thornton, J.M. (2003). Protein–DNA Interactions: The Story so far and a New Method for Prediction. *Comp. Func. Genom.* 4:428–431.
167. Olson, W.K., Gorin, A.A., Xiang, J.L., Hock, L.M., Zhurkin, V.B. (1998). DNA Sequence-Dependent Deformability Deduced from Protein–DNA Crystal Structures. *Proc. Natl. Acad. Sci. USA* 95:11163–11168.
168. Westermann, S., Cheeseman, I.M., Anderson, S., Yates III, J.R., Drubin, D.G., Barnes, G. (2003). Architecture of the Budding Yeast Kinetochore Reveals a Conserved Molecular Core. *J. Biol. Chem.* 163:215–222.
169. Hegeman, J.H., Flieg, U.N. (1993). The Centromere of Budding Yeast. *BioEssays* 15:451–460.
170. Henemuth, B., Marx, K.A. (2006). Protein Deformability Properties of Single Base Mutants of Yeast DNA Kinetochore Protein Binding Sites Correlate with Measured Chromosomal Loss Rates for the Mutants in Yeast Cells. *BMC Mol. Biol.* 7:12–20.
171. Watson, J.D., Crick, F.H.C. (1953). Molecular Structure of Nucleic Acids. *Nature* 171:737–738.
172. Baisnee, P.F., Baldi, P., Brunak, S., Pedersen, A.G. (2001). Flexibility of the Genetic Code With Respect to DNA Structure. *Bioinformatics* 17:237–248.
173. Witten, I.H., Frank, E. (2000). Data Mining: Practical Machine Learning Tools and Techniques with Java Implementations. San Francisco: Morgan Kaufmann.
174. http://www.microarray.org/sfgf/jsp/home.jsp (accessed May 27, 2005).
175. Schafer, J.L. (1997). Analysis of Incomplete Multivariate Data. Monographs on Statistics and Applied Probability. Vol. 72. London, UK: Chapman & Hall/CRC.
176. Jen, J. (2005). Distinguishing Lung Tumors Based Upon Small Numbers of Genes Using Flow-Through-Chips. National Cancer Institute, Bethesda, MD.
177. McCarthy, J.F., Marx, K.A., Hoffman, P.E., Gee, A.G., O'Neil, P., Ujwal, M.L., Hotchkiss, J. (2004). Applications of Machine Learning and High-Dimensional Visualization in Cancer Detection, Diagnosis and Mangement. In: Umar, A., Kapetanovic, I., Khan, J., eds. *The Applications of Bioinformatics in Cancer Detection, Ann. N.Y. Acad. Sci.* 1020:239–262.
178. Meyerson, M. Dana-Farber Cancer Institute, http://research.dfci.harvard.edu/meyerson-lab/lungca/data.html (accessed 2003).
179. Affymetrix, Inc., Santa Clara, CA., http://www.affymetrix.com (2003).
180. Grinstein, G., Jessee. C.B., Hoffman, P., O'Neil, P., Gee, A. (2002). High-Dimensional Visualization Support for Data Mining Gene Expression Data. In: Grigorenko, E.V., ed. *DNA Arrays, Technologies and Experimental Strategies.* New York, NY: CRC Press.
181. Marx, K.A., O'Neil, P., Hoffman, P., Ujwal, M.L. (2003). Data Mining the NCI Cancer Cell Line Compound GI_{50} Values: Identifying Quinone Subtypes Effective against Melanoma and Leukemia Cell Classes. *J. Chem. Inf. Comput. Sci.* 43:1652–1667.
182. Gad-el-Hak, M. (2002). *The MEMS Handbook.* Boca Raton, FL: CRC Press.
183. Patolsky, F, Zheng, G., Hayden, O., Lakadamyali, M., Zhuang, X., Lieber, C.M. (2004). Electrical Detection of Single Viruses. *Proc. Natl. Acad. Sci. USA* 101:14017–14022.
184 Cooper, M.A., Dultsev, F.A., Minson, T., Ostanin, V.P., Abell, C., Klenerman, D. (2001). Direct and Sensitive Detection of a Human Virus by Rupture Event Scanning. *Nat. Biotechnol.* 19:833–837.
185. Hahm, J.L., Lieber, C.M. (2004). Direct Ultrasensitive Electrical Detection of DNA and DNA Sequence Variations using Nanowire Nanosensors. *Nanoletters* 4:51–54.

186. Barone, P.W., Baik, S., Heller, D.A., Strano, M.S. (2005). Near-Infrared Optical Sensors Based on Single-Walled Carbon Nanotubes. *Nat. Mater.* 4:86–92.
187. Bowden, M, Brittain, S., Evans, A.G., Hutchison, J.W., Whitesides, G.M. (1998). Spontaneous Formation of Ordered Structures in Thin Films of Metals Supported on an Elastomeric Polymer. *Nature* 393:146–149.
188. Jackman, R.J., Brittain, S.J., Adams, A., Prentiss, M.G., Whitesides, G.M. (1998). Design and Fabrication of Topologically Complex Three-Dimensional Microstructures. *Science* 280:2089–2091.
189. Zhang, Y., Seeman, N.C. (1994). Construction of a DNA-Truncated Octahedron. *J. Am.. Chem. Soc.* 116:1661–1669.
190. Winfree, E., Liu, F., Wenzler, L.A., Seeman, N.C. (1998). Design and Self-Assembly of Two-Dimensional DNA Crystals. *Nature* 394:539–544.
191. Winfree, E. (1996). On the Computational Power of DNA Annealing and Ligation. In: Lipton, R.J., Baum, E.B., eds. *DNA Based Computers.* Prov. Rhode Island: American Mathematical Society, pp. 199–221.
192. Park, S.J., Lazarides, A.A., Storhoff, J.J., Pesce, L., Mirkin, C.A. (2004). The Structural Characterization of Oligonucleotide -Modified Gold Nanoparticle Networks Formed by DNA Hybridization. *J. Phys. Chem. B* 108:12375–12380.
193. Fu, A., Micheel, C.M., Cha, J., Chang, H., Yang, H., Alivisatos, A.P. (2004). Discrete Nanostructures of Quantum Dots/Au with DNA. *J. Am. Chem. Soc.* 126:10832–10837.
194. Dirks, R.M., Pierce, N.A. (2004). Triggered Amplification by Hybridization Chain Reaction. *Proc. Natl. Acad. Sci. USA* 101:15275–15278.
195. http://stoddart.chem.ucla.edu/ (accessed April 22, 2005).
196. van Delden, R.A., Mecca, T., Rosini, C., Feringa, B.L. (2004). A Chiroptical Molecular Switch With Distinct Chiral and Photochromic Entities and its Application in Optical Switching of a Cholesteric Liquid Crystal. *Chem. Eur. J.* 10:61–70.
197. Ecker, D.J., Sampath, R., Blyn, L.B., Eshoo, M.W., Ivy, C., Ecker, J.A., Libby, B., Samant, V., Sannes-Lowery, K.A., Melton, R.E., Russell, K., Freed, N., Barrozo, C., Wu, J., Rudnick, K., Desai, A., Moradi, A., Knize, D.J., Robbins, D.W., Hannis, J.C., Harrell, P.M., Massire, C., Hall, T.A., Jiang, Y., Ranken, R., Drader, J.J., White, N., McNeil, J.A., Crooke, S.T., Hofstadler, S.A. (2005). Rapid Identification and Strain-Typing of Respiratory Pathogens for Epidemic Surveillance. *Proc. Natl. Acad. Sci. USA* 102:8012–8017.
198. Petricoin, E.F., Liotta, L.A. (2004). SELDI-TOF-Based Serum Proteomic Pattern Diagnostics for Early Detection of Cancer. *Curr. Opin. Biotechnol.* 15:24–30.

2

Optical Methods of Single Molecule Detection and Applications in Biosensors

Xiaomeng Wang, Melissa Massey, Paul A.E. Piunno, and Ulrich J. Krull

CONTENTS

2.1 Introduction

The development of highly sensitive and selective instruments are opening new inroads for the detection and monitoring of increasingly small amounts of material. The ultimate in such instrumentation is that which allows investigation of the properties of single molecules (1–7). The advent of specialized instrumentation techniques, such as the scanning tunneling microscopy (STM) and atomic force microscopy (AFM), has been a crucial technical advance that has permitted the area of nanotechnology to move forward. These latter methods are commonly used for investigations of molecules at solid planar interfaces (6,7), while optical methods are more commonly used when ensembles of molecules to be interrogated are located in a three-dimensional medium (1–5). However, various iterations of optical methods, including laser scanning confocal microscopy (CM), total internal reflection microscopy (TIRM), and near-field scanning optical microscopy (NSOM), are also able to provide information with sufficient spatial resolution for investigation of the location and behavior of individual molecules on surfaces (8–10).

The ability to use optical methods is often based on the fact that one single molecule can be repetitively cycled between its ground state and excited states many times in 1 s when using a wavelength that is resonant with this transition (11). The consequence is the ability to collect a large number of photons for detection and analysis from a single molecule, resulting in a sufficiently large signal-to-noise ratio for analytical interpretation.

The first papers to suggest practical measurement of single molecules by an optical method was published by Hirschfeld in 1976 (12), and reported the use of TIRM to detect a single antibody molecule labeled with nearly 100 fluorophores. Single fluorophore detection was subsequently reported using a flow cytometry method at room temperature (2,13). In 1993, using NSOM, the first optically generated image of a single molecule was obtained at room temperature (14). These were all important steps, culminating in the ability today to use techniques such as TIRM to produce two- and three-dimensional images of targets associated with emission by single fluorescent molecules (3–5). The advent of powerful small laser sources and semiconductor detectors with excellent amplification has made it possible for biosensor technology to advance and implement optical strategies for single molecule detection.

A fundamental question to pose at the outset is why detection and study of a single molecule is important in the areas of bioanalytical chemistry and biosensing? Most experiments typically involve a large number of molecules. The typical static ensemble measurements only yield average values based on collections of molecules that may actually be in different states, and such static measurements tend to report about only the final stages of reaction paths or molecular conformations (15). However, there is much to be learned from the study of intermediate steps during complicated reaction processes that involve many molecules, as there is substantial diversity and distribution of states. Dynamic processes would require the synchronization of all of the individual molecules in a large ensemble, and this is difficult to achieve and impossible to maintain (16). Therefore, single molecule dynamics and kinetics are seldom accessible. In contrast, single molecule detection offers observation of individual states or conformations as correlated with time. Biosensors that are designed to detect and study selective binding interactions can achieve new levels of confidence in determination of selective binding interactions by accessing information about individual states rather than ensemble averages. Furthermore, there continues to be a demand for detection of very small numbers of targets in complicated biological samples, such as the early detection of pathogens, or markers associated with genetic mutations that may signal the onset of cancers.

This chapter reviews instrumentation and procedures for single molecule detection by optical methods, and applications of such methods in bioanalytical chemistry and biosensor development.

2.2 The Principle of Single Molecule Detection Using Optical Methods

The fluorescence process of a single molecule begins with an electronic excitation caused by absorption of a photon, followed by internal relaxation to the lowest energy state within the excited state, and then radiative decay to the ground state. The lifetime of a fluorescence cycle is dominated by the radiative and nonradiative relaxation processes, and tends to fall in a range of approximately 10 ns–10 ps. Multiple excitation and emission cycles are possible at rates of about 10^7–10^8 per second when using a 1.0 mW laser beam focused to the diffraction limit (11). Fluorescence intensity and the time for emission are related to several deactivation channels. These can involve processes such as collisional deactivation, photochemical reactions leading to photodestruction of the molecule, energy transfer, and electron transfer. Photochemical reactions include photooxidation, photoionization, photodissociation, and photoisomerization. Another process that is not commonly a significant energy loss pathway in bulk solution, but that does influence solids and immobilized systems, is intersystem crossing from the excited singlet state to the triplet state, resulting in phosphorescence rather than fluorescence emission (11).

2.2.1 Origins of Fluorescence

One of the fundamental requirements of an optical biosensor is the generation of a photoluminescent signal indicative of target analyte binding. Fluorescence emission that is designed to be indicative of selective reactions can be detected and converted into an analytical signal.

Fluorescence is a type of photoluminescence that is generated when a photon of specific energy is absorbed by a molecule wherein an electron within the molecule is promoted from the ground state to an electronically excited state. The excited state can exist as either a singlet or triplet state (17). The singlet excited state occurs when the electron is paired with another electron of opposing spin (17). This is a quantum-mechanically allowed transition, and fluorescence emission occurs as a result of the spin-paired excited electron returning to the ground state (17). The triplet excited state occurs when the excited electron is of the same spin as the second electron in the pair, and upon relaxation, the electron must undergo a "spin-flip" to return to the singlet ground state (17). These processes are represented in the Jablonski diagram.

Within each of the energy levels of the excited state, there are several nonradiative vibrational energy levels (17). The relaxation of a molecule from an excited vibrational state to the lowest energy excited vibrational state occurs much faster than the radiative relaxation of the molecule from the excited electronic state to the ground state; hence, fluorescence occurs from the thermally equilibrated excited state (17). Because some energy is lost through nonradiative processes associated with the vibrational levels present within each of the excited and ground electronic states, there is a shift to longer wavelengths (lower energy) for the fluorescence emission with respect to wavelength of light required for excitation. This is referred to as a Stokes' shift. Stokes' shifts are also encountered due to collisional processes between molecules, solvent effects and reactions, which occur while the molecule is in the excited state (17). Stokes' shift measurements can reveal important information about a molecule. For example, when the excited state dipole moment is higher than the ground state dipole moment, the Stokes' shift increases as a function of increasing solvent polarity (17).

2.2.2 Measurement of Fluorescence

Fluorescence intensity is defined by an extension of the Beer–Lambert law:

$$F = \Phi K I_0 (1 - 10^{\varepsilon bC}) \qquad [2.1]$$

where F is fluorescence, Φ is the quantum yield, K is the instrument response coefficient, I_0 is the intensity of the excitation radiation, ε is the molar absorptivity of the fluorophore, b is the pathlength, and C is the concentration of fluorescent species. If the concentration of the species is relatively low (below 0.01 M), then the equation can be approximated as

$$F = 2.303 K I_0 \varepsilon b C \Phi \qquad [2.2]$$

When considering the use of fluorophores as an analytical tool, the quantum yield of fluorescence must be taken into account as it directly relates to the intensity of the fluorescence signal generated from the population of fluorophores under interrogation. The quantum yield of fluorescence is defined by:

$$\Phi = \frac{k_r}{k_r + \Sigma k_{nr}} \qquad [2.3]$$

where k_r represents the decay lifetime (rate) of the radiative processes and k_{nr} represents the decay lifetime of the nonradiative processes. Upon inspection of Equation 2.3, as the number of nonradiative processes and their lifetimes decrease, the quantum yield approaches unity and as such, the quantum yield represents the fraction of excited fluorophores that relax by fluorescence (radiative) emission.

The quantum yield of fluorescence is related to the lifetime of the excited state by the following relationship:

$$\Phi_F = {k_r}/{k_r + k_{nr}} = k_r \tau_s \qquad [2.4]$$

where τ_s represents the excited state lifetime or fluorescence decay time of the singlet state. This parameter defines the time window in which dynamic processes can be observed (17). Several parameters influence quantum yield, and hence lifetime.

2.2.3 Parameters Affecting Fluorescence

2.2.3.1 Molecular Structure: (Example—Degree of Conjugation)

A significant number of fluorophores are aromatic. The π conjugation of aromatics leads to a shift of the fluorescence excitation and emission maxima to longer wavelengths and results in an increase in the fluorescence quantum yield as the degree of conjugation increases (17).

2.2.3.2 Substituent Effects: (Example—Electron-Donating and Electron-Withdrawing Groups)

Electron-donating substituents on the parent ring system lead to an increase in molar absorptivity and a shift in both the fluorescence excitation and emission spectral profiles (17). Electron-withdrawing substituent effects are difficult to predict and depend on whether the carbonyl group is an aldehyde, ketone, or carboxylic acid. It also depends on the nature of aromatic carbonyl compound and solvent polarity. Nitroaromatics are usually phosphorescent, and if they are fluorescent, the fluorescence is usually not observed due to internal conversion caused by the charge-transfer efficiency of the excited state (17).

2.2.3.3 Heterocyclic Compounds

Fluorescence of aromatics that contain one or more heterocyclic nitrogens (azarenes) depends heavily on solvent polarity (17). Related heterocycles containing oxygen and sulfur can be similarly compared, and in the case where heterocycles contain N, O, or S atoms with single bonds to carbon, the quantum yield is relatively high owing to the orientation of the ? electron system of the rings and the nonbonding orbital of the N, O, or S atom (17). Several fluorophores of practical use are heterocyclic and contain either an N, O, or S atom or some combination of the three. Coumarins, rhodamines, pyronines, fluoresceins, cyanine, and oxazine classes are examples of some of the more common classes of heterocyclic fluorophores (17).

2.2.3.4 Temperature

An increase in temperature will decrease the quantum yield of fluorescence, as the nonradiative thermal processes owing to the increase in random molecular motion and increased average kinetic energy such as collision and intramolecular vibrations and rotations are enhanced as temperature increases. If the nonradiative processes are now higher, this will contribute to a lower quantum yield of detectable fluorescence.

2.2.3.5 Charge Transfer and Internal Rotation

Photoinduced intramolecular charge transfer occurs as a result of the excitation of an electron from a lower lying molecular orbital to a higher energy state molecular orbital (usually HOMO to LUMO) transition (17). If the excited-state molecular orbital is sufficiently separated in space from the ground-state molecular orbital, this causes an instantaneous change in the dipole moment of the fluorescent molecule (17). This change is not in equilibrium with the surrounding solvent molecules and has an enhanced effect as polarity of the solvent increases (17). In a fluid medium, the solvent molecules rotate within the excited-state lifetime of the fluorophore such that an organized solvent shell that is in equilibrium with the fluorophore is developed (17). This leads to a relaxed intramolecular charge transfer state (17). The relaxation caused by the solvent organization leads to an increased red shift of the fluorescence emission of fluorophores as the solvent polarity is increased (17). The intramolecular charge transfer may also be accompanied by internal rotation within the fluorophore under certain conditions (17). In some cases, the internal rotation and intramolecular charge transfer processes accompanied by an increase in solvent polarity will stabilize the excited state and lead to fluorescence emission. However, in other cases the internal rotation can occur without solvent relaxation and can lead to a twisted intramolecular charge transfer species that is not fluorescent (17). Photoisomerization can also occur where rotation about double bonds takes place and leads to nonradiative deexcitation in nonpolar solvents. However, photoisomerization in polar media is inefficient owing to the stabilization of the intramolecular charge transfer species, which leads to a radiative fluorescence emission (17).

2.2.3.6 Solvent Viscosity/Twist Angle: (Example—the Cyanine Dyes)

Cyanine molecules possess two resonance structures where there is a charged and an uncharged nitrogen atom in each resonance structure. This results in a symmetry of charge distribution when both of the isomers are considered (17). This leads to complexity in terms of the photophysics of relaxation. The fluorescence quantum yield and decay are dependent upon both twist angle and solvent viscosity (17). The nonradiative decay rate depends on twist angle, and consequently quantum yield depends on solvent viscosity/rigidity of surrounding media (17). For example, thiazole orange, an unsymmetric cyanine dye, has a double bond structure that allows intramolecular rotation to occur. It also contains two nitrogen atoms that create charge symmetry due to resonance. Thus, thiazole orange is weakly fluorescent in solution. However, it is highly fluorescent in viscous or rigid media, which makes it an excellent choice as an intercalator indicative of DNA hybridization.

Bridging rotational bonds between two ring structures with an oxygen atom can also restrict the bridging bond rotations (17). Thus, reduction in rotation is achieved, which is the main mode of reducing bond rotation between bridging bonds of the substituted phenyl rings in rhodamine dyes (17). This allows rhodamine dyes to be fluorescent in polar media and exhibit high quantum yields in polar solvents as a result (17). The fluorescence emission from fluorophores is therefore shown to be very sensitive to the structure of fluorophores as well as the immediate environment. This allows them to be excellent candidates for use as fluorescent probes in applications such as detection of nucleic acid hybridization. The characteristics of fluorescence (spectrum, quantum yield, lifetime) can provide a wealth of information about the environment close to the fluorophore, and thus fluorescence spectroscopy becomes a powerful investigative tool that provides spatial and temporal information with excellent sensitivity.

2.2.3.7 Quenching

Since quantum yield is essentially a measure of the fraction of photons emitted to the number of photons absorbed, a theoretical upper limit for Φ would be equal to unity. This value is never achieved for molecules due to nonradiative energy losses as well as a phenomenon that contribute to fluorescence quenching. The main photophysical processes that contribute to fluorescence quenching are the following: collisions with atoms/other molecules; electron transfer, excimer formation, exciplex formation; proton transfer, and energy transfer (17). Fluorescence energy transfer is an important mode of fluorescence quenching and is discussed in an individual section. Fluorescence quenching can also be an effective way to interrogate the environment near a fluorophore both quantitatively and qualitatively. Several molecular deactivation processes that lead to the quenching of fluorescence are outlined below.

2.2.3.8 Dynamic Quenching: (Example—Stern–Volmer Relation)

In the initial treatment of fluorescence quenching, the quenching phenomenon is assumed to be time-independent and, in essence, does not consider the role of diffusion-limited processes. This leads to a fluorescence decay, which is a single exponential profile, and the Stern–Volmer relation can then be represented using the following equation:

$$\frac{\Phi_0}{\Phi} = \frac{I_0}{I} = 1 + k_q \tau_0 [\mathrm{Q}] = 1 + K_{\mathrm{SV}} [\mathrm{Q}] \qquad [2.5]$$

where I_0 and I are the fluorescence intensities in the absence and presence of quencher, respectively, Φ_0 and Φ are the quantum yield terms in the absence and presence of quencher, respectively, K_{SV} is the Stern–Volmer constant, k_q represents the rate of decay of the fluorophore in the presence of quencher, and τ_0 is the excited-state lifetime of the fluorophore in the absence of quencher. In a plot of I_0/I vs. [Q] (Stern–Volmer plot), K_{SV} can be extracted as the slope if the plot yields a linear relationship of I_0/I vs. [Q] (17).

In reality, quenching of fluorescence is a time-dependent process, as a result of being a diffusion-controlled process (17). During the excited-state lifetime of the fluorophore, the quencher can approach the fluorophore to meet the proximity condition required to induce a quenching phenomenon. In this case, the k_q rate constant must be replaced by a rate constant that is a function of time, $k_1(t)$. In the diffusion-controlled model, k_q is equal to the diffusion rate constant k_1, where k_1 can be represented by the following equation:

$$k_1 = 4\pi\left(N_a/1000\right)R_c D \quad [2.6]$$

where N_a represents Avogadro's number, R_c is the distance of closest approach of the fluorophore and quencher (in cm), and D is the diffusional coefficient (in cm^2s^{-1}) which can be expressed by the Stokes–Einstein equation:

$$D = D_F + D_Q = \frac{kT}{f\pi\eta}\left(\frac{1}{R_F} + \frac{1}{R_Q}\right) \quad [2.7]$$

where D_F and D_Q are the translational diffusion coefficients of fluorophore and quencher, respectively, k is the Boltzmann constant, T is temperature, R_F and R_Q are the radii of the fluorophore and the quencher, respectively, η represents the viscosity of the medium, and f is a coefficient of boundary condition (17). This leads to a variation in the Stern–Volmer relationship:

$$\frac{I_0}{I} = \frac{1 + 4\pi N R_C D \tau_0 [Q]}{Y} \quad [2.8]$$

where Y is the "error function" and serves as a multiplying factor for the transient term that leads to deviations in the linearity of the Stern–Volmer plot (17).

2.2.3.9 *Static Quenching*

The argument for dynamic quenching suggests that the concentration of the quencher is not in large excess with respect to the fluorophore and therefore must approach the fluorophore to a distance where quenching would occur (17). If the quencher is in large enough excess with respect to the fluorophore, Q can be sufficiently close to F such that the interaction between F and Q is significant and no approach of F to Q (or vice versa) is necessary because of the high probability of the excited F and Q being in close proximity to each other (17). This type of quenching is called static quenching and refers to two different possible models.

2.2.3.10 *Perrin/Sphere of Effective Quenching*

If Q is located within a defined "sphere of quenching" of volume V_q, it has the ability to quench F. The constraint applied on this model is that the F and Q are in fixed positions brought upon by being in a viscous medium or a rigid matrix (17). A quencher that lies outside of the sphere of effective quenching does not have an effect on the fluorescence emission of the fluorophore in that the fluorescence decay after pulse excitation is not affected because the quencher is too far away from the fluorophore to reach a proximity where it would be effective during the excited-state lifetime of the fluorophore. The number of quenchers in V_q can be expressed by a Poisson distribution:

$$P_n = \frac{\langle n \rangle^n}{n!} \exp(-\langle n \rangle) \qquad [2.9]$$

where P_n is the probability of finding Q in the quenching sphere volume dimensions and n represents the mean number of quenchers that occupy the volume of the quenching sphere:

$$V_q : \langle n \rangle = V_q N_a [\mathrm{Q}] \qquad [2.10]$$

Similarly, the probability of Q not being in the quenching sphere can also be defined as a Poisson distribution:

$$P_0 = \exp(-\langle n \rangle) = \exp(-V_q N_a [\mathrm{Q}]) \qquad [2.11]$$

Therefore, fluorescence intensity would be proportional to P_0 because this population would represent the distribution of fluorophores not affected by the quencher, as it lies outside the sphere of effective quenching (17). This leads to the following expression of Perrin's model for quenching in terms of intensity:

$$\frac{I_0}{I} = \exp(V_q N_a [\mathrm{Q}]) \qquad [2.12]$$

A plot of $\ln(I_0/I)$ vs. [Q] will give the value of V_q where the typical sphere radius is ~10 Å (17). In the Perrin model of the presence of a sphere of effective quenching, the quencher and the fluorophore do not form a complex, but are rather just in very close proximity to one another.

Another form of static quenching exists where the fluorophore and quencher form a 1:1 complex that is nonfluorescent.

2.2.3.11 Formation of a Ground-State Nonfluorescent Complex (Preassociation)

The equilibrium of a complex formed between and F and Q can be expressed as:

$$\mathrm{F} + \mathrm{Q} \leftrightarrow \mathrm{FQ} \qquad [2.13]$$

Analogous to the Perrin model for static quenching, the fluorescence decay lifetime after time-resolved pulse excitation for unassociated (uncomplexed) fluorophore is unaffected while the overall fluorescence intensity of the solution decreases due to the presence of quencher in solution (17). The expression for the above equilibrium situation can be written in terms of an association constant, K_s:

$$K_s = \frac{[\mathrm{FQ}]}{[\mathrm{F}][\mathrm{Q}]} \qquad [2.14]$$

and according to the law of conservation of mass:

$$[\mathrm{F}_0] = [\mathrm{F}] + [\mathrm{FQ}] \qquad [2.15]$$

In turn, this allows the complex ([FQ]) to be replaced such that:

$$\frac{[\mathrm{F}]}{[\mathrm{F}_0]} = \frac{1}{1+K_s[\mathrm{Q}]} \qquad [2.16]$$

If fluorescence intensity is proportional to the concentration of the fluorophores, then this equation can be inverted and expressed in terms of intensity:

$$\frac{I_0}{I} = 1+K_s[\mathrm{Q}] \qquad [2.17]$$

Static and dynamic quenching can be encountered simultaneously as well. Cases where there are multiple interactions lead to deviations in the linearity of Stern–Volmer plots for the F and Q systems involved as well as complexities in combining mathematical relationships of both dynamic and static expressions.

In addition to the static and dynamic quenching models, several other modes of fluorescence quenching exist such as photoinduced electron transfer, excimer/exciplex formation, and photoinduced proton transfer. Transfer of energy from a donor chromophore to an acceptor species is also a mode of fluorescence quenching. This transfer is referred to as fluorescence resonance energy transfer (FRET).

2.2.3.12 Fluorescence Resonance Energy Transfer

When FRET occurs, the relaxation energy of the donor species is absorbed by the acceptor such that the fluorescence emission of the donor species is effectively quenched by the acceptor provided that there is sufficient spectral overlap between the fluorescence emission and the acceptor absorbance wavelengths. In this donor–acceptor relationship, the fluorescence emission of the donor is quenched by the acceptor. The acceptor, which absorbs the relaxation energy emitted by the donor species, then undergoes relaxation processes, and this emission of radiation can either be in the visible range of the electromagnetic spectrum or lie beyond the detection wavelength range for typical fluorescence detection instrumentation.

Energy transfer from a donor to an acceptor where the donor and the acceptor are two different molecules can be represented by the simple equation:

$$\mathrm{D}^* + \mathrm{A} \rightarrow \mathrm{D} + \mathrm{A}^* \qquad [2.18]$$

The energy transfer between D and A can be either radiative or nonradiative (17). FRET follows the nonradiative transfer mode and will be discussed exclusively below.

In nonradiative transfer, there is no emission of photons from the donor at distances shorter than the wavelength of excitation (17). This quantum mechanical photophysical process occurs when there is sufficient spectral overlap between the emission spectrum of the donor and the absorption spectrum of the acceptor. This overlap ensures that there are a sufficient number of vibronic transitions which are equal in energy in both the donor and the acceptor species such that the transitions are in resonance (17).

The energy transfer between the donor and the acceptor is due to several interactions between the two species. The interactions can be due to either short-range intermolecular orbital overlap or long-range coulombic interactions (17). Long-range coulombic interactions are long-range dipole–dipole interactions, and are also referred to as the Förster mechanism of interaction (17). This coulombic term refers to the energy transfer process where the excited electron occupying the LUMO of the donor relaxes and an electron in the HOMO of the acceptor are excited to the LUMO of the acceptor simultaneously (17). In this case, long-range interaction occurs over distances up to 8–10 nm (in comparison to short-range interactions which are a few tens of Angstroms) (17).

Förster developed a series of expressions that define this long-range dipole–dipole transfer of energy from a donor to an acceptor, and as a consequence, FRET is sometimes referred to as Förster resonance energy transfer.

The dipole–dipole energy transfer rate constant can be defined as:

$$k_T^{dd} = k_D \left[\frac{R_0}{r}\right]^6 = \frac{1}{\tau_D^0}\left[\frac{R_0}{r}\right]^6 \qquad [2.19]$$

where k_D represents the emission rate constant of the donor in the absence of energy transfer, is the excited-state lifetime of the donor in the absence of energy transfer (absence of the acceptor), r is the separation distance between the donor and the acceptor, and R_0 is the Förster distance, where the transfer to acceptor and decay of the excited donor are both equally probable (17). In other words, the transfer efficiency is 50% at the Förster distance. The Förster distance can be defined using the following equation:

$$R_0 = 0.2108\left[\kappa^2 \Phi_D n^{-4} \int_0^\infty I_D(\lambda)\varepsilon_A(\lambda)\lambda^4 d\lambda\right]^{1/6} \qquad [2.20]$$

where κ^2 is the orientation factor of the transition dipole moments of the donor and the acceptor, Φ_D^o is the fluorescence quantum yield of the donor in the absence of acceptor, n is the refractive index of the medium in the wavelength region of spectral overlap, and the integral expression represents the overlap integral between the donor emission and the acceptor absorption spectrum (17). This equation can be simplified by allowing $J(\lambda)$ to represent the overlap integral expression:

$$R_0 = 8.79 \times 10^{-5}\left[n^{-4}\Phi_D \kappa^2 J(\lambda)\right]^{1/6} \text{Å} \qquad [2.21]$$

The orientation factor, κ, can then be defined using the angles involved in the orientation of the transition dipoles of the donor and acceptor relative to one another:

$$\kappa^2 = \cos\theta_T - 3\cos\theta_D \cos\theta_A = \sin\theta_D \sin\theta_A \cos\varphi - 2\cos\theta_D \cos\theta_A \qquad [2.22]$$

The values of κ^2 can range from 0 to 4 where zero represents a perpendicular transition moment orientation between donor and acceptor and four represents parallel transition moments between the two species (17):

The transfer efficiency of the fluorescence energy transfer can be written in terms of the Förster distance and the separation distance between the donor and the acceptor:

$$E = \left[1/(1+(r/R_0)^6)\right] \qquad [2.23]$$

From this expression, it is clear that a $1/r^6$ distance dependence exists for FRET.

Fluorescence resonance energy transfer and the information that can be gathered from spectroscopic measurements using FRET pairs is a very valuable tool for determining distances within biomolecules and for biomolecular associations and assemblies. One such biomolecular association would be the hybridization of nucleic acid strands. When determination of distances within or between molecules is desired, FRET can be used as a "spectroscopic" or "molecular" ruler in the distance range of 1–10 nm (17). There are several spectroscopic experiments that can be carried out to determine information on parameters such as FRET distance, orientation factor, and the distance between donor and acceptor. The donor–acceptor distance can be determined by both steady state and time-resolved fluorescence measurements (17). The steady-state measurement of three different parameters can be taken to determine r: decrease in donor fluorescence, comparison of the absorption spectrum and the excitation spectrum (through observation of the acceptor fluorescence), or enhancement of acceptor fluorescence (17). In the time-resolved regime, decay of donor fluorescence and increase in the acceptor fluorescence (if the acceptor fluoresces in the wavelength region capable of being detected by the instrumentation used for the experiment) can be used to determine a value of r (17). One limitation of using these methods is that the distance between donor and acceptor is assumed to be constant; however, if the donor and acceptor are linked by a flexible chain such as a nucleic acid strand, the donor–acceptor distance is not always constant. Similarly, if the donor and acceptor are linked to two different nucleic acid strands, there is still a degree of uncertainty in the location of the donor and acceptor relative to one another. In this case, the method used for the estimation of a distribution of donor–acceptor distances is based on the measurement of donor fluorescence decay and the shape the decay profile takes as a result of these variable interactions between the donor and the acceptor (17). Time-resolved donor fluorescence decay gives the best approximation to the orientation and distance between the donor–acceptor pair; however, at best, this is only an average distance where the distance distribution has contributions from both the distance and the orientation parameters (17).

Viscosity of the medium that the donor–acceptor pair experiences also affects the dynamics of energy transfer because of the dependence of diffusion on viscosity (17). The distance that the donor and acceptor diffuse toward each other during the excited-state lifetime of the donor is dependent upon viscosity (17). If the donor and acceptor are unable to diffuse a significant distance toward one another during the excited-state lifetime of the donor, then a static limit exists between the donor and the acceptor. As a consequence, two modes of energy transfer are possible: the Perrin model of the sphere of effective quenching can approximate the energy transfer process, and if the distance between donor and acceptor is less than $2R_0$, then the energy transfer occurs via Förster transition dipole interactions (17). If the mean distance the donor and acceptor diffuse relative to each other is larger than the mean separation distance, r, during the excited-state lifetime of the donor species, then once again two modes of energy transfer are possible (17). If energy transfer occurs via a collisional process, then the Stern–Volmer model for kinetics is used to define the donor–acceptor system; if the Förster mechanism is considered, then a rapid diffusion limit exists (17). The energy transfer from the donor to acceptor is transferred to acceptors that lie in the $<2R_0$ distance range, as well as any acceptors that happen to diffuse to within that distance during the excited-state lifetime of the donor species (17). If the mean distance diffused by the donor and acceptor relative to each other during the excited-state lifetime is approximately equal to the distance r between the donor and the acceptor, then a very complex case is encountered and requires an intricate set of mathematical expressions to approximate values for FRET (17). Restrictions in terms of orientation geometry also bring about deviations from the "ideal" model of Förster decay (17).

2.2.4 Signal-to-Noise and the Practical Issues of Single Molecule Detection

Perhaps the main practical concern encountered in the use of fluorescence techniques is the improvement of the signal-to-noise ratio. This problem becomes a fundamental obstacle that must be addressed in the case of single molecule detection. Raman and Rayleigh scattering can produce significant background contributions, as can fluorescence from the environment surrounding the target molecule (e.g., the solid substrate in the case of an immobilized single molecule), and fluorescence from the detection optics. To ameliorate these problems, the fluorophore, buffer, embedding matrix, and rejection filter are first selected carefully with due consideration for how each variable would be controlled when simply optimizing an experiment for detection of a large number of molecules. Optical isolation of a single molecule is often the next step as an effective method to further improve signal-to-noise ratio. This approach attempts to assure that only one molecule is in resonance with the excitation radiation within an illuminated spot, and ideally results in a situation in which the average number of molecules in the excitation area is less than one (15).

There are two common experimental conditions that are used to reduce the number of molecules that are interrogated. The first and easiest approach is to use extremely dilute sample solutions to deposit or deliver target molecules to a microscopic area where interrogation occurs. Individual molecules in an excitation area (or volume) yield separate bursts of emission corresponding to different time periods when such molecules are located within the excitation region of the microscope optics. A second approach is to use a selective spectroscopic excitation or emission protocol to excite or detect only one molecule in mixture with others in the sample area or volume (16). This method requires that different molecules in the excitation zone must have sufficiently different excitation or emission properties so that a signal-to-noise advantage can be achieved.

Several other conditions should be considered to ensure that a fluorescence signal actually originates from one single molecule (16). Molecules that do not readily aggregate are preferred for single molecule detection. An emitter should have a substantial absorption and emission dipole so that spectroscopic properties can be assessed when the molecule is immobilized. Fluorescence emission from single molecules can exhibit on and off behavior due to "blinking" or photobleaching. Blinking is a photophysical property that offers distinctive emission levels, and this characteristic is peculiar to certain single molecule emitters (and quantum dots). Determination of fluorescence intensity using different concentrations of fluorophore can be used to correlate the intensity anticipated from individual molecules.

2.3 Methods of Detection

Several optical methods have been used for single molecule detection. Point detection schemes include confocal (18) and near-field scanning optical microscopies (19). Wide-field detection schemes include epifluorescence illumination with lamps (20) and defocused laser excitation (3), and total internal reflection (TIR) methods (21,22). Other techniques include two-photon fluorescence detection (23) and imaging (24), and surface enhanced Raman spectroscopy (SERS) (25,26). Figure 2.1 suggests the physical principles that represent the basis for different instrumental methods (e.g., see [27–37]). The methods that are presented in this article are meant to be representative but not all inclusive, and will focus on near-field microscopy, CM, wide-field epi-illumination microscopy, and TIR.

2.3.1 Fluorescence Correlation Spectroscopy

Fluorescence correlation spectroscopy (FCS) measures the time-dependent fluctuating fluorescence intensity in a defined area or volume of optical interrogation as the number of fluorescent molecules changes due to diffusion, or due to transition between fluorescent and nonfluorescent states (15,28,29). The name of this fluorescence method is given by the mathematical process of signal analysis. The light emitted from the sampling zone is proportional to the number of fluorescent molecules present, and the fluorescence intensity signal will change as a result of any chemical and photophysical reactions, and

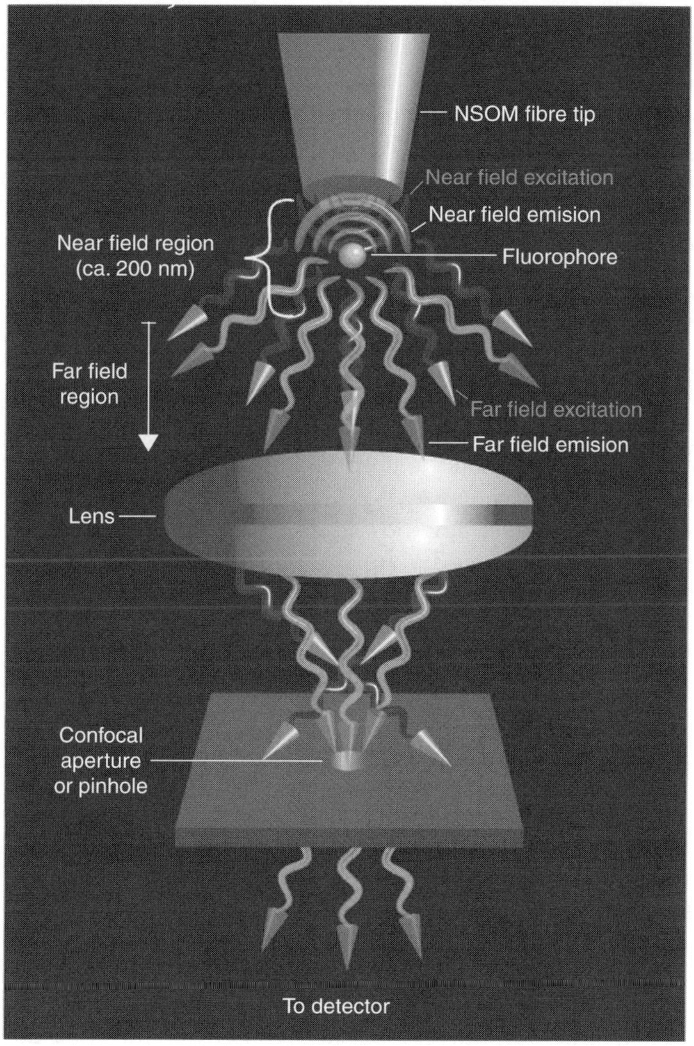

FIGURE 2.1
Physical principles that provide the basis for a number of different single molecule detection methods. An NSOM tip based on a tapered optical fiber that terminates with a diameter less than that of the excitation radiation is surrounded by a metal cladding. An evanescent field (wave) surrounds the fiber tip, decaying with distance in the "near-field" region. A fluorescent molecule (fluorophore) within the near-field region is excited into emission by the evanescent field, and the emitted radiation scatters into the environment. The emitted radiation is a photon in the "far-field", and can be collected by a suitable microscope. A lens serves to focus the emitted light onto a suitable photodetector. If a pin hole is placed before the detector, it is possible to adjust the depth of focus (focal plane depth) by restricting the angles of radiation that reach the detector, and this represents the principle of operation of a confocal microscope.

conformational alterations of these fluorescent molecules (15). By temporally autocorrelating the fluorescence intensity, the signal fluctuations can be quantified in terms of strength and duration. For a small number of molecules present in the detection volume, the fluctuations in the light intensity are large considering the amplitude and decay rate. FCS has been used as a tool to study single molecule behavior where the sample volume is on the order of nano- or pico-liters (29). Very low focal volume and a high quantum yield of the fluorophore are desired (29).

Rigler and Mets (30) were one of the first teams to use FCS to study single molecules. FCS has been applied to problems such as determination of rate constants of chemical reactions, protein folding, DNA transcription, and singlet-to-triplet crossing rate determinations (11,29). Applications of FCS in the area of biosensor development have included the determination of the mobility and function of "receptors," such as association and dissociation constants of intermolecular reactions. For example, when a short-labeled nucleic acid probe associates with a larger unlabeled molecule, the diffusion rate of this labeled probe becomes slower and the diffusion time is longer (29). According to the diffusion characteristics, the binding fraction could be calculated from the correlation mathematics (29).

There are a number of other optical methods that can make use of stochastic and correlation mathematics. Single molecules to be detected in flow cell when a dilute liquid sample flows through a confined volume that is excited by laser beam. Fluorescence is collected at 90 or 180 degrees with respect to the direction of the laser beam (11), and the fluorescence signal corresponds to single molecules passing across the excitation beam. This method can be used in rapid DNA sequencing, optical sizing of DNA fragments, rare-event detection, and ultrasensitive screening of combinatorial chemical libraries. This technique is operated as a compromise that limits large sample volume throughput and high efficiency for single molecule detection. An alternative to use of flowing streams or large volumes of static solution is to detect single molecules in microdroplets. One single molecule can be confined in one picoliter-sized droplet. An electrodynamic trap is used to levitate the droplet so that the molecule can be interrogated by a laser excitation for a longer period (11). The time required for levitation and fluorescence measurement tends to be too long to allow this methodology to be practical for single molecule counting and sorting.

2.3.2 Near-Field Scanning Optical Microscopy

Understanding of the organization and behavior of many basic molecular systems requires an in situ analysis of conformational dynamics at the single-molecule level. NSOM provides the ability to study single biomolecules without the ensemble averaging effects of traditional far-field optical spectroscopy (Figure 2.1). NSOM overcomes the Rayleigh limit of optical resolution and operates to provide spatial resolution in the range of 20–200 nm. The method relies on excitation using a standing evanescent field from a small probe tip, and is sensitive and well suited to study the dynamics studies of protein structure and conformational change (31,32). When a localized source of radiation having an intensity profile that is much smaller than its wavelength is placed within a fraction of one wavelength distance to the interested surface, then the excitation area will be similar to the size of the localized source (33). The source of radiation is usually a tapered optical fiber where the sides are coated with metal to prevent light leakage (16). Finally, the optical signal is detected by using an objective lens in either the transmission or collection mode. As one example, an aluminum-coated tapered optical fiber with a subwavelength aperture (5–10 nm) has been used as a waveguide for excitation using optical laser excitation, and can be raster-scanned at nanometer distances across a sample (34). Shear-force feedback (11), as found in related methods such as AFM, was used to maintain the tip at a constant distance from the sample, resulting in collection of a spatially resolved dataset that provided topographic reconstruction with nanometer resolution. Fluorescence that was emitted by individual molecules was collected by an oil immersion lens of high numerical aperture, and signal was developed using a high-sensitivity avalanche photodiode detector (34).

High-resolution NSOM has allowed the imaging of a fluorophore-labeled actin network associated with intact fibroblast cells, and the results were deemed to be far superior to those of the best far-field confocal images. Biomembrane imaging remains one of the frontiers of near-field research (35). NSOM has been used to determine the dipole orientation of individual fluorophores (16). The orientations of individual emitting dipoles were determined by modeling the field distribution, simulation of the absorption intensity with polarized excitation, and comparison with data obtained at two different excitation polarizations (16). NSOM has also been used to study dynamics, spectroscopy, and resonance energy transfer of single molecules (11).

Near-field scanning optical microscopy suffers from several limitations. There is unavoidable silica Raman scattering from optical fiber probes. A diffuse luminescence emission is always generated from the aperture of the fiber probe. The preparation of the probe is difficult to reproduce and sometimes the sample is perturbed by the presence of the probe. It is interesting to note that single-molecule experiments have not yet been reported at the highest theoretically possible resolution that should be achievable with near-field method (33). This may be due to the weak interaction between a single molecule and the radiation field, and therefore a greater power field may be required (while still maintaining room temperature conditions).

2.3.3 Far-Field Confocal Microscopy

Laser scanning confocal fluorescence microscopy is particularly attractive due to its combination of relatively large fluorescence (signal) capture and concurrent reduction of background signal (noise). In CM, pinhole plates are used to define and limit the depth of focus as shown in Figure 2.1, rejecting "out-of-focus" radiation from areas beyond the focal plane (37). Typically a laser source is reflected from a dichroic mirror (sometimes dichroic beam splitter), into a large numerical aperture microscope objective that can focus excitation radiation to a spot of approximately a 1 μm diameter on the sample. Fluorescence from the sample is collected by the same microscope objective and passed by the dichroic mirror through a pinhole located before a photon detector. The combination of high spatial and depth resolution defines an excitation volume of 0.5–1.0 fL in the sample (25,26). A cylindrically shaped excitation volume is typically about 0.5 μm in diameter and 2 μm in height, which is determined by the diffraction limit for the wavelength of visible light used, and the spherical aberration of the microscope objective.

Some of the first articles describing confocal fluorescence microscopy to detect single molecules were published by Rigler and coworkers (30,38). They recorded the fluorescence from single rhodamine-6G molecules in water diffusing through a 0.24-fL volume. The signal-to-noise ratio was determined to be about 1,000. With the development of photon-counting systems, Nie et al. (39,40) applied confocal fluorescence microscopy to study fluorescently labeled proteins and DNA fragments directly in real time. Furthermore, with the improvement of laser excitation power, it has even possible to trap and manipulate single macromolecules, such as DNA (11), either in microchannels or using an approach termed "optical tweezers" (Figure 2.2). A number of research groups have successfully applied these techniques to detect and identify single molecules (41–46). As one example, the binding of green fluorescence protein (GFP)-labeled XPA protein to a cyanine-labeled DNA substrate was assessed quantitatively by simultaneous detection of both fluorophores using confocal fluorescence microscopy. Colocalization of cyanine and GFP signals within one diffraction-limited spot indicated the formation of a single complex of XPA with DNA (8). The use of microcapillary, microstructure, or monomodal waveguide techniques have been provided for improved detection efficiencies, and have reduced the need to operate using very dilute sample concentration conditions (11).

The advantages of far-field CM include reduction of background interference, unlimited laser throughput (mW to W), three-dimensional sectioning capability, noninvasive detection, high sensitivity, and experimental simplicity. However, because a confocal laser beam probes only a single small excitation volume at a time, it is time-consuming to scan large sample areas of

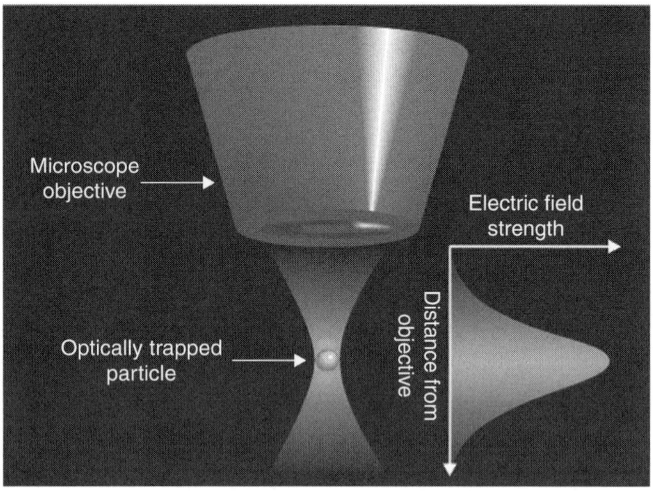

FIGURE 2.2
The principle of "optical tweezers". A high-intensity optical beam can be shaped to form a physical zone that has variation of electric field intensity. The gradient of electric field intensity can be used to trap particles.

interest. In cases where substantial areas must be interrogated it may be possible to use wide-field epi-illlumination as an alternative method for single molecule detection.

2.3.4 Wide-Field Epi-Illumination Microscopy

A conventional mode of fluorescence excitation in wide-field microscopy is epi-illumination, which is perhaps the most straightforward way to image single molecules at the diffraction limit. The optical excitation system consists of a laser source, defocusing optics, a high-performance dichroic beamsplitter, and an oil-immersion objective with low autofluorescence, where the excitation light and emission light travel through the same optics (11,37,47). Such instruments often use a multichannel charge-coupled device (CCD) camera or an intensified CCD (ICCD) as a detector. The imaged area is typically about 100×100 μm, but it also depends on the illumination field and the CCD-active area.

This technique has been used to achieve single molecule detection for numerous different applications. Single myosin molecules that are labeled with one or two copies of a fluorescent dye have been studied using this method (21). In this experiment, the large myosin molecules are adsorbed spontaneously onto a glass slide surface. Schmidt et al. (3) have measured the random two-dimensional motion of single fluorophore-labeled lipid molecules, which moved slowly in a lipid membrane. These examples indicate that wide-field epi-illumination can be used to investigate dry surfaces or targets in aqueous media. The use of a sensitive Peltier-cooled video-rate CCD camera has provided defocused images of single molecules so that orientation was observed (48). A comparison of experiment results with theoretical dipole orientation calculations was accomplished, which used a series of five images taken with increasing defocusing values ranging from 0 to 1.2 mm in steps of 300 nm. A significant correlation between the fluorescence from images and the patterns that were calculated for a parallel dipole orientation was seen (48). Owing to the exposure time of 3 s, only a few molecules survived all exposures without photobleaching or blinking.

The method of wide-field epi-illumination can reduce the amount of time to accumulate a statistically significant number of observations, but there are technical limitations. While single molecules are detected when they are in focus, there is often overwhelming background interference generated from both in-focus and out-of-focus excitation volumes. The reduction of some background scatter can be achieved by the use of high-quality optics. Excitation can also be separated from emission radiation by appropriate design of optics, for example the use of external reflection (16). TIRM is an effective wide-field alternative to epi-illumination to realize this design advantage.

2.3.5 Total internal Reflection Microscopy

Total internal reflection microscopy is a wide-field technique that can be used to observe the interface between two media with different refractive indexes, such as glass and water (49). The experiment is based on a near-field optical effect. An excitation laser beam enters a waveguide and is trapped in a medium of high refractive index, which is surrounded by a medium of low refractive index if the beam is launched at an incident angle greater than the critical angle of TIR as defined by Snell's Law (50). A standing (i.e., nonpropagating) evanescent field appears at the interface of the two media and the intensity exponentially decays into the medium of lower refractive index. The wavelength of the radiation in the evanescent field is the same as in the waveguide, and the exponential decay of intensity limits the excitation depth to a range of a few wavelengths, with most excitation energy being located close to the surface of the waveguide (few hundred nanometers of visible radiation). The evanescent wave allows for excitation of single molecules on (or near) the surface of the waveguide. While the total energy in the evanescent field is often a very small fraction of that pumped into the waveguide, the optical configuration substantially reduces background signals based on scatter and emission from bulk solution. Finally, emission from single molecules is collected by a microscopy objective and is typically imaged using an ICCD camera.

Hirschfeld is regarded as the pioneer of TIRM to detect single molecules. Early work involved the detection of targets such as a single antibodies that were each labeled with tens of fluorophores. With the advent of more sophisticated detection and sample manipulation technologies, single fluorophores can be detected. Examples include detection of labeled myosin and kinesin molecules (21,51,52), and visualization of cell signaling proteins including peptide hormones, membrane receptors, small G proteins, cytoplasmic kinases as well as small signaling compounds after labeling (36). A three-dimensional image of a single Nile red dye molecule confined in nanometer-sized pores of polyacrylamide gels has been obtained by TIRM (53). Although TIRM has been used to study single molecules that can freely move in solution, it is clear that more information can be obtained if the molecules are immobilized (15).

A variety of geometries for TIR excitation of fluorescence have been used near a dielectric interface in wide-field microscopy. Strategies include use of prisms, optical fibers, gaps between optical slides, and use of multimodal and monomodal waveguides (54). The largest signal-to-noise ratio can be obtained by using prism-TIR, while the largest total number of detected photons can be obtained using prism-less through-objective TIR (54). No matter which configuration is used, single molecule detection with TIR excitation seems more effective than that with wide-field epi-fluorescence because of lower optical background contributed by out-of-focus excitation volumes and due to interfacial excitation (9,55–58).

2.4 Applications of Single Molecule Detection

Single molecule detection has the potential to explore nanoenvironments at a molecular level. Single molecule allows exploration of heterogeneity and dynamic state changes without synchronization in condensed phases (15). DNA sequencing, DNA fragment sizing, protein conformations, and molecular dynamics have been addressed by single molecule detection methods. Note that one important distinction to always consider when developing quantitative analyses is that there are different statistical models that are applied to ensemble averages and to individual events (stochastic processes). In the extreme, a bioanalytical or biosensing system based on single molecule detection must invoke an element of time and the concept of "on" and "off" switching. Theoretically, detection of a single molecule of target in a sample by a single molecular receptor in a measurement system only requires that the two species eventually collide in an associative interaction. Practically, this can take time periods of years if sample volumes are large and processes are diffusion controlled. This implies the need for rethinking of the common definitions in analytical chemistry of limit-of-detection and sensitivity so as to accommodate stochastic processes.

2.4.1 DNA Sequencing and DNA Fragment Sizing

DNA sequencing methods often involve incorporation of fluorescently tagged nucleotides by the use of a polymerase chain reaction (PCR). In subsequent steps, the labeled DNA strand is anchored to a surface in a flowing stream, exonuclease enzymes are used to release the tagged nucleotides from the free end of the DNA, and then the DNA sequence is determined after detection and identification of the cleaved nucleotides by their fluorescent tags (11). Single molecule detection has the potential for sequencing about 10^4 bases with high speed, and this is attractive in comparison to gel-based sequencing technologies where detection is often limited to lengths of less than 1,000 bases and can be sensitive to molecular conformation. Keller and coworkers (59) were the first to use this novel method for rapid DNA sequencing, and several other research groups have also used such single molecule detection methods for DNA sequencing (60–65).

It has been established that the number of fluorescent dye molecules that are incorporated by the DNA fragment of interest is related to its physical properties, such as size, surface area, volume, and reactivity (15). In the case of intercalating fluorescent dyes that have no sequence preference and are uniformly distributed along the DNA, the fluorescence intensity is linearly related to the fragment length of DNA. There are two single molecule detection methods that are used for rapid DNA fragment size analysis by measurement of fluorescent intensity from intercalators. One method introduces a solution that contains DNA fragments into a sheath flow cuvette, and DNA fragments can then be detected by a downstream laser beam. The smallest detectable fragment size is approximately 1,000 base pairs (66). The second approach is to stretch DNA molecules on a glass surface and then cut them with restriction enzymes. The relaxation of strands at cleaved sites form micrometer-wide gaps, and these can be seen using an optical microscope. The integrated fluorescence signals of individual fragments reflect the initial fragment size of the DNA sequence. Note that a single molecule detection scheme based on intercalating dyes cannot achieve single-base pair resolution because DNA intercalates at a density of about 3–5 base pairs per dye molecule (16).

2.4.2 Single-Pair Fluorescence Resonance Energy Transfer

Fluorescence resonance energy transfer has been widely used in structural biology, biochemistry, and polymer science for measuring separation distances (15). In FRET, energy of a donor fluorescence molecule is transferred to an acceptor by a through-space dipole interaction. Because FRET is strongly dependent on distance, the process can be used as a tool to evaluate distances, as well as to signal changes of distances between acceptor–donor species.

If the acceptor and donor are in close proximity to one another, the acceptor can absorb the emission energy of the donor in a nonradiative mechanism as long as the pair is within the Förster distance for that particular set of donor and acceptor species. The donor and acceptor can both be fluorescent dyes where the wavelength of excitation for the acceptor is in the overlap integral of the fluorescence emission wavelength for the donor. Alternatively, the acceptor can be a "dark absorber" which does not emit fluorescence radiation such that the only signal that is observed is from the donor when it moves sufficiently far enough away from the acceptor species. There are several solution-based assays that operate upon this principle, which have been developed in recent years, and "Molecular Beacons" are one such example, which have been used for single molecule detection.

Molecular beacons (MB) are single-stranded probe oligonucleotides that possess a donor and an acceptor bound to the terminal ends of the sequence. A specified number of bases at the 3′ and 5′ ends of the sequence (usually 4–7 bases) are complementary to one another allowing the MB to adopt a hairpin or "stem-loop" conformation. This conformation allows the donor and acceptor to be in close proximity to one another and quenching of the donor species by the acceptor species results. The MB maintains this conformation until a target sequence of sufficient complementarity to the MB is introduced which allows the system to overcome the energetics of binding of the hairpin structure, which is then dissociated, and binds to its complementary sequence. This binding event elongates the MB sequence which increases the distance between the donor and acceptor species beyond the Förster distance and the fluorescence of the donor species is restored. Several combinations of donor/acceptor pairs can be considered including two fluorophores, a fluorophore and a quencher that does not emit fluorescence radiation, a fluorophore and an intercalator, a fluorophore and a gold particle, and a fluorophore and quantum dot (Figure 2.3). Quantum dots are small semiconductor particles of approximately 10 nm dimension, which have very large quantum yield and very narrow emission spectra. The narrow spectral emission band and the lack of any significant photobleaching suggest that quantum dots can be excellent labels for

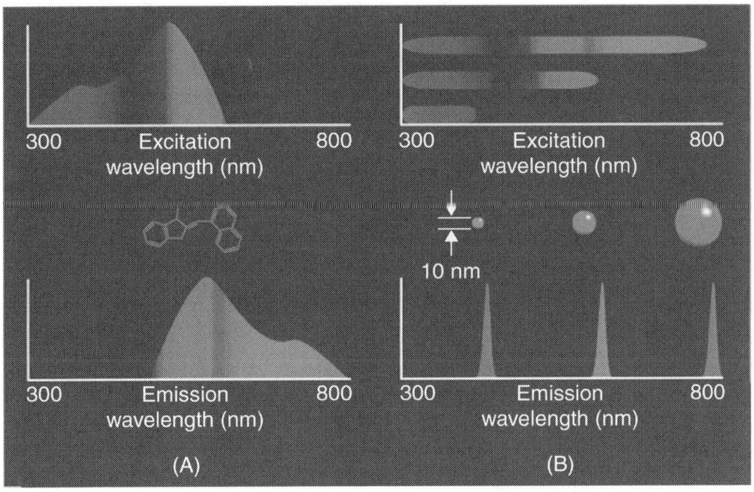

FIGURE 2.3
Aspects for consideration of selection of donor–acceptor pairs when developing energy transfer quenching systems. Panel A shows a representation of the broad band molecular excitation and emission that might be observed for a fluorophore. The width of the spectral excitation band offers the opportunity for selection of a wavelength that could provide selectivity for targeting a specific fluorophore in a mixture. The broad molecular emission band is often undesirable, in that there can be spectral overlap with the absorption band of the same molecular species, and any quenching partner would ideally need to have an absorption profile that would mirror the emission band. Panel B is representative of the spectroscopy of quantum dots. Excitation can be done using wavelength bands, or even full spectral excitation across the entire visible region. Emission is confined to relatively narrow bands, and these can be tuned to appear at different wavelengths by selection of the diameter of the quantum dot (size scale range of a few nanometers and higher). The narrow band of emission provides advantages in selection of operating wavelengths to maximize energy transfer and to avoid background spectral interference (when using narrow band excitation).

tracking single molecules, either directly or as a member of a donor–acceptor pair. There are several mechanisms available for the acceptor to quench the fluorescence of the donor species including static quenching, dynamic quenching, and contact mediated quenching (17). The length and nature of the linker species involved in tethering the donor and acceptor moieties to the 3′ and 5′ ends of the oligonucleotide are also factors that must be considered when optimizing an MB system. Tan et al. (67) has designed MB species with limits of detection of 1.7×10^{-10} M using a simple spectrometer detection scheme, and has used MB probes inside cellular cytoplasm as miniature biosensors.

Fluorescence resonance energy transfer has been used to investigate conformations of RNA based on energy transfer between the donor and acceptor (68,69). The advantage of single-molecule FRET is that it provides for study of time-dependent phenomena. In contrast to FRET from ensembles, which rely on signal averaging, single-molecule FRET measures only one donor–acceptor complex at a time without synchronization (15), and the observations can be uniquely attributed to a single species. Single-molecule FRET has been applied to study immobilized molecules at interfaces, such as for investigation of RNA and protein folding (70–72), and has also found applications in cases where molecules diffuse freely through an excitation laser beam (73–75). Excellent review articles have been published by Selvin, and by Ha (68,76).

2.4.3 Single-Molecule Electrophoresis

Single-molecule electrophoresis measures the different electrophoretic velocities of each molecule and this can be used to identify components of mixtures of a sample. The method has been applied to analyze mixtures of nucleic acids and mixtures of proteins and has the potential to be applied to the area of organic and inorganic chemical analyses. The detection instrumentation uses a laser beam that is split. The two beams are focused into a capillary cell to yield two 5-μm spots, which are separated by 250 μm from each other. An external electric field is applied between the two ends of the capillary cell and the individual molecules are detected as they pass through each of the two laser beams. Each laser beam has a separate time correlated single-photon detector under computer control, and the electrophoretic velocity can be calculated from the time required for each molecule to travel the distance between the two beams (77). The computer then produces a histogram of electrophoretic velocities, which shows a statistical maximum for each molecule present in the sample. A comparison of measured velocities with the velocity characteristics of a calibration species allows tentative identification of the molecules in a sample. The electrophoretic velocity of a single molecule is determined by its size, shape, ionic charge, and the chemical environment of the solution in which it is contained (78). The electrophoretic velocity, therefore, can provide a unique signature for different molecular species.

The method combines the advantages of bulk solution capillary electrophoresis and single molecule detection. It is possible to develop automation, speed, reproducibility, and high sensitivity by optimizing separation and choosing appropriate detection technology. Single-molecule electrophoresis has the potential to be used in fluorescence immunoassay, hybridization, and DNA fingerprinting techniques without the need for extensive DNA amplification by using PCR or other methods (77).

2.4.4 Single Molecule Detection in the Study of Dynamics

Two different time scales are encountered when studying the dynamics of single molecules. A slow scale of milliseconds to hundreds of seconds provides time for processes such as spectral fluctuation and diffusion. The scale of picoseconds to nanoseconds is a time frame associated with excited-state processes. In 1994, Trautman et al. (79) reported that the emission spectra of single molecules dispersed on a film of PMMA changed with time. In the same year, Ambrose et al. and Xie et al. (80,81) each reported the observation that there were sudden jumps in the emission intensity from single rhodamine molecules dispersed on a glass surface. Furthermore, Xie et al. modulated the polarization of excitation light to examine whether the single-molecule spectral jumps were due to molecular reorientation on the surface. This represented the beginning of direct monitoring of dynamics of single molecules (15). Many other experiments have subsequently been reported and cover topics such as the Brownian motion of single chromophores in a lipid membrane (3) and spectral fluctuations of single sulforhodamine-101 molecules (82,83). Radiative and nonradiative energy transfer, charge transfer, electron transfer, and other processes have been studied by single molecule detection (15). Lu and Xie (84) reported the single-molecule chemical kinetics for interfacial electron transfer from cresyl violet molecule to indium tinoxide nanoparticles. Importantly, it was demonstrated that electron transfer data obtained from a single molecule detection system showed single exponential kinetics, compared with the observation of multiexponential kinetics from an ensemble average of experiments using the same chemical system. The reason for this contradiction lies in site heterogeneity of single dye molecules with their environments.

2.4.5 Single Molecule Detection in Biomolecular Dynamics

Conformational states and changes between states are fundamental to most biochemical processes that play a role in biological function. Often there are many states and many conformations that are weakly separated in energy. Such states cannot be easily distinguished by ensemble-average experiments. Single molecule detection techniques have been used in a broad range of applications in life sciences and biotechnology, including molecular motors, DNA transcription, enzyme reactions, protein dynamics, and cell signaling.

Individual enzyme reaction activities based on different conformational states is an area of substantial interest. A review by Greulic (85) reports techniques for determination of the activity of a single enzyme molecule. Xue and Yeung (86) have confined single lactate dehydrogenase molecules along a capillary to detect individual molecular activities, and the catalytic activities among different enzymes have been found to vary by up to a factor of four. This result suggests the presence of different enzyme conformational states. Fluorescence resonant energy transfer has been used to observe ribozyme folding intermediates (71). Kinosita and coworkers (87) have observed the discrete and stochastic 120° rotation steps of the g subunit in F1-ATPase. The 120° rotation step is composed of a 90° rotation step corresponding to the binding of ATP and a 30° rotation step corresponding to the hydrolysis of ATP.

Using single molecule detection technique, it is possible to probe the heterogeneity in reaction rates among nearly identical molecules, and also to study the extent to which activity is related to the previous reaction history. Such a "molecular memory effect" has been observed for flavoenzyme when it catalyzes the oxidation of cholesterol by oxygen (88,89). This experiment was done using a confocal microscope with enzymes immobilized in an agarose gel. This experiment provided evidence that the memory effect was due to slow conformational fluctuations of the enzyme molecules.

Besides enzyme reactivity studies, single molecule detection method has also been widely used in the area of protein dynamics and protein folding. The fluorescence of GFP from the jellyfish *Aequorea victoria* shows repeated on/off cycles on a time scale of seconds (90). Wazawa et al. (91) also observe similar spectral fluctuation of a single fluorophore, tetramethylrhodamine conjugated to the myosin S1, on the same time scale. The phenomena observed from these experiments may be due to slow conformational changes of proteins (89). Single pair FRET has also been used for such studies (92,93). Taking advantage of the distance dependence of FRET, the intermediate states during protein-folding processes have been monitored at different denaturant concentrations. The fluorescence polarization method has also been used to determine the motion of proteins (94–98).

Besides enzyme reactivity and protein dynamics, various types of biomolecule measurements have been made at single-molecule level. These nanometer-sized biomolecules use low energy to work efficiently and accurately in our everyday life. The single molecule detection measurements have provided a clue to understand the mechanism of these processes (89). Though using recent advanced ensemble measurements, the structure of the biomolecules and the roles they play have already been determined, the dynamic properties of these molecules are averaged and could not be observed. Therefore, single molecule detection technique made it possible for us to observe the structure change and the interaction between these biomolecules dynamically. Furthermore, this technique has allowed us to record the behavior of individual molecules in real time.

2.5 Conclusions

Optical single molecule detection techniques have been widely developed and applied in physical chemistry, material science, and the biosciences (102). Compared with the ensemble average experiment, observations of single molecule structure and function provide advantages in understanding reactivity and dynamics. The optical approach permits such interrogation without contacting the molecule, although there must be consideration of the influence and perturbations caused by spectroscopic labels in cases where these are added. Other single molecule methods that do not make use of optical probing have been described, and include surface force microscopies, electrochemical wiring to single molecules, and lipid membranes that can detect the transit of single molecules such as DNA oligonucleotides through artificial pores.

Optical methods that have been adopted for single molecule detection include FCS, NSOM, CM, TIR, and wide-field epi-illumination. These methods have already been used to provide much insight into particular chemical and biochemical systems of interest. A number of other methods are suitable for single molecule analysis, including frequency modulation spectroscopy and two-photon excitation. It is clear that substantial advantage can be gained by integration of methods, such as NSOM and AFM. This combination can provide molecular information by fluorescence detection and nanometer-resolved topographic information by AFM (11,99–101).

References

1. Ambrose, W.P., Goodwin, P.M., Keller, R.A. (1994). Alterations of single molecule fluorescence life times in near-field optical microscopy. *Science* 265(5170): 364–367.
2. Shera, E.B., Seitzinger, N.K., Davis, L.M., Keller, R.A., Soper, S.A. (1990). Detection of single fluorescent molecules. *Chem. Phys. Lett.* 174: 553–557.
3. Schmidt, T., Schutz, G.J., Baumgartner, W., Gruber, H.J., Schindler, J. (1996). Imaging of single molecule diffusion. *Proc. Natl. Acad. Sci. U.S.A.* 93: 2926–2929.

4. Schutz, G.J., Pastushenko V., Kahr, H., Pragl, B., Knaus, H.G., Schindler, H. (2000). 3D-imaging of single molecules on living cells. *Biophys. J.* 78(1): 446A–446A.
5. Hibino, K., Watanabe, T.M., Kozuka, J., Iwane, A.H., Okada, T., Kataoka, T., Yanagida, T., Sako, Y. (2003). Single- and multiple-molecule dynamics of the signaling from H-Ras to cRaf-1 visualized on the plasma membrane of living cells. *Chem. Phys. Chem.* 4(7): 748–753.
6. Binning, G., Quate, C.F., Gerber, C. (1986). Atomic force microscope. *Phys. Rev. Lett.* 56: 930–933.
7. Binnig, G., Rohrer, H. (1982). Scanning tunneling microscopy. *Helvetica Physica Acta* 55(6): 726–735.
8. Segers-Nolten, G.M.J., Wyman, C., Wijgers, N., Vermeulen, W., Lenferink, A.T.M., Hoeijmakers, J.H.J., Greve, J., Otto, C. (2002). Scanning confocal fluorescence microscopy for single molecule analysis of nucleotide excision repair complexes. *Nucleic Acids Res.* 30: 4720–4727.
9. Funatsu, T., Harada, Y., Tokunaga, M., Salto, K., Yanagida, T. (1995). Imaging of single fluorescent molecules and individual ATP turnovers by single myosin molecules in aqueous solution. *Nature* 374: 555–559.
10. Moyer, P.J., Paesler, M. (1996). *Near-field optics: Theory, instrumentation, and applications,* New York, Wiley Publishers.
11. Nie, S., Zare, R.N. (1997). Optical detection of single molecules. *Ann. Rev. Biophys. Biomol. Struct.* 26: 567–596.
12. Hirschfeld, T. (1976). Optical microscopic observation of single small molecules. *Appl. Opt.* 15: 2965–2966.
13. Nguyen, D.C., Keller, R.A., Jett, H., Martin, J.C. (1987). Detection of single molecules of phycoerythrin in hydrodynamically focused flows by laser-induced fluorescence. *Anal. Chem.* 59: 2158–2161.
14. Betzig, E., Chichester, R.J. (1993). Single molecules observed by near-field scanning optical microscopy. *Science* 262: 1422–1425.
15. Kumbhakar, M., Nath, S., Mukherjee, T., Mittal, J.P., Pal, H. (2004). Single-molecule detection in exploring nanoenvironments: An overview. *J. Photochem. Photobiol. C: Photochem. Rev.* 5: 113–137.
16. Michalet, X., Weiss, S. (2002). Single-molecule spectroscopy and microscopy. *Comptes Rendus Physique* 3: 619–644.
17. Valeur, R. (2002). *Molecular fluorescence principles and applications,* Weinheim, Wiley, VCH Verlag, GmbH.
18. Pawley, J.B. (1995). *Handbook of biological confocal microscopy,* New York, Plenum Press
19. Betzig, E., Trautman, J.K. (1992). Near-field optics—microscopy, spectroscopy, and surface modification beyond the diffraction limit. *Science* 257: 189–195.
20. Goulian, M., Simon, S.M. (2000). Tracking single proteins within cells. *Biophys. J.* 79: 2188–2198.
21. Ruckstuhl, T., Seeger, S. (2004). Attoliter detection volumes by confocal total-internal-reflection fluorescence microscopy. *Opt. Lett.* 29: 569–571.
22. Paige, M.F., Bjerneld, E.J., Moerner, W.E. (2001). A comparison of through-the-objective total internal reflection microscopy and epifluorescence microscopy for single-molecule fluorescence imaging. *Single Mol.* 2: 191–201.
23. Mertz, J., Xu, C. (1995). Single-molecule detection by two-photon-excited fluorescence. *Opt. Lett.* 20: 2532–2534.
24. Sanchez, E.J., Novotny, L., Holtom, G.R., Xie, X.S. (1997). Room-temperature fluorescence imaging and spectroscopy of single molecules by two-photon excitation. *J. Phys. Chem.* 101: 7019–7023.
25. Kneipp, K., Harrison, G.R., Emory, S.R., Nie, S.M. (1999). Single-molecule Raman spectroscopy—Fact or fiction? *Chimia* 53(1–2): 35–37.
26. Kneipp, K., Wang, Y., Kneipp, H., Perelman, L.T., Itzan, I. (1997). Single molecule detection using surface-enhanced Raman scattering (SERS). *Phys. Rev. Lett.* 78: 1667–1670.
27. http://www.physik.uni-stuttgart.de/SMG/boersch/PDF/single_molecule_spectroscopy-01.pdf (accessed January 30, 2005).
28. Maiti, S., Haupts, U. (1997). Fluorescence correlation spectroscopy: Diagnostics for sparse molecules. *Proc. Natl. Acad. Sci. U.S.A.* 94: 11753–11757.

29. Brown, R., Milton, M., Gallop, J. (2001). *Review of techniques for single molecule detection in biological applications,* Teddington, Middlesex, UK, National Physics Laboratory, (http://www.npl.co.uk/biotech/smdtechniquesfinal.pdf, accessed January 30, 2005).
30. Rigler, R., Mets, U., Widengren, J., Kask, P. (1993). Fluorescence correlation spectroscopy with high count rate and low background—analysis of translational diffusion. *Eur. Biophys. J. Biophys. Lett.* 22(3): 169–175.
31. Moyer, P.J., Kammer, S.B. (1996). High-resolution imaging using near-field scanning optical microscopy and shear force feedback in water *Appl. Phys. Lett.* 68(24): 3380–3382.
32. Vickery, S.A., Dunn, R.C. (2001). Combining AFM and FRET for high resolution fluorescence microscopy. *J. Microsc.* 202: 408–412.
33. Trautman, J., Ambrose, W.P. (1997). *Single-molecule optical detection, imaging and spectroscopy, near-field optical imaging and spectroscopy of single molecules*, Berlin, VCH publishers.
34. Vickery, S.A., Dunn, R.C. (1999). Scanning near-field fluorescence resonance energy transfer microscopy. *Biophys. J.* 76: 1812–1818.
35. de Lange, F., Cambi, A., Huijbens, R., de Bakker, B., Rensen, W., Garcia-Parajo, M., van Hulst, N., Figdor, C.G. (2001). Cell biology beyond the diffraction limit: Near-field scanning optical microscopy. *J. Cell Sci.* 114(23): 4153–4160.
36. Sako, Y., Uyemura, T. (2002). Total internal reflection fluorescence microscopy for single-molecule imaging in living cells. *Cell Struct. Funct.* 27: 357–365.
37. http://www.biophys.leidenuniv.nl/Research/FvL/SMTutorial.pdf (accessed January 30, 2005).
38. Edman, L., Mets, U., Rigler, R. (1996). Conformational transitions monitored for single molecules in solution. *Proc. Natl. Acad. Sci. U.S.A.* 93(13): 6710–6715.
39. Nie, S., Chiu, D.T., Zare, R.N. (1994). Probing individual molecules with confocal fluorescence microscopy. *Science* 266: 1018–1021.
40. Nie, S., Chiu, D.T., Zare, R.N. (1995). Real-time detection of single molecules in solution by confocal fluorescence microscopy. *Anal. Chem.* 67: 2849–2857.
41. Lee, Y.H., Maus, R.G., Smith, B.W., Winefordner, J.D. (1994). Laser-induced fluorescence of a single molecule in a capillary. *Anal. Chem.* 66: 4142–4149.
42. Ishii, Y., Ishijima, A., Yanagida, T. (2001). Single molecule nanomanipulation of biomolecules. *Trends Biotechnol.* 19(6):211–216.
43. Lyon, W.A., Nie, S. (1997). Confinement and detection of single molecules in submicrometer channels. *Anal. Chem.* 69: 3400–3405.
44. Becker, W., Hickl, H., Zander, C., Drexhage, K.H., Sauer, M., Siebert, S., Wolfrum, J. (1999). Time-resolved detection and identification of single analyte molecules in microcapillaries by time-correlated single-photon counting (TCSPC). *Rev. Sci. Instrum.* 70(3): 1835–1841.
45. Zander, C., Drexhage, K.H., Han, K.T., Wolfrum, J., Sauer, M. (1998). Single-molecule counting and identification in a microcapillary. *Chem. Phys. Lett.* 286: 457–465.
46. Wagner, G.B., Mathis, H., Schmidt, K., Kalusche, G., McCaskill, J. (1997). Single molecule detection in microstructures. *Nucleosides Nucleotides* 16(5–6): 635–642.
47. Adachi, K., Kinosita, K., Ando, T. (1999). Single-fluorophore imaging with an unmodified epi-fluorescence microscope and conventional video camera. *J. Microsc.* 195: 125–132.
48. Bohmer, M., Enderlein, J. (2003). Orientation imaging of single molecules by wide-field epi-fluorescence microscopy. *J. Opt. Soc. Am.* 20(3): 554–559.
49. Axelrod, D., Burghardt, T.P., Thompson, N.L. (1984). Total internal reflection fluorescence. *Ann. Rev. Biophys. Bioeng.* 13: 247–268.
50. Stout, A.L., Axelrod, D. (1989). Evanescent field excitation of fluorescence by epi-illumination microscopy. *Appl. Opt.* 28: 5237–5242.
51. http://www.m-boersch.org/ (accessed January 30, 2005).
52. Vale, R.D., Funatsu, T., Pierce, D.W., Romberg, L., Harada, Y., Yanagida, T. (1996). Direct observation of single kinesin molecules moving along microtubules. *Nature* 380: 451–453.
53. Dickson, R.M., Norris, D.J., Tzeng, Y.L., Moerner, W.E. (1996). Three-dimensional imaging of single molecules solvated in pores of poly(acrylamide) gels. *Science* 274: 966–969.
54. Ambrose, W.P., Goodwin, P.M., Nolan, J.P. (1999). Single-molecule detection with total internal reflection excitation: Comparing signal-to-background and total signals in different geometries. *Cytometry* 36: 224–231.

55. Schwille, P. (2003). TIR-FCS: Staying on the surface can sometimes be better. *Biophys. J.* 85(5): 2783–2784.
56. Tokunaga, M., Kitamura, K., Saito, K., Iwane, A.H., Yanagida, T. (1997). Single molecule imaging of fluorophores and enzymatic reactions achieved by objective-type total internal reflection fluorescence microscopy. *Biochem. Biophys. Res. Commun.* 235: 47–53.
57. Axelrod, D. (2001). Total internal reflection fluorescence microscopy in cell biology. *Traffic* 2(11): 764–774.
58. Drexhage, K.H. (1974). Interactions of light with monomolecular dye layers. In: Wolf, E., ed. *Progress in optics.* Vol. XII. Amsterdam, North-Holland, pp. 163.
59. Jett, J., Keller, R.A., Martin, J.C., Marrone, B.L., Moyzis, R.K. (1989). High-speed DNA sequencing—and approach based upon fluorescence detection of single molecules. *J. Biomol. Struct. Dyn.* 7: 301–309.
60. Eigen, M., Rigler, R. (1994). Sorting single molecules—application to diagnostics and evolutionary biotechnology. *Proc. Natl. Acad. Sci. U.S.A.* 91: 5740–5747.
61. Werner, J.H., Cai, H., Jett, J.H., Reha-Krantz, L., Keller, R.A., Goodwin, P.M. (2003). Progress towards single-molecule DNA sequencing: A one color demonstration. *J. Biotechnol.* 102(1): 1–14.
62. Goodwin, P.M., Cai, H., Jett, J.H., Keller, R.A., Semin, D.J. (1996). DNA sequencing by single molecule detection. *Progr. Biophys. Mol. Biol.* 65: SMI02–SMI02.
63. Goodwin, P.M., Cai, H., Jett, J.H., Ishaug-Riley, S.L., Machara, N.P., Semin, D.J., Van Orden, A., Keller, R.A. (1997). Application of single molecule detection to DNA sequencing. *Nucleosides Nucleotides* 16: 543–550.
64. Stephan, J., Dorre, K., Brakmann, S., Winkler, T., Wetzel, T., Lapczyna, M., Stuke, M., Angerer, B., Ankenbauer, W., Foldes-Papp, Z., Rigler, R., Eigen, M. (2001). Towards a general procedure for sequencing single DNA molecules. *J. Biotechnol.* 86(3): 255–267.
65. Keller, R.A., Ambrose, W.P., Goodwin, P.M., Jett, J.H., Martin, J.C., Wu, M. (1995). Single molecule fluorescence analysis in solution. *Appl. Spectrosc.* 50: 12A–32A.
66. Haab, B.B., Mathies, R.A. (1995). Single-molecule fluorescence burst detection of DNA fragments separated by capillary electrophoresis. *Anal. Chem.* 67: 3253–3260.
67. Tan, W.H., Wang, K.M., Drake, T.J. (2004). Molecular beacons. *Curr. Opin. Chem. Biol.* 8(5): 547–553.
68. Ha, T. (2001). Single-molecule fluorescence resonance energy transfer. *Methods* 25: 78–86.
69. http://www.m-boersch.org/PDF/single-molecule-FRET-01.pdf (accessed January 30, 2005).
70. Zhuang, X., Kim, H., Pereira, M.J.B., Babcock, H.P., Walter, N.G., Chu, S. (2002). Correlating structural dynamics and function in single ribozyme molecules. *Science* 296: 1473–1476.
71. Zhuang, X.W., Bartley, L.E., Babcock, H.P., Russell, R., Ha, T.J., Herschlag, D., Chu, S. (2000). A single-molecule study of RNA catalysis and folding. *Science* 288: 2048.
72. Rhoades, E., Gussakovsky, E., Haran, G. (2003). Watching proteins fold one molecule at a time. *Proc. Natl. Acad. Sci. U.S.A.* 100: 3197–3202.
73. Schuler, B., Lipman, E.A., Eaton, W.A. (2002). Probing the free-energy surface for protein folding with single-molecule fluorescence spectroscopy. *Nature* 419: 743–747.
74. Deniz, A.A., Laurence, T.A., Beligere, G.S., Dahan, M., Martin, A.B., Chemia, D.S., Dawson, P.E., Schultz, P.G., Weiss, S. (2000). Single-molecule protein folding: Diffusion fluorescence resonance energy transfer studies of the denaturation of chymotrypsin inhibitor 2. *Proc. Natl. Acad. Sci. U.S.A.* 97: 5179–5184.
75. Talaga, D.S., Lau, W.L., Roder, H., Tang, J.Y., Jia, Y.W., DeGrado, W.F., Hochstrasser, R.M. (2000). Dynamics and folding of single two-stranded coiled-coil peptides studied by fluorescent energy transfer confocal microscopy. *Proc. Natl. Acad. Sci. U.S.A.* 97: 13021–13026.
76. Selvin, P.R. (2000). The renaissance of fluorescence resonance energy transfer. *Nat. Struct. Biol.* 7: 730–734.
77. Castro, A., Shera, E.B. (1995). Single-molecule detection—applications to ultrasensitive biochemical analysis. *Appl. Opt.* 34(18): 3218–3222.
78. http://www.lanl.gov/p/p21/smd/sme.shtml (accessed January 30, 2005).
79. Trautman, J.K., Macklin, J.J., Brus, L.E., Betzig, E. (1994). Near-field spectroscopy of single molecules at room temperature. *Nature* 369: 40–42.

80. Ambrose, W.P., Goodwin, P.M., Martin, J.C., Keller, R.A. (1994). Single-molecule detection and photochemistry on a surface using near-field optical excitation. *Phys. Rev. Lett.* 72: 160–163.
81. Xie, X.S., Dunn, R.C. (1994). Probing single-molecule dynamics. *Science* 265: 361–364.
82. Lu, H.P., Xie, X.S. (1997). Single-molecule spectral fluctuations at room temperature. *Nature* 385: 143–146.
83. Xie, X.S., Trautman, J.K. (1998). Optical studies of single molecules at room temperature. *Ann. Rev. Phys. Chem.* 49: 441–480.
84. Lu, H.P., Xie, X.S. (1997). Single-molecule kinetics of interfacial electron transfer. *J. Phys. Chem.* 101: 2753–2757.
85. Greulich, K.O. (2004). Single molecule techniques for biomedicine and pharmacology. *Curr. Pharm. Biotechnol.* 5(3): 243–257.
86. Xue, Q., Yeung, E.S. (1995). Differences in the chemical reactivity of individual molecules of an enzyme. *Nature* 373: 681–683.
87. Yasuda, R., Noji, H., Yoshida, M., Kinosita, K., Itoh, H. (2001). Resolution of distinct rotational substeps by submillisecond kinetic analysis of F-1-ATPase. *Nature* 410: 898–904.
88. Lu, H.P., Xun, L., Xie, X.S. (1998). Single-molecule enzymatic dynamics. *Science* 282: 1877–1882.
89. Wazawa, T., Ishii, Y., Funatsu, T., Yanagida, T. (2000). Spectral fluctuation of a single fluorophore conjugated to a protein molecule. *Biophys. J.* 78(3): 1561–1569.
90. Dickson, R.M., Cubitt, A.B., Tsien, R.Y., Moerner, W.E. (1997). On/off blinking and switching behaviour of single molecules of green fluorescent protein. *Nature* 388: 355–358.
91. Wazawa, T., Ishii, Y., Funatsu, T., Yanagida, T. (2000). Spectral fluctuation of a single fluorophore conjugated to a protein molecule. *Biophys. J.* 78(3): 1561–1569.
92. Ha, T., Tiong, A.Y., Liang, J., Caldwell, B., Dentz, A.A., Chemla, D.S., Schultz, P.G., Weiss, S. (1999). Single-molecule fluorescence spectroscopy of enzyme conformational dynamics and cleavage mechanism. *Proc. Natl. Acad. Sci. U.S.A.* 96: 893–898.
93. Zhuang, X., Rief, M. (2003). Single-molecule folding. *Curr. Opin. Struct. Biol.* 13: 88–97.
94. Sase, I., Miyata, H., Ishiwata, S., Kinosita, K. (1997). Axial rotation of sliding actin filaments revealed by single-fluorophore imaging. *Proc. Natl. Acad. Sci. U.S.A.* 94: 5646–5650.
95. Warshaw, D.W., Hayes, E., Gaffney, D., Lauzon, A., Wu, J., Kennedy, G., Trybus, K., Lowey, S., Berger, C. (1998). Myosin conformational states determined by single fluorophore polarization. *Proc. Natl. Acad. Sci. U.S.A.* 95: 8034–8039.
96. Bopp, M.A., Sytnik, A., Howard, T.D., Cogdell, R.J., Hochstrasser, R.M. (1999). The dynamics of structural deformations of immobilized single light-harvesting complexes. *Proc. Natl. Acad. Sci. U.S.A.* 96: 11271–11276.
97. Harms, G.S., Sonnieitner, A., Schutz, G.J., Gruber, H.J., Schmidt, T. (1999). Single-molecule anisotropy imaging *Biophys. J.* 77: 2864–2870.
98. Ha, T., Laurence, T.A., Chemla, D.S., Weiss, S. (1999). Polarization spectroscopy of single fluorescent molecules. *J. Chem. Phys.* 103: 6839–6850.
99. Bustamante, C., Rivetti, C. (1996). Visualizing protein–nucleic acid interactions on a large scale with the scanning force microscope. *Ann. Rev. Biophys. Biomol. Struct.* 25: 395–429.
100. Hansma, H.G., Hoh, J.H. (1994). Biomolecular imaging with the atomic force microscope. *Ann. Rev. Biophys. Biomol. Struct.* 23: 115–139.
101. You, H.X., Lowe, C.R. (1996). Progress in the application of scanning probe microscopy to biology. *Curr. Opin. Biotechnol.* 7: 78–84.
102. Kulzer, F., Orrit, M. (2004). Single-molecule optics. *Ann. Rev. Phys. Chem.* 55: 585–611.

3

Nanoscale Optical Biosensors and Biochips for Cellular Diagnostics

Brian M. Cullum

CONTENTS

3.1 Introduction to Nanosensors and Biochips

Within the last few decades, controlled fabrication and manipulation of nanometer-scale molecular components and assemblies, from either a "bottom-up" or "top-down" approach, allowed for rapid developments in the field of nanosensors. As described in several other chapters of this book, nanoscale assemblies have resulted in the development of a wide variety of biological and biologically modified receptors for sensor recognition, as well as provided unique methods of signal transduction capable of sensitively detecting analytes in the smallest of environments. Employing these new developments, the fields of chip-based sensing (i.e., biochips, etc.) and nanosensing have seen a recent explosion of research.[1–11] The small-size scales of these sensors and their ability to probe previously inaccessible samples, and locations within those samples with a high degree of spatial resolution, have opened a great many doors in a variety of different fields. While such sensors have been applied to samples as widespread as embryos[12] and micropores in synthetic structures,[13] perhaps one of the most dramatically impacted areas of scientific research has been the field of cellular analyses. Owing to the small-size scale of these sensors, and the often noninvasive nature of the signal transduction technique employed, they have been able to probe individual living cells, both extracellularly and intracellularly for various lengths of time, providing the first true measurements of cellular pathways and metabolic processes without the potential variability experienced from multicell averaging. Employing nanosensors and biochips (e.g., gene chips), cellular analyses ranging from medical diagnostics[14–16] to the decoding of the human genome[9,17,18] have been reported. This chapter will provide a highlight into some of the major achievements in these fields over the last two decades.

Because of their potential for providing an enhanced understanding of cellular responses to various stimuli, several reviews have already been devoted to the subject of optical nanosensors and biochips, despite their short existence.[1–4,19–21] This chapter will provide a look at the evolution of biochips and nanosensors from the beginning to the present (biosensors capable of probing subcellular compartments of individual cells) and discuss their application to biological measurements, as well as future directions in nanosensing. On account of the vast number of publications in this field over the last few years, this chapter is not meant to provide a comprehensive review, but a critical review of optical-based nanosensors and biochips, describing significant achievements and advancements that have occurred in these fields.

Biosensors, generally defined, are devices that consist of a bioreceptor and a signal transduction mechanism capable of producing a measurable signal in the presence of an analyte of interest. A bioreceptor is a biochemical species (e.g., an antibody, an oligonucleotide, and an enzyme) or a living biological system (e.g., cells, tissue, or whole organisms) that utilizes typical biochemical recognition mechanisms (e.g., lock-and-key fit) for analyte recognition. Upon interaction of the analyte with the bioreceptor the transducer converts the resulting change into a measurable signal (e.g., optical emission, and electrical current). Figure 3.1 illustrates the conceptual principle of the overall biosensing process.

Owing to the wide array of biosensors that exist, they are often classified into different categories based upon either the transduction method that they employ or the type of biological receptor used to provide specificity. Common classes of signal transduction techniques include (1) optical measurements (i.e., luminescence, absorption, and surface plasmon resonance); (2) electrochemical measurements; and (3) mass-sensitive measurements (i.e., surface acoustic wave, and microbalance). Bioreceptor-based classification of biosensors also involves several different categories, with the most common forms of biorecognition being (1) antibody or antigen interactions, (2) nucleic acid interactions, (3) enzymatic interactions, (4) cellular interactions (i.e., microorganisms and proteins), and (5) interactions using biomimetic materials (i.e., synthetic bioreceptors). These bioreceptors are generally contained in a biosensitive layer doped with the bioreceptor molecules or covalently attached to the surface of the sensor. When several transducers or transducer–bioreceptor combinations are arrayed onto individual integrated circuit microchips, these biosensors are often referred to as biochips. In general, biochips consist

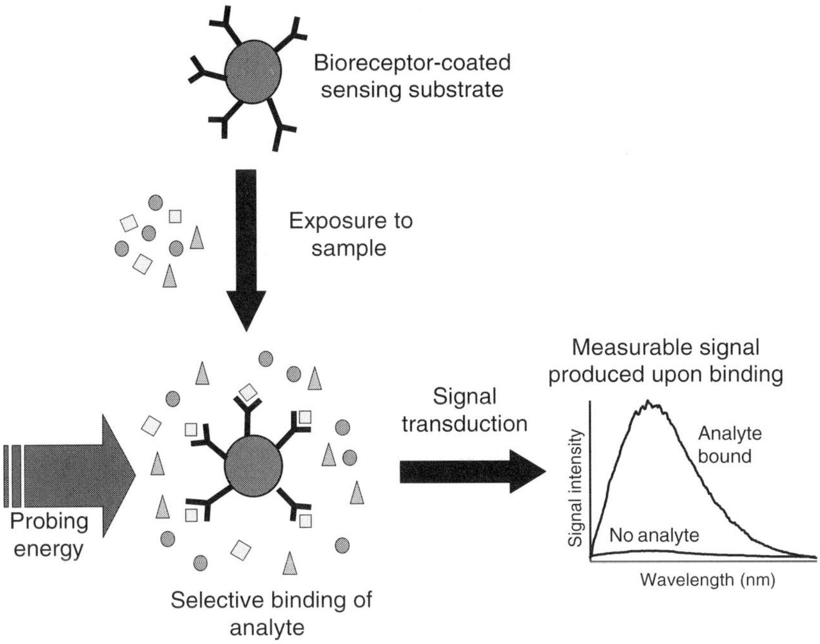

FIGURE 3.1
Schematic diagram depicting the fundamental concept of a biosensor. A bioreceptor is used to provide specific chemical recognition, followed by probing of this interaction to provide a measurable signal.

of an array of individual biosensors that can be individually monitored and used to detect multiple analytes simultaneously.

3.1.1 Beginnings of Optical Nanosensor

Since the first integrated macroscopic biosensors were reported in the early 1960s,[22] there has been a constant desire to reduce the size of these sensors to take advantage of significant enhancements in sensitivity, response time, and many other important characteristics while still retaining the specificity and sensitivity derived from using biological receptor molecules or organisms. In the early 1990s, the fields of chemical and biochemical sensing experienced several simultaneous advances that allowed for the realization of optical nanosensors. Chief among these advances was the development of tapered probes for near-field-scanning optical microscopy (NSOM).[23] NSOM is a relatively new technique, employing light sources and detectors that are smaller than the wavelength of light used for imaging, allowing for chemical images of a sample to be taken with nanometer-scale (i.e., 10–50 nm) spatial resolution.[23–26] One common method for performing such measurements is to construct an excitation probe with dimensions that are smaller than the wavelength of light that is being used for sample interrogation, effectively providing a light source with subwavelength dimensions.[27,28] The first such excitation probe was developed by Betzig et al.[27] by tapering a single-mode optical fiber to dimensions of approximately 20 nm, thereby confining the excitation radiation to the fiber tip. Using such tapered optical fibers, Betzig and coworkers[24,25,29,30] were able to reconstruct chemical images of a known pattern, with signal enhancements of greater than 10^4 over previous NSOM analyses that employed apodized detectors. Since then, tapered fiber-optic probes have been used by many researchers for the analysis of a wide variety of samples, including the imaging of individual strands of deoxyribonucleic acid (DNA) via near-field surface-enhanced resonance Raman spectroscopy.[31,32] With the advent of fiber-optic nanoprobes, NSOM is an area of research that holds a great deal of promise for biological analyses.

3.1.1.1 Fiber-Optic Tapering

Fabrication of tapered fiber-optic probes for use in NSOM, and later for fiber-optic nanosensors, is typically performed through one of the two different methods: heated pulling or chemical etching. Employing these methods can result in the fabrication of fibers with various sizes and taper angles, depending upon the conditions of the tapering process. The most commonly employed procedure involves the use of a laser-based micropipette puller. In this procedure, a CO_2 laser heats the fiber while it is stretched along its major axis by a tension device (see Figure 3.2), thereby tapering it, until finally the fiber is pulled into two pieces.

After the pulling process is complete, each piece has one large end and one end with nanometer-scale dimensions. Using this process, tips have been generated with diameters ranging in size from 20 nm to >500 nm, simply by changing the heating temperature and the applied pulling tension.[12,13,27,33,34] The entire process, using a micropipette puller, takes only seconds, producing two very reproducible tips, making it a very fast and reliable process. An example of a fiber-optic probe tip fabricated via this process is shown in the scanning electron micrograph (SEM) image in Figure 3.3.

In addition to the heated pulling process, optical fibers with nanometer-scale tips have also been produced via chemical etching. While chemical etching tends to produce a

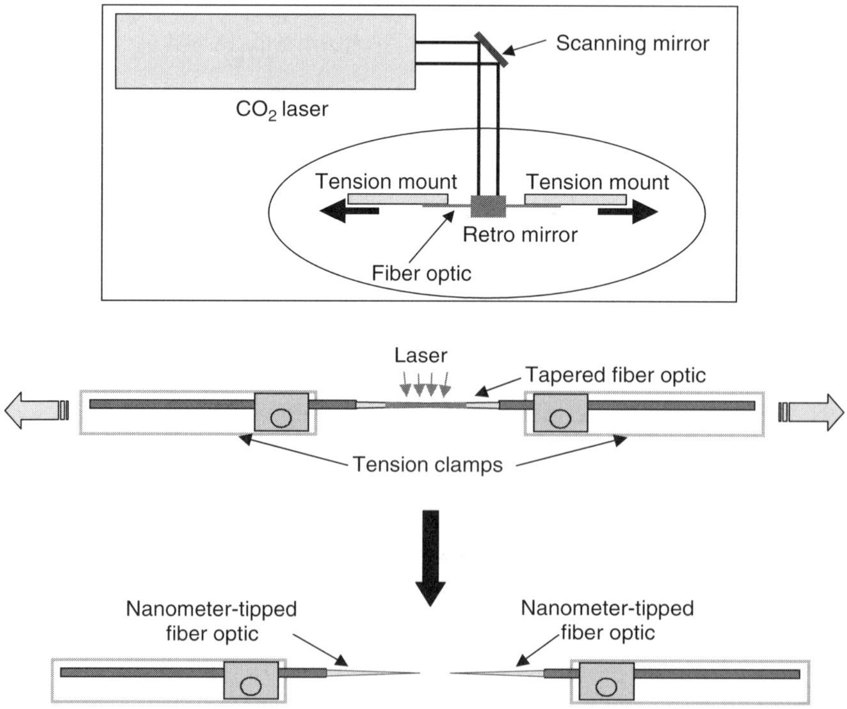

FIGURE 3.2
Diagram depicting the heated pulling process for the fabrication of nanometer diameter fiber-optic probes. After placing a polished fiber in the laser-based micropipette puller, the fiber is heated and pulled apart into two individual fiber probes.

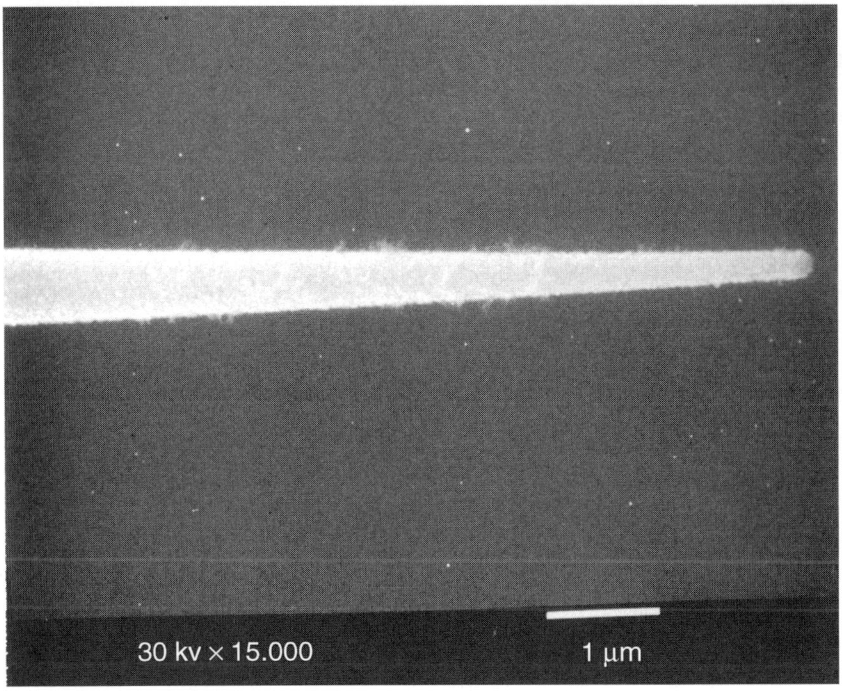

FIGURE 3.3
Scanning electron micrograph of a tapered fiber-optic probe fabricated via a heated pulling process.

greater variability in fiber tips from batch to batch, it does not require any special equipment and can provide tips with much greater taper angles than are generally produced via heated pulling processes. Two different types of chemical etching have been reported in the literature: (1) Turner etching[35,36] and (2) tube etching.[37] In the first method, Turner etching, a silica fiber is placed in the meniscus between hydrofluoric acid (HF) and an organic overlayer, causing the HF to etch the silica while the organic overlayer protects the region of fiber suspended in it. By changing the position of the fiber in this meniscus during the etching process, as well as various other parameters, it is possible to accurately control the taper angle and degree of taper of these fiber-optic tips. However, due to the dual chemical nature of the etchant solution, environmental parameters such as vibrations can cause significant batch-to-batch variations in the characteristics of the fiber tips.

To overcome the issues associated with this dual chemical etchant, a process known as tube etching was developed. In this process, a silica core fiber having a polymer cladding is placed into a solution of HF and the silica core is allowed to dissolve while the polymer cladding remains unaffected. As the silica core is etched it rises above the surface of the HF solution. At this point, capillary action draws HF up the walls of the polymer cladding where it reaches the silica core of the fiber again and then drains back into the solution below, forming a smooth tip with a large taper angle. The entire etching process takes approximately 2 hours. To remove the surrounding polymer cladding and expose the silica tip for use, the fibers can either be placed in a suitable organic solvent to dissolve the cladding or burned off using a high-temperature flame. While this process can produce more reproducible tapered fiber-optic probes than Turner etching, difficulties associated with submerging multiple fibers to the same depth in the HF solution tend to make it less reproducible than fibers produced via the heated pulling process.

3.2 Submicron and Nanoscale Sensors and Biosensors

3.2.1 Fiber-Optic Nanosensors

Shortly after the development of tapered optical fibers for NSOM, they began to be applied to the field of chemical sensing, allowing for the probing of specific chemicals in highly localized areas ($\lesssim$50 nm from the fiber tip). Owing to this high degree of localization, monitoring of concentration gradients of chemicals and their spatial inhomogeneity in submicroscopic environments via various spectroscopic techniques became possible. In particular, these submicron sensors or nanosensors provide a means of obtaining highly localized analyses of chemical species within subcellular compartments of individual cells and multicell organisms.

3.2.1.1 Fiber-Optic Chemical Nanosensors

The first such optical nanosensor was reported by Kopelman and coworkers[12,13] in 1992 for the measurement of pH. In this work, the heated pulling process was used to taper multimode and single-mode fibers to diameters of 100–1000 nm at the tip. Following the tapering process, the walls of the fiber were coated with aluminum in a vacuum evaporator to ensure total internal reflection over the pulled region of the fiber. During aluminum deposition, the fibers were placed with their tips facing away from the metal source, thereby shadowing them from the evaporating metal and leaving the tips uncoated (see Figure 3.4).

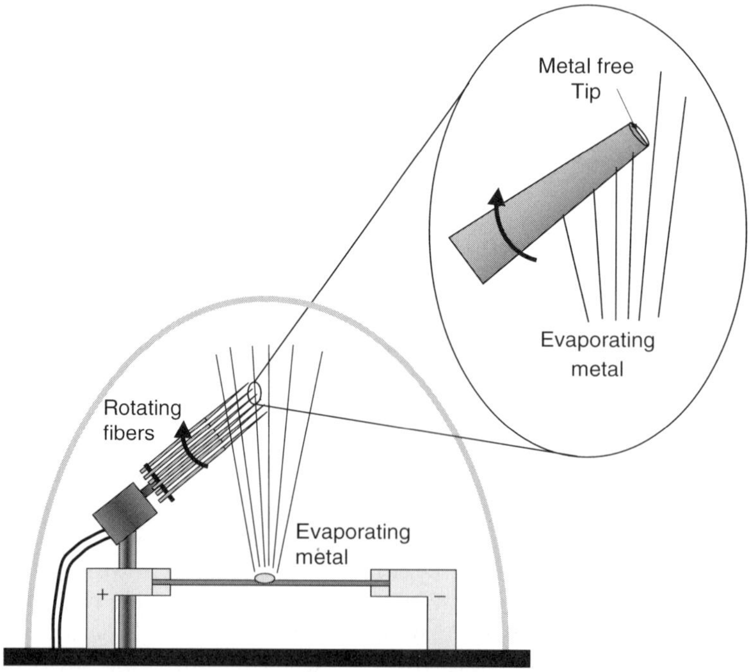

FIGURE 3.4
Schematic diagram depicting the shadowing of a tapered fiber-optic probe tip from the evaporating metal used to coat the walls of the fiber.

The uncoated silica surface on the tip then allowed for the binding of the pH-sensitive fluorescent dye acrylofluoresceinamine via a variation of a common photopolymerization process often used in the construction of larger chemical sensors.[38,39] However, since the tips of these fibers were smaller than the wavelength of light used to initiate the photopolymerization process, the crosslinking of the dye-doped polymer solution was restricted to the near field of the fiber. This near-field polymerization process resulted in a sensor with a sampling volume greater than six orders of magnitude smaller than the conventional fiber-optic chemical sensors, making it ideal for subcellular measurements. This reduced sampling volume resulted in a 100-fold faster response time than conventional fiber-optic chemical sensors fabricated via similar processes. After the development of this first fiber-optic-based chemical nanosensor, several additional sensors have been reported for a wide variety of chemical species, including pH[40–43] and various other ions[44,45] as well as nonionic species.[46]

Application of these fiber-optic nanosensors to biological analyses was first reported for the monitoring of pH in rat embryos.[12] In this study, nanosensors similar to the ones described above were inserted into the extraembryonic space of a rat conceptus and pH measurements were performed, resulting in a means of monitoring pH differences with minimal damage to the surrounding visceral yolk sac. In a similar study using the same type of sensor, indirect measurements of nitrite and chloride levels in the yolk sac of rat conceptuses were also performed.[47] Such minimally invasive analyses provide a great deal of promise for biological measurements and may aid in furthering our understanding of the effect that environmental factors play in embryonic growth.

As the field of optical-fiber-based nanosensors has evolved, the size of the environment capable of being probed has become smaller and smaller, allowing for the monitoring of chemical species inside individual living cells. The first report of such single-cell analyses using opto-chemical fiber-optic nanosensors employed nanosensors fabricated in a process similar to that described above. These fiber-optic nanosensors were then employed to measure sodium ion (Na^+) concentrations in the cytoplasmic space inside a single mouse oocyte.[1] In another application of fiber-optic nanosensors to biological measurements, calcium-ion-sensitive nanosensors were used to measure calcium ion fluctuations in vascular smooth muscle cells while the cells were being stimulated.[48]

3.2.1.2 Fiber-Optic Nano-Biosensors

Owing to the complexity of biological systems and the number of possible interferences present in cellular environments, added specificity is often required. Like their larger counterparts, conventional fiber-optic biosensors, biological receptor molecules (i.e., antibodies, enzymes, etc.) are often used in conjunction with tapered fiber-optic probes to provide added specificity. The different types of bioreceptor molecules that have been used for the fabrication of fiber-optic nano-biosensors include antibodies, oligonucleotides, and enzymes, thereby allowing for the detection of a wide array of analytes.

3.2.1.2.1 Antibody-Based Fiber-Optic Nano-Biosensors

The first such fiber-optic nano-biosensor was reported in 1996 by Vo-Dinh et al.[33] In this work, antibody-based nano-biosensors were developed for benzo[*a*]pyrene tetrol (BPT), a DNA adduct of the carcinogen benzo[*a*]pyrene (BaP). As a biomarker for exposure to BaP, the measurement of BPT is of great interest for health monitoring. These immuno-based nano-biosensors were fabricated in a simple four-step process. In the first step, 600-μm-diameter multimode fiber optics was tapered to 40 nm at the tip. Following tapering, the fibers were coated with 200 nm of silver to prevent light leakage and enhance the delivery of the excitation light to the end of the fiber. Once the metal was coated, the fibers were silanized to allow for covalent attachment of the antibodies to the fiber tips. Covalent attachment of the antibodies was performed by first activating the silanized fiber tip with a solution of carbonyl diimidazole, followed by incubation of the activated fiber with the antibody of interest at 4°C for 3 days. After antibody attachment, the biosensors were tested for retention of antibody activity as well as their sensitivity and absolute detection limits. From these analyses, it was found that the antibodies had retained greater than 95% of their original binding affinity for BPT and that the absolute detection limit for BPT was approximately 300 zmol (zepto = 10^{-21}).

Shortly after the development of this first antibody-based fiber-optic nano-biosensor, they were applied to in vitro measurements within individual living cells.[3,34,49–51] In one such study, nano-biosensors for BPT were prepared and used to quantitatively measure intracellular concentrations of BPT in the cytosol of two different cell lines: (1) rat liver epithelial cells and (2) human mammary carcinoma cells.[34] Unlike previous intracellular analyses that employed relatively large cells (i.e., mouse oocytes and neurons) the cells analyzed in this study were approximately 10 μm in diameter, demonstrating that fiber-optic nanosensors and nano-biosensors could be used to analyze cells, the size of typical mammalian somatic cells (i.e., 10–15 μm), without destroying them (see Figure 3.5). This study also demonstrated that the insertion of a fiber-optic nano-biosensor into a cell and the subsequent spectroscopic measurement had little effect on the cell's routine function. This was demonstrated by inserting the fiber-optic nano-biosensor into the cytoplasm of a cell beginning to undergo mitosis and monitoring division of the cell into two identical daughter cells over the next 2 h.

3.2.1.2.2 *Protein-Based Fiber-Optic Nano-Biosensors*

Since the development and application of these first antibody-based fiber-optic nano-biosensors, several other nano-biosensors, employing different biological receptor molecules, have been fabricated and applied to the analysis of various biologically relevant species in individual cells.[52,53] These have included nano-biosensors for the detection of nitric oxide, employing the nonenzymatic protein cytochrome *c* and fluorescently labeled cytochrome *c*,[52] as well as enzymatic-based nano-biosensors for the indirect detection of glucose and glutamate. By employing enzymes as bioreceptors, not only is a high degree of specificity achieved, but also the catalytic activity of the enzymes can amplify the species being measured, providing extremely sensitive analyses. One such enzymatic fiber-optic nano-biosensor

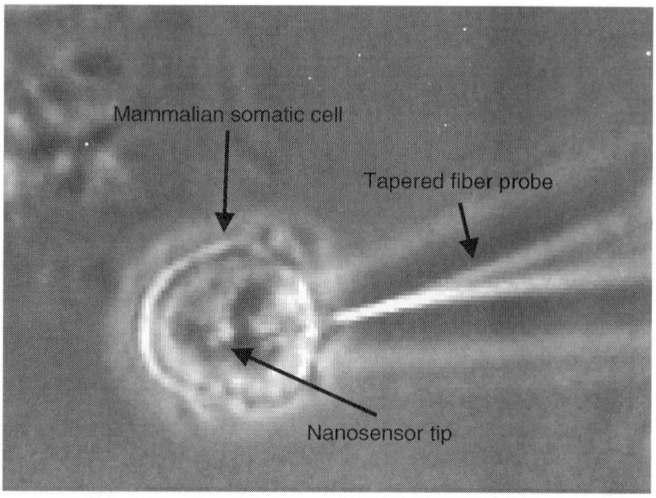

FIGURE 3.5
Microscopic image of a fiber-optic nanosensor being used to probe the chemical cytosol of a rat liver epithelial cell.

was developed for the indirect detection of glucose.[54] In this work, Kopelman and coworkers photopolymerized the enzyme glucose oxidase and the oxygen-sensitive indicator tris(1,10-phenanthroline)ruthenium chloride in an acrylamide polymer onto the tapered end of a nanofiber. When this nano-biosensor is in the presence of glucose, the enzyme catalyzes the oxidation of glucose into gluconic acid, thereby consuming oxygen. The resulting changes in oxygen levels are then measured by monitoring the fluorescence of the oxygen-sensitive indicator dye. Using glucose oxidase as an enzymatic bioreceptor, these nano-biosensors were capable of absolute detection limits of approximately 10^{-15} mol and a sensitivity 5 to 6 orders of magnitude greater than the previous glucose optodes.[54] In another example of enzymatic-based nano-biosensors, the enzyme glutamate dehydrogenase was attached to the end of a tapered fiber optic by Tan et al.[53] When glutamate was present, the glutamate dehydrogenase reduced NAD^+ to NADH and the fluorescence of the NADH was measured. Employing such an enzymatic-based nano-biosensor, the continuous monitoring of glutamate levels released from an individual cell were monitored, providing a powerful new tool for the field of neurophysiology.

3.2.1.2.3 Molecular Beacon-Based Fiber-Optic Nano-Biosensors

To detect the presence of oligonucleotides (i.e., ribonucleic acid (RNA) and DNA), in intracellular environments, a class of fiber-optic nano-biosensors has been developed that employs a bioreceptor molecule known as a molecular beacon.[55] Molecular beacons are hairpin-shaped oligonucleotide probes that rely on the complementarity of nucleic acids (i.e., adenine:thymine, cytosine:guanine, etc.) to provide the molecular recognition for a specific oligonucleotide sequence.[56] In their native form molecular beacons form a stem-loop structure, in which a fluorophore (e.g., fluorescein isothiocyanate (FITC)) on one end of the stem is in close proximity to a fluorescence-quenching moiety (e.g., dimethylaminophenylazobenzoic acid (DABCYL)) on the other end (see Figure 3.6).

In this state, when the fluorophore on the beacon is excited, energy transfer takes place between the excited fluorophore and the quencher, by either direct energy transfer or fluorescence resonance energy transfer (FRET), resulting in minimal fluorescence emission. However, in the presence of an oligonucleotide sequence complementary to the loop

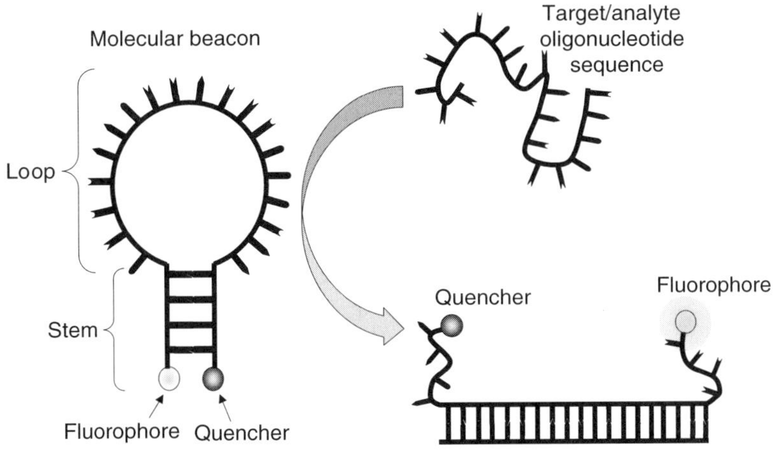

FIGURE 3.6
Schematic depiction of a molecular beacon. In its native form, the molecular beacon exists in a hairpin structure in which a quenching species minimizes any fluorescent emission. After binding the complementary oligonucleotide sequence, the quencher and fluorophores are separated resulting in fluorescence emission.

sequence of the molecular beacon, the loop sequence binds to the complementary strand, causing the stem to unzip and creating a spatial separation between the fluorophore and the quencher. This spatial separation in turn generates a dramatic increase in fluorescence intensity of the molecular beacon, with some reports of enhancement exceeding 200-fold.[56] Owing to their inherent sensitivity and specificity, molecular beacons have recently been used to fabricate fiber-optic nano-biosensors for single-stranded DNA and RNA.[55,57] In this work, molecular beacons were attached to the silica core of a tapered fiber-optic nano-biosensor through biotin–avidin linkages. By binding the molecular beacons to the tapered tip of the fiber optic in this method, it was shown that a strong attachment could be achieved while not degrading the activity of the beacon.

3.2.1.3 Fiber-Optic Nano-Imaging Probes

While fiber-optic nanosensors and nano-biosensors have proven to be useful tools for the spatially localized, quantitative analysis of chemical species in individual living cells, they are limited in the number of locations that can be probed simultaneously. In most cases only one sensor can be inserted into a typical mammalian somatic cell at a time, without causing significant damage or altering the normal cellular function of the cell.[58] This inability to monitor multiple locations on or within an individual living cell simultaneously can dramatically limit the applicability of fiber-optic nanosensors and nano-biosensors for many types of cellular analyses. To overcome this limitation, two different types of fiber-optic nano-imaging probes, comprised of thousands of individual fibers arranged in a coherent bundle, have recently been developed by a couple of research groups and used for chemical imaging.[59,60] These nano-imaging probes represent the latest advance in fiber-optic-based nanosensors, and are capable of obtaining measurements at thousands of different locations simultaneously with nanometer-scale spatial resolution.

Fabrication of these fiber-optic nano-imaging probes is performed through either a chemical etching process or a heated pulling process, similar to single-fiber-based nanosensors and nano-biosensors. In the first of these fabrication methods, chemical etching is performed on commercially available coherent imaging bundles having as many as 30,000 individual fibers. The individual fiber elements comprising the imaging bundle are etched in HF to form an array of silica tips, with the core of each fiber element corresponding to a nanotip.[61] After the array of silica tips is formed from this etching process, a layer of gold is sputtered over the entire surface to prevent light from escaping from the tapered sides of each fiber element. Once the gold overlayer is deposited on the fiber tips a polymer coating is electrochemically deposited over the gold and heat-cured to expose the gold only at the tips of the individual fiber elements. Lastly, the exposed gold on the tips of the coated fiber elements is dissolved in aqua regia to produce free silica surfaces for spectroscopic analyses.

As with the single-core fiber-optic nanosensors and nano-biosensors, nano-imaging probes have also been fabricated via a heated pulling process.[59] When fiber-optic nano-imaging probes are fabricated via this process, the resulting fiber probe has a flat tip, with each of the individual fiber elements comprising the bundle having been tapered equally. This can be seen from the microscopic image shown in Figure 3.7a, which shows the side-on view of a tapered fiber-optic nano-imaging probe produced on a CO_2 laser-based micropipette puller that has been specially optimized for producing nano-imaging probes. To produce the flat surface on these nano-imaging probes, which is necessary for imaging, the fiber is allowed to cool slightly before the final hard pull of the micropipette puller cleaves the bundle into two individual pieces. In addition to providing a flat surface, this tapering process also ensures that each fiber element in the bundle is tapered equally. By changing the heating temperature and the initial pull strength of the micropipette puller,

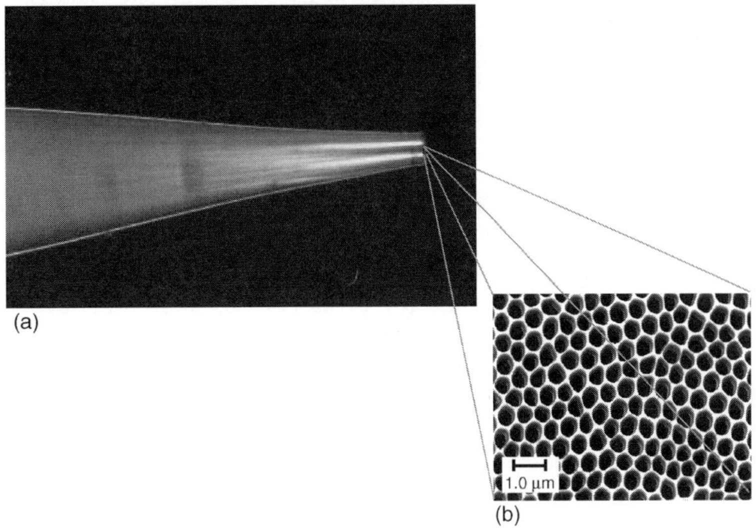

FIGURE 3.7
(a) Microscopic image of a tapered fiber optic imaging bundle, demonstrating the flat edge produced from the heated pulling process. (b) Scanning electron micrograph of a tapered fiber-optic imaging bundle that has been etched via hydrofluoric acid.

fibers can be accurately and reproducibly tapered to diameters ranging from the initial 4 μm diameter to 150 nm before cross talk is observed.[59] The untapered end of these fiber probes can then be imaged onto a charge-coupled device (CCD), allowing images of the tapered end of the fiber and any sample in contact with it to be imaged with resolutions of approximately 150 nm.

To provide chemical information about the species imaged using fiber-optic nano-imaging probes fabricated via the heated pulling process, a roughened layer of silver has been deposited on these tapered bundles to allow surface-enhanced Raman scattering (SERS) images to be obtained.[59] To produce this roughened metal surface, tapered imaging bundles were placed in a solution of HF, thereby etching the core of each individual fiber element faster than the cladding surrounding it. This can be seen in the SEM shown in Figure 3.7b. Once the fiber elements were etched, silver was deposited on the surrounding claddings, via vacuum evaporation, to produce an efficient SERS substrate with a highly ordered and controllable spacing. Using such SERS imaging substrates, images of homogenous SERS active substances with variability of less than 2% across the surface of the probe have been produced, providing a probe capable of obtaining chemical images of samples with spatial resolutions of 150 nm.

With the two different types of fiber-optic imaging probes that have been developed, it is possible to obtain quantitative measurements of chemical species at many different locations simultaneously. Although these fiber-optic nano-imaging probes are too large to be inserted into individual living cells, they can be used to monitor cellular membranes without damage.

3.2.2 Implantable Nanosensors

While fiber-optic nanosensors and fiber-optic nano-imaging probes have had and are having large impacts on the fields of cellular biology and biochemistry, significant advances and variations in optical nanosensors are constantly being made. These advances range from the use of more selective bioreceptors to the development of different types of optical-based nanosensors and nano-biosensors capable of being implanted into individual cells. One such advance in the last decade has been the development of nanoparticle-based opto-chemical sensors and biosensors, with nanometer-scale sizes in all three dimensions. The small dimensions of these sensors allow for the implantation of thousands of them within an individual cell, allowing for the monitoring of many locations simultaneously. Since the first such nanoparticle-based nanosensor was developed approximately a decade ago, many different such sensors have been developed for single-cell analyses, based upon different bioreceptors and transduction schemes. However, despite the large number of nanoparticle-based sensors developed, each type can be classified into one of four different classes: (1) quantum-dot-based nano-biosensors, (2) polymer-encapsulated nanosensors (known as probes encapsulated by biologically localized embedding (PEBBLEs)), (3) phospholipid-based nanosensors, and (4) SERS-based nanosensors.

3.2.2.1 Quantum-Dot-Based Nano-Biosensors

One of the earliest classes of nanoparticle-based nanosensors that were developed employ semiconductor nanoparticles, known as quantum dots,[62–65] attached to various biological receptor molecules such as antibodies,[66,67] oligonucleotides[66–69] and many other species[66,67,70] that can be used to translocate these quantum dots to the location of interest on or within a cell.[71–75] The quantum dots used in these early nanosensors and the most commonly employed to date are ZnS particles capped with CdSe. By varying the diameter of these quantum dots the wavelength of maximum luminescence emission is shifted, with nanometer variations in diameter resulting in shifts of tens of nanometer in the emission, maximum.[63] Additionally, by using quantum dots instead of conventional organic dyes, these sensors exhibit much more intense emission and a greater degree of photostability, which is important when trying to monitor changes in chemical or biochemical species concentrations over time. This high degree of photostability is a direct result of reduced photobleaching and photodegradation of the luminescent material, as the quantum dots are much more robust than most organic dyes. While these nanosensors represent an excellent tool for labeling or staining and obtaining qualitative information about cellular reactions, it is often difficult or impossible to employ them in a conventional "on/off" sensing modality, as the quantum dots exhibit luminescence emission whether or not they are bound to the analyte of interest. Additionally, these quantum-dot-based nanosensors and nano-biosensors represent a potential problem for cellular analyses in terms of their biocompatibility, since the materials used to fabricate these quantum dots (e.g., CdSe) are often toxic. Because of the toxicity issues associated with Cd and Se as well as other materials that have been used to fabricate quantum dots, a significant effort is currently underway in many research groups to develop quantum dots composed of more biocompatible materials. Another means of overcoming the potential toxicity issues associated with quantum-dot-based nanosensors that is currently being investigated is to thoroughly coat the surface of the quantum dots with biological receptor molecules. This prevents the semiconductor core of the sensor from interacting with the cellular environment, thus allowing quantum-dot-based nano-biosensors to perform long-term monitoring of cellular reactions or processes. Recent results in this area are showing a great deal of promise in terms of overcoming the toxicity of these nanoparticle-based nanosensors.[76]

3.2.2.2 Polymer-Encapsulated Nanosensors

Another early class of nanoparticle-based nanosensors that has already had a large impact on the fields of cellular biology and biochemistry is known as PEBBLEs.[77–85] These nanosensors, developed by Kopelman et al. over the last decade, are comprised of fluorescent indicator dyes that are embedded in 20–200 nm diameter polymer or sol–gel spheres.[78,79] By encapsulating fluorescent indicator dyes that are sensitive to particular ionic species (i.e., H^+, and Ca^{2+}, etc.) in a polymer matrix, the dye is protected from cellular degradation by proteins or enzymes, while still allowing ions, which are much smaller in size, to pass into the matrix and react with the dye. Additionally, the polymer coating protects the cell from any potential toxic effects of the dye.

Since they were first developed in 1996, PEBBLEs have been used to monitor fluxes in the concentration of many different species within individual cells, including pH, Ca^{2+}, NO, O_2, and Zn^{2+}.[78–80,83–86] Upon fabrication of the particular type of PEBBLE sensor desired for a measurement, they are inserted into a cell typically via nano-injection or gene-gun insertion. Owing to the small size of the particles, hundreds to thousands of these PEBBLE sensors can be inserted into individual cells without affecting routine cellular function. This allows for sensors to be present throughout the cell and various subcellular compartments, providing a means for imaging cellular response at many locations within the cell simultaneously. Once the sensors have been injected into the cell, the entire cell is illuminated and the resulting fluorescence signal from the indicator dye is measured over the autofluorescent background and correlated to the amount of analyte species present.

Owing to the potential for intense autofluorescence from cellular materials following excitation of the implanted nanosensors with ultraviolet or blue light, detection limits for the analyte of interest from such analyses can be relatively high. Because of this potential problem, Kopelman and coworkers have recently developed a variation of these PEBBLE nanosensors that can be magnetically modulated at a known frequency. These magnetically modulated nanosensors, known as magnetically modulated optical nanoprobes (MagMOONs), are fabricated by either creating the polymer-encapsulated sensor around an aspheric magnetic nanoparticle, or coating a spherical nanoparticle on one side with a thin layer of a magnetic material.[87,88] This aspheric magnetic particle is then inserted into the cell of interest, and a magnetic field is applied. As the magnetic field is applied, the particles begin to rotate at a rate that is related to the strength and location of the applied field. When these MagMOONs are optically excited, the resulting fluorescence emission, which can be related to the presence of the analyte of interest, is modulated. This modulated luminescent signal can then be recovered from the continuous autoluminescence signals of the cell via conventional techniques such as lock-in amplification, dramatically improving the detection limit of the MagMOON over the PEBBLE sensors that require energetic ultraviolet or blue excitation wavelengths.

3.2.2.3 Phospholipid-Based Nanosensors

A third class of implantable nanosensor that has focused on the need for biocompatibility is the phospholipid-based nanosensors. These nanoparticle-based sensors, developed by Rosenzweig and coworkers,[89–95] rely on the concept of encapsulation of the indicator dye, similar to PEBBLEs; however, they employ biological components for the encapsulation instead of polymers or silica. Encapsulating fluorescent indicator dyes inside biological materials not only reduces any toxic effects to the cell and biodegradative effects to the dye but it also allows for common biological mechanisms to be employed for cellular uptake of the sensors, as compared with potentially intrusive methods such as nano-injection.[95]

Within the class of phospholipid-based nanosensors, two distinct subclasses exist: (1) liposome sensors[93,94] and (2) lipobead sensors.[89,91] The former of these two subclasses, liposome-based nanosensors, employ fluorescent indicator dyes encapsulated in the internal aqueous compartment of a liposome. By encapsulating the dye in the aqueous media of the liposome, it is allowed to retain its solution-based characteristics such as spectral emission profile, Stokes shift and response time, which can be varied slightly when embedded in other media such as a polymer. In addition, the surrounding lipid membrane allows particular analytes to pass into the interior of the sensor and react with the indicator dye, while preventing the dye from exiting and being degraded by the cell. Presently, liposome-based nanosensors have been developed and applied to the measurement of pH[93] and molecular oxygen[94] in various cellular environments, demonstrating the high degree of sensitivity as well as specificity of such sensors.

A recent variation to these liposome-encapsulated nanosensors has been to first partially immobilize the fluorescent indicator dye of interest in a polystyrene nanoparticle, prior to encapsulation in a phospholipid membrane.[89,91,96,97] This modification in the nanosensor fabrication allows the dyes to be partially embedded in the polystyrene bead, providing a much more stable sensor that is less susceptible to biological degradation, while still retaining much of the solution-based characteristics of the dye. This second subclass of phospholipid-based nanosensors, known as lipobead nanosensors, also allows the fluorescent indicator dye to be partially embedded in the phospholipid membrane, increasing the accessibility of the dye to the analyte species. This dramatically increases the potential number of analytes that can be investigated over other encapsulated nanosensors such as the PEEBLEs that require the analyte to travel through tiny pores to interact with the indicator dye.[89,91]

3.2.2.4 SERS-Based Nanosensors

In addition to the luminescent-based nanosensors and nano-biosensors that have been developed over the past decade, a fourth class of implantable nanosensors and nano-biosensors have been developed that rely on SERS to provide information about the environment being probed. Unlike their luminescent counterparts, these nanosensors take advantage of the rich molecular information provided by Raman spectroscopy and the localized sensitivity provided by the roughened metal surface necessary for SERS to provide a sensitive and selective detection scheme. In addition, the narrow linewidths associated with Raman and SERS spectroscopies allow for the multiplexed monitoring of many different species simultaneously.[98–100] While metallic colloidal nanoparticles have been used for SERS analyses for several decades, the first true nanoparticle-based SERS nanosensors and nano-biosensors have been around for a much shorter period of time.

Some of the earliest reported SERS nano-biosensors developed were reported by Mirkin and coworkers,[99,100] in which oligonucleotide sequences were attached to metallic nanoparticles. These oligonucleotide-labeled nanoparticles were then used to monitor the binding of complementary strands of dye-labeled oligonucleotides, resulting in an SERS signal from the particular dye used for labeling. Since a different dye was attached to each different oligonucleotide sequence being investigated and the SERS spectra of each type of dye molecule is unique, this resulted in a form of molecular "barcoding" that could be used to interrogate individual strands of DNA or RNA. By taking advantage of the narrow linewidths, these nanosensors were capable of probing many different oligonucleotide sequences simultaneously.

More recently, antibody-labeled SERS nanoparticles that are capable of being implanted into individual cell for the direct detection of specific protein species have been developed by Cullum and coworkers. Unlike previous nanoparticle-based SERS biosensors, these sensors do not require a washing step or dye label to identify the presence of the analyte. These nano-biosensors are fabricated by coating a thin layer of a SERS-active metal such as silver or gold onto a silica nanoparticle followed by attachment of the antibody of interest through a thiol linkage.[101,102] Once bound, the sensors are implanted into a cell by either nano-injection or a pressured-injection technique such as a gene gun, and positioned at the location of interest via optical tweezers. Once in place within the cellular compartment of interest, antigens bind to the antibodies on the sensor either bringing the antigen close enough to the metal surface for direct SERS monitoring, or causing conformational changes in the antibody that can be monitored via the resulting SERS spectrum. By optimizing the surface coverage of the antibody on the nano-biosensor, exposure of the metal surface to other potential species within the cellular environment is dramatically reduced, resulting in minimal interferences.[102] In addition to SERS-based nano-biosensors such as those described above, similar nanosensors have recently been developed for real-time monitoring of pH in microscopic environments, in which organic molecules are attached to the surface of a SERS-active nanoparticle and the protonation of those organic molecules is monitored via SERS spectroscopy.[103] Based upon the sensitivity and selectivity of SERS-based nanosensors and nano-biosensors and their ability to monitor many different species simultaneously, they should see a great deal of application to cellular analyses in the future.

3.3 Biochip and Chip-Based Biosensor Arrays

While fiber-optic and implantable nanosensor and nano-biosensor technologies have provided the scientific community with unprecedented methodologies for quantitatively determining the presence and location of cellular species ranging from ions to oligonucleotides (e.g., DNA and RNA) to proteins, knowledge of the intracellular location of these species is not always necessary. In addition, it is sometimes necessary to monitor thousands of different analytes from a cell or tissue sample simultaneously. Analyses such as these have resulted in the advent of biosensor arrays, with individual receptor sites having micrometer- or nanometer-scale dimensions. Such arrays, are often referred to as "gene chips," "DNA chips," or "protein chips," depending upon the analyte being probed, or the more general terms of "biochip" or "microchip array."[17,104–106] The underlying technologies necessary for biochips were developed in the early 1980s and since that time these devices have found a large number of applications,[11,105–107] spawning the fields of genomics and proteomics. Owing to the rapid growth and widespread application of biochip technologies, several good reviews have been written on the subject and their application to various fields over the years.[6–11]

In general, biochips can be described as consisting of a highly ordered array of microscopic- or submicroscopic-sized spots of biological receptor molecules on a substrate of one form or another (see Figure 3.8) and an array transducer for location-specific measurements. Although transduction can occur through many different means (e.g., optical and electrochemical), this chapter focuses solely on optical transduction schemes, which account for the vast majority of biochip-based arrays. Fabrication of high-density bioreceptor arrays used for biochip analyses is typically performed using robotic spotting systems, capable of delivering microliter to nanoliter volumes of the bioreceptor of interest to the appropriate substrate. The most common substrates employed for such analyses are microscope slides; however, membrane materials (e.g., zeta-probe membrane) have also been used depending upon the sample that is being analyzed and the degree of nonspecific binding of the sample to the array substrate.[108] Once an array has been produced with each spot corresponding to a different particular analyte, the sample solution of interest is placed on the surface of the array, and the different bioreceptors are allowed to bind their corresponding analyte. After binding, the array is typically washed off to remove any unbound sample materials from the analysis area, followed by reaction with an optically measurable marker, such as a fluorescent dye, that binds solely to the bioreceptor spots that have reacted with analyte.

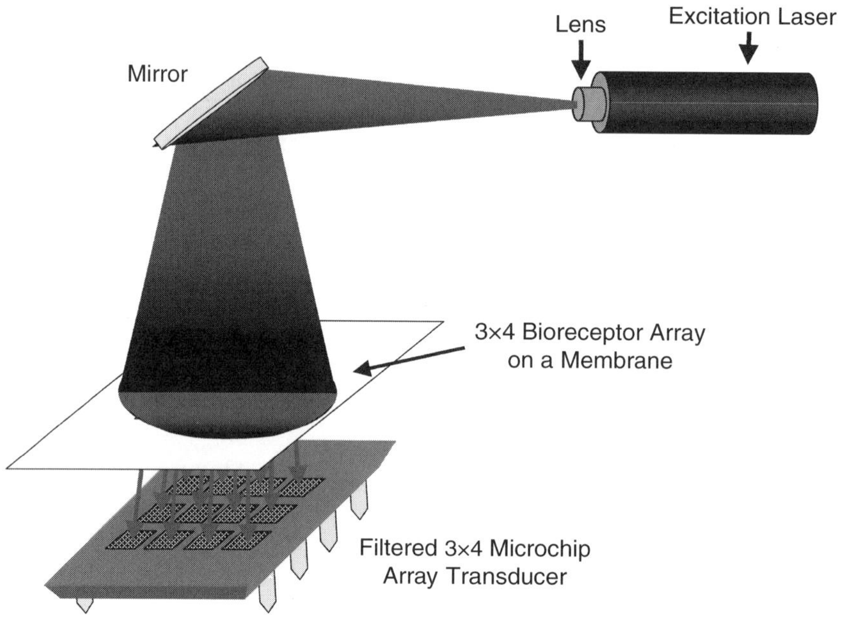

FIGURE 3.8
(See color insert following page 330) Schematic diagram of the basic concept of a biochip-based sensor. A light source is used to excite an array of different bioreceptors and the resulting optical emission is detected using an array detector.

After a final wash, to remove any unbound excess marker, the array is interrogated, and the resulting signal is measured from each location in the array simultaneously. The locations that exhibit signal are then related back to the bioreceptor spotted in that location, and the identity of the analyte is determined. By performing assays in this fashion, a vast amount of analytes (e.g., DNA, RNA, and proteins) can be measured simultaneously with the same excitation and detection scheme, resulting in a rapid and relatively inexpensive means of diagnosing the presence of particular genes or biomarkers.

3.3.1 Oligonucleotide-Based Biochips

One of the early impetuses for the development of biochip arrays was the decoding of the human genome. In this work, high-density arrays of thousands to hundreds-of-thousands of oligonucleotide strands were placed on a suitable substrate and monitored for subsequent binding of a complementary sequence of single-stranded DNA. Since this time, similar analyses have been performed for diagnostic purposes on various cellular and tissue samples. In one such diagnostic analysis, 96,000 different oligonucleotide sequences were synthesized by polymerase chain reaction (PCR) of RNA and patterned on a substrate by Hacia et al.,[109] allowing for the screening of the entire 5.53-kb coding region of the hereditary ovarian cancer gene, breast cancer 1 early onset (BRCA1), in a single analysis. Such analyses can be used for the screening of genetic dispositions to certain forms of disease such as cancer.

3.3.2 Antibody-Based Biochips

In addition to oligonucleotide bioreceptors, many other types of biological receptor molecules have been employed in the development of biochip arrays. In one such case, antibodies were immobilized on the surface of a microscope slide in a medium-to-low-density array, and used for the sensitive and simultaneous detection of several different biological toxins, including ricin, staphylococcal enterotoxin B (SEB), and yersinia pestis.[110] Attachment of these antibodies was accomplished using a photoactivated optical adhesive that was placed in small wells on the microscope slide. Once the captured antibodies were immobilized on the slide, toxins were introduced, followed by a second fluorescently labeled antibody. Upon binding of the fluorescently labeled secondary antibodies, the array was excited and the resulting fluorescence image was captured on a CCD camera and correlated to the various toxins. From this work it was found that the various toxins could be detected at concentrations ranging from 5 to 25 ng/mL.

3.3.3 Aptamer-Based Biochips

A third class of bioreceptor that has recently been used in the development of biochip arrays is a relatively new type of bioreceptor known as an aptamer.[111] Aptamers are synthetic oligonucleotide sequences that fold into three-dimensional structures and are capable of binding proteins and other biological species with affinities approaching those of antibodies.[112,113] By employing aptamers as biological recognition elements in a biochip array, it has been possible to develop proteomic assays that require minimal washing steps while providing a high degree of sensitivity and specificity.[113] In addition, to providing a high degree of sensitivity and specificity, aptamers also offer the advantage of being able to be synthetically fabricated and amplified via PCR, as well as the ability to be placed in denaturing conditions and later be reconstituted. When employing aptamer arrays, analyte proteins from a sample are placed in contact with aptamers immobilized in an array format on a suitable substrate, and then washed with a fluorescent dye that selectively attaches to proteins. Once the dye label has been attached to the immobilized proteins, fluorescence images of the arrays are obtained using a CCD or other imaging device, and related to the presence and concentration of analyte proteins, similar to other types of biochips.

3.3.4 Integrated Multifunctional-Based Biochips

When performing complex biological diagnostics, the need to monitor more than just one type of analyte is often essential. To this end, biochip arrays have been developed that employ a mixture of different types of bioreceptors (e.g., antibodies and oligonucleotides) on the same substrate array. Biochip arrays formed using a mixture of bioreceptor molecules are often termed "multifunctional biochips." One early example of a multifunctional biochip was demonstrated by Vo-Dinh and coworkers,[114] in which DNA and antibodies were immobilized on a single-sensing array and used to detect various biomarkers or disease end points. This work also demonstrated the development of one of the earliest, self-contained bioreceptor–transducer array formats, in which a complementary metal oxide semiconductor (CMOS)-based photodiode sensor array with on-chip amplifiers, discriminators, and logic circuitry were employed for the detection of optical signals corresponding to analyte binding.[115,116] Unlike most previous biochip detection systems, which employed optical microscopes or similar microscopic imaging systems with CCD detectors, this integrated biochip system fit onto a single integrated circuit microchip, creating a device with a small footprint that could be mass produced inexpensively. Based upon the concept of a fully integrated and disposable biochip microarray and the various bioreceptor molecules that have been used for biochip fabrication, many different commercial variations of this CMOS-based chip design have been fabricated and are commercially available, ranging from low- to high-density arrays for many different applications.

3.4 Conclusions

Unlocking the vast amounts of information present within individual cells and tissue cultures has historically involved extensive biological procedures such as gel electrophoresis, Southern and Northern blotting, mass spectrometry and many others, resulting in cellular population analyses of a single chemical species that could take a long time. However, with the advent of nanosensors and biochips and mankind's increased control over materials on the nanometer scale, significant advances have been made in the field of cellular and biological analyses. With the development of intracellular nanosensors (i.e., fiber-optic nanosensors and particle-based nanosensors), the ability to monitor individual chemical species in a site-specific fashion within an individual cell has already provided and continues to provide unprecedented information about cellular and biological systems. Employing these nanosensors, it is now possible to monitor, in real time, the effects of a particular stimulus on cellular response, and correlate this response to parameters such as growth phase of the cell. Such analyses could potentially provide insight into cell-to-cell variabilities in response of a particular type of cell to a stimulus. By employing implantable nanosensors that rely on narrow bandwidth spectral profiles, such as the SERS-based nanosensors, the ability to detect hundreds of different chemical species simultaneously within an individual cell may provide important insights into the interrelated interaction of various biological or cellular pathways.

In addition to the vast amount of knowledge that can be gained from site-specific monitoring of cellular and biological reactions using nanosensors, rapidly growing advances in biochip technology and high-density arraying of biological receptor molecules has allowed for the rapid detection of as many as hundreds-of-thousands of different analytes simultaneously. Biochips represent a unique bioanalytical tool that could allow for the screening of potential biological markers to many different diseases or illnesses with minimal sample and low cost, resulting in a reduction in the number and volume of samples to be sent to labs for medical diagnoses. In the last 15 to 20 years, biochip-based sensing has seen a rapid and dramatic growth, in which this technology has developed from a research-based tool to commercially available devices for a wide variety of applications.[6,117] In fact, currently, there are over 15 companies that have been formed worldwide with multiple product lines devoted to biochip sales and this number is continuously growing. Based upon these events, the future for biochip-based analyses looks very promising. With the great deal of diagnostic and predictive medical information that could be gained from a small biological sample (e.g., drop of blood) and the hundreds of thousands of different biomarkers that could be examined in a single biochip-based analysis, biochip-based technologies may one day result in a small integrated device that can be found in a doctor's office or even home for the rapid screening of various illnesses or diseases. Recent advances in the integration of the sensing arrays and transducer arrays used for signal readout have already resulted in commercially available biochips. However, further advances in the integration of sampling systems, inexpensive optical excitation, and signal amplification and discrimination are still needed before reliable and inexpensive devices can be made available to the population. Presently, biochip technologies have revolutionized the fields of genomic and proteomic research and current trends in miniaturization and inexpensive fabrication methodologies offer the promise of small integrated systems capable of providing rapid medical screenings for many different diseases with only a small amount of sample.

References

1. Tan, W. H., Kopelman, R., Barker, S. L. R., and Miller, M. T. (1999) Ultrasmall Optical Sensors for Cellular Measurements. *Anal. Chem.* 71 (17): 606A–612A.
2. Vo-Dinh, T., Cullum, B. M., and Stokes, D. L. (2001) Nanosensors and Biochips: Frontiers in Biomolecular Diagnostics. *Sensors Actuat. B.* 74 (1–3): 2–11.
3. Cullum, B. M. and Vo-Dinh, T. (2000) The Development of Optical Nanosensors for Biological Measurements. *Trends Biotechnol.* 18 (9): 388–393.
4. Cullum, B. M. (2004) Optical Nanosensors and Nanobiosensors, in *Encyclopedia of Nanoscience and Nanotechnology.* New York: Marcel Dekker, pp. 2757–2768.
5. Lu, J. Z. and Rosenzweig, Z. (2000) Nanoscale Fluorescent Sensors for Intracellular Analysis. *Anal. Chem.* 366 (6–7): 569–574.
6. Vo-Dinh, T. (1998) Development of a DNA Biochip: Principles and Applications. *Sensors Actuat. B* 51: 52–58.
7. Vo-Dinh, T. and Askari, M. (2001) Micro-Arrays and Biochips: Applications and Potential in Genomics and Proteomics. *Curr. Genomics* 2: 399–406.
8. Schena, S., Heller, R. A., Theriault, T. P., Konrad, K., Lachenmeier, E., and Davis, R. W. (1998) Microarrays: Biotechnology's Discovery Platform for Functional Genomics. *Trends Biotechnol.* 16 (7): 301–306.
9. Wallace, R. W. (1997) DNA on a Chip: Serving up the Genome for Diagnostics and Research. *Mol. Med. Today* 3 (9): 384–387.
10. Nuwaysir, E. F., Bittner, M., Trent, J., Barrett, J. C., and Afshari, C. A. (1999) Microarrays and Toxicology: The Advent of Toxicogenomics. *Mol. Carcinog.* 24 (3): 153–161.
11. Ekins, R. and Chu, F. W. (1999) Microarrays: Their Origins and Applications. *Trends Biotechnol.* 17 (6): 217–223.
12. Tan, W. H., Shi, Z. Y., Smith, S., Birnbaum, D., and Kopelman, R. (1992) Submicrometer Intracellular Chemical Optical Fiber Sensors. *Science* 258: 778–781.
13. Tan, W. H., Shi, Z. Y., and Kopelman, R. (1992) Development of Submicron Chemical Fiber Optic Sensors. *Anal. Chem.* 64 (23): 2985–2990.
14. DeRisi, J., Penland, L., Brown, P. O., Bittner, M. L., Meltzer, P. S., Ray, M., Chen, Y. D., Su, Y. A., and Trent, J. M. (1996) Use of a cDNA Microarray to Analyse Gene Expression Patterns in Human Cancer. *Nat. Genet.* 14 (4): 457–463.
15. Kononen, J., Bubendorf, L., Kallioniemi, A., Barlund, M., Schraml, P., Leighton, S., Torhorst, J., Mihatsch, M. J., Sauter, G., and Kallioniemi, O. P. (1998) Tissue Microarrays for High-Throughput Molecular Profiling of Tumor Specimens. *Nat. Med.* 4 (7): 844–848.
16. Afsahri, C. A., Nuwaysir, E. F., and Barrett, J. C. (1999) Application of Complimentary DNA Microarray Technology to Carcinogen Identification. *Cancer Res.* 59 (19): 4759–4766.
17. Blanchard, A. P. and Hood, L. (1996) Sequence to Array: Probing the Genome's Secrets. *Nat. Biotechnol.* 14 (13): 1649–1652.
18. Wang, D. G., Fan, J. B., Siao, C. J., Berno, A., Young, P. C., Sapolsky, R., Ghandour, G., Perkins, N., Winchester, E., Spencer, J., Kruglyak, L., Stein, L., Hsie, L., Topaloglou, T., Hubbel, E., Robinson, E., Mittmann, M., Morris, M. S., Shen, N. P., Kilburn, D., Rioux, J., Nusbaum, C., Rozen, S., Hudson, T. J., Lipshutz, R., Chee, M., and Lander, E. S. (1998) Large-Scale Identification, Mapping and Genotyping of Single-Nucleotide Polymorphisms in the Human Genome. *Science* 280: 1077–1083.
19. Lu, J. Z. and Rosenzweig, Z. (2000) Nanoscale Fluorescent Sensors for Intracellular Analysis. *Fresenius J. Anal. Chem.* 366 (6–7): 569–575.
20. Vo-Dinh, T. and Cullum, B. M. (2000) Biosensors and Biochips: Advances in Biological and Medical Diagnostics. *Fresenius J. Anal. Chem.* 366 (6–7): 540–551.
21. Cullum, B. M. and Vo-Dinh, T. (2003) Nanosensors: Development and Application of Nanosensors for Cellular Diagnostics, in *Biomedical Photonics Handbook*, 1 ed., Vo-Dinh, T., Ed., Boca Raton, FL: CRC Press, pp. 60-1–60-28.

22. Clark, L. C. J. and Lyons, C. (1962) Electrode Systems for Continuous Monitoring in Cardiovascular Surgery. *Ann. Acad. Sci.* 102: 29–45.
23. Betzig, E., Harootunian, A., Lewis, A., and Isaacson, M. (1985) Near-Field Scanning Optical Microscopy (NSOM)—Investigation of Radiation Transmitted through Sub-Wavelength Apertures. *Biophys. J.* 47 (2): A407–A407.
24. Betzig, E., Lewis, A., Harootunian, A., Isaacson, M., and Kratschmer, E. (1986) Near-Field Scanning Optical Microscopy (NSOM)—Development and Biophysical Applications. *Biophys. J.* 49 (1): 269–279.
25. Durig, U., Pohl, D. W., and Rohner, F. (1986) Near-Field Optical-Scanning Microscopy. *J. Appl. Phys.* 59 (10): 3318–3327.
26. Pohl, D. W. (1984) Scanning Near-Field Optical Microscopy, in *Advances in Optical and Electron Microscopy*, Sheppard, C. J. R. and Mulvey, T., Eds., London: Academic Press.
27. Betzig, E., Trautman, J. K., Harris, T. D., Weiner, J. S., and Kostelak, R. L. (1991) Breaking the Diffraction Barrier—Optical Microscopy on a Nanometric Scale, *Science* 251(5000): 1468–1470.
28. Betzig, E. and Chichester, R. J. (1993) Single Molecules Observed by Near-Field Scanning Optical Microscopy. *Science* 262 (5138): 1422–1425.
29. Betzig, E., Isaacson, M., and Lewis, A. (1987) Collection Mode Near-Field Scanning Optical Microscopy. *Appl. Phys. Lett.* 51 (25): 2088–2090.
30. Lieberman, K., Harush, S., Lewis, A., and Kopelman, R. (1990) A Light-Source Smaller Than the Optical Wavelength. *Science* 247 (4938): 59–61.
31. Zeisel, D., Deckert, V., Zenobi, R., and Vo-Dinh, T. (1998) Near-Field Surface-Enhanced Raman Spectroscopy of Dye Molecules Adsorbed on Silver Island Films. *Chem. Phys. Lett.* 283 (5–6): 381–385.
32. Deckert, V., Zeisel, D., Zenobi, R., and Vo-Dinh, T. (1998) Near-Field Surface Enhanced Raman Imaging of Dye-Labeled DNA with 100-nm Resolution. *Anal. Chem.* 70 (13): 2646–2650.
33. Alarie, J. P. and VoDinh, T. (1996) Antibody-Based Submicron Biosensor for Benzo a Pyrene DNA Adduct. *Polycyc. Aromat. Compounds* 8 (1): 45–52.
34. Cullum, B. M., Griffin, G. D., Miller, G. H., and Vo-Dinh, T. (2000) Intracellular Measurements in Mammary Carcinoma Cells Using Fiber-Optic Nanosensors. *Anal. Biochem.* 277 (1): 25–32.
35. Hoffmann, P., Dutoit, B., and Salathe, R. P. (1995) Comparison of Mechanically Drawn and Protection Layer Chemically Etched Optical Fiber Tips. *Ultramicroscopy* 61 (1–4): 165–170.
36. Turner, D. R. (1984) Patent #4,469,554.
37. Stockle, R., Fokas, C., Deckert, V., Zenobi, R., Sick, B., Hecht, B., and Wild, U. P. (1999) High-Quality Near-Field Optical Probes by Tube Etching. *Appl. Phys. Lett.* 75 (2): 160–162.
38. Munkholm, C., Parkinson, D. R., and Walt, D. R. (1990) Intramolecular Fluorescence Self-Quenching of Fluoresceinamine. *J. Am. Chem. Soc.* 112 (7): 2608–2612.
39. Munkholm, C., Walt, D. R., Milanovich, F. P., and Klainer, S. M. (1986) Polymer Modification of Fiber Optic Chemical Sensors as a Method of Enhancing Fluorescence Signal for pH Measurement. *Anal. Chem.* 58 (7): 1427–1430.
40. Song, A., Parus, S., and Kopelman, R. (1997) High-Performance Fiber Optic pH Microsensors for Practical Physiological Measurements Using a Dual-Emission Sensitive Dye. *Anal. Chem.* 69 (5): 863–867.
41. Tan, W. H., Shi, Z. Y., and Kopelman, R. (1995) Miniaturized Fiberoptic Chemical Sensors with Fluorescent Dye-Doped Polymers. *Sensors Actuat. B* 28 (2): 157–163.
42. Samuel, J., Strinkovski, A., Lieberman, K., Ottolenghi, M., Avnir, D., and Lewis, A. (1994) Miniaturization of Organically Doped Sol-Gel Materials—A Microns-Size Fluorescent pH Sensor. *Mat. Lett.* 21 (5–6): 431–434.
43. McCulloch, S., and Uttamchandani, D. (1995) Development of a Fibre Optic Micro-Optrode for Intracellular pH Measurements. *IEEE Proc. Optoelec.* 144 (162): 146–154.
44. Koronczi, I., Reichert, J., Heinzmann, G., and Ache, H. J. (1998) Development of a Submicron Optochemical Potassium Sensor with Enhanced Stability Due to Internal Reference. *Sensors Actuat. B* 51 (1–3): 188–195.
45. Bui, J. D., Zelles, T., Lou, H. J., Gallion, V. L., Phillips, M. I., and Tan, W. H. (1999) Probing Intracellular Dynamics in Living Cells with Near-Field Optics. *J. Neurosci. Meth.* 89 (1): 9–15.

46. Barker, S. L. R., Clark, H. A., Swallen, S. F., Kopelman, R., Tsang, A. W., and Swanson, J. A. (1999) Ratiometric and Fluorescence Lifetime-Based Biosensors Incorporating Cytochrome C ′ and the Detection of Extra- and Intracellular Macrophage Nitric Oxide. *Anal. Chem.* 71 (9): 1767–1772.
47. Barker, S. L. R., Thorsrud, B. A., and Kopelman, R. (1998) Nitrite- and Chloride-Selective Fluorescent Nano-Optodes and in in Vitro Application to Rat Conceptuses. *Anal. Chem.* 70 (1): 100–104.
48. Barker, S. L. R. and Kopelman, R. (1998) Development and Cellular Applications of Fiber Optic Nitric Oxide Sensors Based on a Gold-Adsorbed Fluorophore. *Anal. Chem.* 70 (23): 4902–4906.
49. Cullum, B. M. and Vo-Dinh, T. (2000) Optical Nanosensors and Biological Measurements. *Biofutur* 205: A1–A6.
50. Vo-Dinh, T. G., G. D.; Alarie, J. P.; Cullum, B. M.; Sumpter, B. and Noid, D. (2000) Development of Nanosensors and Bioprobes. *J. Nanopart. Res.* 2: 17–23.
51. Vo-Dinh, T., Alarie, J. P., Cullum, B. M., and Griffin, G. D. (2000) Antibody-Based Nanoprobe for Measurement of a Fluorescent Analyte in a Single Cell. *Nat. Biotechnol.* 18 (7): 764–767.
52. Barker, S. L. R., Kopelman, R., Meyer, T. E., and Cusanovich, M. A. (1998) Fiber-Optic Nitric Oxide-Selective Biosensors and Nanosensors. *Anal. Chem.* 70 (5): 971–976.
53. Cordek, J., Wang, X. W., and Tan, W. H. (1999) Direct Immobilization of Glutamate Dehydrogenase on Optical Fiber Probes for Ultrasensitive Glutamate Detection. *Anal. Chem.* 71 (8): 1529–1533.
54. Rosenzweig, Z. and Kopelman, R. (1996) Analytical Properties and Sensor Size Effects of a Micrometer-Sized Optical Fiber Glucose Biosensor. *Anal. Chem.* 68 (8): 1408–1413.
55. Liu, X. J., Farmerie, W., Schuster, S., and Tan, W. H. (2000) Molecular Beacons for DNA Biosensors with Micrometer to Submicrometer Dimensions. *Anal. Biochem.* 283 (1): 56–63.
56. Tyagi, S. and Kramer, F. R. (1996) Molecular Beacons: Probes That Fluoresce Upon Hybridization. *Nat. Biotechnol.* 14 (3): 303–308.
57. Liu, X. J. and Tan, W. H. (1999) A Fiber-Optic Evanescent Wave DNA Biosensor Based on Novel Molecular Beacons. *Anal. Chem.* 71 (22): 5054–5059.
58. Cullum, B. M., Griffin, G. D., and Vo-Dinh, T. (2001) Nanosensors for Analysis of a Single Cell. *SPIE Proc.* 4254: 35–40.
59. Hankus, M., Gibson, G., Chandrasekharan, N., and Cullum, B. M. (2005) Surface Enhanced Raman Scattering (SERS)—Nanoimaging Probes for Biological Analysis. *SPIE Proc.* 5588: 106–117.
60. Pantano, P. and Walt, D. R. (1997) Toward a Near-Field Optical Array. *Rev. Sci. Instrum.* 68 (3): 1357–1359.
61. Walt, D. R. (2005) Near Field Arrays, http://ase.tufts.edu/chemistry/walt/index.htm.
62. Alivisatos, A. P. (1996) Semiconductor Clusters, Nanocrystals and Quantum Dots. *Science* 271: 933–937.
63. Sooklal, K., Cullum, B. M., Angel, S. M., and Murphy, C. J. (1996) Photophysical Properties of ZnS Nanoclusters with Spatially Localized Mn^{2+}. *J. Phys. Chem.* 100 (11): 4551–4555.
64. Chan, W. C. W. and Nie, S. (2003) Luminescent Quatum Dots as Advanced Biological Labels, in *Biomedical Photonics Handbook*, Vo-Dinh, T., Ed., Boca Raton, FL: CRC Press, pp. 58-1–58-14.
65. Dahan, M., Laurence, T., Pinaud, F., Chemla, D. S., Alivisatos, A. P., Sauer, M., and Weiss, S. (2001) Time-Gated Biological Imaging by Use of Colloidal Quantum Dots. *Opt. Lett.* 26 (11): 825–827.
66. Hermanson, G. T. (1996) *Bioconjugate Techniques* New York, Academic Press.
67. Evident Technologies Product Line (2004), *www.evidenttech.com*.
68. Mitchell, G. P. and Mirkin, C. A. (1999) Programmed Assembly of DNA Functionalized Quantum Dots. *J. Am. Chem. Soc.* 121: 8122–8127.
69. Fu, A., Micheel, C. M., Cha, J., Chang, H. P., Yang, H., and Alivisatos, A. P. (2000) Discrete Nanostructures of Quantum Dots/Au with DNA. *J. Am. Chem. Soc.* 126: 10832–10833.
70. Mattoussi, H., Mauro, J. M., Goldman, E. R., Anderson, G. P., Sundar, V. C., Mikulec, F. V., and Bawendi, M. G. (2000) Self-Assembly of CeSe-ZnS Quantum Dots Bioconjugates Using an Engineered Recombinant Protein. *J. Am. Chem. Soc.* 122: 12142.

71. Chan, W. C. W. and Nie, S. M. (1998) Quantum Dot Bioconjugates for Ultrasensitive Nonisotopic Detection. *Science* 281: 2016–2018.
72. Lyon, W. A. and Nie, S. M. (1997) Single Molecule Methodologies for DNA Analysis. *Abstr. Am. Chem. Soc.* 213: 49-PHYS.
73. Taylor, J. R., Fang, M. M., and Nie, S. M. (200) Probing Specific Sequences on Single DNA Molecules with Bioconjugated Fluorescent Nanoparticles. *Anal. Chem.* 72 (9): 1979–1986.
74. Zhang, C. Y., Ma, H., Nie, S. M., Ding, Y., Jin, L., and Chen, D. Y. (2000) Quantum Dot-Labeled Trichosanthin. *Analyst* 125 (6): 1029–1031.
75. Zhang, C. Y., Ma, H., Ding, Y., Jin, L., Chen, D. Y., Miao, Q., and Nie, S. M. (2001) Studies on Quantum Dots-Labeled Trichosanthin. *Chem. J. Chin. Univ.* 22 (1): 34–37.
76. Hoshino, H. (2004) Applications of T-lymphoma Labeled with Fluorescent Quantum Dots to Cell Tracing Markers in Mouse Body. *Biophys. Biochem. Res. Comm.* 314: 46–53.
77. Sasaki, K., Shi, Z. Y., Kopelman, R., and Masuhara, H. (1996) Three-Dimensional pH Microprobing with an Optically-Manipulated Fluorescent Particle. *Chem. Lett.* 2: 141–142.
78. Clark, H. A., Hoyer, M., Philbert, M. A., and Kopelman, R. (1999) Optical Nanosensors for Chemical Analysis inside Single Living Cells. 1. Fabrication, Characterization, and Methods for Intracellular Delivery of PEBBLE Sensors. *Anal. Chem.* 71 (21): 4831–4836.
79. Clark, H. A., Kopelman, R., Tjalkens, R., and Philbert, M. A. (1999) Optical Nanosensors for Chemical Analysis inside Single Living Cells. 2. Sensors for pH and Calcium and the Intracellular Application of PEBBLE Sensors. *Anal. Chem.* 71 (21): 4837–4843.
80. Sumner, J. P., Aylott, J. W., Monson, E., and Kopelman, R. (2002) A Fluorescent PEBBLE Nanosensor for Intracellular Free Zinc. *Analyst* 127 (1): 11–16.
81. Clark, H. A., Hoyer, M., Philbert, M. A., and Kopelman, R. (1999) Optical Nanosensors for Chemical Analysis inside Single living cells. 1. Fabrication, Characterization, and Methods for Intracellular Delivery of PEBBLE Sensors. *Anal. Chem.* 71 (21): 4831–4836.
82. Clark, H. A., Kopelman, R., Tjalkens, R., and Philbert, M. A. (1999) Optical Nanosensors for Chemical Analysis inside Single Living Cells. 2. Sensors for pH and Calcium and the Intracellular Application of PEBBLE Sensors. *Anal. Chem.* 71 (21): 4837–4843.
83. Clark, H. A., Barker, S. L. R., Brasuel, M., Miller, M. T., Monson, E., Parus, S., Shi, Z. Y., Song, A., Thorsrud, B., Kopelman, R., Ade, A., Meixner, W., Athey, B., Hoyer, M., Hill, D., Lightle, R., and Philbert, M. A. (1998) Subcellular Optochemical Nanobiosensors: Probes Encapsulated by Biologically Localised Embedding (PEBBLEs). *Sensors Actuat. B* 51 (1–3): 12–16.
84. Xu, H., Aylott, J. W., Kopelman, R., Miller, T. J., and Philbert, M. A. (2001) A Real-Time Ratiometric Method for the Determination of Molecular Oxygen inside Living Cells Using Sol-Gel-Based Spherical Optical Nanosensors with Applications to Rat C6 Glioma. *Anal. Chem.* 73 (17): 4124–4133.
85. Xu, H., Aylott, J., and Kopelman, R. (2000) Sol-gel PEBBLE Sensors for Biochemical Analysis inside Living Cells. *Abstr. Am. Chem. Soc.* 219: 97-ANYL.
86. Xu, H., Yan, F., Monson, E. E., and Kopelman, R. (2003) Room-Temperature Preparation and Characterization of Poly (Ethylene Glycol)-Coated Silica Nanoparticles for Biomedical Applications. *J. Biomed. Mat. Res.* 66A (4): 870–879.
87. Anker, J. N., Behrend, C., and Kopelman, R. (2003) Aspherical Magnetically Modulated Optical Nanoprobes (MagMOONs). *J. Appl. Phys.* 93 (10): 6698–6700.
88. Anker, J. N. and Kopelman, R. (2003) Magnetically Modulated Optical Nanoprobes. *Appl. Phys. Lett.* 82 (7): 1102–1104.
89. Ji, J., Rosenzweig, N., Jones, I., and Rosenzweig, Z. (2001) Molecular Oxygen-Sensitive Fluorescent Lipobeads for Intracellular Oxygen Measurements in Murine Macrophages. *Anal. Chem.* 73 (15): 3521–3527.
90. Ji, J., Rosenzweig, N., Griffin, C., and Rosenzweig, Z. (2000) Synthesis and Application of Submicrometer Fluorescence Sensing Particles for Lysosomal pH Measurements in Murine Macrophages. *Anal. Chem.* 72 (15): 3497–3503.
91. McNamara, K. P., Nguyen, T., Dumitrascu, G., Ji, J., Rosenzweig, N., and Rosenzweig, Z. (2001) Synthesis, Characterization, and Application of Fluorescence Sensing Lipobeads for Intracellular pH Measurements. *Anal. Chem.* 73 (14): 3240–3246.
92. Nguyen, T., McNamara, K. P., and Rosenzweig, Z. (1999) Optochemical Sensing by Immobilizing Fluorophore-Encapsulating Liposomes in Sol-Gel Thin Films. *Anal. Chim. Acta* 400: 45–54.

93. McNamara, K. P., Rosenzweig, N., and Rosenzweig, Z. (1999) Liposome-Based Optochemical Nanosensors. *Mikrochim. Acta* 131 (1–2): 57–64.
94. McNamara, K. P. and Rosenzweig, Z. (1998) Dye-encapsulating Liposomes as Fluorescence-Based Oxygen Nanosensors. *Anal. Chem.* 70 (22): 4853–4859.
95. Rosenzweig, Z., Rosenzweig, N., and Blagoi, G. (2004) FRET-Based Luminescence Sensors for Carbohydrates and Glycoproteins Analysis, *SPIE Proc.* 5588: 1–8.
96. McNamara, K. P. and Rosenzweig, Z. (1999) Nanosized Phospholipid-Coated Particles as Biological Sensors. *Abstr. Am. Chem. Soc.* 218: 26-ANYL.
97. DeCoster, M. A., Nguyen, T., Bazan, N. G., and Rosenzweig, Z. (2002) Adherence of Lectin-Coated Micro Lipobeads to Cortical Neurons in Cell Culture. *FASEB* 16 (4): A37–A37.
98. Cullum, B. M., Mobley, J., Stokes, D. L., Allain, L. R., Stratis, D. N., Wintenberg, A. L., Maples, R. A., and Vo-Dinh, T. (2002) Development of a Portable Raman Spectrometer for Medical Diagnostics. *SPIE Proc.* 4615: 82–90.
99. Cao, Y. C., Jin, R., and Mirkin, C. A. (2002) Nanoparticles with Raman Spectroscopic Fingerprints for DNA and RNA Detection. *Science* 297: 1536–1540.
100. Nam, J.-M., Thaxton, C. S., and Mirkin, C. A. (2003) Nanoparticle-Based Bio-Bar Codes for the Ultrasensitive Detection of Proteins. *Science* 301: 1884–1886.
101. Li, H. and Cullum, B. M. (2003) Development and Optimization of SERS-Based Immuno-Nanosensor for Single Cell Analyses. *SPIE Proc.* 5261: 142–154.
102. Li, H., Sun, J., and Cullum, B. M. (2005) Nanosphere-Based SERS Immuno-Sensors for Protein Analysis. *SPIE Proc.* 5588: 19–31.
103. Talley, C. E., Jusinski, L., Hollars, C. W., Lane, S. M., and Huser, T. (2004) Intracellular pH Sensors Based on Surface-Enhanced Raman Scattering. *Anal. Chem.* 76 (23): 7064–7068.
104. Marshall, A. and Hodgson, J. (1998) DNA Chips: An Array of Possibilities. *Nat. Biotechnol.* 16 (1): 27–34.
105. Schena, M., Shalon, D., Davis, R. W., and Brown, P. O. (1995) Quantitative Monitoring of Gene-Expression Patterns with a Complimentary-DNA Microarray. *Science* 270: 467–471.
106. Vo-Dinh, T. (2003) Biochips and Microarrays: Tools for the New Medicine, in *Biomedical Photonics Handbook*, Vo-Dinh, T., Ed., Boca Raton, FL: CRC Press, pp. 51-1–51-29.
107. Chee, M., Yang, R., Hubbell, E., Berno, A., Huang, X. C., Stern, D., Winkler, J., Lockhart, D. J., Morris, M. S., and Fodor, S. P. A. (1996) Accessing Genetic Information with High-Density DNA Arrays. *Science* 274 (5287): 610–615.
108. Allain, L. R., Stratis-Cullum, D. N., and Vo-Dinh, T. (2004) Investigation of Microfabrication of Biological Sample Arrays Using Piezoelectric and Bubble-Jet Printing Technologies. *Anal Chim. Acta* 518: 77–85.
109. Hacia, J. G., Woski, S. A., Fidanza, J., Edgemon, K., Hunt, N., McGall, G., Fodor, S. P., and Collins, F. S. (1998) Enhanced High Density Oligonucleotide Array-Based Sequence Analysis Using Modified Nucleoside Triphospohates. *Nucl. Acids Res.* 26: 4975–4988.
110. Wadkins, R. M., Golden, J. P., Prisiolas, L. M., and Ligler, F. S. (1998) Detection of Multiple Toxic Agents Using a Planar Array Immunosensor. *Biosens. Bioelectron.* 13 (3-4): 407–415.
111. Kirby, R., Cho, E. J., Gehrke, B., Bayer, T., Park, Y. S., Neikirk, D. P., McDevitt, J. T., and Ellington, A. D. (2004) Aptamer-Based Sensor Arrays for the Detection and Quantification of Proteins. *Anal. Chem.* 76 (14): 4066–4075.
112. Potyrailo, R. A., Conrad, R. C., Ellington, A. D., and Hieftje, G. M. (1998) Adapting Selected Nucleic Acid Ligands (Aptamers) to Biosensors. *Anal. Chem.* 70 (16): 3419–3425.
113. Baskerville, S., Zapp, M., and Ellington, A. D. (1999) Anti-Rex Aptamers as MIMICS of the Rex-Binding Element. *J. Virol.* 73 (6): 4962–4971.
114. Vo-Dinh, T., The Multi-Functional Biochip (1999) *Sixth Annual Biochip Technol. Chips Hits '99.*
115. Vo-Dinh, T., Alarie, J. P., Isola, N., Landis, D., Wintenberg, A. L., and Ericson, M. N. (1999) DNA Biochip Using a Phototransistor Integrated Circuit. *Anal. Chem.* 71 (2): 358–363.
116. Vo-Dinh, T., Isola, N., Alarie, J. P., Landis, D., Griffin, G. D., and Allison, S. (1998) Development of a Multiarray Biosensor for DNA Diagnostics. *Instrumen. Sci. Technol.* 26 (5): 503–514.
117. Affimetrix Inc. (2005) http://www.affymetrix.com.

4

Conducting Polymer Nanowire-Based BioFET for Label-Free Detection

Adam K. Wanekaya, Wilfred Chen, Nosang V. Myung, and Ashok Mulchandani

CONTENTS

4.1 Introduction

The use of biological recognition elements, such as receptors, binding proteins, antibodies, and DNAs, as sensing elements, is particularly attractive because of their highly sensitive and specific nature. Conventional bioaffinity-based sensors involve the use of a label or tag to generate the necessary signal to monitor the binding event. An attractive and highly desirable alternative to this strategy would eliminate the tagging step and, instead, would rely on detection of the change of an inherent property of the analyte or receptor or the molecular aggregates formed upon binding. The advantages of label-free detection include a simple homogeneous assay format without separation and washing steps and rapid near real-time response. The development of label-free sensing has the potential to impact basic biological research as well as screening in medical and bioterrorism applications.

Label-free sensing using planar (two-dimensional (2-D)) thin-film as the gate of a field effect transistor (FET) has been demonstrated. The direct label-free detection of biomolecules by FETs is based on the change of conductance or a related property due to depletion or accumulation of carriers within the transistor structures from the binding of the charged biomolecule to the gate. The selectivity and sensitivity of the FET can be improved by modifying the gate of the FET with (bio)receptors/(bio)recognition molecules such as antibodies, antigens, and oligonucleotides [1]. While successful demonstration of the label-free detection of biomolecules such as proteins and DNA using planar films has been reported, the limited sensitivity of planar devices has precluded them from having a large impact.

Field effect transistors fabricated from one-dimensional (1-D) nanostructured materials are based on a similar framework but are more sensitive because, unlike planar FETs, they avoid the reduction in conductance changes caused by lateral current shunting. When used as the gate of FET device, 1-D nanostructures offer significant advantages over 2-D thin-film planar gates. First, binding to the surface of 1-D nanostructures can lead to depletion or accumulation of carriers in the "bulk" of the nanometer diameter structure versus only the surface region of a planar device (Figure 4.1), giving rise to large resistance/conductance changes to the point that single-molecule detection is possible. Also, as shown in Figure 4.2, real-time monitoring can be achieved when the binding between the receptor and the target is reversible. Second, the direct conversion of chemical information into an electronic signal can take advantage of the existing microelectronic technology and lead to miniaturized sensor devices. Finally, the small size of the nanostructures makes it possible for the development of high-density arrays of individually addressable nanostructures for simultaneous analysis of a range of different species and massive redundancy to reduce false positives/negatives [2]. Also, 1-D nanostructures, such as nanowires, nanobelts, nanosprings, and carbon nanotubes (CNTs), are extremely attractive for nanoelectronics because they can function both as devices and as the wires that access them.

Until recently the advancement of 1-D nanostructures has been slow because of many difficulties associated with the synthesis and fabrication of these materials with well-controlled dimensions, morphology, phase purity, and chemical composition. Three classes of 1-D nanostructured materials, namely CNTs, silicon nanowires (Si NWs), and lately conducting polymer nanowires (CP NWs), have shown profound performance in device fabrication in general and in label-free detection technology in particular.

4.1.1 Scope and Overview of the Chapter

Recently, extensive reviews covering the synthesis of Si NWs and CNTs and their applications for label-free detection have been published [3–7]. We will therefore limit the scope of this chapter to biosensors based on CP NWs. We begin by outlining the methods used to synthesize and fabricate CP NWs. We highlight the successes and limitations of various methods

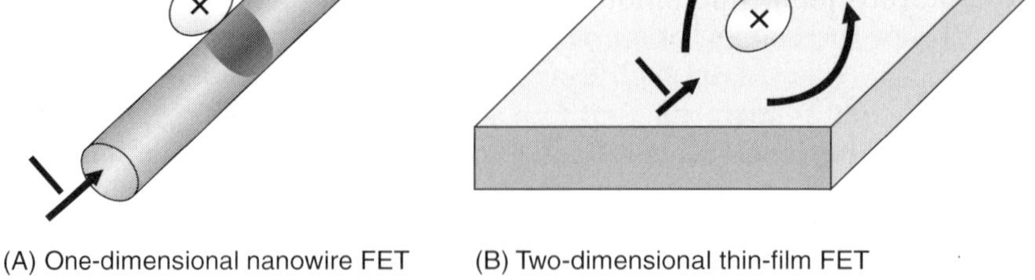

FIGURE 4.1
Advantages of one-dimensional (1-D) nanowire-based field effect transistor (FET) (A) over two-dimensional (2-D) thin-film FET (B). Binding to 1-D nanowire leads to depletion or accumulation in the bulk of the nanowire as opposed to only the surface in 2-D thin film case.

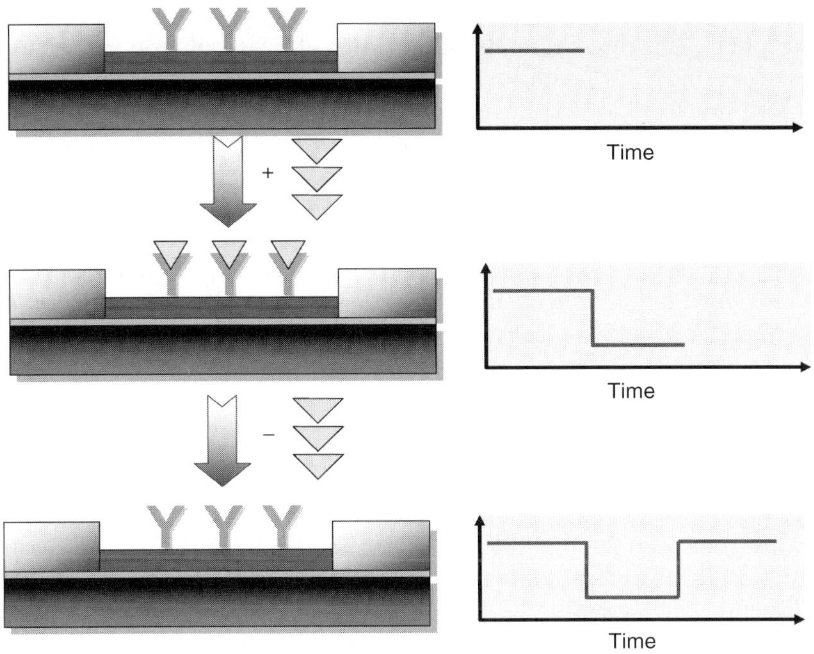

FIGURE 4.2
Schematic of nanowires-based detection. Nanowire device modified with antibodies (*Left*). Specific binding of targets on the nanowires produces a response change (*Right*) characteristic of the surface charge of the targets on the nanowire. When the target unbinds from the surface, the response return to the baseline value.

from different laboratories and discuss how we have attempted to address these limitations. We then continue to discuss the functionalization and assembly of these 1-D nanostructures. We demonstrate how they have been used for label-free detection of various biological molecules and how such devices have enabled the achievement of extremely high sensitivity and low detection limits. Finally, we conclude by highlighting some of the challenges we face in the 1-D nanostructured materials research arena and also predict the direction toward which future research on this area might be directed.

4.2 Fabrication of One-Dimensional Conducting Polymer Nanostructures

The ability to synthesize, pattern, and write conducting polymers (CPs) is one of the target areas for fabrication of organic electronic devices. Several techniques such as photochemical lithography [8,9], micromoulding [10], electropolymerization using modified electrodes [10], deposition using scanning electrochemical microscopy [11], and printing techniques [12] have been successfully applied to fulfill the micrometer and submicrometer dimension. There is, however, increased demand for miniaturization of CPs to the nanometer scale. Among the methods that have been used to fabrication CPs to the nanometer scale are electrochemical dip-pen lithography, mechanical stretching [13], electrospinning [14,15], and polymer-templated electrochemical synthesis [16].

4.2.1 Dip-Pen Nanolithography

Invented by Mirkin and coworkers, Dip-Pen nanolithography (DPN) is a scanning probe nanopatterning technique in which an atomic force microscopy (AFM) tip is used to deliver molecules to a surface via a solvent meniscus, which naturally forms in the ambient atmosphere. This direct-write technique offers high-resolution patterning capabilities for a number of organic [17,18], semiconducting [19], or metallic materials [19,20] in nanoscale dimensions. Lim and Mirkin [21] have reported the preparation of self-doped polyaniline (PANI) and doped polypyrrole (PPY) lines down to 310 and 290 nm widths, respectively, at writing speeds of about 0.8 μm/s. The ionically charged CPs were used as the "ink" for writing on oppositely charged substrates. Electrostatic interactions between the water-soluble ink materials and charged substrates provided a significant driving force for the generation of stable DPN patterns on the silicon substrates. Liu and coworkers [22] have reported the writing of poly 3,4-ethylenedioxythiophene down to 30 nm by polymerization of 3,4-ethylenedioxythiophene at the AFM tip/substrate interface in what they term as electrochemical-DPN.

4.2.2 Mechanical Stretching

In this procedure, CP NWs are fabricated by electrochemical polymerization of the corresponding monomer onto a sharp scanning tunneling microscope (STM) tip that is held at a small distance (between 20 and 100 nm) from an electrode [13]. The STM tip is coated with an insulation layer except a few nm^2 at the end so that the growth of PANI is localized at the tip end. After forming a CP NW between the STM tip and the electrode, the diameter of the nanowire is further reduced by stretching it with the STM tip. Highly conductive PANI with diameter of about 20 nm has been reported using this method [13].

4.2.3 Electrospinning

This method uses a microfabricated scanned tip as an electrospinning source. The tip is dipped in a polymer solution to gather a droplet as a source material. A voltage applied to the tip causes the formation of a Taylor cone, and at sufficiently high voltages, a polymer jet is extracted from the droplet. By moving the source relative to a surface, acting as a counter-electrode, oriented nanofibers can be deposited and integrated with microfabricated surface structures. In addition to the uniform fiber deposition, the scanning tip electrospinning source can produce self-assembled composite fibers of micro- and nanoparticles aligned in a polymeric fiber [23–26].

4.2.4 Template-Directed Methods

Like other 1-D nanostructures, 1-D CP nanostructrures can be synthesized by template-directed methods. Porous anodic alumina oxide (AAO), track-etched porous polymer membranes, and mica are the three types of templates that are commonly used and electrochemical deposition is usually the technique of choice. Electrochemical deposition is accomplished by coating one face of the template with an inert conducting film (e.g., gold and platinum) and using this inert conducting film as the anode. The polymer is then electrochemically synthesized within the pores of the membrane. The length of the nanowires is determined by the current density and deposition time. The diameter of nanowire is determined by the pore diameter of template. CPs show preferential deposition along the walls of the polycarbonate membrane resulting in nanotubule structures due to solvophobic interactions. These tubules close up as the deposition time is increased and eventually results in nanowires [27,28]. Chemical template synthesis can be similarly accomplished by simply immersing the membrane into a solution of the desired monomer and its oxidizing agent. Following nanowire electrodeposition, the conducting film used for electrochemical polymerization and the template are then dissolved using appropriate acids or bases. Organic solvents may be used to dissolve polymer templates. Figure 4.3 depicts a schematic of the template-based synthesis process. It has been shown that template-directed CP nanotubes and nanowires have higher electronic conductivities than bulk samples of the same material. Furthermore, the mechanism of electronic conduction in these template-directed nanostructures can be varied at will by changing their diameter [29].

He and coworkers [30] synthesized PANI nanowires by stretching, aligning, and immobilizing double-stranded λ-DNA on a thermally oxidized Si chip by the molecular combing method [31,32]. Then the DNA templates were incubated in protonated aniline monomer solution to emulsify and organize the aniline monomers along the DNA chains. Finally, the aligned aniline monomers were polymerized enzymatically by adding horseradish peroxidase (HRP) and H_2O_2 successively to form PANI/DNA nanowires.

Simmel and coworkers [33] also synthesized PANI nanowires templated by DNA using three methods. They found that DNA templating worked best for PANI formed by oxidative polymerization of aniline with ammonium persulfate, both in solution and on templates immobilized on a chip. DNA was also good template for PANI formed by enzymatically catalyzed polymerization utilizing HRP. However, immobilization of these structures between contact electrodes was compromised by extensive protein adsorption to the surface. A photo-oxidation method using ruthenium *tris*(bipyridinium) complexes as photo-oxidant resulted in less uniform PANI /DNA structures than the other methods.

4.2.5 Individually Addressable Single Nanowires

Although several examples of uses of these methods of synthesis and fabrication of 1-D nanostructures have been reported, many properties of these nanostructures and their fabrication methods had significantly limited their full-scale development, particularly for high-density arrays [34–37]. For example,

(1) the harsh conditions, highly concentrated sodium hydroxide or phosphoric acid, required to dissolve most templates in template-synthesized CP NWs might not be suitable for many biological applications.
(2) Functionalization/modification for incorporation of specific sensing capabilities can only be performed as postassembly processes.
(3) Nanodevice fabrication requires complex postsynthesis assembly using sophisticated manipulating tools.

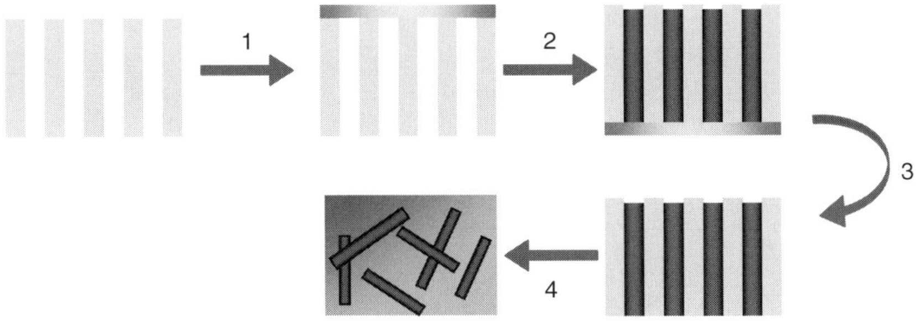

FIGURE 4.3
Schematic of template-based synthesis of conducting polymer nanowires (CP NWs). 1) Deposition of inert conducting thin film on one side of template 2) Polymerization of the monomers within the pores of the template 3) Etching of inert conducting thin film 4) Dissolution of template.

(4) Limitation of dopants and inability to functionalize materials other than silicon limit the diversity of semiconductor NWs.
(5) Single-walled CNTs can exhibit either metallic or semiconducting behavior depending on the diameter and helicity [38]. The inability to control whether NTs are semiconducting or metallic makes specific device fabrication largely a random event.
(6) Synthesis of Si NWs and CNTs require extreme conditions followed by purification before functionalization with appropriate receptor and device fabrication/assembly.

In an attempt to address some of the above-mentioned challenges, we recently reported a facile technique for synthesis of high aspect ratio (100 nm wide by up to 13 μm long), dendritefree, single and multiple individually addressable CP NWs by electrodeposition within channels between two electrodes on the surface of silicon wafers [39]. The electrodes were fabricated on a (100)-oriented silicon wafer with a chemical vapor deposition grown 1 μm thick insulator layer of low-stress Si_3N_4 film. A Cr adhesion layer and a ~3000 Å thick Au contact layer were then deposited, with contacts being patterned by lift-off. Thermally evaporated SiO_2 was then deposited on the wafer at room temperature followed by a coat of PMMA photoresist. Using e-beam lithographic patterning, the deposited SiO_2 and PMMA were selectively opened with reactive ion etching to form electrolyte channels. The widths of the electrolyte channels were fixed at 100, 200, 500, or 1000 nm [40].

Figure 4.4 shows a cross-section of the silicon wafer. The deposition and growth of the nanowire were based on well-known electrochemical oxidative polymerization, starting with monomers and dopants. Multiple channels were etched between gold electrode pairs

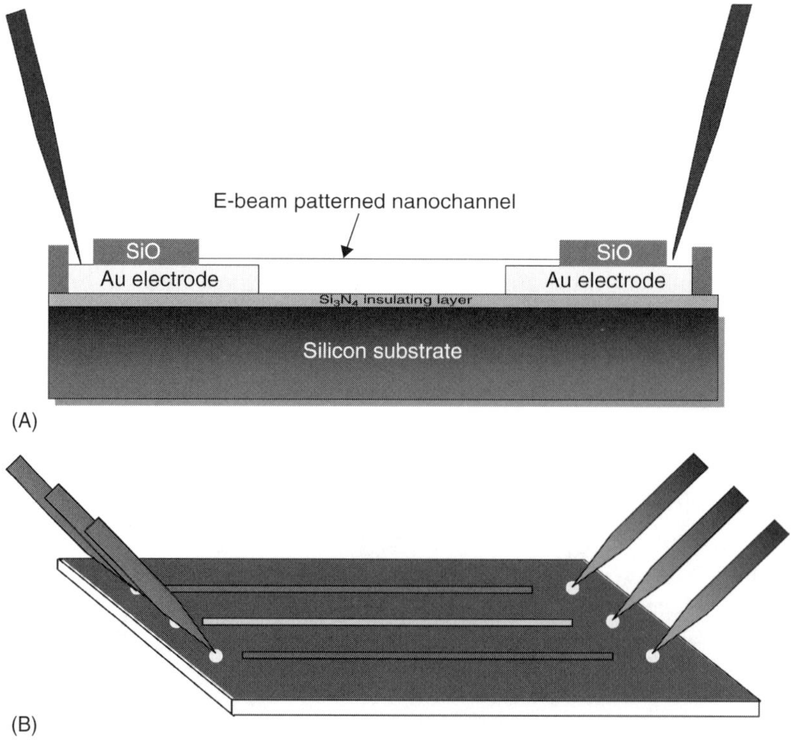

FIGURE 4.4
Schematic of e-beam lithographically patterned electrolyte channels with defined channel dimensions, showing cross-sectional (A) and top-view (B) details.

in the form of arrays to provide formation of any number of individual nanowires. The procedure is a single-step deposition process for each nanowire, and multiple-nanowire arrays of different materials can be deposited on the same wafer sequentially. PPY and PANI were used as models for demonstration.

Figure 4.5 shows the SEM image of a 100 nm wide and 4 μm long electrochemically grown PANI nanowire. As observed, the nanowire was continuous, well-confined, and nondendrite, spanning the entire length and making contact between the two electrodes. The ability to make individually addressable nanowires in high density was evaluated by making two 200 nm wide by 2.5 μm long PPY nanowires separated by 6 μm sequentially as shown in Figure 4.6.

A complementary approach to the one reported above is based on CP nanojunctions formed by bridging two electrodes separated with a gap of 1–100 nm with polymer and was reported by Tao's group [41]. The process is started with an array of electrode (Au or other materials) pairs, 1–100 nm apart, fabricated on an oxidized Si chip using either optical lithography or electron beam lithography (Figure 4.7a). The gap between the Au electrodes is then reduced down to ~1 nm by first electrochemically depositing Au onto the electrodes (Figure 4.7b). This causes an increase in the current flowing across the gap, due to quantum

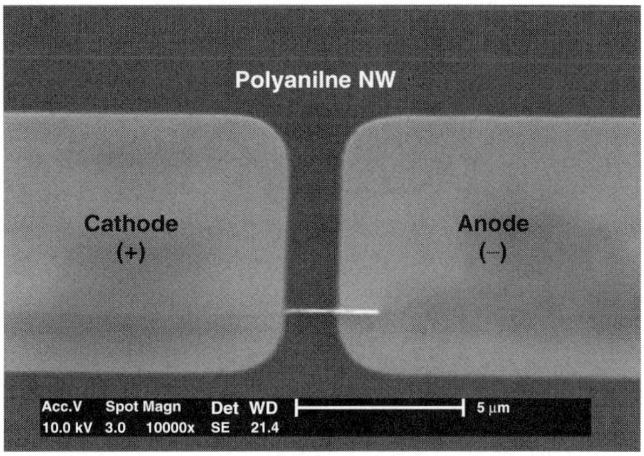

FIGURE 4.5
SEM image of a 100 nm wide by 4 μm long polyaniline (PANI) nanowires [39]. (From Ramanathan, K., Banger, M. A., Yun, M., Chen, W., Mulchandani, A., Myung, N. V. (2004). Individually addressable conducting polymer nanowires array. *Nano Lett.* 4:1237–1239. With permission.)

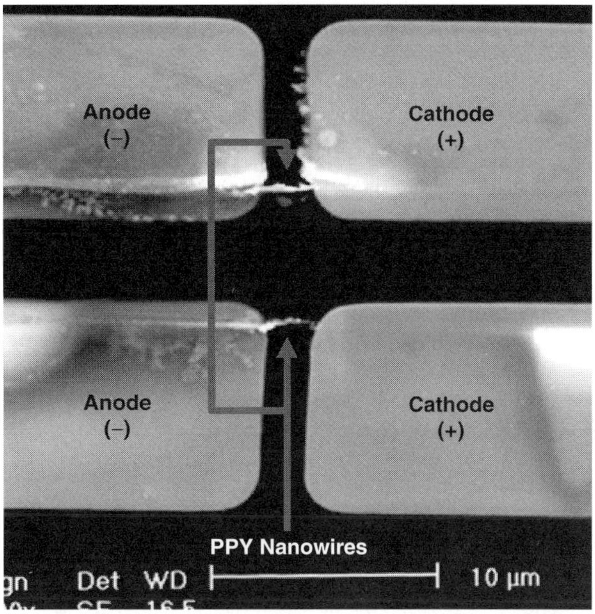

FIGURE 4.6
SEM image of two 200 nm wide by 2.5 μm long polypyrrole (PPY) nanowires separated by 10 μm deposited one at a time [39]. (From Ramanathan, K., Banger, M. A., Yun, M., Chen, W., Mulchandani, A., Myung, N. V. (2004). Individually addressable conducting polymer nanowires array. *Nano Lett.* 4:1237–1239. With permission.)

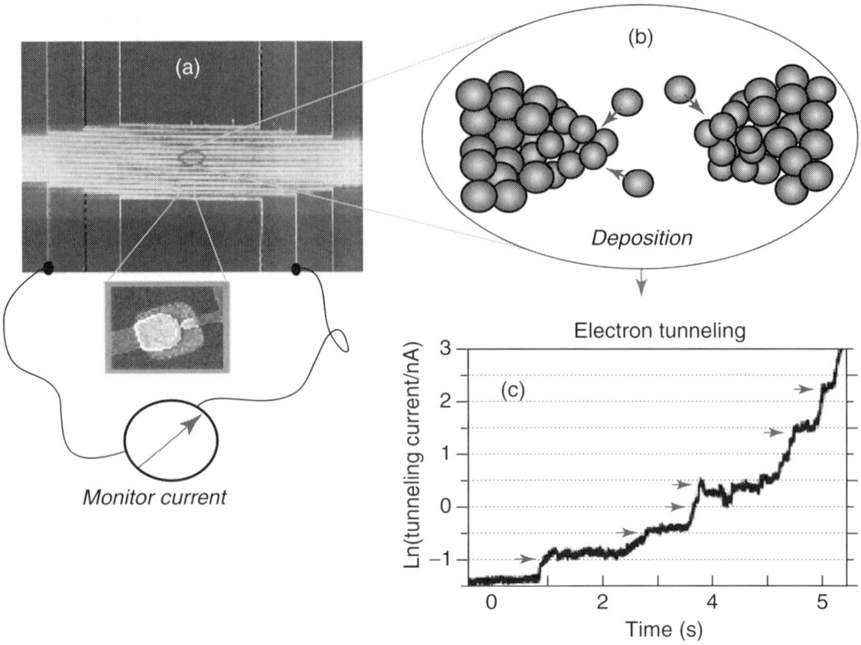

FIGURE 4.7
Fabrication of nanoelectrode arrays. (a) Scanning electron microscopy image of an array of Au electrode pairs on an oxidized Si chip. The inset shows a pair of electrodes (a big and a small one) separated with a gap of a few nm. (b) The gap between the electrodes in each pair decreases as one electrochemically deposits atoms onto the electrodes, and the process is monitored by the current across the gap. (c) When the gap is reduced to a few nanometers, electron tunneling takes place between the electrodes. The tunneling current increases in discrete steps, reflecting the discrete nature of atoms. Adapted from Li, C. Z., He, H. X., Tao, N. J. (2000). Quantized tunneling current in the metallic nanogaps formed by electrodeposition and etching. *Appl. Phys. Lett.* 77: 3995–3997. With permission.

tunneling effect (Figure 4.7c). Because the tunneling current is extremely sensitive to the gap width, the width can be controlled using the tunneling current as a feedback signal. Subsequently, taking advantage of the reversibility of the electrochemical process, Au atoms are etched away from the electrodes to enlarge the gap to desired dimension. Once the nanoelectrodes separated with nanometer-scale gaps are fabricated, the gaps are bridged with CPs (e.g., PANI and PPY) to form nanojunctions by cycling the potential of the nanoelectrodes in a solution containing monomer (Figure 4.8) [41]. The potential cycling converts the monomers into polymers and deposits the polymers into the gap. A controlled amount of polymers is deposited in the gap by controlling the cycling time. Unlike the nanowires produced using preformed channel, the length to width ratio of the nanojunction is not well controlled and the aspect ratio is small. However, since the conductance path defined by the separation between the electrodes is small, this nanojunction approach is particularly suitable for polymers that are poorly conductive under physiological conditions, or polymers that lose much of their conductivity upon attachment of receptor groups.

4.3 Functionalization of Conducting Polymers

Most of the research in the functionalization of CPs has been based on thin-films. However, the same techniques should also be applicable in the functionalization of CP NWs. Various functional groups can be immobilized within the CPs matrix inducing specific properties in

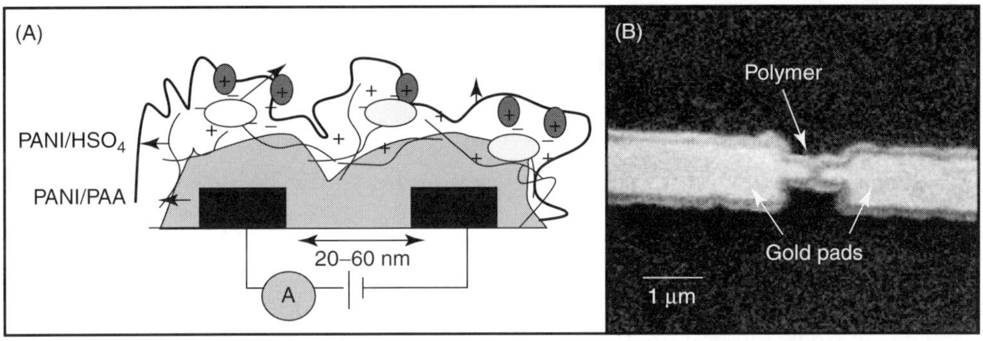

FIGURE 4.8
(A) PANI and poly(acrylic acid) (PAA) nanojunction sensor. (B) SEM image of PANI–PAA films deposited on gold pads with 20–60 nm gaps [41]. (From Forzani, E. S., Zhang, H., Nagahara L. A., Ishamshah, A., Tsui, R., Tao, N. J. (2004). A conducting polymer nanojuntion sensor for glucose detection. *Nano Lett.* 4:1785–1788. With permission.)

the polymer. This process offers attractive possibilities to develop materials that show behaviors resulting from the combination of properties due to CPs and those due to the functional groups. The functionalization of CPs can be carried out in three ways: before, during, and after the polymerization process. The fourth procedure is an entrapment technique where the target material is immobilized during electrochemical polymerization processes.

The first technique involves covalently linking a specific group to the starting monomer and subsequently preparing the functionalized polymer [42]. For example, the hydrogen attached to the nitrogen on the pyrrole molecule can easily be substituted with a specific group. This method can be adopted only if the specific group is stable during the polymerization. A good application of this technique was demonstrated by Cosnier who synthesized conducting PPY films functionalized by biotin groups [43,44]. The strategy involved the conversion of pyrrole to the *N*-biotinylated pyrrole (Figure 4.9) followed by its electropolymerization and the subsequent functionalization of the resulting conducting PPY film by avidin and, finally, its coupling using biotinylated biomolecules. This method combines the advantages of a spatially localized deposition with those of the avidin–biotin coupling scheme, in particular the accessibility of the immobilized biomolecule.

Other pyrrole derivatives are also possible but cationic derivatives of pyrrole N-substituted with alkylammonia groups are interesting because they can be used in the immobilization of enzymes. The derivatives are amphiphilic molecules as they contain both polar and nonpolar zones that generate stable electroactive cationic films. Electrochemical polymerization of these surfactants can be carried out in aqueous media without added electrolyte, which aids the incorporation of biomolecules with a negative charge during formation of the polymer. The monomer and biomolecule are immobilized together by adsorption on the electrode surface before the electropolymerization step. This method has permitted the immobilization of different enzymes [43,45,46] with an increased enzyme loading compared with the entrapment technique.

The disadvantage of N-substituted pyrroles is the slower rate of polymerization and a marked drop in the conductivity of the PPY matrix due to nonplanarity of the PPY chain induced by the substituents. An alternative to the N-substitution is derivatization of the pyrrole ring in the 3 or the 4 positions, which normally requires lower anodic potentials for its oxidation and has higher electronic conductivities as demonstrated by Garnier and coworkers [47].

In the second route, some of the specific anions can be electrostatically incorporated simultaneously during the electropolymerization. In this way, the functionalization is

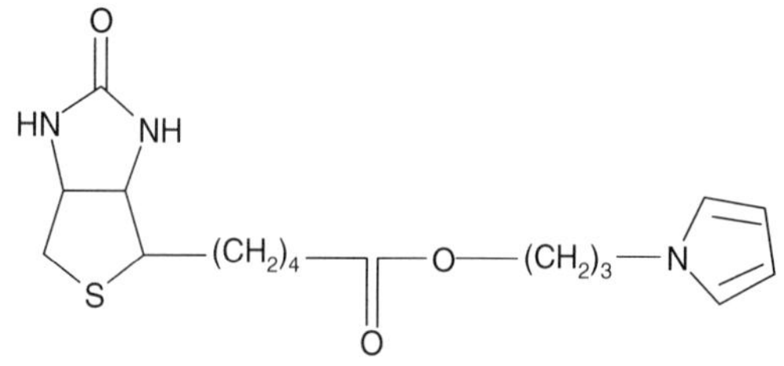

FIGURE 4.9
Structure of biotinylated pyrrole.

obtained if the doping anion is irreversibly captured in the polymer matrix. The incorporation of anionic complexing ligands to the CPs comes under this category [48]. PPY functionalized with biomolecules and other receptors such as 2,6-pyridinedicarboxylic acid and ethylenediaminetetraacetic acid have been prepared using this method [49]. We also have used this procedure to incorporate 2(2-pyridylazo)chromotropic acid anion ($PACh^{2-}$) as a dopant within PPY and utilized it for the selective determination of lead in water down to 10 ng/mL [50]. In this technique, an aqueous solution of pyrrole containing the disodium salt of 2(2-pyridylazo)chromotropic acid was electropolymerized at constant potential. Since $PACh^{2-}$ was the only anion present in the solution, it was incorporated in the growing cationic PPY film as the charge-balancing counter anion. The $PACh^{2-}$ acted both as an electrolyte and a dopant. Wang and Musameh [51] also recently demonstrated a one-step preparation route of amperometric enzyme electrodes based on incorporating a CNT dopant and glucose oxidase (GO*x*) enzyme within an electropolymerized PPY film [51]. The CNT dopant retained its electrocatalytic activity to impart high sensitivity upon entrapment within the PPY network. Such simultaneous incorporation of CNT and GO*x* thus imparts biocatalytic and electrocatalytic properties onto amperometric transducers and represents a simple and effective route for preparing enzyme electrodes.

The third procedure is the postpolymerization functionalization method. In this case, the functionalization is performed after the polymerization. An appropriate functional group in a polymer is allowed to covalently bind to another functional group of a specific molecule. This approach requires the electrosynthesis of CPs possessing reactive entities used as anchoring points to graft the functional groups. A good example is the recent postpolymerization functionalization of poly(N-subsituted pyrrole) film [52]. The β-ferrocene ethylamine used as redox probe was immobilized via a chemical coupling on the surface of a preformed PPY film bearing activated ester groups on to which biotin entities were immobilized.

The fourth way involves entrapment of target molecules within CPs. This route involves the application of an appropriate potential to the working electrode soaked in a solution containing the target molecule, the monomer, and dopant. It allows spatially controlled deposition of the molecules in CP matrices [53–56]. Moreover, the entrapment of the molecules occurs without any chemical reaction that could affect their activity. Electrochemical polymerization ensures that films are prepared in a rapid one-step procedure. Further, this method enables the exact control of the thickness of the polymer based on the measurement of the electrical charge passed during the electrochemical polymerization. However, in some cases, physical entrapment in polymer films and steric hindrances may drastically reduce the catalytic activity and flexibility of some immobilized biomolecules such as enzymes [57,58].

4.3.1 Functionalization of Conducting Polymer Nanowires

The functionalization of CP NWs has recently generated a lot of interest due to the fact that polymeric nanowires are characterized with higher conductivity than polymeric macroscopic films, and their electronic conductance is strongly influenced by minor surface perturbations. The following examples demonstrate some of the procedures that have been used to functionalize CP NWs. Most functionalization processes involving CP nanowires could be termed as noncovalent as the molecule of interest is usually immobilized within the CP matrix during the polymerization procedure. For example, we immobilized biological molecules within CP NWs by entrapping avidin during the polymerization process [59]. In this process, avidin, streptavidin decorated Cd–Se quantum dots (Aqd) and biotin decorated quantum dots functionalized PPY nanowires were synthesized/fabricated within 100- and 200-nm wide by 3-μm long channels between gold electrodes on a silicon substrate fabricated as previously mentioned. As observed earlier for unfunctionalized CP NWs without biomolecules, well-confined, dendritefree, and high aspect ratio nanowires were realized by our methodology (Figure 4.10A). Energy-dispersive X-ray (EDX) analysis confirmed the presence of biomolecule through the presence of quantum dot Cd (Figure 4.10B and Figure 4.10C), while AFM phase imaging analysis verified that biomolecules were uniformly and well distributed and present on the surface of the polymer (Figure 4.11). The latter is necessary for a rapid response and the detection of large analytes and particles like viruses. The study was performed on a film equivalent to the thickness/depth of the channel used for nanowire fabrication. Moreover, the procedure is safe and fast, and environmentally benign reagents are used in the process.

Mallouk and coworkers [60] have demonstrated the synthesis of segmented biologically functionalized CP NWs using template method. Au/(PPY)/Au segmented nanowires loaded with avidin and streptavidin in the polymer component was synthesized as the first step toward the development of nanoscale biosensors and assemblies. Avidin and streptavidin were introduced into the nanowires by entrapment during PPY polymerization. The high specificity of the association between these proteins and biotin was used to monitor protein incorporation and accessibility in the CP segments of the nanowires as a function of the conditions of synthesis.

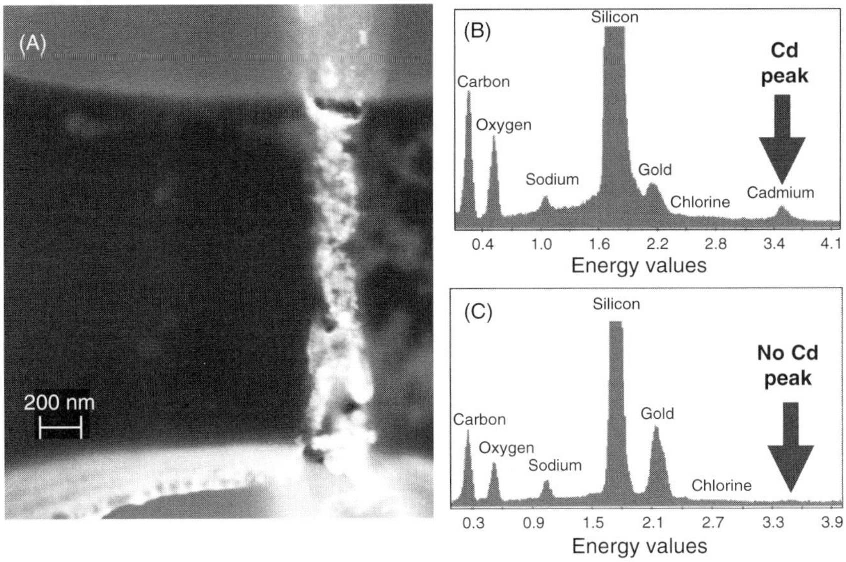

FIGURE 4.10
(A) SEM image of an Aqd-embedded PPY nanowire (200 nm wide). The energy-dispersive X-ray (EDX) analysis of PPY nanowire with embedded Aqd (B) and without embedded Aqd (C) [59]. (From Ramanathan, K., Bangar, M., Yun, M., Chen, W., Myung, N., Mulchandani, A. (2005). Bioaffinity sensing using biologically functionalized conducting-polymer nanowire. *J. Am. Chem. Soc.*, 127:496–497. With permission.)

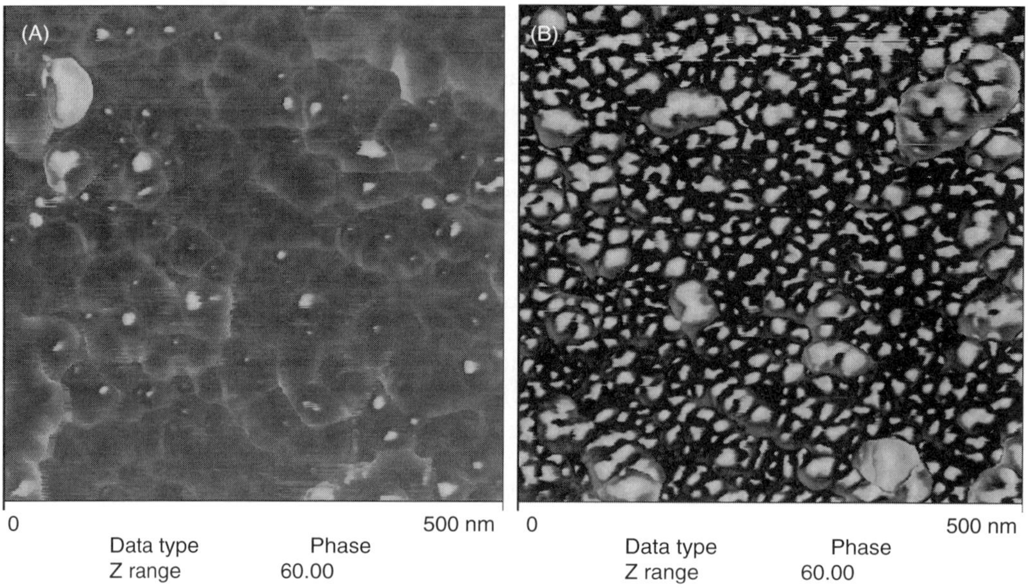

FIGURE 4.11
Atomic force microscopy (AFM) phase images for a 50 nm thick film on an area for (A) PPY and (B) PPY with entrapped Aqd [59]. (From Ramanathan, K., Bangar, M., Yun, M., Chen, W., Myung, N., Mulchandani, A. (2005). Bioaffinity sensing using biologically functionalized conducting-polymer nanowire. *J. Am. Chem. Soc.*, 127:496–497. With permission.)

Wang recently reported the functionalization of PPY nanowires with CNTs using a template-directed electrochemical synthetic route [61]. PPY was electrodeposited in the pores of a host membrane in the presence of shortened and carboxylated CNT dopants that served as charge-balancing counterions. The electron flow within these composite nanowires was enhanced by the CNTs.

Tao and coworkers [41] incorporated biomolecules in the CP nanojunction using both a presynthesized monomer-biomolecule conjugate and postpolymerization functionalization. In the former, small peptides (gly-gly-his and hexa-his) conjugated to aniline and aniline copolymerized to form the nanogap junction while in the latter the enzyme GOx was covalently attached through its lysine $-NH_2$ to the dopant poly(acrylic acid) –COOH groups via the condensation reaction.

4.4 Assembly/Fabrication of Nanowire-Based Field Effect Transistor Sensor Device

The conducting polymer nanowires synthesized in the majority of the techniques presented above require assembly to fabricate functional sensor device. The most widely used method of assembling the nanowires into functional sensor device is based on dispersing a dilute suspension of nanowires onto silicon oxide followed by e-beam lithography to form the source and drain electrodes. An improvement over this random dispersion is the method of fluidic alignment of the nanowires and then patterning the electrodes by e-beam lithography. Dielectrophoresis and magnetically assisted assembly are two other powerful techniques that can be potentially employed for assembling CP NWs into FET sensor devices. We recently demonstrated assembling, positioning, and spatial manipulating of nanowires by integrating ferromagnetic ends on nanowires and using the magnetic interaction between ferromagnetic ends and electrodes [62]. Hundred percent alignment of nanostructures to the imposed magnetic fields was achieved by applying low external magnetic field of 0.02 T. In addition, directional controllability of magnetic assembling technique was demonstrated by assembling nanostructures with angles from 45° to 135° with respect to the electrodes.

4.5 Label-Free Sensing using Nanowire-Based Field Effect Transistor

While there have been several reported demonstrations of the application of the CP NWs as FET devices and chemical sensors, there has been only a couple of reports elucidating the application of biologically functionalized CP NWs for label-free sensing. Using avidin-functionalized PPY nanowires, we illustrated the label-free detection of biotin–DNA (20 mer oligo) conjugate and antiavidin IgG. Figure 4.12 shows the response of an avidin-functionalized PPY nanowire when challenged with biotin–DNA. The resistance of the functionalized nanowire increased as a function of concentration, while a control nanowire without embedded avidin showed no response. The biosensor response was very rapid and as low as 1 nM of biotin–DNA was detected in a few seconds. Similar response was observed with antiavidin IgG. The nanomolar sensitivity of the present biosensor, while lower than that for Si NW and CNT, was obtained with nanowires that were not optimized with respect to the conductivity. A sensitivity of potentially single-molecule detection is possible by adjusting the nanowire's conductivity to a value closer to the lower end of a semiconductor. Other demonstration of label-free sensing using biologically functionalized CP comes from Tao's group in which PANI functionalized with gly-gly-his and hexa-his was used for the detection of Cu^{2+} and Ni^{2+} (Figure 4.13).

4.6 Conclusions and Future Perspectives

One-dimensional nanostructures and most of the applications derived from these materials are still in their infancy. Hence, several issues including their chemical/thermal/mechanical stability need to be addressed before these materials can be utilized to their full potential. This is of crucial importance because 1-D nanostructures are known to be both less thermally and

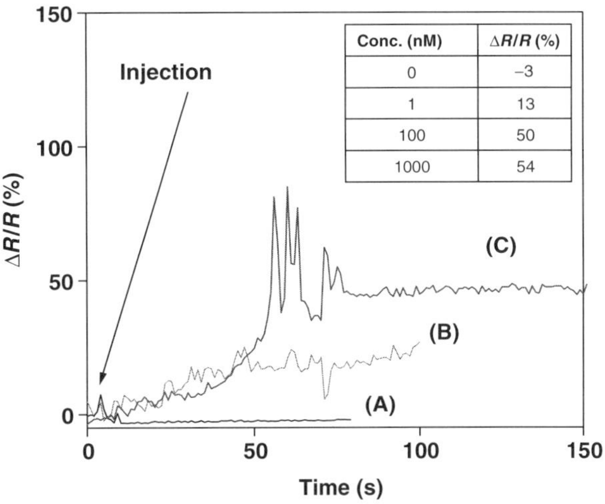

FIGURE 4.12
Electrical responses of an unmodified nanowire (A) to 100 nM biotin–DNA (single stranded) and avidin-embedded PPY (200 nm) nanowires to 1 nM (B) and 100 nM (C) biotin–DNA. The responses were recorded on two separate PPY-avidin nanowires. PPY nanowire containing entrapped avidin were grown using 25 nM pyrrole in 10 mM NaCl and of avidin [59]. (From Ramanathan, K., Bangar, M., Yun, M., Chen, W., Myung, N., Mulchandani, A. (2005). Bioaffinity sensing using biologically functionalized conducting-polymer nanowire. *J. Am. Chem. Soc.*, 127:496–497. With permission.)

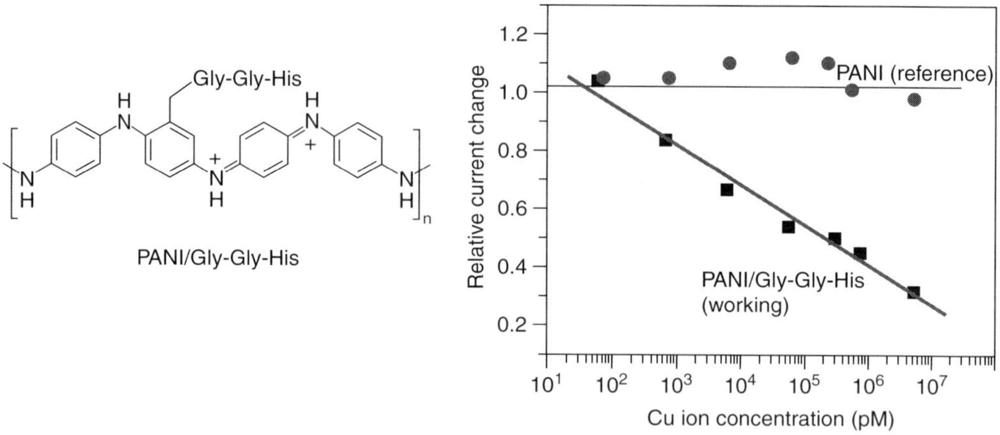

FIGURE 4.13
Left: Covalent attachment of peptides to PANI chains. *Right*: Conductance (current) of a 50 nm-working junction (with peptide-modified PANI) and a 50 nm-reference junction (without peptide) vs. Cu2+ concentration. The bias voltage across the junctions is 0.1 V, and the solution is drinking water containing Cu2+ of various concentrations. Adapted from Zhang, H., Boussaad, S., Ly, N., Tao, N. J., C. Z., He, H. X., Tao, N. J. (2004). Magnetic field assisted assembly of metal/polymer/metal junction sensors. *Appl. Phys. Lett*. 84: 133–135. With permission.

mechanically stable than their bulk cousins. Another challenge is posed by chemically synthesized nanowires which are known to assemble into complex structures.

Even though there has been tremendous advancement in fabrication techniques of 1-D nanostructures, there are still difficulties associated with the fabrication of these nanostructures with well-controlled and consistent dimension, morphology, phase purity, and chemical composition. Practical routes to large quantities of 1-D nanostructures from a diversified range of raw materials rapidly and at reasonable cost is still a big challenge.

Environmental and health questions have been raised. At the present time products derived from these emerging technologies are viewed by the public in a more demanding perspective from the standpoint of safety and environmental impact. The potential for exposure to some of these nanostructures will increase as the quantity and types used in society grow. The studies reported by Hunter and coworkers [63] provided a first insight into the in vivo toxicity of a specific type of manufactured SWNTs.

In spite of the numerous challenges, 1-D nanostructured materials offer unlimited research opportunities. These techniques exceed capabilities of other methods such as PCR [64] and micromechanical devices [65].

Acknowledgments

The authors thank DOD/DARPA/DMEA and US EPA for supporting studies on individually addressable biologically functionalized CP NWs array for label-free immunosensors.

References

1. Schasfoort, R. B. M., Kooyman, R. P. H., Greve, J. (1990). A new approach to ImmunoFET operation. *Biosens. Biolelectr.* 5:103–124.

2. Cui, Y., Duan, X., Huang, Y., Lieber, C. M. (2003). Nanowires as building blocks for nanoscale science and technology. In: Wang, Z. L., ed. *Nanowires and Nanobelts Materials, Properties and Devices, Vol. 1: Metals and Semiconductor Nanowires*. Boston, MA: Kluwer Academic Publishers, pp. 3–68.
3. Sadik, O. A. (1999). Bioaffinity sensors based on conducting polymers: Short Review. *Electroanalysis* 11:839–844.
4. Patolsky, F., Lieber, C. M. (2005). Nanowire nanosensors. *Materials Today* April 2005:20–28.
5. Wang, J. (2005). Carbon nanotubes based electrochemical biosensors: A review. *Electroanalysis* 17:7–14.
6. Xia, Y. N., Yang, P. D., Sun, Y. G., Wu, Y. Y., Mayers, B. Gates, B., Yin, Y. D., Kim, F., Yan, Y. Q. (2003). One-dimensional nanostructures: Synthesis, characterization, and applications. *Adv. Mater.* 15:353–389.
7. Law, M., Goldberger, J., Yang, P. (2004). Semiconductor nanowires and nanotubes. *Annu. Rev. Mater. Res.* 34:83–122.
8. Persson, S. H. M., Dyreklev, P., Inganas, O. (1996). Patterning of poly(3-octylthiophene) conducting polymer films by electron beam exposure. *Adv. Mater.* 8:405–408.
9. Beh, W. S., Kim, I. T., Qin, D., Xia, Y., Whitesides, G. M. (1999). Formation of patterned microstructures of conducting polymers by soft lithography, and applications in microelectronic device fabrication. *Adv. Mater.* 11:1038–1041.
10. Huang, Z., Wang, P.-C., MacDiarmid, A. G., Xia, Y., Whitesides, G. (1997). Selective deposition of conducting polymers on hydroxyl-terminated surfaces with printed monolayers of alkylsiloxanes as templates. *Langmuir* 13:6480–6484.
11. Kranz, C., Ludwig, M., Gaub, H. E., Schuhmann, W. (1995). Lateral deposition of polypyrrole lines by means of the scanning electrochemical microscope. *Adv. Mater.* 7:38–40.
12. Rogers, J. A., Bao, Z., Baldwin, K., Dodabalapur, A., Crone, B., Raju, V. R., Kuck, V., Katz, H., Amundson, K., Ewing, J., Drzaic, P. (2001). Paper-like electronic displays: Large-area rubber-stamped plastic sheets of electronics and microencapsulated electrophoretic inks. *Proc. Natl. Acad. Sci. U.S.A.* 98:4835–4840.
13. He, H. X., Li, C. Z., Tao, N. J. (2001). Conductance of polymer nanowires fabricated by a combined electrodeposition and mechanical break junction method. *Appl. Phys. Lett.* 78:811–813.
14. Doshi, J., Reneker, D. H. (1999). Electrospinning processes and applications of electrospun fibers. *J. Electrost*, 35:151–160.
15. Reneker, D. H., Yarin, A. L., Fong, H., Koombhongse, S. (2000). Individually addressable conducting polymer nanowires array. *J. Appl. Phys.* 87:4531–4547.
16. Jerome, C., Jerome, R. (1998). Electrochemical synthesis of polypyrrole nanowires. *Angew. Chem. Int. Ed.* 37:2488–2490.
17. Piner, R. D., Zhu, J., Xu, F., Hong, S. H., Mirkin, C. A. (1999). Dip-pen nanolithography. *Science* 283:661–663.
18. Ivanisevic, A., Mirkin, C. A. (2001). "Dip-Pen" nanolithography on semiconductor surfaces. *J. Am. Chem. Soc.* 123:7887–7889.
19. Li, Y., Maynor, B. W., Liu, J. (2001). Electrochemical AFM "Dip-Pen" nanolithography. *J. Am. Chem. Soc.* 123:2105–2106.
20. Maynor, B. W., Li, Y., Liu, J. (2001). Au "Ink" for AFM "Dip-Pen" nanolithography. *Langmuir* 17:2575–2578.
21. Lim, J-H., Mirkin, C. A. (2002). Electrostatically driven Dip-Pen nanolithography of conducting polymers. *Adv. Mater.* 14:1474–1477.
22. Maynor, B. W., Filocamo, S. F., Grinstaff, M. W., Liu, J. (2002). Direct-writing of polymer nanostructures: polythiophene nanowires on semiconducting and insulating surfaces. *J. Am. Chem. Soc.* 124:522–524.
23. Kameoka, J., Craighead, H. G. (2003a). Fabrication of oriented polymeric nanofibers on planar surfaces by electrospinning. *Appl. Phys. Lett.* 83:371–373.
24. Kameoka, J., Orth, R., Yang, Y., Czaplewski, D., Mathers, R., Coates, G. W., Craighead, H. G. (2003b). A scanning tip electrospinning source for deposition of oriented nanofibres. *Nanotechnology* 14:1124–1129.
25. Kameoka, J., Orth, R., Ilic, B., Czaplewski, D., Wachs, T., Craighead, H. G. (2002). An electrospray ionization source for integration with microfluidics. *Anal. Chem.* 74:5897–5901.

26. Liu, H., Kameoka, J., Czaplewski, D. A., Craighead H. G. (2004). Polymeric nanowire chemical sensor. *Nano Lett.* 4:671–675.
27. Martin, C. R. (1996). Membrane-based synthesis of nanomaterials. *Chem. Mater.* 8:1739–1746.
28. Parthasarathy, R. V., Martin, C. R. (1994). Template-synthesis of polyaniline microtubules. *Chem. Mater.* 6:1627–1632.
29. Spatz, J. P., Lorenz, B., Weishaupt, K., Hochheimer, H. D., Menon, V. P., Parthasarathy, R. V., Martin, C. R., Bechtold, J., Hor, P.-H. (1994). Observation of crossover from 3-dimensional to 2-dimensional variable-range hopping in template-synthesized polypyrrole and polyanilne. *Phys. Rev. Lett.* 50:14888–14892.
30. Ma, Y. F., Zhang, J. M., Zhang, G. J., He H. X. (2004). Polyaniline nanowires on Si surfaces fabricated with DNA templates. *J. Am. Chem. Soc.* 22:7097–7101.
31. Bensimon, A., Simon, A., Chiffaudel, A., Croquette, V., Heslot, F., Bensimon, D. (1994). Alignment and sensitive detection of DNA by a moving interface. *Science* 265:2096–2098.
32. Nakao, H., Hayashi, H., Yoshino, T., Sugiyama, S., Otobe, K., Ohtani, T. (2002). Development of novel polymer-coated substrates for straightening and fixing DNA. *Nano Lett.* 2:475–479.
33. Nickels, P., Dittmer W. U., Beyer, S. Kotthaus, J. P., Simmel, F. (2004). Polyaniline nanowire synthesis templated by DNA. *Nanotechnology* 15:1524–1529.
34. Cui, Y., Wei, Q., Park, H., Lieber, C. M. (2001). Nanowire nanosensors for highly sensitive and selective detection of biological and chemical species. *Science*, 293:1289–1292.
35. Kong, J., Franklin, N. R., Zhou, C. W., Chapline, M. G., Peng, S., Cho, K. J., Dai, H. (2000). Nanotube molecular wires as chemical sensors. *Science* 287:622–625.
36. Li, C. Z., He, H. X., Bogozi, A., Bunch, J. S., Tao, N. J. (2000). Molecular detection based on conductance quantization of nanowires. *Appl. Phys. Lett.* 76:1333–1335.
37. Favier, F., Walter, E. C., Zach, M. P., Benter, T., Penner, R. M. (2001). Hydrogen switches from electrodeposited palladium mesowire arrays. *Science* 293:2227–2231.
38. Odom T. W., Huang, J.-L., Kim, P., Lieber C. M. (1998). Atomic structure and electronic properties of single-walled carbon nanotubes. *Nature* 391:62–64.
39. Ramanathan, K., Banger, M. A., Yun, M.. Chen, W., Mulchandani, A., Myung, N. V. (2004). Individually addressable conducting polymer nanowires array. *Nano Lett.* 4:1237–1239.
40. Yun, M., Myung, N. V., Vasquez, R. P., Menke, E., Penner, R. M. (2004). Electrochemically grown wires for individually addressable sensor arrays. *Nano Lett.* 4:419–422.
41. Forzani, E. S., Zhang, H., Nagahara L. A., Ishamshah, A., Tsui, R., Tao, N. J. (2004). A conducting polymer nanojuntion sensor for glucose detection. *Nano Lett.* 4:1785–1788.
42. Cai, Q., Khoo, S. B. (1994). Poly(3,3′-diaminobenzidine) film on a gold electrode for selective preconcentration and stripping analysis of selenium(IV). *Anal. Chem.* 66:4543.
43. Cosnier, S. (1998). Fabrication of amperometric biosensors by entrapment of enzymes in functionalized polypyrrole films. *Can. J. Chem. Eng.* 76:1000–1007.
44. Cosnier, S., Lepellec, A. (1999). Poly(pyrrole-biotin): A new polymer for biomolecule grafting on electrode surfaces. *Electrochim. Acta* 44:1833–1836.
45. Cosnier, S., Gondran, C., Watelet, J. C. (2001). A polypyrrole-bienzyme electrode (salicylate hydroxylase-polyphenol oxidase) for the interference-free determination of salicylate. *Electroanalysis* 13:906–910.
46. Mousty, C., Galland, B., Cosnier, S. (2001). Electrogeneration of a hydrophilic cross-linked polypyrrole film for enzyme electrode fabrication. Application to the amperometric detection of glucose. *Electroanalysis* 13:186–190.
47. Delabouglise, D., Roncali, J., Lemaire, M., Garnier, F. (1989). Control of the lipophilicity of polypyrrole by 3-alkyl substitution. *J. Chem. Soc., Chem. Commun.* 475–477.
48. Shiu, K. K., Chan, O. Y., Pang, S. K. (1995). Factors affecting the electroanalytical behavior of polypyrrole-modified electrodes bearing complexing ligands. *Anal. Chem.* 67:2828–2834.
49. Wallace, G. G., Lin, Y. P., J. (1988). Preparation and application of conducting polymers containing chemically active counter ions for analytical purposes. *Electroanal. Chem.* 247:145–156.
50. Wanekaya, A. K., Sadik, O. (2002). Electrochemical detection of lead using overoxidized polypyrrole films. *J. Electroanal. Chem.* 537:135–143.
51. Wang, J., Musameh, M. (2005). Carbon-nanotubes doped polypyrrole glucose biosensor. *Anal. Chim. Acta* 539:209–213.

52. Calvo-Munoz, M.-L., Bile B. E.-A., Billon, M., Bidan, G. (2005). Electrochemical study by a redox probe of the chemical post-functionalization of N-sustituted polypyrrole films: Application of a new approach to immobilzation of biotinylated molecules. *J. Electroanal. Chem.*, 578:301–313.
53. Bartlett, P. N., Cooper, J. M. (1993). A review of the immobilization of. enzymes in electropolymerized films. *J. Electroanal. Chem.* 363:1–12.
54. Schuhmann, W. (1995). Conducting polymers and their application in amperometric biosensors. *Mikrochim. Acta* 121:1–29.
55. Trojanowicz, M., Krawczynski, T. K. V. (1995). Electrochemical biosensors based on enzymes immobilized in electropolymerized films. *Mikrochim. Acta* 121:167–181.
56. Cosnier, S. (1997). Electropolymerization of amphiphilic monomers for designing amperometric biosensors. *Electroanalysis* 9:894–902.
57. Coche-Guérente, L., Deronzier, A., Mailley, P., Moutet, J.-C. (1994). Electrochemical immobilization of glucose oxidase in poly(amphiphilic pyrrole) films and its application to the preparation of an amperometric glucose sensor. *Anal. Chim. Acta* 289:143–153.
58. Cosnier, S., Galland, B., Innocent, C. (1992). A novel biosensor elaboration by electropolymerization of an adsorbed amphiphilic pyrrole-tyrosinase enzyme layer. *J. Electroanal. Chem.* 328:361–366.
59. Ramanathan, K., Bangar, M., Yun, M., Chen, W., Myung, N., Mulchandani, A. (2005). Bioaffinity sensing using biologically functionalized conducting-polymer nanowire. *J. Am. Chem. Soc.*, 127:496–497.
60. Hernandez, R., Richter L., Semancik, S., Stranick S., Mallouk, T. E. (2004). Template fabrication of protein-functionalized gold-polypyrrole-gold segmented nanowires. *Chem. Mater.* 16:3431–3438.
61. Wang, J., Dai, J., Yarlagadda, T. (2005). Carbon nanotube-conducting-polymer composite nanowires. *Langmuir* 21:9–12.
62. Hangarter, C. M., Myung, N. V. (2005). Magnetic alignment of nanowires. *Chem. Mater.* 17:1320–1324.
63. Lam, C., James J. T, McCluskey, R., Hunter, R. L. (2003). Pulmonary toxicity of carbon nanotubes in mice 7 and 90 days after intratracheal instillation. *Toxicol. Sci.* 72:44–44.
64. Mackay, I. M., Arden K. E., Nitsche A. (2002). Real-time PCR in virology. *Nucleic Acids Res.* 30:1292–1305.
65. Gupta, A., Akin, D., Bashir, R. (2004). Single virus particle mass detection using microresonators with nanoscale thickness. *Appl. Phys. Lett.* 84:1976–1778.

5

Smart Biosensor Functions—A Machine Learning Perspective

George K. Knopf

CONTENTS

5.1 Introduction

From a historic perspective biosensors have been largely regarded as a subgroup of chemical sensors. These devices utilize a highly selective biomolecular recognition component with a high degree of affinity to the sample, or analyte, being investigated. The constituent components of the sensor system transform one physical quantity as identified by the bioreceptor element or biochemical reaction into another more easily observed secondary signal proportional to a single analyte or a related group of analytes (1,2). A variety of signal parameters such as changes in electrical resistance, conductance, potential difference, ion concentration, current, optical properties, pH, and oxygen consumption can be measured using the appropriate transducer. The magnitude or frequency of the measured transducer signal will often change with respect to the presence of the analyte or concentration of the substance being measured. The basic stages of a biosensor system are summarized in Figure 5.1.

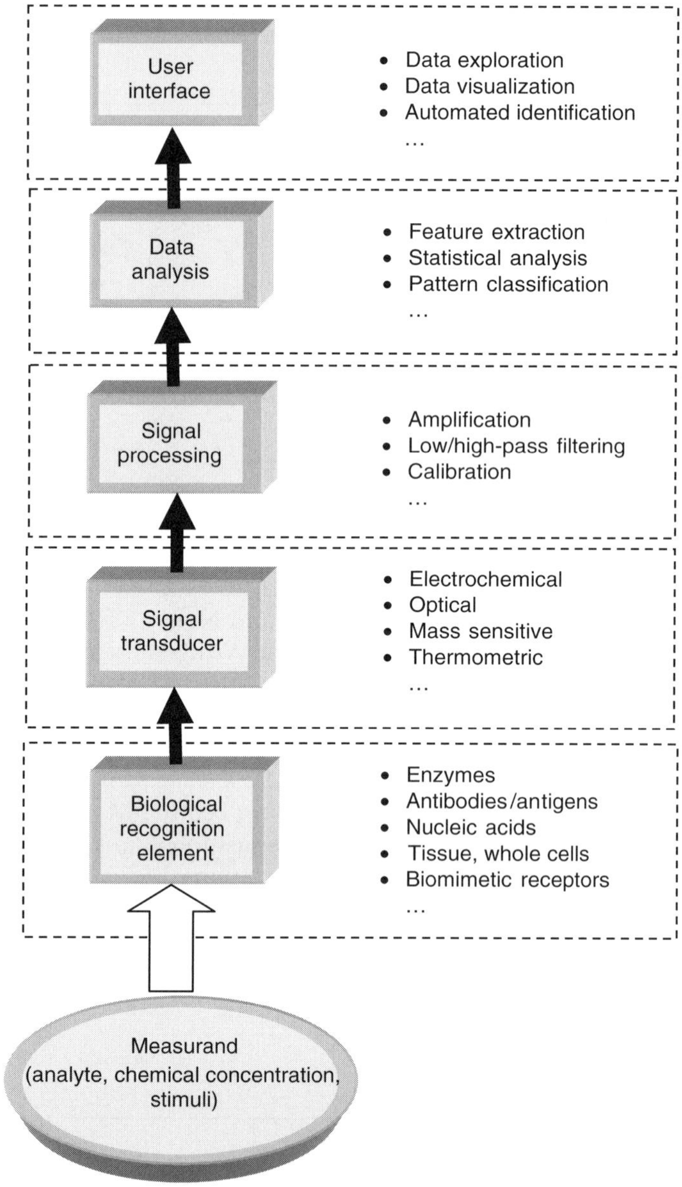

FIGURE 5.1
The functional blocks of a typical biosensor.

A biosensor with successful commercial appeal must exhibit performance capabilities and cost advantages over other currently available technologies (3). The advantages of one type of biosensor over another are often related to detection speed, system portability, or the required level of user skill. In recent years, however, the functionality and design of the biosensor have greatly changed with the introduction of new "intelligent" materials and rapid advances in computing technology and microfabrication techniques. Modern biosensors are now integrated devices that closely connect the biomolecular recognition element with a physical transducer to create an electrical or optical signal that is proportional to the concentration of a specific chemical or biological agent being measured. These microsystems are biosensors that include detecting elements, transducers, and signal processing on a single integrated chip.

In general it is not a single technology or specific material that makes a biosensor smart or intelligent, but the synergistic interaction between the constituent components that comprise the sensory system. Seamless functional integration requires the development team to understand fundamental scientific and design principles derived from chemistry, physics, biology, material science, electronics and optics, and informatics. These seemingly diverse branches of science, engineering, and information science all provide unique perspectives on the meaning of "intelligent biosensors," their design, and potential applications. To further increase a sensor's specificity, selectivity, and overall functionality many commercial biosensors must incorporate the computing power of microprocessors and microcontrollers. These information-intensive biosensor systems exploit numeric algorithms and new analytical techniques being developed in areas of adaptive signal processing, machine learning, pattern recognition, and artificial intelligence (AI).

The field of intelligent biosensors has borrowed terms, concepts, and technologies from a variety of scientific and engineering disciplines. Consequently, the words *smart* and *intelligent* can mean different things to different groups of researchers (4). For example, the IEEE 1451 smart transducer interface standard defines "smartness" in terms of on-board data storage/processing capability and the existence of an interfaced/integrated analog or digital sensor (5). Clearly, this is a very narrow interpretation of a smart device. In other areas of electrical engineering, a smart device is defined as a system with integrated computing power that allows self-calibration, nonlinear correction, offset elimination, failure detection, communication, and decision-making ability (6,7). A general definition of a smart or intelligent sensor is a sensing device that includes signal-processing capability integrated into it to reduce physical deficiencies in the intrinsic hardware. The signal processing can be either analog or digital. Intelligence can result in the raw signal data from the sensor being significantly transformed or even reconfigured as meaningful information.

The terminology used in biosensor technology is further influenced by the disciplines of chemistry and material science. Biosensor systems comprised of biological macromolecules, assemblies of these molecules, and living cells exhibit some intelligent properties. Most of these smart biosensors are created using highly specific materials. The materials are termed *intelligent* because they are selective and respond to only the desired receptor signal. Intelligent properties include template-based self-assembly, self-multiplication, self-repair, self-degradation (selective), redundancy, self-diagnosis, learning, and prediction/notification. The basic philosophy is that intelligent material properties incorporated into biosensors would enable the systems to respond in real time to environmental changes and be capable of integrating multiple functions such as signal feedback and pattern.

A common feature of all smart materials and smart sensors is *adaptation*. This may be in the form of adapting to environmental stimuli, adaptive signal processing algorithms, adaptive control, and intelligent decision-making. Although several distinct elements comprise a sensor system, it is when these elements are effectively integrated and function collectively that the sensor truly exhibits "smartness." The integrated structure will perform in a way that is similar to biological systems, adapting to the circumstances and exploiting available energy as efficiently as possible. As the system becomes more complex, the capabilities of the sensor are increased with the inclusion of structural condition monitoring and health maintenance. The sensor must act in some way to optimize or improve its performance. The fundamental mechanism is the inherent ability to adjust its internal parameters to changing environmental conditions. The capability of modifying behavior to external influences is a defining attribute of many living cells or all biological organisms.

At the higher organizational level, adaptive sensor signal processing and sophisticated pattern recognition exist. The ability to acquire diverse sensory inputs, extract relevant features, and rapidly comprehend the underlying meaning of the stimuli is the key to a truly smart biosensor. Similar to a living organism, these attributes cannot be realized in a biosensor by only selecting a "smart material," but requires an integrated system that is greater than the sum of the individual parts. Sophisticated biosensor designs now incorporate hardware and software technologies that enable adaptation, learning from past experience, and stimulus identification. These systems draw from the disciplines of computer science and electrical engineering. The computational algorithms may be embedded in physical sensors using programmable microcontrollers or implemented off-line on computer workstations where the captured information is graphically displayed for human analysts.

In this chapter, the role of machine learning and how it can be applied to biosensor calibration and pattern classification is discussed from an engineering and informatics perspective. In this context, artificial neural networks (ANNs) are introduced as a viable AI tool for analyzing large volumes of multidimensional data from single sensors or sensor arrays. Finally, the computational mechanism of two commonly used ANNs, the radial basis function (RBF) network and the self-organizing feature map (SOFM), are summarized and their role in enabling smart functions is illustrated.

5.2 The Role of Machine Learning in Developing Smart Biosensor Functions

The discipline of *machine learning* is concerned with extracting meaningful information from sensor data using statistical, analytical, and connectionist algorithms. The various algorithms used to perform machine learning tasks can be organized into a taxonomy that is based on the desired system output. These learning algorithms include: supervised, unsupervised, semisupervised, reinforcement, transduction, inductive transfer, and approximate inference techniques. Pattern recognition is a subarea of research within machine learning that seeks to classify data (patterns) based on either a priori information or on statistical information extracted directly from the dataset.

The functional capabilities of an individual biosensor or a sensor array can be significantly enhanced by taking advantage of pattern recognition algorithms to help categorize and analyze the large number of features extracted from the acquired signals. Pattern analysis involves sensing biomolecules from one or more sites, preprocessing the acquired transducer data, extracting relevant features that enable classification, clustering features into groups with similar characteristics called classes, and interpreting the underlying meaning of the final class assignment (8).

5.2.1 Pattern Recognition

Pattern recognition is the process of classifying the analyte, detecting chemical composition, or estimating the pattern's descriptive parameters (9). The goal of *classification* is to assign the sensor signal pattern, or feature vector, to a predefined class. *Detection* strives to determine the temporal or spatial occurrence of a specific instance of the desired pattern. In contrast, *parameter estimation* involves determining specific parameters that assist the analyst in defining a given pattern. In large datasets, parameter estimation is used to either remove noise and redundant information in the original data or measure the similarity between patterns. Data compression and data fusion algorithms are often used to reduce the amount of data vectors for interpretation or analysis, while pattern discrimination techniques help establish measures of similarity between two or more patterns.

An essential step in pattern analysis is the transformation of measured signals from the transducer to a limited selection of features that best represent meaningful patterns in the signal (10). Feature vectors (also called patterns) contain all salient information necessary to describe the input analyte or chemical substance. Statistical and syntactic approaches are often used to extract suitable features from the transducer output. These techniques use various criteria for identifying which features are best for the classification. An optimal feature selection technique must satisfy some criterion that minimizes a measure of representation error. In addition, the feature vector used to describe the pattern produced by the signal(s) is often a reduced dimensional representation. The primary objective of dimensionality reduction is to satisfy practical constraints imposed by the computing software and hardware. Furthermore, the classification result is often more accurate when the pattern embedded in the raw signal is simplified through reduction with only the most important or critical features being represented. The choice and number of required features for a classification task are problem specific and governed by the level of permissible information loss.

Classification algorithms are used to assign the extracted feature vectors to predefined classes (11). A class is a set of feature vectors that have been grouped based on a similarity measure. The metric used to define "similarity" depends on the objective of the classification task. Often the distribution of classes within a feature space is determined using a clustering algorithm. Most pattern classification algorithms are statistical, syntactic, or connectionist. An example of a statistical algorithm is the minimum distance classifier, which assigns a feature vector to a class based on its Euclidean distance from the class prototypes located in the feature space. Other approaches involve linear discriminate operators, entropy criteria methods, and maximum likelihood estimation classifiers.

Syntactic techniques, in contrast, classify feature vectors based on formal grammars and symbolic descriptors. The low-level pattern structures, called primitives, correspond to simple features in the sensor signal. Each primitive may require multiple attributes defined by statistical metrics to describe it. The sensor signal is decomposed into primitives by means of formal grammars. Each class of patterns is represented by a distinct grammar that can be expressed in symbolic form as sentences. Two basic approaches to performing syntactic pattern recognition are to define distinct grammars that reflect class structure, and use graphs to match relational descriptions.

Connectionist algorithms are largely ANN approaches that establish classes based on training a densely interconnected network of simple computing units called neurons (12). These approaches are preferred in pattern analysis applications where there only exists a minimal amount of a priori class information. The network parameters are adjusted by presenting a collection of input–output examples and iteratively adjusting the internal parameters to provide the desired output. Connectionist approaches can efficiently determine nonlinear class boundaries in multidimensional feature spaces.

5.2.2 Artificial Intelligence and Neural Networks

Artificial intelligence has emerged as a major tool for solving complex problems arising from ill-defined or massive amounts of data. Common AI tools include Boolean logic, deductive and inductive reasoning, soft or fuzzy logic, expert systems, knowledge-base systems, genetic algorithms, support vector machines, and ANNs. Neural network structures (12–15) have gained popularity in recent decades as a viable mechanism for embedding computational intelligence into sensors and enabling the analysis of large-volume high-dimensional datasets. A neural network is defined as a computing system made up of numerous simple, highly interconnected processing elements which process information by responding to external inputs. The various classes of ANNs are identified by the type of neuron (static, dynamic, Gaussian), structure of the network (feed forward, feedback, clustering map), and the learning algorithm (supervised, unsupervised, competitive) employed. A large number of neural networks have been described in the literature including adaptive linear neuron, multilayered perceptron, backpropagation network, RBF network, SOFM, Hopfield net, bidirectional associative memory, Boltzmann machine, adaptive resonance theory, and neocognitron to name a few.

The functional capabilities of most neural networks used in practice can be viewed from three different perspectives (14). The first is that the fundamental feedforward network structure can be trained to implement any Boolean logic function. The second perspective is that the same network architecture can approximate any nonlinear function and, therefore, be used to establish a nonlinear mapping between sensor inputs and outputs. The third perspective is that the neural network can learn to partition multidimensional input (or feature) spaces for pattern classification problems. In terms of biosensor data, functional approximation and pattern classification are the two main applications of neural networks.

In general, the methodology does not require explicit knowledge of the network structure and permits the internal system parameters to be learned through a process of repetitive exposure to experimentally acquired examples of input–output vector pairs. Most neural network algorithms separate the training phase from the final operation phase. It is during the training phase that the system develops an acceptable mapping function or data partitions. The major drawbacks are the model structure hidden in a "black box" and the computational time necessary to optimize the parameters through training.

Neural networks can be applied to adaptive signal processing and control (16), or used to perform complex chemical analysis and analyte identification (17). The most direct application of these nonlinear maps to analytical chemistry involves creating nonlinear, multivariate calibration models where the spectral intensities at different wavelengths are used as the input variables and the corresponding concentrations of the analyte species in the same sample are the output attributes.

Simple feedforward neural networks have been used successfully to establish a calibration standard for a number of sensors including an optical fiber pH sensor (18–20). The network implemented by Taib and Narayanaswamy (18) was tested with noisy data and produced an average error of 0.2 pH units for additive noise levels of up to 13% of the input signal. Raimundo and Narayanaswamy (21) used a similar neural network to perform multivariate calibration on an optical fiber chemical sensor for determining the ammonia and relative humidity in air. Ferreira et al. (22) developed an alcohol fermentation control system based on biosensor measurements interpreted by feedforward neural networks. The neural network established a correlation between the experimentally acquired glucose and sucrose measurements. The nonlinear mapping determined by a similar training procedure was also used to successfully model the forward and inverse dynamics of a nonisothermic, continuously stirred tank reactor (23).

The pattern classification capability of neural networks has also been extensively studied for analyzing large volumes of sampled data or high-dimensional data vectors. A multipoint optical evanescent wave U-bend sensor that exploits a network pattern recognition system was described by Lyons et al. (24). The signals generated from the optical time domain reflectometer sensing fiber was complex due to the cross-coupling effects arising from the various interfering parameters. However, the authors determined that the neural network used to perform the signal analysis resulted in 100% correct classification for all test data analyzed. Blank and Brown (25) also investigated how neural networks can be applied to problems in analytical chemistry by comparing the performance of a feedforward network with more well-established chemometric techniques that exploit matrix regression methods. Bachmann et al. (26) also used a neural network approach to perform chemometric data analysis on the outputs from an amperometric multielectrode biosensor.

The effectiveness of feedforward neural networks for identifying single molecules, according to their fluorescence lifetime, was reported by Bowen et al. (27). For a nonideal single molecule fluorescence data sample, the neural network approach was performed better than the maximum likelihood estimator. Neural networks have also been used for processing signals from microbial sensors (28,29). Lobanov et al. (28) describe how it is possible to use sensors based upon whole microbial cells to selectively measure ethanol and glucose in mixtures.

One active area of pattern recognition and identification research in recent years is the development of "electronic noses" for qualitative classification of various kinds of chemical stimuli (30); Figure 5.2. A comprehensive overview of electronic nose design are provided by Göpel et al. (31) and Ziegler et al. (32). The role of neural networks and genetic algorithms in enhancing the performance of electronic noses is discussed in the paper by Kermani et al. (33).

The two main components of an electronic nose are the sensing system and the automated pattern recognition system (30). The combination of broadly tuned biosensors coupled with sophisticated information processing algorithms enables the electronic nose to be a powerful instrument for odor and toxin analysis. The sensing system may be an array of biochemical sensors where each individual sensor measures a different property of the sensed analyte, or it can be single sensor that produces an array of measurements for each sample. The desired odorant (or volatile chemical compound) presented to the sensor array produces a signature

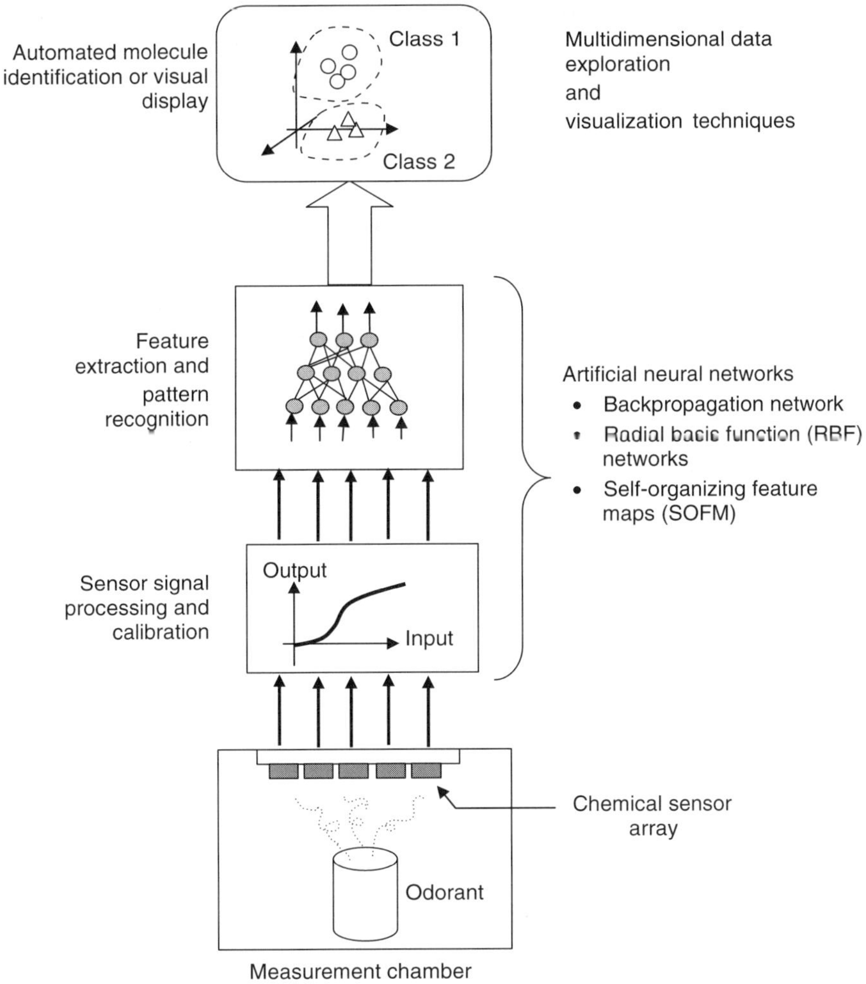

FIGURE 5.2
Flow diagram showing the components of an electronic nose that utilizes artificial neural networks for performing sensor calibration, pattern classification, and multidimensional data visualization.

or characteristic pattern of the odorant. By presenting different odorants to the sensor array it is possible to build up a database of signatures. The goal of many of these systems is to train or configure the pattern recognition system to produce unique clusters or classifications of each odorant so that automated identification can occur. In this manner, ANNs have become a significant tool in developing adaptive, intelligent electronic noses.

Among the many applications of electronic noses where neural network pattern classification has been used are in the area of food analysis (34), including determining the quality of wheat (35), monitoring the ripeness of bananas (36) and other fruits (37), evaluating the freshness of eggs (38), and identifying spoiled beef (39).

An aroma identification system based on an array of semiconductor tin dioxide gas sensors and a feedforward neural network to discriminate the five different aromatic species is described by Brezmes et al. (40). Daqi et al. (41) use an electronic nose with modular RBF network classifiers to recognize multiple fragrant materials. An electronic nose in combination with a neural network to detect the presence of *Mycobacterium tuberculosis* in sputum samples of patients is presented by Pavlou et al. (42). The system is developed for rapid qualitative analysis for screening patient samples and the clinical diagnosis of TB patients. Dutta et al. (43) use an array of 32 polymer carbon black composite sensors and several neural networks to help identify two species of *Staphylococcus aureus* bacteria. The object-oriented data clustering approach is a combination of several techniques including principal component analysis, fuzzy C-means, and the SOFM.

5.3 Biosensor Data Analysis Using Artificial Neural Networks

This section will explore how artificial neural networks can be used to enhance sensor performance and assist the scientist with complex data analysis by first determining the nonlinear relationship between experimentally observed biosensor inputs and outputs, and then performing pattern classification and multidimensional data visualization tasks. The RBF network is used to determine a calibration standard for a simple bR photocell. A variant unsupervised clustering network called the SOFM is used to partition and classify high-dimensional medical data. The concept of self-organization is extended to the task of associating patterns from several sensors. Finally, the process of mapping high-dimensional data vectors onto a colorized deformable spherical SOFM is described as mechanism for visualizing complex numeric datasets and searching for hidden patterns.

5.3.1 Sensor Calibration by Functional Approximation

The signals generated by the biosensor transducer must be linked or correlated in some manner with the biochemical identifier (target) to generate predictable and reproducible results with only small deviations from the expected values. The target may be the presence or absence of an analyte, the concentration level of analytes, or a specific chemical composition. The relationship between the transducer signal (output) and biochemical identifier (input) is often described as a *calibration standard* or *function* (44). The process of biosensor calibration may be viewed as developing a mapping function between the inputs and outputs. Typically, the calibration model is mathematically represented as

$$y = F(x,w) \tag{5.1}$$

where y is the sensor output vector, x is the vector that represents the inputs to the biosensor (e.g., biochemical identifier), and w is the state vector representing the internal parameters of the sensor model $F(\cdot)$. The model F and parameters w are designed using a parameter-estimation process that is optimized through an iterative procedure, such as a gradient search algorithm, to minimize errors. However, most real-world sensors are inherently nonlinear and characterizing the relationship between the inputs and measured transducer output over a wide range of operation is often difficult. For simplicity, a limit range of operation is selected to insure linearity and dynamic stability. Unfortunately, this can greatly restrict the biosensor's functionality.

Static neural networks such as back propagation neural networks and RBF networks (16) implement nonlinear mapping operations of the form $y = F(u)$ where u is a function of the inputs. Typically, u is a measure of similarity between the external inputs x and internal system parameters, w, and can be written as

$$u = w \otimes x \tag{5.2}$$

where $\otimes$ represents a mathematical operator that performs similarity measure between the two vectors. The two most common types of similarity measures are: the scalar product and the distance metrics (8). The individual neurons that comprise most artificial neural networks described in the technical literature employ a scalar product. One popular exception is the RBF network that uses an Euclidean distance measure for describing how an input vector relates to the weights. In nearly all cases, both the inputs and outputs to the system are often normalized such that they are also bound over the range [0, 1] or [−1, 1]. The key characteristics of these neural network algorithms are the mathematical model used to describe the network structure, its capabilities in terms of approximating continuous functions or pattern discrimination, adaptive learning algorithms, and the inherent complexity of learning. Since the RBF network is easier to relate to more traditional approximation and pattern clustering algorithms, it will be described and used for biosensor calibration.

5.3.1.1 Radial Basis Function Network

A RBF network is a two-layer computing structure whose output is a linear combination of weighted basis (or kernel) functions as shown in Figure 5.3. The basis functions produce a localized response to the input vector $x^p = \left[x_1^p, x_2^p, \ldots, x_n^p\right]^T$ where n is the number of elements, $p = 1, 2, \ldots P$, and P is the total number of data vectors used for training. In this manner, the constituent basis units of the first-layer w_j^1 generate a significant nonzero response only when the input vector falls within a localized region of the feature (input) space. Typically, the basis function is a Gaussian kernel defined by

$$u_j^1 = \exp\left[-\frac{\sum_{j=1}^{N1}(x^p - w_j^1)^2}{2(\sigma_j)^2}\right] \tag{5.3}$$

where $j = 1, 2, \ldots N_1$, u_j^1 is the output of the jth node in layer 1, x^p is the input pattern vector, w_j^1 is the weight vector for the jth node in the first layer (which is the center of the Gaussian kernel function for node j), σ_j^2 is the normalization parameter for the jth node, and N_1 is the number of nodes in the first layer. The outputs from the nodes in the first layer are bound over the range of [0, 1] by Eq. (5.3) such that if the input vector is close to a particular Gaussian center its response will be large. The name "radial basis function" is derived from the radially symmetric Gaussian kernels that are often used to compute the node output. Consequently, the basis node's response will be identical for any input x that lies at a fixed radial distance from the center of the kernel, w_j^1. However, a variety of

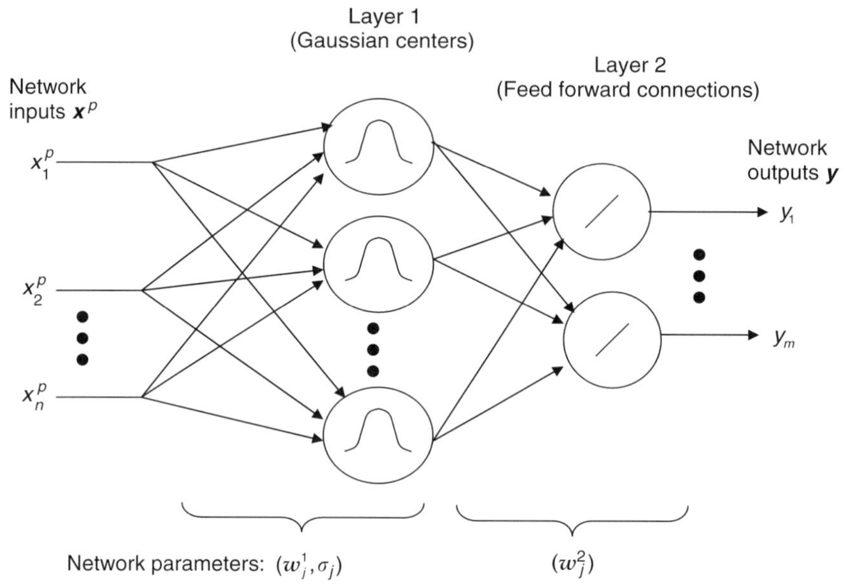

FIGURE 5.3
The computational structure of a radial basis function (RBF) network. The input is the data vector $\boldsymbol{x}^p$ and the output is the response given by vector y.

different definitions for the basis function have been proposed in the recent literature but most nonsymmetrical approaches are application dependent (14).

Each node in the second layer of the RBF network computes a weighted linear summation of the response produced by the basis functions in the previous layer, and its output is given by

$$y_i = \sum_{i=0}^{N_2} u_j^1 w_i^2 \tag{5.4}$$

where y_i is the output response of the ith node in layer 2, w_j^2 is the weight vector for this node, u_i^1 is the output from the ith neuron in the first layer, and N_2 is the number of output nodes. The second layer of the RBF network also uses a bias input of $u_0^1 = 1$ to adjust the approximation curve or classification boundary. Thus, the overall network performs a nonlinear transformation from $\Re^{N_1}$ to $\Re^{N_2}$ by forming a linear combination of the nonlinear basis functions. Theoretically, the RBF network can form an arbitrarily close approximation to any continuous nonlinear mapping operation. Since the individual basis functions of the RBF network cover small localized regions, the accuracy of the mapping is related to the number of functions used in the network.

A variety of different training algorithms are used to determine the location of the Gaussian centers (w_j^1), normalization parameters (σ_j), and weights associated with the output layer (w_i^2). Most algorithms begin by separating the problem into two stages: *unsupervised* clustering to determine Gaussian centers and normalization parameters, and *supervised* least-mean square (LMS) adaptation of weights in the output layer. There are a number of different clustering algorithms that can be used but the most popular choice is the K-means clustering algorithm (14). The normalization parameters are obtained once the K-means algorithm has completed the process of determining the kernel centers, w_j^1. These values represent the spread of data associated with each node in layer 1 (i.e., Gaussian center). The most common technique for problems with large datasets is to make them equal to the average distance between the cluster centers and the training patterns. This can be mathematically given by

$$\sigma_j^2 = \frac{1}{M_j} \sum_{x \in \Theta_j} (x^p - w_j^1)^2 \tag{5.5}$$

where Θ_j is the set of training patterns grouped with cluster center w^1_j, and M_j is the number of training patterns assigned to the group Θ_j.

The adaptation of the weights in the output layer occurs only when the parameters of the basis function (w^1_j, σ_j) have been determined. The nodes in the output layer are computed using a LMSs learning algorithm that utilizes error correction. The training data for this process is a set of input–output pairs $(x^p, \boldsymbol{d})$ where is $\boldsymbol{d}$ is the desired output given x^p. Based on the previously determined basis function parameters the output of the ith node, the node output, is calculated using Eqs. (5.3) and (5.4). Typically, the weight adaptation for the feed forward weights of the ith node in layer 2 is

$$w^2_i(new) = w^2_i(old) + \mu(y_1 - d_i)u^1_j \tag{5.6}$$

where μ is the learning rate which determines the step size for changing the value of the weight after each iteration.

5.3.1.2 Multivariate Calibration Surface for a bR Photocell

Among the numerous biomaterials being investigated for engineering applications, bacteriorhodopsin (bR) has received a great deal of attention as a "smart material" for accomplishing a variety of complex functions. bR is a retinal protein found in the cell membrane of *Halobacterium salinarum*. Absorption of light by the retinal chromophore of bR causes a photoinduced isomerization, followed by a complex photochemical cycle through several spectrally distinct intermediate states. During the photocycle, bR pumps a proton from the cytoplasmic side to the extracellular side of the cell membrane, resulting in a charge displacement and an electrochemical potential across the membrane. This proton gradient is used by the bacterium to generate energy by the synthesis of ATP from ADP.

A simple bR-based photocell was recently constructed by placing a thin film of bR between two indium tin oxide (ITO) electrodes as shown in Figure 5.4. The purple membrane (PM) that contained the bR molecules was isolated from strains 14 of *Halobacterium halobium* according to the method described in Wang et al. (45). The PMs were orientated and electrophoretically deposited onto the ITO electrode that was connected to ground. The cytoplasmic side of the bR membrane was attached to the ITO conductive surface; the extracellular side of the bR membrane was contacted with the second ITO electrode that was the input of the measurement circuit. A polyester thin film was used as the spacer separating the two ITO electrodes. The diameter of the active area of the photocell is about 5 mm.

The signal generated by the bR photocell corresponds to the charge displacement inside the protein and can be detected as a photovoltage across the membrane or a photocurrent through the membrane. The photovoltage generated by the dried bR film can be modeled as an equivalent RC circuit (45,46) as shown in Figure 5.5. The experimental results reveal that both wavelength and intensity of the incident light influence the amplitude of the signal. Under illumination of a laser light with a fixed wavelength, the photocell signal as shown in Figure 5.6 is approximately linear except near the saturating intensities. The saturation of light intensity is related to the physical properties of the bR photocell such as thickness, area, and optical density of the bR film. This property was studied by using a tunable argon or krypton laser system with five selectable wavelengths ranging from 476

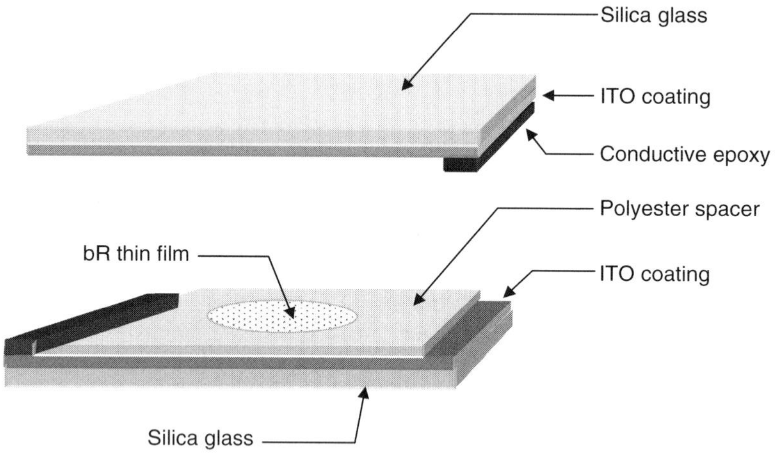

FIGURE 5.4
The ITO electrode/bR film/ITO electrode structure of the dried bR-based photocell.

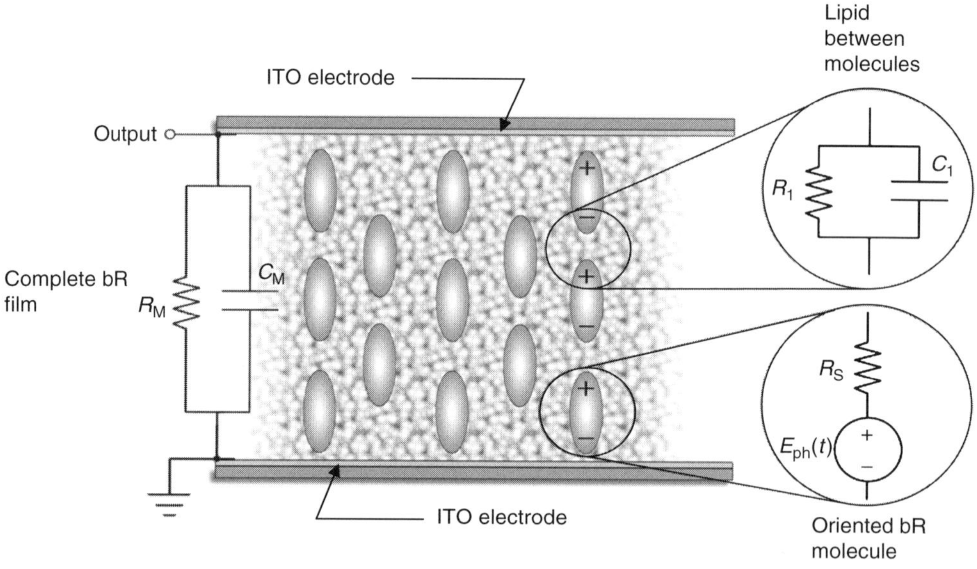

FIGURE 5.5
Schematic diagram of the physical bR photocell and equivalent RC circuit for the photovoltage generated.

nm (blue) to 568 nm (red) (Melles Griot, 35KAP431-220). The output power for individual wavelength can be tuned over a wide range. Furthermore, it was observed that the bR photocells can response to extremely small change of light intensity (46,47).

To illustrate how a neural network can adaptively determine a multivariate calibration function, consider the example of the bR-based photocell described above. Recall that the photovoltage response of the device is proportional to both the wavelength and intensity of the incident light source. For biosensor calibration purposes, the RBF network is used to model the relationship between photovoltage and light intensity of the bR-based photocell for five wavelengths 476, 488, 520, 530, and 568 nm. The input–output dataset is first

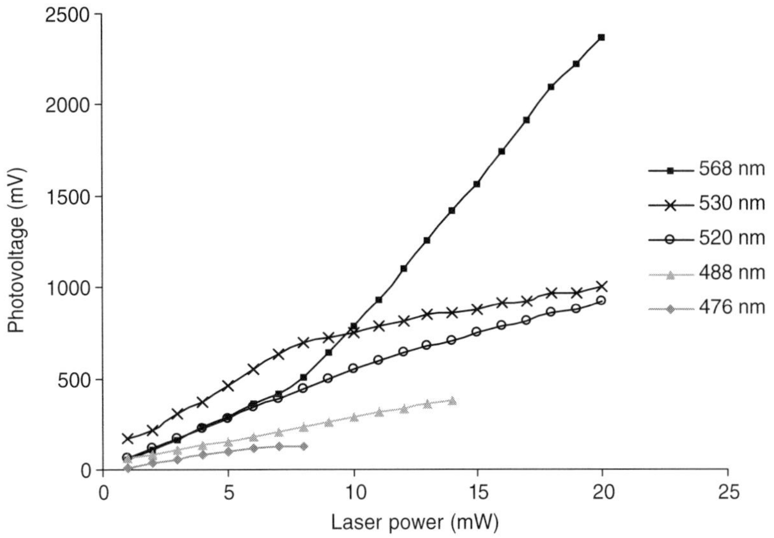

FIGURE 5.6
The intensity-dependent photoelectric response of bR photocell measured using a tunable argon or krypton laser system with selectable wavelengths at 476, 488, 520, 530, and 568 nm. The power of the laser system is in the range of 0.65 mW to 37.4 mW.

separated into training and testing data. Typically, 10–14% of the sampled data pair is randomly removed from the dataset for testing and validation. For this very limited dataset only 10% were removed for testing. The dual-input single-output system is modeled using a RBF network with two external inputs, 25 basis nodes in layer 1, and only one feedforward node in layer 2.

During the initial learning phase, the basis nodes in the first layer are fed with a randomly selected input vector x^p from the training data. Once the parameters of the basis function were determined, the nodes in the second layer were adapted using an LMS algorithm. The learning rate is set to 0.1, and the number of cycles through the training dataset for the LMS adaptation of weights in layer 2 is 10,000. Upon completion, it is possible to test the resultant model with the remaining test data. The continuous calibration surface displayed in Figure 5.7 for the bR photovoltage was generated from 21×21 (wavelength, light power) evenly spaced points along the 3D functional surface. The mean-squared error in this situation is 0.092.

5.3.2 Pattern Analysis

Many of the pattern classification algorithms exploit supervised or unsupervised learning methods to assign data samples (8) to the classes. Supervised techniques are used in the analysis of biosensor data where individual classes of analytes are already known. The class labeled data is used to train the pattern recognition system to identify specific biosensor inputs or develop a response among the many true and false possibilities. These classification techniques often test sample features for statistical significance using *K*-nearest neighbor, logical regression, support vector machines, and artificial neural networks (9–11). Once the supervised learning phase is complete, the pattern recognition system must be tested and evaluated for satisfactory performance. The process of validation involves determining the expected error rate by introducing a new unknown dataset to

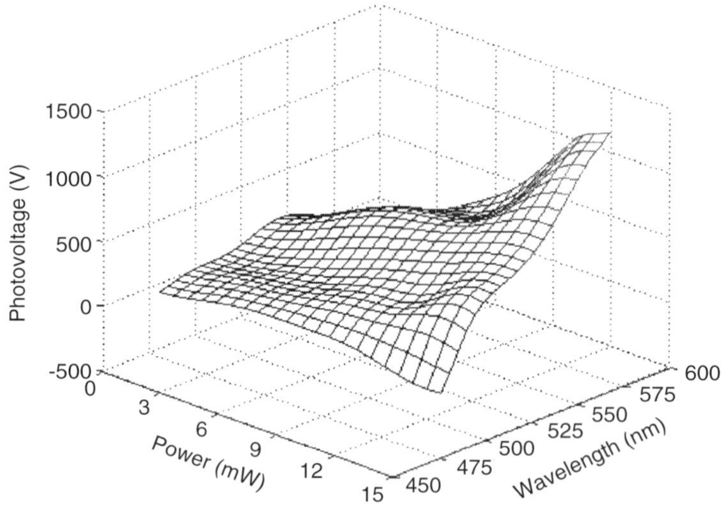

FIGURE 5.7
Continuous calibration surface for the bR photocell generated by 2 × 25 × 1 RBF network. The inputs are wavelength and intensity, and the output is photocell response.

the trained classifier. One validation technique is based on the principle of "leave-one-out." In this case, a small number of data vectors are randomly removed from the original dataset and used to test the classifier model. This is essential the approach used for testing a trained neural network. In some implementations, the process data is repeated until every data point has been left out once for testing the classifier and the estimated accuracy for each test is then averaged for final performance assessment.

Alternatively, unsupervised pattern classification has become a popular approach for exploratory data analysis where class assignment is largely unknown or when it is necessary to reduce the volume or dimensionality of the sensor data into a manageable size for meaningful analysis. These techniques are referred to as data grouping, clumping, and cluster analysis (8). The primary goal is to identify groups of similar data vectors that exist in a common feature space by projecting them into a lower dimensional space that reflects the novel attributes in the original data.

Most unsupervised classification techniques search for clusters that are disjoint or mutually exclusive, as opposed to clusters that overlap due to clumping. Once identified, it is possible to reduce the number of elements necessary for visual display by restricting the view to only the clusters and not the original data. The mapping of large numbers of data vectors into a limited number of clusters provides the analyst with an overview of the data structure, and permits the observer to retain proper context while reducing visual complexity. Most clustering algorithms seek to find a balance between the number of clusters and the number of data vectors assigned to each cluster. A small number of clusters will permit fast information processing but the information content that is visible to the analyst will become very low. Critical details for some applications can be lost in this transformation and "lossless" techniques must be used.

Two exploratory clustering methods commonly used to identify groups of data vectors that form patterns in the feature space are *K*-means clustering algorithm and Kohonen's SOFM. *K*-means is a clustering algorithm that finds groups of similar vectors that are represented by their respective cluster center, which is computed as the mean of the data vectors assigned to it (8). The SOFM learns patterns by forming cluster centers through an unsupervised *K*-means clustering algorithm. However, the cluster centers generated by the SOFM are ordered on the map such that neighboring clusters in the *N*-dimensional feature space are also neighbors on the lower-dimensional 2D map.

5.3.2.1 Self-Organizing Feature Map

The SOFM is an unsupervised clustering algorithm that searches for hidden patterns, or relationships between data vectors, by establishing an ordered topology for the cluster centers in the feature space (15,17,48). Associations among the cluster centers and consequently those among groups of similar data vectors are reflected in the relative position of the nodes in the lattice of the SOFM. The lattice of ordered cluster centers is also called a map. For pattern classification applications, the two-layer SOFM structure shown in Figure 5.8 is often used where the first layer performs topologically ordered clustering and the second layer computes a linear weighted summation of cluster centers in an attempt to derive a decision or assign a class label. The method is similar to the RBF network described in the previous section except that the position of the Gaussian kernel centers in the feature space for the SOFM is determined by the predefined lattice. Data vectors that reside in a local region of the original *N*-dimensional feature space should be assigned to the same or neighboring nodes in the 2D map. The node assignment is a function of the number of nodes in the lattice and the distance measure used for clustering.

The SOFM in the first layer of the pattern classifier incorporates a competitive self-learning algorithm that develops a topological order of the data by exploiting redundancies between randomly acquired points. The self-organizing map arranges the weights

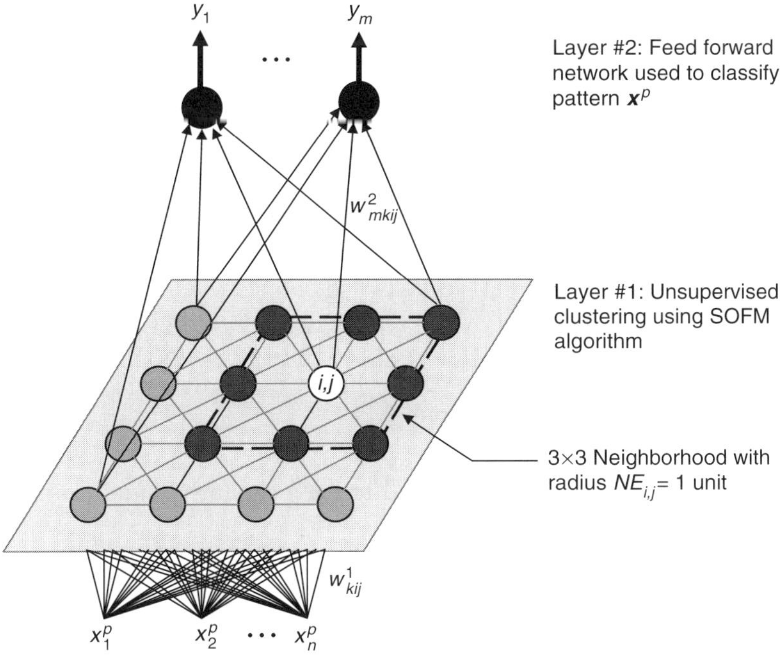

FIGURE 5.8
Architecture of a two-layer pattern classifier where the first layer is a 2D SOFM and the second layer is a linear weighted summation operator. The input feature vector is x^p and the class assignment is given by y.

assigned to the constituent nodes, w^1_{ij}, such that they reflect some physical characteristic of the external input x^p where $p = 1, 2, \ldots P$, and P is the total number of data points used for training. The weight vector for each node is equivalent to the cluster center in the feature space. Every input vector x^p is connected to each node in the network lattice or map by a weight vector, w^1_{ij} . Initially, the weights are set to small random values located in the n-dimensional data space. A node located at (i, j) in the network has an associated variable neighborhood operator that covers an area of the map given by $(2NE_{ij} + 1) \times (2NE_{ij} + 1)$ where NE_{ij} is the decreasing neighborhood radius. The weight adaptation algorithm is designed such that the selected node, and all nodes residing in its neighborhood are simultaneously updated. The size of the neighborhood is decreased as training progresses until it consists of only the winning unit.

For each randomly selected x^p, an error or difference, D^p_{ij} , is computed between the input vector and the weight vectors for all the cluster units in the network

$$D^p_{ij} = \Phi(\delta_{ij}) \sum_{k=1}^{n} (x^p_k - w^1_{kij})^2 \tag{5.7}$$

where x^p_k is the kth input, w^1_{kij} is the weight from the kth input to the $(i, j)^{\text{th}}$ cluster unit, $i = 1, 2, \ldots$ I, and $j = 1, 2, \ldots$ J. A count-dependent nondecreasing function, $\Phi(\delta_{ij})$, is used to prevent cluster under utilization.

During the adaptation or learning process, weights of the cluster unit $(i,j)^*$ nearest to the input vector x^p, called the winning unit or neuron, and all the units residing within the specified neighborhood, NE_{ij*}, are updated using

$$w^1_{ij*}(new) = w^1_{ij*}(old) + \alpha\left(x^p - w^1_{ij*}(old)\right) \tag{5.8a}$$

where

$$\alpha = 0.001 + \mu\left(\frac{NE_{ij*}}{NE_{\text{initial}}}\right) \tag{5.8b}$$

μ is a predefined learning rate, and NE_{initial} is the initial neighborhood size in terms of number of units. The radius of the topological neighborhood, NE_{ij*}, is gradually reduced. Initially, the nodes are uniformly distributed in the planar lattice. As the training process progresses, the weights assigned to the nodes get organized in a natural manner and the cluster centers move to reflect similarity in clusters.

The adaptation of the weights in the second layer used to identify the class occurs only when the parameters w^1_{ij} of the SOFM in layer 1 have been determined. The nodes in the output layer are computed using a LMSs learning algorithm that utilizes error correction. The training data for this process is a set of input–output pairs (x^p, c_m) where is c_m is the desired class assignment for training vector x^p. Once more, the weight adaptation for the feedforward weights of the ith node in layer 2 is

$$w^2_m(new) = w^2_m(old) + \mu\left(w^{1m}_{ij} - c_m\right) \tag{5.9}$$

where μ is the learning rate which determines the step size for changing the value of the weight after each iteration, w^{1m}_{ij} is the nearest cluster to the input vector x^p, and m is the identifier of the output neuron or class.

The SOFM performs relatively well in noisy data because the number of clusters is fixed, and adaptation stops after training has been completed. However, for datasets with a limited number of feature vectors the results may depend on the presentation order of the input data. Another common problem with this type of competitive learning algorithm is that all cluster units are not equally utilized during the training process. Often during training only a few cluster units will be updated and this leads to unwanted distortions and misclassifications (34).

5.3.2.2 ***Pattern Classification***

Breast cancer data from the machine learning repository maintained by the Computer Science Department at the University of California, Irvine (49), will be used to illustrate how an SOFM can be used to map high-dimensional feature vectors into a 2D space. The publicly available dataset consists of 683 fine needle aspirate (FNA) tissue samples provided by Dr. William H. Wolberg (50–52), University of Wisconsin Hospitals, Madison, and is preclassified into two categories: malignant and benign tissue samples. Every FNA tissue sample is a vector consisting of nine attributes: clump thickness, uniformity of cell size and shape, marginal adhesion, single epithelial cell size, bare nuclei, bland chromatin, and normal nucleoli and mitoses, all of which need to be considered to make a diagnosis. Each attribute is assigned a value on a scale of 1–10 by the pathologist, 10 indicating a very high probability of malignancy (50).

Data vectors are randomly selected and presented to the 10×10 SOFM network. An initial neighborhood of 3 U radius is selected and the network is trained for approximately 1,000 cycles. Upon completion of training the SOFM, a second adaptation algorithm based on LMSs is used to adjust the weights of layer 2. A learning rate of 0.1 is used and the tolerance for correct classification was set to 0.0001. The weights of the second layer w_m^2 were trained for 10,000 iterations. The separation boundary shown on the resultant map (Figure 5.9) was based on the output of the classification nodes in layer 2. All test data vectors were correctly classified except the one false match identified by (X) in the figure.

The success of the unsupervised clustering technique is however dependent upon the signal signatures or features used for clustering. A poorly selected feature set can result in numerous incorrect classifications regardless of the effectiveness of the clustering algorithm employed. This will always be a problem because of the garbage-in garbage-out principle of information processing. The computational time required to achieve a usable result is significant and does not lend itself to real-time applications. The problems arising from selecting appropriate feature sets or incomplete feature sets must still be

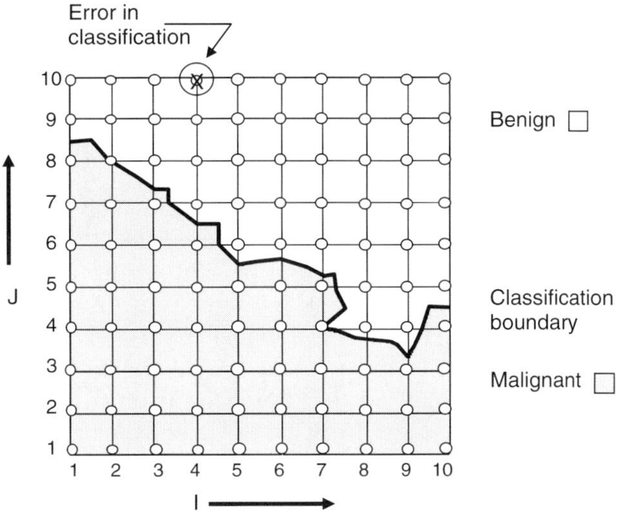

FIGURE 5.9
The decision boundary for the Wisconsin breast cancer training data as determined by 10×10 SOFM lattice. The data is separated into two classes: malignant and benign. The 50 test data vectors were correctly classified except the one shown at (4,10) as (X).

investigated. An important positive attribute of the technique is that no assumption is made about the underlying data structure. If there is no correlation between the features in the dataset then the method results in as many cluster units as feature vectors.

5.3.2.3 Pattern Association

A single biosensor is largely used to determine the presence of a single chemical compound or the concentration level of one substance contained in a sample. This approach to sensor design ensures high specificity for each detector but, unfortunately, provides low analyte selectivity. The problem of low selectivity is especially prevalent in biosensors that utilize whole cells, organelles, or tissues. To compensate for these limitations it is possible to use *N*-sensors to detect *M*-substances. The information can originate from different sensory devices during a single period of time, or from a single bioreceptor over an extended period of time. The synergistic use of data acquired from several very selective and highly sensitive biological and chemical detectors is the basic design principle behind many electronic noses (31,32) and tongues (53).

A variety of mathematical and statistical techniques can be used to "optimally" combine different sources of sensory information into a single representational format (54,55). The process, also known as data fusion, can take place either at the signal, object property, or symbolic level. Signal-level fusion can be used in real-time applications and can be treated as another step in the overall processing of the signal. Property and symbol-level fusion can provide additional features to enhance recognition and interpretation capabilities. The different levels are distinguished by the type of information provided to the system, how the information is modeled, the degree of the sensor registration required for fusion, the methods used for fusion, and the means by which the fusion process can improve the "quality" of the information provided to the observer or the automated analyzer.

Two important advantages of using information from multiple biosensors include *data redundancy* and *complementary information*. Combining redundant data from several similar bioreceptors increases system accuracy and reliability by reducing the uncertainty associated with the operating fidelity of any single sensor. In addition, different types of sensors monitoring the same physical phenomenon can provide complementary information that is difficult or impossible to acquire with a single type of sensor. The complementary information can either represent redundant data for enhanced system reliability or new information used to increase the analyte discrimination capabilities. One reoccurring problem in integrating redundant or complementary information from multiple biosensors is that of associating features, or signatures, acquired by one biological element with those simultaneously captured by another sensor or the same sensor at another instance of time (Table 5.1).

A constraint imposed on feature association is that corresponding feature elements from the various sensors represent the same type of information. The match matrix is often generated using an optimization technique such as genetic algorithms, simulated annealing, competitive neural networks, or the SOFM (56). The key issue in these approaches has been to maximize data compression while retaining relevant information on the novel input. In essence, the task involves finding a more abstract, compressed, feature set by going from one representation level to another. Evidence of hidden redundancy and measures of similarity are used to guide the feature recognition algorithms. The Kohonen SOFM is one method that has been used for dimensionality reduction and multisensor correlation. Figure 5.10 is an illustration of a multiple biosensor system that uses a 3D SOFM (56) to correlate the features from three separate detectors. The SOFM has been widely used for classification purposes and several researchers have presented ways in which the correlated data is presented on the SOFM to reflect the classification and the

TABLE 5.1

Biological Recognition Elements that Are Commonly Used in Biosensor Designs

Biological Element	Mechanism for Recognition
Antibodies and antigens	Based on the specific and high-affinity antibody–antigen binding interactions. Tracers such as fluorescent molecules, enzymes, and radioisotopes are used to generate a detectable signal.
Biomimetric receptors	Genetically engineered molecules such as single-chain antibody fragments. RNA and DNA aptamers. Artificial membranes.
Enzymes	Catalytic transformation of an analyte to induce a signal or generate a product that can be detected by the transducer. Monitoring enzyme characteristics that change upon direct interaction with the analyte.
Nonenzymatic proteins	Class of proteins that contain one or more selective binding sites and produce a signal through a transmembrane ion channel or a secondary messenger enzyme activation system.
Nucleic acids	Detection of DNA–ligand interactions (e.g., monitoring pollutants). Detection of specific DNA sequences by hybridization.
Whole cells	A substance-dependent increase or inhibition of microorganism respiration (e.g., changes in bacteria and fungi respiration to pollutants).

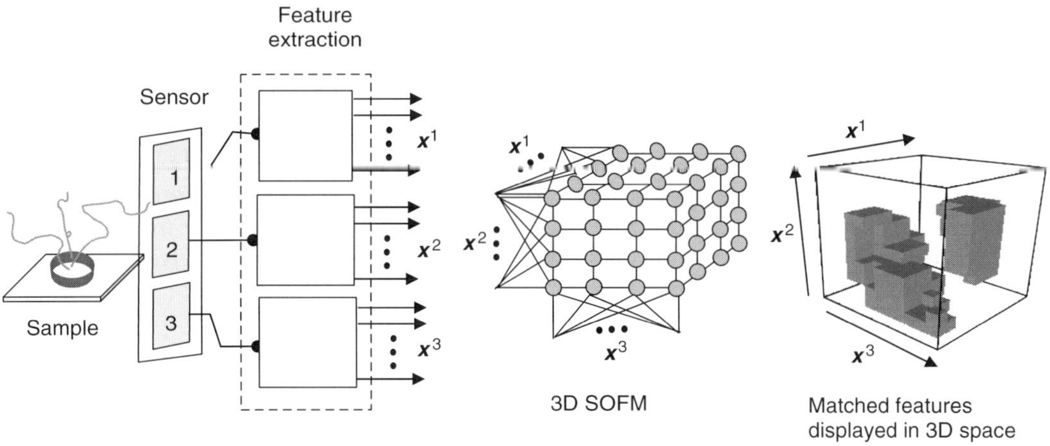

FIGURE 5.10
Feature associate determined using a 3D SOFM. At each instant in time, the three biosensors generate a triplet (x^1, x^2, x^3).

relationships between the classes. The SOFM lattice not only influences how data is reorganized during weight adaptation but also provides a mechanism for displaying the desired relationships between the numeric data vectors that have been mapped onto the SOFM space.

5.3.2.4 Scientific Data Visualization

Data visualization involves the exploitation of statistical, computer graphics, and geometric modeling techniques to transform numeric or symbolic datasets into graphical plots and displays that help reduce the difficulty of interpreting the behavior of complex physical systems. These techniques are used to correlate huge volumes of data, reduce the quantity and dimensionality of the original data to a manageable size, and graphically encode as much information as possible into the visual display. Key information embedded within the data is represented in the display space using color difference, shape parameters, relative proximity of graphical icons, and movement of the icons over time. Problems in visualization arise from the various ways in which different levels of numeric and symbolic data can be presented to a human observer, and how the individual's perception of these graphical representations affect the final interpretation of the underlying information.

Over the past decade two branches of visualization research have evolved: *scientific data* and *information visualization*. Techniques that generate graphical displays from raw or transformed experimental data are usually referred to as scientific data visualization. Raw data are measured signals acquired directly by the transducer while transformed experimental data, also known as analytical data, is created using mathematical tools that modify or combine single or multimodal raw data into new entities for enhanced interpretation. In scientific data visualization a variety of mathematical techniques are used to transform multidimensional multivariate datasets into simple graphical objects called *glyphs* that provide scientists and engineers with a clearer understanding of the underlying system behavior. Information visualization, in contrast, focuses on the graphical display of document databases and information spaces. Clearly, the functionality and presentation of information by the visualization application tool depends upon the tasks to be performed by the end user and the type of graphical cues needed to represent the data.

The SOFM (15) has been investigated as a visualization tool because of its ability to internally order and correlate data without making any assumptions on the underlying relationships present in the data (57–59). While a majority of the glyph-based methods are directed towards generating graphical forms for individual data vectors, the SOFM clusters the data based on inherent similarities and presents information on the data by analyzing relationships between the clusters. Knopf and Sangole (58,60) describe how the spherical SOFM can be utilized to represent arbitrary numeric data vectors as a closed 3D form. Local surface distortions and colors are introduced to the spherical lattice to display the local statistical variance or distance metrics between neighboring units. The shape transformation mechanism is summarized in Figure 5.11. The nodes of the tessellated sphere represent cluster units and each is defined by a weight vector w_{ijk}.

The relative similarity, dissimilarity, or variation amongst the vectors assigned to a cluster unit or between units in a common map can be displayed on the spherical SOFM lattice using a metric or measure-of-information. The selection of this metric is application dependent and defined by the observer such that it quantifies a desired characteristic of the dataset to be displayed on the 3D glyph (60). Common metrics include the Euclidean distance between the cluster unit and the center of the entire map, and the magnitude of each unit's weight vector. The standard deviation of all data vectors assigned to a particular cluster unit is another measure of information that can be used. In this manner, each unit on the spherical lattice is assigned a scalar value based on a metric that reflects the type of information required by the observer. The computed metric is then mapped onto the various visual dimensions of the graphical primitive to generate a unique, reproducible color-coded closed surface of the original dataset. Deformation is introduced in the spherical SOFM during vector projection using a technique similar to the extended Gaussian image mapping method that generates spherical representations of objects in an

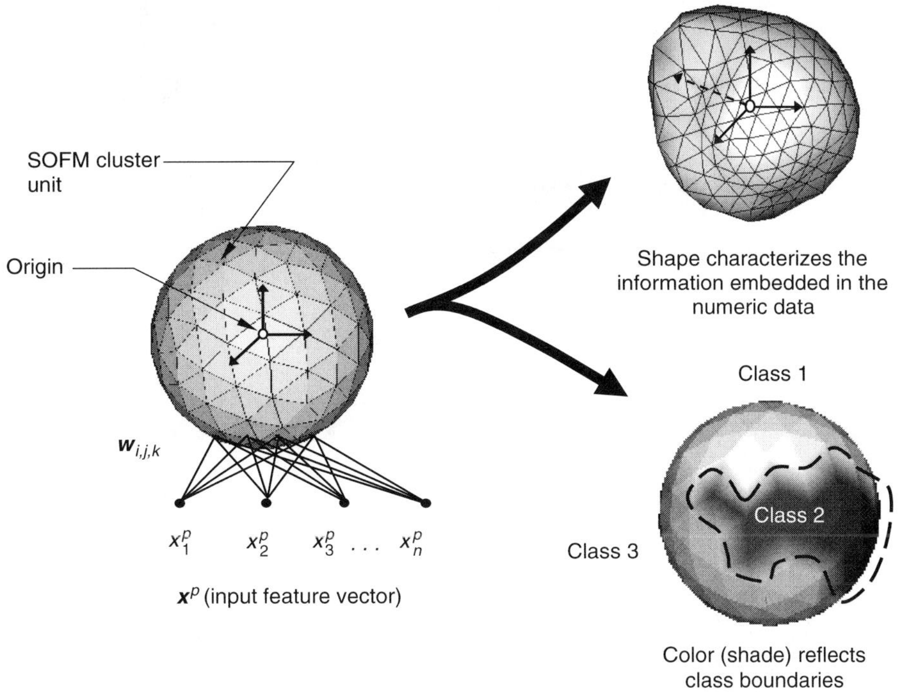

FIGURE 5.11
Spherical SOFM used to enhance visualization and assist in data exploration.

independent coordinate system (59,60). Correspondingly, color is added to reflect a measure of information that relates the weight vectors of the cluster units. The color values assigned can be scaled to lie in the visible range of the color spectrum, and thus every node will have a color vector associated with it. The resulting color-coded deformed spherical representations can provide a comprehensive visual representation of different aspects of information embedded in the numeric data.

The process of multidimensional data clustering for enhanced scientific data visualization is now illustrated using the same Wisconsin breast cancer data (48) described in Section 5.3.2.2. Again, the dataset consists of 683 FNA tissue samples preclassified into two categories: malignant and benign tissue samples. Every FNA tissue sample is a vector consisting of nine attributes that need to be considered to make a correct diagnosis. Each attribute is assigned a value between 1 and 10 by the pathologist, with 10 indicating a very high probability of malignancy.

The training data vectors are randomly selected and presented to a spherical SOFM network that consists of 642 cluster units uniformly distributed on a tessellated sphere. An initial neighborhood of 3 units radius is selected and the network is trained for approximately 200 cycles. On completion of the training process, the spherical lattice is deformed and color-coded to reflect information about the physical phenomenon. The spherical lattice is color-coded using prior information on the two categories of samples, red indicating malignant tissue and yellow representing benign tissue samples. Figure 5.12 presents the same colorized glyph from two separate viewpoints. All the malignant samples contribute to the elevated red region in the shape. A misclassified tissue sample is identified in the right view. A slightly above average value (6.0) for clump thickness and a value of 10 for the

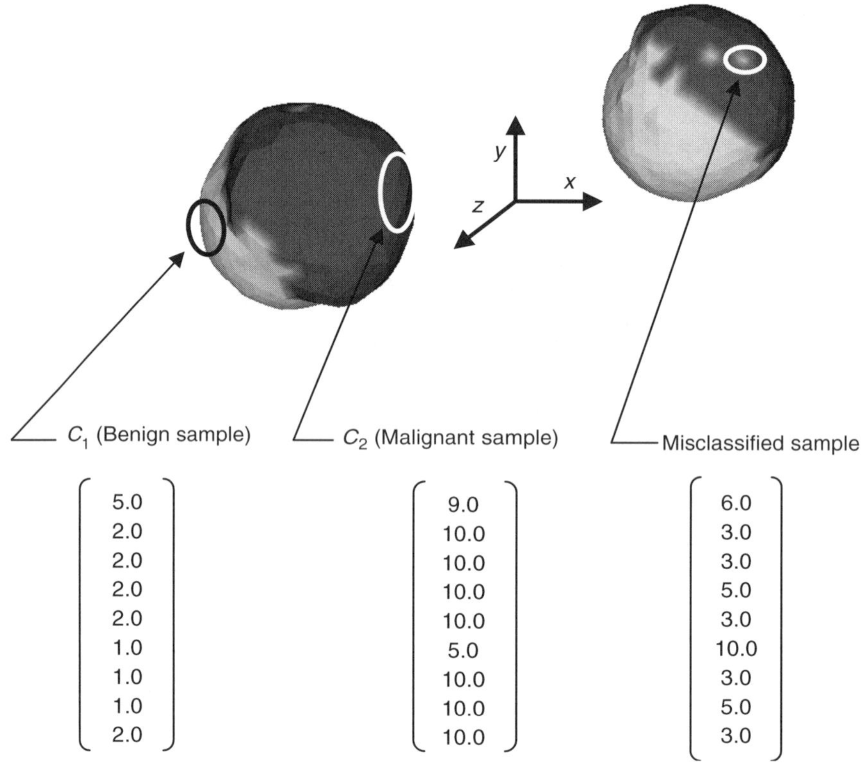

FIGURE 5.12
(See color insert following page 330) Two views of the same 3D colored glyph used to cluster 683 nine-dimensional data vectors into two classes (benign and malignant). Vectors assigned to the benign class (C_1) are represented by the yellow facets, while those associated with the malignant class (C_2) are shown as red. Visual identification of a misclassified data vector based on the color cue is shown in the right view.

epithelial cell size in the feature vector appear to have resulted in classifying it as a malignant-type tissue. By maintaining a record of the inputs assigned to every cluster unit, it is possible to trace those that contribute to significant changes in surface characteristics, either shape or color, in the graphical form thereby indicating the presence of distinct attributes as compared to the rest of the input vectors in the dataset.

5.4 Conclusions

Smart and intelligent biosensors utilize highly selective biological recognition elements to accurately and rapidly identify specific stimuli, analytes, or biochemical reactions. Although the concept of a biosensor has been around over several decades, it is only recently with the introduction of new biomaterials, microfabrication technologies, and advanced computing algorithms that it became possible to incorporate some semblance of intelligence into the device. This chapter briefly examined a variety of key issues associated with biosensor design and presented an overview as to how AI and pattern recognition algorithms can enhance system performance and assist the analyst with interpreting the acquired data (Table 5.2). In this regard, several adaptive connectionist computing

TABLE 5.2

Design Parameters for Selecting a Transducer for a Biosensor

Parameter	Definition
Sensitivity	Ratio of change of sensor output, Δy, to change in the input, Δx: $S = \frac{\Delta x}{\Delta y}$. Maximum sensitivity is normally a design goal as long as other key parameters, such as linearity and accuracy, are not sacrificed.
Linearity	The constancy of the ratio of output to input. If relationship is linear then $y = Kx$ where K (also called gain) is a constant. If the sensor is nonlinear then the input–output relationship is given by an equation: $y = f(x)$.
Working range	Difference between the maximum and minimum values that can be measured by the biosensor or sensor element. A large working range is often preferred.
Accuracy	Difference between the measured and the actual values. It can either be defined as an absolute value which is the maximum error within the working range, or a relative value which is the ratio of the maximum error to the range of measurement.
Repeatability	Difference in value between two successive measurements under the same operating environment.
Resolution	Minimum change of input that can be detected at the output of the biosensor. In some circumstances it is preferred to define resolution as the number of incremental measurements within the range from minimum to maximum value.
Output	Voltage signal is preferred because microcontrollers and computers are being used to automatically gather the data. If the proposed biosensor output signal is not a voltage then it must be converted to an electrical signal with additional hardware, increasing the source of errors.
Response time	Time required for a change in the input to be observed as a stable output change (also known as *settling time*). In most cases, the response time is measured from the start of an input change to the time when the output has settled to the specified range. The response time of the sensor system is determined by its time constant.
Bandwidth	Under ideal conditions, a transducer would produce identical amplitude responses for input signals with the same amplitude, independent of their constituent frequencies. However, the output of a physical transducer is dependent on the amplitude and frequency of an input signal because of the transducer's inherent bandwidth limitations. This characteristic is often described as the transfer function in the frequency domain.

architectures with self-learning capabilities were reviewed. The discussion focused on the general principles of these artificial neural networks and how these algorithms can be applied to biosensor calibration, pattern classification, analyzing patterns generated by multiple detectors, and representing complex numeric datasets for enhanced scientific visualization. Illustrations of function approximation for calibration and unsupervised clustering were achieved using RBF networks and SOFMs, respectively. Although this overview focused on algorithms taken from the neural network literature, other more classical functional approximation or innovative pattern classification techniques may also have worked. The choice depends on the quality of biosensor data and application goals being addressed. All these issues play a greater role in capturing and analyzing data and signals from modern biosensors.

References

1. Powner, E.T., Yalcinkaya, F. (1997). Intelligent biosensors. *Sensor Review*. 17(2): 107–116.
2. Collings, A.F., Caruso, F. (1997). Biosensors: recent advances. *Rep. Prog. Phys.* 60: 1397–1445.
3. Griffiths, D., Hall, G. (1993). Biosensors—what real progress is being made? *Trends Biotechnol.* 11(4): 122–130.
4. Gaura, E., Newman, R.M. (2004). Smart, intelligent and cogent MEMS based sensors. In: *Proceedings of the 2004 IEEE International Symposium on Intelligent Control*. Taipei, Taiwan, 431–436.
5. Rankey, P.G. (2002). Smart sensors. *Sensor Review*. 22(4): 312–318.
6. Powner, E.T., Yalcinkaya, F. (1995). From basic sensors to intelligent sensors: definitions and examples. *Sensor Review*. 15(4): 19–22.
7. Yamasaki, H. (1996). What are the intelligent sensors. In: *Intelligent Sensors*, Yamasaki, H., ed., Amsterdam: Elsevier, pp. 1–17.
8. Duda, R.O., Hart, P.E., Stork, D.G. (2000). *Pattern Classification*. Second Edition. New York: John Wiley.
9. Ciaccio, E.J., Dunn, S.M., Akay, M. (1993). Biosignal pattern recognition and interpretation systems. Part 1 of 4: fundamental concepts. *IEEE Eng. Med. Biol.* 12(3): 89–97.
10. Ciaccio, E.J., Dunn, S.M., Akay, M. (1993). Biosignal pattern recognition and interpretation systems. Part 2 of 4: methods for feature extraction and selection. *IEEE Eng. Med. Biol.* 12(4): 106–113.
11. Ciaccio, E.J., Dunn, S.M., Akay, M. (1993). Biosignal pattern recognition and interpretation systems. Part 3 of 4: Methods of classification. *IEEE Eng. Med. Biol.* 12(3): 89–97.
12. Haykin, S. (1994). *Neural Networks: A Comprehensive Foundation*. New York: Macmillan.
13. Fausett, L. (1994). *Fundamentals of Neural Networks*. Englewood Cliffs, N.J.: Prentice Hall.
14. Hush, D.R., Horne, B.G. (1993). Progress in supervised neural networks. *IEEE Signal Process. Magazine*. 10(1): 8–39.
15. Kohonen, T. (1997). *Self-Organizing Maps*. New York: Springer.
16. Gupta, M.M., Knopf, G.K. (1994). *Neuro-Vision Systems: Principles and Applications*. New York: IEEE Press.
17. Zupan, J., Novic, M., Ruisanchez, I. (1997). Kohonen and counter propagation artificial neural networks in analytical chemistry. *Chemometr. Intell. Lab. Syst.* 38: 1–23.
18. Taib, M.N., Narayanaswamy, R. (1997). Multichannel calibration technique for optical-fibre chemical sensor using artificial neural network. *Sensors Actuators* 38–39: 365–370.
19. Suah, F.B.M., Ahmad, M., Taib, M.N. (2003). Applications of artificial neural network on signal processing of optical fibre pH sensor based on bromophenol blue doped with sol-gel film. *Sensors Actuators* 90: 182–188.
20. Suah, F.B.M., Ahmad, M., Taib, M.N. (2003). Optimisation of the range of an optical fibre pH sensor using a feed-forward artificial neural network. *Sensors Actuators* 90: 175–181.
21. Raimundo, I.M., Narayanaswamy, R. (2001). Simultaneous determination of relative humidity and ammonia in air employing an optical fibre sensor and neural network. *Sensors Actuators* 74: 60–68.
22. Ferreira, L.S., De Souza, M.B., Folly, R.O.M. (2001). Development of an alcohol fermentation control system based on biosensor measurements interpreted by neural networks. *Sensors Actuators* 75: 166–171.
23. Zupan, J., Novic, M., Gasteiger, J. (1995). Neural networks with counter-propagation learning strategy used for modelling. *Chrometr. Intell. Lab. Syst.* 27: 175–187.
24. Lyons, W.B., Flanagan, C., Lochmann, S., Ewald, H., Lewis, E. (2001). A multipoint optical evanescent wave U-bend sensor system based on artificial neural network pattern recognition. In: *Advanced Photonic Sensors and Applications II*, Asundi, A. K., Osten, W., Varadan, V.K., eds., Proceedings of SPIE Vol. 4596: 320–328.
25. Blank, T.B, Brown, S.D. (1993). Nonlinear multivariate mapping of chemical data using feed-forward neural networks. *Anal. Chem.* 65: 3081–3089.
26. Bachmann, T.T., Leca, B., Vilatte, F., Marty, J.-L., Fournier, D., Schmid, R.D. (2000). Improved multianalyte detection of organophosphates and carbamates with disposable multielectrode

biosensors using recombinant mutants of *Drosophila* acetylcholinesterase and artificial neural networks. *Biosen. Bioelectr.* 15: 193–201.

27. Bowen, B.P., Scruggs, A., Enderlein, J., Sauer, M., Woodbury, N. (2004). Implementation of neural networks for the identification of single molecules. *J. Phys. Chem.* 108: 4799–4804.
28. Lobanov, A.V., Borisov, I.A., Gordon, S.H., Greene, R.V., Leathers, T.D., Reshetilov, A.N. (2001). Analysis of ethanol–glucose mixtures by two microbial sensors: application of chemometrics and artificial neural networks for data processing. *Biosens. Bioelectr.* 16: 1001–1007.
29. Bulsari, A., Saxen, H. (1994). Using feed-forward neural networks for estimation of microbial concentration in a simulated biochemical process. *Biosens. Bioelectr.* 9: 105–109.
30. Keller, P.E. (1999). Overview of electronic nose algorithms. In: *Proceedings of the International Joint Conference on Neural Networks*. Vol. 1. pp. 309—312.
31. Göpel, W., Ziegler, C., Breer, H., Schild, D., Apfelbach, R., Joerges, J., Malaka, R. (1998). Bioelectronic noses: a status report Part I. *Biosens. Bioelectr.* 13(3–4): 479–493.
32. Ziegler, C., Göpel, W., Hammerle, H., Hatt, H., Jung, G., Laxhuber, L., Schmidt, H.-L., Schütz, S., Vögtle, F., Zell, A. (1998). Bioelectronic noses: a status report. Part II. *Biosens. Bioelectr.* 13: 539–571.
33. Kermani, B.G., Schiffman, S.S., Nagle, H.T. (1999). Using neural networks and genetic algorithms to enhance performance in the electronic nose. *IEEE Trans. Biomed. Eng.* 46(4): 429–439.
34. Di Natale, C., Macagnano, A., Davide, F., D'Amico, A., Paolesse, R., Boschi, T., Faccio, M., Ferri, G. (1997). An electronic nose for food analysis. *Sensors Actuators* 44: 521–526.
35. Evans, P., Persaud, K.C., McNeish, A.S., Sneath, R.W., Hobson, N., Magan, N. (2000). Evaluation of a radial basis function neural network for determination of wheat quality from electronic nose data. *Sensors Actuators* 69: 348–358.
36. Llobet, E., Hines, E.L., Gardner, J.W., Franco, S. (1999). Non-destructive banana ripeness determination using a neural network-based electronic nose. *Measure. Sci. Technol.* 10: 538–548.
37. Brezmes, J., Llobet, E., Vilanova, X., Saiz, G., Correig, X. (2000). Fruit ripeness monitoring using an electronic nose. *Sensors Actuators* 69: 223–229.
38. Dutta, R., Hines, E.L., Gardner, J.W., Udrea, D.D., Boilet, P. (2003). Non-destructive egg freshness determination: an electronic nose based approach. *Measure. Sci. Technol.* 14: 190–198.
39. Panigrahi, S., Balasubramanian, S., Gu, H., Logue, C., Marchello, M. (2006). Neural-network-integrated electronic nose system for the identification of spoiled beef. *LWT.* 39: 135–145.
40. Brezmes, J., Ferreras, B., Llobet, E., Vilanova, X., Correig, X. (1997). Neural network based electronic nose for classification of aromatic species. *Anal. Chim. Acta.* 348: 503–509.
41. Daqi, G., Shuyan, W., Yan, J. (2004). An electronic nose and modular radial basis function network classifiers for recognizing multiple fragrant materials. *Sensors Actuators* 97: 391–401.
42. Pavlou, A.K., Magan, N., Jones, J.M., Brown, J., Klatser, P., Turner, A.P.F. (2004). Detection of *Mycobacterium tuberculosis* (TB) in vitro and in situ using electronic nose in combination with a neural network system. *Biosens. Bioelectr.* 20: 538–544.
43. Dutta, R., Morgan, D., Baker, N., Gardner, J.W., Hines, E.L. (2005). Identification of *Staphylococcus aureus* infections in hospital environment: electronic nose based approach. *Sensors Actuators* 109: 355–362.
44. Golden, J.P., Anderson, G.P., Cao, L.K., Ligler, F.S. (1994). Calibration methods for an evanescent wave fiber optic sensor. In: *Annual International Conference of the IEEE Engineering in Medicine and Biology—Proceedings*. Vol. 16(2). pp. 822–823.
45. Wang W.W., Knopf G.K., Bassi A. (2005). Photoelectric properties of a detector based on dried bacteriorhodopsin film. *Biosens. Bioelectr.* 21: 1309–1319.
46. Wang, W.W., Knopf, G.K., Bassi, A. (2005). Protein-based photocell for high-speed motion detection. In: *2005 IEEE International Conference on Control Applications*. pp. 731–736.
47. Wang, W.W., Knopf, G.K., Bassi, A. (2005). Implementation of bidirectional motion detection using protein based photoreceptors. In: *IEEE/ASME Advance Intelligent Mechatronics (AIM 2005)*. pp. 72–75.
48. Knopf, G.K., Sangole, A. (2004). Interpolating scattered data using 2-D self-organizing feature maps. *J. Graph. Models* 66: 50–69.
49. Blake, C.L., Merz, C.J. (1998). *UCI Repository of Machine Learning*, University of California, Department of Information and Computer Science, Irvine, CA. [http://www.ics.uci.edu/~mlearn/MLRepository.html].

50. Wolberg, W.H., Mangasarian, O.L. (1990). Multisurface method of pattern separation for medical diagnosis applied to breast cytology. *Proc. Natl. Acad. Sci.* 87: 9193–9196.
51. Mangasarian, O.L., Street, W.N., Wolberg, W.H. (1995). Breast cancer diagnosis and prognosis via linear programming. *Operations Res.* 43(4): 570–577.
52. Lee, Y.-J., Mangasarian, O.L., Wolberg, W.H. (2003). Survival-time classification of breast cancer patients. *Computat. Optimization Appl.* 25: 151–166.
53. Gutés, A., Ibáñez, A.B., Céspedes, F., Alegret, S., del Valle, M. (2005). Simultaneous determination of phenolic compounds by means of an automated voltammetric electronic tongue. *Anal. Bioanal. Chem.* 382: 471–476.
54. Abidi, M.A., Gonzalez, R.C. (1992). *Data Fusion in Robotics and Machine Intelligence.* Boston: Academic Press.
55. Hall, D.L. (2004). *Mathematical Techniques in Multi-Sensor Data Fusion.* Boston: Artech House.
56. Knopf, G.K., Sangole, A. (2000). Trinocular data registration using a three-dimensional self-organizing feature map. In: *2000 IEEE Conference on SMC*. pp. 2863–2868.
57. Gross, M. (1994). *Visual Computing: The Integration of Computer Graphics, Visual Perception and Imaging*. Berlin: Springer—Verlag.
58. Gross, M., Seibert, F. (1993). Visualization of multidimensional image data sets using a neural network. *Visual Comput.* 10: 145–159.
59. Sangole A., Knopf G.K. (2002). Representing high-dimensional data sets as closed surfaces. *Inform. Visualization J.* 1: 111–119.
60. Sangole, A., Knopf, G.K. (2003). Visualization of randomly ordered numeric data sets using spherical self-organizing feature maps. *J. Comput. Graph.* 27: 963–976.

6

Neuronal Network Biosensors

Guenter W. Gross and Joseph J. Pancrazio

CONTENTS

6.1 Introduction

Neuroactive compounds can alter performance and behavior of organisms over a wide range: from life-threatening catastrophic failure of organ systems that rely on neural input to subtle but debilitating changes in cognitive processes and motor coordination. Neuroactive compounds include neurotoxins, several of which are considered potential biological weapons, as well as neurotoxicants, where a risk may arise via environmental or occupational exposure. Evaluation of a vast number of potentially neuroactive compounds, which range from candidate therapeutics to putative environmental toxicants, presents a significant burden to the pharmaceutical industries, the military, and the academia. As supplements to current methods, which primarily rely on animal neurobehavioral testing, in vitro test systems using either cell lines or high-yield primary cell cultures are receiving increased attention. Data accumulated so far indicate that in vitro systems are well suited for rapid, inexpensive, and quantitative screening of toxicants under well-defined and controlled experimental conditions. The key to effective in vitro models is the ability to capture physiologically relevant activity. In the case of neuronal systems, electrophysiological activity is paramount relative to physiological function and behavior. A major component of such activity, the suprathreshold action potential dynamics, can now be captured from many neurons comprising the network and processed in real time. If cultured on glass plates with transparent thin film indium–tin oxide (ITO) conductors, high-power microscopy and associated fluorescence techniques can be applied in parallel with electrophysiological monitoring (as shown in Figure 6.1). The latter includes discriminated action potential data where waveshape changes can be used to study pharmacological effects on voltage-gated channels.

Neuronal network biosensors (NNBS), as described in this chapter, are positioned between cell-based and tissue-based biosensors. The NNBS systems consist of functional, spontaneously active networks that are formed on substrate-integrated microelectrode arrays (MEAs). This approach provides extensive multisite electrophysiological data in terms of action potential patterns and waveshapes, as well as simultaneous cytological information through high-power microscopy and fluorescence. The dissociation and seeding of embryonic neural tissue on appropriately prepared surfaces of MEAs produces an intimate cellular contact with that surface, resulting in strong cell-surface and cell-electrode coupling if the ratio of cell mass to adhesion area is small (see Figure 6.2). Such coupling provides long-term morphological stability and excellent signal-to-noise ratios (SNRs). In fact, the NNBS has the capacity to provide long-term electrophysiological data over periods ranging from weeks to months, which constitutes a far greater performance window than virtually any other in vitro electrophysiological assay.

The Center for Network Neuroscience has placed value in generating cultures that contain all cell types present in the parent tissue in approximately the same ratios. Although the developing embryonic circuitry is disrupted during the cell dissociation process, the

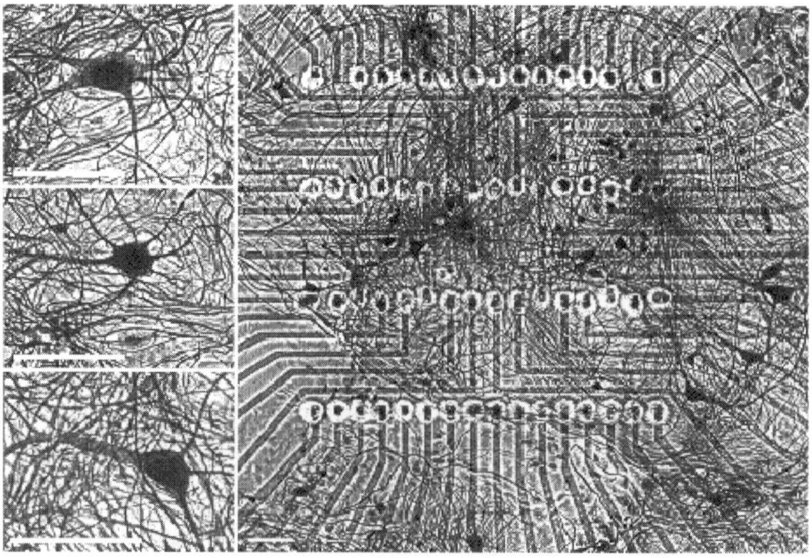

FIGURE 6.1
Neuronal network on the 1 mm^2 recording matrix of an MEA. Indium–tin oxide conductors connect this area with amplifier contacts at the edges of a 5 × 5 cm glass plate. The electrode array is insulated with 2 μm of polysiloxane resin perforated with laser shots at the end of the conductors. Bodian stain with fixation 96 days after seeding. Bar: 50 μm.

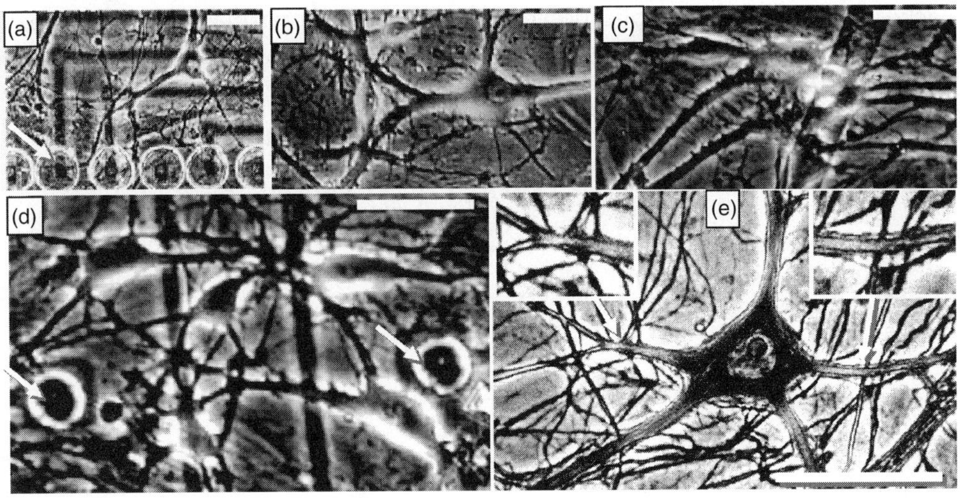

FIGURE 6.2
(a–d) Phase contrast micrographs of living multipolar neurons on microelectrode arrays. Electrode recording craters in A and D are identified by white arrows. (e) Neurofilament antibody stain of multi-polar neuron. Inserts show synaptic boutons. (bar; 40 μm, spinal cord culture, 48 d.i.v.). White bars represent 40 μm.

networks reorganize in a nonrandom manner [1] to form dynamic systems where the overall spontaneous activity patterns and, moreover, the pharmacological response profiles are reproducible. Extensive pharmacological data have shown that all basic synaptic mechanisms as well as key receptors and channels are retained in culture. In fact, even tissue specificities, characteristic of the parent tissue region, are being documented in culture [2, 3, 4]. Although responses from such networks must be considered as "cell culture correlates" to the changes observed in animals, such in vitro responses are repeatable and can be quantified. Given this level of data, it is not surprising that serious efforts are now underway to use such systems for investigations of neurotoxicity and cytotoxicity, and for exploiting their properties for environmental toxicology, drug development, and even in defense scenarios as broadband biosensors. Judging from toxicological investigations performed till date, NNBS respond to a diverse array of compounds that attack neural life support systems or alter the behavior or performance of mammals. These compounds include bacterial toxins, metabolic poisons, toxic metals, neuropharmacological compounds, hallucinatory drugs, and epileptogenic agents. The recognition that neurobehavioral changes can have profound impacts on operational readiness of deployed forces drives the military-related interests in broadly sensitive biosensors for neuroactive compounds [5, 6].

It is important to emphasize that pharmacological and toxicological investigations address primarily receptor-dependent mechanisms. Response measurements to global exposure of neural tissue to specific test compounds require stable, spontaneous activity, but not necessarily specific circuitry. Numerous recent articles have demonstrated the histotypic responses of primary cultures where EC_{50} values virtually overlap with those obtained from animal experiments. Such cultures develop in vitro, stabilize, and are incorporated into recording chambers without injury to the existing circuitry. This avoids problems of deafferentation and trans-synaptic degeneration. The virtual monolayer allows rapid, uniform pharmacological exposure of all cells and the extraction of massive spike-pattern information. Added to the optical accessibility and general longevity, primary cultures should no longer be considered a second choice to brain slices. These experimental systems have matured and provide a rich environment for a large number of investigative approaches and for specific applications in the domains of pharmacology, toxicology, and drug development.

The aim of this chapter is to describe and summarize the experimental basis for the suggested applications with emphasis on biosensors. Once the utility of these systems is validated, technical improvements can quickly adapt NNBS procedures for a great variety of applications. Prototype systems of compact portable recording have already been realized [7, 8] and are briefly described herein. Dual network recording on the same MEA, providing a "twin" network as a control, is already a routine procedure; and eight-network recording modules, served by a liquid handling robot, are under development at the Center for Network Neuroscience (CNNS).

6.2 Methods

6.2.1 Cell Culture

The cell culture procedures are well established and generate robust, spontaneously active networks that can withstand 50 g of tangential impact and allow shipping to other laboratories. Many experiments performed at the U.S. Naval Research Laboratory have relied on shipment of intact living networks grown on MEAs at the CNNS [9, 8]. In addition to the spinal cord and frontal cortex networks discussed in this chapter, other neuronal cell systems such as auditory cortex [10, 2] and brain stem [11] have been used and grown on MEAs with nearly identical methods. The only major difference between spinal and cortical cell culture development is the use of a glycine-free medium for spinal tissue to avert glycine-induced inhibition of spinal circuit activity. Cultures are prepared according to the basic pioneering methods established in the laboratory of Dr. Phillip Nelson [12]. Central nervous system (CNS) tissues are obtained from embryonic HSD:IRC mice (Harlan, Sprague, Dawley, Indianapolis, IN). Spinal cord tissue is dissected at 14–15 days' gestation and cortical tissue at 16–17 days. The tissue is dissociated enzymatically (papain) and mechanically (via trituration), and seeded at a concentration of $0.2–0.5 \times 10^6$ cells per milliliter onto electrode arrays coated with polylysine and laminin to promote cell adhesion. The cell pool is seeded as 0.1 ml droplets (approximately 20,000–50,000 cells or 2,800–7,100 cells per square millimeter). Assuming a neuron to nonneuron cell ratio of 1:10, this procedure provides a neuronal density of approximately 300–700 neurons per square millimeter. After cell adhesion (~2 h), 1 ml of medium is added, confined by a silicon gasket. Spinal cord and cortical cultures are incubated in minimal essential medium (MEM) and Dulbecco's modified minimal essential medium (DMEM), respectively, each supplemented with 5% horse serum and 5% fetal bovine serum for a week. Thereafter, the cells are fed twice a week using half-media changes with the respective media containing only 5% horse serum. Typically, 80% of the cultured networks exhibit robust spike activity on at least 50% of the 64 recording sites. Mean SNRs are generally in a range from 3–6 with several channels exceeding SNRs of 10. This level of activity has proven to be suitable for reproducible pharmacological studies. Thereafter, the cells are fed twice a week using half-media changes with the respective media containing only 5% horse serum. Moreover, pooled tissues from 10 embryonic mouse spinal cords, for example, can seed more than 200 MEAs. Similarly, specific brain regions, such as frontal cortex, can easily seed more than 50 MEAs. This provides a heretofore unprecedented efficiency in tissue utilization and could clearly serve to reduce the number of animals presently necessary for neurotoxicological investigations.

6.2.2 Multichannel Recording

The techniques used to fabricate and prepare passive MEAs and the associated recording chambers have been described and are well established [13, 14, 15]. A typical laboratory-based workstation for single and dual array recording is shown in Figure 6.3. Both chambers are interconvertible from constant-medium-volume configurations to closed-flow configurations. Dual arrays feature two recording areas formed by 32 microelectrodes separated by 2.5 cm on the same 5×5 cm MEA plate with identical amplifier contact geometries (Figure 6.4). Cruciform electrodes were incorporated to provide three deinsulation sites per conductor (or channel) to increase the electrode yield (percent of electrodes with activity of 2:1 SNR or higher). Although this is indeed an advantage at low cell densities (<300 neurons per square millimeter), it causes excessive pickup at high cell densities, making the clean separation of waveforms in real time more difficult. The two separate networks however provide twin networks that are matched for age and maintenance schedule and allow selection of one control and one experimental culture. This is not yet high throughput, but it paved the way for eight-network MEAs (described in Section 6.5).

In the past decade, relatively sophisticated commercial hardware and software have emerged for multichannel extracellular recording (Plexon Inc., Dallas; Multichannel Systems, Reutlingen, Germany). The Plexon digital display is shown in Figure 6.5. It allows simultaneous scanning of (1) discriminated activity on a specific channel of interest, (2) time-stamp patterns from all the units discriminated,, and (3) all waveshapes discriminated. A CNNS population vector display, presently based on LabView, is shown in Figure 6.6. It depicts the average network spike and burst production per selected time base (usually per minute) and can display the evolution of activity for continuous recordings over a period of days. Expansion of the *X*-axis between any two time points allows detailed scrutiny of minute-to-minute activity changes. Such displays of simple activity variables are invaluable to real-time monitoring and data interpretation, but can also be used for automated data analyses using stable activity levels as reference. The dashed horizontal line shows an activity recovery to near reference values and stability of network spike and burst production over a 5-h period. The fluctuations of the minute means about the episode mean (dashed line) provides a standard deviation (box) from which all substance-induced deviations are measured.

Burst identification is generally achieved via integration with time constants of approximately 70 ms. For digitized burst identification, a two-threshold method is used (Figure 6.7), whereby the first threshold (T1) identifies the beginning of a putative burst, which is then

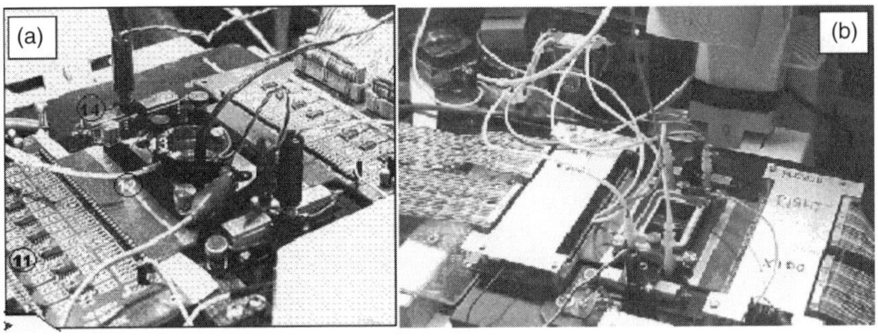

FIGURE 6.3.
Workstations for multichannel recording from networks in vitro combined with light microscopy. Chambers and preamplifiers (32 per side) are situated on the stage of an inverted microscope allowing simultaneous electrophysiological and optical monitoring. (a) Open chamber with heated cap (to prevent condensation) and CO_2 line for pH control. (b) Dual closed chamber served by two-channel peristaltic pump.

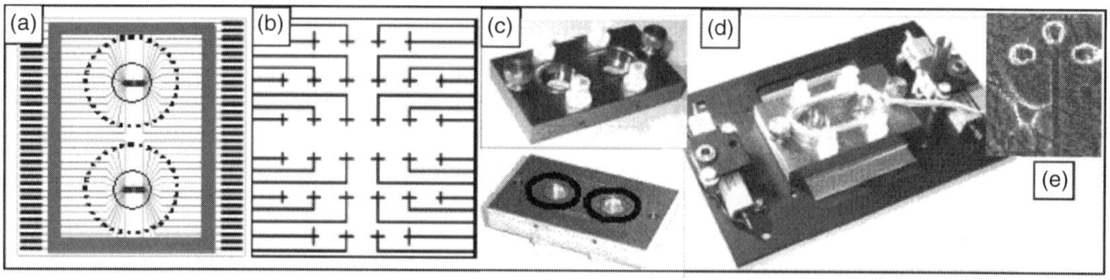

FIGURE 6.4
Dual-array MEAs compatible with the regular amplifier contact geometry. (a) 5 × 5 cm plate showing recording site, contact circle of 'O'-rings (dashed lines), and gasket holding medium during network development. The inner circles represent the cell adhesion areas. (b) Electrode configuration of one recording site showing 32 cruciform electrodes separated by 200 μm laterally and between rows. (c) Top and bottom views of recording chambers. (d) Assembled chamber and base plate with heating power resistors. (e) Multipolar neuron near cruciform electrode terminals.

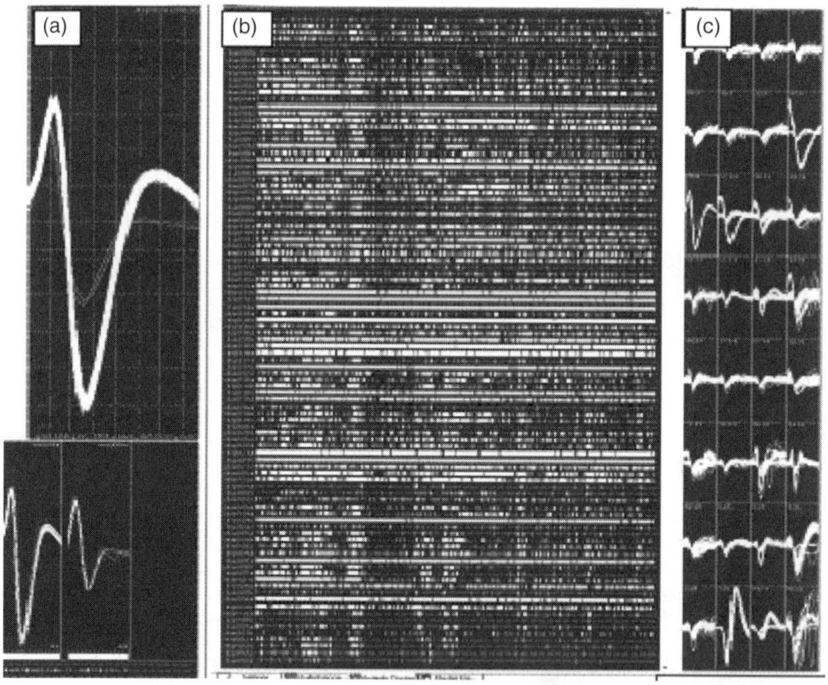

FIGURE 6.5
Plexon multichannel data analysis display. (a) A selected channel shows action potential waveshapes on that channel and allows setting of templates for real time discrimination (b) Separation into logical timestamp channels. (c) All discriminated units, selected by template matching algorithms, are shown in the right panel.

confirmed by a second, higher threshold (T2). In this manner small bursts of 3–4 spikes can be eliminated if emphasis is to be placed on larger bursts. Burst identification is very much an operational process and depends on the type of activity pattern generated by the cultures.

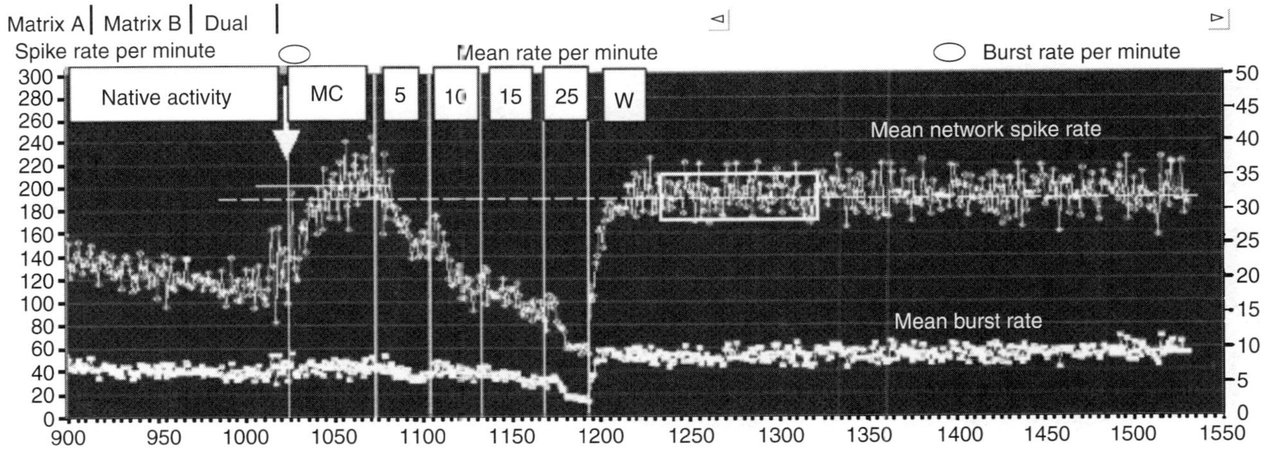

FIGURE 6.6
Population vector display of burst and spike rates for a 10.7-h experimental episode (single network, constant medium bath). Each data point represents an average across all discriminated units per minute. Native activity is increased by a medium change (MC), followed by a verapamil titration (final concentration in μM) and a further medium change (W) with return to the MC reference activity level. Final activity is stationary in time with an activity fluctuation that can be described in terms of an episode mean and standard deviation (box).

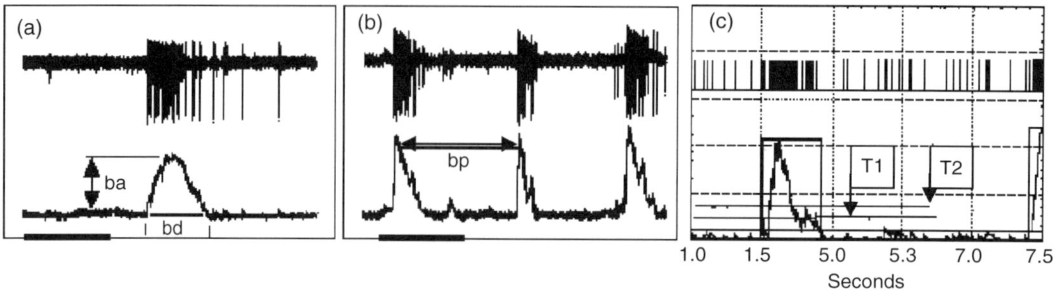

FIGURE 6.7
Burst detection and quantification using integration. (a) Analog channel activity, (b) digital channel activity, and (c) discriminated unit activity showing three activity variables: burst amplitude (ba), burst duration (bd), and burst period (bp). Burst rate (usually per minute) and area under the burst outline are also extracted. Digital burst detection is based on a two-threshold method where the first threshold (T1) identifies the start of a burst only if T2 is reached in 100 ms thereafter.

6.3 Network Response Profiles

6.3.1 Pharmacological Dose Responses

Experimental networks do not have to display identical activity patterns to allow quantitative comparisons between networks. If each network is treated as an individual and all changes in activity are referenced to the native activity of that culture, then dose response reproducibility can be achieved. This is illustrated in Figure 6.8, which depicts the response of a frontal cortex network to ethanol [16].

In this example, a slow, partial activity recovery complicates the determination of response plateau values. Because of the high ethanol volatility, ethanol was added to the wash bottle that humidified the CO_2–air mixture at the same concentration as that in the recording chamber. Ethanol concentrations in the bath medium were measured with a diagnostic kit (Sigma Diagnostics Inc., St. Louis, MO) and confirmed that concentrations were stabilized. The activity recovery was not considered as linked to ethanol evaporation. A similar activity rebound was seen with low concentrations of anandamides [17]. Table 6.1 shows the overlap between cell culture and animal experiments.

Figure 6.9 shows results from an analysis of novel pharmacological compounds designed to inhibit acetylcholinesterase [18]. A series of seven symmetrical, bifunctional alkylene-linked bis-thiocarbonate compounds were tested, all of which had been characterized biochemically as mixed inhibitors of AChE, with K_i's in the 10^{-7}–10^{-6} M range. The neuronal network assays revealed high repeatability for each compound but, surprisingly, diverse effects among these closely related structures. Six of the seven compounds produced changes in network activity at concentrations from 10–350 μM. Three were excitatory, two were biphasic (excitation followed by inhibition), and one was solely inhibitory. Two of the inhibitory compounds produced irreversible inhibition of activity and were considered to be functionally neurotoxic in the absence of cytotoxicity. Only two compounds showed activity profile changes consistent with that of a classic AChE inhibitor. This study demonstrated that spontaneously active neuronal networks reflect the interactions of multiple neurotransmitter and receptor systems, and can reveal unexpected side effects due to secondary binding that can have dominant effects on neurophysiological function.

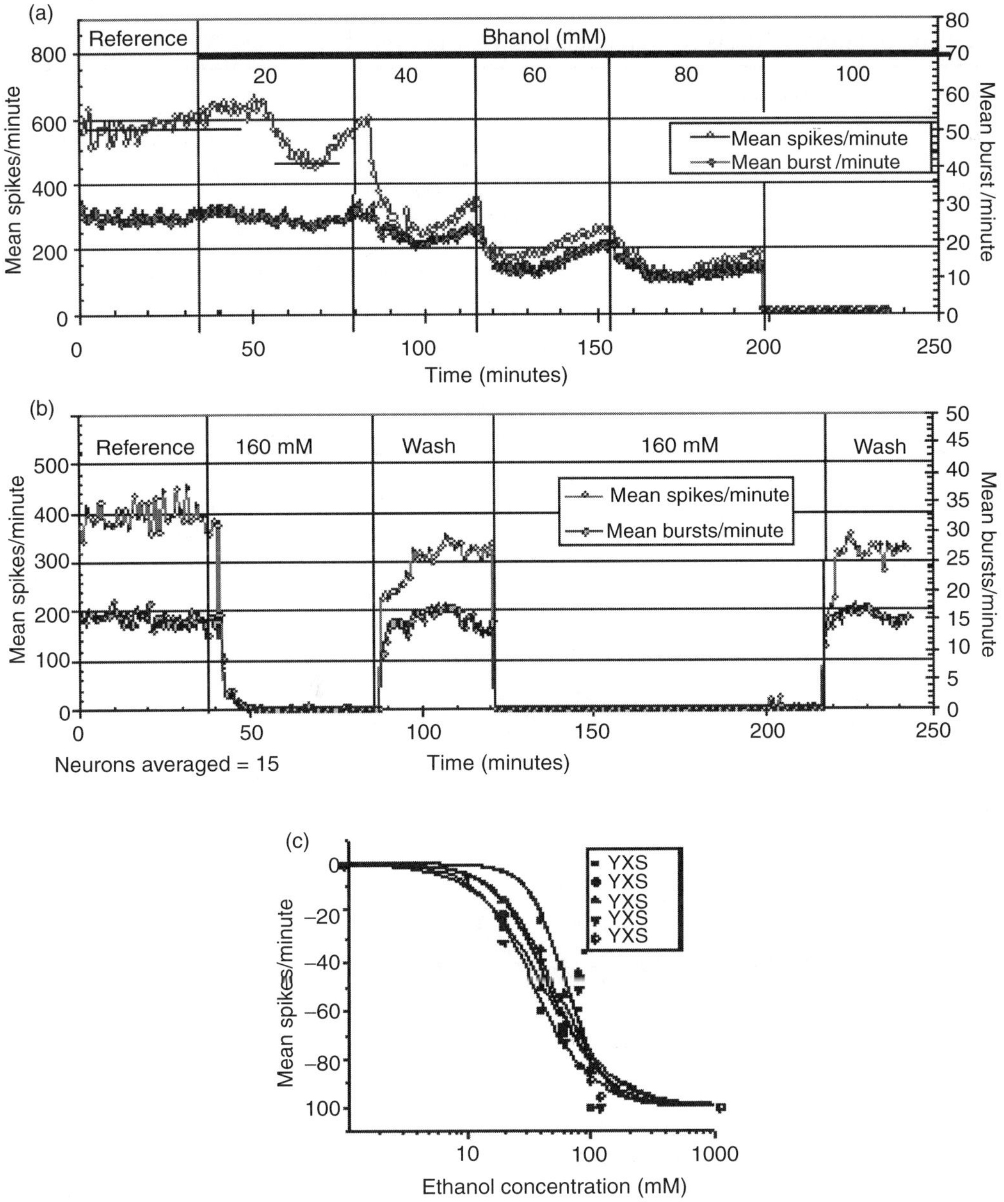

FIGURE 6.8
(a) Frontal cortex network response to increasing concentrations of ethanol. Each data point represents the per-minute average of spike (open circles) and burst (closed symbols) production for all discriminated units recorded. Percent decreases are calculated from a mean reference activity preceding all applications of test substances. In the case of slow spontaneous recovery of activity, the maximum decrease attained at a particular test substance concentration is used for response calculations. (b) Reproducibility of recovery after exposure to high concentrations of ethanol. (c) Concentration–response curves for five different frontal cortex networks; EC_{50}: 48.8 ± 5.4 mM (redrawn from Xia and Gross, 2003).

Recent work has shown that dissociation constants can also be determined using spontaneously active primary cultures on MEAs [19]. A summary for the titration of muscimol in the presence of its antagonist gabazine is presented in Figure 6.10. Muscimol is a potent and specific gamma-aminobutyric acid (GABA) analog [20]. Because there is no re-uptake

TABLE 6.1
Comparison of Ethanol Effects on Mammals and FC Cultures

Mammal	Concentration	Effect
Humans	5–15	Slight impairment of attention and judgment
Humans	10–22	Impairment of speech and balance
Humans	15–30	Significant sedation
Rats	20	Sedation
FC cultures	10	First measurable change in activity pattern (slight excitation)
FC cultures	15	First overt change in spike and burst production (inhibition)
Humans	30–55	Mental confusion
Mice	40	Loss of righting reflex
FC cultures	48.8	EC_{50} for spike production
Humans	100	Coma and death
Mice	120	Coma, hypothermia
FC cultures	100–140	Cessation of all spontaneous activity

Note: FC = frontal cortex.

Source: From Charness, M.E., Simon, R.P., and Greenberg, D.A. (1989). Ethanol and the nervous system. *The New England Journal of Medicine* 321: 442-454; Little, H.J. (1990) Mechanisms that may underlie the behavioral effects of ethanol. *Progress in Neurobiology* 36: 171–194. With permission.

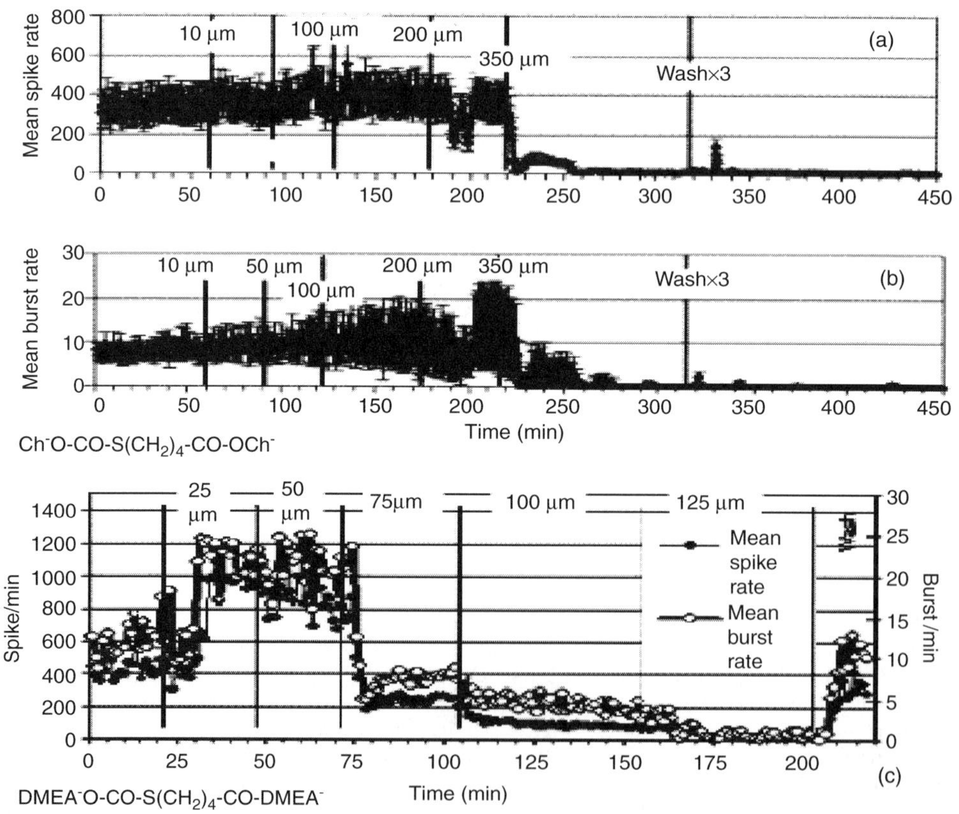

FIGURE 6.9
Irreversible inhibition of network activity by a choline-based compound with a six-carbon spacer. (a) Mean spike rates per minute and burst rates per min (b) with standard deviations per minute show a small increase in activity until activity shut-off at 350 μM. The SD for the burst rate also increases until catastrophic network failure. Inhibition was not reversible by three medium changes. Frontal cortex culture, 28 d.i.v. (c) Biphasic response of the four-carbon spacer, DMEA residue bis-thiocarbonate showing excitation at 25–50 μM followed by inhibition and activity recovery after a single medium change at 200 min (redrawn from Keefer et al., 2001).

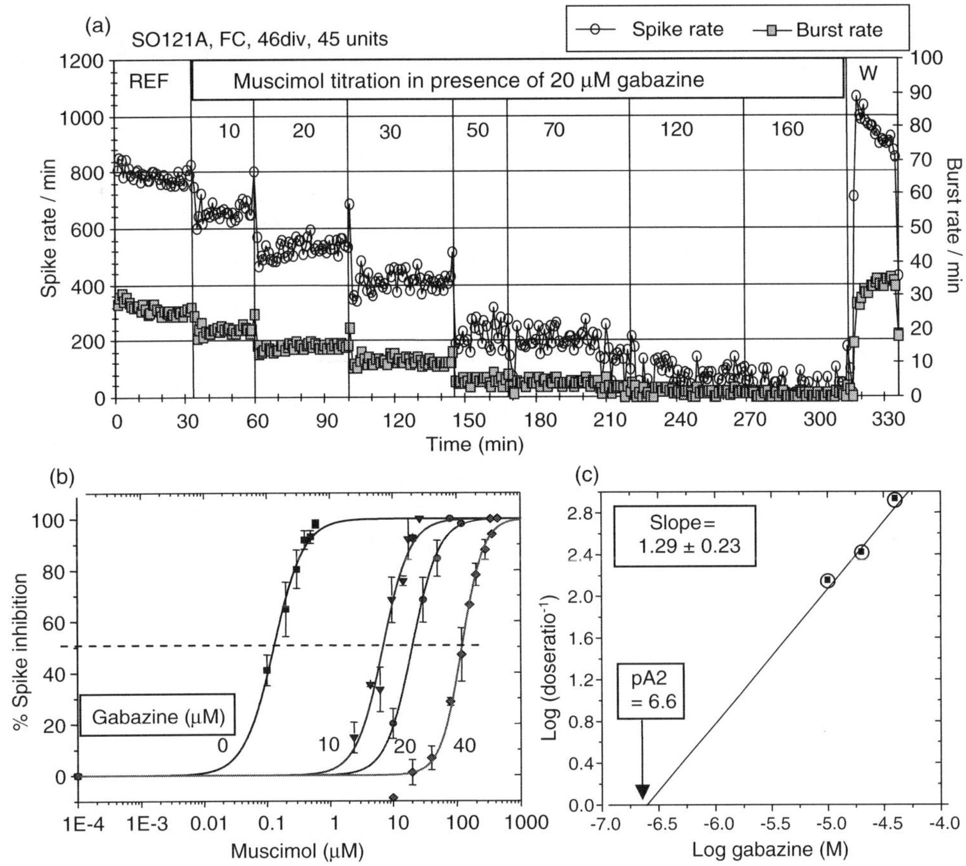

FIGURE 6.10
Quantification of receptor interactions for the GABA antagonist gabazine. (a) Muscimol titration in the presence of 20 μM gabazine. Each data point represents average spike rate (left ordinate) and burst rate (right ordinate) for all discriminated units (45) in bins of 1 min. Note level response activity plateaus from which percent decrease from reference activity can be determined. (b) Concentration–response curves of the pooled data in the presence of 0, 10, 20, 40 μM gabazine. Vertical bars represent SD of percent spike activity decrease. The mean IC_{50} for muscimol in the presence of gabazine are 20.1, 37.2, and 120.4 μM. (c) Schild plot of log (dose ratio-1) vs. log gabazine concentration. The *X*-intercept represents a pA_2 value (–log of K_B) of 6.6. The dissociation constant K_B is 0.25 μM (redrawn from Oli-Rijal, 2005).

mechanism for muscimol, its concentration in medium remains constant and it is well suited for quantitative pharmacology of the $GABA_A$ receptor. Gabazine, [2-(3′-carboxy-2′-propyl)-3-amino-6-p-methoxyphenylpyridazinium bromide], is an aryl-aminopyridazine derivative of gamma-aminobutyric acid (GABA) and represents a potent, selective, competitive, and reversible $GABA_A$ antagonist [21, 22]. Muscimol titration in the presence of increasing concentrations of gabazine shifts the concentration-response curves (CTCs) to the right (higher concentrations of muscimol), but with identical efficacies and with no measurable change in slopes (Figure 10a). A Schild plot [23]) with linear regression predicts an *X*-intercept of 0.23 μM. This intercept is the dissociation constant K_B for the antagonist gabazine. The theoretical slope of 1 identifies a substance as a competitive antagonist. The experimental slope of 1.29 ± 0.23 is not different from unity and identifies gabazine as a competitive antagonist of muscimol. The K_B and the pA_2 value (–log K_B) of 6.63 agree well with published data [24, 25] and validate primary cultures as useful, histotypic biosensor candidates.

6.3.2 Responses to Toxins

Given the histotypic responses observed with pharmacological studies, it is not surprising that the same is seen with toxins. Extensive data have been gathered with tetrodotoxin, botulinum toxin A, tetanus toxin, and the nerve gas hydrolysis products methyl phosphonate, pinacolyl methylphosphonate, and isopropyl methyl phosphonate. Whereas tetrodotoxin and botulinum toxin revealed rapid and slow (respectively) activity shut-off in the nanomolar and picomolar ranges (unpublished), the nerve gas hydrolysis products were found non-toxic, as millimolar concentrations were required for partial activity decreases [5]. Methyl phosphonate had no effect on network spike production up to 5 mM.

Figures 6.11 and 6.12 show network responses to tetanus and botulinum toxins. The tetanus toxin data show tissue specificity with an expected greater effect on spinal networks. Both the time course of activity changes and the eventual involvement of all the neurons in the network agree well with the published data. Williamson et al. [26] investigated the effect of this toxin on the release of radiolabeled glycine and glutamate, followed over time intervals corresponding to the early phase of convulsant activity through the later phase of electrical quiescence. Approximately 90 min after toxin application, the release of glycine was blocked and glutamate release increased to twice the normal. Blockage of

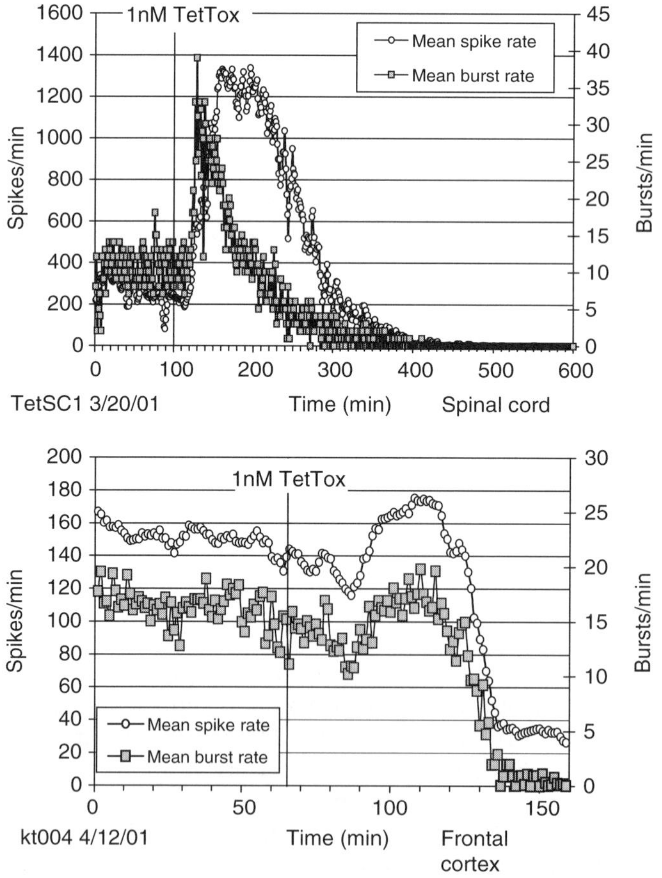

FIGURE 6.11
Responses of spinal and frontal cortex networks to tetanus-toxin (1 nM). Whereas the spinal network shows strong excitation after a 30-min delay followed by gradual decline to catastrophic failure in 200 min, the frontal cortex network responds with weak excitation after 25 min followed by catastrophic network failure in 70 min.

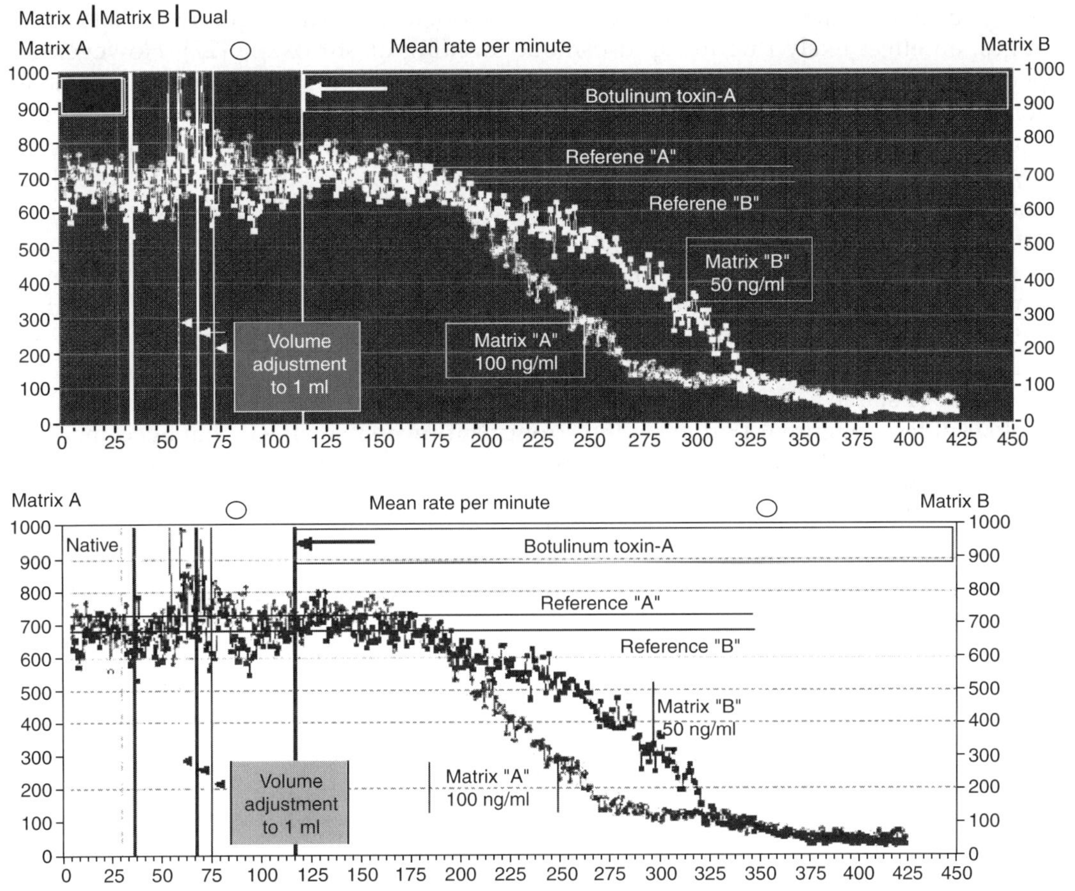

FIGURE 6.12
Dual network array experiment showing responses to 100 ng/ml (Matrix A) and 50 ng/ml (Matrix B) botulinum toxin A. Time to 50% activity decay is 1.9 and 2.9 h for the high and low concentrations, respectively. The concentrations respond to approximately 660 and 320 picomoles. The lowest concentration at which responses were repeatable was 2 ng/ml. Real-time computer display (CNNS NACTAN program).

glutamate release occurred after 150–210 min. These results agree well with the evolution of activity changes depicted in Figure 6.11 where burst rates reach a maximum at 35 min followed by a spike rate maximum at 60–100 min. Activity is lost (>90%) at 240 min.

6.3.3 Life Support Stability and Medium Biochemistry

The stability of the native activity and the repeatability of the responses are directly related to the stability of the culture life support provided. Temperature, pH, and osmolarity of the medium must be held as constant as possible. Small changes of pH (7.3–7.5) are tolerated, but may affect activity in a subtle manner that has not yet been quantitatively described. Temperature will directly affect the episode means as well as standard deviation of the minute means. Although any temperature between 35 and 38°C can be used as reference, this value must be kept within ±0.2°C during the course of a compound evaluation. One interesting feature of the networks is that they tolerate exposure to hypothermic conditions at room temperature (20–22°C), even showing in many cases spontaneous activity although substantially reduced from that observed at physiologic temperatures. Like other cells, neurons osmoregulate, but are vulnerable to sudden osmolarity gradients and rapid changes, an effect likely due to rapid osmotic swelling or shrinkage [27]. However, for experiments lasting 8–12 h, exposure to high-osmolarity conditions for experiments involving seawater dilutions have been successfully performed for marine toxin detection at the U.S. Naval Research Laboratory. In addition to these common environmental variables, changes in medium biochemistry associated with long-term culture and recording, spanning days and weeks, represent a still challenging and relatively unexplored domain. Here, accumulation of ammonia, changes in concentration of neuroactive lipids, the presence of serum albumin, and even subtle effects of metal ions can all impact the stability and response reliability of the network sensor. This is the reason why the native medium is often changed to a reference medium, of which a large pool is available, if several medium changes are anticipated. The network responds to the new medium, but reaches a new level of stable activity that determines the reference level for all subsequent pharmacological tests. Figure 6.13 depicts the strong, but reversible effects that aged medium can have on activity. The reduction of activity was attributed to the rise of ammonium ions resulting from breakdown of glutamine (see Table 6.2).

Confirmation of glutamine problems for long-term experiments with slowly aging medium is shown with a dual array, closed-chamber experiment where 10 ml of the same medium was circulated at 50 μl/min (Figure 6.14). Loss of units occurred after 8 days in

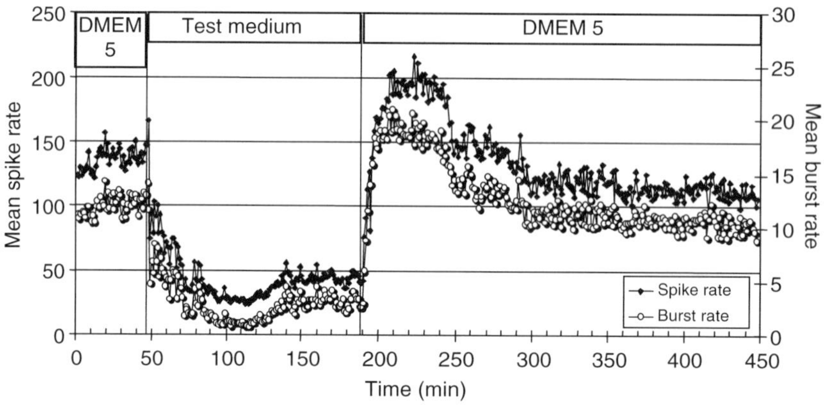

FIGURE 6.13
Activity reduction in mean network spike and burst rates due to a medium change to an aged medium maintained in a closed container for 20 days at 37°C. Recovery to original activity levels was achieved by a return to fresh DMEM-5 medium (Dulbecco's modified Eagles medium with 5% heat-inactivated horse serum).

TABLE 6.2
Ionic and Metabolic Components of Fresh and 20-Day-Old Medium (All Concentrations in mM)[a]

Medium Component	Fresh	20 Days
Glucose	25.3	25.3
Glutamine	3.79	0.15
Glutamate	0.00	0.09
Ammonium	0.37	1.50
Sodium	151.2	150.8
Potassium	5.19	4.99

[a]Measured with Nova Bioprofile 100 Analyzer.

normal medium but was extended beyond 30 days in modified medium with a glutamine dipeptide, which is known to limit glutamine degradation [28].

The long-term activity fluctuations in terms of spike and burst production are shown in the highly time-condensed record of Figure 6.15. A frontal cortex network was used for four morphine applications (between 50 and 250 min) and then left unattended for 72 h in a constant-volume bath medium. Life support was provided by temperature control to 36 ± 0.1°C, 10% CO_2 in humidified air, and 30 µl/h water addition (syringe pump) to compensate for evaporation (empirically determined). The graph is representative of a general slow, gradual decay in activity (43% for bursting in 72 h) and shorter periods of minor fluctuations. The SD of the minute means also increases after the 3000-min mark (arrow). The long-term changes are attributed to alterations in the medium biochemistry, as medium replacement (not shown here) generally raises the activity to original levels, and often above those levels. Short-term fluctuations may be caused by pH changes, as that parameter is difficult to monitor continuously. Even short-term electronic noise cannot be ruled out. Alternatively, it may be generic to spontaneous network activity. Regardless, it is readily apparent that the NNBS approach allows long-term assessment of detailed electrophysiological function that extends far beyond the time window permitted by any other in vitro neuronal methodology based on dissociated cells.

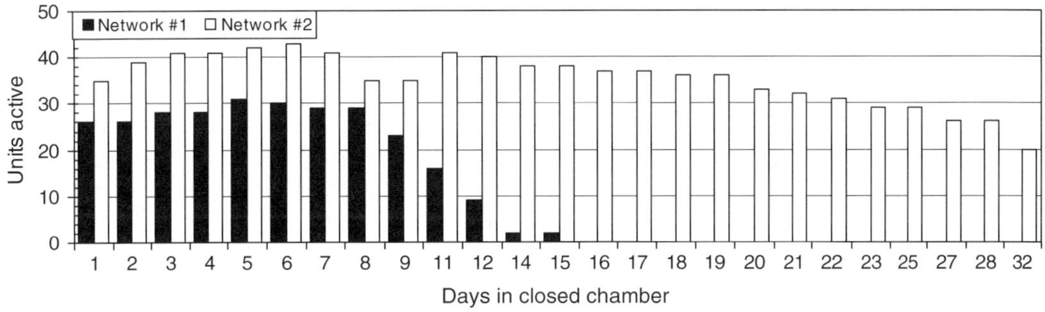

FIGURE 6.14
Long-term twin network experiment with dual recording array. The culture with stable glutamine (network 2) performed better than the one in normal medium. Frontal cortex networks 69 d.i.v. at the beginning of experiment.

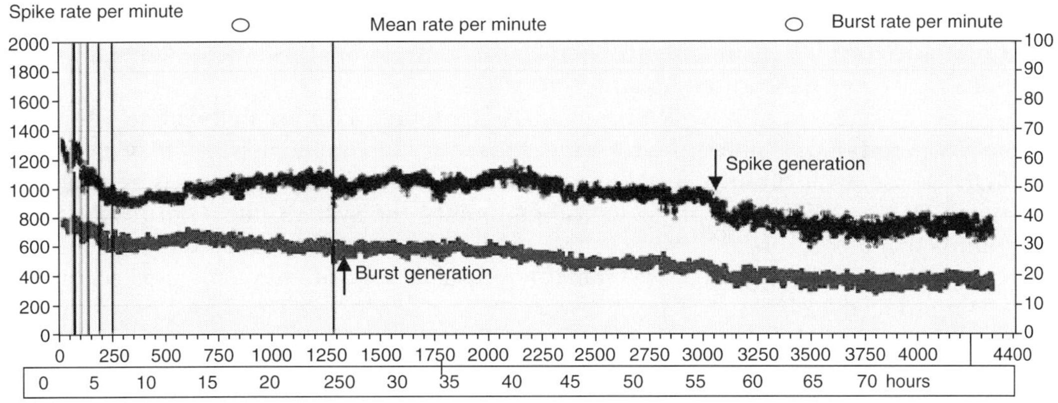

FIGURE 6.15
Continuous 75-h record of frontal cortex culture in a constant-volume bath medium showing spike and burst production averaged over 58 discriminated units per minute. Pharmacological manipulation between 50 and 250 min was followed by 70 h of unattended recording except for an osmolarity check (10 µl removed) at 1250 min (vertical line).

One reason for scrutiny of such an activity record is that it represents the realistic activity domain in which an environmental biosensor is required to operate. Response patterns must exceed both the short-term and long-term fluctuations. A totally level line with almost no minute-to-minute fluctuations cannot be achieved, even in flow chambers. The fact that some tests have to be performed when only small amounts of substances are available is another reason for trying to understand network behavior in a small, constant-volume medium bath. It is clear that compounds producing very slow changes that are also small in magnitude require experimental repetition and stable control cultures.

Although we are only considering here two activity variables, mean spike and burst rates per minute, these metrics capture a great deal of network dynamics and complexity. For example, it is possible for a network to change from spiking into bursting behavior without a major change in the mean spike rate, but it is very difficult to change time-stamp patterns in such a manner that neither the spike rate nor the burst rate is noticeably affected. Other measures that have value in describing changes in network behavior include spike waveforms [9] and measures of network synchronization [29]. With the exception of some recent work [30], efforts with nonlinear dynamical methods have not yielded new significant activity parameters.

6.4 The NNBS as a Broadband Water Toxicity Sensor

With increased concern over the security of water sources and supplies, the capacity for the NNBS to act as a "broadband physiological biosensor" is being explored. In contrast to the existing sensor paradigms, such as immunoassays or molecular-biology-based measures, the NNBS does not require a predetermined molecular signature and responds to any compound, known and unanticipated, capable of affecting general cellular metabolism and organization as well as specific neural functions in mammals. There is a trade-off relative to the types of information generated by the NNBS versus structure-based assays. While both approaches are capable of detection, structure-based assays yield quantitative data regarding concentration within limits of specificity. The strength of the NNBS approach, like other function-based assays, is broad specificity where quantitation of concentration and identification of the analyte are more difficult. Nevertheless, the data derived from the NNBS are rich and it is reasonable to expect that it will be possible to classify compounds based on reproducible functional effects on classes on neuronal receptors, which in some situations may drive the choice of countermeasures.

Unlike olfactory-based cells, which have also been proposed for use in threat detection [31, 32], the NNBS responds in concentration ranges that affect mammalian nervous system function. In that sense, an attribute of the NNBS is that it is not overly sensitive to any and all compounds; in fact, by relying on physiological endpoints, the system filters out responses to compounds that are at concentrations below an effective level. One can envision situations where such physiologically based detection can reduce false positives in threat detection scenarios. A reproducible observation from demonstrations of cell- or tissue-based detection systems, including the NNBS, has been a markedly low number of false-positive events [33].

6.4.1 Detection, Classification, and Identification

The simplest approach to detection is to monitor network spike production averaged across all channels per 30 s or 1 min (minute means). As in all pharmacological and toxicological experiments performed so far, changes in activity are always measured relative to the reference activity of the same network. The reference state is characterized by its stability and variability. Stability refers to a level, linear evolution of minute means in time, preferably with a zero or very shallow slope. This is called the network reference mean and is assigned a standard deviation (SD). Variability refers to the positive and negative excursions of the minute means about the reference mean. A high variability of the minute means yields a large SD and low sensitivity. A steep slope indicates failure of life support and a permanent or temporary loss of the network for sensing. Shallow slopes can be tolerated. A detection event occurs when the mean network activity drops out of the ± 1 SD reference state for a period of several minutes. Effective toxicants affect neural tissues over a long period of time, and a single movement of a minute mean outside the SD box is generally insufficient for reliable detection. This is illustrated in Figure 6.16 with a network response to tetrodotoxin. Transient response decreases appear to be caused by a lack of proper mixing of the toxicant with the medium.

Detection in terms of multiples of the reference SD (often called *z*-score) immediately allows classification into two road categories: inhibition and excitation. If burst analysis is included in the detection variables, then it is also possible to read out "disinhibition," a response to the blocking of inhibitory circuitry that leads to burst coordination and burst oscillation. All compounds that cause epileptiform activity are in the "disinhibitory" category. Figure 6.17 shows a pattern transition to trimethylolpropane phosphate (TMPP) where the spike rate is minimally affected but a major reorganization of spikes into bursts occurs. Hence, although spike rate is a dominant variable, it is necessary but not sufficient for detection.

Figure 6.18 summarizes a simple numbering scheme that can be used for an initial compound identification. It takes into consideration classification into four categories: excitatory, inhibitory, biphasic, and disinhibitory, as well as the speed of the response, the reversibility of the response, and the manner in which the activity dies. Biphasic responses are almost always excitation followed by inhibition whereas disinhibition generates coordinated burst oscillation. Response speed is somewhat qualitative and has been given 5 numbers with "5" being the fastest response such as is seen with tetrodotoxin and "1" the slowest as is seen with botulinum toxin. The characterization of activity death is not trivial as different affected mechanisms create different activity decay profiles. These categories are presently (1) burst period death (BPD), (2) burst duration death (BDD), (3) spike amplitude death (SAD), and cytotoxic death (CTD). This vector is open and will allow addition of other variables or replacement of variables if they are not compatible with automated detection schemes. Using this approach, the inhibitory transmitter GABA, or its analog muscimol, would be identified by the number sequence 1, 5, 1, 1. These molecules are reversibly inhibitory, act very fast, and create a decrease in burst period without appreciably affecting the burst duration. Breakaway bursts, even if the period is as long as minutes, generally look

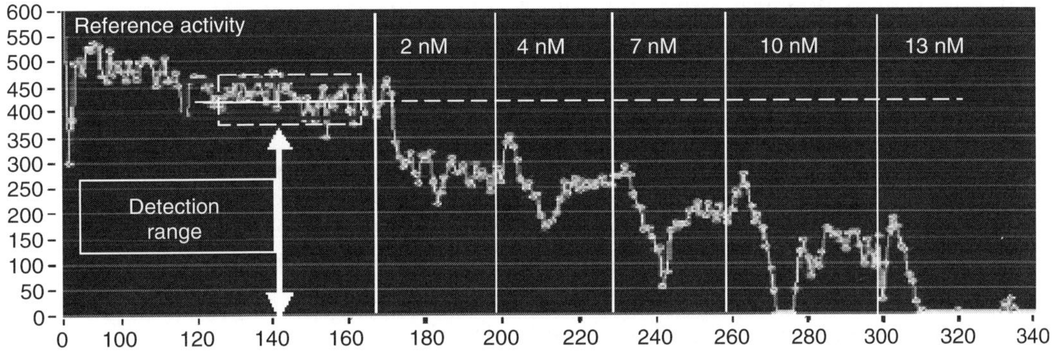

FIGURE 6.16
Response of a frontal cortex culture to tetrodotoxin in terms of network spike production. The dashed horizontal line represents the network reference mean and the dashed box the ±1SD detection reference.

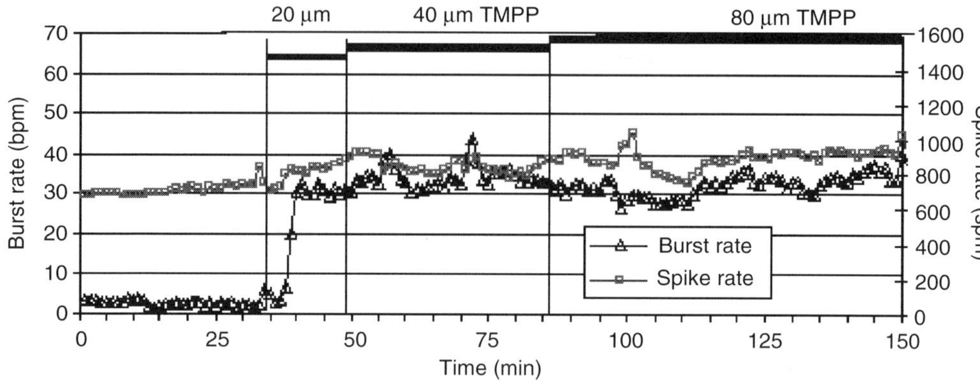

FIGURE 6.17
Organization of activity pattern from spiking with minimal embedded bursting to extensive bursting without a major change in network spike production. TMPP is a bicuculline analog and generates epileptiform network response patterns.

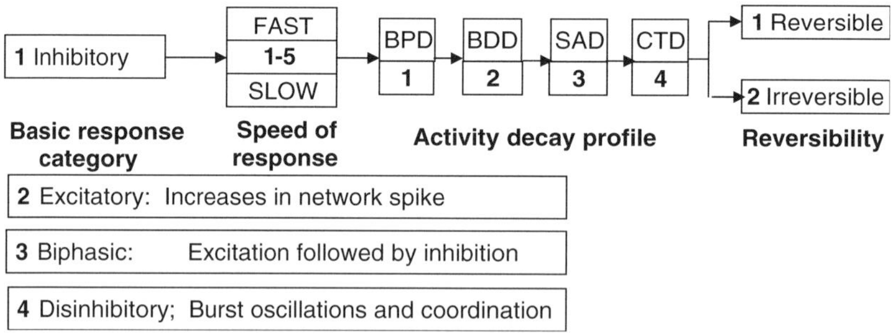

FIGURE 6.18
A number scheme for the classification of response profiles and possible identification of unknown compounds based on a matching of number sequences representing major features of activity changes. For explanation, see text.

normal and also retain some of the cellular burst-shape characteristics. Botulinum toxin would be identified by the sequence 1, 1, (1, 2) 2. Here, both the burst period and duration are affected as the network activity dies. Finally, the $GABA_A$ receptor blocker bicuculline can be identified by 4, 4, 0, 1 as it generates disinhibition, is relatively fast, does not generate burst activity decay, and is reversible. Application of this scheme will undoubtedly generate some new categories and lead to a simple classification system that will allow rapid comparisons of unknown compound responses with a library of established response profiles.

Several other methods may prove useful to achieve quantitative discrimination of classes of compounds based on mechanisms of action, including artificial neural networks, linear discriminant analysis, and self-organizing maps. In fact, artificial neural networks have been used with hippocampal slice recordings [34] and are likely to prove useful for the NNBS. An alternative approach is based on binary decision tree classifiers where qualitative aspects of the NNBS response are used as inputs. Characteristic changes in the response profile can be used to perform classification. This approach has been successful using fish chromatophores, another cell-based biosensor paradigm under consideration for water toxicity testing [35].

6.4.2 Sample Introduction and Resistance to Chlorine

Whether used as a remote detector or as a test platform in a laboratory facility, it is desirable to introduce water samples in the most direct manner to the culture medium without changing osmolarities. However, if a toxicant is present in the sample at low concentrations, a high dilution may make the substance no longer detectable. Concentration steps are time consuming and may interfere with the efficacy of the toxicant. Therefore, the most direct and rapid method is to mix samples with preconcentrated medium. Doubly concentrated medium has already been tested for long-term stability and utility, allowing a 50/50 mixture of medium and test sample. This approach was used to test a series of samples during a function-based assay workshop entitled Eilatox–Oregon, where the NNBS successfully detected a range of environmental samples [7]. A mixture of three parts of aqueous test sample with one part of concentrated medium is also possible, although long-term medium storage has not been explored. With this approach the test sample need only be filtered if microorganisms are to be excluded although short-term tests of acute toxicity can be performed in the presence of microorganism. Clearly, if the microorganisms are the target of the scrutiny, they will grow very well in culture medium and can be explored optically and with a variety of staining methods.

One of the major misconceptions concerning the use of mammalian cells as detection elements is the presumed lack of tolerance to chlorine, which is commonly used in water purification. For water purification, chlorine levels reach 2–4 ppm and must be considered a potential interferent in the analysis of water toxicity. The University of North Texas (UNT) group has thoroughly examined this issue with respect to the NNBS, finding virtually no alteration in the pharmacological responsiveness of the networks in the presence of high levels of chlorine. Figure 6.19 shows the effect of a synaptic disinhibitor, TMPP, in the absence and presence of 10 ppm chlorine, a concentration well above that which is used in water purification. In fact, networks grown and maintained in DMEM (frontal cortex) and MEM (spinal cord) with 5% serum withstand acute additions of up to 20 ppm chlorine without suffering damage. This protection was linked to serum concentrations and specifically to serum albumin [36].

6.4.3 Portable Sensor Units

Efforts to develop a portable cell-based biosensor have been ongoing for nearly a decade [37, 38, 8]. The major interest in this area has been the recognition that current sensor paradigms for environmental threats are incapable of detecting unanticipated threats, raising the need for a broad-spectrum detection capability [6]. The initial design philosophy centered on

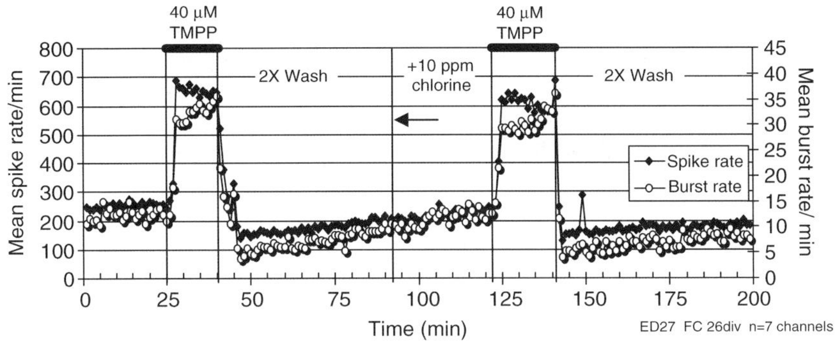

FIGURE 6.19
Stable baseline spike and burst rates increase over twofold upon addition of 40 μM TMPP. Two full medium changes with DMEM-5 restore baseline activity (wash). Treatment with 10 ppm chlorine (trichlor) at 92 min (arrow) does not produce a marked effect, and a subsequent TMPP application under 10-ppm chlorine elicits the stereotypical TMPP response in this culture.

instrumentation: developing portable amplifiers and electrode arrays suitable for extracellular recording with limited demonstrations with spontaneously active cardiac cells. More recently, the NRL has worked closely with the UNT to develop the first portable recording system for cultured neuronal networks as shown in Figure 6.20 [8]. It was through the recognition that the elements of the engineering design had to be driven by the practical issues of handling and maintaining cultured neuronal networks. As a result, this prototype system has been designed specifically to accommodate the UNT neuronal network format in a manner conducive to cell-culture handling, while providing robust control of temperature and flow conditions as well as low noise recording of microvolt-level extracellular potentials. Recording levels are virtually identical to those achieved with laboratory-based systems.

The robustness of the cultured networks led to a concept of use where the biological component could be shipped to end users and treated as modules for the portable system. Moreover, with the use of a standard, the capacity to compare results derived between the laboratory and field settings can be realized. This has been a major issue limiting the adoption of cell-based biosensors for environmental threat detection. No advance in engineering will circumvent the need to build a library of response profiles for characterization of system sensitivities or eventually threat classification.

The other aspects of the portable recording system are driven by the need for simplicity: feedback-temperature-control unit that inserts stainless steel housing into the culture module; fluidic interconnects where media is prewarmed to ensure temperature stability; and a software user interface that allows recording to be initiated immediately without delineation of units. In fact, our experience has been that unit discrimination for environmental threat detection has had only marginal influences on the estimations of EC_{50} values (Figure 6.21) and often has had little impact on threat detection. The user interface displays neuronal network activity at multiple time scales: microsecond resolution to display the digitized waveforms over three user-selected and modifiable channels, per-second resolution where supra-threshold spikes are counted and displayed as a false color

FIGURE 6.20
Portable system for field use tailored to monitoring neuronal extracellular potentials. The system incorporates standardized microelectrode arrays developed by and in use at the University of North Texas. Wherever possible, off-the-shelf components have been utilized for system design and fabrication. Key components of the system include low-noise amplifier and filter boards, a thermal control system with integrated fluidics to minimize perturbations in temperature under flow conditions. Data collection and processing can be accomplished in real time using a personal computer running a graphical user interface for data acquisition and control (not shown).

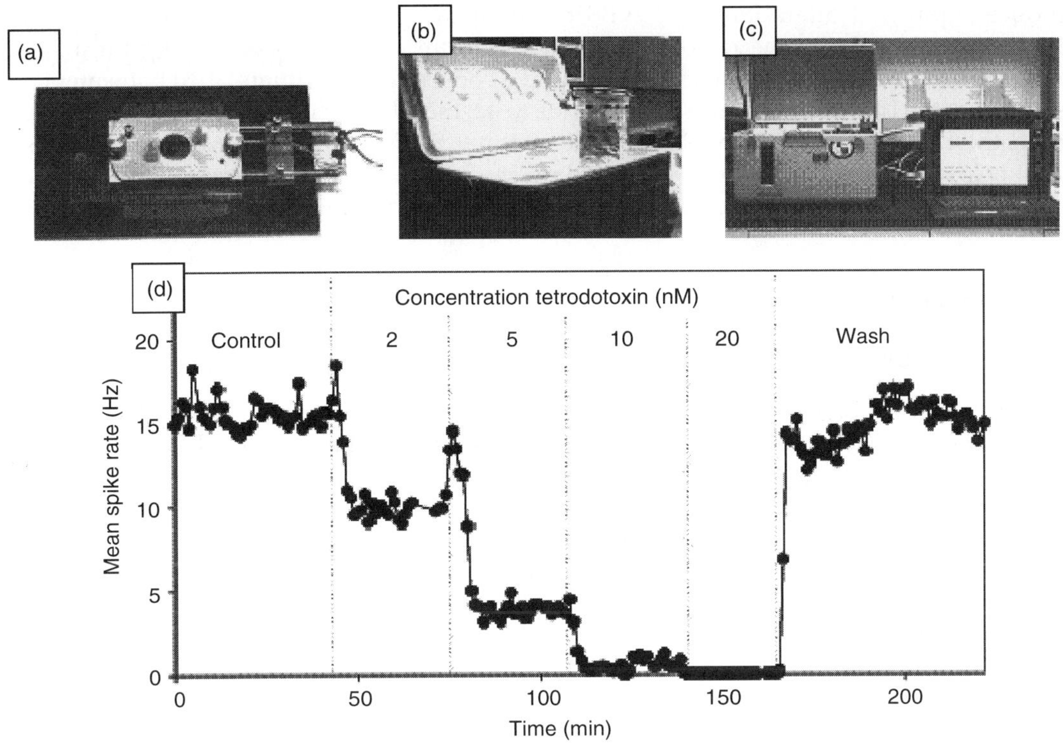

FIGURE 6.21
Concentration-dependent effects of tetrodotoxin for a typical spinal cord network using the NRL portable system after shipping of network. The pharmacological sensitivity to tetrodotoxin for the portable recording system is similar to that achieved with laboratory-based recording systems. (a) Shipping and recording chamber; (b) shipping containers; (c) portable recording assembly; (d) Concentration-response profile of a culture shipped from Texas to Washington, DC.

or "heat" plot, and a per-minute display of mean spike rate across the network. These are indeed simple measures that are not intended to be exhaustive, but are sufficient to allow a user insight into the status of the network. Off-line analysis can then be performed to achieve spike sorting and measure other parameters such as burst dynamics, synchronization, and waveform amplitude.

A typical recording medium used for monitoring NNBS activity in the portable system consists of MEM supplemented with (in mM) 25 glucose, 40 HEPES, and 26 $NaHCO_{3;}$, and is allowed to equilibrate for 4 h in air before adjustment of the pH to 7.4. There is a slow drift toward higher pH levels; however, the networks tolerate this drift very well over the 1-to-2-day time frame. Based on this protocol for aqueous-phase sample introduction under constant flow conditions, minimal variation in mean spike rate is observed, consistent with temporal stability, such that changes of >10% are readily distinguished. With this particular formulation, the NRL group has observed stability in neuronal network function exceeding 2 days of continuous perfusion. As mentioned earlier, the system has been successfully used off-site at a bioassay workshop [7] in Oregon and recently at the National Oceanographic and Atmospheric Administration in South Carolina. During these demonstrations, sensitivity has been shown for a multiplicity of neuroactive compounds including tetrodotoxin, titysustoxin, brevetoxin-2, brevetoxin-3, arsenic, cyanide, cadmium, mercury, strychnine, and the organophosphate phosdrin.

To assess the potential utility of the NNBS approach for detection of threats in aqueous phase, one can compare the concentration-dependent inhibition by various threats to those levels that induce lethality. For example, consider cadmium and strychnine, which induce significant alterations in NNBS behavior at approximately 1 μM and 600 nM, respectively. Applying the rat oral LD_{50} for cadmium chloride of 88 mg/kg for a 70-kg human, a molecular weight of 183 g/mol, and a conservative estimate of 10 l of water consumption, the LD_{50} for cadmium corresponds to 4.8 mM. For similar calculations for strychnine, the LD_{50} corresponds to 190 μM. The fact that the NNBS shows profound responses well below these levels suggests potential utility for function-based environmental threat detection. Similar relationships can be observed for a wide range of toxicants.

6.5 Development of Multinetwork Platforms for High Throughput

A major limitation of present techniques for laboratory-based analyte screening is the low throughput and lack of data analysis automation. Although the CNNS at UNT has recently moved to simultaneous recording from two separate networks (32 electrodes each), this approach is still far from what is needed for efficient applications of network dynamics. Statistical evaluations of interculture repeatability, of multiple drug and multiple concentration applications, and of the temporal evolution of toxic responses, are tedious and far too slow for present industrial and even research utility. The eight-network module under development is an important step to increasing the throughput of this approach (Figure 6.22).

6.6 Summary

There are substantial advantages provided by the NNBS approach: (1) the cultured networks provide both electrophysiological and morphological data; (2) the chemical environment of the networks and the concentrations of the toxicants can be controlled with high precision; (3) the simplicity of the tissue allows repeatable measurements resulting in reproducible EC_{50} and dissociation constants where major shifts from these constants imply changes in receptor densities, properties, or linkages to chemical pathways; (4) large, stable SNR allow waveshape analysis for voltage-gated ion channel toxicology and pharmacology; (5) the robustness of the cultures enables shipping of nonfrozen networks to provide high-quality cell cultures

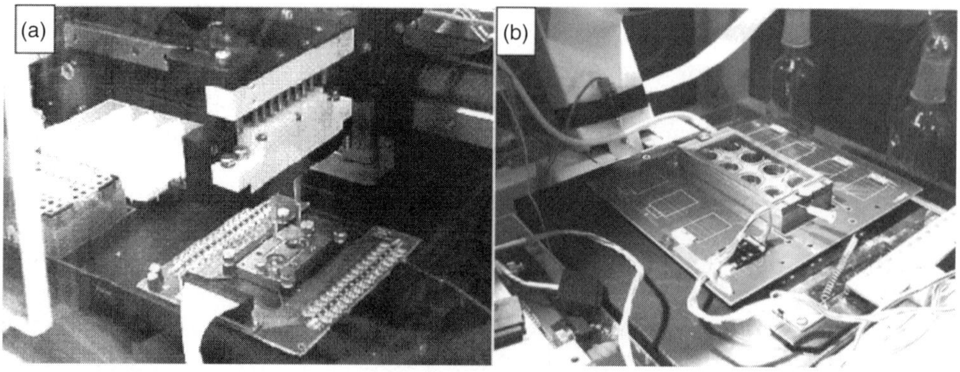

FIGURE 6.22
Emerging system for parallel recording from multiple networks. (a) Two-network open chamber served by a liquid handling robot in an environmental chamber. (b) Eight-network module with experimental preamplifier (256 channels; Plexon Inc., Dallas) on microscope stage.

for end users; (6) noncytotoxic, but functionally toxic compounds are often reversible and allow repeated studies with the same networks; (7) on glass plates with transparent ITO conductors high-power microscopy, access can be achieved making simultaneous electrophysiological, optical, and fluorescent studies possible; (8) one pregnant mouse can seed over 400 MEAs, thus reducing the number of experimental animals, at least, in the compound screening stages; and (9) the networks can serve as the biological component for a portable recording system for environmental threat detection.

Current sensor paradigms for environmental threats are incapable of detecting unanticipated threats, raising the need for a broad-spectrum detection capability. The number of novel chemical compounds that have been or can be synthesized to be toxic to mammals is large and is continually growing. Unknown compounds will most likely not be detected by conventional in-line scanning with existing instruments unless they appear in such large concentrations as to trigger a follow-up testing in animals or cellular systems. As instruments cannot measure toxicity, living tissues must be used for broadband scanning. However, it is difficult to use living tissue in a similar manner as electronic monitoring instruments. In-line water monitoring, a relatively simple task for instruments, becomes a major technical problem when using cell-based systems. However, if it is considered vital to have a broadband monitoring capability, then it is technically possible to route supply water in such a manner that only tested supply water enters a secure central holding tank. Such testing can be accomplished with high confidence using neuronal networks on MEAs. With the development of multinetwork systems, a larger number of tests can be performed in parallel, providing greater reliability, accuracy, and throughput. The eight-network platform described in this chapter takes a step in that direction. The automated use of multiple eight-network modules is anticipated and a further increase of networks per plate is feasible. However, water-testing stations using such networks require a reliable supply of functional living tissue on MEAs. Thus, the first step has been taken with the shipment of living networks to end users. Yet, such a supply service would be greatly simplified by an ability to freeze networks, with network thawing and activation at the water-testing stations. This has not yet been accomplished and will require further research.

Acknowledgments

This research effort was supported by the Defense Advanced Research Projects Agency (DARPA) Tissue-Based Biosensor and Activities Detection Technology Programs, by two grants from the Texas Advanced Technology Program, and by the Charles Bowen Memorial Endowment to the CNNS. The authors recognize the intellectual and funding support by Dr. Alan S. Rudolph in his role as a Program Manager at the DARPA. The opinions and assertions contained herein are the private ones of the authors and are not to be construed as official or reflecting the view of the National Institutes of Health (NIH).

References

1. Stephens, G.J., Bettencourt, L.M.A., and Gross, G.W. (2004) Motifs in the functional connectivity of primary neuronal cultures. Society for Neuroscience Abstract # 420.15.
2. Gopal, K.V. and Gross, G.W. (2004) Unique responses of auditory cortex networks *in vitro* to low concentrations of quinine. *Hearing Res.* 192: 10–22.

3. Gross, G.W. and Gopal, K.V. (2006) Emerging histiotypic properties of cultured neuronal networks. In: Advances in Network Electrophysiology using Microelectrode Arrays (M. Taketani and M. Baudry, eds), Springer, pp. 193–214.
4. Xia, Y., Gopal, K.V., and Gross, G.W. (2003) Differential acute effects of fluoxetine on frontal and auditory cortex networks *in vitro*. *Brain Res.* 973: 151–160.
5. Pancrazio, J.J., Keefer, E.W., Wu Ma, Stenger, D.A., and Gross, G.W. (2001) Neurophysiologic evaluation of chemical agent hydrolysis products on cortical neurons *in vitro*. *NeuroToxicology* 22: 393–400.
6. Stenger, D.A., Gross, G.W., Keefer, E.W., Shaffer, K.M., Andreadis, J.D., Ma, W., and Pancrazio, J.J. (2001) Detection of physiologically active compounds using cell-based biosensors. *Trends Biotechnol.* 19: 304–309.
7. O'Shaughnessy, T.J., Gray, S.A., and Pancrazio, J.J. (2004) Cultured neuronal networks as environmental biosensors. *J. Appl. Toxicol.* 24:379–385.
8. Pancrazio, J.J., Gray, S.A., Shubin, Y.S., Cuttino, D.S., Shaffer, K.M., Eisemann, K., Curran, A., Zim, B., Gross, G.W., and O'Shaughnessy, T.J. (2003) A portable microelectrode array recording system incorporating cultured neuronal networks for neurotoxin detection. *Biosens. Bioelectron.* 18: 1339–1347.
9. Kulagina, N.V., O'shaughnessy, T.J., Ma, W., Ramsdell, J.S., and Pancrazio, J.J. (2004) Pharmacological effects of the marine toxins, brevetoxin and saxitoxin, on murine frontal cortex neuronal networks. *Toxicon* 44:669–676.
10. Gopal, K.V. and Gross, G.W. (1996) Auditory cortical neurons in vitro: Initial pharmacological studies. *Acta Otolaryngologica* 116: 697–704.
11. Keefer, E.W. and Gross, G.W. (2000) Spiking properties of respiratory-related pacemaker cells in vitro. Society for Neuroscience Abstracts V26, Abst. # 347.7.
12. Ransom, B.R., Neale, E.A., Henkart, M., Bullock, P.N., and Nelson, P.G. (1977) Mouse spinal cord in cell culture. *J. Neurophysiol.* 40: 1132–1150.
13. Gross, G.W., Wen, W., and Lin, J. (1985). Transparent ITO patterns for extracellular, multisite recording in neuronal cultures. *J. Neurosci. Meth.* 15: 243–252.
14. Gross, G.W. and Kowalski, J.M. (1991) Experimental and theoretical analyses of random network dynamics. In: Neural Networks, Concepts, Application and Implementations, Vol. 4 (P. Antognetti and V. Milutinovic, eds), Prentice Hall, N.J., pp. 47–110.
15. Gross, G.W. (1994) Internal dynamics of randomized mammalian neuronal networks in culture. In: Enabling Technologies for Cultured Neural Networks (D.A. Stenger and T.M. McKenna, eds), Academic Press, N.Y., pp. 277–317.
16. Xia, Y. and Gross, G.W. (2003) Histiotypic electrophysiological responses of cultured neuronal networks to ethanol. *Alcohol* 30: 167–174.
17. Morefield, S. I., Keefer, E.W., Chapman, K. D., and Gross, G.W. (2000) Drug evaluations using neuronal networks cultured on microelectrode arrays. *Biosens. Bioelectron.* 15: 383–396.
18. Keefer, E.W., Norton, S.J., Boyle, N.A.J., Talesa, V., and Gross, G.W. (2001b) Acute toxicity screening of novel AChE inhibitors using neuronal networks on microelectrode arrays. *NeuroToxicology* 22 (1): 3–12.
19. Oli-Rijal, S. (2005) Determination of dissociation constants for GABAA receptor antagonists using spontaneously active neuronal networks *in vitro*. MS Thesis, University of North Texas, December 2005.
20. Johnston, G.A., Curtis, D.R., De Groat W.C., and Duggan, A.W. (1968) Central actions of ibotenic acid and muscimol. *Biochem. Pharmacol.* 17: 2488–2489.
21. Chambon, J.P., Duggan, A.W., Heaulme, M., Restle, S., Schlichter, R., Biziere, K., Wermuth, C.G. (1985) An arylaminopyridazine derivative of gamma-aminobutyric acid (GABA) is a selective and competitive antagonist at the $GABA_A$ receptor site. *Neurobiology* 82: 1832–1836.
22. Heaulme, M., Chambon, J.P., Leyris, R., Wermuth, C.G., and Biziere, K. (1986) Biochemical characterization of the interaction of three pyridazinyl GABA derivatives with the $GABA_A$ receptors sites. *Brain Res.* 384: 224–231.
23. Arunlakshana, O. and Schild, H.O. (1959) Some quantitative use of drug antagonists. *Br. J. Pharmacol.* 14: 48–58.

24. Hamann, M., Desarmenien, M., Desaulles, E., Bader, M.F., and Feltz, P. (1987) Quantitative evaluation of the properties of a pyridazinyl GABA derivative (Sr 95531) as a GABAA competitive antagonist. An electrophysiological approach. *Brain Res.* 442: 287–296.
25. McCabe, R.T., Wamsley, J.K., Yezuita, J.P., and Olsen, R.W. (1988) A novel $GABA_A$ antagonist [H-3]SR95531—Microscopic analysis of binding in the rat-brain and allosteric modulation by several benzodiazepine and barbiturate receptor ligands. *Synapse* 2: 163–173.
26. Williamson L.C., Fitzgerald, S.C., and Neale, E.A. (1992) Differential effects of tetanus toxin on inhibitory and excitatory neurotransmitter release from mammalian spinal cord cells in culture. *J. Neurochem.* 6: 2148–215.
27. Somjen, G.G. (2002) Ion regulation in the brain: implications for pathophysiology. *Neuroscientist* 8: 254–267.
28. Arii, K., Kai, T., and Kokuba, Y. (1999) Degradation kinetics of L-alanyl-L-glutamine and its derivatives in aqueous solution. *Eur. J. Pharmaceut. Sci.* 7: 107–112.
29. Selinger, J.V., Pancrazio, J.J., and Gross, G.W. (2004) Measuring synchronization in neuronal networks for biosensor applications. *Biosens. Bioelectron.* 19: 675–683.
30. Ignaccolo, M., Grigolini P., and Gross, G.W. (2004) Towards the timely detection of toxicants. *Chaos, Solitons and Fractals* 20: 87–93.
31. Ko, H.J. and Park, T.H. (2005) Piezoelectric olfactory biosensor: Ligand specificity and dose-dependence of an olfactory receptor expressed in a heterologous cell system. *Biosens. Bioelectron.* 20: 1327–1332.
32. Minic, J., Persuy, M.A., Godel, E., Aioun, J., Connerton, I., Salesse, R., and Pajot-Augy, E. (2005) Functional expression of olfactory receptors in yeast and development of a bioassay for odorant screening. *FEBS* 272: 524–537.
33. Pancrazio, J.J., McFadden, P.N., Belkin, S., and Marks, R.S. (2004b) EILATox-Oregon biomonitoring workshop: Summary and observations. *J. Appl. Toxicol.* 24: 317–321.
34. Gholmieh, G., Courellis, S., Fakheri, S., Cheung, E., Marmarelis, V., Baudry, M., Berger, T. (2003) Detection and classification of neurotoxins using a novel short-term plasticity quantification method. *Biosens. Bioelectron.* 18: 1467–1478.
35. Mojovic, L., Dierksen, K.P., Upson, R.H., Caldwell, B.A., Lawrence, J.R., Trempy, J.E., and McFadden, P.N. (2004) Blind and naive classification of toxicity by fish chromatophores. *J. Appl. Toxicol.* 24: 355–361.
36. Dian, E. (2004) Application of Neuronal Network Biosensors for use as Biological Sensors in Water Toxicology and Lipid Signaling. Ph.D. Dissertation, University of North Texas, June 2004.
37. Gray, S.A., Kusel, J.K., Shaffer, K.M., Shubin Y.S., Stenger D.A., and Pancrazio J.J. (2001) Design and demonstration of an automated cell-based biosensor. *Biosens. Bioelectron.* 16: 535–542.
38. Kovacs, G.T.A. (2003) Electronic sensors with living cellular components. *Proc. IEEE* 919: 15–929.
39. Beaumont, K., Chilton, W.S., Yamamura, H.I., and Enna, S.J. (1978) Muscimol binding in rat brain association with synaptic GABA receptors. *Brain Res.* 148(1): 153–162.
40. Keefer, E.W., Gramowski, A., and Gross, G.W. (2001a) NMDA receptor dependent periodic oscillations in cultured spinal cord networks. *J. Neurophysiol.* 86: 3030–3042.
41. Little, H.J. (1999) The contribution of electrophysiology to knowledge of the acute and chronic effects of ethanol. *Pharmacol. Therapeutics* 84, 333–353.
42. Marom, S. and Shafaf, G. (2002) Development, learning and memory in large random networks of cortical neurons: Lessons beyond anatomy. *Q. Rev. Biophys.* 35: 63–87.
43. Pancrazio, J.J., Kulagina, N.V., Shaffer, K.M., Gray, S.A., and O'Shaughnessy, T.J. (2004a) Sensitivity of the neuronal network biosensor to environmental threats. *J. Toxicol. & Env. Health A* 67: 809–818.

Part II

Material Design and Selection I

This section focuses specifically on a discussion of materials that form the basis of the biological sensing layer in biosensors. In Chapter 7, Amarjeet Bassi reviews some examples of approaches that combine either the "passive" materials or stimuli-responsive materials within the biological sensing component. This chapter discusses the application of such materials that make up the sensing layer in biosensors and presents the state of the art in these areas. First some recent studies on cantilever biosensors are discussed. Then other types of "smart" materials including dendrimers, conducting films and gels, and stimuli-responsive materials are considered. Several techniques have been utilized in biosensor design for forming the biological sensing layer. The deposition of single layers on the biosensor surface offers the opportunity of careful control and design of surface properties. In Chapter 8, Professor Shin-ichiro Suye and Haitao Zheng describe the layer-by-layer (LBL) adsorption technique based on electrostatic force between polyelectrolytes and proteins. This method can produce thin multicomponent layers by alternate adsorption in cationic and anionic polyelectrolytes. The characterization of the layers using QCM and AFM is described. In Chapter 9, Professor Claudio Nicolini, Manuela Adami, and Christina Paternolli have summarized the present state of the art on nanostructuring of sensing organic matrices and their applications in biosensor design for health and environment. Several applications are discussed such as Potentiometric Stripping Analyzer for metal detection using neural networks, nanotube biosensors for HCl detection, cytochromes on nanoparticles or as thin films, bacteriorhodopsin thin films for anesthetic detection, and nanocomposites for gas sensing. In Chapter 10, Professor Shekhar Bhansali and others discuss the concept of a bioengineered interface that interacts directly with skin layers using microfluidics-based sensing systems. There is sufficient background on the structure of skin and its layers. The design of MEMS devices that interface with the skin is discussed next. Techniques for information gathering such as electrochemical microelectrodes are covered. Several issues involved in microelectrode design, testing, and implementation are covered. The microfluidic system contains sensor arrays integrated with local computing and communication systems. Thus, this "smart" biosensor technology can provide real-time information on a minimally invasive platform.

7

Material Design and Selection for Smart Biosensors

Amarjeet S. Bassi

CONTENTS

7.1 Introduction

A biosensor consists of a biological detection or sensing layer in close proximity to a signal transducer. The sensing layer contains biomolecules immobilized on the biosensor surface. Biosensors utilize biomolecules with specific recognition and binding for the analyte of interest. The interaction of the analyte with the sensing layer produces a signal, which is converted by the transducer into a measurable response. The biomolecules can be incorporated into the sensing layer in a biosensor in different ways. The key design goals in constructing a biosensor are rapid response, sensitivity, selectivity, and long-term stability or reuse. These criteria can be met by patterning the biosensor surface with a high concentration of appropriate biomolecules (enzymes, antibodies, or other types) in such a way which minimizes mass-transfer limitations or partitioning effects and enhances stability of the biomolecules. The surface should also selectively interact with the environment to filter out interferences and nonspecific molecules from the analyte of interest. One approach in such design is to improve the performance of the sensing layer using "smart" materials.

Biosensors can incorporate two types of "smart" materials. These are: (a) "passive" materials which have unusual properties that can enhance significantly the functional properties of the sensing layer such as providing faster response and stabilization of the biological components and (b) stimuli-responsive materials (StRMs) which themselves respond to changes in the environment and in some cases may act as switches to turn the biosensor response on or off. The ultimate goal may be to achieve a level of biomimetics that approaches the sensing capabilities of living cells. A classification of the materials used for biosensor design is presented in Table 7.1. Figure 7.1 shows some examples of these materials, which include conducting polymers, smart fabrics and textiles, dendrimers, carbon nanotubes, and pH-, temperature- or photo-responsive polymers or hydrogels. Other materials that can be potentially utilized or are being reported for novel

TABLE 7.1

Classification of "Smart" Materials in Biosensor Design

Material Type	Examples	Advantages
Passive Materials	*Inorganic Materials*	
	Conducting organic and polymeric materials	High electrical conductivity Thin films
	Dendrimers	Custom designed with specific properties Star-shaped or many other configuration
	Ferromagnetic particles	Can be applied in label-free detection
Stimuli Responsive Materials	Shape-sensitive alloys	High-temperature applications Long term stability and robustness
	PH-sensitive polymers	Operate under mild conditions
	Temperature-sensitive polymers	Versatile
	Photosensitive polymers	Easy processing and incorporation
	Rheological fluids	High-pressure applications

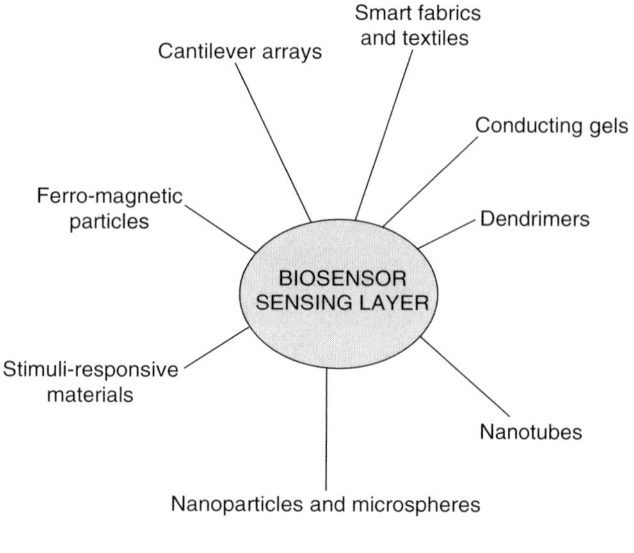

FIGURE 7.1
Examples of "smart" materials and technologies in biosensor design.

biosensor designs include shape-sensitive alloys, rheological fluids, magnetic particles, or technologies such as cantilever arrays.

Several approaches are underway which combine either the "passive" materials (Table 7.1) or StRMs with the biological sensing component. This chapter discusses the application of those materials that make up the sensing layer in biosensors, and presents the state of the art in these areas. First some recent studies on cantilever biosensors are discussed. Then other types of "smart" materials including dendrimers, conducting films and gels, and StRMs are considered.

7.1.1 Cantilever Arrays

Microcantilevers have been used in the atomic force microscope. Their application in microbiosensors represents an example of "smart" biosensor technology. These sensors are arrays of microfabricated flat rods (1 μm thick and less than 100 μm), which can be coated or patterned with biological molecules (e.g., single-stranded DNA, antibodies, or single-chain antibody fragments) for detection and binding with specific analytes of interest. These devices are now commercially available. Ziegler (1) has reviewed the theory, design, and future prospects of cantilever sensors. Figure 7.2 is a schematic representation of the principle of cantilever biosensor.

Cantilever biosensors have the advantages of local, high resolution of the response signal and label-free molecular recognition measurements on a portable device. Also, the very small size of the sensor can allow for sensitive response and minute levels of reagent consumption. When a biomolecular interaction such as ligand–analyte binding (antibody–antigen or enzyme–substrate complex) is combined with a microcantilever operation, a highly sensitive biosensor signal is generated. The microcantilevers are usually placed in piezoelectric cantilever holders and vibrated using external AC voltage. Microcantilevers

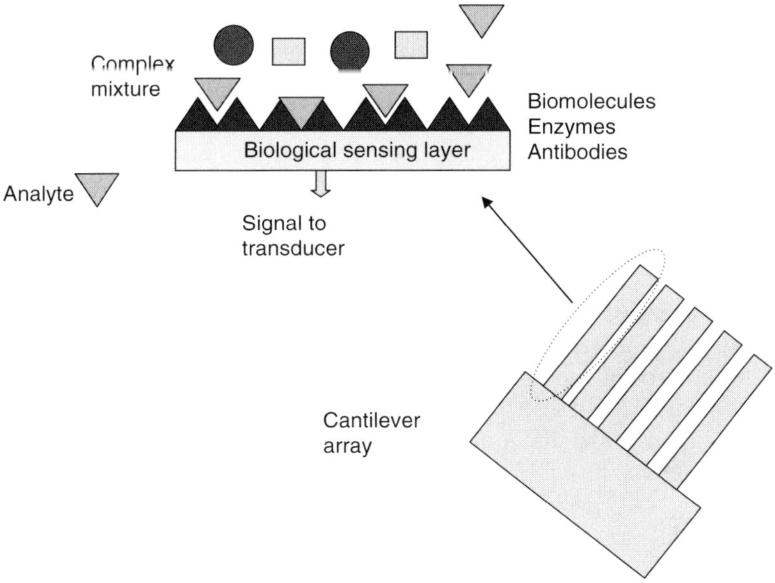

FIGURE 7.2
Schematic representation of a cantilever array biosensor. Microfabricated silicon (flat) rods are coated with appropriate biomolecule or molecules. Specific interactions are picked up due to ligand–analyte binding resulting in specific response.

can also undergo mechanical bending due to a differential surface stress caused by adsorption. The signals can be detected using optical detection with diode lasers. Thus, there are specific biosensor signals the result of which can be a shift in resonance frequency, deflection due to mass loading, or damping effects. Microcantilevers are very sensitive and can detect forces in the pico-Newton range and displacements in the Angstrom range. Thus, they can also be employed to detect force, pressure, or viscosity. Specific metal-coated cantilevers are also very sensitive to temperature changes. Microcantilevers can also operate in a liquid medium making them practical for biosensor design.

The resonance frequency of a microcantilever rod is shown by Equation 7.1.

$$\nu = \frac{1}{2\pi}\sqrt{\frac{\kappa}{M}} \tag{7.1}$$

Where υ is the resonance frequency, κ is the spring constant of the oscillating cantilever, and M is the effective mass of the cantilever. For commercial silicon nitride cantilevers for example, the spring constant κ can be 0.03–0.06 N/m and M can be (0.14–0.18) * mass of the individual rod or beam. A second method used is mechanical displacement, and Equation 7.2 below describes the relationship between mechanical displacement z and differential surface stress δ_s.

$$z = \frac{3L^2(1-\lambda)}{Yt^2}\delta_s \tag{7.2}$$

where z is the displacement, L and t are the length and thickness of the cantilever, λ is the Poisson's ratio, and Y is the Young's modulus.

Recently, Calleja et al. (2) have reported on the fabrication of cantilever arrays for biosensors using materials other than silicon. The cantilevers were made by spin coating a photosensitive polymer that had a Young's modulus much lower than that of silicon. The spring constant, resonant frequency, and other mechanical properties of the cantilevers were investigated as a function of the dimensions and the medium. The devices were then applied for the adsorption of ss-DNA. It was demonstrated by the authors that this cantilever biosensor had six times the sensitivity of commercial silicon nitride cantilevers.

Nugaeva et al. (3) demonstrated the application of micromechanical cantilever arrays for the capture and detection of fungal spores. The authors utilized either microfabricated uncoated or gold-coated silicon cantilevers. These were functionalized with concanavalin A (Con A), fibronectin, or immunoglobulin G. The proteins were utilized to bind to the spores of either *Aspergillus niger* or *Saccharomyces cerevisiae*. The authors found that binding resulted in shifts in resonance frequencies of the cantilever arrays. The immunoglobulin G-functionalized cantilever surfaces produced the best response, and very low levels of spores could be detected in several hours compared to days with conventional approaches. The proposed techniques may have potential applications as mold or odor sensors or in medical and agricultural diagnostics, food- and water-quality monitoring.

Recently Shu et al. (4) describe an interesting mechanical motion-based biosensor using a DNA molecular motor. The DNA motors were integrated with an array of microfabricated silicon cantilevers. The forces exerted by the precise conformational changes were probed via differential measurements using an in situ reference cantilever coated with a nonspecific sequence of DNA. Both the direction and amplitude of motion of the cantilevers were a function of buffer pH and ionic strength. The authors showed that changing the pH in a controlled fashion produced compressive stress repeatedly and the process was reversible. Thus, their device converts biochemical energy from conformational changes into micromechanical work.

7.1.2 Nanotubes

There is a great deal of interest in both multiwalled and single-walled carbon nanotubes (CNTs) for biosensors. Figure 7.3 illustrates the structure of carbon nanotubes.

The main advantages of CNTs are the high surface area to volume ratio, very high electrical conductivity and mechanical strength. The high surface area allows the possibility of increased biomolecular packing density on the biosensor surface. Only two recent examples are discussed here as this aspect is dealt with in greater detail elsewhere in the text. Li et al. (5) have described multiwalled carbon nanotubes (MWCNT) containing covalently linked glucose oxidase (GOD) with high-function density for use as a biosensing interface. The GOD–MWCNT composites were found to be highly water soluble. Electrochemical characterization of the GOD–MWCNT composites that were modified on a glassy carbon electrode shows that the covalently linked GOD retained its bioactivity for the oxidation of glucose. The biosensor response was linear up to 40 mM and had a minimum detection limit of 30 μM. Huang et al. (6) describe an alternative technique to functionalize CNTs for biosensor applications. They utilized electroactive and photoactive thionine for cross-linking to self-assembled multilayers containing CNTs via the alternate layer deposition technique. The resulting multilayer system was found to electrocatalyze the oxidation of NADH.

Gold surfaces have also been extensively applied in biosensor design such as for surface assembled monolayers. Recently the production of gold nanotubes has been reported. Barreca et al. (7) have produced hollow 1-D gold nanostructures with controlled morphology by RF-sputtering of gold into porous polycarbonate or polyester scaffolds at low temperatures. The scaffolds were removed by a postsynthesis membrane etching by oxygen plasmas or in solution. This allowed the preparation of freestanding Au nanotubes, which maintained the original morphology.

Nanostructured titanium dioxide (TiO_2) nanotubes have been reported, which can be prepared by anodizing titanium in a hydrofluoric acid aqueous solution (8). The materials are nontoxic and have good biocompatibility. In the recent study (8), TiO_2 nanotubes were

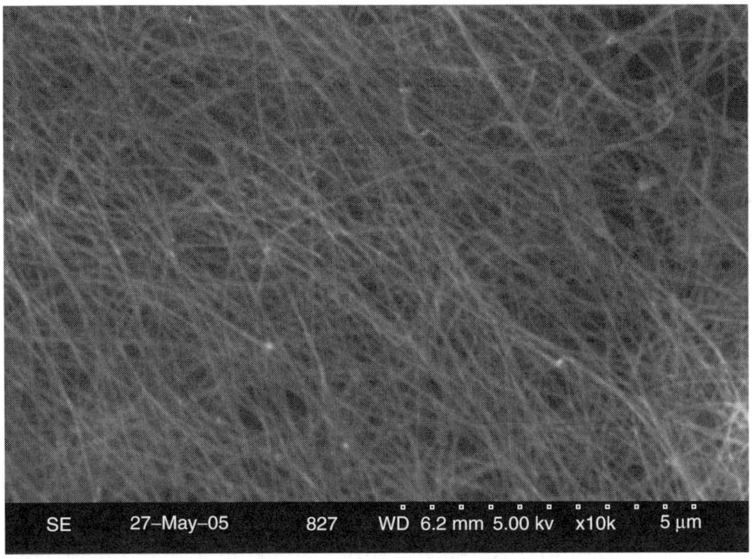

FIGURE 7.3
Illustration of carbon-nanotubes prepared by the chemical vapor deposition method.

prepared by anodic oxidation of titanium as follows: Titanium foil (0.80 mm thick) was ultrasonically cleaned in distilled water and in acetone for 20 min. Next, after rinsing thoroughly the clean titanium foil electrodes (3 × 10 mm, geometrical area ca. 0.6 cm^2) was first etched in an 18% HCl solution at 85°C for 10 min. The etched titanium foil was washed again and then subjected to anodization in a 1:8 acetic acid–water solution containing 1.0 vol% hydrofluoric acid at 20 V for 45 min. This resulted in TiO_2 nanotube arrays on the Ti substrate. The nanotubes were then applied to adsorb peroxidase enzyme. Immobilized thionine on the surface was found to be electrochemically reduced. This study demonstrates a method for biosensor design using a material with high surface area and low electrical conductivity.

7.1.3 Ferromagnetic Particles

Recently magneto-electronics has shown great potential for biosensor design (9). This approach is based on the interaction of magnetically labeled biomolecules with a magnetic field sensor. These are termed as magneto resistive sensors. Magnetic nanoparticles coated with biochemical ligands are being applied in such sensors for numerous biological and medical applications. In particular biomagnetic biosensors can provide simple design and rapid analysis. One of the approaches applied for magnetic bead labeled detection is based on the principle of giant magneto-resistance (GMR) (10,11). When an external magnetic field is applied to layers of alternating magnetic and nonmagnetic thin films (e.g., Fe and Cr), this leads to a decrease in resistance of the thin films. This effect known as GMR occurs due to the scattering of the electrons in the layers, which is influenced by the orientation of the electron spin with respect to the magnetization direction of the magnetic layers. Application of the external applied field aligns the magnets of all the layers in one direction, and thus the scattering for electrons of one of the spins is reduced leading to the reduction in electrical resistance. This detection principle has advantages over established methods, and the sensitivity is promising. Several challenges that need to be met include the design of suitable magnetic nanoparticles and their functionalization. Sensitive GMR magnetic field sensors can be combined with suitable detection protocols such as antibody- or antigen-labeled magnetic immunosensors.

In a separate approach for biosensor design magnetic beads have been utilized for the immobilization of biomolecules. Recently, a phenol biosensor was developed based on the immobilization of tyrosinase on the surface of modified magnetic $MgFe_2O_4$ nanoparticles (12). The tyrosinase was first covalently immobilized to core-shell ($MgFe_2O_4$–SiO_2) magnetic nanoparticles, which were modified with amino group on its surface. The resulting magnetic bionanoparticles were attached to the surface of carbon paste electrode with the help of a permanent magnet. The immobilization matrix provided a good microenvironment for the retention of the bioactivity of tyrosinase. Phenol was determined by the direct reduction of biocatalytically generated quinone species at −150 mV versus SCE. The resulting phenol biosensor had high tyrosinase loading of the immobilization matrix.

Chung et al. (13) have demonstrated another method utilizing a substrateo-free biomagnetic sensing, which depends on the magnetic susceptibility of ferromagnetic particles suspended in a liquid. The binding of the particles to the intended analyte modifies the magnetic relaxation of these particles. This scheme has several advantages: (i) it requires only one binding event; (ii) there is an inherent check of integrity; and (iii) the signal contains additional information about the size of the target analyte.

In another recent report, magnetite (Fe_3O_4) nanoparticles (average size 12 nm) modified with electroactive Prussian blue (PB) were synthesized (14). The magnetic properties of the sample were investigated by low-field alternating current susceptibility and superconducting quantum interference device measurement; the results indicated super paramagnetic properties at room temperature. The PB-modified nanoparticles were then immobilized on the surface of glassy carbon electrode and applied to construct a sensor. The sensor has potential to be applied as a mediator-less biosensor for the redox enzyme-based reactions. The above studies indicate the versatile applications of magnetic sensing. Thus magnetic biosensors are rapidly coming on-stream.

7.1.4 Smart Polymer Composites, Fabrics, and Textiles

A recent interesting area of biosensor design is the utilization of flexible substrates such as plastic film, which can lead to shape-selective or wearable biosensors (15). In particular, the combination of polymers with carbon nanotubes has been applied to form composites with interesting properties. Reinforced epoxy composites have been fabricated with multiwalled nanotubes (16). Fangming et al. (17) used a coagulation method to induce better dispersion of single-walled nanotubes (SWNTs) in a polymer matrix to produce SWNT/poly(methyl methacrylate) PMMA composites. These composites had superior elastic modulus, electrical conductivity, and thermal stability. Chang et al. (18) used in situ polymerization technique to fabricate nanotube-polypyrrole nanocomposites with better conductivity than polypyrrole alone. These were made by using ultrasonic energy to disperse the nanotubes in the solvent and incorporate them into composites. In another study (19) nanotube-based composites were fabricated using an electrospinning technique. The addition of CNTs led to an improvement in strain-sensing capabilities of the sensors.

Smart textiles can be produced by in situ polymerization of the conducting polymers such as polypyrrole or polyaniline in the presence of the textile. The simultaneous polymerization and deposition of the polymer on the fabric avoid the need for further processing. The resulting fabrics are conductive and they have found applications in medical monitoring. Brady et al. (15) report on the development and characterization of polypyrrole–polyurethane foam-based material. Sawicka et al. (20) have prepared electrospun biocomposite nanofibers and applied them in a biosensor for urea. Nanocomposite fibers of urease and polyvinylpyrrolidone were first prepared by the electrospinning technique. Fast response and sensitivity to low concentrations of urea and a versatile design were the key advantages noted.

7.1.5 Conducting Films and Gels

Conducting materials such as polyaniline, polythiophene, and polypyrrole have been previously quite frequently applied in biosensor design and investigations. The main advantages of these materials are: (i) they possess the electrical conductivity similar to metals and the mechanical properties of polymers; (ii) they form thin films that can be applied to entrap biomolecules while minimizing the mass transfer resistance, thus leading to very small response times; and (iii) electropolymerization can be applied to form the films. Thus, these materials can easily be applied in the design of electrochemical biosensors.

New advances in conducting film applications have been reported including the preparation of hybrid 3D structures. For instance, Ionescu et al. (21) report on the preparation of novel pyrrole-alginate gels. In situ electrochemical polymerization of the linked pyrrole groups was carried out at 0.93 V, which led to the formation of a composite polypyrrole-gel matrix. This gel was found to be superior in terms of enzyme retention as well as increased alginate stability in the presence of phosphate anions unlike the natural alginate gel. Thus, such conducting gels can be applied for enzyme retention and electrochemical reactions. Biosensors based on GOD-based composite alginate were also developed by these authors.

Li et al. (22) describe the syntheses and electropolymerization of a series of new 3-alkylthiophene derivatives functionalized with the maleimide, azide, and anhydride. Despite the nucleophilic or electrophilic nature of the reactive groups, the synthesized thiophene monomers exhibit rather good polymerizability, and the reactive groups withstand the polymerization conditions and are correctly incorporated into the resulting electroactive polymers. The reactivity of the pendant reactive groups of the resulting polymers to attach biomolecules was also investigated.

Syritski et al. (23) have described a technique to monitor ion transport in electrochemically synthesized inherently conducting polymer poly(3,4-ethylenedioxythiophene) (PEDOT) induced by the redox switching of the polymer. Mass changes were monitored by an electrochemical quartz crystal microbalance and scanning electrochemical microscopy. Direct evidence of the exchange capability anions entrapped in the electropolymerized with Cl^- anion from the solution was found.

There have been several reports on new conducting films or mediators in the literature. Ngamna et al. (24) have studied the properties of poly(2-methoxyaniline-5-sulfonic acid) (PMAS), a water-soluble polymer, as a potential mediator for biosensors. Poly(L-Lysine) (PLL) was added to the PMAS to make it water insoluble. Uniform and stable films were fabricated on ITO-coated Mylar using 0.1% (w/v) PMAS and 0.05% (w/v) PLL. Immobilization of enzyme horseradish peroxidase (HRP) onto the PMAS/PLL film was also investigated. Amperometric responses were observed upon the addition of hydrogen peroxide. HRP-immobilized concentration was optimized, and a linear response range between 0.01 and 0.1 mM H_2O_2 was observed.

Similarly Kros et al. (25) have produced electroactive membranes and investigated the stability of PEDOT and polypyrrole as the electroactive component in amperometric biosensors. PEDOT and polypyrrole were first synthesized in the pores of a track-etch membrane by immersing the membrane into a solution of the corresponding monomer and ferric chloride. Amperometric and resistance measurements, FT-IR spectroscopy, ESR spectroscopy, and scanning electron microscopy were applied in the investigations. The results indicated that polypyrrole was unsuitable as an electroactive component. However, PEDOT was found to be superior for long-term applications in these novel electroactive membranes.

Interpenetrating networks (IPNs) of polypyrroles/hydroxyethylmethacrylate hydrogels have recently been prepared for enzyme entrapment in amperometric biosensor applications (26). The amperometric responses for hydrogen peroxide oxidation were evaluated. Measurements were made of the amperometric responses via H_2O_2 oxidation for each biosensor. The biosensors were found to be more stable and sensitive compared to the conventional approach. Conducting polymers films and 3D gels will continue to play an important role in biosensor design configurations in the future.

7.1.6 Dendrimers

Dendrimers are 3D tree-like macromolecules. Figure 7.4 is a cartoon representation of a dendrimeric structure. In a dendrimer, the branches are interlinked polymerized chains of molecules, each of which generates new chains, all of which congregate to a core. Each radial concentric region that is formed around the core is called a generation.

Since dendrimers are chemically synthesized, each structure can be tailored or tuned for specific applications. The advantages offered are structural homogeneity and porosity and a controllable composition with multiple homogeneous chain ends.

Modification or selection of appropriate chain ends can lead to specific properties such as electric charge. The entire dendrimer can function as a polyelectrolyte. Control of the

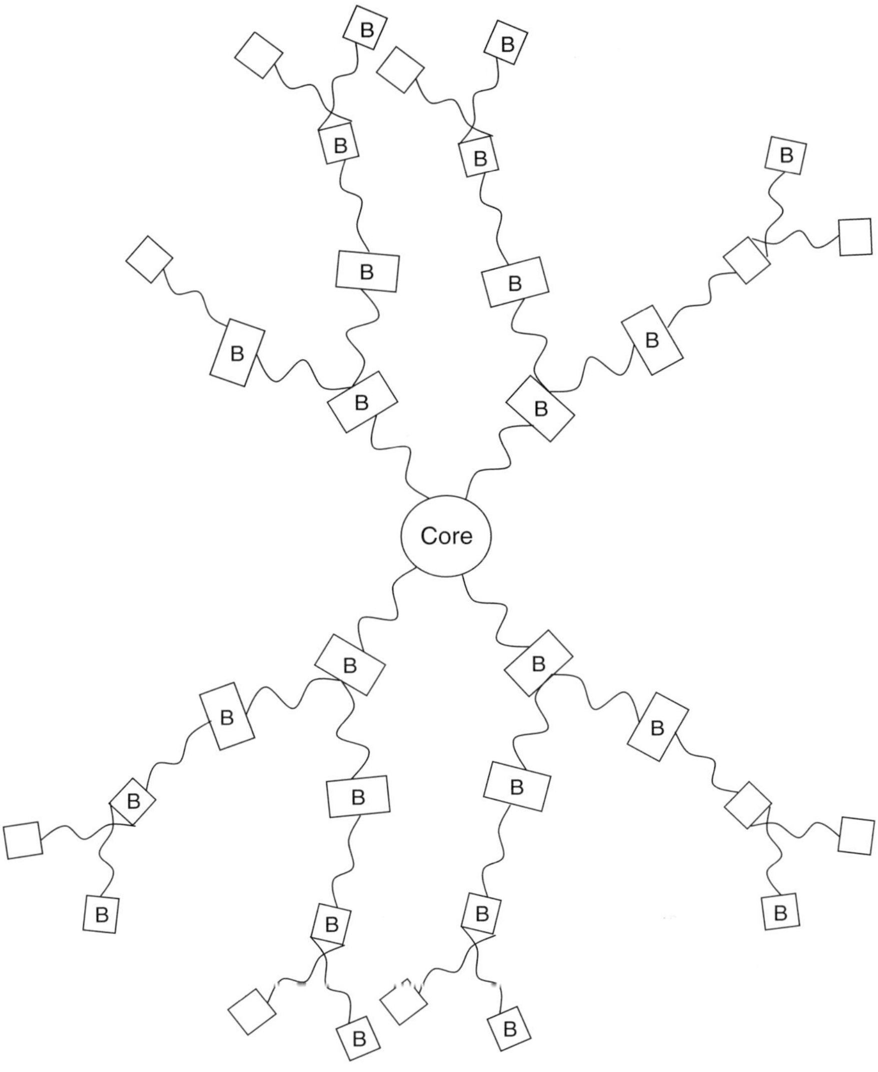

FIGURE 7.4
Cartoon representation of a dendrimeric star-shaped structure. Repeating blocks B emanate from a core to form the structure. Only one type of representation is shown, and other structures are possible.

external size and internal architecture can lead to the design of interior cavities or channels with properties different from those on the exterior surface. This enables the possibility of dendrimers to enclose guest molecules for targeted drug delivery or gene therapy. Thus, these highly branched dendritic macromolecules are being applied to solubilize drugs, in DNA biosensors, and in the delivery of targeted oligonucleotide sequences.

Constable et al. (27) have reported on dendrimers containing a metallic core of ruthenium, iron, or cobalt. These macromolecules open possible applications in electrochemical biosensing. Wendland and Zimmerman (28) produced dendrimers with hollow cores. Dendrimers with hydrophobic or hydrophilic interiors have also been reported [Maciejewski (29), Newkome et al. (30)].

Yoon and Kim (31) and others have reported on the synthesis and electrochemistry of several families of silicon- and nitrogen-based dendritic molecules modified with organometallic moieties as surface functional groups (32–34). The synthesis of poly(propyleneimine) dendrimers containing both ferrocene and cobaltocenium units (33). Such dendritic macromolecules containing a controlled number of redox-active organometallic units at the core, within the branches or at the periphery of the dendritic structure can also be utilized for many practical applications such as catalysts, electron transfer, ion sensors, or in electronics devices. Ashton et al. (35) also describe a new family of globular-shaped fourth-generation dendrimers constructed with 5-hydroxyisophthalic acid and diethanolamine. Both the convergent and divergent methods were used in the synthesis.

Recently, dendrimers have been adsorbed onto gold substrates (36). These layers revealed a high mechanical stability and can be functionalized without any loss of dendrimers from the surface. The large area of dendrimers allows an increase in the number of immobilized biomolecules, which can lead to greater sensitivities. Multiarrayed enzyme films composed of a GOD) and of fourth-generation dendrimers have been reported (G4) (37). Further improvement in the sensitivity of this biosensor has been obtained by modification of the dendrimer molecules with ferrocene carboxaldehyde (37). The biosensors showed good stability, and a 32% modification of the surface amines on the dendrimers to ferrocenyls was found to be optimum. Yang et al. (38) report on a novel route to prepare bioreactive surfaces on gold by the self-assembly of third-generation hydroxyl-terminated dendron thiols (G3-OH) and subsequent bridging reactions using generation two amine-terminated dendrimers (G2-NH_2). It has been shown that G3-OH dendron thiols form a stable and uniform self-assembled monolayer on gold and have potential applications for DNA microarrays and protein chips.

Chang et al. (39) describe the detection of live *Pseudomonas aeruginosa* using a sensing film containing a fourth-generation hydroxy-terminated polyamidoamine (PAMAM) dendrimer (i.e., G4-OH) and SYTOX Green fluorescent nucleic acid stain. The films were prepared on disposable plastic and placed on optical fibers. In the presence of PAMAM-OH (G4-OH) in water, the bacterial cell becomes permeable to the SYTOX dye and the fluorescence is significantly enhanced. The fluorescence increased with the bacteria concentration and the intensity was greater compared to controls. The dendrimers stabilized the sensing film. After drying and desiccation, the SYTOX Green/PAMALM-OH films were still able to quantitatively detect *P. aeruginosa* in water. It is evident that dendrimers offer a great potential for the design of "smart" biosensing layers and this area continues to be of great interest for biosensors.

7.1.7 Nanoparticles and Microspheres

It is interesting to note that silicon technology has revolutionized the concepts of intelligent or smart sensing, since the sensing technology and detection circuits can be incorporated on single chips. Sensing technologies based on radiation or light (40), mechanical (41), thermal signals (42,43), magnetic signals (44), or chemical signal detection (45) have been previously reported. Silicon-based MOSFET devices have been applied for chemical detection and as humidity sensors (46). Recently, Nather et al. (47) have described pH- and penicillin-sensitive electrolyte–insulator–semiconductor sensors ($At/Si/SiO_2/Ta_2O_5$ and penicillinase), which were integrated into a commercial flow-injection analysis system. The FIA set-up contained miniature flow-through cells on silicon wafers. High throughput, minimum dispersion, fluid consumption, and sensitivity and reproducibility of response were reported by the authors. One disadvantage of using silicon-based integrated systems in biosensors is the corrosive environment they may have to operate in leading to poor lifetimes. Several challenges that still need to be overcome in the design of smart integrated biosensors include improving the signal-to-noise ratio;

nonlinearity of response; cross-sensitivity; and off-set and signal drift. More efforts need to be devoted to these aspects and integrated silicon-based biochip design is an area of active and ongoing research. Alternative approaches such as nanoparticle technologies are therefore being considered. Nanoparticles have been prepared using not only silicon, but also gold, carbon, and ferromagnetic particles. Nanoparticles provide the advantages of the inorganic material as well as very high surface to volume ratios.

In a recent study, a water-soluble gold nanoparticles/polyaniline nanocomposite has been reported (48). The particles were prepared using 3-aminophenylboronic acid (3-APBA) as reductive and protective reagent. Poly(vinyl alcohol) (PVA) was used as disperser based on the covalent bond interaction between the boronic acid and diol group. The particle size was found to be a function of the concentration of PVA. The nanocomposite particles were found to detect glucose based on the competitive reaction of 3-APBA and PVA, and that of 3-APBA.

In an alternative approach, Chinnayelka and McShane (49) present a design of biosensors based on "smart" hollow microspheres. These spheres were prepared using a layer-by-layer approach by self-assembly of Con A/dextran into multilayer films, followed by polymer multilayers. First dissolvable resin microparticles were coated with fluorescein isothiocyanate (FITC)–dextran + tetramethylrhodamine isothiocyanate (TRITC)–Con A multilayers. Next polyelectrolyte multilayers were added, and the core particles were then dissolved to yield hollow capsules. For biosensing, as glucose was added to the matrix, FITC–dextran was displaced by TRITC-labeled Con A. The basic transduction principle was the change in resonance energy transfer efficiency from FITC to TRITC.

7.2 Stimuli-Responsive Materials

Stimuli-responsive materials (StRMs) respond in a controlled and often reversible fashion to changes in the environment such as ionic strength, pH, and temperature, nature of the solvent, or light. The materials may respond by changing shape, polarity, charge, solubility, or viscosity. StRMs are finding extensive applications in drug delivery, biosensors, actuators, bioseparations, or other areas. The applications and characteristics of StRMs have previously been reviewed (50,51). Here only some aspects of this very popular area of research are presented with the focus on biosensor applications.

7.2.1 pH-Sensitive Materials

Materials that are pH sensitive contain ionizable functional groups in which charge is generated in response to pH. This causes electrostatic repulsion and corresponding swelling of the material. Examples include natural materials such as the polysaccharides chitosan, alginate, and κ-carrageenan, which respond to changes in pH, Ca^{2+}, and K^+, respectively (52). Synthetic polymers or gels that are pH sensitive include polyethyleneimine, polylysine, and poly-*N*, *N*-dimethyl aminoethyl methacrylamide, polyacrylic acid, and polymethacrylic acid (PMAA). Other pH-sensitive materials include PMAA and PEG copolymers (53), poly(acrylic acid-co-octyl acrylate) (54), poly(methacrylic acid-co-ethacrylic acid) P(MA-co-EA) (55), 4-amino-*N*-(4, 6-dimethyl-2-pyrimidinyl) benzene sulfonamide-*N*, *N*-dimethylacrylamide (56), and poly-*N*-acryloyl-*N*-propylpiperazine (PNANP) (57).

A genetically engineered silk-elastin-like protein-based copolymer (SELP) with an amino acid repeat sequence of [(GVGVP)(4)GKGVP(GVGVP)(3)(GAGAGS)(4)](12) also has been made (58). The influence of environmental conditions such as pH, temperature, and ionic strength on the equilibrium swelling ratio of physically cross-linked networks of these gels was investigated. The effects of gelation cure time and initial polymer concentration on the equilibrium swelling ratio and soluble fraction of the hydrogels were also studied. It was found that the soluble fraction linearly correlated with the initial polymer concentration at higher gelation times. Soluble fraction results suggest that final hydrogel water content may be controlled by both initial polymer concentration and gelation time. Differentially charged analogs of block copolymers containing repeating sequences from silk (GAGAGS) and elastin (GVGVP) were synthesized using genetic engineering techniques by replacing a valine residue with glutamic acid. The sensitivity to pH and temperature was examined at various polymer concentrations, ionic strengths, and polymer lengths. The polymers transitioned from soluble to precipitate state over narrow temperature ranges. The transition temperature *T–t* (the temperature at which half-maximal spectrophotometric absorption was observed) increased with increasing pH up to pH 7.0 and leveled off above this value for the Glu-containing polymer (17E)(1*t*). *T–t* was independent of p*I* for the Val-containing polymer (17V)(1*t*). It decreased with increasing ionic strength, polymer concentration, and polymer length for both polymers. These results suggest that by substituting charged amino acids for neutral amino acids at strategic locations in the polymer backbone and by control of the length of silk-elastin-like block copolymers using genetic engineering techniques, it is possible to precisely control sensitivity to pH, temperature, and ionic strength. Equilibrium swelling studies demonstrated that these hydrogels are relatively insensitive to environmental changes such as pH, temperature, and ionic strength. Over the concentration range studied, it was found that an increase in gelation time at 37°C resulted in lower hydrogel weight equilibrium swelling ratios, which corresponds to less soluble polymer released postgelation. Together, these results have implications for the controlled delivery of bioactive agents from silk-elastin-like hydrogels.

Cai et al. (59) have applied a magneto-elastic transducer-based biosensor using a pH-sensitive gel (Figure 7.5). The magneto-elastic transducer vibrates in a time-varying magnetic field at a resonance frequency the value of which inversely depends on sensor mass loading. The glucose biosensor was made by coating with a pH-sensitive polymer and a GOD layer on the transducer surface. The GOD-catalyzed oxidation of glucose produced gluconic acid, causing shrinking of the pH-responsive polymer and decrease in mass. The changes in magnetic flux were remotely detected using a pickup coil. The sensitivity of the glucose biosensor was a function of ionic strength. The response was linear up to 10 mM, and the detection limit was 0.6 mM.

A miniature conductimetric pH sensor based on the measurement of the conductivity of a pH-responsive hydrogel has been described (60). The microfabricated sensor consisted of electrode arrays coated with a photo-lithographically patterned hydrogel membrane.

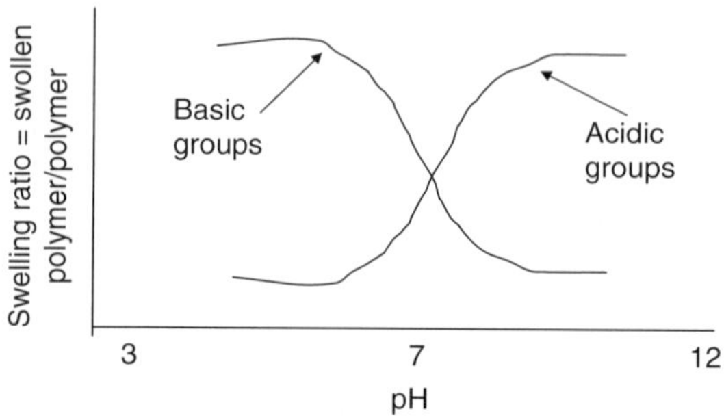

FIGURE 7.5
Swelling behavior of pH-sensitive hydrogel materials.

The hydrogel sensing layer responded to the pH of the solution in which it is immersed leading to changes in ionic mobility across the gel. The sensor could operate in high-ionic strength solutions, which was a big advantage. The response time was 5 min and was operated for more than 6 weeks with a 50% loss of sensitivity over 3 weeks.

Simonian et al. (61) have produced biosensor platforms for detection of pesticides. Organophosphate hydrolase enzymes were integrated with several different transduction platforms including conventional pH electrodes, fluoride ion-sensitive electrodes, and pH-responsive fluorescent dyes. Detection limit for most systems was found to be in the low parts per million concentration range.

7.2.2 Thermosensitive Materials

Several investigations have been carried out in the literature on the temperature-responsive polymer poly(*N*-isopropylacrylamide) (PNIPAAm). Park and Hoffman were among the first to report on the temperature-dependent swelling and collapse of PNIPAAm (62). Following this several copolymers were investigated with *N*-isopropylacrylamide (NIPAAm) including acrylic acid and methacrylic acid, 2-methyl-2-acrylamidopropane sulfonic acid, trimethyl-acrylamidopropyl ammonium 3-methyl-1-vinylimidazolium iodide, sodium acrylate, sodium methacrylate, and 1-(3-sulphopropyl)-2-vinyl-pyridinium betaine (63–65). The NIPAAm-methacrylic acid polymer is both pH and temperature responsive (55). Bioactive "smart" protein–polymer conjugates were also synthesized by polymerizing from defined initiation sites on proteins (66). Cysteine 34 of bovine serum albumin (BSA) and Cys-131 of T4 lysozyme V131C were modified. Polymerization of NIPAAm from the protein macroinitiators resulted in thermosensitive BSA–PNIPAAm and lysozyme–PNIPAAm in greater than 65% yield. The resultant conjugates were characterized by gel electrophoresis and size exclusion chromatography (SEC) and easily purified by preparative SEC. Lytic activities of the lysozyme conjugates were determined by two standard assays and compared to that of the unmodified enzyme prior to polymerization; no statistical differences in bioactivity were observed (66).

In another study, NIPAAm gels seeded with ferromagnetic materials were prepared, which showed magnetic-field-sensitive swelling–deswelling transition. The likely reason was the heating of magnets in a magnetic field. This led to the collapse of temperature-sensitive NIPAAm (67). Similarly, Yoshida et al. (68) demonstrated swelling–deswelling transition controlled by redox reaction using NIPAAm seeded with tris (2,2′-bipyridyl) ruthenium (II). The gel was found to undergo contraction and shrinking repeatedly as ruthenium changed from +2 to +3 oxidation states.

Thermal-responsive polymer gels have been applied in immunoassays (69). Miyata et al. (70) first attached rabbit IgG to *N*-succinimidylacrylate and subsequently copolymerized this with acrylamide, along with *N,N*′-methylene bisacrylamide as a cross-linker. The resulting hydrogel was labeled antigen–antibody semiIPN. When an antigen to IgG was introduced in the network, the gel swelled. Removal of the antigen resulted in gel shrinkage. Thus, a shape memory response occurred on exposure to the antigen, which was found to be reversible.

Hoffman (71) has designed "intelligent" polymer gels by conjugating them near a protein's ligand binding site. On exposure to an appropriate stimulus or trigger, the protein-gel network collapses and blocks the ligand binding. This was demonstrated using PNIPAAm and streptavidin. Temperature was applied as the trigger to prevent the binding of biotin. Similarly, Ding et al. (72) have conjugated PNIPAAm to a site near a genetically engineered streptavidin. They demonstrated that the temperature-induced collapse of the polymer released bound biotin. Similarly, using NIPAAm–acrylic acid copolymers pH can also be utilized for biotin release. This brings about the release of biotin. In this case the polymer is conjugated to a genetically engineered streptavidin at a site near its biotin binding site.

7.2.3 Photo-Responsive Materials

Photo-responsive chromophores have been incorporated into polymer chains to produce photo-responsive polymers. Azobenzene has been one of the more common chromophores utilized. The molecule can undergo changes in shape and polarity upon irradiation. Polymeric chains containing this molecule show reversible photoisomerization (73). Junge and McGrath (74) produced benzyl aryl ether dendrimers, which underwent reversible *cis*- or *trans*-isomerization on exposure to UV light.

Desponds and Freitag (75) have prepared biopolymeric conjugated molecules of biotin with isopropylacrylamide, *N*-acryloxysuccinimide, and (3-minopropyloxy)azobenzene. These bioconjugates were photosensitive due to the presence of the chromophore. Irradiation with UV light (330 nm) or visible light (440 nm) caused oscillation of the chromophore between *cis* and *trans* states. The bioconjugates were applied to capture avidin from solution. The proposed technique may lead to optically switchable biosensors and bioaffinity ligands.

Bioresponsive hydrogel-based microlenses have recently been reported for protein detection (76). Stimuli-responsive poly(*N*-isopropylacrylamide-co-acrylic acid) (pNIPAm-co-AAc) microgels were functionalized using biotin. Silane-modified glass substrate was coated with arrays containing both pNIPAm-co-AAc microgels (as an internal control) and biotinylated pNIPAm-co-AAc microgels. The microlens were applied for avidin capture and detection. The optical properties of the lens changed with the binding events and the process was reversible. This study is an example of label-free detection of proteins or small molecules.

Analyte-dependent swelling/shrinking properties of ultrathin polymer layers are an appropriate means for the detection of various analytes (77). Optical metal nanoclusters can be used to determine the change of the layer's thickness, which is shown by a change in the color of the chip. By using different cross-linking agents and different polymers (biological or artificial as well) it was possible to design various sensitive layers showing different swelling or shrinking behaviors. Sensitivity on various analytes could be observed, since the different types of polymers employed differed in structure, functional groups, or biorecognitive properties.

Leclerc and Ho (78) describe the thermochromic, ionochromic, and affinity chromic properties of various neutral conjugated polymers. In particular, the recent development of cationic, water-soluble, chromic polythiophenes has allowed the easy, rapid, specific, and ultrasensitive detection of various nucleic acids and proteins in aqueous media.

7.3 Conclusions and Future Prospects

Significant progress has been made in membrane protein engineering over the last 5 years, based largely on the redesign of existing scaffolds (79). Engineering techniques that have been employed include direct genetic engineering, both covalent and noncovalent modification, unnatural amino acid mutagenesis, and total synthesis aided by chemical ligation of unprotected fragments. Combinatorial mutagenesis and directed evolution remain, by contrast, underemployed. Techniques for assembling and purifying heteromeric multisubunit pores have been improved. Progress in the de novo design of channels and pores has been slower. But, we are at the beginning of a new era in membrane protein engineering based on the accelerating acquisition of structural information, a better understanding of molecular motion in membrane proteins, technical improvements in membrane protein refolding, and the application of computational approaches developed for soluble proteins. In addition, the next 5 years should see further advances in the applications of engineered channels and pores, notably in therapeutics and sensor technology. Li et al. (80) have been investigating applications of nanopore membranes in analytical chemistry—specifically in membrane-based bioseparations, in electroanalytical chemistry, and in the development of new approaches to biosensor design. Membranes that have conically shaped pores (as opposed to the more conventional cylindrical shape) may offer some advantages for these applications. They describe a plasma-etch method that converts cylindrical nanopores in track-etched polymeric membranes into conically shaped pores. The key advantage of the conical pore shape is a dramatic enhancement in the rate of transport through the membrane, relative to an analogous cylindrical pore membrane. They have demonstrated this by measuring the ionic resistances of the plasma-etched conical pore membranes.

Goryll et al. (81) present a method to fabricate an aperture in a silicon wafer that can be used to suspend a freestanding lipid bilayer membrane. The design offers the feature of scalability of the aperture size into the submicron range. Lipid bilayer membranes formed across the aperture in the oxidized silicon substrate show a gigaohm sealing resistance. The stability of these membranes allowed the insertion of a nanometer-sized ion channel protein (OmpF porin) and the measurement of voltage-dependent gating that can be expected from a working porin ion channel.

References

1. Ziegler, C. (2004). Cantilever-based biosensors. *Anal. Bioanal. Chem.* 379 (7–8): 946–959.
2. Calleja, M., Nordstrom, M., Alvarez, M., Tamayo, J., Lechuga, L.M., Boisen, C. (2005). A highly sensitive polymer-based cantilever-sensors for DNA detection. *Ultramicroscopy* 105 (1–4): 215–222.
3. Nugaeva, N., Gfeller, K.Y., Backmann, N., Lang, H.P., Duggelin, M., Hegner, M. (2005). Micromechanical cantilever array sensors for selective fungal immobilization and fast growth detection. *Biosen. Bioelectron.* 21(6): 849–856.
4. Shu, W.M., Liu, D.S., Watari, M., Riener, C.K., Strunz, T., Welland, M.E., Balasubramanian, S., McKendry, R.A. (2005). DNA molecular motor driven micromechanical cantilever arrays. *J. Am. Chem. Soc.* 127(48): 17054–17060.
5. Li, J., Wang, Y.B., Qiu, J.D., Sun, D., Xia, X.H. (2005). Biocomposites of covalently linked glucose oxidase on carbon nanotubes for glucose biosensor. *Anal. Bioanal. Chem.* 383(6): 918–922.
6. Huang, M.H., Jiang, H.Q., Qu, X.H., Xu, Z., Wang, Y.L., Dong, S.J. (2005). Small molecules as cross-linkers: fabrication of carbon nanotubes/thionine self-assembled multilayers on amino functionalized surfaces. *Chem. Commun.* 44: 5560–5562.
7. Barreca, D., Gasparotto, A., Maragno, C., Tondello, E. (2005). Synthesis of gold nanotubes by sputtering of gold into porous materials. *J. Nanosci. Nanotechnol.* 5(11): 1883–1886.
8. Liu, S.Q., Chen, A.C. (2005). Coadsorption of horseradish peroxidase with thionine on TiO_2: nanotubes for biosensing. *Langmuir* 21(18): 8409–8413.
9. Graham, D.L., Ferreira, H.A., Freitas, P.P. (2004). Magnetoresistive-based biosensors and biochips. *Trends Biotechnol.* 22(9): 455–462.
10. Baselt, D.R., Lee, G.U., Natesan, M., Metzger, S.W., Sheehan, P.E., Colton, R.J. (1998). A biosensor based on magnetoresistance technology. *Biosens. Bioelectron.* 13: 731–739.
11. Megens, M., Prins, M. (2005). Magnetic biochips: a new option for sensitive diagnostics. *J. Magnet. Magnetic Mat.* 293(1): 702–708.

12. Liu, Z.M., Liu, Y.L., Yang, H.F., Yang, Y., Shen, G.L., Yu, R.Q. (2005). A phenol biosensor based on immobilizing tyrosinase to modified core-shell magnetic nanoparticles supported at a carbon paste electrode. *Anal. Chim. Acta* 533(1): 3–9.
13. Chung, S.H., Hoffmann, A., Guslienko, K., Bader, S.D., Liu, C., Kay, B., Makowski, L., Chen, L. (2005). *J. Appl. Phys.* 97(10): 10R101 Part 3.
14. Zhao, G., Feng, J.J., Zhang, Q.L., Li, S.P., Chen, H.Y. (2005). Synthesis and characterization of Prussian Blue modified magnetite nanoparticles and its application to the electrocatalytic reduction of H_2O_2. *Chem. Mater.* 17(12): 3154–3159.
15. Brady, S., Lau, K.T., Megill, W., Wallace, G.G., Diamond, D. (2005). The development and characterization of conducting polymeric sensing devices. *Synth. Met.* 154: 25–28.
16. Carrara, S., Bavastrello, V., Ricci, D., Stura, E., Nicolini. C. (2005). Improved nanocomposite materials for biosensor applications investigated by electrochemical impedance spectroscopy. *Sens. Actuat. B-Chem.* 109(2): 221–226.
17. Fangming, D.U., Fischer, J.E., Winey, K.I. (2003). Coagulation method for preparing single wall carbon nanotubes/poly(methyl methacrylate) composites and their modulus, electrical conductivity and thermal stability. *J. Polym. Sci. Part B: Polym. Phys.* 41: 3333–3338.
18. Chang, B.H., Liu, Z.Q., Sun, L.F., Tang, D.S., Zhou, W.Y., Wang, G., Qian, L.X., Xie, S.S., Fen, J.H., Wan, M.X. (2000). Conductivity and magnetic susceptibility of nanotube/polypyrrole nanocomposites. *J. Low Temp. Phys.* 119(1/2): 41–48.
19. Laxminarayana, K., Jalili, N. (2005). Functional nano-tube based textiles: pathway to next generation fabrics with enhanced sensing capabilities. *Textile Res. J.* Sept.: 670–681.
20. Sawicka, K., Gouma, P., Simon, S. (2005). Electrospun biocomposite nanofibers for urea biosensing. *Sens. Actuat. B-Chem.* 108(1–2): 585–588.
21. Ionescu, R.E., Abu-Rabeah, K., Cosnier, S., Marks, R. (2005). Improved enzyme retention from an electropolymerized polypyrrole-alginate matrix in the development of biosensors. *Electrochem. Comm.* 7(12): 1277–1282.
22. Li, G.T., Bhosale, S., Tao, S.Y., Bhosale, S., Fuhrhop, J.H. (2005). Conducting polythiophenes with a broad spectrum of reactive groups. *J. Polym. Sci. Part A-Polym. Chem.* 43(19): 4547–4558.
23. Syritski, V., Gyurcsanyi, R.E., Opik, A., Toth, K. (2005). Synthesis and characterization of inherently conducting polymers by using scanning electrochemical microscopy and electrochemical quartz crystal microbalance. *Synth. Metals* 152(1–3): 133–136.
24. Ngamna, O., Morrin, A., Moulton, S.E., Killard, A.J., Smyth, M.R., Wallace, G.G. (2005). An HRP based biosensor using sulphonated polyaniline. *Synth. Metals* 153(1–3): 185–188.
25. Kros, A., Sommerdijk, N.A., Nolte, R.J.M. (2005). Poly(pyrrole) versus poly (3,4-ethylenedioxythiophene): implications for biosensor applications. *Sens. Actuat. B-Chem.* 106(1): 289–295.
26. Brahim, S., Narinesingh, D., Guiseppi-Elie, A. (2002). Bio-smart materials: kinetics of immobilized enzymes in *p*(HEMA)/*p*(pyrrole) hydrogels in amperometric biosensors. *Macromol. Symp.* 186: 63–73.
27. Constable, E.C., Harverson, P., Oberholzer, M. (1996). Convergent and divergent approaches to metallocentric metallodendrimers. *J Chem. Soc. Chem. Commun.* 15: 1821–1822.
28. Wendland, M.S., Zimmerman, S.C. (1999). Synthesis of cored dendrimers *J. Am. Chem. Soc.* 121(6): 1389–1390.
29. Maciejewski, M. (1982). Concepts of trapping topologically by shell molecules. *J. Macromol. Sci. Chem.* A17(4): 689–703.
30. Newkome, G.R., Moorfield, C.N., Vogtle, F. (1996). *Dendritic Molecules: Concepts, Synthesis, Perspectives*. VCH, Weinheim.
31. Yoon, H.C., Kim, H.S. (2000). Multilayered assembly of dendrimers with enzymes on gold: thickness controlled biosensing layer. *Anal. Chem.* 72: 922–926.
32. Casado, M.A., Hack, V., Camerano, J.A., Ciriano, M.A., Teiel, C., Oro, L.A. (2005). Unprecedented hybrid scorpionate/phosphine ligand able to be anchored to carbosilane dendrimers. *Inorgan. Chem.* 44(25): 9122–9124.
33. Diaz, I., Garcia, B., Alonso, B., Casado, C.M., Moran, M., Losada, J., Perez-Pariente, J. (2003). Ferrocenyl dendrimers incorporated into mesoporous silica: new hybrid redox-active materials. *Chem. Mater.* 15(5): 1073–1079.
34. Takada, K., Diaz, D.J., Abruna, H.D., Cuadrado, I., Casado, C., Alonso, B., Moran, M., Losada, J. (1997). Redox-active ferrocenyl dendrimers: thermodynamics and kinetics of adsorption, insitu

electrochemical quartz crystal microbalance study of the redox process and tapping mode AFM imaging *J. Am. Chem. Soc.* 119(44): 10763–10773.

35. Ashton, P.R., Anderson, D.W., Brown, C.L., Shipway, A.N., Stoddart, J.F., Tolley, M.S. (1998). The synthesis and characterization of a new family of polyamide dendrimers. *Chem.-A Euro. J.* 4(5): 781–795.
36. Svobodova, L., Snejdarkova, M., Toth, K., Gyurcsanyi, R.E., Hianik, T. (2004). Properties of mixed alkanethio-dendrimer layers and their applications in biosensing. *Bioelectrochemistry* 63(1–2): 285–289.
37. Yoon, H.C., Hong, M.Y., Kim, H.S. (2000). Functionalization of a poly(amidoamine) dendrimer with ferrocenyls and its application to the construction of a reagentless enzyme electrode. *Anal. Chem.* 72(18): 4420–4427.
38. Yang, M., Tsang, E.M.W., Wang, Y.A., Peng, X.G., Yu, H.Z. (2005). Bioreactive surfaces prepared via the self-assembly of dendron thiols and subsequent dendrimer bridging reactions. *Langmuir* 21(5): 1858–1865.
39. Chang, A.C., Gillespie, J.B., Tabacco, M.B. (2001). Enhanced detection of live bacteria using a dendrimer thin film in an optical biosensor. *Anal. Chem.* 73(3): 467–470.
40. Hartel, V. (1978). *Optoelectronics.* Mc-Graw Hill, New York.
41. Noorlag, D.J.W. (1982). Quantitative analysis of effects causing non-linear position response in position sensitive photodetectors. *IEEE Trans. Electron. Devices ED.* 29: 158—161.
42. Miejer, G.C.M. (1982). A low power easy to calibrate temperature transducer. *IEEE J. Solid State Circuits* SC-17: 609–613.
43. van Herwaarden, A.W. (1984). The Seeback effect in silicon Ics. *Sens. Actuat.* 6: 245–254.
44. Zaeren, V., Middelhoek, S. (1982). Magnetic field sensor based on a two collector transistor structure. *Sens. Actuat.* 2: 251–261.
45. Saaman, A.A. Bergveld, P. (1985). A classification of chemically sensitive semi-conductor devices. *Sens. Actuat.* 7: 75–87.
46. Lundstrom, I. (1981). Hydrogen sensitive MOS structures, Part 1, principles and applications. *Sens. Actuat.* 1: 403–426.
47. Nather, N., Rolka, D., Poghossian, A., Koudelka-Hep, M., Schoning, M.J. (2005). Two microcell flow-injection analysis (FIA) platforms for capacitive silicon-based field-effect sensors. *Electrochim. Acta* 51(5): 924—929.
48. Ma, Y., Li, N., Yang, C., Yang, X.R. (2005). One-step synthesis of water-soluble gold nanoparticles/polyaniline composite and its application in glucose sensing. *Colloid. Surf. A-Phycicochem. Eng. Aspects* 269(1–3): 1–6
49. Chinnayelka, S., McShane, M.J. (2004). Glucose-sensitive nano-assemblies comprising affinity-binding complexes trapped in fuzzy micro-shells. *J. Fluoresens.* 14(5): 585–595.
50. Jeonga, B., Gutowska, A. (2002). Lessons from nature: stimuli-responsive polymers and their biomedical applications. *Trends Biotechnol.* 20(7): 305–311.
51. Roy, I., Gupta, M.N. (2003). Smart polymeric materials: emerging biochemical applications. *Chem. Biol.* 10(12): 1161–1171.
52. Torres-Lugo, M., Peppas, N.A. (1999). Molecular design and in vitro studies of novel pH sensitive hydrogels for the oral delivery of calcitonin. *Macromolecules* 32: 6646–6651.
53. Thomas, J.L., You, H., Tirrell, D.A. (1995). Tuning the response of a pH-sensitive membrane switch. *J. Am. Chem. Soc.* 117: 2949–2950.
54. Philippova, O.E., Hourdet, D., Audebert, R., Khokhlov, A.R. (1997) pH-responsive gels of hydrophobically modified poly(acrylic acid). *Macromolecules* 30: 8278–8285.
55. Park, S.Y., Bae, Y.H. (1999). Novel pH sensitive materials containing sulfonamide groups. *Macromol. Rapid Commun.* 20: 269–273.
56. Luo, L.B., Kato, M., Tsuruta, T., Kataoka, K., Nagasaki, Y. (2000). Stimuli sensitive material gels that stiffen upon swelling. *Macromolecules* 33(14): 4992–4994.
57. Gan, L.H., Gan, Y.Y., Doon, G.R. (2000). Poly(*N*-acryloyl-*N*-propylpiperazine): a new stimuli responsive material. *Macromolecules* 33(21): 7893–7897.
58. Dinerman, A.A., Cappello, J., Ghandehari, H., Hoag, S.W. (2002). Swelling behavior of a genetically engineered silk-elastinlike protein polymer hydrogel. *Biomaterials* 23(21): 4203–4210.
59. Cai, Q.Y., Zeng, K.F., Ruan, C.M., Desai, T.A., Grimes, C.A. (2004). A wireless, remote query glucose biosensor based on a pH-sensitive polymer. *Anal. Chem.* 76(14): 4038–4043.

60. Sheppard, N.F., Lesho, M.J., Mcnally, P. (1995). Micro-fabricated conductimetric pH sensor. *Sens. Actuat. B-Chem.* 28(2): 95–102.
61. Simonian, A.L., Flounders, A.W., Wild, J.R. (2004). FET-based biosensors for the direct detection of organophosphate neurotoxins. *Electroanalytes* 16(22): 1896–1906.
62. Park, T.G., Hoffman, A.S. (1994). Estimation of temperature-dependent pore size in poly(*N*-isopropylacrylamide) hydrogel beads. *Biotechnol. Prog.* 10: 82–86.
63. Lee, W.-F., Shieh, C.-H. (1999). pH-thermoreversible hydrogels. II. Synthesis and swelling behaviours of *N*-isopropylacrylamide-co-acrylic acid-co-sodium acrylate hydrogels. *J. Appl. Polym. Sci.* 73: 1955–1967.
64. Oya, T., Enoki, T., Grosberg, A.Y., Masamune, S., Sakiyama, T., Takeoka, Y., Tanaka, K., Wang, G.Q., Yilmaz, Y., Feld, M.S., Dasari, R., Tanaka, T. (1999). Reversible molecular adsorption based on multipoint interaction by shrinkable gels. *Science* 286:1543–1545.
65. Lee, S.B., Song, S.C., Jin, J.I., Sohn, Y.S. (1999). A new class of biodegradable thermosensitive materials. 1. Synthesis and characterization of poly(organophosphazenes) with methoxy poly(ethylene glycol) and amino acid ester side groups. *Macromolecules* 32: 2188–2193.
66. Heredia, K.L., Bontempo, D., Ly, T., Byers, J.T., Halstenberg, S., Maynard, H.D. (2005). In situ preparation of protein—"Smart" polymer conjugates with retention of bioactivity. *J. Am. Chem. Soc.* 127(48): 16955–16960.
67. Kato, N., Takizawa, Y., Takahashi, F. (1997). Magnetically driven chemomechanical device with poly(*N*-isopropylacrylamide) hydrogel containing gamma-Fe_2O_3. *J. Intell. Mater. Syst. Struct.* 8(7): 588–595.
68. Yoshida, R., Yamaguchi, T., Kokufuta, E. (1999). Molecular design of self-oscillating polymer gels and their dynamic swelling–deswelling behaviours. *J. Intell. Mater. Syst. Struct.* 10(6): 451–457.
69. Monji, N., Hoffman, A.S. (1987). A novel immunoassay system and bioseparation process based on thermal phase separating polymers. *Appl. Biochem. Biotechnol.* 14: 107–120.
70. Miyata, T., Asami, N., Urugami, T. (1999). A reversibly antigen-responsive hydrogel. *Nature* 399: 766–769.
71. Hoffman, A.S. (2000). Bioconjugation of intelligent polymers and recognition proteins for use in diagnostics and affinity separations. *Clin. Chem.* 46: 1478–1486.
72. Ding, Z., Long, C.J., Hayashi, Y., Bulmus, E.V., Hoffman A.S., Stayton, P.S. (1999). Temperature control of biotin binding and release with a streptavidin-poly(*N*-isopropylacrylamide) site-specific conjugate. *Bioconjug. Chem.* 10: 395–400.
73. Izumi, A., Nomura, R., Masuda, T. (2000). Synthesis of a new class of *n*-dopable and photoluminescent conjugated polymers having phenazine units in the main chain. *Macromolecules* 33(24): 8918–8920.
74. Junge, D.M., McGrath, D.V. (1997). Photo-responsive dendrimers. *Chem. Commun.* 9: 857–858.
75. Desponds, A., Freitag, R. (2005). Light-responsive bioconjugates as novel tools for specific capture of biologicals by photoaffinity precipitation. *Biotechnol. Bioeng.* 91(5): 583–591.
76. Kim, J., Nayak, S., Lyon, L.A. (2005). Bio-responsive hydrogel microlenses. *J. Am. Chem. Soc.* 127(26): 9588–9592.
77. Lepek, M., Pittner, F. (2004). Novel transducers for nano-optical biosensor chips based on biological and synthetic polymers with analyte-dependent swelling/shrinking behaviour. *J. NanoSci. Nanotechnol.* 4(1–2): 106–114.
78. Leclerc, M., Ho, H.A. (2004). Affinity-chromic polythiophenes: a novel bio-photonic tool for high-throughput screening and diagnostics. *Synletters* 2: 380–387.
79. Bayley, H., Jayasinghe, L. (2004). Functional engineered channels and pores—(Review). *Mol. Membr. Biol.* 21(4): 209–220.
80. Li, N.C., Yu, S.F., Harrell, C.C., Martin, C.R. (2004). Conical nanopore membranes. Preparation and transport properties. *Anal. Chem.* 76(7): 2025–2030.
81. Goryll, M., Wilk, S., Laws, G.M., Thornton, T., Goodnick, S., Saraniti, M., Tang, J., Eisenberg, R.S. (2003). Silicon-based ion channel sensor. *Superlat. Microstruct.* 34(3–6): 451–457.

8

Molecular Film Assembled Detection System for Biosensors by Layer-by-Layer Adsorption

Shin-ichiro Suye and Haitao Zheng

CONTENTS

8.1 Introduction

Enzymatic biosensors have been extensively applied in clinical, food, and environmental areas as an analytical instrument with fast detection speed, high selectivity and sensitivity. The construction of functional protein- or mediator-containing layers is important for biosensors because the immobilization procedure of enzyme and mediator should be beneficial to electron transfer between enzymes and electrode surface by facilitating redox mediator activity [1]. Modification of a solid surface has become a major challenge in the area of materials science. This modification allows the adjustment and control of parameters such as binding and frictional properties, biocompatibility, conductivity, and adhesion of a solid surface. The deposition of single layers on the surface offers the opportunity to manipulate surface properties at the molecular level, and multilayer films, consisting of organic compounds on a solid surface, have been the focus of materials science for more than several decades. In all research works, the control of layer thickness, sequence, density, and roughness is critical. Various approaches have been utilized to immobilize protein on a modified electrode surface. Although some achievements have been made by the Langmuir–Blodgett method and molecular self-assembly (chemisorption from solution), this method is not widely used for protein immobilization, because water-soluble proteins seem difficult to spread on the air–water surface, and are denatured easily at the interface [2].

Recently, the layer-by-layer (LBL) adsorption technique based on electrostatic force between polyelectrolytes and proteins has received great attention and interest. A novel technique proposed by Decher and Kunitake, LBL, has been introduced to fabricate multilayer films, and it represents a promising preparation method because no complicated instruments or chemical reactions are involved and ultrathin multicomponent architecture can be constructed simply by alternate adsorption in cationic and anionic polyelectrolytes [3,4].

8.2 Preparation of Layer-by-Layer Film on a Solid-State Device

LBL method is based on the electrostatic force between polyanions and polycations. Generally, in the LBL adsorption process, the substrate (after proper pretreatment) is immersed in a beaker containing either a polycationic or polyanionic solution for a given period of time (usually a few minutes), and then washed in a solution of approximately the same pH to remove nonadsorbed molecules. The substrate is then immersed in another solution containing the oppositely charged polymer and washed again. By repeating the alternate adsorption steps in oppositely charged polyelectrolyte solutions, a multilayer structure can be formed. A schematic illustration of the formation of multilayer structures composed of polyelectrolytes or other charged molecular or colloidal objects is shown in Figure 8.1.

In fact, the LBL method is dependent on charge neutralization and resaturation between polycations and polyanions, and the procedure of LBL can be regarded as a template controlled growth process. First, the solid surface acts as a template, and after the adsorption, it is refunctionalized and acts as a new template for the following adsorption. The multilayer film grows according to the above steps, and the effective electrostatic force between the oppositely charged layers is critical. For this reason, an ideal adsorption is not only the adsorption of one layer but also the overcompensation of the electric charges of previous layers and the supply of additional charges to facilitate further adsorption.

Unlike the chemisorption method, multilayer film could be fabricated on various kinds of inorganic substrates such as glass [5], quartz [6], silicon wafers [7], metals [8], and others

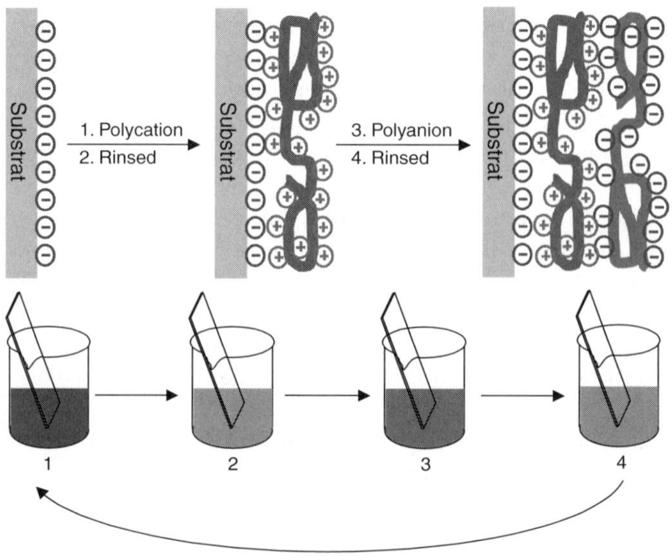

FIGURE 8.1
(See color insert following page 330)
Multilayer assembly of oppositely charged polyelectrolytes by layer-by-layer (LBL) method.

after proper pretreatment. The LBL alternating assembly of oppositely charged polyelectrolytes provides a method to construct nanoscale films with various kinds of substances in a preset sequence. A wide range of materials, for example polyelectrolytes [9,10], dyes [11,12], ceramics [13,14], and others, have been used in the fabrication of multilayer structures by the LBL method, which makes it possible to incorporate substances with desired chemical or physical properties. Because the macromolecules are adsorbed from aqueous solution in the LBL procedure, it opens up the possibility to immobilize biomolecules such as proteins [4,15,16] and DNA [17,18] in their active state.

The method of LBL adsorption can be easily manipulated. A concrete example of LBL adsorption of enzymes and polymerized mediator on a carbon electrode is shown as follows [19]. A negative or positive charge was first introduced onto a base material such as a carbon (PG) or gold electrode. The introduction of a carboxy group to a carbon electrode is achieved by electrochemical oxidation in nitric acid solution containing 2.5% $K_2Cr_2O_7$. A potential of 2.2 V vs. Ag/AgCl was applied to a bare electrode dipped into the above solution for 10 s. In the assembly of multilayer structures by the LBL method, PEI (polyethyleneimine) was used as polycation, and glucose-6-phosphate dehydrogenase (G6PDH), diaphorase (DI), and sodium alginate (Alg-Na) as polyanions. The modified electrode was first immersed in solution (5.0 mg mL^{-1}) for 20 min, and then thoroughly rinsed with water. It was then immersed in Alg-Na solution (5.0 mg mL^{-1}) for another 20 min and rinsed again to construct one layer of mediator. The multilayer mediator film was constructed by repeating the above procedure, and the electrode was referred to as PG/PEI-Fc_{n1}. The multilayers of DI and G6PDH were immobilized by repeated alternate adsorption in PEI solution (1.0 mg mL^{-1}) and enzyme solution (1.0 mg mL^{-1}) under 4°C.

8.3 Properties of Layer-by-Layer Film

Data about the structure of these multilayer assembly constructed using LBL methods have been evaluated by various kinds of instrumental techniques, such as quartz crystal microbalance (QCM), ultraviolet adsorption, atomic force microscopy (AFM), and scanning electron microscopy [4,17,20–25]. It has been found that a linear mass increase occurred with the layer number, and many factors such as salt concentration [22], adsorption time, dry conditions [21], pH [26], and charge density [27] affected the formation of multilayer films. It has also been demonstrated that an ultrathin organized multilayer structure could be formed utilizing the LBL procedure. The LBL procedure was also found to allow the adsorption amount and thickness of layers to be controlled, which made it possible to build up desired multilayer architectures with multiple components [25].

8.3.1 Characterization of Multicomponent Films by Quartz Crystal Microbalance

Mass-controlled LBL sequential adsorption technique was analyzed using QCM to monitor the immobilization procedure [6,16,22,24,28–30]. It has been proven that LBL method is a simple and elegant method which facilitates nanoscale preparation of ultrathin films with defined composition and uniform thickness.

The Sauerbrey equation describes the relationship between frequency shift and area mass density for one cycle of adsorption (Eq. 8.1), where F_0 is the fundamental resonant frequency (9×10^6 Hz), μ_q is the shear modulus of the quartz (2.947×10^{11} g·cm^{-1}·s^{-2}), ρ_q is the density of the quartz (2.648 g·cm^{-3}), and A is the electrode area (0.16 cm^2).

$$\Delta F = \frac{2F_0^2}{(\mu_q \rho_q)^{1/2}} \frac{\Delta m}{A} \tag{8.1}$$

By taking into account the characteristics of quartz resonators, the adsorbed mass can be calculated from the frequency shift utilizing Eq. (8.2), which indicates that 1 Hz frequency shift corresponds to a mass increase of about 0.9 ng.

$$\Delta F = -1.83 \times 10^8 \frac{\Delta m}{A} \tag{8.2}$$

The QCM data was also used to estimate the thickness, d (cm), of multilayer films by means of Eq. (8.3), where ρ is the density of adsorbed materials (1.2 g cm^3 for polyelectrolytes and 1.3 g cm^3 for enzymes) [4] and A is the area of the QCM electrode. As adsorption occurred on both sides of QCM electrode, total mass increase should be divided by 2 [16].

$$d = \frac{\Delta m}{2A\rho} \tag{8.3}$$

When the assembly of multilayer structure was accompanied by alternate adsorption in oppositely charged polyelectrolyte and enzyme solutions, for example at pH 7.0, PEI and PEI-Fc are positively charged, and Alg-Na, DI, and G6PDH are negatively charged. The multilayer formation of the LBL method was investigated by QCM. A gold electrode modified with 3-MPA self-assembled monolayer (SAM) was alternately adsorbed through sequential immersion in PEI/Alg-Na, PEI-Fc/Alg-Na, PEI/DI, and PEI/G6PDH solution. The final assembly of the multilayer film was referred to as Au/(PEI/Alg-Na)$_2$(PEI-Fc/Alg-Na)$_2$(PEI-DI)$_2$(PEI-G6PDH)$_2$. After each adsorption step, the frequency of the Au electrode was measured and frequency shifts were recorded (Figure 8.2). A negative linear progression of frequency shift was found to be dependent on adsorption cycles between four pairs of oppositely charged polyelectrolytes and enzymes, which indicated that multilayer assembly could be successively carried out by the LBL method. From Eq. (8.3) and QCM data, the

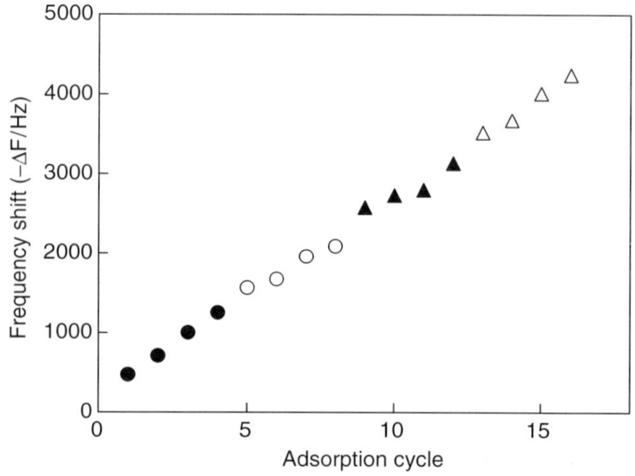

FIGURE 8.2
Frequency shift ($-\Delta F$) due to cycles of alternate adsorption in PEI/Alg-Na (●), PEI-Fc/Alg-Na (○), PEI/DI (▲), and PEI/G6PDH (△) solutions. (From Zheng, H., Okada, H., Nojima, S., Suye, S., Hori, T. (2004) Layer-by-layer assembly of enzymes and polymerized mediator on electrode surface by electrostatic adsorption. *Sci. Technol. Adv. Mater.* 5:371–376.)

thickness of DI and G6PDH layers were estimated to be 5.8 and 5.0 nm, and these estimates are close to the enzyme diameters (DI: 6.4 nm and G6PDH: 6.3 nm), which were calculated from molecular weight by assuming the enzymes as being rigid, spherical molecules [15]. Lvov et al. [4] reported that the thickness of most of the protein layers was from 22 to 57 nm. This suggests that most of the enzymes are assembled into monolayers. However, typically, immobilization of enzymes is very complex, as pointed out by Lvov, and in some cases, the aggregation of enzymes during the LBL procedure resulted in a multilayer conformation in one adsorption step [4]. In our work, QCM data showed that monolayer enzyme adsorption occurred, if the interface was considered to be a plane after immobilization [19].

8.3.2 Atomic Force Microscopy

To identify the immobilization by LBL and clarify the surface smoothness after alternate adsorption, the interface was viewed by AFM, and observed images are shown in Figure 8.3. Figure 8.3A is the image of a bare gold electrode surface, and Figure 8.3B is the image of the surface after alternate adsorption in PEI and DI solution. A relatively smooth surface is found on a bare gold electrode, and after one bilayer immobilization, it has been changed to a rough surface, which confirmed that the LBL absorption steps resulted in enzyme aggregation.

8.4 Application of Layer-by-Layer Adsorption for Biosensors

Organized multilayer architectures containing proteins or other biomolecules have attracted great interest because of their broad range of applications in biotechnology, especially in bioanalytical areas such as biosensors [31].

Enzyme immobilization on electrode surface is one of the most important factors in the design and construction of biosensors because the immobilization procedure should not

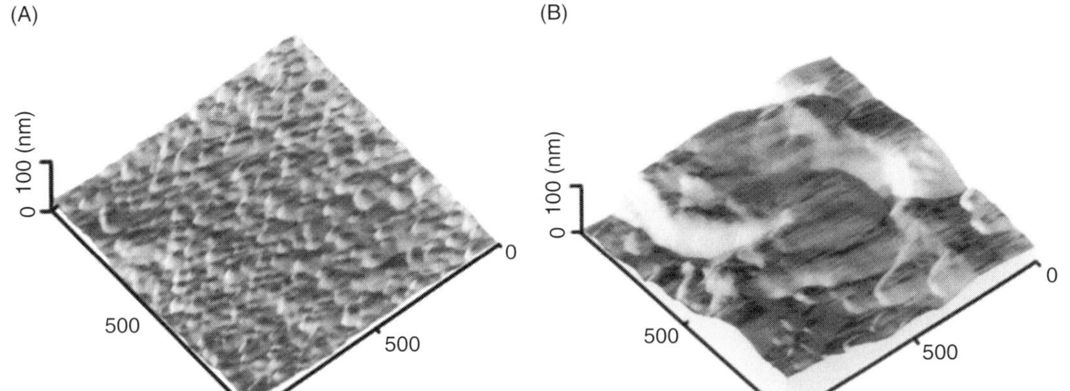

FIGURE 8.3
Atomic force microscopy (AFM) images of bare Au electrode (A) and after one bilayer adsorption of PEI/DI on Au electrode first assembled by 3-MPA. (From Zheng, H., Okada, H., Nojima, S., Suye, S., Hori, T. (2004) Layer-by-layer assembly of enzymes and polymerized mediator on electrode surface by electrostatic adsorption. *Sci. Technol. Adv. Mater.* 5:371–376.)

only facilitate the stabilization of the enzymes but also ensure diffusion of substrate and efficient electron transfer [32]. Many methods have been employed in the immobilization of enzymes, but it is difficult to control the amount and spatial distribution of these biocatalysts. Compared with these methods, the LBL method presents great advantages. For instance, the amount of enzyme immobilized can be controlled by adjusting the number of layers, and as all the adsorption steps are conducted in aqueous solution and no chemical reactions are involved, enzymes can be immobilized without being denatured.

Appropriate pH values relative to the isoelectric point of enzymes to be immobilized result in net positive or negative charges as required, to facilitate the LBL method. Several works have been devoted to the immobilization of enzymes by LBL method [4,16,33–36]. A monomolecular layer was formed after each adsorption step, and the enzymes incorporated in these multilayer films showed long-term storage stability and relatively high thermostability after heat treatment [33]. The applications of LBL adsorption for biosensors are listed in Table 8.1.

8.5 Conclusions

The LBL technique has many potential applications in the biosensor area. Alternate adsorption of various enzymes and synthetic or natural polyions on an electrode surface has been used to fabricate biosensor systems. Finely controlled multilayer films were constructed that were composed of enzymes and redox polyelectrolytes, and the enzymes retained their activities in a multilayer assembly formed by using the LBL method.

TABLE 8.1

The Application of Layer-by-Layer Adsorption for Biosensors

Substrate	Electrode	Polyanion	Polycation	Reference
Fructose	Au	Fructose dehydrogenase Horseradish peroxidase (HRP)	Poly[(vinylpyridine)Os(bpy)$_2$Cl	[36]
Glucose	ITO	Glucose oxidase Prussian Blue	Poly(allylamine)hydrocloride	[37]
Cholesterol	Au	Cholesterol oxidase Poly(styrenesulfonate)	Microperoxidase	[38]
Catechol	Glassy carbon	Polyphenol oxidase Poly(styrenesulfonate) IgG	Chitosan	[39]
Choline	Pt	Choline oxidase	Poly(ethyleneimine) Poly (diallyldimethylammonium chloride)	[40]
Cholesterol	ITO	Cholesterol oxidase Cholesterol esterase Poly(styrenesulfonate)	Poly(ethyleneimine)	[41]
DNA	Carbon Streptavidin	Algnic acid Polyethyleneimine)-ferrocene	Poly(ethyleneimine)	[42]

References

1. Chaubey, A., Malhotra, B. D. (2002). Mediated biosensors. *Biosens. Bioelectron.* 17:441–456.
2. Ishii, T., Muramatsu, M. (1971). Spread ability of ovalbumin monolayer at air–water interface. Effects of additives to spreading solutions. *Bull. Chem. Soc. Jpn.* 44:679–681.
3. Decher, G. Fuzzy. (1997). Nanoassemblies: toward layered polymeric multicomposites. *Science* 277:1232–1337.
4. Lvov, Y., Ariga, K., Ichinose, I., Kunitake, T. (1995). Assembly of multicomponent protein films by means of electrostatic layer-by-layer adsorption. *J. Am. Chem. Soc.* 117:6117–6223.
5. Santos, J. P., Welsh, E. R., Gaber, B. P., Singh, A. (2001). Polyelectrolyte-assisted immobilization of active enzymes on glass beads. *Langmuir* 17:5361–5367.
6. Li, W., Wang, Z., Sun, C., Xian, M., Zhao. M. (2000). Fabrication of multilayer films containing horseradish peroxidase and polycation-bearing Os complex by means of electrostatic layer-by-layer adsorption and its application as a hydrogen peroxide sensor. *Anal. Chim. Acta* 418:225–232.
7. Yang, X., Johnson, S., Shi, J., Holesinger, T., Swanson, B. (1997). Polyelectrolyte and molecular host into self-assembly to multilayer thin films: An approach to thin film chemical sensors. *Sens. Auctuators B* 45:87–92.
8. Mizutani, F., Sato, Y., Yabuki, S., Hirata, Y. (1996). Enzyme ultra-thin layer electrode prepared by the co-adsorption of poly-L-lysine and glucose oxidase onto a mercaptopropionic acid-modified gold surface. *Chem. Lett.* 251–252.
9. Lvov, Y., Decher, G., Möhwald, H. (1993). Assembly, structural characterization and thermal behavior of layer-by-layer deposited ultrathin films of polyvinylsulfonate and polyallylamine. *Langmuir* 9:481–486.
10. Lvov, Y., Haas, H., Decher, G., Möhwald, H., Kalachey, A. (1993). Assembly of polymeric molecular films onto the plasma treated glass. *J. Phys. Chem.* 97:12835–12840.
11. Ariga, A., Lvov, Y., Kunitake, T. (1997). Assembling alternate dye-polyion molecular films by electrostatic layer-by-layer adsorption. *J. Am. Chem. Soc.* 119:2224–2231.
12. Lvov, Y., Yamada, S., Kunitake, T. (1997). Nonlinear optical effects in layer-by-layer alternate films of polycations and azobenzene-containing polyanion. *Thin Solid Films* 300:107–112.
13. Lvov, Y., Ariga, K., Ichinose, I., Kunitake, T. (1996). Formation of ultrathin mutilayers and hydrated gel from montmorillonite clay and linear polycations. *Langmuir* 12:3038–3044.
14. Lvov, Y., Ariga, K., Onda, M., Ichinose, I., Kunitake, T. (1997). Alternate assembly of ordered multilayers of SiO_2 and other nanoparticles and polyions. *Langmuir* 13:6195–6203.
15. Lvov, Y., Ariga, K., Ichinose, I., Kunitake, T. (1996). Molecular film assembly via layer-by-layer adsorption of oppositely charged macromolecules (linear polymer, protein, and clay) and of Concanavalin A/glycogen pair. *Thin Solid Films* 284:797–801.
16. Li, Z., Hu, N. (2002). Assembly of electroactive layer-by-layer films of myoglobin and lonomer poly(ester sulfonic acid). *J. Colloid Interf. Sci.* 254:257–265.
17. Sukhorukov, G. B., Möhwald, H., Decher, G., Lvov, Y. (1996). Assembly of polyelectrolyte multilayer films by consecutively alternating adsorption of polynucleotides and polycations. *Thin Solid Films* 284–285:220–223.
18. Jin, Y., Shao, Y., Dong, S. (2000). Direct electrochemistry and surface plasmon resonance characterization of alternate layer-by-layer self-assembled DNA-myoglobin thin films on chemically modified gold surfaces. *Langmuir* 19:4771–4777.
19. Zheng, H., Okada, H., Nojima, S., Suye, S., Hori, T. (2004). Layer-by-layer assembly of enzymes and polymerized mediator on electrode surface by electrostatic adsorption. *Sci. Technol. Adv. Mater.* 5:371–376.
20. Ram, M. K., Salerno, M., Adami, M., Faraci, P., Nicolini, C. (1999). Physical properties of polyaniline films: assembled by the layer-by-layer technique. *Langmuir* 15:1252–1259.
21. Lvov, Y., Ariga, K., Onda, M., Ichinose, I., Kunitake, T. (1999). A careful examination of the adsorption step in the alternate layer-by-layer assembly of linear polyanion and polycation. *Colloids Surf. A* 146:337–346.
22. Shiratori, S., Yamada, M. (2000). Nano-scale control of composite polymer films by mass-controlled layer-by-layer sequential adsorption of polyelectrolytes. *Polym. Adv. Technol.* 11:810–814.

23. Baba, A., Kaneko, F., Advincula, R. C. (2000). Polyelectrolyte adsorption processes characterized in situ using the quartz crystal microbalance technique: alternate adsorption properties in ultrathin polymer films. *Colloid Surf. A* 173:39–49.
24. Clark, S. L. Hammond, P. T. (2000). The role of secondary interactions in selective electrostatic multilayer deposition. *Langmuir* 16:10206–10214.
25. Okayama, Y., Ito, T., Shiratori, S. (2001). Optimization of the feedback constant control for the mass controlled layer-by-layer sequential adsorption technique for polyelectrolyte thin films. *Thin Solid Films* 393:132–137.
26. Kim, B., Bruening, M. L. (2003). pH-Dependent growth and morphology of multilayer dendrimer/poly(acrylic acid) films. *Langmuir* 19:94–99.
27. Schoeler, B., Kumaraswamy, G., Caruso, F. (2002). Investigation of the influence of polyelectrolyte charge density on the growth of multilayer thin films prepared by the layer-by-layer technique. *Macromolecules* 35:889–897.
28. Forzani, E. S., Otero, M., Pérez, M. A., Teijelo, M. L., Calvo, E. J. (2002). The structure of layer-by-layer self-assembled glucose oxidase and $Os(Bpy)_2ClPyCH_2NH$-poly(allylamine) multilayers: Ellipsometric and quartz crystal microbalance studies. *Langmuir* 18:4020–4029.
29. Calvo, E. J., Etchenique, R., Pietrasanta, L., Wolosiuk, A. (2001). Layer-by-layer self-assembly of glucose oxidase and Os(Bpy)2ClPyCH2NH-poly(allylamine) bioelectrode. *Anal. Chem.* 73:1161–1168.
30. Decher, G., Lehr, B., Lowack, K., Lvov, Y., Schmitt, J. (1994). New nanocomposite films for biosensors: layer-by-layer adsorbed films of polyelectrolytes, proteins or DNA. *Biosens. Bioelectron.* 9:677–684.
31. Schuhmann, W. (2002). Amperometric enzyme biosensors based on optimized electron-transfer pathways and non-manual immobilization procedures. *Rev. Mol. Biotech.* 82:425–441.
32. Onda, M., Ariga, K., Kunitake, T. (1999). Activity and stability of glucose oxidase in molecular films assembled alternately with polyions. *J. Biosci. Bioeng.* 87:69–75.
33. Forzani, E. S., Solís, V. M., Calvo, E. J. (2000). Electrochemical behavior of polyphenol oxidase immobilized in self-assembled structures layer by layer with cationic polyallylamine. *Anal. Chem.* 72:5300–5307.
34. Caruso, F., Schüler, C. (2000). Enzyme multilayers on colloid particles: Assembly, stability, and enzymatic activity. *Langmuir* 16:9595–9603.
35. Forzani, E. S., Pérez, M. A., Teijelo, M. L., Calvo, E. J. (2002). Redox driven swelling of layer-by-layer enzyme-polyelectrolyte multilayers. *Langmuir* 18:9867-9873.
36. Narváez, A., Suárez, G., Popescu, I. C., Katakis, I., Domínguez. E. (2000). Reagentless biosensors based on self-deposited redox polyelectrolyte-oxidoreductases architectures. *Biosen. Bioelectron.* 15:43–52.
37. Ferreira, M., Fiorito, P. A., Oliveira, Jr., O. N., Córdoba de Torresi, S. I. (2004). Enzyme-mediated amperometric biosensors prepared with the layer-by-layer (LbL) adsorption technique. *Biosens. Bioelectron.* 19:1611–1615.
38. Gobi, K. V., Mizutani, F. (2001). Layer-by-layer construction of an active multilayer enzyme electrode applicable for direct amperometric determination of cholesterol. *Sens. Actuators B Chem.* 80:72–277.
39. Coche-Guérente, L., Desbrières, J., Fatisson, J., Labbé, P., Rodriguez, M. C., Rivas, G. Physicochemical characterization of the layer-by-layer self-assembly of polyphenol oxidase and chitosan on glassy carbon electrode. *Electrochim. Acta* 56:2865–2877.
40. Shi, H., Song, Z., Huang, J., Yang, Y., Zhao, Z., Anzai, J., Osa, T., Chen, Q. Effects of the type of polycation on the amperometric response of choline biosensors prepared by a layer-by-layer deposition technique. *Mater. Sci. Eng. C* 25:433–435.
41. Ram, M. K., Bertoncello, P., Ding, H., Paddeu, S., Nicolini, C. (2001). Cholesterol biosensors prepared by layer-by-layer technique. *Biosens. Bioelectron.* 16:849–856.
42. Suye, S., Zheng, H., Okada, H., Hori, T. (2005). Assembly of alternating polymerized mediator, polymerized coenzyme, and enzyme modified electrode by layer-by-layer adsorption technique. *Sens. Actuators B Chem.* 108:671–675.

9

Nanostructured Organic Matrices and Intelligent Sensors

Claudio Nicolini, Manuela Adami, and Cristina Paternolli

CONTENTS

9.1 Introduction

Over the last decade, (1–59) the Biophysics Institute of the University of Genova and the Fondazione Elba in close cooperation with the Scientific and Technological Park of the Elba Island and with numerous leading international companies (STM, ABB, Edison, FIAT, and Elsag–Bailey) have been quite active in developing neural sensors (32,34), based on organic (13,17) and biological (6,8,14) materials and utilizing a wide range of transducers (3,6): amperometric (6,8), potentiometric (32,34,42), conductimetric (3), nanogravimetric (31), spectrophotometric (14), and fluorometric (43). Several immobilizing techniques were also employed over the time, ranging from self-assembly (8,11), to layer by layer (18,20) and Langmuir–Blodgett (1,2,13,16). This chapter intends to summarize the present state of the art with emphasis on the recent nanostructuring of sensing organic matrices and to critically assess the potential industrial relevance of intelligent sensors for a wide variety of applications for health and environment.

9.2 Intelligent Metal Sensing: The Psa Analyzer

In 1976, a new technique derived from polarographic methods, the potentiometric stripping analysis (PSA), was proposed. This chemical method (60) is very sensitive for the detection of metal ions in aqueous samples and consists of two steps: the ions are first electrolytically concentrated by depositing them on a working electrode (preconcentration or plating stage), and then a metal stripping phase follows, during which no control is performed on the potential. The latter step is accomplished, usually with a chemical oxidant in solution (Hg^{++} or dissolved oxygen). After the preconcentration phase, the analytical signal (the potential of the working electrode) is recorded as a function of time and then utilized to obtain quantitative information about the metal ions in solution.

A recent paper (61) demonstrates the possibility to design and realize a simple and low-cost instrument for monitoring environmentally significant metals by implementing the PSA technique (Figure 9.1), connecting a cheap electronic device to a computer. Such analyzer, suitable for heavy metal analyses (Figure 9.2), is based on the constant-current potentiometric stripping analysis and on a very simple electrochemical cell. The software

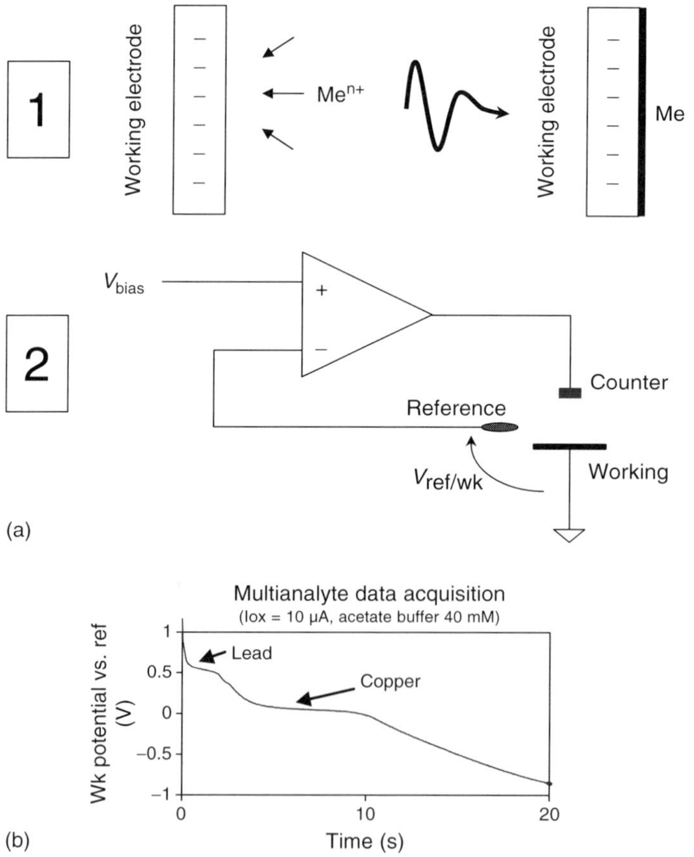

FIGURE 9.1
Schematic description of a PSA experiment. In (a), part 1, the electrochemical process taking place during the first step is illustrated: the working electrode polarization causes the deposition of the metal ions on it; part 2 shows the circuital schematics to drive this phase. Part (b) describes, in the same manner, the second step of the analytical technique.

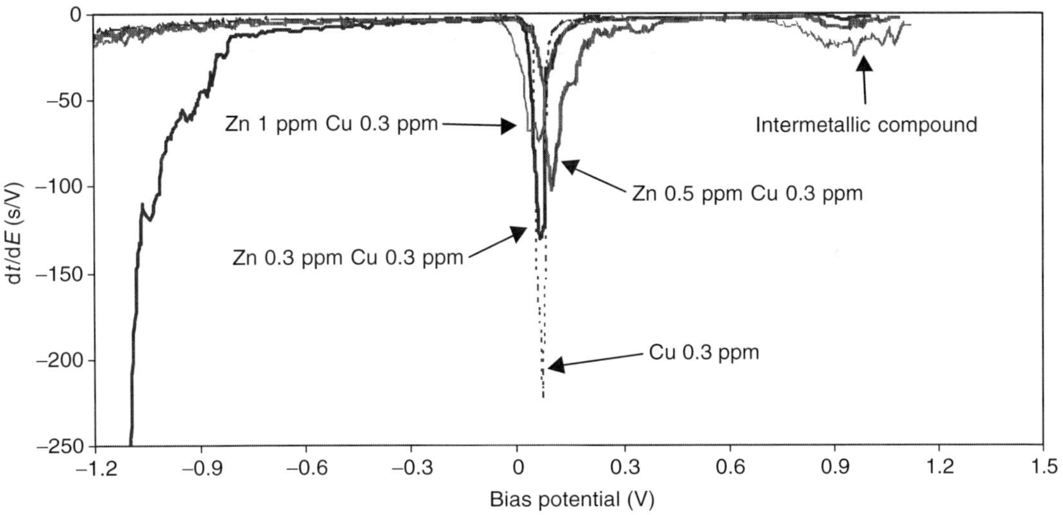

FIGURE 9.2
Examples of curves obtained with our PSA analyzer with synthetic samples containing lead, copper, cadmium, and zinc. The figure shows the linear behavior between peaks area and concentration for all the analytes we tested. The curves are obtained under the experimental conditions specified in Adami et al. (2005).

tools, which are user-friendly and easy to use, allow the detection of the analytical signals and extraction of the information data. With synthetic sample we obtained enough sensitivity to propose a first-screening apparatus but, when the sample is a real matrix, it can contain some different components, in particular some different metal ions, each one acting as a possible interferant. To overcome this problem, we tried to integrate in the proposed system a neural network algorithm, the multilayer perceptron (MLP), which appears to be able to reveal the real concentrations of the different metal ions, after a severe training (Figure 9.3). The PSA analyzer is based on the schematic block diagram proposed in Figure 9.1(a). A computer drives the apparatus and a suitable software interface allows even nontechnical personnel, without any specific hardware knowledge, to perform the experiments. The analyzer drives a glassy-carbon working electrode, a calomel reference electrode, and a platinum counter electrode; each user can utilize its proper electrochemical cell and its proper electrodes. The system is designed to perform the two-step process described in an article by Adami, Sartore, and Nicolini (61): during the first stage the working electrode polarization causes the deposition of the reduced metal ions on it. A potentiostat (62,63) drives the electrochemical cell through the counter electrode ensuring the proper biasing voltage $V_{ref/wk}$. In the second step (Figure 9.1(b)), a constant oxidizing current (1–100 μA) is fed into the working electrode and the metal is forced to move again into the solution, producing a potential drop between reference and working electrodes. The system becomes a galvanostat and the potential drop $V_{ref/wk}$ is now recorded as a function of time. The instrument software consists of two distinct parts: a low-level driver, which manages the data transfer to and from the interface electronics, and a high-level application, which provides a user-friendly interface with data acquisition and analysis options. The software tools are entirely written in C under the LabWindows CVI 5.5 environment (National Instruments). It operates under Microsoft Windows 95/98/NT/2000. The data acquisition software implements an algorithm to drive the experimental procedure, by setting the proper parameters on the experimental windows (61). The data analysis software transforms and plots the acquired data $E(t)$ in the inverse derivative form dt/dE (64,65) as a function of the recorded potential (Figure 9.3): those portions of the stripping potentiogram exhibiting a plateau are evidentiated, in the derivative

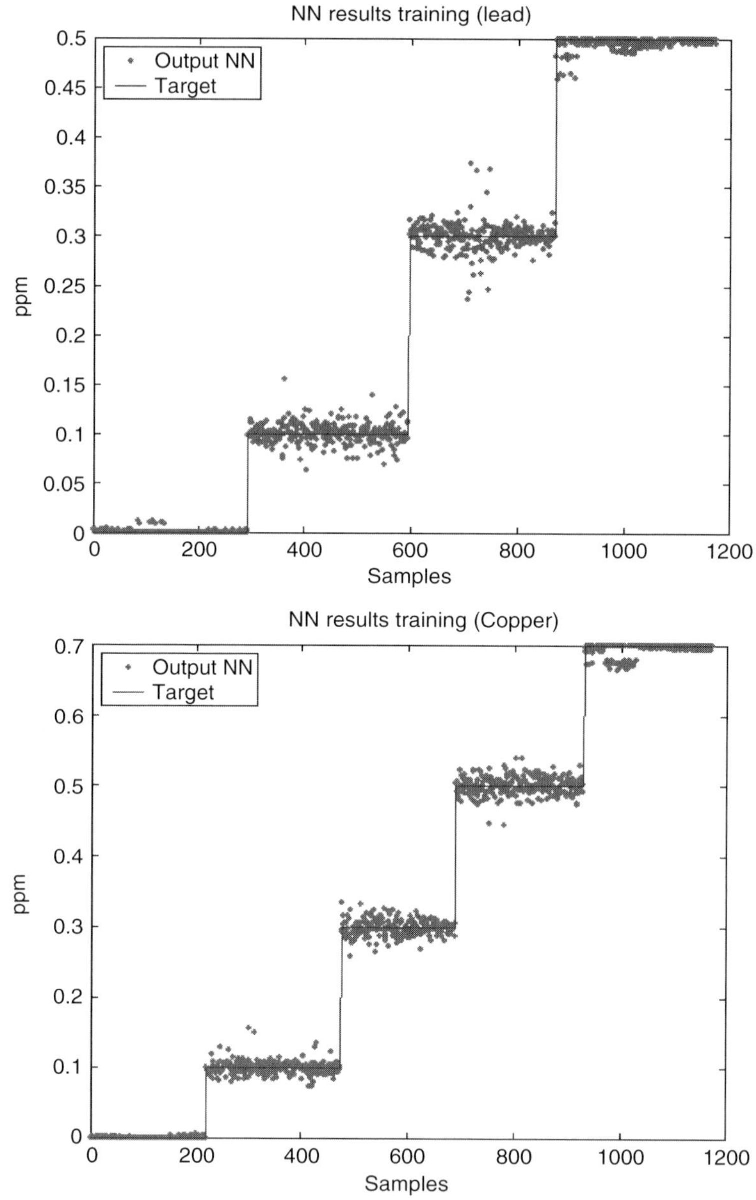

FIGURE 9.3
Neural Network response after the training phase. These plots report the expected and calculated ion concentrations for different samples on a test set for either lead or copper.

potentiogram, as peaks whose area is a linear function of the concentration of the metal in solution. The software automatically computes the inverse function and integrates the peaks within appropriate, user-specified intervals, relative to specific metal ions. In particular, for lead the integration limits are 0.42–0.62 V, for copper –0.1– 0.1 V, for cadmium 0.7–0.9 V, and for zinc 1–1.2 V. The software then recovers the concentration of an unknown sample by means of preacquired calibration curves. The neural network processing software is an MLP with one hidden and one output layer. The learning is achieved with a back propagation algorithm. The input data are the coefficients of the polynomial curve fitting the acquired data, i.e., the potential–time plot.

From the results described earlier it appears that sophisticated software implementation, such as those based on neural network (NN) algorithm could be utilized for recovering an analytical signal derived from a multisensing application. Many examples were proposed for gas sensing but not so many for ion screening in solution. We are optimizing our NN for a better discrimination of metal traces, by changing some parameters and some strategies and we hope that a software solution will eliminate the need to introduce modifications in the proposed analyzer, such as the use of a modified working electrode (with a mercury layer or with some chemical agents, such as chelant compounds or resins) or some sample treatments or more sophisticated electronics.

9.3 Liquid Sensing Using Polymer and Nanocomposite Matrices

Novel nanomaterials for use in bioassay applications represent a rapidly advancing field. Various nanostructures have been investigated to determine their properties and possible applications in biosensors. These structures include nanotubes, nanofibers, nanorods, nanoparticles, and thin films. Of these, nanoparticles are the best-studied structures. The biosensors based on different kinds of nanostructures are discussed in this review. A gas sensor, fabricated by selective growth of aligned carbon nanotubes (CNTs) by pulsed plasma on Si3N4/Si substrates patterned by metallic platinum, is presented for inorganic vapor detection at room temperature. Poly(o-anisidine) (POAS) deposition onto the CNTs device was shown to impart higher sensitivity to the sensor. Upon exposure to HCl the variation of the CNTs sensitivity is less than 4%, while the POAS-coated CNTs devices over a higher sensitivity (i.e., 28%). The extended detection capability to inorganic vapors is attributed to direct charge transfer with electron-hopping effects on intertube conductivity through physically adsorbed POAS between CNTs.

The CNTs thin film was grown using a radio frequency pulsed plasma-enhanced chemical vapor deposition (RF-PECVD) system. Prior to the nanotube growth, a Si_3N_4/Si substrate was patterned with platinum film (60 nm thick) by vacuum deposition through shadow masks, containing rectangular stripes 30 μm wide and a back-deposited thin film platinum heater commonly used in gas-sensor applications. A thin film (3 nm) of Ni catalyst was deposited onto the Si_3N_4/Si substrates using thermal evaporation. The CNTs film was obtained by pulsed RF-PECVD deposition on Si_3N_4/Si substrates as reported previously (66).

POAS was chemically synthesized by oxidative polymerization of the monomers by using $(NH_4)_2S_2O_8$ under controlled conditions. For the synthesis, the monomer, previously distilled, was dissolved in HCl 1M aqueous solution. The oxidant was thus added slowly maintaining the temperature at 0–4°C by means of an ice bath. The reaction was continued for about 12 h. The dark green precipitate emeraldine-salt form recovered from the reaction vessel was filtered and washed with distilled water. The precipitate was subsequently treated with a diluted solution of NH_4OH for 6 h to obtain the crude materials in the undoped form (blue in color). The precipitate recovered after filtration was washed again with water and treated first with methyl alcohol and then with ethyl ether for 24 h to remove the oligomers constituted by few monomer units. The material obtained from the last filtration was heated at 80°C in a vacuum system to remove the traces of solvents. The final product contained the emeraldine base form of POAS that was completely soluble in chloroform.

Figure 9.4 shows two distinct groups of *I*–*V* curves resulting from nanotubes and polymer POAS-coated nanotubes. Each group consists of results from several sensors. It can be seen from Valentini et al. (3) that the conductivity (slope) is almost the same for

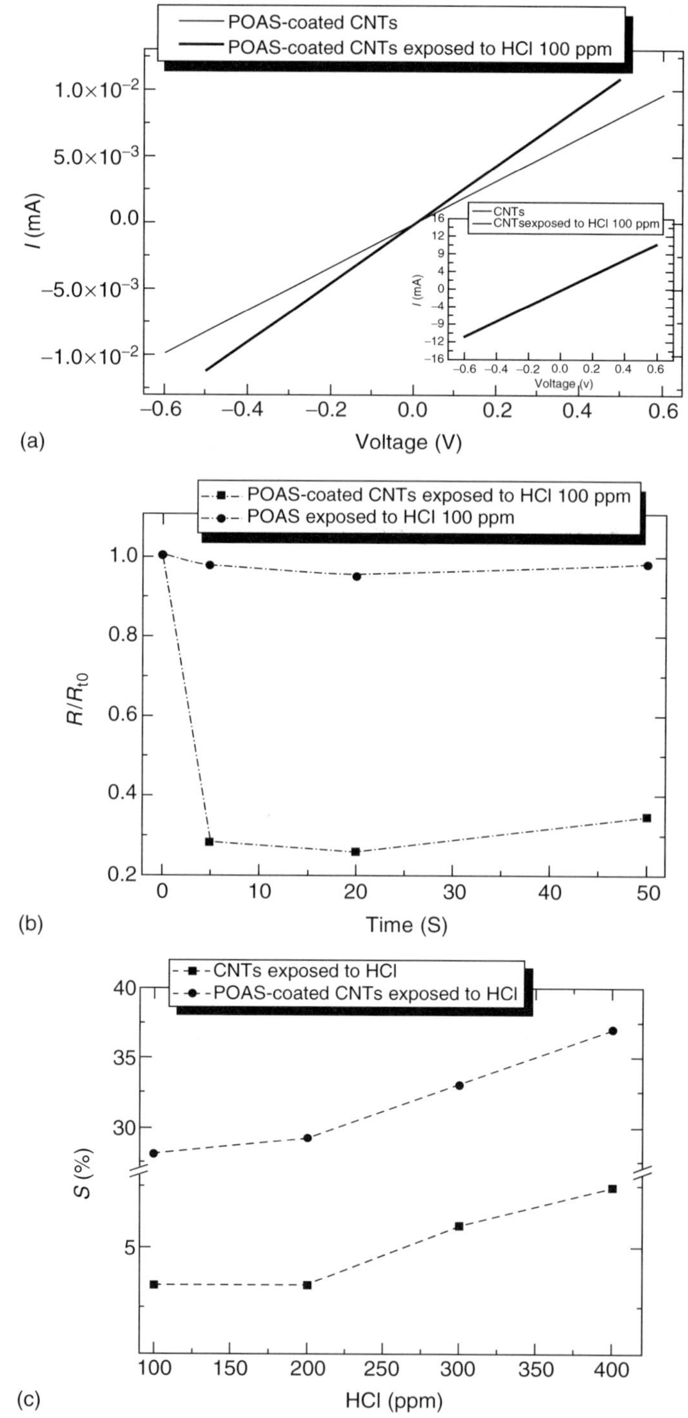

FIGURE 9.4
(a) *I–V* curves at room temperature of POAS-coated devices to HCl vapor. The inset shows *I–V* curves of the CNTs device at room temperature to HCl 100 ppm. (b) The time-dependent change of the normalized resistance (R_{t0} is the initial resistance of the sample) of POAS-coated device and CNT film at room temperature to HCl 100 ppm. (c) Sensitivity vs. HCl concentrations of POAS-coated device and CNT .lm at room temperature.

CNTs and POAS-coated CNTs, but it varies for the device when exposed to HCl. In particular, it shows that the conductance of nanotubes slightly increases, while the POAS-coated device shows a significant increment in conductance. It is interesting to note that the response transient of POAS-coated device is a few seconds, while a sensor based on resistance changes of POAS exhibits a poorer response time. The circumstance to maintain .lm resistance below 1 kΩ, which is significantly lower with respect to that reported for POAS sensors (about 50 MΩ), makes POAS-coated CNTs film integration in electronic circuitry easier and cheaper, since lower direct current (DC) voltages are required to drive the sensor response. The doping process of polyanilines is always associated with conformational modifications of the polymer chains owing to the local distortions created by the addition of H^+ ions to the basic sites and usually provides stable systems. It means that the conducting polymer in the doped form can be maintained in this state for long periods of time till the material reacts with basic reagents and strongly changes its chemical–physical properties. In other words, the reversibility of the process is not spontaneous.

If we define sensor sensitivity (S) as the ratio $S = [(R_A - R_G) = R_A] \times 100$, where R_A represents the resistance in air and R_G the resistance in vapor, the gas sensitivity increases from $S = 3.0$–27.9%. It reveals that by selecting proper polymer functionalization, sensor sensitivity to HCl may be improved.

In conclusion, CNTs thin films prepared by pulsed RF-PECVD demonstrated their potentiality as a new class of materials for HCl detection for environmental applications. Moreover, polymer functionalization enhances the sensitivity to these devices. The advances made here shall pave the way for future work in developing CNTs sensor arrays for highly sensitive and specific molecular detection and recognition in gases and in solutions.

9.4 Nanostructured Cytochrome Matrices for Liquid Sensing

All known cytochromes are heme-proteins, which are involved in the electron transfer process. P450 is the most versatile biological catalyst ever known (67,68). Cytochrome P450 is known to catalyze a great deal of reactions such as hydroxylation, O-dealkylation, and N-oxidation. Recombinant cytochrome P450s were employed, since recombinant proteins are more suitable for technological applications. In fact, they represent the homologous population of molecules with a controlled sequence, which can be ad hoc modified by site-specific mutagenesis, and they can undergo cheap mass production and furthermore have a high level of purity.

The presence of various important substrates for cytochrome P450s shows the possibility of utilizing the interaction between these enzymes and some specific substances for biosensor applications. In this field, the immobilization of the enzymes onto solid support is crucial. Several kinds of deposition techniques have been used over the last few years to produce sensitive elements based on protein molecules. In the previous studies, we used the layer-by-layer technique to immobilize P450 fusion protein (11,14), and the spreading technique to develop P4501A2 nanostructures (6). By comparing the results to the ones obtained by utilizing the Langmuir–Blodgett (LB) technique, it was found that in all cases the latter is the most suitable method for producing sensitive cytochrome matrixes. Moreover, this type of immobilization preserves the functionality of the molecules under the same working conditions of the biosensors (e.g., room temperature and dry conditions). Another work was concerned with an investigation of electron transfer between cytochrome P450scc (CYP11A1) and gold nanoparticles immobilized on rhodium–graphite electrodes. Thin films of gold nanoparticles were deposited onto the rhodium–graphite electrodes by drop casting. Cytochrome P450scc was deposited onto both gold nanoparticle-modified and bare rhodium–graphite electrodes. Cyclic voltammetry indicated enhanced activity of the enzyme at the gold nanoparticle-modified surface. The role of the nanoparticles in mediating electron transfer to the cytochrome P450scc was verified using ac impedance spectroscopy (69).

The amperometric responses were obtained by adding aliquots of clozapine (40 μM in methanol) in a working stirred system of 10 mM K-phosphate buffer, pH 7.4 (2 ml). Equilibration time was 5 min for each measurement. The cytochrome was electrochemically reduced at the potential of –600 mV vs. Ag/AgCl.

Protein studies utilizing electrochemical techniques such as cyclic voltammetry have provided insight into the functional properties of redox-active centers.

In several studies, we performed an electrochemical study of the cytochrome P450scc (15) and defined the best electrochemical parameters for this enzyme, which are also used for cytochrome P4501A2 and 2B4.

Electrodes with gold nanoparticles and P450scc were used for measurements of cholesterol concentration in solution.

Aliquots of a 100 μM cholesterol stock solution in 0.3% sodium cholate were added to the analyzed solution and the steady-state current was recorded. The steady-state current as a function of cholesterol concentration is shown in Figure 9.5. The plot shows a sensitivity of the biosensor of 0.13 $\mu A\ \mu M^{-1}$ and a detection range between 10 and 70 μM of cholesterol. This indicates that gold nanoparticles mediate electron transfer to the P450scc better than riboflavin (sensitivity 13.8 $nA\ \mu M^{-1}$ and detection limit 300 μM). The current response of the cytochrome P4501A2 electrode, in a function of clozapine concentration, was measured by chronoamperometry (6). This analysis was performed using the screen-printed thick film electrodes: a three-electrode standard configuration was used, where

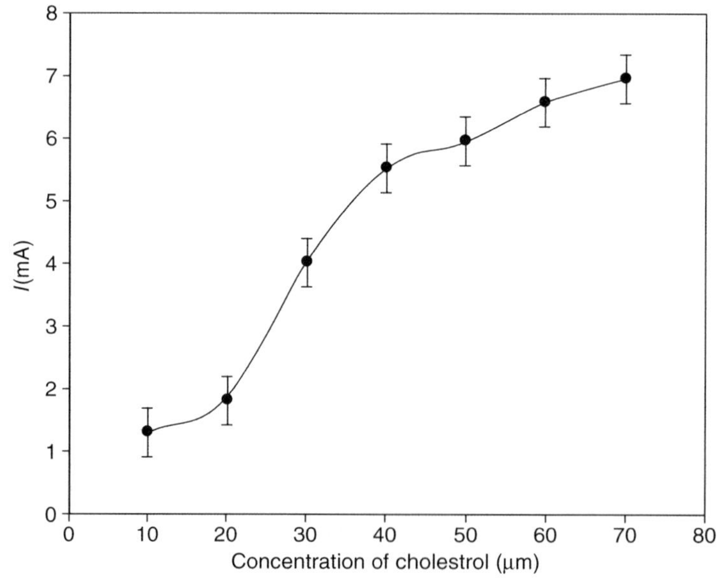

FIGURE 9.5
The amperometric response of screen-printed rhodium–graphite Au–P450scc electrode to increasing cholesterol concentration. About 100 μM stock solution in 0.3% sodium cholate, measured when fractions of 10 μM of cholesterol has been repeatedly added. The total volume of electrolyte was 60 μl. The current was measured at the potential −400 mV (vs. Ag/AgCl).

platinum wire acted as the counter electrode and a screen-printed electrode served as both the reference and the working one. The screen-printed rhodium–graphite electrodes were produced using a modified protocol according to (70,71). The area of the working electrode was 2 × 10 mm.

The chronoamperometry was employed to verify the possibility of producing an amperometric sensor to detect clozapine using cytochrome P4501A2. Aliquots of clozapine (40 μM in methanol) were added to the reaction mixture to obtain an amperometric response curve. Figure 9.6 shows the resulting current as a consequence of constant potential. It is observed that raising the clozapine concentration increased the response current. These results indicate that it is possible to detect the concentration of clozapine, free in solution, in the 20–300 nM range. This sensitivity is sufficient for routine measurements, since the therapeutic range of clozapine in plasma is between 0.16 and 1.83 μM (50–600 ng/ml) (72–74).

The results underline that the LB technique can be used to immobilize molecules of cytochrome P450 onto solid supports in a stable and reproducible way and to employ heme-protein structures as sensitive elements in biosensors. Nevertheless, to detect clozapine by electrochemical method we have found that, since we are dealing with a low-grade cytochrome, it appears more advantageous to use cytochrome P4501A2 entrapped in gel matrix. The data demonstrated that utilizing both types of systems (LB film and gel matrix) the presence of styrene, cholesterol, and clozapine employing cytochrome P4502B4, P450SCC, and P4501A2, respectively, could be properly detected. These results are encouraging for their possible future applications in the medical and ecological fields.

In conclusion, due to the very attractive pleiotropic properties of the heme-enzymes, P450s cytochrome isoforms confirm their potential to be used as sensing elements for a wide range of organic substances. It is worth noticing, however, that even if the LB or LS films appear to yield the best stable working conditions, the sensor technology may yield optimal results also with other immobilization techniques, such as solution casting improving cholesterol sensing to the range 11.5–1.2 mg/dl without gold nanoparticles (8) and to the range 2.7–0.4 mg/dl with gold nanoparticles (68) in presence of 200 μg of P450scc on the electrode.

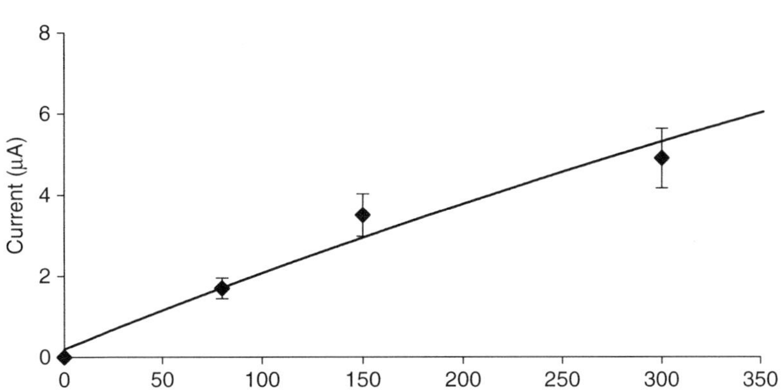

FIGURE 9.6
Current response in function of the concentration of clozapine (20 nM in methanol) added to cytochrome P4501A2 gel-matrix, utilizing s.p.e. in 10 mM K-phosphate buffer, pH 7.4. The working electrode was poised at $U = -600$ mV.

9.5 Nanostructured Bacteriorodhopsin Matrices for Liquid Sensing

Starting from the effects of volatile anesthetics on the structure of bacteriorhodopsin (bR) in the purple membrane (PM) in solution this work tries to investigate the interaction of ether and hydrocarbon-type anesthetic vapors with self-assembled bR thin films. A dedicated constant flux chamber has been built to maintain the sample in a rather constant atmosphere of anesthetics, during the absorption measurements. The kinetics of absorption and desorption have been determined (Figure 9.7). The results obtained have been compared with those of bR in solution (36).

The suspensions of PM have been obtained by dilution of the stock suspension (5 mg ml^{-1}) with distilled water. The concentration was checked spectrophotometrically; in particular, the peak of proteins (at 280 nm) should not show the characteristic scattering signal. The concentration chosen for the experiments was 0.3 mg ml^{-1}.

In the case of bR in a solution, the anesthetic solution was just added to the bR solution and stirred for 10 min. This was done to equilibrate the system. In the case of chloroform, not mixable with water, the cuvette was shaken to promote the contact of PM with the solvent. The self-assembled samples (drop-coating multilayers) were formed by spreading 1 ml of the original stock suspension of PM over the flat quartz surface (75).

This operation had to be carried out carefully to obtain a homogeneous distribution of material over the surface. Moreover, the time of drying, by exposure to the ambient atmosphere, was important to have a good sample. After 12 h the self-assembled thick sample containing 5 mg of PM patches, 80% of which is bR (MW 26,000) assembled over a surface of 1.5 cm^2, was ready. In our case, the bR film was exposed to the anesthetic vapors, coming to the reaction chamber through tubes from the separate vessel containing the anesthetic in liquid form (saturated steam). The amount of the incoming anesthetic vapors was qualitatively monitored, injecting by a syringe different amounts of vapor proceeding from the heated vessel.

In Figure 9.8(a), a self-assembled film of bR in PM was treated with diethylether vapors: (a) the absorption spectra of the film, pure and treated with diethylether for different periods of time, is shown; (b) the decrease in the absorbance at 570 nm was recorded during the injection of successive amounts of vapors of anesthetic. In Figure 9.8(a), the absorbance maximum of the PM-thick film appears at 563 nm and the intensity of absorbance decreases with the treatment with anesthetic. At the same time, a broad band occurs at about 400 nm during the treatment and disappears after relaxation. It is difficult to give an interpretation of such a band, which cannot be precisely attributed to the anesthetic-induced bR480 and bR380 forms; further experiments will probably clarify the role of this band. Nevertheless, this figure reveals either the functionality of the protein or the reversibility of the process qualitatively comparable with that obtained in solution. Zones A, B, C, D are evident in Figure 9.8(b), each of them corresponds to a progressive increase of the anesthetic pressure zone A is more wide than the others, this could be due to a higher pressure of the vapor injected in the cell or an effect of saturation of the sample that makes the successive binding in the other zones more difficult. After the decrease of the absorbance of 0.1 units, compressed air was injected in three steps (regions 1, 2, 3), obtaining almost the total recovery of the structure after 15 min.

The work pointed out the possibility of using the self-assembled bR films as sensitive layers for optical biosensors for anesthetics. The important point is that the optical properties of the layer are reversible for not only ether-type anesthetics but also hydrocarbon-type such as chloroform. It allows one to consider such films as good candidates for biosensors of continuous monitoring of anesthetics.

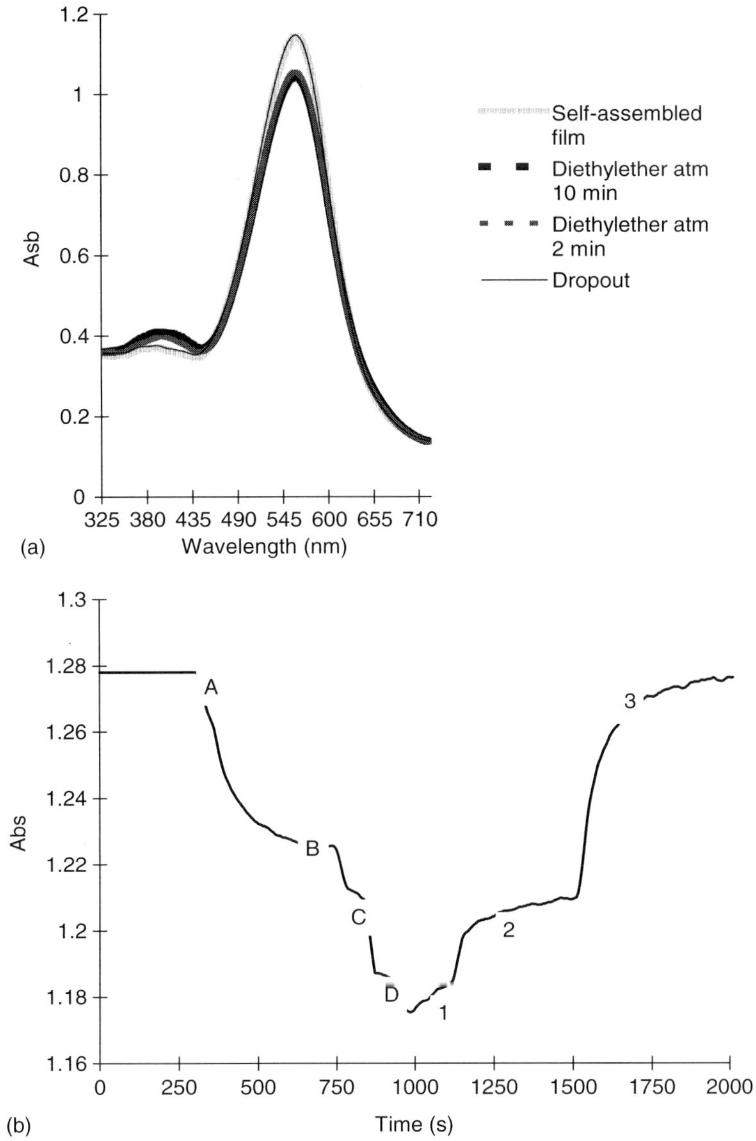

FIGURE 9.7
(a) The absorption spectra of the film, pure and treated with diethylether, for different periods of time and after relaxation. (b) The absorbance profile at 570 nm of a self-assembled film of bacteriorhodopsin upon exposition to vapors of diethyl ether by increasing pressure (A, B, C, D) and upon exposition to a compressed air stream for the desorption of the anesthetic (regions 1, 2, 3).

9.6 Organic Nanostructured Matrices for Gas Sensing

A nanocomposite of multiwalled carbon nanotubes (MWNTs) embedded in poly(2,5-dimethylaniline) (PDMA) was synthesized by oxidative polymerization. The conducting polymer (PDMA) was chemically synthesized by oxidative polymerization of the monomers by using $(NH_4)_2S_2O_8$ under controlled conditions. For the synthesis, a quantity of 100 mg of accurately weighed MWNTs was dispersed in 0.5 M HCl aqueous solution by

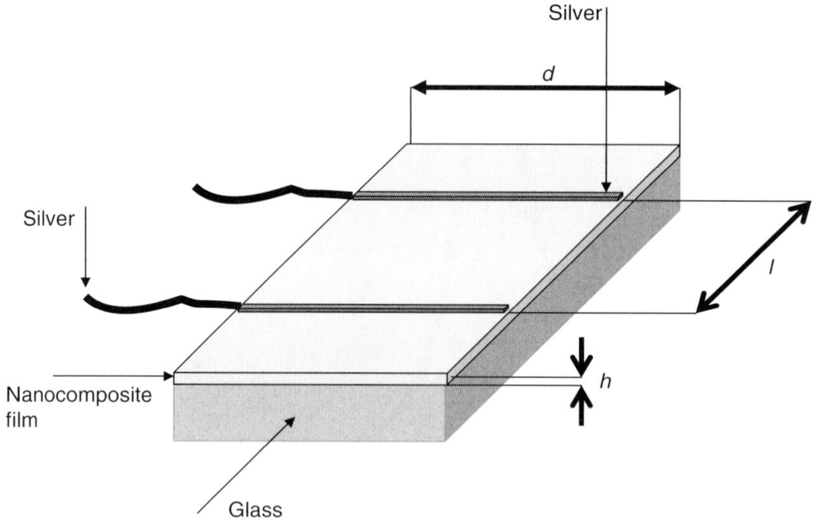

FIGURE 9.8
Schematic of devices employed for the determination of the specific resistance.

means of ultrasonic equipment. The final product contained the emeraldine base (EB) form of PDMA–MWNTs nanocomposite material. The EB form was completely soluble in chloroform. Langmuir–Scahefer (LS) films were obtained by spreading solutions of PDMA–MWNTs nanocomposite and dissolving 5 mg of the material in 100 ml of chloroform. The related compression isotherms were recorded in a manner similar to our earlier investigation on POAS–MWNTs nanocomposite (76). The cyclic voltammetry was useful in determining the oxidation states of PDMA–MWNTs nanocomposite. The steric hindrance and the simultaneous presence of MWNTs in the polymeric matrix are responsible for the decreased number of redox states and also affect the stability of the nanocomposite ES form, giving a possible explanation for the quick undoping process revealed during the experiments. This phenomenon was discovered by observing the quick color change, from green to violet, of the films of material deposited on a substrate for further characterizations. The nanocomposite (PDMA–MWNTs) showed a progressive spontaneous undoping process along the time associated with the instability of the doping agent, constituted by HCl, inside the polymeric matrix. The study of the undoping process revealed that this phenomenon is related to the synergetic effect of the steric hindrance of the substituents on the aromatic rings and the presence of MWNTs inside the polymeric matrix. The conducting properties connected to a doping–undoping equilibrium in the presence of the doping agent were also investigated. To verify whether or not the deposited films could be affected by basic vapors in the environment, the various samples were transferred in a sealed container. This simple experiment showed that the undoping process was spontaneous and did not need the presence of basic chemicals to be carried out. Since they had shown a spontaneous undoping process, the samples were tested very quickly to obtain accurate values of the specific resistance. To investigate the electrical properties of the material focused on industrial prospective, other configurations were tested. The deposition of thin films of material with adequate thickness (several microns) was performed by spin-coating technique. This method allowed us to fabricate thicker films in a very short period of time.

The fabricated devices were connected to the electrometer by means of silver wires and silver paint, as shown in the schematic of Figure 9.8. In Figure 9.9, the *V*/*I* characteristics

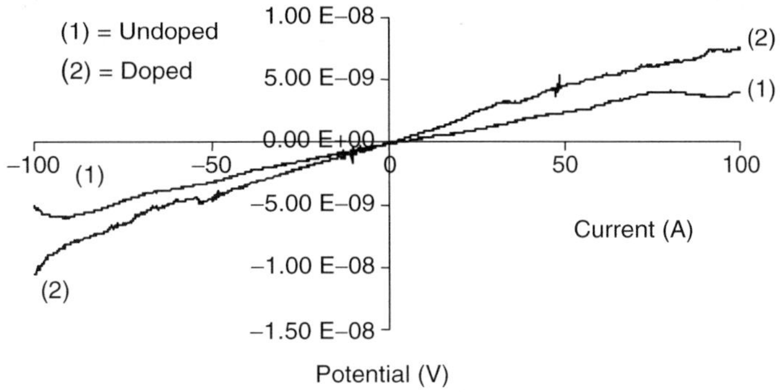

FIGURE 9.9
V/*I* characteristics of PDMA–MWNTs nanocomposite in both the undoped (curve 1) and doped (curve 2) forms. The material showed a quasi-linear behavior.

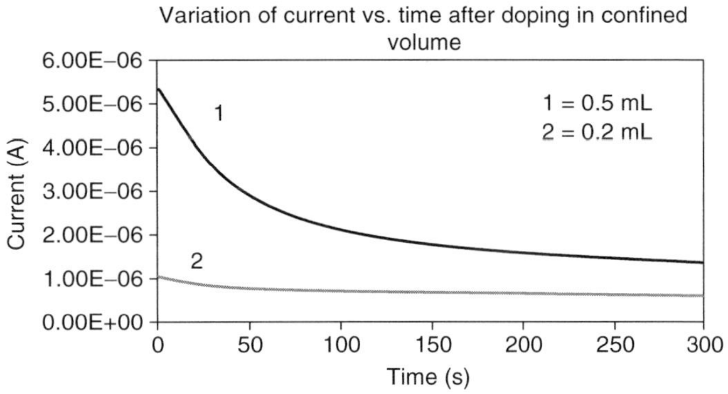

FIGURE 9.10
Response of the sensor device to HCl vapors.

of PDMA–MWNTs nanocomposite in the undoped and doped forms evidenced by curves 1 and 2, respectively, are illustrated. The instability of the doping process allowed us to fabricate a spontaneous reversible sensor for acid vapors (7) by setting up a comparative potentiometric circuit and engineering the sensitive element directly on the circuit board. To test the response of the fabricated device after exposing to HCl vapors, an experimental set-up was realized as following. The device was placed inside a container of 120 ml and sealed by means of a parafilm layer, letting the wires out through it to be connected to the electrometer. Volumes of 0.2 and 0.5 ml of saturated vapors of HCl were injected through the parafilm membrane. Current versus time measurements at a fixed bias of 10 V were then carried out. The relative results obtained from the experiments are shown in Figure 9.10. It is observed that the system tends to return to the initial conditions of resistance after about 3 min since the injection of HCl vapors. This behavior is the consequence of the instability of EB/ES equilibrium, which can be shifted to the stable formation of the ES form only by means of a continuous dynamic inflation of acid vapors. Curve 1 represents the saturation level of the sensor device valuable in a concentration of 4200 ppm of HCl vapors, since higher levels of concentration gave the same results. Curve 2 shows the data obtained for an intermediate concentration between the absence and the saturation of vapors, and the device evidenced a response in function of HCl vapors concentration. The experiment was carried out by quickly injecting the acid vapors into the system containing the sensor device; and the acquisition of data started immediately after the complete injection (7).

It will be the future goal of our research to study other nanocomposites synthesized by using different monomers and concentrations of MWNTs inside the polymeric matrix for a wide range of gas-sensor applications.

9.7 Conclusions

This chapter has shown, with few key examples that recently appeared in the literature by our group, the significant potential of nanostructured organic matrices for developing intelligent sensors for gas and liquid having potential impact on health care and environment control (19,22,29,38).

Acknowledgments

We are particularly grateful to other members of the Nanoworld Institute and Fondazione Elba active in this field as shown by the list of publications, namely Sandro Carrara, Valter Bavastrello, and Enrico Stura. This work was supported by FISR-MIUR grant on Molecular Electronics and Lithography to Fondazione Elba and Nanoworld Institute-CIRNNOB of the University of Genova and by a FIRB-MIUR grant on Organic Nanoscience and Nanotechnology to the Nanoworld Institute, namely to the Fondazione EL.B.A and to the Centro Interuniversitario di Ricerca sulle Nanotecnologie e Nanoscienze Organiche e Biologiche of Genova University. This work is also supported by a research grant from MIUR-PNR on Biocatalysis.

References

1. Valentini, L., Bavastrello, V., Armentano, I., D'Angelo, F., Pennelli, G., Nicolini, C., and Kenny, J. M. (2004). Synthesis and electrical properties of CdS Langmuir–Blodgett multilayers nanoparticles on self-assembled carbon nanotubes. *Chem. Phys. Lett.* 392:214–219.
2. Ghisellini, P., Paternolli, C., Antonini, M., and Nicolini, C. (2004). P450scc mutant nanostructuring for optimal assembly. *IEEE Trans. Nanobioscience* 3:121–128.
3. Valentini, L., Bavastrello, V., Stura, E., Armentano, I., Nicolini C., and Kenny, J. M. (2004). Sensors for inorganic vapor detection based on carbon nanotubes and poly(o-anisidine) nanocomposite material. *Chem. Phys. Lett.* 383:617–622.
4. Bavastrello, V., Carrara, S., Ram, M. K., and Nicolini, C. (2004). Optical and electrochemical properties of poly(ortho-toluidine)-multi walled carbon nanotubes composite Langmuir-Schaefer films. *Langmuir* 20:969–973.
5. Bertoncello, P., Notargiacomo, A., and Nicolini, C. (2004). Synthesis, fabrication and characterizatoin of poly[3-3'(vinylcarbazole)] (PVK) Langmuir-Schaefer films. *Polymer* 45:1659–1664.
6. Antolini, M., Ghisellini, P., Paternolli, C., and Nicolini, C. (2004). Electrochemical study of the engineerized cytochrome P450scc interaction with free and in lipoproteins cholesterol. *Talanta* 62:945–950.
7. Bavastrello, V., Stura, E., Carrara, S., Erokhin, V., and Nicolini, C. (2004). Poly(2,5-dimethylaniline-MWNTs nanocomposite: a new material for conductometric acid vapors sensor. *Sens. Actuators, B* 98:247–253.
8. Shumantseva, V., De Luca, G., Bulko, T., Carrara, S., Nicolini, C., Usanov, S. A., and Archakov, A. (2004). Cholesterol amperometric biosensor based on cytochrome P450scc. *Biosens. Bioelectron.* 19:971–976.
9. Pechkova, E. and Nicolini, C. (2003). *Proteomics and Nanocrystallography*, New York, USA, Kluwer Academic Publisher, pp. 1–212.
10. Pastorino, L., Disawal, S., Nicolini, C., Lvov, Y. M., and Erokhin, V. V. (2003). Complex catalytic colloids on the basis of firefly luciferase. *Biotechnol. Bioeng.* 84:286–291.

11. Paternolli, C., Ghisellini, P., and Nicolini, C., (2002). Development of immobilization techniques of cytochrome P450-GST fusion protein. *Colloid Surf. B* 23:305–311.
12. Ghisellini, P., Paternolli, C., Chiossone, I., and Nicolini C. (2002). Spin state transitions in Langmuir–Blodgett films of recombinant cytochrome P450scc and adrenodoxin. *Colloid Surf. B* 23:313–318.
13. Ding, H., Bertoncello, P., Ram, M. K., and Nicolini, C. (2002). Electrochemical investigations on MEH-PPV/C_{60} nanocomposite Langmuir-Schaefer films. *Electrochem. Commun.* 4:503–505.
14. Paternolli, C., Ghisellini, P., and Nicolini, C. (2002). Pollutant sensing layer based on cytochrome P450. *Mat. Sci. Eng. C.* 22:155–159.
15. Nicolini, C., Erokhin, V., Ghisellini, P., Paternolli, C., Ram M. K, and Sivozhelezov V, (2001). P450scc engineering and nanostructuring for cholesterol sensing. *Langmuir* 17:3719–3726.
16. Ram, M. K., Bertoncello, P., and Nicolini, C., (2001). Langmuir-Schaefer films of processable poly(o-ethoxyaniline) conducting polymer: fabrication and characterization as sensor for heavy metals. *Electroanalysis* 13:574–581.
17. Ding, H., Ram, M. K., and Nicolini, C. (2001). Nanofabrication of organic/inorganic hybrid of TiO_2 with substituted phthalocyanine or polythiophene. *J. Nanosci. Nanotechnol.* 1:207–213.
18. Ram, M. K., Bertoncello, P., Ding, H., Paddeu, S., and Nicolini, C. (2001). Cholesterol biosensors prepared by layer-by-layer technique. *Biosen. Bioelectron.* 16:849–856.
19. Nicolini, C., Erokhin, V., and Ram M. K., (2001). Supramolecular layer engineering for industrial nanotechnology. In *Nano-surface chemistry*, Rosoff, M. (ed.), Marcel Dekker New York, pp. 141–212.
20. Shumyantseva, V. V, Bulko, T. V, Usanov, S. A., Schmid, R. D., and Nicolini, C. (2001). Construction and characterization of bioelectrocatalytic sensors based on cytochromes P450. *J. Inorg. Biochem.* 87:185–190.
21. Ram, M. K., Adami, M., Paddeu, S., and Nicolini, C. (2000). Nanoassembly of glucose oxidase on the in situ self-assembled electrochemical characterizations. *Nanotechnology* 11:112–119.
22. Ram, M. K. and Nicolini, C. (2000). Thin Conducting Polymeric Films and Molecular Electronics. In *Recent Research Development in Physical Chemistry*, Kerala, India, TransWorld Publishing 4:219–258.
23. Nicolini, C., Erokhin, V., Paddeu, S., Paternolli, C., and Ram, M. K. (1999). Toward bacteriorhodopsin based photocells. *Biosen. Bioelectron.* 14:427–433.
24. Nicolini, C., Erokhin, V., Paddeu, S., and Sartore, M. (1998). Towards light-addressable transducer bacteriorhodopsin based. *Nanotechnology* 9:223–227.
25. Erokhin, V., Carrara, S., Amenitch, H., Bernstorff, S., and Nicolini, C. (1998). Semiconductor nanoparticles for quantum devices. *Nanotechnology* 9:158–161.
26. Paddeu, S., Ram, M. K., Carrara, S., and Nicolini, C. (1998). Langmuir–Schaefer films of poly(o-anisidine conducting polymer for sensors and displays. *Nanotechnology* 9:228–236.
27. Ram, M. K., Carrara, S., Paddeu, S., Maccioni, E., and Nicolini, C. (1997). Poly(o-anisidine Langmuir–Schaefer films: fabrication and characterization. *Langmuir* 13:2760–2765.
28. Ram, M. K., Maccioni, E., and Nicolini, C. (1997). The electrochromic response of polyaniline and its copolymeric systems. *Thin Solid Films* 303:27–33.
29. Nicolini, C. (1997). Protein monolayer engineering: principles and application to biocatalysis. *Trends Biotechnol.* 15:395–401.
30. Erokhin, V., Facci, P., Carrara, S., and Nicolini, C. (1997). Fatty acid based monoelectronic device. *Biosen. Bioelectron.* 12:601–606.
31. Nicolini, C., Erokhin, V., Facci, P., Rossi, A., Guerzoni, S., and Paschkevitch, P. (1997). DNA Based multiquartz sensor. *Biosen. Bioelectron.* 12:613–618.
32. Adami, M., Zolfino, I., Fenu, S., Nardelli, D., and Nicolini, C. (1996). Potentiometric alternating biosensing toxicity tests on cell population. *J. Biochem. Biophys. Methods* 32:171–181.
33. Facci, P., Erokhin, V., Carrara, S., and Nicolini, C. (1996). Room temperature single-electron junction. *Proc. Natl. Acad. Sci. U S A* 93:10556–10559.
34. Adami, M., Sartore, M., and Nicolini, C. (1996). Potentiometric and nanogravimetric biosensors for drug screening and pollutants detection. *Food Technol. Biotechnol.* 34:125–130.
35. Fanigliulo, A., Accossato, P., Adami, M., Lanzi, M., Martinoia, S., Paddeu, S., Parodi, M. T., Rossi, A., Sartore, M., Grattarola, M., and Nicolini, C. (1996). Comparison between a LAPS and a n FET-based sensor for cell-metabolism detection. *Sens. Actuators, B Chem.* 32:41–48.

36. Maccioni, E., Radicchi, G., Erokhin, V., Paddeu, S., Facci, P., and Nicolini, C. (1996). Bacteriorhodopsin thin film as a sensitive layer for an anaesthetic sensor. *Thin Solid Films* 284–285:898–900.
37. Erokhin, V., Facci, P., Carrara, S., and Nicolini, C., Monoelectron phenomena in nanometer scale particles formed in LB films. *Thin Solid Films* 284–285:891–893.
38. Nicolini, C., Supramolecular architecture and molecular bioeletronics. (1996). *Thin Solid Films* 284–285:1–5.
39. Tronin, A., Dubrovsky, T., Radicchi, G., and Nicolini, C. (1996). Optimisation of IgG Langmuir film deposition for application as sensing elements. *Sens. Actuators, B* 34:276–282.
40. Piras, L., Adami, M., Fenu, S., Dovis, M., and Nicolini, C. (1996). Immunoenzymatic application of a redox potential biosensor. *Anal. Chimica. Acta* 335:127–135.
41. Troitsky, V. I, Sartore, M., Berzina, T. S., Nardelli, D., and Nicolini, C. (1996). Instrument for depositing Langmuir–Blodgett films composed of alternating monolayer using a protective layer of water. *Rev. Sci. Instrum.* 67:4216–4223.
42. Adami, M., Sartore, M., and Nicolini, C. (1995). PAB: a newly designed and integrated potentiometric alternating biosensor. *Biosen. Bioelectron.* 10:155–167.
43. Paddeu, S., Fanigliulo, A., Lanzi, M., Dubrovsky, T., and Nicolini, C. (1995). LB-based PAB immunosystem: activity of immobilized urease monolayer. *Sens. Actuators, B Chem.* 25:876–882.
44. Nicolini, C., Adami, M., Dubrovsky, T., Erokhin, V., Facci, P., Paschkevitch, P., and Sartore, M. (1995). High-sensitive biosensor based on LB technology and on nanogravimetry. *Sens. Actuators, B Chem.* 24:121–128.
45. Adami, M., Alliata, D., Del Carlo, C., Martini, M., Piras, L., Sartore, M., and Nicolini, C. (1995). Characterization of silicon transducers with Si_3N_4 sensing surfaces by AFM and PAB systems. *Sens. Actuators, B Chem.* 24–25:889–893.
46. Nicolini, C., Sartore, M., Zunino, M., and Adami, M. (1995). A new instrument for the simultaneous determination of pH and redox potential. *Rev. Sci. Instrum.* 66:4341–4346.
47. Nicolini, C., Lanzi, M., Accossato, P., Fanigliulo, A., Mattioli, F., and Martelli, A. (1995). A silicon-based biosensor for real-time toxicity testing in normal versus cancer liver cells. *Biosen. Bioelectron.* 10:723–733.
48. Diaspro, A., Radicchi, G., and Nicolini, C. (1995). Polarized light scattering: a biophysical method for studying Bacterial cells. *IEEE Trans. Biomed. Eng.* 42:1038–1043.
49. Erokhin, V., Facci, P., Carrara, S., and Nicolini, C. (1995). Observation of room temperature mono-electron phenomena on nanometre-sized CdS particles. *J. Phys. D: Appl. Phys.* 28:2534–2538.
50. Facci, P., Radicchi, G., Erokhin, V., and Nicolini, C. (1995). On the mobility of Immunoglobulines G in Langmuir-Blodgett films. *Thin Solid Films* 269:85–89.
51. Paddeu, S., Antolini, F., Dubrovsky, T., and Nicolini, C. (1995). Langmuir–Blodgett film of glutathione S-transferase immobilised on silanized surfaces. *Thin Solid Films* 268:108–113.
52. Sartore, M., Adami, M., Bousse, L., Afeman, D., Mostarshed, S., and Nicolini, C. (1994). Investigation of carrier transport through silcon wafers by AC photocurrent measurements. *J. Appl. Phys.* 75:4000–4008.
53. Adami, M., Piras, L., Lanzi, M., Fanigliulo, A., Vakula, S., and Nicolini, C. (1994). Monitoring of enzymatic activity and quantitative measurements of substrates by means of a newly designed silicon-based potentiometric sensor. *Sens. Actuators, B* 18–19:178–182.
54. Dubrovsky, T., Vakula, S., and Nicolini, C. (1994). Preparation and immobilization of Langmuir Blodgett films of antibodies conjugated to enzyme for potentiometric sensor application. *Sens. Actuators* 22:69–73.
55. Adami, M., Sartore, M., Rapallo, A., and Nicolini, C. (1992). Possible developments of a potentiometric biosensor. *Sens. Actuators B* 7:343–346.
56. Sartore, M., Adami, M., Nicolini, C., Bousse, L., Mostarshed, S., and Hafeman, D. (1992). Minority carrier diffusion length effects on light-addressable potentiometric sensor (LAPSdevices). *Sens. Actuators A* 32, 431–439.
57. Sartore, M., Adami, M., and Nicolini, C. (1992). Computer simulation and optimization of a light addressable potentiometric sensor. *Biosen. Bioelectron.* 7:57–68.
58. Sartore, M., Adami, M., Baldini, E., Rossi, A., and Nicolini, C. (1992). New measuring principles for LAPS devices. *Sens. Actuators B* 9:25–36.

59. Nicolini, C., Adami, M., Antolini, F., Beltram, F., Sartore, M., and Vakula, S., (1992). Biosensors: a step to bioelectonics. *Phys. World* 5:30–34.
60. Estela, J. M., Tomás, C., Cladera, A., and Cerdà, V. (1995). Potentiometric stripping analysis: a review. *Crit. Rev. Anat. Chem.* 25:91–141.
61. Adami, M., Sartore, M., and Nicolini, C. (2005). A Potentiometric stripping analyzer for multianalyte screening. *Sens. Actuators, B Chem.* submitted.
62. Bard, A. J. and Faulkner, L. R. (1980). *Electrochemical Methods: Fundamentals and Applications.* Wiley, New York.
63. Jagner, D., Sahlin, E., Axelsson, B. and Ratana-Ohpas, R. (1993). Rapid method for the determination of copper(II) and lead(II) in tap water using a portable potentiometric stripping analyzer. *Anal. Chimica. Acta* 278:237–242.
64. Lo Coco. F., Monatti, P., Rizzotti, S., and Ceccon, L. (1999). Determination of lead in oil products by derivative potentiometric stripping analysis. *Anal. Chimica. Acta* 386:41–46.
65. Valentini, L., Armentano, I., Kenny, J. M., Lozzi, L., and Santucci, S. (2003). Pulsed plasma-induced alignment of carbon nanotubes. *Mater. Lett.* 57:3699.
66. Gonzalez, F. J. (1992). Human cytochromes P450: problems and prospects. *Trends Pharmacol. Sci.* 13:346–352.
67. Wrighton, S. A. and Stevens, J. C., (1992). The hepatic cytochromes P450 involved in drug metabolism. *Crit. Rev. Toxicol.* 22:1–21.
68. Shumyantseva, V. V., Carrara, S., Bavastrello, V., Riley, D. J., Bulko, T. V., Skryabin, K. G., Archakov, A. I., and Nicolini, C. (2005). Direct electron transfer between cytochrome P450scc and gold nanoparticles on screen-printed rhodium–graphite electrodes. *Biosen. Bioelectron.* 21:217–222.
69. Kulys, J. and D'Costa, E. J. (1991). Printed amperometric sensor based on TCNQ and cholinesterase. *Biosen. Bioelectron.* 6:109–115.
70. Bachmann, T. T. and Schmid, R. D. (1999). A disposable multielectrode biosensor for rapid simultaneous detection of the insecticides paraoxon and carbofuran at high resolution. *Anal. Chimica. Acta* 401:95–103.
71. Haring, C., Barnas, C., Saria, A., Humpel, C., and Fleischhacker, W. W., (1989). Dose-related plasma levels of clozapine. *J. Clin. Psychopharmacol.* 9:71–72.
72. Buur-Rasmussen, B. and Brøsen, K. (1999). Cytochrome P450 and therapeutic drug monitoring with respect to clozapine. *Eur. Neuropsychopharm.* 9:453–459.
73. Rahden-Staron, I., Czeczot, H., and Szumilo, M. (2001). Induction of rat liver cytochrome P450 isoenzymes CYP 1A and CYP 2B by different fungicides, nitrofurans, and quercetin. *Mutat. Res./Genet Toxicol. Environ. Mutagen.* 498:57–66.
74. Grindlay, J. E. and Bailyn, C. D. (1993). Birth of millisecond pulsars in globular clusters. *Nature* 336:48–50.
75. Bavastrello, V., Ram, M. K., and Nicolini, C. (2002). Synthesis of multiwalled carbon nanotubes and poly(o-anisidine) nanocomposite material: fabrication and characterization of its Langmuir–Schaefer films. *Langmuir* 18:1535–1541.

10

The Challenge of Human Skin—Engineering the Biotic/Abiotic Interface

Puneet Khanna, Steven Hoath, Rod Smallwood, and Shekhar Bhansali

CONTENTS

10.1 Introduction

Technological advances in molecular biology, tissue engineering, and medicine have been paralleled by rapid and profound gains in microelectronics, smart materials research, and information processing. The confluence of these emerging disciplines promises to bridge the biotic (organism) and abiotic (environment) realms in unanticipated ways. The dynamic coupling of organism and environment draws attention to those boundary structures that naturally allow flexible and adaptive sensory communication while simultaneously providing the physical barrier required for life outside the womb or ocean.

This review focuses on the unique biological boundary of the body manifested by the surface of the human skin. Specifically, attention is drawn to the outermost layer of the skin, the stratum corneum (SC), which as the terminally differentiated (dead) product of epidermal differentiation forms the ultimate biopolymeric interface with the environment. The SC, viewed through the lens of a materials engineer, is a multifunctional nanocomposite with all the hallmarks of a smart material. The strategic location of this highly organized, thin (~20 μm), bioengineered film places it in an optimal location to mediate biotic/abiotic interactions. Improved sensor systems for noninvasive detection of physiologically relevant mechanical, electrical, or magnetic signatures at the surface of the body (data beyond current electrocardiography or electroencephalography) require bridging to the body via this complex but assessable biological structure.

The iterative development of such extended, noninvasive, bioengineered interfaces should not ignore the body's own exquisite material solution to biotic/abiotic interface design. Seamless coupling to the body surface requires a better understanding of the molecular architecture of the SC as a polymer film and the functional unit organization of the epidermis. This challenge amounts to nothing less than an initial understanding of the boundary conditions of the human body. This will necessitate integration of broad advances in a wide variety of fields viz. microsystems, clinical medicine, cell and molecular biology, and information technology. Such extended interface systems would be typically freestanding, multifunctional, multimode, microfluidics-based sensor systems (Figure 10.1) coupled to human skin for real-time monitoring at global (physiological) and local (molecular) levels.

The scope of systems containing bioengineered microinterfaces goes beyond sampling and extraction of high-quality distributed physiological data. Advanced systems hold the

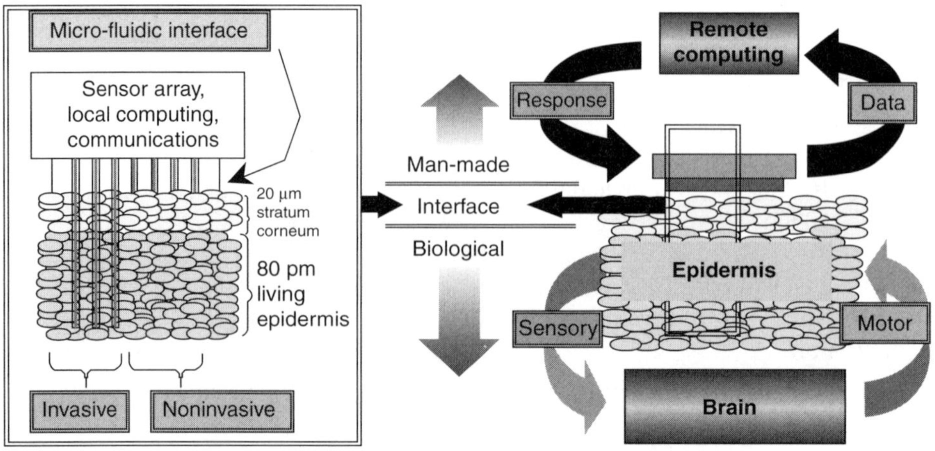

FIGURE 10.1
Bioengineered interface between analogous biological and MEMS-based systems.

potential to incorporate both diagnostic as well as therapeutic functionality. Such systems would utilize information in the same manner as the body uses the skin surface for sensorimotor feedback control [2–8].

Development of bioengineered skin-interfacing systems require interdisciplinary research combining diverse fields such as tissue engineering, molecular biology, anatomy, physiology, medical physics, biology, mechanical engineering, microfluidics, chemical processes, microfabrication, nanotechnology, and signal processing. This chapter reviews the development of these interface systems from the MEMS perspective.

10.2 Skin-Interface Systems: Concept

When a foreign object is introduced in the body, the body rejects it by forming scar tissue around it, thus isolating it from the rest of the body. An interface between the sensing system and the skin needs to ensure that the body does not initiate this defense mechanism. Such ability would make it possible for the interface material to act as a seamless bridge that performs transfer of signals and samples to the sensing system. Systems with such interfaces would have the potential to integrate detection, diagnostics, and treatment of adverse medical and physiological states affecting the human body. An example of major drivers for such systems is diabetes. There are numerous active efforts underway to develop a seamless glucose sensing and insulin delivery systems that can function over extended durations. Other systems could enable diagnosis of adverse states like heat prostration, exhaustion, dehydration, hemorrhagic shock, and anxiety. This approach can potentially result in physiologically and clinically relevant information far in advance of changes in central measures like blood pressure, arterial blood gases, or acid–base status.

Positioning sensors strategically on the body surface and incorporating developments in Computer Science and Engineering, viz. distributed computing algorithms and low-power wireless communication, can allow observation of the target biological processes in the body in real time, similar to electrocardiography and electroencephalography. Figure 10.2 depicts the schema of distributed smart sensor arrays on the human body.

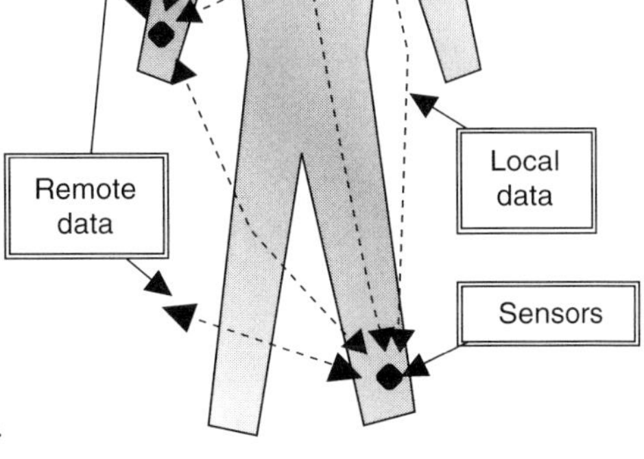

FIGURE 10.2
Schema of distributed smart sensor arrays on the human body.

Essentially, the bioengineered interfaces will facilitate (a) characterization of the quantitative boundary conditions of the human body; (b) determination of variation of those conditions with disease and environmental change; and (c) putting into effect suitable therapeutic response, as needed.

10.3 Anatomy of Skin

To develop a long-lasting interface, the nuances involved in interfacing with skin need to be understood. Toward that goal, a brief review of the skin is discussed below.

10.3.1 General Structure

Skin is the largest organ of the human body and accounts for ~16% of the body weight. It serves multiple purposes, which can be broadly classified into four functions—(1) Protection: physical, biological, against UV light, and from dehydration; (2) Thermoregulation: control of body temperature; (3) Metabolic functions: synthesis of vitamin D, excretion of waste products, and storage of fat; and (4) Sensory functions: pressure, heat, cold, contact, pain, as well as integral interaction with vision (pigmentation patterns) and smell (pheromone release). It is evident from its range of functionality that the skin hosts complex physiological, biomechanical, and biochemical processes, all occurring simultaneously.

The skin possesses a layered structure comprising various kinds of tissues. It icomprises two layers: the dermis and the epidermis. The epidermis is the outer layer of the skin. Being a tough and waterproof layer, it protects the inner regions. The underlying dermis is a thicker layer and is responsible for imparting strength and elasticity to the skin. Beneath the dermis is the subcutaneous layer, which is not part of the skin itself. It is a layer of tissue made of protein fibers and adipose tissue (fat). It contains glands, sensory receptors, and other skin structures. Figure 10.3 depicts the cross-section of the skin [1].

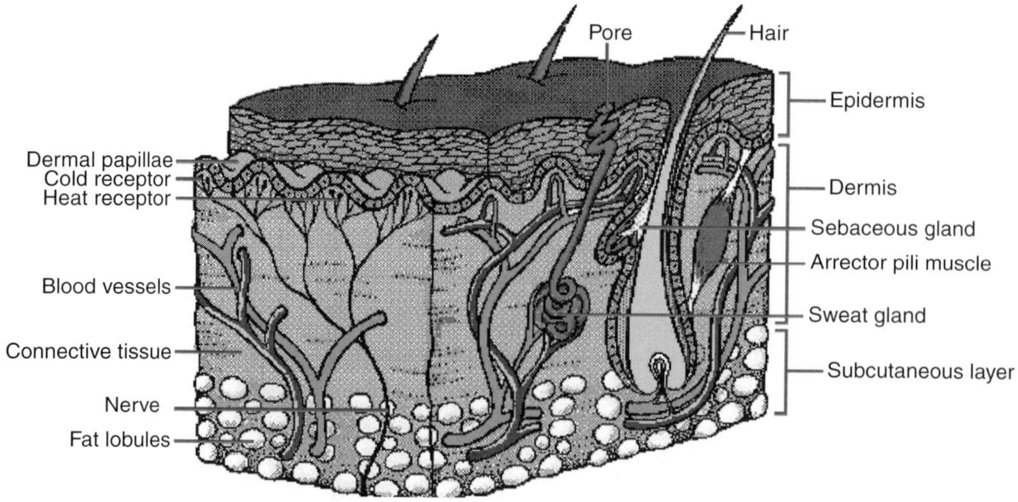

FIGURE 10.3
Cross-section of skin [1]. (From "Structure of the Skin" *Microsoft® Encarta® Encyclopedia*. http://encarta.msn.com© 1993–2004 Microsoft Corporation. All rights reserved, With permission.)

The protection provided by the epidermis forms the first barrier against harmful foreign substances. Most of the cells (~90%) in the epidermis are "keratinocytes." They produce a tough, fibrous, and intracellular protein called keratin; hence the name. The keratinocytes are stacked in layers. The youngest cells occupy the lower layers while older cells are present in the upper ones. The lower layer cells multiply continually, while the older upper layer keratinocytes constantly slough off. By the time the cells move up to the uppermost layer of the epidermis, they are dead and completely filled with the keratin. From bottom to top, the layers are named stratum basale, stratum spinosum, stratum granulosum, and SC. Figure 10.4 shows the cross-section of the epidermis.

Various skin properties are routinely measured at the broadest classification level whereby considering the entire epidermis as a single functional entity. Such information, though essential, is incomprehensive, in that it does not account for the heterogeneity of the epidermis. It is imperative to possess precise knowledge of the distinct properties of each cutaneous component to design interfaces that interact individually with the cuticle layers. SC in particular dominates various design considerations and is often times the principal cause for development of micromachined interfaces.

10.3.2 Stratum Corneum

The outermost layer of the epidermis, SC, is a thin, flexible, high-impedance biopolymer composed of interconnected "dead" cells called corneocytes. This complexly organized, anucleate, 15–20-μm-thick, biopolymeric structure is essential to life and serves to couple the organism to the environment. This structure is particularly well developed in humans who lack a protective mantle of fur. It is the major barrier to all environmental insults (climatic, toxic from xenobiotics, or microbial) [10]. Despite its resilience, it can bind a certain amount of water to stay soft, smooth, and pliable even in a dry environment, thus allowing free-body movement without cracking or scaling of the skin surface.

The underlying hypothesis in the development of bioengineered systems is that SC can be a portal to the biological and physiological processes. It plays a major role in modulating sensory signals for visual and tactile perception. Also, the hydration status of SC is a key determinant of the mechanical, electrical, optical, and chemical properties of the skin. The properties of this layer serve as a health status indicator and valuable insights into the

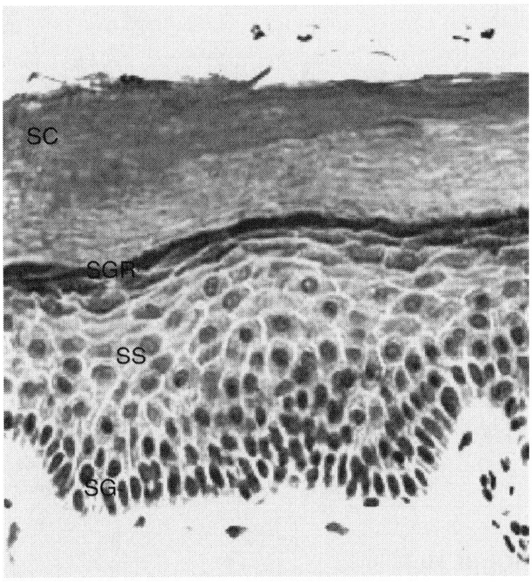

FIGURE 10.4
Cross-section of epidermis [9]. (From Swanson, J. R., http://www.meddean.luc.edu/lumen/MedEd/medicine/dermatology/melton/skinlsn/epider.htm. Accessed May 15, 2005. With permission.)

human biosystem can be attained by characterizing the quantitative boundary conditions of the human body and determining variation of those conditions with disease and environmental change.

As a secondary consideration, the condition of this layer serves vital social and cultural functions as it is a major determinant of our appearance and is thus of central interest to the cosmetic industry. The multifaceted significance of SC necessitates a multidisciplinary approach toward its analyses. Current researchers include dermatologists, cosmetic scientists, physiologists, neonatologists, pharmaceutical scientists, and bioengineers.

10.4 System Architecture

The conceptual microsystem, proposed in Figure 10.1 involves the transfer of electrical or physical (cellular/fluidic) entities. Currently accepted approaches to collect these samples include disruption in skin structure through needle puncture, electroporation or vaporization, or removal of SC through gels or tapes. Micromachined electrodes and needles (microelectrodes or microneedles) can enable collection of the same information, with significantly less trauma to the tissue, and have the potential of even eliminating it.

The microneedles and microelectrodes, when incorporated into the interface, will enable the development of a generic front end to the microsystem. As illustrated in Figure 10.5, such an interface can take care of most applications involving skin interaction, whether diagnostic or therapeutic in nature. The rest of the system architecture would be highly application specific thus obviating the need for a rigid architecture. The subsystems can be broadly classified by the type of entity transferred: physical (fluid transfer) or electrical (charge transfer), and the system may incorporate any or both types of interactions.

The interface system may be either a stand-alone diagnostic or therapeutic system, or have both capabilities. An active feedback-controlled system in which the therapy depends upon the diagnosis constitutes a "smart" system. The data collected may either

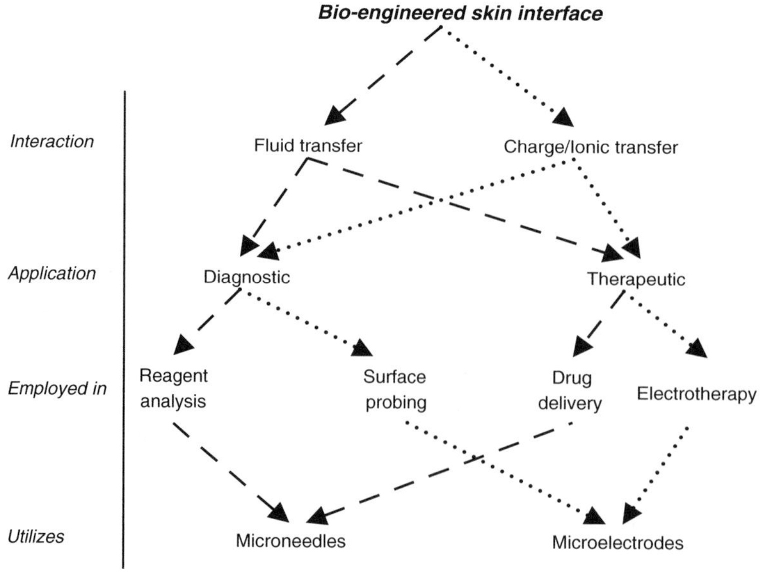

FIGURE 10.5
Application tree of skin-interface system.

be used locally or be relayed further—in which case the system would incorporate an onboard data communication capability—to transmit raw or significant data. Advanced sensor systems would follow a node-based approach for a global-level monitoring. They shall include low-power wireless communication protocols within multiple nodes, distributed computing algorithms and protocols at these nodes.

The MEMS devices, which interact with the skin, would depend upon the application and will generally be microfluidics-based (for fluid transfer systems), or IC-based (for charge transfer systems). Fluid transfer systems can be expected to be more complex and will have multiple component levels. Figure 10.6 depicts a schematic representation of a typical system.

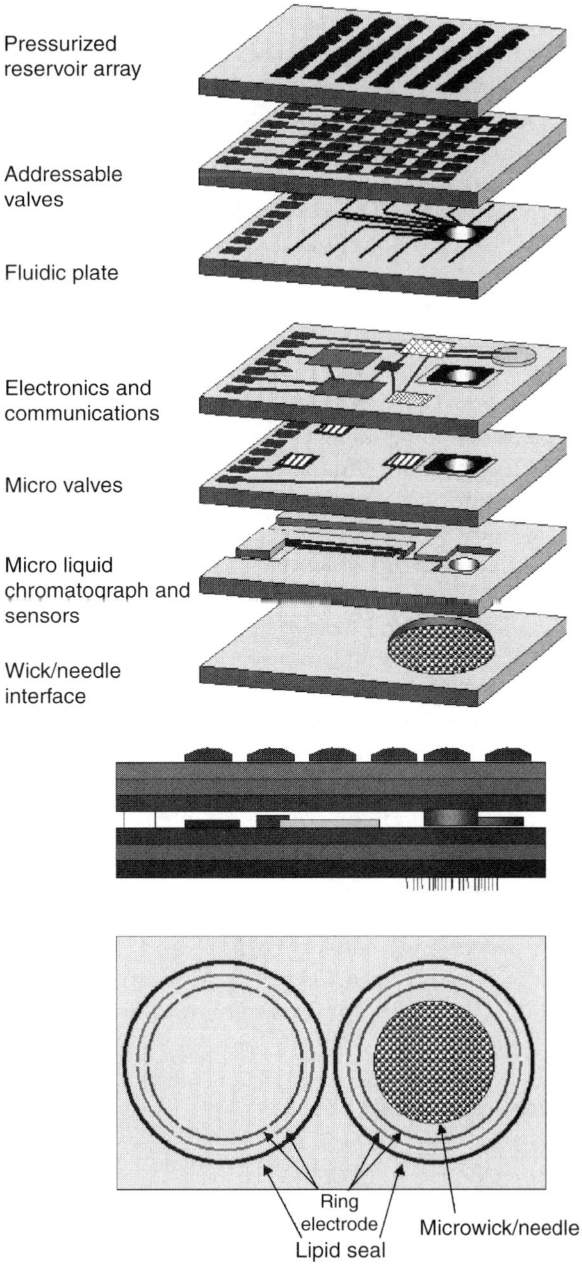

FIGURE 10.6
Elements of the prototypical MEMS system.

10.5 Design of the System

Despite architectural differences, the tasks involved in designing the systems have certain standard objectives. These objectives establish the foundation of knowledge and capabilities required to build the specific system. They include: (a) understanding the cellular and molecular biology of the skin–environment interface, that is, the epidermal barrier; and (b) developing tools, techniques, and protocols to nonintrusively or semiintrusively collect relevant physiological information using an array of micro/nanoprobes, sensors, and analysis protocols. For a global-level system, it includes: (c) using distributed computing, imaging, and wireless communication to analyze, correlate, and understand the response of the skin at various body sites to a single-controlled stimulus and across a group of sensors at a population level. Fluid transfer systems entail: (d) developing models to understand microfluidic coupling between SC and MEMS-based tools and on-chip immunoassays and sensors. Therapeutic systems include: (e) developing therapeutic application methodologies and protocols to deliver required medication based upon computational algorithms.

Each of the objectives mentioned above encompasses numerous tasks, which comprise varied undertakings in biology, medicine, chemistry, information technology, and microtechnology. The first task includes interface testing on various surfaces, such as bioengineered cellular films, bioengineered skin, excised human skin, and the intact human body. Relevant target measurements such as bioimpedance, immunoassay, and analyte detection would have to be conducted. This would assist in identifying key analytes to be monitored, and determining the baseline and deviation standards. Information/models and images would be used to enable extrapolation of data from the tissue-level microenvironment, to an understanding of meaningful physiological responses at the level of the organism. For reagent analysis systems, immunoassays and analyte detection protocols have to be developed. Sensor/analyzer designs have to be built accordingly by the MEMS designers. Deciding upon the physical design and establishing fabrication procedures of the individual micro/nanostructures would form the most vital component of system development. The biggest challenge would lie in integration of the various microdevices (such as the electrodes, sensors, microchromatographs, and fluidic components). To ease the development process, the microinterface systems can be expected to have a modular design, that is each subsystem would be functionally independent. However, subsystem development cannot be independent of other tasks, and continuous collaboration between MEMS and biotechnology/information technology groups would have to be a standard feature of the design process. The information technology tasks could include development of data processing capability and a data transfer interface. Depending upon whether the skin-interface unit is a stand-alone node or part of a global sensing system, the tasks would include creation of low-power wireless communication protocols within multiple nodes, distributed computing algorithms and protocols at these nodes, and transmission of raw or significant data collected from individual nodes. Data management tasks would comprise signal conditioning, post processing, sensor fusion, and feedback control to manage and optimize the device/skin interface. Data depiction front-ends would have to be created to facilitate easier data interpretation, for example providing the data as a 3-D image composite.

10.6 Significance of Skin-Interface Systems

Bioengineered systems would comprise application-specific MEMS devices coupled to the skin via microelectrodes or microneedles. Systems may incorporate elements such as lab-on-a-chip devices, wireless communication capability, complex information processing capability, data archiving, etc. There is no limit to the versatility of smart skin systems. This section highlights few significant applications that could potentially revolutionize the biomedical industry.

While such systems offer immense scope and opportunity, *it should be noted that some of the applications involve safety-critical systems. Also, certain systems intend (or at least provide the option) to remove human intervention from their functionality, thus demanding more stringent precision requirements. Such features open up an assortment of technical, ethical, and legal issues. A discussion on these issues is beyond the scope of this chapter. However, they would need to be addressed when this technology is put in application.*

10.6.1 Bioimpedance and Bioelectricity Measurement

Bioimpedance measurements at a macroscale have routinely been used for analyzing tissue quality and health monitoring in wide-ranging applications including the freshness of fish, analysis of skin [5,6,11], and electrocardiogram (ECG). Bioimpedance measurements give information about electrochemical processes in the tissue and can hence be used for characterizing the tissue or for monitoring physiological changes. The difference in the bioimpedance characteristics of different tissues and their condition (healthy or unhealthy) form the basis of electrical impedance tomography, a standard technique in medicine (currently being used to measure pulmonary ventilation and gastric emptying).

Determination of skin conductance enables the evaluation of the hydration state of the upper portion of SC [12]. The technique is now being used for many new applications and methods, for example, in areas like body composition, cell micromotion, organ viability, and skin pathology. When coupled with long-term monitoring capability, it can be used for searching heart arrhythmia or epileptic spikes. Along with finding new applications, basic research is going on to provide a better understanding of such basic questions as—is living tissue a purely ionic conductor; are there also local electronic or semiconductor mechanisms; what is the conduction mechanism, for example, in keratinized tissue or DNA [13]?

10.6.2 Electrochemical Sensing

Potentiometry using microelectrodes is a highly sensitive technique of electrochemical detection capable of detecting zepto molar concentrations in nanoliter solution volume. Potentiometric sensors respond to the presence of analyte by a change in electrochemical potential. The most widely used sensors in this category are the ion-selective electrodes (ISE). Instrumentally, an ISE is designed to observe the potential of a membrane that interacts selectively with a particular chemical, for example pH meter. The selectivity of this kind of sensor depends entirely upon the membrane design. The number of chemicals that can be determined by this method are limited only by one's ability to design a membrane with sufficient sensitivity and selectivity for the analyte in question. Integrated sensor arrays are capable of detecting ions, properties of fluids, and dissolved gases that are collected from the skin (from sweat to interstitial fluids). Some of the analytes that can be detected via this method are pH, Na^+, K^+, O_2, CO_2, and NH_3 [14].

10.6.3 Programmed Drug Delivery

Advanced drug-delivery systems will be able to deliver fixed quantities of drug at certain time intervals automatically. Such systems would comprise a microneedle interface attached to a MEMS device consisting of a drug reservoir, a microchip, and micropumps.

The microchip would initiate micropumps automatically to inject-controlled quantities of drug at specified intervals or on demand. Such systems would eliminate human errors such as skipped medications or incorrect dosage. Another variation of the same systems could have the ability to continuously inject drugs at a constant rate for extended periods of time.

10.6.4 Automated Feedback-Controlled Therapy

Natural evolution of smart skin-interface systems would ultimately lead to a combination of diagnostic and therapeutic systems. The interface would extract physiological data, and medication would be provided on the basis of the response. The diagnostic part might be either potentiometry-based electronic sensing or more complicated lab-on-a-chip device which would extract physiological samples via microneedles. Most prevalent system of such a type is expected to be one for diabetic patients. Microneedles would remove biofluids for glucose-level measurements and supply insulin as required.

10.6.5 Electrotherapeutic Pain Relief

Transcutaneous electrical nerve stimulation (TENS) [13] provided via surface microelectrodes can be used for pain relief. The electrical pulses stimulate the body's own mechanism for obtaining pain relief. The method works using either of the following physiological mechanisms—(1) Gating theory: Pain perception is regulated by a gate mechanism in the synapses, which is controlled by separate nerve fibers. Stimulation with pulses of high frequency (50–200 Hz) activates the gate mechanism, providing pain relief; (2) Endorphins: The human body uses natural forms of morphine called endorphins for pain relief, and their secretion can be provoked with low-frequency (2–4 Hz) stimulation; (3) Vasodilation: This effect is usually associated with pain in cold extremities and results in increase in temperature due to increased blood flow. The same effect can be elicited using electrical stimulation.

10.6.6 Iontophoresis

Iontophoresis is a method to accelerate transcutaneous drug delivery by application of DC current. If the drug is in ionic form, the migration velocity may be increased either by application of long-term DC voltages of less than 5 V, or by using high-voltage pulses up to 200 V decaying in about 1 ms. The effect is due to creation of aqueous pathways in SC. Various therapeutic agents can be introduced in the skin in this manner, such as anesthetic agents, antibodies, or metals such as silver or zinc (for treatment of ischemic ulcers).

10.7 Technological Merit of Skin-Interface Systems

Micromachined electrodes and needles form the technological foundation for bioengineered skin interfaces. As mentioned earlier, these enabling components will be a common denominator in design of these systems. The choice of other components would be purely application driven. The physical characteristics of microneedles/microelectrodes provide unique technological advantages to these systems as compared to traditional invasive methodologies. Next-generation therapeutic methods using advances in genomics, proteomics, pharmacogenomics, etc., have all come into existence due to the capabilities provided by these microcomponents.

Tissue interaction is usually done using macroscale components such as electrodes or hypodermic needles (henceforth referred to as macroelectrodes or macroneedles). Such components offer immense utilitarian advantages in their being easy to fabricate and handle. They are irreplaceable in appropriate problem identification and drug-delivery applications, which warrant the use of large-sized interface tools. However with the advent new technologies and discoveries, these conventional tools are being fast replaced with their micromachined counterparts.

10.7.1 Benefits from Microelectrodes

Traditional noninvasive techniques used in ECG and EKG record the fluctuating bioelectrical and biophysical properties of the epidermis. Usually macroelectrodes are pasted onto the skin in the vicinity of the organ. Unfortunately this method does not provide cellular-level information or interaction, and thus endows limited knowledge of the biosystem. They average the activity in many neurons and cannot establish precise activity in a particular region. *Furthermore, the skin/sensor interface is not intimate enough due to the inherent noise component introduced by the high resistivity of* SC. Therefore, such methods are extremely crude at best and only provide an unspecific diagnosis of the relevant organ.

Transdermal measurement techniques using microelectrodes have recently come to the limelight due to the inherent benefits that accompany their small size. They are increasingly being used for well-established as well as experimental electrode applications, such as ECG, electroencephalogram, electro impedance tomography [15], and blood flow measurement [16].

Some of the technological advantages obtained from microelectrodes are:

1. *High spatial resolution and precision*—The small size translates to closer probes and thus higher spatial sensitivity. This provides the opportunity to monitor localized effects, as opposed to the broad "system level only" macroelectrode monitoring.
2. *High temporal resolution*—The small area of microelectrodes leads to lower diffusion capacitance and thus a smaller RC time constant. This translates to a faster response and higher refresh rate (in range of microseconds).
3. *Enhanced mass transport properties*—When performing electrochemical analysis, the rates of mass transport to and from the electrode surface are much higher for microelectrodes. Furthermore, time-independent/steady-state mass transport rates are much easier to establish [14]. This is because of the different shapes of diffusion layer formed around a microelectrode and a macroelectrode. The diffusion layer is hemispherical for microelectrodes and planar for macroelectrodes. To maintain the rate of reaction, the diffusion layer does not need to extend as far into solution in the case of microelectrodes, due to the hemispherical shape. Thus, the population of electroactive species affected by a microelectrode exceeds that of conventional electrodes. This translates to better mass transport properties. Figure 10.7 illustrates this phenomenon.
4. *Reduced iR effects*—Lower area of the electrodes means low current requirement. This allows for measurement to be done in high-resistivity media. Highly resistive media can cause severe 'iR' distortions in macroelectrodes, due to the high current in its case.
5. *Increased deployment options*—Beyond surface-level skin testing, microelectrodes can by deployed in small or hard to reach areas within the human body such as

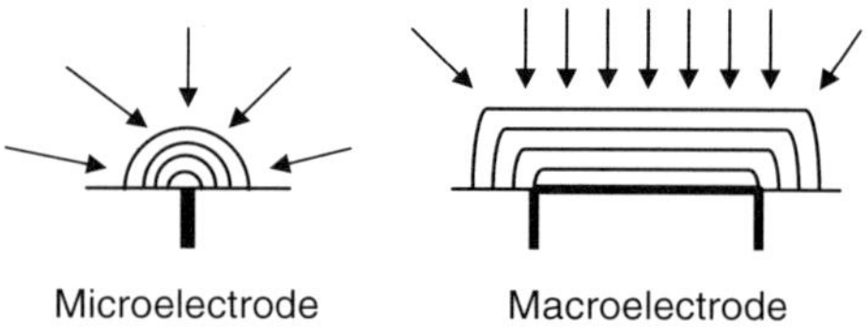

FIGURE 10.7
Hemispherical diffusion in microelectrodes as opposed to planar transport in macroelectrodes [14]. (From Imisides, M. D., John, R., and Wallace, G. G., Microsensors based on conducting polymers. *Chemtech*, 1996, **26**(5): 19–25.)

in pulmonary ventilation and gastric emptying. Furthermore, they are minimally intrusive and thus can be deployed to make measurement in sensitive areas that cannot afford deep probing (e.g., brain).

Although the benefits of microelectrodes do remove the disadvantages of macroelectrodes, one has to bear in mind that microelectrodes can tend to be too focused. For example, in the case of recording electrical activity of the brain, while taking single neuron measurements using microelectrodes, neuronal interactions may be neglected. One may not be able to differentiate between different neurological conditions or behavioral states. In such cases, microelectrode measurements supplement macroelectrode analysis. Optimum diagnosis is obtained by resolving the "system level" data along with "component level" data.

10.7.2 Benefits from Microneedles

In typical clinical and biomedical applications, drug injection or biofluid sampling is done using hypodermic needles. Though utilitarian, this method causes undesirable pain and excessive tissue trauma. This is particularly troublesome when frequent administration of drugs is necessary (such as administration of insulin). Furthermore, control of drug delivery is highly approximate using macroneedles.

A painless alternative to macroneedles, which has lately started to become popular, is the transdermal patch. It is simply a piece of plastic with an adhesive on it with the drug being suspended in the adhesive. The patch is stuck on the skin and the drug passively diffuses slowly to lower layers. However, this method of drug delivery poses certain challenges. The effectiveness of the adhesive is directly affected by the concentration of the drug within the adhesive. Thus, if a larger dose of the drug is required, either it has to be reapplied more frequently or a larger patch has to be used. Increasing the concentration of the drug does not help if the adhesive bond with the skin is weak [17]. Furthermore, it is a slow method and not very precise, and is thus not suitable to many applications. The barrier properties of SC pose unique obstacles to drug application by providing a very low topical bioavailability [18]. The "route" taken by the externally applied drug is tortuous—around the dead, dense corneocytes. The permeant diffuses in the intercellular channels, which contain structured lipid bilayers. Thus, the problem is compounded, as molecules have to cross, sequentially, lipophilic and hydrophilic domains.

Microneedles are increasingly being sought out as an alternative to the macroneedle and the transdermal patch as it combines their advantages while eliminating the disadvantages of both. In recent times, numerous types of microneedles have been fabricated and used for transdermal drug delivery [19–23], vaccine delivery [24], fluid analysis and sampling [25], dialysis [26], and cellular DNA delivery [27], among other applications. Figure 10.8 shows an array silicon dioxide microneedles fabricated at the University of South Florida.

Vital technological advantages obtained by using microneedles are:

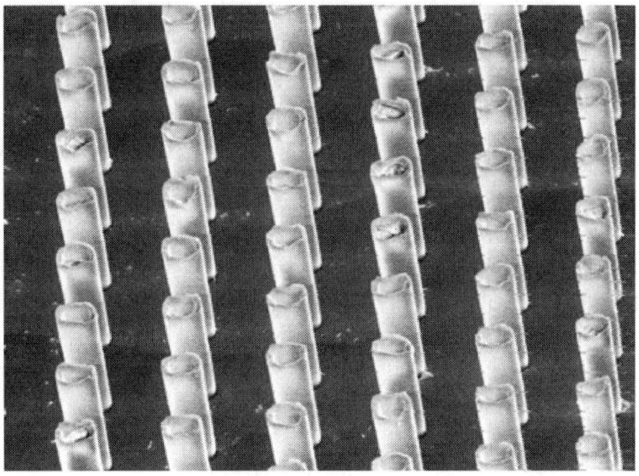

FIGURE 10.8
Silicon dioxide microneedle array.

1. *Increased Permeability*—The self-sealing properties of SC are poorly understood. It is probable that SC seals holes spontaneously up to a given magnitude, but beyond that limit it cannot close itself. This property can be used by microneedles to increase the permeability of skin by orders of magnitude, making it possible to diffuse macromolecules across the skin. Thus in some cases, the microneedles may not themselves deliver the drug. Unmedicated microneedles may simply puncture the skin and then medicines may be smeared on and left to be rapidly absorbed through the holes.
2. *Application customization*—Microneedles can be fabricated in various shapes (cylindrical, pyramidal, and polyhedral) or sizes and may be hollow or solid. Such geometrical flexibility allows for fabrication of microneedle arrays customized for the type and volume of drug and the duration of use. This feature also translates to flexibility in method of drug delivery. For example, medicines may be delivered the way a regular hypodermic needle does so, or microneedles may be coated with medication which would be absorbed on contact.
3. *Material flexibility*—Microneedles can be fabricated using a variety of materials such as silicon, glass, carbon nanotubes, biodegradable polymers, and metals. The choice of material would depend on the application. For example, metallic microneedles are the most popular because they have high strength and biocompatibility. Polymer needles have the advantage of being biodegradable, and thus if the tips snap off during insertion, the body would simply absorb the remains.
4. *Minimal medical training requirements*—Microneedles are expected to become a boon for immunization programs in developing countries as they could be applied by people with minimal medical training. Drugs could similarly be administered at home instead of presently compulsory clinical visits. The significance of such capability is further enhanced due to possibility of mass vaccination or antidote administration in bioterrorism incidents. The cause is further helped by the fact that microneedle fabrication promises to be less expensive than any other reported technology.
5. *Site specific therapy*—Highly targeted drug administration to individual cells is possible using microneedles. A potential application is to modulate the immune system by administering drugs precisely into the area where special immune cells reside. Langerhans cells, for example, are the major immunomodulatory cells that modify delivery of HIV particles from the epidermis to the lymph

nodes—these cells are located only 40–50 μm deep in the epidermis. Vaccine delivery directly to these cells provides a target of mass vaccination strategies.

6. *Precise control of drug quantities*—The small volumes of drug passing through the microneedles, either through diffusion or through auxiliary pumping systems, make the delivery process highly controllable.

Microneedles are presently at various stages of development, although they are yet to be introduced clinically. They will surely revolutionize drug-delivery systems and provide more options and possibilities to patients. The day is not far when administering of polypeptides and even proteins will not be via injection or infusion.

10.8 Designing the Interface Microtools

Microneedle and microelectrode design is the most critical part of designing the bioengineered system. It involves simultaneous consideration of multiple parameters. There is always present some form of trade-off, for example ease of skin penetration vs. ease of fluid flow and probe strength vs. biocompatibility. There is no specific optimum design; it varies with each application. The combinatorial possibilities are vast, and the parametric interdependency is highly complex. Every application will have a few structural or functional demands that bear higher priority than others. Fortunately, not every application needs a new microprobe design. Most applications can be grouped into classes having broadly similar requirements. Thus, generic probes can be designed with specifications that provide enough latitude for parametric variations within the class.

10.8.1 Functional Requirements or Constraints

Bioengineered interfaces will have to satisfy diverse functional requirements, a large majority of them being either electrical or mechanical in nature. For the sake of discussion, these requirements are thus broadly classified here as: (1) electrical parameters, (2) mechanical parameters, and (3) other parameters.

10.8.1.1 Electrical Parameters

Electrical considerations take prominence for sensing applications. The foremost aim for such applications is optimizing accuracy and reliability of retrieved data. One of the first factors that affect electrode design is whether it is used in galvanic or nongalvanic contact with tissue. In galvanic contact, the electrode is the site of shift from electronic to ionic conduction. The ionic conduction may be in an applied coating of conduction liquid or in natural tissue fluids surrounding the electrode. Galvanic contact generally relates to highly environmentally sensitive probing. It entails cleaning procedures or surface preparation. The electrode array should thus be designed accordingly (spacing and material). Nongalvanic contact is often characteristic of a thin isolating layer of Teflon or glass. Coupling is by capacitive or inductive fields, or by electromagnetic waves. Such electrodes might employ high voltages and thus have to be designed such. Also, they are prone to stray capacitances etc. and the array design has to compensate accordingly. Depending on the application, probes may have to satisfy various electrical specifications, which might include parameters such as electrical conductivity, resistance, effective electrode area, capacitive properties, inductive properties, electromagnetic properties, half-cell potentials, polarizability, and noise levels, among many others.

10.8.1.2 Mechanical Parameters

Bioengineered interfacing entails insertion of microelectrodes or microneedles into the upper epidermal layers of the skin, and thus the probes have to be able to withstand the applied pressure and puncture the skin without fracture. Optimum design demands characterization of various mechanical features such as mechanical strength, buckling characteristics, fracture point, elasticity, angular strength and ease of penetration. Microneedles have to be further characterized for their capability to withstand fluidic pressure at the sidewalls and the inlet pressure for a particular flow rate.

10.8.1.3 Other Parameters

Microcomponents also have to satisfy certain requirements that are neither electrical nor mechanical in nature. Most significant of them is biocompatibility of the interface with the skin. Interaction of nonbiocompatible sensors or electrodes with cells and tissue is known to trigger "foreign body response" and foul the electrodes and systems. Biologically, such a system constitutes a breach in the integrity of the organism and would be responded to as a wound. Thus, it becomes difficult to determine whether one is sampling "intracellular" fluid from impaled cells or "extracellular" fluid from needles placed between cells. The most common approach to minimize biofouling is surface passivation of the functional materials.

Microfluidic properties of microneedles are also important design factors. The design has to take into account fluid flow characteristics for the passing fluid and also environmental flow resistance. There should be no clogging of the fluid at orifices or along the needle lumens. Also capillary forces have to be sufficient to ensure fluid delivery or extraction.

Other common parameters affecting microcomponent design include adhesive interactions (adhesion of analytes to the probe, unwanted adhesion of the probe to the skin, or other structural parts), life duration, chemical inertness, and biodegradability.

10.8.2 Design Variables

Common variables that control the component characteristics can be broadly classified into: (1) Material Selection, (2) Geometric Considerations, and (3) Array Layout.

10.8.2.1 Material Selection

There is no one best material available for the fabrication of microelectrodes or needles. The choice varies according to the application. Metallic microprobes are popular due to strength and biocompatibility; they have the lowest probability of causing allergic reactions. They are relatively easy to fabricate, common methods being micromolding or electroforming. Within metals, there is a wide range of material choices.

Silver is one of the best materials available for current carrying bioapplications. Silver electrodes are usually coated with AgCl using electrolytic deposition. As body fluids contain Cl^-, AgCl forms a nonpolarizable electrode capable of passing currents with less overvoltage. However, in this case, biocompatibility may be compromised. Platinum is the metal most preferred for biocompatibility-sensitive applications due to its inertness. However, it can be highly polarizable by DC currents in a physiological environment. Nevertheless, it is extremely suited to DC potential reading applications under strict restriction of zero DC currents. Pure noble metals are too soft for practical use, and thus alloys with other noble metals are often used to improve mechanical properties. Iridium alloys are quite common, as they also provide catalytic properties in some cases. Other metals are also used for various reasons. Stainless steel is used due to its high strength, noncorrosive nature, and low price. They are however unsuitable for low-noise, small-signal measurements. Tin and lead alloys are used for low-noise properties. Also they have low melting points, which translates to fabrication advantages in that they can be easily formed or molded. Nickel is used for its flexibility. However, it may give allergic skin reactions. In applications such as electrotherapy or skin iontophoresis, the pharmaceutical or bactericidal properties of the metal are of prime importance; thus, iron, aluminum, or zinc might be considered [13].

Nonmetallic materials have other advantages. Glass microelectrodes are best suited for recording DC or slowly varying DC potentials. Silicon dioxide microneedles are popular due to their ease of fabrication and good biocompatibility characteristics. However, strength and reliability may not be sufficiently high for certain applications. Composite material microneedles, such as those made of oxide–nitride–oxide stack might be used instead. Another nonmetallic material being investigated for microelectrodes is carbon because it is X-ray translucent but is still an electronic conductor. Combinations of carbon and rubberlike materials may be used because they are soft and thus highly adaptable to anatomical variations. Polymers such as polyaniline, polypyrrole, polythiophenes, and their derivatives [28] are becoming highly popular because of their multiple advantages. Since they are biodegradable, microneedles can be used without danger of infection due to needle tips breaking off. Conductive polymers are often used for microelectrodes due to their selective interaction properties with proteins, enzymes, and antibodies, thus making them ideal for certain biochemical sensing applications. Furthermore, polymers are highly flexible too.

10.8.2.2 Geometrical Considerations

Geometry of the microprobe affects most of the electrical, mechanical, or other characteristics highlighted earlier. Common geometrical considerations are probe length, probe width, area/perimeter of cross-section, shape, thickness of sidewalls (for microneedles), tip geometry, and taper angle.

Each geometrical variable can affect a number of parameters, and thus the optimum choice of geometry is often a trade-off between efficiencies. For example, sharper needle tips can be expected to require less force for insertion, but the reduced penetration force comes at the expense of reduced strength near the tip [29].

Probe length determination is particularly critical as the probe must be sufficiently deep to allow reliable test results, and for drug delivery or sample extraction, they have to reach the blood capillaries. The possibility of probe fracture and buckling increase as probe length increases. Also, to be painless, the probes cannot be so long as to touch the nerve endings. Increasing cross-sectional area increases strength but may compromise certain electrical characteristics [30]. For microneedles, constriction of needle lumen size is limited by the size of species being transferred. Finally, biofouling properties always impose critical constraints, it being less intensive for less invasive geometries.

10.8.2.3 Array Layout

Microprobes are most often used in an array formation. Thus, device functionality depends not only on their individual, but also on their "team" efficiency. Array size, density, and configuration are most important considerations.

The electrochemical behavior of microelectrode arrays varies strongly with center-to-center spacing; the larger the spacing the more sigmoidal are the recorded results, which translates to better sensitivity. Also, the temporal response improves with increased center-to-center spacing, with current–time transients reaching a steady-state response more quickly using loosely packed arrays [30]. However, decreased density compromises the spatial resolution and also results in higher area requirement.

10.9 Evaluation of Microstructure Interaction with Skin

It is evident that the design of the bioengineered interfaces involves optimization of varied performance parameters. This process is not an easy task. The physiological and structural complexity of the skin hinders accurate biophysical or biochemical modeling. Thus, simulation alone cannot arrive at the finalized design. Testing the interface mechanism has to be included within almost each design iteration. This increases the criticality of testing procedure, and care must be given to precise calibration of test equipment.

10.9.1 Microneedle Testing

Testing methods are being evaluated at various research facilities. For example, at the University of South Florida, DRIE-based SiO_2 microneedle arrays have been employed for experimental testing to examine the significance of geometry on transdermal testing. Reliability testing based on measurement of buckling force and skin penetration force is performed before confirming use of the needles for applications like drug delivery and fluid extraction. Measurements are carried out via testing module utilizing an accurate load transducer and real-time viewing capability. Preliminary characterization is established on skin-like polymer before real-time experimentation on excised split-skin samples. Mechanical tests are also performed on the polymer to correlate its properties with skin. Even though the polymer does not behave like skin since its properties are isotropic and elastic unlike skin, this material can be used for preliminary characterization of microneedles before real-time testing on cadaver skin. An important part of skin-interface testing is fracture and penetration testing. It is necessary to validate the ability of the microneedles to penetrate skin without fracturing. This is performed by quantifying the penetration force in real-time microneedle insertion tests done on split-thickness skin and isolated SC (Figure 10.9).

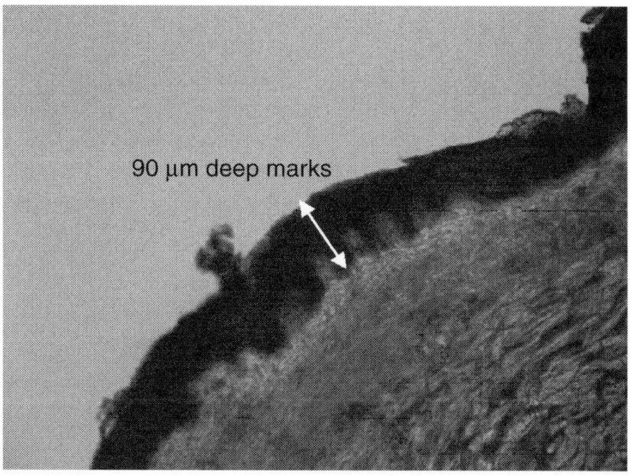

FIGURE 10.9
Optical microscope image (20× magnification) illustrating split thickness. Skin section with 90-μm-deep, 20-μm-wide needle marks into epidermis.

The point of fracture is visually captured by the camera with the data acquisition system displaying a sharp peak in force–time graph, interpreted as "needle failure point."

10.9.2 Microelectrode Testing

Most of the mechanical constraints tested for microneedles also apply to microelectrodes. However, as is obvious from the nature of application, testing of electrical parameters is of prime importance here. There may be various methodologies to test a microelectrode array depending on its design. One of them is highlighted here as an example.

Microelectrode testing is performed at the University of South Florida using a precision Impedance Analyzer in conjunction with a Probe Station. The microelectrode array (Figure 10.10) is placed on the probe station, and the impedance analyzer provides a small voltage potential while measuring the resultant radio frequency current. Four terminal sources are configured to create a bipolar measurement analysis. The most important step is the calibration of the instrument to eliminate/null any parasitic capacitance and resistance from additional cables attached to the probe tips. Phase compensations are also done to allow signal stability over the specified frequency range. A vital consideration is to avoid tissue heating during testing. The oscillating voltage should be chosen such that the maximum current output is $\lesssim 2\ \mu A$.

10.10 Conclusion

Skin interaction systems containing bioengineered interfaces will revolutionize the biomedical industry in the coming decades. They promise exceptional capabilities in diverse transdermal diagnostic and therapeutic applications. Development of customized instrumentation will enable real-time electrical, mechanical, and electrochemical measurements on the skin surface and also provide novel methods of drug delivery.

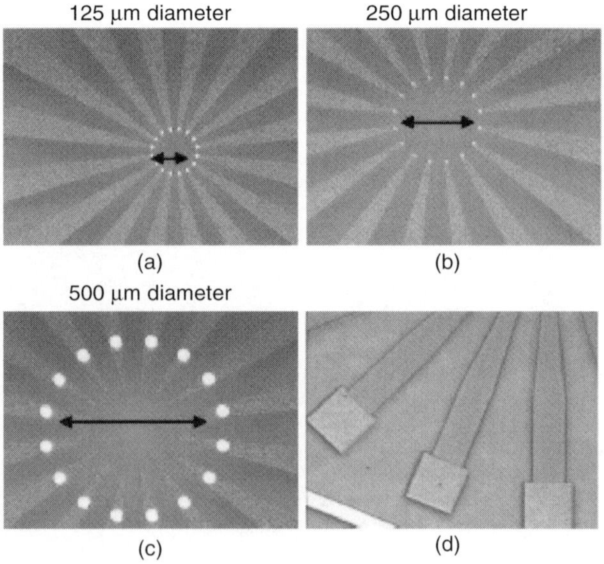

FIGURE 10.10
Microelectrode arrays with electroplated tips, (a) 125-diameter configuration, (b) 250-diameter configuration, (c) 500-diameter configuration, and (d) exposed bonding pads.

The skin interfaces benefit from the unique properties of microneedles and microelectrodes. It is now evident that these components offer multifaceted advantages over their conventional counterparts. However, successful clinical implementation of such systems would entail overcoming numerous challenges in areas of design, fabrication, electronics integration, deployment, testing, and data interpretation, among many others. The interdependencies between the various design variables and the obtained characteristics are highly complex, and warrant multidisciplinary research initiatives. Optimum utilization of this technology depends upon our capacity to identify and resolve, qualitatively and quantitatively, the various parametric entities involved.

References

1. Structure of the Skin, *Microsoft® Encarta® Encyclopedia*. http://encarta.msn.com© 1993–2004 Microsoft Corporation. All rights reserved.
2. Elias, P., Stratum corneum architecture, metabolic activity and interactivity with subjacent cell layers. *Exp. Dermatol.*, 1996, **5**: 191–201.
3. Elias, P., The stratum corneum revisited. *J. Dermatol.*, 1996, **23**: 756–758.
4. Hoath, S. B., Donnelly, M. M., and Boissy, R. E., Sensory transduction and the mammalian epidermis. *Biosens. Bioelectron.*, 1990, **5**(5): 351–366.
5. Kell, D. B., The principles and potential of electrical admittance spectroscopy: An introduction. In: *Biosensors; Fundamentals and Applications*. Turner, A. P. F., Karube, I., and Wilson, G. S., (eds.) Oxford University Press, 1987: pp. 427–468.
6. Nicander, I., Ollmar, S., Eek, A., Lundh, R. B., and Emtestam, L., Correlation of impedance response patterns to histological findings in irritant skin reactions induced by various surfactants. *Br. J. Dermatol.*, 1996, 134: 221–228.
7. Rawlings, A. et al., Stratum corneum moisturization at the molecular level. *J. Invest. Dermatol.*, 1994, 103: 731–741.
8. Sinclair, B., Everything's great when it sits on a chip – A bright future for DNA arrays. *Scientist*, 1998, 13: 18–20.
9. Swanson, J. R., http://www.meddean.luc.edu/lumen/MedEd/medicine/dermatology/melton/skinlsn/epider.htm (Accessed May 15, 2005).
10. Marks, R. and Leveque, J.-L., *The Essential Stratum Corneum: Introduction*. London, UK: Martin Dunitz Ltd., 2002.
11. Ollmar, S., Making electronic biopsies into a viable future for non-invasive diagnostics with electrical impedance. *Med. Biol. Eng. Comput.*, 1999, **7**(2): 116–117.
12. Elsner, P., Berardesca E., and Maibach, H. I., Bioengineering of Skin: *Water and Stratum Corneum*. Boca Raton, FL, US: CRC Press, 1994.
13. Grimnes, S. and Martinsen, O. G., *Bioimpedance and Bioelectricity Basics*. London, UK: Academic Press, 2000.
14. Imisides, M. D., John, R., and Wallace, G. G., Microsensors based on conducting polymers. *Chemtech*, 1996, **26**(5): 19–25.
15. Griss, P. et al., MicroMachined electrodes for biopotential measurements. *JMEMS*, 2001, **10**: 10–16.
16. Perchon, S. et al., Micro blood flow measurement using thermal conductivity micro needles: A new CMOS compatible manufacturing process onto porous silicon. *IEEE*, 2000: 184–187.
17. Morrow, T., Transdermal patches are more than skin deep. *Managed Care*, April 2004.
18. Marks, R., and Leveque, J.-L., *The Essential Stratum Corneum: Crossing the Barrier*. London, UK: Martin Dunitz Ltd., 2002.
19. Stoeber, B., and Liepmann D., Fluid injection through out of plane micro needles. In *First Annual International IEEE-EMBS, Special Topic Conference on Micro Technologies in Medicine and Biology*. 2000, Lyon, France: IEEE.

20. McAllister, D. V. et al., Micro needles for transdermal delivery of molecules. *IEEE*, 1999: 836.
21. Gardeniers, J. G. E. et al., Silicon micro machined hollow micro needles for transdermal liquid transfer. *IEEE*, 2002: 141–144.
22. Liu, K. et al. Electroplated micro needle array for BIO medical applications. In *Proceedings of SPIE – The International Society for Optical Engineering*, 2001.
23. Henry, S. et al., Micromachined needles for the transdermal delivery of drugs. *IEEE*, 1998: 494–498.
24. Griss, P. and Stemme, G., Novel, side opened out-of-plane micro needles for microfluidic transdermal interfacing. *Micro Electro Mechanical Systems, 2002. The fifteenth IEEE International Conference on 20–24 Jan.* 2002: 467–470.
25. Brazzle, J., Papautsky, I., and Frazier, A. B., Micro machined needle array for drug delivery or fluid extraction. *Engineering in Medicine and Biology Magazine. IEEE*, 1999, 18(6): 53–58.
26. Zahn, J. D., Trbotich, D., and Liepmann, D., Microfabricated microdialysis microneedles for continuous medical monitoring. Microtechnologies in Medicine and Biology. 1st Annual International Conference on 12–14 Oct. 2000. IEEE, 2000: 375–380.
27. McAllister, D. V., Allen, M. G., and Prausnitz, M. R., Microfabricated microneedles for gene and drug delivery. *Ann. Rev. BioMed. Engg.*, 2000, 2: 289–313.
28. Sung, J.-H., Kim, S.-J., and Lee, K.-H., Fabrication of microcapacitors using conducting polymer microelectrodes. *J. Power Sources*, 2003, 124(1): 343–350.
29. Chandrasekaran, S. and Frazier, A. B., Mechanical characterization of surface micromachined hollow metallic microneedles. In *IEEE EMBS Special Topic Conference on Microtechnologies in Medicine and Biology*. May 2002. Madison, WI.
30. Sandison, M. E. et al., Optimization of the geometry and porosity of microelectrode arrays for sensor design. *Anal. Chem.*, 2002, **74**: 5717–5725.

Part III

Material Design and Selection II

The focus in this section is on porous silicon and carbon nanotube technology. These materials are being applied in the design of new generations of biochips and sensors. In Chapter 11, Dr. Benjamin Miller covers the use of porous silicon as the base material in the design of smart "label-free" sensors. Porous silicon can be made by electrochemical etching silicon wafers to form refractive index detectors, or band gap photonic materials. These in turn can function as "smart bandages" and "smart dust." A very good example presented is the alternative approach to the Gram stain test. In Chapter 12, Xueliang Sun and Chen-zhong Li focus on nanotube technologies. The processing techniques used to control the nanostructure of carbon nanotubes and nanowires are first described. The factors that govern the growth mechanisms of these nanomaterials are presented. Methods by which carbon nanotubes and nanowires can be functionalized with appropriate agents for biosensor applications are discussed. This section is followed by a review of interesting new applications of these materials in electrochemical biosensors, field-effect-transistor-based biosensor and nanofabrication of these biosensors. In Chapter 13, Dr. Won-Yong Lee focuses on carbon nanotube-based electrochemical biosensors. The chapter presents recent advances in the use of CNT for the construction of electrochemical biosensors. Common preparation methods for CNT-modified electrodes and their analytical characteristics are next presented. Applications of these CNT-modified amperometric biosensors based on oxidase or dehydrogenase are provided. DNA biosensors based on the enhanced detection of the product of the enzyme label or of guanine base in the target DNA is also given. Finally, the functionalization of the field-effect transistors (FETs) with biomolecules is discussed.

11

Porous Silicon in Biosensing Applications

Benjamin L. Miller

CONTENTS

11.1 Introduction

The detection of harmful microorganisms has undergone a considerable metamorphosis since Hans Joachim Gram, a Danish physician, first observed, in the early 1860s, the capacity of crystal violet to cause a certain class—but not another—of bacterium to acquire a purplish color.[1] In particular, the explosive growth in our knowledge of the molecules that make us and the organisms that might do us harm, through the parallel fields of genomics,[2] proteomics,[3] and metabolomics,[4] has made improved detection methods both possible and a necessity.

Microbiological analysis since the days of Gram has relied primarily on visual examination of cultures, cell staining, or antibody- or nucleic-acid-based assays that require significant sample manipulation and "labeling" of the molecular target with a secondary antibody, a fluorescent tag, or a radiolabel. Such methods have been enormously successful in centralized diagnostic laboratories. However, with their requirements for expensive reagents and operation by trained personnel, they are problematic as point-of-care diagnostic aids, and completely unsuitable for field-use as biosensors. Therefore, the need for "label-free," versatile, and simple-to-use sensing technologies has become starkly clear. This chapter will cover the use of porous silicon (PSi) as the base material for a number of such "label-free" sensors. Even though a significant amount of development work remains to be done for PSi to be widely used in the field, the doctor's office, and the clinic, studies by several groups (including ours) indicate that it shows considerable promise.

Our discussion on the development and current "state of the art" of PSi in biosensing proceeds roughly chronologically. For the sake of brevity, we focus primarily on efforts at the University of Rochester and by the Sailor and Ghadiri groups at the University of California, San Diego (UCSD) and Scripps, respectively. However, we stress that many other groups have made significant contributions to the development of this field.

11.2 Origins and Definitions

The initial discovery of PSi and early studies on its properties have been discussed at length elsewhere,[5] and will not be covered in this chapter. Formed by anodic oxidation in a solution containing hydrofluoric acid, PSi is exactly what its name implies: a mesh-like structure formed by partial dissolution of a silicon wafer (Figure 11.1). Porosity of the material produced, defined as a percentage, is dependent on a number of factors including the dopant level and type in the wafer, the concentration of HF in the etchant, and the amount of current passed through the etch cell. The production of PSi is notable in that conditions are chosen such that etching almost exclusively occurs only at the interface between unmodified silicon and the etchant solution; once a porous structure has been

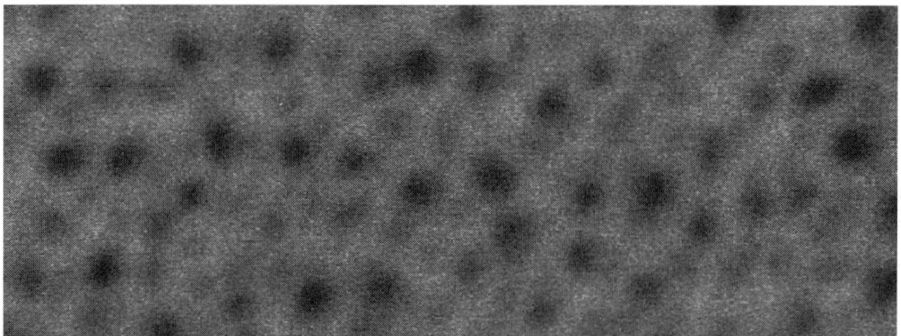

FIGURE 11.1
Scanning electron micrograph (SEM) of porous silicon.

formed, it is generally unreactive through the rest of the etch cycle. By varying the etch current as a function of time, it is therefore possible to produce complex-layered structures of precisely defined thickness and porosity. As we shall discuss later, these layered structures are important in the development of optical sensors.

Regardless of the precise device structure, PSi sensors function by transducing a change in the bulk refractive index of the material effected by infiltration of analytes into the porous matrix (Figure 11.2). In some respects, this makes PSi sensors somewhat like other refractive index-based (or "mass sensing") techniques such as surface plasmon resonance (SPR)[6] and reflective interferometry (RI).[7] However, the three-dimensional nature of the transducing surface that is unique to PSi results in both interesting opportunities and challenges.

The size of the pores produced is just as important as the ability to produce structures of specific porosity. Generally, PSi is described as falling into one of the three primary categories: *microporous silicon* (pore diameter <10 nm), *mesoporous silicon* (10–50 nm), and *macroporous silicon* (50–10 μm). The bulk of studies on PSi sensors have, till date, focused on mesoporous silicon; microporous silicon is too small to permit infiltration of most analytes of interest, and until recently macroporous silicon could not be produced with a sufficient level of control to permit effective optical sensing. Recent work however suggests that the manufacturing challenges associated with macroporous silicon can be overcome, as we will discuss later in this chapter.

The seminal discovery that led to much of our early work on PSi sensors was by Canham, who observed room-temperature photoluminescence from mesoporous material.[8] However, the first "proof of concept" demonstrations of PSi-based biosensing relied on the reflectivity spectra of single-layer PSi devices. Building on initial work toward understanding the optical reflection spectra of PSi,[9] Sailor, Ghadiri, and coworkers observed subtle yet detectable binding-dependent shifts in the Fabry–Perot fringes

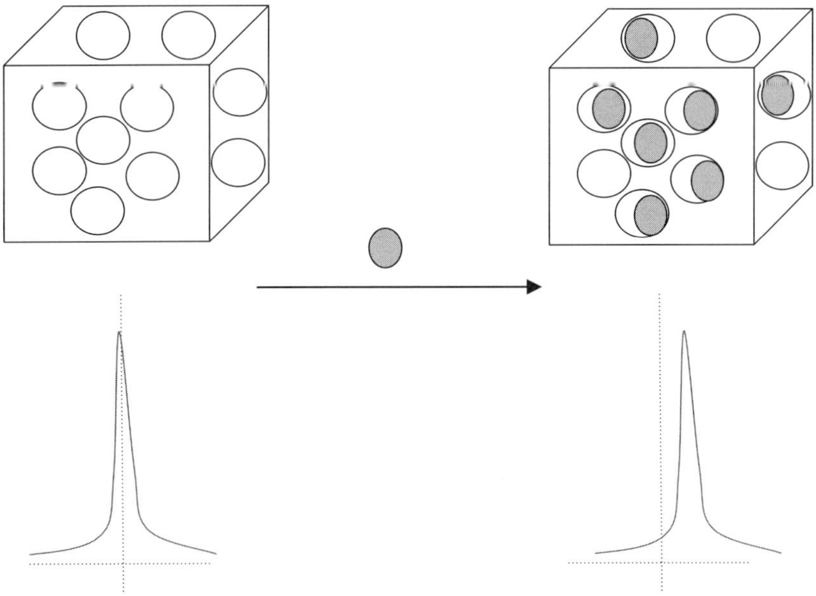

FIGURE 11.2
Basic operation of a porous silicon sensor: infiltration of an analyte into the porous structure causes a change in the bulk refractive index of the material, which in turn causes a change (for example, a redshift) in its optical properties.

produced by these devices.[10] Importantly, femtomolar sensitivity for DNA was reported, and the device was also adaptable to the detection of other types of molecular targets. In the subsequent work, Sailor and coworkers examined reversible binding of immunoglobulin G (IgG) to a PSi surface covalently modified with protein A.[11] Derivatization of the surface with protein A was found to reduce oxidative degradation of the PSi significantly.

11.3 PSi Biosensors: Early Work at Rochester

In 1999, in collaboration with Philippe Fauchet's research group and initially also in collaboration with Prof. Lewis Rothberg, we began to examine the properties of an alternative device structure in the context of biosensing: the microcavity optical resonator. Such a device consists of two Bragg mirrors (produced by repeating alternate high- and low-porosity layers) on either side of a thick "active layer" of PSi (Figure 11.3). This geometry provides a solution to the observation that controlling the size distribution of the nanocrystalline silicon structures responsible for the material's luminescence is exceedingly difficult. Instead of the large (>100 nm full width at half maximum (FWHM)) luminescence bands typically observed with single-layer structures, microcavity resonators produce luminescence spectra with line widths typically FWHM less than 3 nm. Building on methods developed for the production of multilayer PSi structures,[12] Fauchet's laboratory had initially described the fabrication of PSi microcavity devices in 1999.[13] In that instance, the microcavity electroluminescence was used to produce what is essentially an all-PSi light-emitting diode (LED). Importantly, this chapter verified that large changes in porosity could be effected during the etch process (for highly doped *P*-type silicon wafers) by changing the current density as a function of time. Since etching only occurs at the interface between unmodified silicon and the etchant solution, structures remaining behind are not modified further. This is not necessarily the case for alternative etchant solutions or alternative substrate materials. Subsequent analysis of the geometry

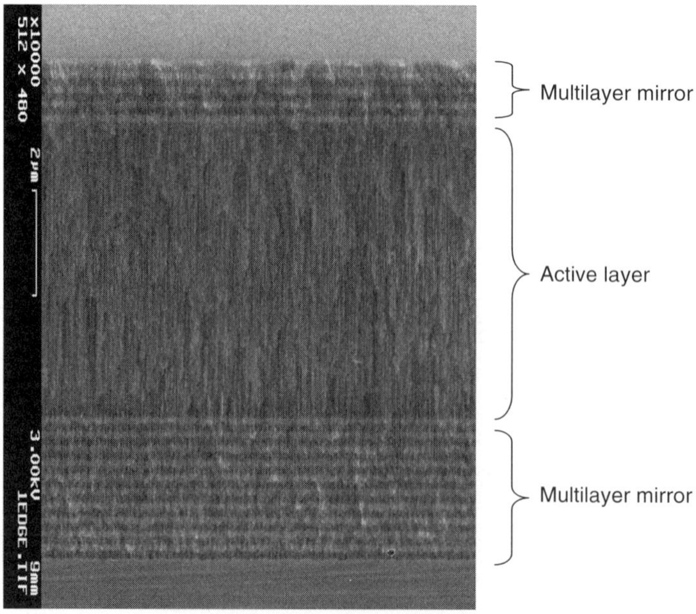

FIGURE 11.3
SEM of a porous silicon microcavity. Note that this particular example is an asymmetric microcavity (different number of mirror periods on either side of the active layer), which potentially increases the rate of diffusion of analytes into the sensor.

according to a photonic model indicated that one can view such PSi microcavities as a one-dimensional band-gap structure.[14,15] The narrow FWHM displayed by these microcavity devices suggested that they might provide a greater sensitivity to changes in the local refractive index over other device configurations (single-layer or multilayer stack), making them ideal for biosensing.

Our initial efforts focused on the detection of nucleic acid targets using microcavities derivatized with synthetic oligonucleotides.[16,17] After etching each microcavity, exposure to high-temperature oxidation was used to both stabilize the underlying structure and provide a reactive surface for probe immobilization. Treatment with (glycidoxypropyl)trimethoxysilane followed by an amino-terminated DNA oligonucleotide provided the sensor in its final form. As shown in Figure 11.4, incubation of this sensor with a solution of the complementary oligonucleotide, followed by a rinsing and drying procedure, causes a strong shift in the visible photoluminescence spectrum. Importantly, no analogous shift is observed in the presence of noncomplementary DNA. As a first attempt at detecting a "real" biological target, we also incubated the sensor with a solution of lambda bacteriophage. Lambda bacteriophage is a 48,502-bp double-stranded DNA virus that, as its name implies, infects bacteria.[18] It was gratifying to observe a strong redshift in the luminescence following incubation with this infectious agent. Preliminary experiments at the time suggested that the sensor was capable of detecting DNA at a very low concentration. However, it should be noted that the sensitivity of the detection will depend on the size of the targeted DNA, since larger targets will produce a larger redshift regardless of the size of the probe sequence. Significant work remains to be done in this area as well with regard to the effect of salt on both the stability of

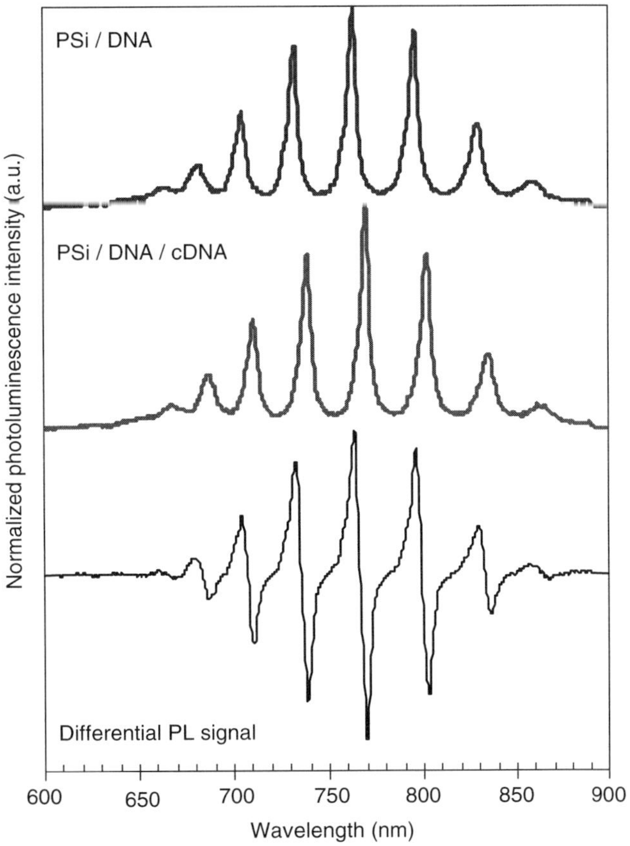

FIGURE 11.4
Top: Photoluminescence spectrum for 50 μM of DNA attached to a porous silicon microcavity structure. Middle: The same chip following exposure to a 1-μM solution of complementary DNA. A 7-nm redshift is observed, consistent with DNA binding. The difference spectrum is shown at the bottom.

the underlying PSi matrix and the reproducibility of the DNA detection. Salt is unfortunately absolutely required for DNA–DNA hybridization; however, a potential alternative is to use immobilized peptide-nucleic acid (PNA) probes instead, since their ability to bind DNA is significantly less salt dependent.[19] We will discuss issues of PSi stability in the presence of salt and other "environmental" solutes in greater detail later in this chapter.

After producing an initial demonstration of the suitability of PSi microcavities for biosensing, we next turned our efforts to an examination of the suitability of nonnucleic acid probes.[20] Because sensing with PSi is "label-free," meaning that no extraneous radionuclides or fluorophores are required, this substantially broadens the range of molecular targets and probes that can be addressed with the technique. Parallel to our work with PSi, we had developed a synthetic receptor for lipid A. Lipid A is the conserved head group of lipopolysaccharide, a key diagnostic molecule for Gram-(–) bacteria[21] as well as one of the primary causative agents of bacterial sepsis.[22] The receptor, dubbed tetratryptophan-ter-cyclopentane (TWTCP) (1),[23] was immobilized on PSi microcavities using a procedure analogous to that described above for DNA immobilization. Initial results proved disappointing: the devices produced in this manner were completely unable to detect either lipid A or Gram-(−) bacteria. However, on reflection, it was hypothesized that immobilization of the probe molecules in this manner produced devices in which the binding surface was occluded either via reaction of all four amino-terminated "arms" of the receptor with the surface, or simply by crowding or aggregation of the receptor. Incorporating a large (10:1) molar fraction of glycine methyl ester as a "spacer" molecule into the probe immobilization step remedied the problem, yielding devices that were capable of detecting both lipid A and a series of Gram-(–) bacteria (Figure 11.5). No signal was produced on incubation of the sensor with Gram-(+) bacteria, as one would expect based on their limited (and essentially inaccessible) amount of lipid A. Thus, we had demonstrated a PSi-based complement to the venerable "Gram stain," a 150-year-old microbiological staining technique that is still in use.

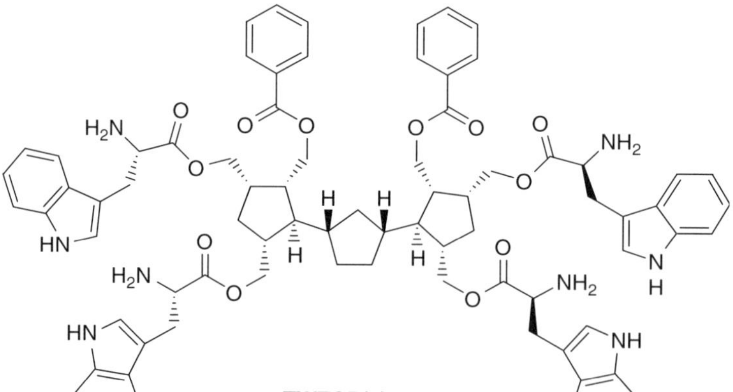

TWTCP(1)

11.4 From Empirical Observations to Predictable Behavior: Understanding the Properties of Mesoporous Silicon Biosensors in Greater Detail

While initial results from our laboratories on mesoporous silicon microcavities provided an indication that such devices could function as biosensors, substantial work remained

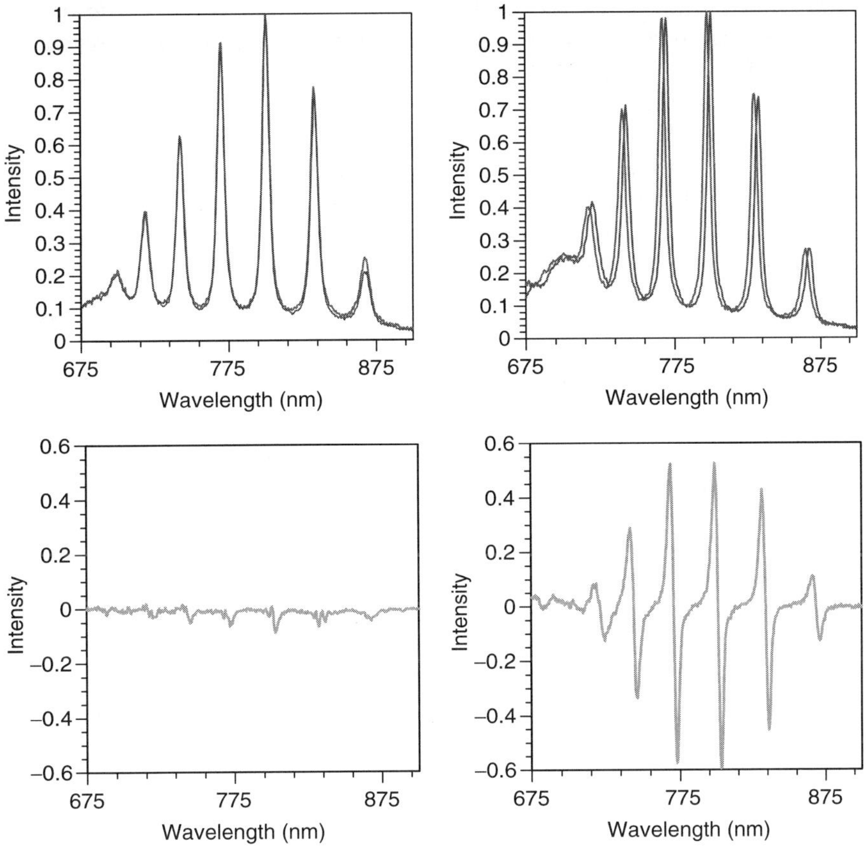

FIGURE 11.5
Photoluminescence spectra of a porous silicon microcavity biosensor in the presence and absence of bacterial cell lysates. Top: overlays of spectra for sensor alone following derivatization with TWTCP and glycine methyl ester with sensor photoluminescence following incubation with cell lysates from Gram-(+) bacteria (*Bacillus subtilis*, left) or Gram-(−) bacteria (*Escherichia coli*, right). The spectra in the bottom panels in each case are the difference between spectra obtained in the presence and absence of bacterial cell lysates.

(and still remains) to fully understand and control the properties of this material. Two concerns were of paramount importance: first, whether the photoluminescent devices we produced would be capable of harboring larger probe molecules and binding larger targets; and second, whether a secondary assay could be developed to correlate optical biosensor response with the amount of material specifically entrained in the pores (either covalently or noncovalently). Research in both these directions in the Rochester biosensor group was driven primarily by Dr. Lisa DeLouise; what follows in this section is a summary of the results of her research.

With regard to the question of molecular size, this was not as straightforward a problem as it might initially appear. While early efforts had demonstrated that high molecular-weight DNA sequences (such as lambda bacteriophage) could be efficiently captured, oligonucleotides are inherently flexible and rod-shaped. Therefore, their "molecular size" is not necessarily relevant to the problem of immobilizing and capturing peptides, proteins, and antibodies. Initial experiments in our laboratories with the extracellular domains of intimin and tir, two proteins critical to the infectious mechanism of enteropathogenic and enterohemmorhagic *E. coli*,[24] were characterized by poor reproducibility. This suggested that even these two relatively small proteins (12 kDa and 32 kDa, respectively) had difficulty entering the pores and/or became entrained in smaller channels within the PSi matrix in an inactive conformation. Since increasing the pore diameter was an essential prerequisite of moving forward, we examined the effect of dilute solutions of KOH post-etch on the microcavities.

The use of KOH solutions to enlarge pore diameters in mesoporous Si was initially explored by Tinsley–Bown, et al.[25] In that case, a solution of 83 mM KOH in 17% water or 83% ethanol was used to effectively double the diameter of mesopores etched in *n*-type silicon from roughly 50–100 nm for 12 min. For *P*-type silicon microcavities, we found that postetch treatment with a solution of 0.5–1.5 mM ethanolic KOH was effective in expanding the pore diameter by as much as 15% (from roughly 15 nm to 19 nm).[26] Treatment with more concentrated KOH solutions caused complete destruction of the multilayer microcavity structure in a matter of seconds. Pore opening was found to be independent of the starting porosity of the material, as measured by blueshift in the optical reflection spectrum.

Two additional important observations were derived from this study. First, KOH treatment almost instantaneously destroyed the photoluminescence of the microcavities. We presume that this is the result of very rapid dissolution of the manocrystalline silicon particles that are believed to be the source of PSi's room-temperature photoluminescence. Thus, such samples must be measured in reflectivity mode. Second, as one might anticipate, KOH treatment caused degradation of the microcavity Q ("Quality" factor, essentially a measure of the narrowness of the microcavity line width) as a function of length of KOH exposure (Figure 11.6). However, microcavities treated for $\leq$30 min— a time sufficient to enlarge pores— were still of sufficient quality for general sensor use.

11.5 Using Enzyme Assays as a Secondary Monitor of Sensor Performance

Since the properties of PSi are a result of its three-dimensional matrix structure, gaining a detailed understanding of the relationship between the structure of the PSi device and its sensor performance is considerably more difficult than for a planar device (i.e., a surface plasmon resonance (SPR) chip or fluorescence microarray) because the active surface cannot be observed directly. In the biosensing examples discussed so far, one observes a change in the optical spectrum of the device on attachment of a probe molecule or binding of an analyte, and based on control experiments with "nonbinding" analytes one attempts to relate this too; but it is not certain that there is a way to correlate these observations with another measure of the presence and amount of the active, immobilized recognition molecule. In early experiments, we employed radiolabeled oligonucleotides to correlate sensor response with the presence of immobilized probe oligonucleotides.[27] One could potentially employ fluorescence-tagged molecules in a similar manner. More recently, in a series of articles published in 2004 and 2005, we used the enzymatic activity of glutathione-S-transferase (GST) covalently immobilized in PSi as a measure of immobilization capacity of the matrix, and to examine in detail how the optical response of the sensor material correlates with the amount and activity of the immobilized molecules.

PSi has many characteristics analogous to other substrates used for immobilized enzyme bioreactor systems, thus combining this concept with biosensing seemed a natural course of action. GST is an ideal model enzyme for testing such systems, since it is relatively small (25 kDa per monomer, and active as the dimer[28]), robust, and inexpensive. The dimeric, active form of the enzyme is essentially spherical (Figure 11.7), permitting the

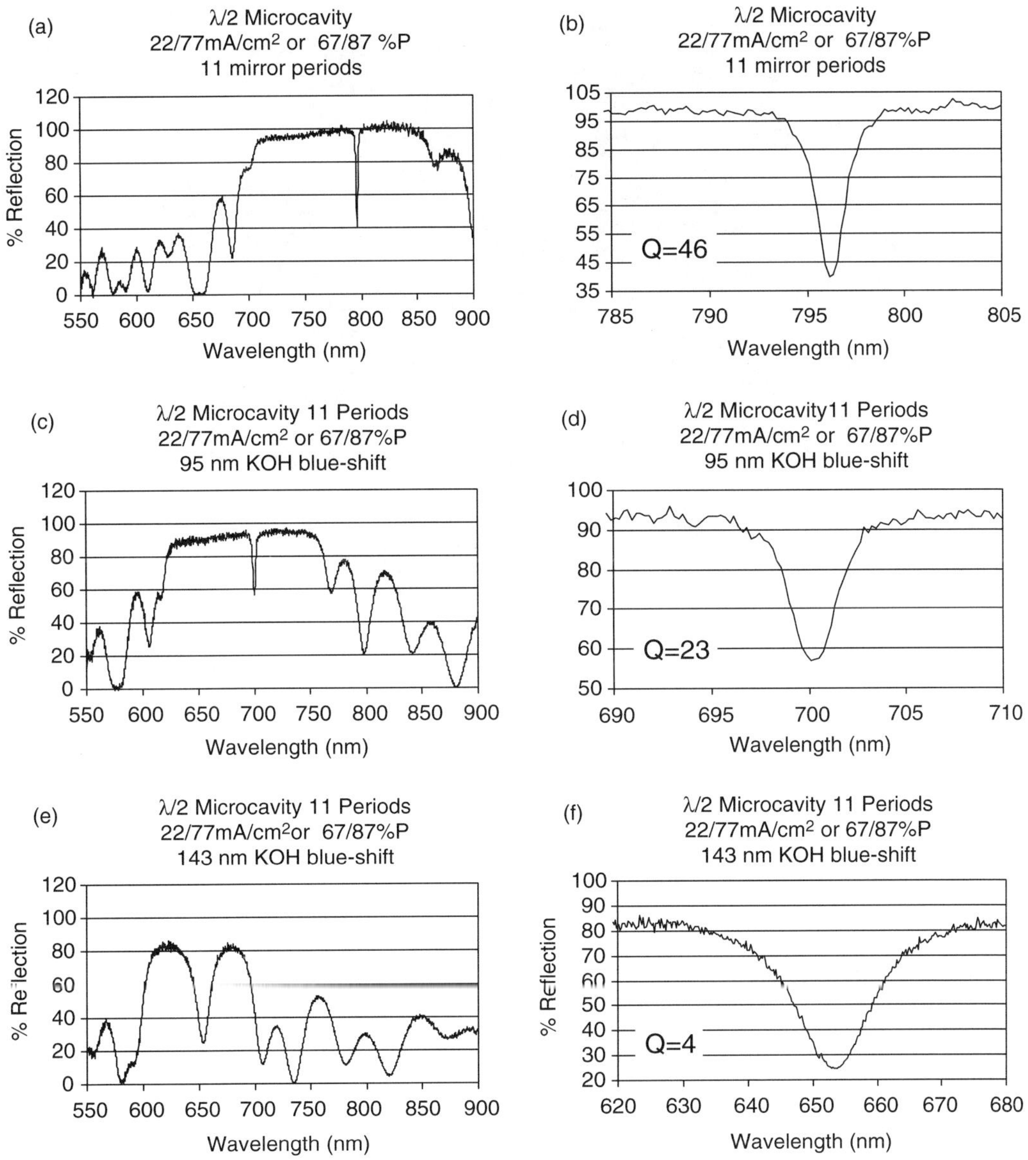

FIGURE 11.6
Optical reflection spectra as a function of 1.5-mM KOH exposure. (a,b) As-etched, (c,d) 30 min, and (e,f) 1 h 45 min. Data illustrates for short exposures, <30 min, the high-quality cavity remains. Longer exposures (>1 h) cause a more severe degradation of cavity quality (Q-factor).

construction of simple geometric models for its incorporation into the PSi matrix. Furthermore, extensive analysis of the performance of GST in a colorimetric assay (employing 1-chloro-2,4-dinitrobenzene (CDNB)) was available in the literature,[29] allowing us to benchmark the performance of our systems against those studied by others.

The first order of business was to employ the enzyme assay as a means of determining the amount of GST that could be immobilized in PSi as a function of sample depth.[30] Single-layer mesoporous devices were etched using standard conditions; thickness varied in the range of 0.03–8.8 μm (as determined by gravimetric and SEM analysis) by varying the etch time. Samples were then subjected to postetch KOH treatment to

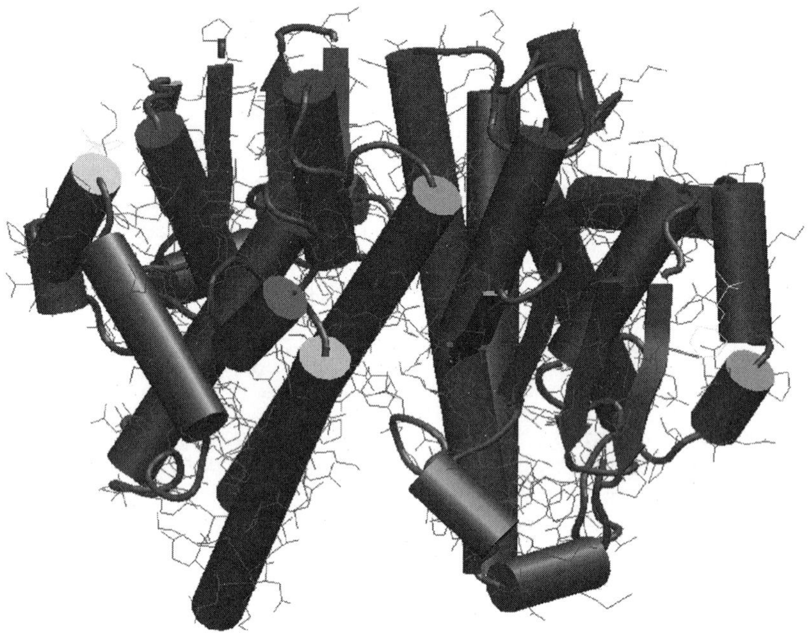

FIGURE 11.7
X-ray crystal structure of the GST Dimer (PDB code 12GS). (From Oakley, A. J., LoBello, M., Nuccetelli, M., Mazzetti, A. P., and Parker, M. W. (1999). The ligandin (non-substrate) binding site of human pi class glutathione transferase is located in the electrophile binding site (H-site). *J. Mol. Biol.* 291: 913.).

enlarge the pores, and stabilized via thermal oxidation. To prepare surfaces for immobilizing GST, PSi samples were first treated with a solution of 3-aminopropyltriethoxysilane (APTS), followed by a solution of glutaraldehyde. This process yields a surface populated with reactive aldehyde groups, capable of trapping amines on the protein (from the amino terminus and lysine side-chains) as imines. Following immobilization of GST, measurement of the enzyme reaction yield on the chip compared to measurements made with free enzyme in solution allowed for the amount of enzyme immobilized to be determined. As expected, we found that the amount of enzyme immobilized increased with increasing depth of the PSi layer (Figure 11.8). Furthermore, the observed immobilization capacity correlated well with a simple geometric model of the PSi matrix in which the pores are approximated as cylinders (Figure 11.9). These results established that immobilization of biomolecules in PSi does indeed proceed in a predictable manner and set the stage for further characterization of the enzymatic and optical performance of multilayer films.

While GST retained its ability to function inside single-layer (single-porosity) PSi films, would the same be true of multilayer films? To examine this question, we prepared $\lambda/2$ microcavity devices consisting of 28 layers of alternating 71 and 84% porosity,[31] and immobilized GST as before. After carefully characterizing the kinetic parameters for free GST in solution as a function of concentration, enzymatic activity of the microcavity-immobilized GST was assessed. Perhaps unsurprisingly, the activity of the immobilized GST was between two- and fourfold lower than that in solution (Table 11.1). Several effects potentially contribute to this observation. First, the immobilization chemistry employed in these studies is not regiospecific: any amine on the surface of the protein has an equivalent probability of reacting with an aldehyde on the PSi surface. Thus, it is possible for the enzyme to be attached in an orientation that blocks access to the substrate-binding site.

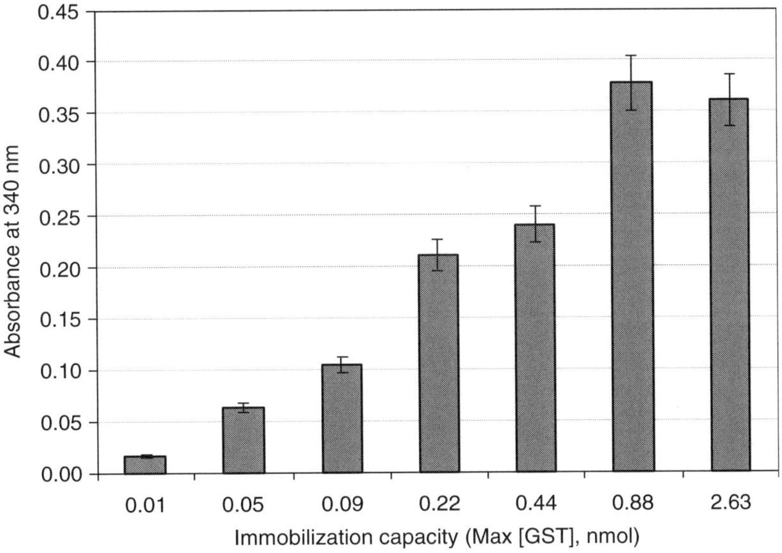

FIGURE 11.8
Conjugation product yield obtained from each of the immobilized enzyme chips. Absorbance values were measured after a 1-min reaction on a colorimetric substrate. The X-axis indicates the immobilization capacity for each chip as calculated using a simple geometric model for PSi.

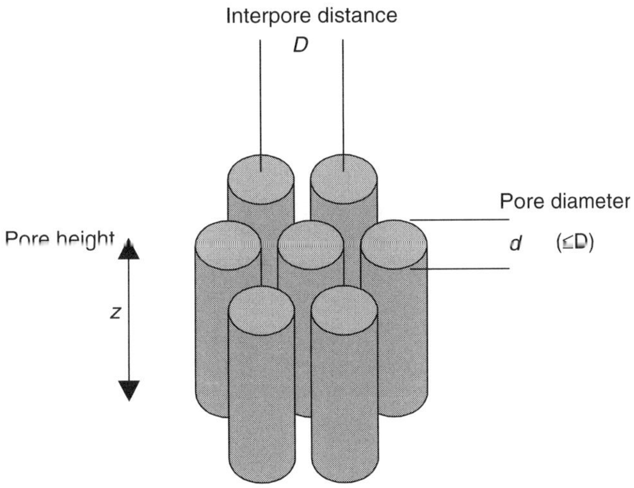

FIGURE 11.9
Simple geometric model for calculating PSi internal volume.

Second, other aspects of the immobilization process, from sample handling to simple contact with the PSi surface, may render some fraction of the enzyme inactive through denaturation or other degradation. Finally, the microenvironment of the PSi matrix itself may contribute to diminished activity, through charge effects (unmodified silicon is anionically charged at neutral pH[32]) or by sterically impeding diffusion. In the context of biosensing, all of these effects can potentially interfere with performance of the sensor, and will require further study.

With an understanding of the immobilization capacity and enzymatic performance of GST-functionalized PSi devices in hand, we could next proceed with correlating these observations with the changes in optical spectra that are the hallmark of PSi biosensors.[33] For a series of devices prepared with increasing concentrations of GST, we were

TABLE 11.1

Comparison of Immobilized vs. Solution GST Enzyme Activity for Fixed Substrate Concentrations

Solution [GST] (nmol)	GST (μg/ml)	Activity (nmol/min)	Specific activity (μmol/(min mg))
Solution Enzyme Reaction			
0.05	33	10.8	4.3
0.10	67	19.4	3.9
0.50	333	49.2	2.0
Immobilized Enzyme Reaction			
0.05	33	2.6	1.0
0.10	67	4.7	0.9
0.50	333	26.2	1.0
1.00	667	48.5	1.0

Note: Glutathione = 3.3 mM; CDNB = 3.3 mM.

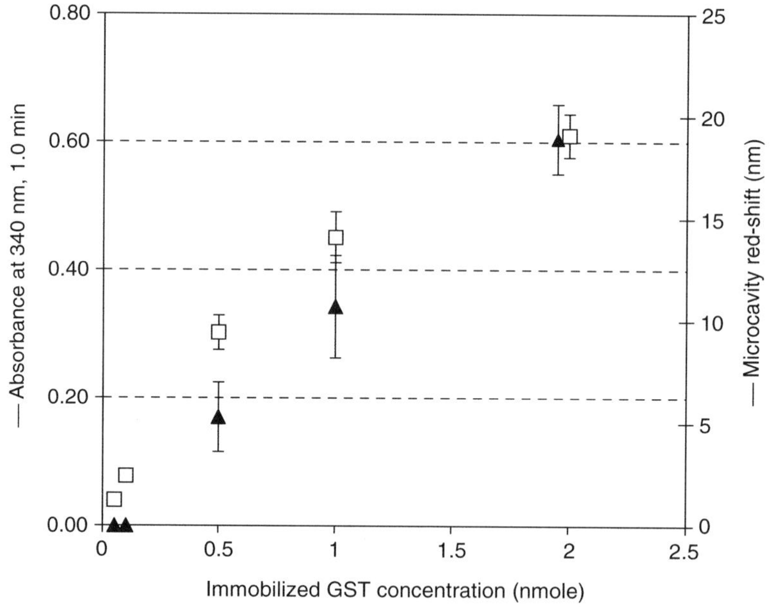

FIGURE 11.10
Correlation of sensor response with enzyme activity. The microcavity detection limit appears to lie between 0.1 and 0.5 nmol of GST.

gratified to observe a strong correlation between enzyme activity and optical response, as measured by redshift in the reflectance spectrum (Figure 11.10). The observation that the optical response is linear, while the enzymatic performance becomes nonlinear at high concentration, is interesting and potentially a reflection of the increased probability of active-site occlusion that occurs when higher concentrations of enzyme are immobilized. Despite being a relatively simple observation this places the use of PSi as a matrix for the construction of biosensors on firmer experimental footing. Simulations conducted during the course of this work suggested that devices tuned to operate closer to the infrared should have improved sensitivity; this assertion remains to be tested.

11.6 Studying the Longevity of PSi under Environmentally and Physiologically Relevant Conditions

Because PSi biosensors function via changes in the amount of bulk material present within the PSi matrix, degradation and dissolution of PSi over time is a major concern. Indeed, PSi has been touted as a delivery vehicle for pharmaceuticals precisely because of its biocompatibility, low toxicity, and ability to dissolve over time in a physiological environment.[34] Fortunately, several methods have been developed to stabilize PSi. These include thermal oxidation (passing the PSi device through a roughly 500-degree furnace in the presence of ambient oxygen), oxidation via aqueous hydrogen peroxide treatment, ozonolysis, thermal carbonization (sample annealing at high temperature in N_2 or acetylene atmosphere), thermal nitridation (sample annealing at upwards of 1100°C in N_2 atmosphere), and dodecene treatment. Björkvist et al. compared four of these treatments in single-layer PSi samples.[35] Brunauer–Emmett–Teller (BET) methodology was employed to examine changes in porosity (and hence surface area) following thermal oxidation, dodecene treatment, thermal carbonization, and thermal nitridation. Thermal oxidation and thermal nitridation were found to cause the greatest changes in porosity (corresponding to an 80% loss in surface area). However, these two treatments were also found to provide the greatest increase in device-shelf life; examining the weight increase of PSi samples following storage at 70°C and 75% relative humidity, these were found to have undergone the smallest change.

The tendency for PSi to degrade rapidly under physiological conditions was recognized early by Canham and coworkers.[36,37] In their 1998 report,[38] Sailor, Ghadiri, and coworkers examined the stability of single-layer PSi structures in 10% EtOH/90% PBS (pH 7.4) solution as a function of surface derivatization. Measured as a time-dependent change in effective optical thickness, "as etched" (hydride-terminated) samples were found to degrade at a rate of 6 nm/min, while ozone oxidation reduced this to 2 nm/min and thermal oxidation reduced it to 1 nm/min. Importantly, functionalization of the ozone-oxidized surface with (2-pyridyldithiopropionamidobutyl)dimethylmethoxysilane reduced the degradation rate to an almost negligible 0.05 nm/min. This suggests that careful choice of probe immobilization chemistry can significantly enhance the stability of the devices produced.

In at least one instance, salt-mediated PSi degradation has been exploited as a "feature," rather than viewed as a problem to be avoided. In 2004, the Ghadiri group reported that hybridization of target DNA to immobilized nucleic acid probes in neutral buffered solution caused a blueshift in the optical-sensor response relative to probe-functionalized samples exposed to solutions lacking target DNA.[39] This observation was interpreted as a hybridization-dependent enhancement in the rate of device dissolution.

11.7 PSi Biosensors in Hydrogels: Toward the "Smart Bandage"

A deficiency of the majority of PSi-based biosensors examined till date is that because the PSi matrix remains attached to the backing Si wafer, diffusion of analyte solution in and out of the matrix is substantially hindered. It has been known for some time that PSi membranes can be obtained as free-standing entities by fracturing them from the backing wafer with a high-current pulse postetch; however, these free-standing membranes are exceptionally fragile. The "Smart Dust" methodology of Sailor and coworkers (described below) is a highly creative solution to this problem; another solution is to transfer the PSi membrane to a backing material that readily permits diffusion of analytes. The Rochester group has recently begun exploring the latter concept, by transferring PSi microcavities to a commercial hydrogel material.[40] Microcavities tuned to operate in the visible range (725 nm) were contact laminated to a commercial wound-care product ("Nu Gel," Johnson & Johnson) (Figure 11.11). As one might expect, transfer to the gel causes a substantial redshift in the optical spectrum and degradation in microcavity Q-factor (Figure 11.12), but importantly the optical response of the device produced remains constant for at least 1 year and through repeated cycles of gel hydration and drying. Response of the sensor to changes in solution environment is also maintained, as cycles of exposure to water and increasing concentrations of sucrose yielded a linear [sucrose] or optical response plot, and return to original (control) response value. While this study did not test the ability of the sensor to selectively detect a target analyte (i.e., there was no immobilized molecular recognition element employed), it nonetheless establishes that such devices have potential for applications in general biosensing. More specifically, one can envision applications in which hydrogel-immobilized PSi biosensors could be used to monitor the progress of wound healing, and given the interest in PSi as a drug-delivery matrix, could also actively participate in the wound-healing process.

11.8 Beyond Mesoporous Silicon: Larger-Volume Structures

One obvious limitation of the PSi sensors discussed so far is that, by definition, mesoporous-type silion has an upper limit on the size of molecules that can be targeted given the 10–50-nm pore diameter. The Rochester group has recently demonstrated that *n*-type silicon can yield pores with diameters in the 60–120 nm range through careful choice of doping level (0.01–0.001 ohm-cm) (Figure 11.13).[41] Importantly, as with mesoporous silicon fabrication, porosity may be varied by changing current as a function of time during the etch

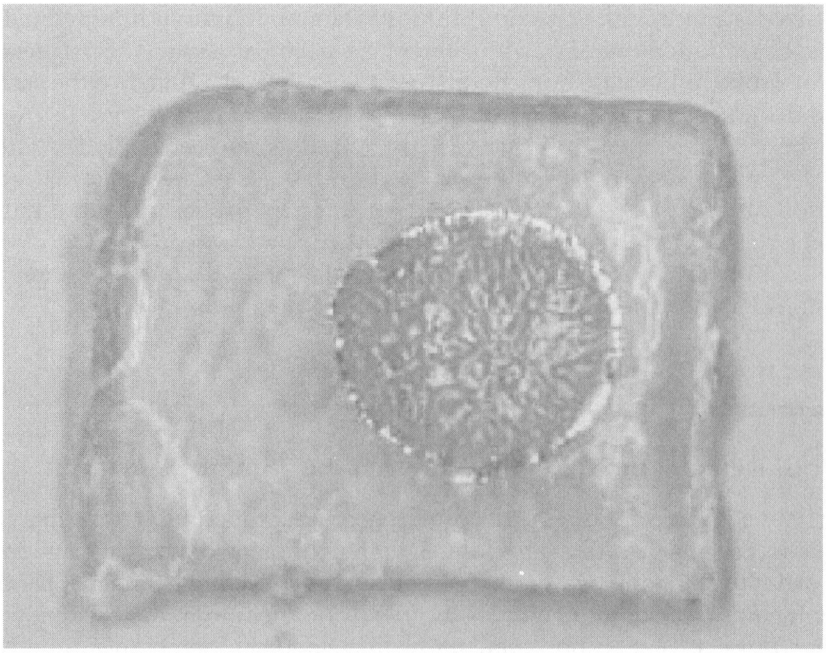

FIGURE 11.11
PSi microcavity sensor after contact-adhesion to NuGel hydrogel.

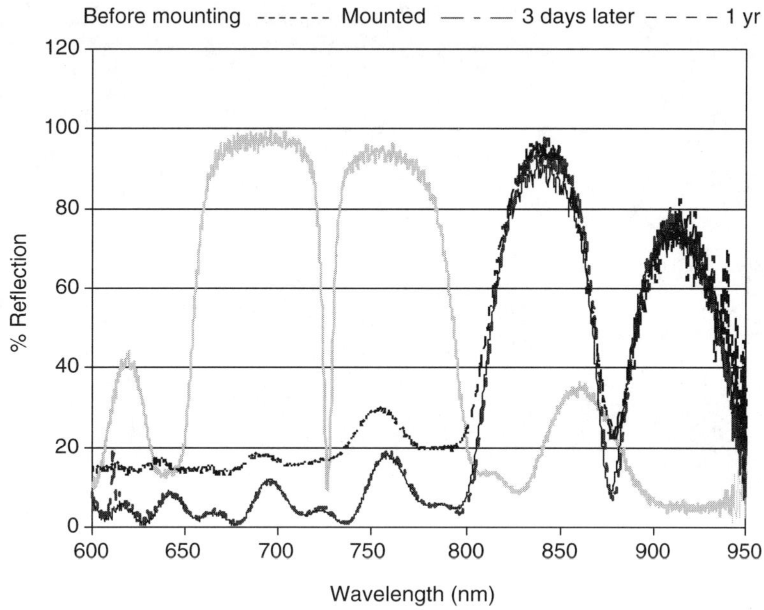

FIGURE 11.12
Reflection spectra for a free-standing 5.2-μm thick microcavity membrane, before and after mounting in a hydrogel sheet.

process, allowing fabrication of microcavities. Two systems were examined to provide an initial proof of concept for biosensing with these devices. First, chips were exposed to varying concentrations of rabbit immunoglobulin G (IgG). IgG is a fairly large protein (150 kDa, or ~17 nm along its largest dimension) that does not infiltrate into mesoporous microcavities well. However, a clear concentration dependence in the redshift of microcavity response was available post-IgG exposure, indicating that infiltration of the protein was facile. Second, biotin was covalently attached to the surface of a macroporous microcavity, and used as a probe for streptavidin. Indeed, as one would expect, chips functionalized with biotin showed a redshift after exposure to streptavidin, while those lacking biotin did not.

11.9 Alternative Sensing Modes and Device Structures

11.9.1 "Smart Dust"

Building on their 1992 observation that ultrasonic treatment of PSi can produce suspensions of luminescent particles,[42] Sailor and collaborators published two reports in 2002 detailing the use of such "smart dust" in chemical[43] and biological[44] sensing. We will focus on the biological application here. PSi samples were etched as rugate filters, or devices in which sinusoidal variation in the refractive index produces a narrow band of high reflectivity (in this case, 11 nm) at a specific wavelength. Control over the precise placement of this high reflectivity band allows for the production of "encoded" particles, conceptually similar to the "smart beads" of Nie[45] or the infrared-encoded polystyrene resins of Fenniri.[46]

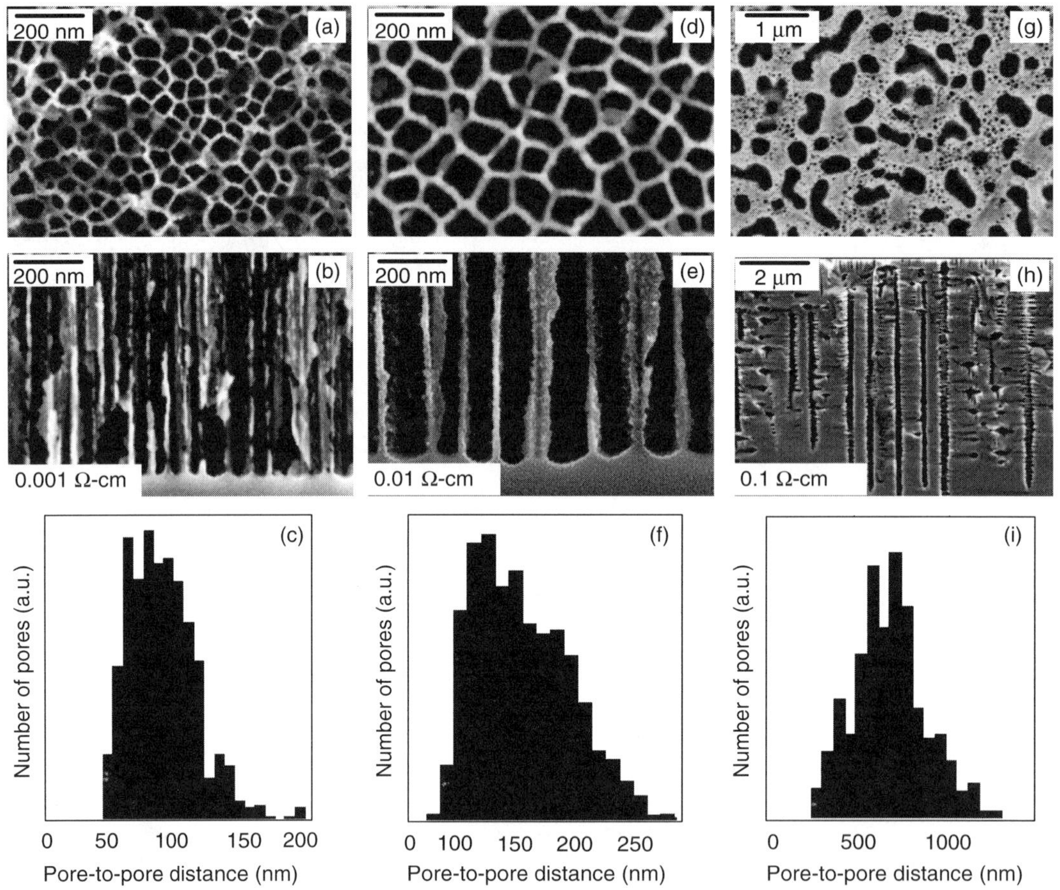

FIGURE 11.13
Top view and cross-sectional SEM images and pore-to-pore distance histograms of macropores etched on *n*-type silicon wafers with different resistivities: (a–c) 0.001 ohm-cm; (d–f) 0.01 ohm-cm; (g–i) 0.1 ohm-cm.

Two sets of encoded particles were examined in the Sailor study (Figure 11.14). The first set, encoded at 750 nm, was treated with a concentrated solution of bovine serum albumin (BSA). The second set, encoded at 540 nm, was treated with a solution of rat albumin, and then with a rabbit antirat-albumin antibody solution. The two sets of particles were then mixed, and exposed to a solution of a fluorescence-tagged goat antirabbit IgG. Importantly, when the reflectance spectra of particles that had taken up the fluorescent antibody were measured, all were found to have maxima at 540 nm. That is, only those particles treated initially with rat albumin and antirat-albumin took up the target fluorescent antibody, and the rugate filter "code" for those particles could be read retrospectively. This is a particularly impressive result, given that the probe molecules were only noncovalently adsorbed onto the particle surface rather than covalently attached. In the context of sensing, one could envision employing this scheme in situations where detection of multiple analytes simultaneously in solution was desirable.[47]

11.9.2 Electrical Sensors as an Alternative Biosensing Mechanism

While the vast majority of studies on PSi as a biosensing platform have focused on optical signaling, changes in the electrical properties of PSi as a function of environment have also

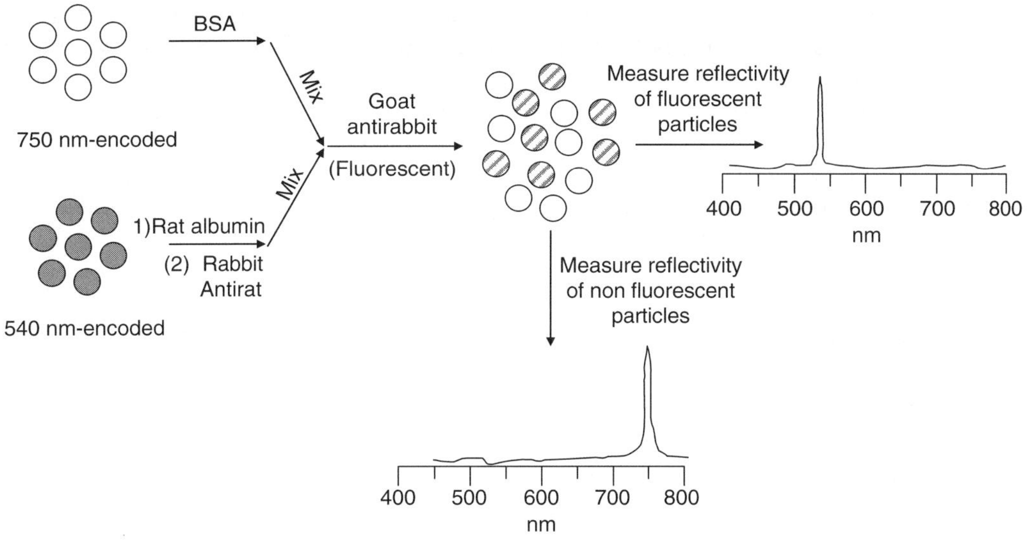

FIGURE 11.14
Sailor "smart dust" experiment.

been observed.[48] Recently, the Fauchet laboratory has begun exploring the potential of employing the electrical properties of macroporous Si in sensing charged molecules such as DNA.[49] Thick layers of macroporous silicon were electrochemically etched in HF or dimethylformamide; the resulting devices were attached to a computer-controlled inductance-capacitance-resistance (LCR) meter. DNA probes were then introduced into the device and allowed to adhere electrostatically, and measurements of changes in impedance and phase angle were made while the sample was still wet. Results indicated that the sensor could reproducibly detect complementary DNA, with good selectivity for the "matched" sequence relative to fully "mismatch" sequence. The results fit well with a "space charge region modulation" (SCRM) model;[50] binding of charged complementary DNA to the electrostatically immobilized probe oligonucleotide alters both the dielectric constant inside the pores, and the charge distribution of the underlying silicon itself. While significant development work remains here to establish the sensitivity and generality of electrical biosensors based on PSi, the potential simplicity of integrating such sensors with microelectronics using standard fabrication techniques makes them highly attractive.[51]

11.10 Conclusions and Prospective

Clearly, significant work remains to be done if PSi biosensors are to evolve from being an area of academic research to a generally available detection (or diagnostic) tool. Standardized methods for producing devices will be important, as will improved methods for device stabilization and probe immobilization. Little is known at this point regarding long-term stability of biopolymer-functionalized devices; for example, does PSi immobilization *enhance* or *retard* degradation of nucleic acids, peptides, and antibodies in ambient conditions? Multianalyte (arrayed) sensors are an important goal, as are optical and electrical sensors integrated with microfluidics, other sample preparation modules, and "on-chip" detectors.

However, the examples described above and ongoing research have provided strong support for the assertion that PSi-based sensors can provide significant value in general biosensing applications, as well as potentially possessing some truly unique attributes. For example, little has been done so far to exploit the "bulk filtration" properties the PSi matrix can provide, as well as potentially carrying out size-selective detection of analytes. Combined with its ease of preparation and the promise of simple integration with commodity microelectronics, we can expect to see PSi-based sensors move beyond the laboratory into the "real world" in the near future.

References

1. Dolley, C. S. (1885). *The Technology of Bacteria Investigation*. Boston, MA: S. E. Cassino and Company.
2. Segal, E.; Friedman, N.; Kaminski, N.; Regev, A.; Koller, D. (2005). From signatures to models: understanding cancer using microarrays. *Nat. Genet.* 37: S38–S45.
3. Liu, H. L.; Hsu, J. P. (2005). Recent developments in structural proteomics for protein structure determination. *Proteomics* 5: 2056–2058.
4. Goodacre, R.; Vaidyanathan, S.; Dunn, W. B.; Harrigan, G. G.; Kell, D. B. (2004). Metabolomics by numbers: acquiring and understanding global metabolite data. *Trends Biotechnol.* 22: 245–252.
5. Canham, L. T., Ed., (1997). *Properties of Porous Silicon*. London, UK: INSPEC: The Institution of Electrical Engineers.
6. Daly, S. J.; Keating, G. J.; Dillon, P. P.; Manning, B. M.; O'Kennedy, R.; Lee, H. A.; Morgan, M. R. A. (2000). Development of surface plasmon resonance-based immunoassay for aflatoxin B1. *J. Agric. Food Chem.* 48: 5097–5104.
7. Lu, J.; Strohsahl. C. M.; Miller, B. L.; Rothberg, L. J. (2004). Reflective Interferometric Detection of Label-Free Oligonucleotides. *Anal. Chem.* 76:4416–4420.
8. Canham, L. T. (1990). Silicon quantum wire array fabrication by electrochemical and chemical dissolution of wafers. *Appl. Phys. Lett.* 57:1046–1048.
9. Doan, V. V.; Sailor, M. J. (1992). Luminescent color image generation on porous silicon. *Science* 256: 1791–1792.
10. Lin, V. S.-Y.; Motesharei, K.; Dancil, K.-P. S.; Sailor, M. J.; Ghadiri, M. R. (1997). A porous silicon-based optical interferometric biosensor. *Science* 278: 840–843.
11. Dancil, K.-P. S.; Greiner, D. P.; Sailor, M. J. (1999). A Porous Silicon Optical Biosensor: Detection of Reversible Binding of IgG to a Protein A-Modified Surface. *J. Am. Chem. Soc.* 121: 7925–7930.
12. Berger, M. G.; Dieker, C.; Thönissen, M.; Vescan , L.; Lüth, H.; Münder, H.; Theiß, W.; Wernke, M.; Grosse, P. (1994). Porosity Superlattices: a new class of Si heterostructures. *J. Phys. D: Appl. Phys.* 27:1333–1336.
13. Chan, S.; Fauchet, P. M. (1999). Tunable, narrow, and directional luminescence from porous silicon light emitting devices. *Appl. Phys. Lett.* 75: 274–276.
14. Lugo, J. E.; Lopez, H. A.; Chan, S.; Fauchet, P. M. (2002). Porous silicon multilayer structures: A photonic band gap analysis. *J. Appl. Phys.* 91: 4966–4972.
15. Lopez, C. (2003). Materials aspects of photonic crystals. *Adv. Mater.* 15: 1679–1704.
16. Chan, S.; Fauchet, P. M.; Li, Y.; Rothberg, L. J.; Miller, B. L. (2000). Porous Silicon Microcavities for Biosensing Applications. *Phys. Stat. Sol. A* 182: 541–546.
17. Chan, S.; Li, Y.; Rothberg, L. J.; Miller, B. L.; Fauchet, P. M. (2001). Nanoscale Silicon Microcavities for Biosensing. *Mat. Sci. Eng. C* 15: 277–282.
18. Hershey, A. D. (1971). *The Bacteriophage Lambda*. New York: Cold Spring Harbor Laboratory.
19. Igloi, G. L. (1998). Variability in the stability of DNA-peptide nucleic acid (PNA) single-base mismatched duplexes: Real-time hybridization during affinity electrophoresis in PNA-containing gels. *Proc. Natl. Acad. Sci. U S A* 95: 8562–8567.

20. Chan, S.; Horner, S. R.; Miller, B. L.; Fauchet, P. M. (2001). Identification of Gram Negative Bacteria using Nanoscale Silicon Microcavities. *J. Am. Chem. Soc.* 123: 11797–11798.
21. Young, L. S.; Martin, W. J.; Meyer, R. D.; Weinstein, R. J.; Anderson, E. T. (1977). Gram-negative rod bacteremia: Microbiologic, immunological, and therapeutic considerations. *Ann. Intern. Med.* 86: 456–471.
22. Ulevitch, R. J.; Tobias, P. S. (1994). Recognition of endotoxin by cells leading to transmembrane signaling. *Curr. Opin. Immunol.* 6: 125–130.
23. Hubbard, R. D.; Horner, S. R.; Miller, B. L. (2001). Highly Substituted *ter*-Cyclopentanes as Receptors for Lipid A. *J. Am. Chem. Soc.* 123: 5810–5811.
24. Kenny, B.; DeVinney, R.; Stein, M.; Reinscheid, D. J.; Fry, E. A.; Finlay, B. B. (1997). Enteropathogenic E. coli (EPEC) transfers its receptor for intimate adherence into mammalian cells. *Cell* 91: 511.
25. Tinsley-Brown, A. M.; Canham, L. T.; Hollings, M.; Anderson, M. H.; Reeves, C. L.; Cox, T. I.; Nicklin, S.; Squirrell, D. J.; Perkins, E.; Hutchinson, A.; Sailor, M. J.; Wun, A. (2000). Tuning the pore size and surface chemistry of porous silicon for immunoassays. *Phys. Stat. Sol. A* 182: 547–553.
26. DeLouise, L. A.; Miller, B. L. (2004). Optimization of mesoporous silicon microcavities for proteomic sensing. *Mat. Res. Soc. Symp. Proc.* 782: A5.3.1–A5.3.7.
27. Chan, S.; Fauchet, P. M.; Li, Y.; Rothberg, L. J.; Miller, B. L. (2000). Porous Silicon Microcavities for Biosensing Applications. *Phys. Stat. Sol. A.* 182: 541–546.
28. Allardyce, C. S.; Mcdonagh, P. D; Lu-Yun, L.; Wolf, C. R.; Roberts, G. C. K. (1999). The role of tyrosine-9 and the C-terminal helix in the catalytic mechanism of Alpha-class glutathione S-transferases. *Biochem. J.* 343: 525–531.
29. Vargo, M. A.; Nguyen, L.; Colman, R. F. (2004). Subunit interface residues of glutathione S-transferase A1-1 that are important in the monomer-dimer equilibrium. *Biochemistry* 43: 3327–3335.
30. DeLouise, L. A.; Miller, B. L. (2004). Quantitative Assessment of Enzyme Immobilization Capacity in Porous Silicon. *Anal. Chem.* 76: 6915–6920.
31. DeLouise, L. A.; Miller, B. L. (2005). Enzyme Immobilization in Porous Silicon: Quantitative Analysis of the Kinetic Parameters for Glutathione-S-Transferases. *Anal. Chem.* 77: 1950–1956.
32. Bousse, L.; Mostarshed, S.; vander Schoot, B.; de Rooij, N. F.; Gimmel, P.; Göpel, W. (1991). Zeta potential measurements of Ta_2O_5 and SiO_2 thin-films. *J. Colloid Interface Sci.* 147: 22–32.
33. DeLouise, L. A.; Kou, P. M.; Miller, B. L. (2005). Cross-Correlation of Optical Microcavity Biosensor Response with Immobilized Enzyme Activity. Insights into Biosensor Sensitivity. *Anal. Chem.* 77: 3222–3220.
34. Canham, L. T.; Stewart, M. P.; Buriak, J. M.; Reeves, C. L.; Anderson, M.; Squire, E. K.; Allcock, P.; Snow, P. A. (2000). Derivatized porous silicon mirrors: Implantable optical components with slow resorbability. *Phys. Stat. Sol. A* 182: 521–525.
35. Björkvist, M.; Salonen, J.; Laine, E.; Niinistö, L. (2003). Comparison of stabilizing treatments on porous silicon for sensor applications. *Phys. Stat. Sol. A* 197: 374–377.
36. Canham, L. T. (1995). Bioactive silicon structure fabrication through nanoetching techniques. *Adv. Mater.* 7: 1033.
37. Canham, L. T.; Stewart, M. P.; Buriak, J. M.; Reeves, C. L.; Anderson, M.; Squire, E. K.; Allcock, P.; Snow, P. A. (2000). Derivatized porous silicon mirrors: Implantable optical components with slow resorbability. *Phys. Stat. Sol. A* 182: 521–525.
38. Janshoff, A.; Dancil, K.-P. S.; Steinem, C.; Greiner, D. P.; Lin, V. S.-Y.; Gurtner, C.; Motesharei, K.; Sailor, M. J.; Ghadiri, M. R. (1998). Macroporous p-Type Silicon Fabry-Perot Layers. Fabrication, Characterization, and Applications in Biosensing. *J. Am. Chem. Soc.* 120: 12108–12116.
39. Steinem, C.; Janshoff, A.; Lin, V. S.-Y.; Völker, N. H.; Ghadiri, M. R. (2004). DNA hybridization-enhanced porous silicon corrosion: mechanistic investigations and prospect for optical interferometric biosensing. *Tetrahedron* 60: 11259–11267.
40. DeLouise, L. A.; Fauchet, P. M.; Miller, B. L.; Pentland, A. P. (2005). Hydrogel Supported Optical Microcavity Sensors. *Adv. Mater.* 17: 2199–2203.
41. Ouyang, H.; Christophersen, M.; Viard, R.; Miller, B. L.; Fauchet, P. M. (2005). Macroporous Silicon Microcavities for Macromolecule Detection. *Adv. Func. Mat.* 15: 1851–1859.

42. Heinrich, J. L.; Curtis, C. L.; Credo, G. M.; Kavanagh, K. L.; Sailor, M. J. (1992). Luminescent colloidal Si suspensions from porous Si. *Science* 255: 66–68.
43. Schmedake, T. A.; Cunin, F.; Link, J. R.; Sailor, M. J. (2002). Standoff detection of chemicals using porous silicon 'smart dust' particles. *Adv. Mater.* 14: 1270–1272.
44. Cunin, F.; Schmedake, T. A.; Link, J. R.; Li, Y. Y.; Koh, J.; Bhatia, S. N.; Sailor, M. J. (2002). Biomolecular screening with encoded porous-silicon photonic crystals. *Nature Mater.* 1: 39–41.
45. Han, M. Y.; Gao, X. H.; Su, J. Z.; Nie, S. (2001). Quantum-dot-tagged microbeads for multiplexed optical coding of biomolecules. *Nature Biotech.* 19: 631–635.
46. Fenniri, H.; Chun, S.; Ding, L.; Zyrianov, Y.; Hallenga, K., (2003). Preparation, Physical Properties, On-Bead Binding Assay and Spectroscopic Reliability of 25 Barcoded Polystyrene-Poly(ethylene glycol) Graft Copolymers. *J. Am. Chem. Soc.* 125: 10546–10560.
47. Sailor, M. J.; Link, J. R. (2005). 'Smart dust': nanostructured devices in a grain of sand. *Chem. Commun.* 2005: 1375–1383.
48. Baratto, C.; Comini, E.; Faglia, G.; Sberveglieri, G.; Di Francia, G.; De Filippo, F.; La Ferrara, V.; Quercia, L.; Lancellotti, L. (2000). Gas detection with a porous silicon based sensor. *Sens. Act. B* 65: 257–259.
49. Archer, M.; Christophersen, M.; Fauchet, P. M. (2004). Macroporous Silicon Electrical Sensor for DNA Hybridization Detection. *Biomed. Microdevices* 6: 203–211.
50. Archer, M.; Christophersen, M.; Fauchet, P. M. (2005). Electrical porous silicon chemical sensor for detection of organic solvents. *Sens. Act. B* 106: 347–357.
51. Archer, M.; Christophersen, M.; Fauchet, P. M.; Persaud, D.; Hirschman, K. D. (2004). Electrical porous silicon microarray for DNA hybridization detection. *Mat. Res. Soc. Symp. Proc.* 782: 385–391.

12

Fundamental Aspects and Applications of Nanotubes and Nanowires for Biosensors

Xueliang Sun and Chen-zhong Li

CONTENTS

12.1 Introduction

Rapid advances in nanoscience and nanotechnology have provided a variety of nanoscale materials with unique optical, electrical, magnetic, or catalytic properties, which have enabled the fabrication of various functional nanoscale devices such as biosensors [1,2]. Integration of biomaterials (e.g., proteins, peptides, or DNA) with nanostructured materials greatly expands the impact of bioelectronics, particularly in biosensors.

Nanostructures (i.e., structures with at least one dimension in the range of 1–100 nm) have attracted steadily growing interest due to their unique, fascinating properties and potential applications complementary to three-dimensional (3D) bulk materials. Dimensionality plays a critical role in determining the properties of materials due to, for example, the different ways in which electrons interact in 3D, two-dimensional (2D), one-dimensional (1D), and zero-dimensional (0D) structures [3]. Compared with 0D nanostructures (so-called quantum dots or nanoparticles) and 2D nanostructures (thin films), 1D nanostructures (including carbon nanotubes (CNTs) and nanowires (NWs)) are ideal as model systems for investigating the dependence of electronic transport, optical, and mechanical properties on size confinement and dimensionality as well as for various potential applications, including composite materials, electrode materials, field emitters, nanoelectronics, and nanoscale sensors.

In comparison with nanoparticles (NPs), however, the integration of 1D nanostructures with biological systems to form functional assemblies has been slow until recently, as it has been hindered by the difficulties associated with the synthesis and fabrication of these materials with well-controlled dimensions, morphology, phase purity, and chemical composition; for example, controlling growth to form semiconducting single-walled carbon nanotubes (SWCNTs). As better techniques are developed, 1D nanostructures should find extensive applications in the construction of novel nanoscale devices such as biosensors, which combine the conductive or semiconductive properties of the nanomaterials with the recognition of biomaterials. Owing to the high surface-to-volume ratio of 1D nanostructures, the variation of their electronic conductance to adsorbed surface species may be sensitive enough for single-molecule detection to become possible.

This chapter is organized as follows: We begin by describing the processing techniques used to control the nanostructure of CNTs and NWs. We pay particular attention to structure control and growth mechanisms of these nanomaterials. We then discuss methods by which CNTs and NWs can be functionalized with appropriate agents for biosensor applications. This section is followed by a review of interesting new applications of these materials in electrochemical biosensors, field-effect transistor (FET)-based biosensor, and nanofabrication of these biosensors.

12.2 Carbon Nanotubes

12.2.1 Introduction

Carbon nanotubes are a class of new, 1D carbon nanomaterials discovered by Iijima in 1991 [4]. The extraordinary mechanical properties and unique electrical properties of CNTs have stimulated extensive research activities worldwide, resulting in many special issues of journals [5–7] and several books [8–11]. Driven by potential applications including nanocomposite materials, nanoelectrode materials, field emitters, nanoelectronics, and nanoscale sensors, significant progress has been made in CNT synthesis and characterization in the past ten years.

Structures comprising one cylindrical tube are called single-walled carbon nanotubes (SWCNTs). SWCNTs have a relatively smaller diameter, as small as 0.4 nm, and can be metallic or semiconducting, depending on their structure. The SWCNT structure is characterized by a vector (m, n) designating the orientation of the graphene layer relative to the tube axis: armchair ($n = m$), zigzag ($n = 0$ or $m = 0$), or chiral (any other n and m). Examples are shown in Figure 12.1A. All armchair SWNTs are metals; those with $n-m = 3k$, where k is a nonzero integer, are semiconductors with a tiny band gap; all others are semiconductors with a band gap that inversely depends on the nanotube diameter [12]. While semiconducting SWCNTs (S-SWCNTs) can be used to build high-performance FETs and sensors, metallic SWCNTs (M-SWCNTs) might be useful for interconnects.

Structures that contain a concentric set of cylinders with a constant interlayer separation of 0.34 nm are called multi-walled carbon nanotubes (MWCNTs) (Figure 12.1B). MWCNTs have relatively large diameter, ranging from a few nanometers to several tens of nanometers, and are conducting materials. The electronic properties of perfect MWCNTs are rather similar to those of perfect SWCNTs because the coupling between the cylinders is weak in MWCNTs.

Carbon nanotubes exhibit various extraordinary properties including mechanical, electrical, thermal, and chemical properties. Their Young's modulus is over 1 TPa and the tensile strength is about 200 GPa. The thermal conductivity can be as high as 3,000 W/mK.

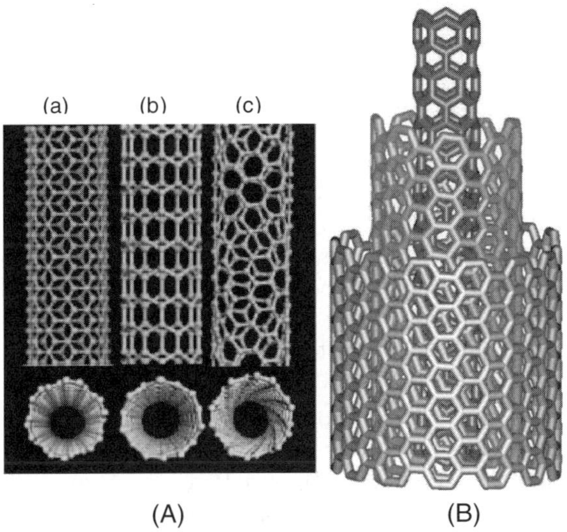

FIGURE 12.1
(See color insert following page 330)
A. Schematic illustrations of the structures of (a) armchair, (b) zigzag, and (c) chiral SWNTs (Reproduced with permission from Baughman, R. H., Zakhidov, A. A., de Heer, W. A. (2002). Carbon Nanotubes—The Route Toward Applications. *Science*, 297, 787–792.). B. Structure of a multi-walled carbon nanotube made up of three shells of differing chirality. (Reproduced with permission from Balasubramanian, K., Burghard, M. (2005). Chemically Functionalized Carbon Nanotubes. *Small*, 1, 180–192.)

CNTs have very high chemical stability and can be chemically functionalized: that is, it is possible to attach a variety of atomic and molecular groups to their ends [12].

The current challenge is to control the synthesis of these nanostructures in ways that allow the potential applications of CNTs to be explored. For example, growing patterned and aligned nanotubes with controlled structure, dimension, and morphology.

12.2.2 Synthesis of Carbon Nanotubes

From a synthesis point of view, there were two significant advances in the early stage after Iijima's first discovery, the MWCNTs. One was the discovery, in 1993, of SWCNTs by Iijima and coworkers [14] as well as by Bethune and coworkers [15]. The other milestone, reached in 1996, was a relatively high-yield synthesis of SWCNTs, which allowed investigations of physical properties to be carried out. Following these advances, considerable effort has been made toward high-quality and large-scale synthesis of SWCNTs and MWCNTs with controlled structures. Clearly, future developments in nanotube-based science and technology will rely on the highly controlled synthesis of nanotube materials.

The principal methods of CNT synthesis are arc discharge, laser ablation, and chemical vapor deposition (CVD). While the first two methods use high-energy input to release the carbon atoms from carbon-containing precursor molecules, CVD relies on carbon atomization via catalytic decomposition of carbon precursors on the surface of transition metal particles. Compared with arc and laser methods that produce powdered samples, the CVD method allows scale-up of production to an industrial level and production of CNTs in a predictive fashion with controlled length, positions, and orientations on substrates. Significant progress has been made by Dai and coworkers [16,17] by developing a novel CVD method to produce high-quality individual SWCNTs on isolated catalyst particles on substrates. This technique yields large numbers of SWCNTs at specific locations and opens up new possibilities for integrated nanotube systems. We will describe recent progress in the CVD process.

During CVD synthesis, it is possible to form several different carbonaceous products, including MWCNTs, SWCNTs, amorphous carbon, and metal particles encapsulated by graphitic shells. Control over the CNT structure and growth is achieved through the variation of a series of experimental parameters such as process temperature, gas mixtures, pressure, flow rates, and catalyst materials. Among these, one of the most important parameters is the specific catalyst material employed [18]. The catalytic materials used are generally transition metals, for example iron, cobalt and nickel or their alloys, at the nanometer scale, either in oxide or metallic forms or as mixtures. The important properties of metals for nanotube formation include their ability to catalytically decompose gaseous carbon-containing molecules, possessing a modest melting temperature, and the solubility of carbon and the carbon diffusion rate in the metal. Various methods have been employed to produce catalytic particles, including the precipitation of metal salts (nitrates, sulfates, and chlorides), metal-organic, or organometallic precursors on a support, followed by drying, calcining and grinding, and sputtering [19]. Extensive mechanistic studies have been conducted to determine the dependence of the CNT diameter and structure on the catalyst particle size. One of the challenges is to predetermine the catalyst size and explore its relationship to the resulting nanotubes. This will be discussed in Section 12.2.3.

Many research groups have contributed to the use of the CVD method to grow high quality and controlled SWCNTs. For example, Dai and coworkers synthesized high-quality SWCNTs from catalyst islands patterned directly on substrates. Further, the SWCNTs thus synthesized bridge adjacent catalytic islands from pillar to pillar [19] (Figure 12.2).

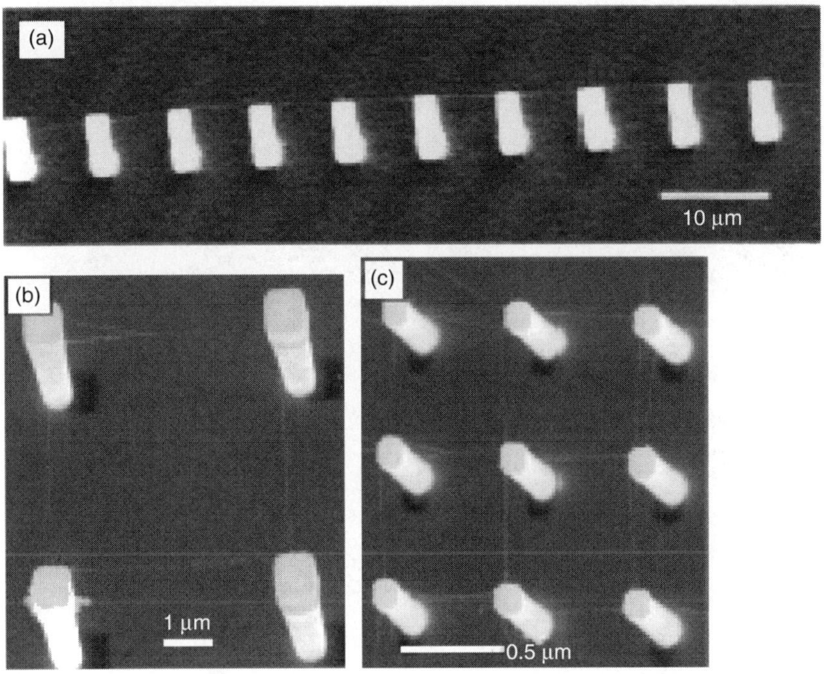

FIGURE 12.2
Directed growth of suspended SWCNT. (a) A nanotube power-line structure. (b) A square of nanotubes. (c) An extensive network of suspended SWCNTs. (From Franklin, N., Dai, H. (2000). An Enhanced CVD Approach to Extensive Nanotube Networks With Directionality. *Adv. Mater.*, 12, 890–894.)

Clearly, the isolated-catalyst CVD method combined with electrode fabrication renders abundant nanotube devices for further studies and provides a route to investigate growth mechanisms and correlations between catalyst particles and nanotube structure (discussed in Section 12.2.3). Liu and his coworkers have developed a highly active Fe/Mo catalyst by sol-gel synthesis for methane CVD growth of SWCNTs [20]. The high surface area and large mesopore volume of this catalyst resulted in a nanotube–catalyst ratio as high as 2:1 (1 g of catalyst yielding 2 g of SWCNTs). This is also an excellent example of the scale-up of high-quality SWCNT synthesis that can be achieved by understanding and optimizing the catalytic materials. Several groups have developed a gas-phase CVD process to grow bulk quantities of SWCNTs [21–23]. Smalley and his coworkers [21] used carbon monoxide as carbon feedstock and iron pentacarbonyl, produced by thermal decomposition, as the catalytic source to achieve bulk production of SWCNTs.

Multi-walled carbon nanotubes have also been controllably obtained by the CVD method in many research groups [24–27]. For example, Ren [24] has grown relatively large-diameter and aligned "forests" of MWCNTs. Dai and coworkers [27] synthesized aligned MWCNTs by CVD; here, the driving force for self-alignment is the Van de Waals interactions between nanotubes, as shown in Figure 12.3.

Another significant advance in the CVD method for growth of CNTs is the use of alcohol. Maruyama and coworkers [28] recently reported the synthesis of highly pure SWCNTs using alcohol. The authors attributed the high quality of the SWCNTs obtained to the effects of OH radicals formed at high temperature, which can efficiently remove amorphous carbon during nanotube growth. More recently, Iijima and coworkers [29] found another method to produce extremely pure SWCNTs (purity 99.98%) by simply adding water to ethylene during the CVD process, as shown in Figure 12.4. In this case,

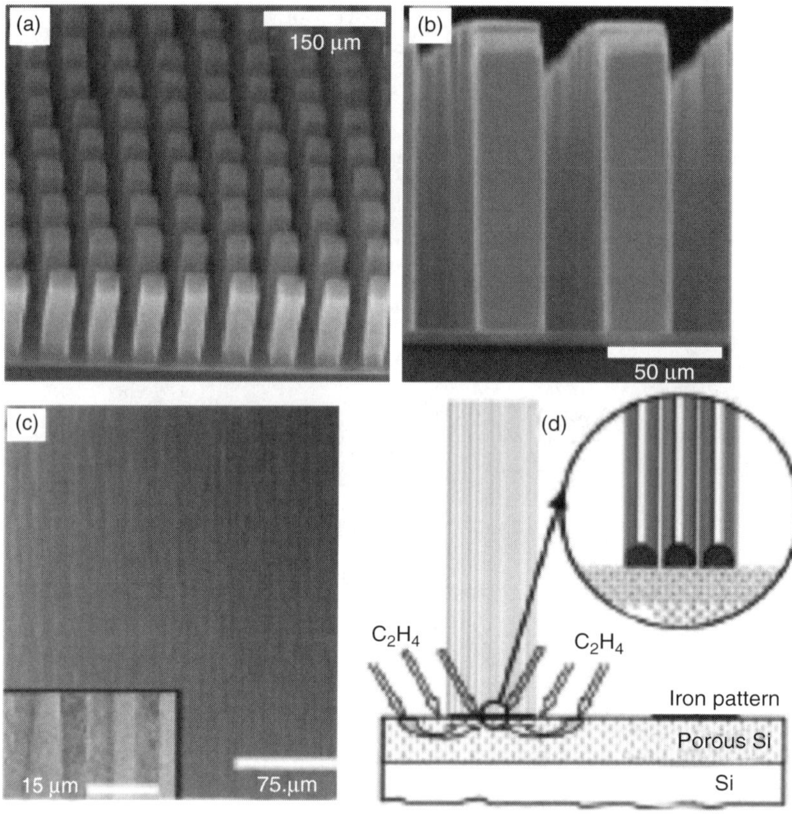

FIGURE 12.3
Self-oriented MWCNTs arrays grown by CVD on a catalytically patterned porous silicon substrate. (a) SEM image of tower structures consisted of aligned nanotubes. (b) SEM image of the side view of the tower. (c) A high-magnification SEM image showing aligned nanotubes in a tower. Inset: TEM image showing two MWCNTs. (d) Schematic diagram of the growth process. (From Fan, S., Chapline, M., Franklin, N., Tombler, T., Cassell, A., Dai, H. (1999). Self-Oriented Regular Arrays of Carbon Nanotubes and Their Field Emission Properties. *Science*, 283, 512–514.)

the SWCNTs/catalyst weight ratio reaches 50,000%. They found that the activity and the lifetime of the catalyst could be enhanced by water. This approach represents a significant advance toward the realization of large-scale and highly pure SWNT material.

12.2.3 Growth Mechanisms and Structure Control

The unique mechanical and electrical properties of CNTs are directly associated with their structure, including their diameter, number of shells, and helicity. A thorough understanding of the formation mechanisms for these nanotubular carbon systems is critical to design procedures for controlling the growth conditions to obtain more practical structures, which might be directly available for nanotechnology including biosensors.

It has been generally accepted that the growth mechanisms of both SWCNTs and MWCNTs in the CVD process can be divided into two distinct classes: base-growth and tip-growth [18]. In the base-growth mode, the catalyst particle remains on the support surface and the nanotube elongates while maintaining a closed-end. In the tip-growth mode, the catalyst particle is lifted off the support surface by the growing nanotube and carried along at the tube end. The operation of these two modes depends on the

FIGURE 12.4
SWNT forest grown with water-assisted CVD. (A) Picture of a 2.5-mm-tall SWNT forest on a 7 by 7 mm silicon wafer. A matchstick on the left and ruler with millimeter markings on the right is for size reference. (B) Scanning electron microscopy (SEM) image of the same SWNT forest. Scale bar, 1 mm. (C) SEM image of the SWNT forest edge. Scale bar, 1 mm. (D) Low resolution TEM image of the nanotubes. Scale bar, 100 nm. (E) High resolution TEM image of the SWNTs. Scale bar, 5 nm. (From Hata, K., Futaba, D. N., Mizuno, K., Namai, T., Yumura, M., Iijima, S. (2005). Water-Assisted Highly Efficient Synthesis of Impurity-Free Single-Walled Carbon Nanotubes. *Science*, 306, 1362–1364.)

interaction between the metal catalyst and the substrate. In the following two sections, we will discuss the growth mechanisms and control of the structure of SWCNTs and MWCNTs, respectively.

12.2.3.1 Single-Walled Carbon Nanotubes

Owing to the direct correlation between the properties of the catalyst particle and the growth mechanism of nanotubes in the CVD method, synthesis using a controllable catalyst on a predictable substrate is the key to controlling nanotube growth chemistry, mechanism, and structure. The understanding of how a catalytic particle initiates growth of individual SWCNTs is a significant step to controlling their structure and properties. Dai [30] have investigated isolated iron NPs for the synthesis of SWCNTs by CVD. Their TEM images revealed the growth process of SWCTs from isolated NPs (Figure 12.5). These results clearly showed that the diameters of SWCNTs are determined from the diameter of catalytic particles. The catalytic particles were found to remain on the substrate; this is a case of the base-growth model. Further, the initial growth of a SWCNT from one catalytic

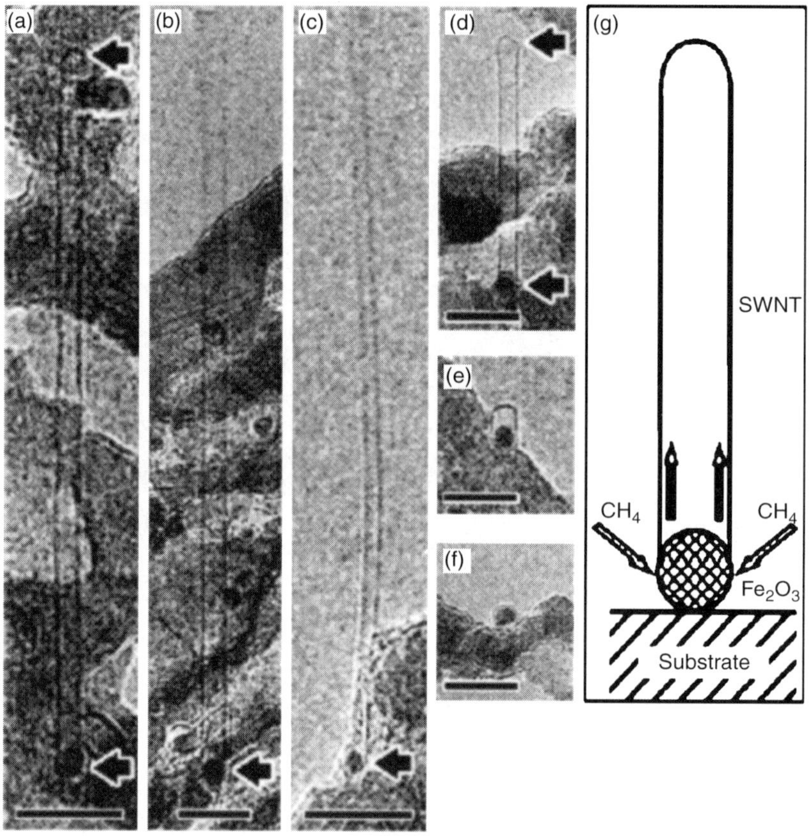

FIGURE 12.5
(a–f) TEM images of SWCNTs grown from discrete nanoparticles, showing particle–nanotube relationship. Scale bars: 10 nm. (a–d) SWCNTs grown from discrete nanoparticles (dark dots at the bottom of the images). The arrows point to the ends of the nanotubes. (e) Image of an ultrashort (~4 nm) nanotube capsule grown from ~2 nm nanoparticle. (f) Image of a nanoparticle surrounded by a single graphitic shell. (g) A schematic model for nanotube growth. (From Dai, H. (2002). Carbon Nanotubes: Synthesis, Integration, and Properties. *Acc. Chem. Res.*, 35, 1035–1044.)

particle, corresponding to f, e, and d in Figure 12.5, can be clearly seen. In fact, the CVD reaction process involves decomposing a hydrocarbon gas such as methane to obtain carbon atoms, which then dissolve into the surface of the catalytic NP on the substrate to form a carbon-metal solid-state solution. As the supersaturation increases, carbon atoms precipitate from the particle, leading to the growth of a nanotube (Figure 12.5g).

In the synthesis of SWCNTs, results have shown that S-SWCNTs and M-SWCNTs coexist in all materials synthesized by arc discharge, laser ablation, and CVD methods. A key challenge is how to separately obtain purely metallic and semiconducting nanotubes. While chemical separation is a possible approach [31,32], preferential growth is also desirable. Recently, the preferential growth of S-SWCNTs (~90%) from Dai's group [33] was obtained in a plasma-enhanced CVD process at 600°C, compared with S-SWCNTs of ~60% for HiPCO and ~30% for laser ablation. Similar observation of SWCNTs with uniform chirality produced from the CVD process has also been reported [34]. More recently, Dai and coworkers [35] reported a correlation between the diameter of SWCNTs and the relative fractions of metallic and semiconducting tubes. Small tubes ($d \approx 1.1$ nm) show higher percentages of S-SWCNTs while large tubes ($d \geq 1.1$ nm) produce higher percentages of M-SWCNTs. The preferential growth of small semiconducting tubes was attributed to a lower free energy for S-SWCNTs, according to a theoretical calculation [35]. This represents great progress toward future control of synthesis resulting in predictable structures on a large scale.

12.2.3.2 Multi-Walled Carbon Nanotubes

The growth mechanisms of MWCNTs remain controversial despite all efforts, both experimental and theoretical. The debate is mainly over whether CNTs exhibit a Russian-doll or scrolled structure [36,37]. Many TEM observations of MWCNTs have shown coaxial structures according to Russian-doll model, which generally consist of multilayered, concentric cylinders of individual graphene sheets (Figure 12.6). Such nanotubes are generally produced either by the electric-arc technique (without the need of any catalyst) or by catalyst-enhanced thermal cracking of hydrocarbon feeds.

Although many people have accepted the Russian-doll model (comprising nested concentric tubes), recent results have also shown the presence of a scroll structure in many tubes. Amelinckx et al. [39] observed a disrupted layer by HRTEM and proposed a mixed model involving both concentric cylinders and a scroll-type structure. Xu and coworkers [40] used high-resolution x-ray diffraction to provide further support for the scroll model. More recently, Sun et al. [41] used HRTEM to observe MWCNTs grown by ohmically heating the catalytic sites; this provided direct evidence for the scroll model. They found that the catalytic particle at the tip exhibits a cork-like shape conforming to the walls of the tube, which also gradually becomes narrower (Figure 12.7A). Such a narrowing of the wall simply corresponds to a decrease in the number of graphitic layers at the lattice fringes as indicated by the arrows (Figure 12.7B). A close observation of the tube structure shows that both the external and internal lattice fringes of the tube wall terminate at more or less regular intervals (Figure 12.7C), suggesting that these MWCNTs have a scrolled structure, as depicted in Figure 12.7E. In this case, the entire nanotube is generated by a single graphene sheet that produces new spirals inside the previous ones. In the scroll model, the Co–Ni particles are continuously lifted up to maintain their initial shape and to create the spiraling structure with a constant diameter. Further, they obtained additional evidence for the growth of MWCNTs according to the scroll model by the deposition of platinum (Pt) NPs

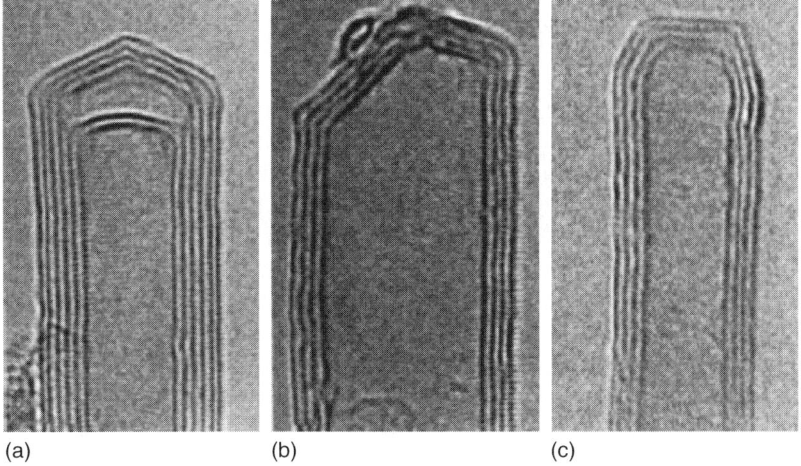

FIGURE 12.6
High-resolution transmission electron microscopy images of concentric MWCNTs with closed caps. (From Iijima, S. (1994). Carbon Nanotubes. *MRS Bull.*, 19, 43–51.)

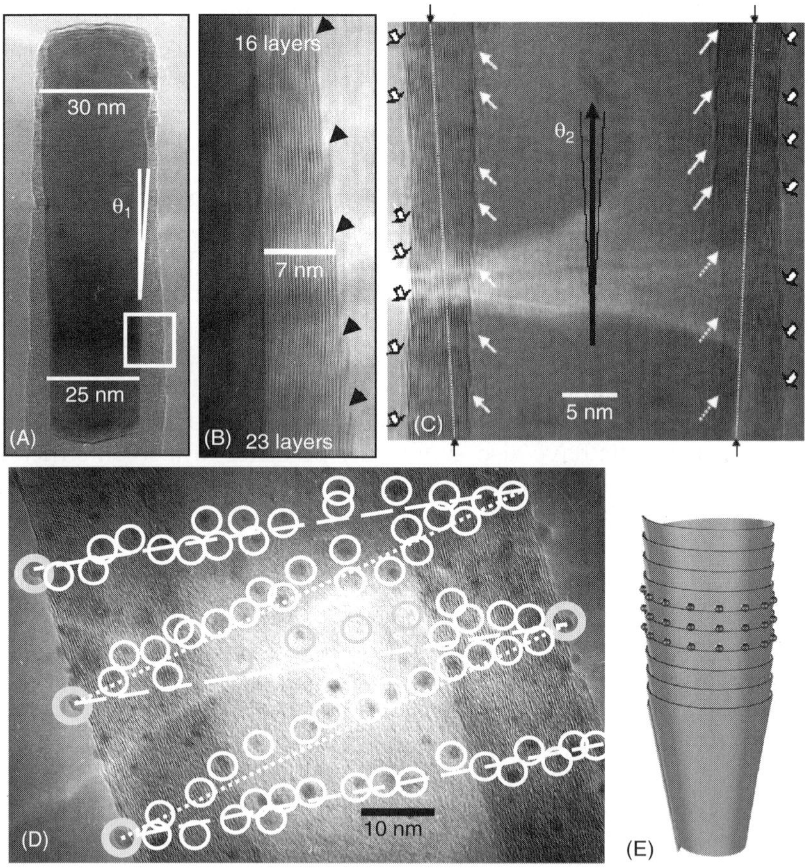

FIGURE 12.7
Microstructure of a scrolled nanotube. (A) TEM image showing a cork-shaped catalyst particle at the tip of a tube, and the presence of an inclined angle θ_1 relative to the tube axis. (B) HRTEM images of the tube at a marked location showing interruption of lattice fringes only at external walls and a reduction of layer number (Note: $\theta_1 = \theta_2$). (C) HRTEM image of the base tube showing the interruption of the external and internal lattice fringes of the nanotube wall and an inclined angle of θ_2 with the tube axis. (D) Decoration of Pt nanoparticles revealing a preferential path along the edges of scrolled structure. (E). 3D model of the scrolled structure and deposited Pt nanoparticles on the edges. (From Sun, X., Li, R., Stansfield, B., Dodelet, J. P., Ménaed, G., Désilets, S., Unpublished work)

on the surface of the tubes. After a slight oxidation treatment of the tubes [42], a preferential deposition of Pt NPs (79 among 112 particles counted on the micrograph in Figure 12.7D) are aligned along the spiraling external edge of the graphene sheet, as shown schematically by the 3D model in Figure 12.7E. Actually, the scroll growth model had been previously proposed for nanofibers [43] and observed in hydrothermal synthesis [44].

Similar to SWCNTs, to further elucidate their growth mechanism, it is essential to understand how nanometer catalyst particles initiate the growth of MWCNTs. So far, there have been few reports and little direct evidence concerning the initiation of the growth of nanotubes due to the extremely fast growth process. Sun et al. [41] observed nanotubes at various stages during their growth by HRTEM measurements of tubes grown from catalyst particles deposited on MWCNTs (see Figure 12.8). Clearly, the initial shape of the catalyst particles deposited on the nanotubes is spherical before growth begins, with a diameter of about 35 nm (Figure 12.8A). In the early stages, the catalyst particle was reshaped to reveal faceting (Figure 12.8B). It can be seen clearly that a thicker graphitic wall appears at the bottom of the catalyst particle, indicating that the initiation of the structure occurs from the

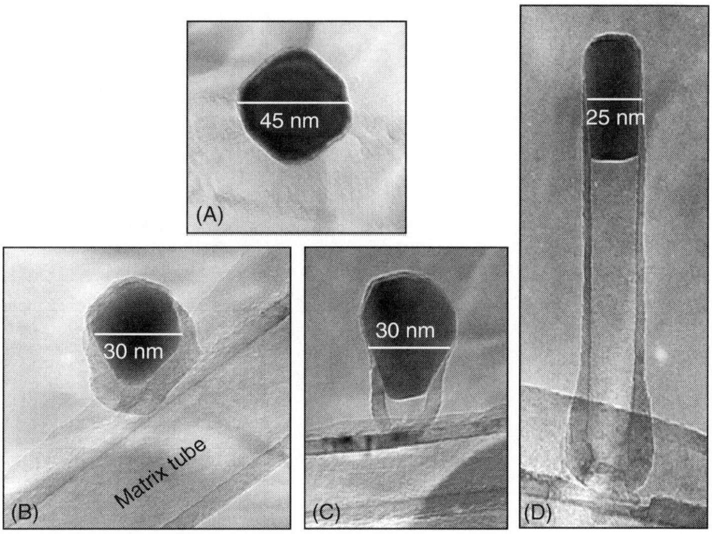

FIGURE 12.8
TEM micrographs of the initiation of MWCNT growth. (A) A catalytic particle before the growth of a nanotube. (B) The graphene layers initiate from the bottom of the catalyst at an inclination with respect to the tube. (C) The catalytic particle is lifted up from the substrate as the tube forms. (D) The length of the tube increases and the catalytic particle becomes cork-shaped. (Reproduced with permission from Sun, X., Li, R., Stansfield, B., Dodelet, J. P., Ménaed, G., Désilets, S. (2006). Controlled Synthesis of Pointed Carbon Nanotubes. *Carbon*, in press.)

bottom. These spherical graphitic layers at the bottom (Figure 12.8D) correspond to the shape of the catalytic particle at the initial stage. Subsequently, the bottom of the catalyst particle separates from the lateral graphitic layers. The catalyst is then lifted up, maintaining a facet at the bottom while becoming round at the top (Figure 12.8C). Continued growth leads to a cork-shape catalyst with uniform walls (Figure 12.8D).

In conclusion, different structures may depend on different synthesis methods. Compared with the nested concentric structure of MWCNTs, scrolled structures appear to have many advantages. The high reactivity of the edges of the graphene sheet facilitates surface functionalization by, for example, –OH, –COOH, or deposition of other metals such as Pt, which is useful for various applications such as biosensors.

12.3 Nanowires

12.3.1 Introduction

Nanowires (NWs), like CNTs, are also 1D and well-defined crystalline structures with a high aspect ratio. NWs have been demonstrated to possess superior electrical, optical, mechanical, and thermal properties. Unlike CNTs, NWs are noncarbon-based materials that can consist of metals, semiconductors, or inorganic compounds. The broader choice of various crystalline materials and doping methods makes the electronic and optical properties of NWs tunable with a high degree of freedom and precision. With the development of new synthesis methods, NWs have attracted increasing attention for sensor applications. *Nature* magazine recently stated that "Nanowires, nanorods, nanowhiskers, it does not matter what you call them, they are the hottest properties in nanotechnology" (*Nature*, 419 (2002) 553). Figure 12.9 shows a well-aligned array of InP NWs grown from gold catalyst particles defined by electron beam lithography [45]. There are a few reviews [46,47] and books [48–51] dealing with the synthesis, characterization, properties, and applications of these materials. A grand challenge is to integrate NWs into assemblies with multifunctions to fabricate nanoscale electronic and photonic devices as well as biosensors. In this section, we will focus on their synthesis, growth process, and properties.

FIGURE 12.9
SEM of InP nanowires grown with MOVPE from gold catalysts defined by electron beam lithography. (From Samuelson, L., Thelander, C., Bjork, M. T., Borgstrom, M., Deppert, K., Dick, K.A., Hansen, A. E., Martensson, T., Panev, N., Persson, A. I., Seifert, W., Skold, N., Larsson, M. W., Wallenberg, L. R. (2004). Semiconductor Nanowires for 0D and 1D Physics and Application. *Physica E.*, 25, 313–318.)

12.3.2 Synthesis of Nanowires

Synthesis methods of NWs include both vapor phase and solution techniques. Compared with physical methods such as nanolithography and other patterning techniques, chemical methods are seen to be more versatile and effective in the synthesis of these NWs. The chemical synthesis includes various routes: thermal evaporation, chemical vapor-phase deposition, metal-organic chemical vapor-phase deposition, arc discharge, laser ablation, sol-gel, and template-based methods. The basic process of 1D nanostructure formation involves two fundamental steps: nucleation and growth. A major challenge in the synthesis of NWs is to control their size, phase purity, crystallinity, and chemical composition. Therefore, a detailed understanding and good control of the nucleation and growth processes at the nanometer scale is required.

12.3.2.1 Vapor Phase Growth of Nanowires

12.3.2.1.1 Vapor–Liquid–Solid Growth

The vapor–liquid–solid (VLS) mechanism was proposed by Wagner [52] in the 1960s, when he studied the growth of millimeter-scale Si whiskers using Au as catalyst. An advantage of the VLS method is that it can produce single-crystalline 1D nanostructures from a rich variety of inorganic materials including elemental semiconductors (Si, Ge, and B), III–V semiconductings (GaN, GaAs, GaP, InP, and InAs), II–VI semiconductors (ZnS, ZnSe, CdS, and CdSe), and oxides (ZnO, MgO, and SiO_2) [47]. A typical VLS process involves the dissolution of gaseous reactants into nanosized liquid droplets of a catalyst metal, followed by the nucleation and growth of single-crystalline rods and then wires. Since the diameter of the NWs is determined by the diameter of the catalyst particles, this method provides an efficient means to obtain uniformly sized NWs. This mechanism has been widely accepted and applied to the growth of various NWs. Indeed, the synthesis of NWs with controlled composition, size, purity, and crystallinity required an increased understanding of the nucleation and growth processes at the nanoscale. In situ high-temperature TEM provided real-time observations of Ge NW growth that confirmed the VLS growth mechanism [53]. Experimental observations suggest that there are three growth stages: metal alloying, crystal nucleation, and axial growth (Figure 12.10).

12.3.2.1.2 Vapor–Solid Growth

The vapor–solid grow method, also called evaporation–condensation growth, is also a popular and simple method to grow oxide NWs. In this process, the vapor species is first generated by evaporation, chemical reduction, and gaseous reaction. The vapor is subsequently transported and condensed onto a substrate. Using this method, various oxide NWs of Zn, Sn, In, Cd, Mg, Ga, and Si have been synthesized. For example, Zhu and

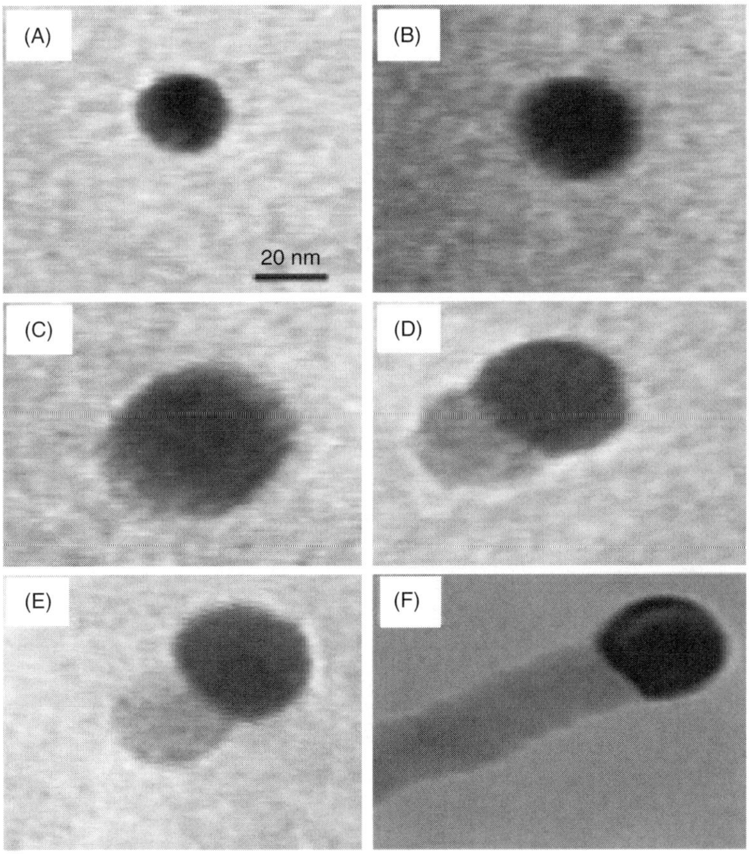

FIGURE 12.10
The birth of a Ge nanowire on a Au nanocluster, as observed using in situ TEM. It is clearly seen that the Au nanocluster started to melt after the formation of Ge–Au alloy, and this was followed by an increase in the liquid droplet size during the Ge vapor condensation process. When the droplet was supersaturated with the Ge component, a Ge nanowire grew out of this droplet of Au–Ge alloy and became longer as time elapsed. (From Wu, Y., Yang, P. (2001). Direct Observation of Vapor–Liquid–Solid Nanowire Growth. *J. Am. Chem. Soc.*, 123, 3165–3166.)

coworkers [54] have grown Si_3N_4, SiC, Ga_2O_3, and ZnO NWs by heating the powders of these materials to elevated temperatures [54]. Wang and coworkers [55] obtained nanobelts by this method.

12.3.2.2 Solution-Based Growth of Nanowires

12.3.2.2.3 Template-Based Synthesis

Template-directed synthesis provides another class of convenient and versatile methods for growing 1D nanostructures. In this approach, the template simply serves as a scaffold within which different materials are synthesized and shaped into a nanostructure with a morphology complementary to that of the template. The templates could be nanoscale channels within mesoporous materials, porous alumina, and polycarbonate membranes. The nanoscale channels are filled using the solution, the sol-gel, or the electrochemical method. The NWs produced are released from the templates by removal of the host matrix [56]. Unlike the polymer membranes fabricated by track etching, anodic alumina membranes (AAMs) containing a hexagonally packed 2D array of cylindrical pores with a uniform size are prepared by anodization of aluminum foils in an acidic medium (Figure 12.11). Several materials have been fabricated into NWs using AAMs in the templating process. NWs of various inorganic materials including Au, Ag, Pt, TiO_2, MnO_2, ZnO, SnO_2, InO_3, CdS, CdSe, CdTe, electronically conducting polymers such as polypyrrole (Ppy), poly(3-methylthiophene), and polyaniline, have been produced, as well as carbon nanotubules.

Although the NWs synthesized using the template method are usually polycrystalline, recent studies have revealed that it is possible to synthesize single crystals of NWs by carefully controlled conditions. For example, Takahashi et al. have demonstrated successful synthesis of single-crystal V205 nonorod arrys by template-based electrodeposition [57]. In recent work, Cao and coworkers [58] discovered that titania NWs obtained from the electrophoretic deposition method became single crystals once their diameters had been reduced below 15 nm.

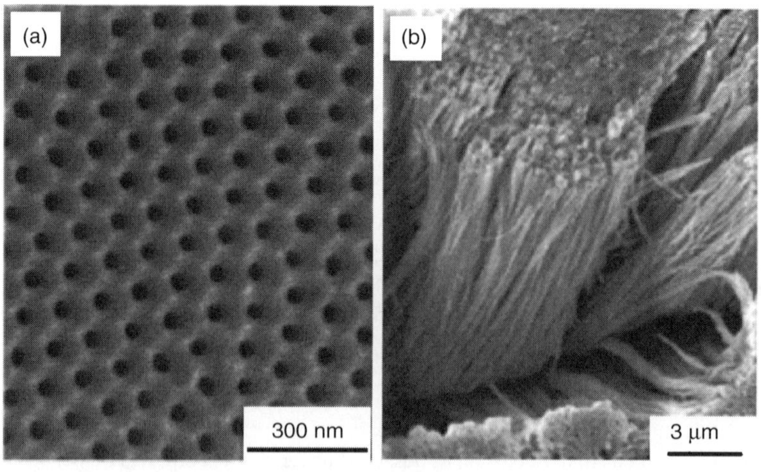

FIGURE 12.11
(a) SEM image of AAM template and (b) Bi_2Te_3 nanowire arrays. (From Jin, C., Xiang, X., Jia, C., Liu, W., Cai, W., Yao, L., Li, L. (2004). Electrochemical Fabrication of Large-Area, Ordered Bi_2Te_3 Nanowire Arrays. *J. Phys. Chem. B.*, 108, 1844–1847.)

12.3.3 Properties of Nanowires

Compared with bulk materials, low-dimensional NWs, with their large surface area and possible quantum confinement effects, exhibit distinct electric, optical, chemical, and thermal properties. Boland and coworkers [59] recently demonstrated the ultrahigh strength of gold NWs. They found that, compared with Au bulk nanocrystalline metals, the Young's modulus of Au NWs is essentially independent of diameter, whereas the yield strength is largest for the smallest diameter wires, with strengths of up to 100 times greater than that of the bulk material. They attributed this to reduced defects and a fewer number of grains across the diameter of NWs. Alivisatos and coworkers [60] showed how semiconducting NWs could be used to enhance the processibility and efficiency of solar cells. They fabricated thin-film photovoltaic devices by blending CdSe NWs with polythiophenes to obtain hybrid materials. They also found that NWs were superior to quantum dots in photovoltaic applications because they could provide a direct path for electrical transport at much lower loading; a power conversion efficiency of as high as 1.7% has been obtained. Based on the extremely high surface-to-volume ratios of NWs, Penner and coworkers [61] fabricated hydrogen sensors with Pd NWs. Since each NW contains many break junctions along their longitudinal axis resulting from hydrogen reduction, the resistance of these NWs exhibited a strong dependence on the concentration of hydrogen gas. Various other applications of semiconducting NWs have recently been exhibited, such as building blocks for assembling a range of nanodevices including FETs, *p-n* diodes, bipolar junction transistors, and various biosensors. The latter will be discussed in detail later on.

12.4 Functionalization of Carbon Nanotubes and Nanowires for Biosensor Development

Integration of biology and materials at the nanoscale has the potential to revolutionize nanobiotechnology. The small size, good charge transport properties, and large active surface area of CNTs and NWs holds great promise for the design of biosensors. The electrical properties of CNTs and NWs are very sensitive to surface charge transfer and changes in the surrounding environment, and drastic changes can be caused by single-molecule interactions [2,62,63]. With the development of nanofabrication technology, the potential of CNTs and NWs as sensing elements and tools for biological applications as well as for chemical analysis has been attracting considerable attention for sensor development [64,65]. To control the chemical and physical properties of CNTs for biosensor applications, it is obviously important to develop reliable methods for CNT or NW functionalization [66–69].

12.4.1 Solubilization and Functionalization of Carbon Nanotubes

It is well known that chemical modification of nanomaterial surfaces is often necessary to generate various functionality and biocompatibility of such nanomaterials [70]. In particular, functionalization of CNTs and NWs with different selected biomolecular components may introduce functional units such as recognition sites, catalytic elements or bioactive groups, which are essential for various applications [71–73]. The recent bloom of chemical modification and functionalization methods [74] has made it possible to conjugate NWs with various biological and bioactive species such as proteins [75–80], enzymes [81–83], peptides [84,85], porphyrin [68,86,87], carbohydrates [88–90], and nucleic acids [32,91–93]. The construction of such functionalized CNTs and NWs has become increasingly important to the assembly of surface-confined architectures or patterns via molecular recognition and self-assembly. Furthermore, by means of site-selective modification (such as CNT tips [94], ends of CNTs [95,96], inner or outer walls [73]), it is possible to modulate precisely the position of the biomolecule, which in turn enables control of their architecture and functionality. To enable easy access for biomolecules to CNTs in the functionalization procedure, it is advantageous to have CNTs solubilized.

12.4.1.1 Solubilization of Carbon Nanotubes

Chemical strategies to solubilize and disperse CNTs in aqueous as well as in organic solvents overcomes a major technical barrier to their facile manipulation and processing in biological environments. Different methods, including oxidative acid treatments [97,98], surfactant [99,100] (such as sodium dodecyl sulphate [101–103], Triton X-100 [104], and tetraoctylammonium bromide[105]) -assisted dispersion and polymer-assisted dispersion, have been widely used to generate relatively stable CNT suspensions. The judicious selection of different methods has made it possible to adjust the solubility properties of these CNTs and to direct their assembly into more complex structure. For example, a concentrated H_2SO_4/HNO_3 mixture has been widely used to generate carboxylic groups on the ends and walls of CNTs via a refluxing/sonication process [106]. Such heavily oxidized CNTs may then be stabilized in aqueous suspensions, and various defects in the structure [73,107] can provide sites for covalent coupling of biomaterials (Figure 12.12) through the creation of amide and ester bonds.

Derivatization of CNTs with functional organic groups will also increase their solubility in organic [110,111] and aqueous solvents. CNTs reacted with an amine-functionalized

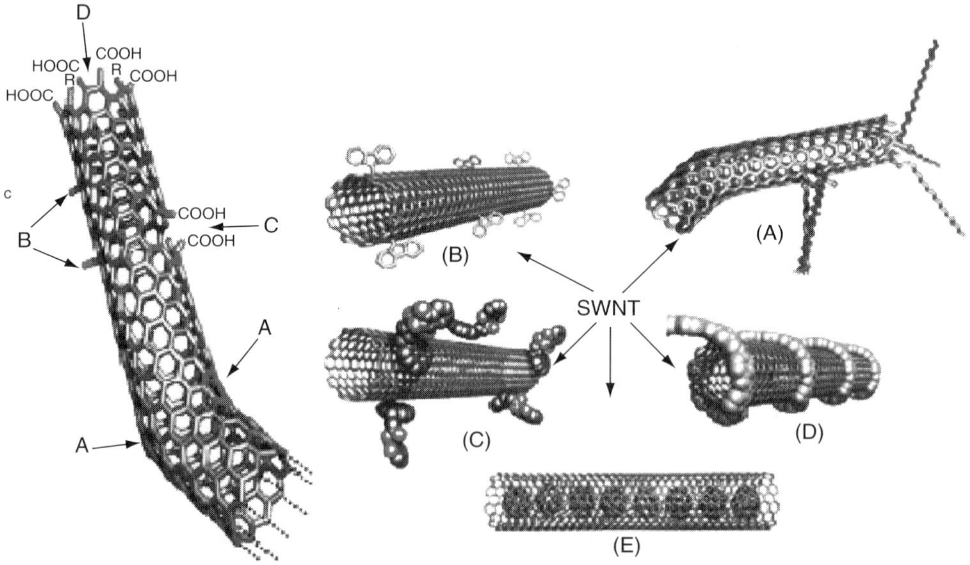

FIGURE 12.12
Left: Typical defects in an SWNT: (A) five- or seven-membered rings in the C framework, instead of the normal six-membered ring, leads to a bend in the tube, (B) sp^3-hybridized defects (R=H and OH), (C) C framework damaged by oxidative conditions, which leaves a hole lined with –COOH groups, and (D) open end of the SWNT, terminated with –COOH groups. Besides carboxy termini, the existence of which has been unambigously demonstrated, other terminal groups such as $-NO_2$, OH, H, and =O are possible. Right: Functionalization possibilities for SWNTs: (A) defect-group functionalization, (B) covalent sidewall functionalization, (C) noncovalent exohedral functionalization with surfactants, (D) noncovalent exohedral functionalization with polymers, and (E) endohedral functionalization with, for example, C_{60}. For methods B–E, the tubes are drawn in idealized fashion, but defects are found in real situations. (From Hirsch, A. (2002). Functionalization of Single-Walled Carbon Nanotubes. *Angew. Chem. Int. Ed.*, 11, 1853–1859.)

crown ether became readily soluble in water and a variety of organic solvents, including acetone and methanol. Such derivatives could be exploited for further covalent and ionic functionalization [112]. Biomaterial modification has been considered as another interesting strategy to prepare SWNT stable aqueous dispersion. Wrapping of CNTs with biopolymers (e.g., helical amylase [113] or DNA [32]) or the self-assembly of biomolecules (e.g., lipid derivatives [102]) on CNT side walls can provide control over the interfacial properties and improve the dispersion and homogeneity of the CNTs in water, thus permitting the assembly of the CNTs into architectures necessary for biosensor applications.

12.4.1.2 Functionalization of Carbon Nanotubes with Biomaterials

Many techniques exist for CNT biofunctionalization including noncovalent and covalent modification. CNTs can be functionalized by various biomolecules via spontaneous adsorption, or be immobilized in a more controllable fashion by pretreatment of CNTs.

12.4.1.2.1 Noncovalent Modification

A variety of proteins such as antibodies [115], enzymes [76], and peptides [85] can nonspecifically bind to the CNTs exterior surface without their covalent coupling. Oligonucleotides can also be nonspecifically bound to the surface and the opened cavity of CNTs. In fact, both computational and experimental evidences showed that nonspecific DNA–CNT interactions in water are from the DNA base stacking on the CNT surface, with the hydrophilic sugar-phosphate backbone pointing to the exterior [114].

12.4.1.2.2 Covalent Binding

Covalent coupling of biomaterials to CNTs is important to meet the specific requirement demanded by biosensor applications. Biofunctional reagents have been used to modify the CNT sidewalls with other biomolecules. For example, amide groups produced by the reaction of the azomethine group and the sidewall of CNTs [116] have been used for the covalent coupling of amino acids [117] and biologically active peptides for immunoassays [118].

The oxidatively introduced carboxyl groups represent useful sites for further modifications, as they enable the covalent coupling of biomolecules through the creation of amide and ester bonds [112] (Figure 12.13). The carboxylic groups on both ends and sidewalls of CNTs can covalently tether various organic dyes, DNA, protein, metallic NPs [108], and even magnetic NPs [109] via carbodiimide coupling or by using heterobifunctional-coupling groups. "Sugar-coated" CNTs generated by derivatized galactoses have been reported as candidates of biosensing materials for pathogen detection [119,120]. The functionalization of CNTs was based on the carbodiimide-activated amidation of the galactose-tethered amino groups with CNT-bound carboxylic acids. The high aspect ratio and large surface area of the CNTs enable the display of abundant sugar arrays, which effectively captured pathogenic *E. coli* via specific adhesion–receptor interactions.

Also, amine-terminated oligonucleotides have been covalently bound to electrochemically generated carboxylic acid functional groups at the sidewalls and at the ends of CNTs [91] (Figure 12.14). The CNT–DNA adduct could then selectively hybridize with complementary stands while minimizing nonspecific interactions with noncomplementary strands. The CNT nanoelectrode arrays functionalized with selective DNA probes on the open end have been considered for promising ultrasensitive DNA biosensors to analyze both poly G-tagged DNA targets and label-free PCR amplicons [121,122]. Importantly, the approach to functionalize sidewalls of CNTs with DNA by growing oligonucleotides using conventional DNA synthesis has made it possible to photolithographically pattern different DNA sequences on the CNTs and to fabricate new types of DNA biosensors [123].

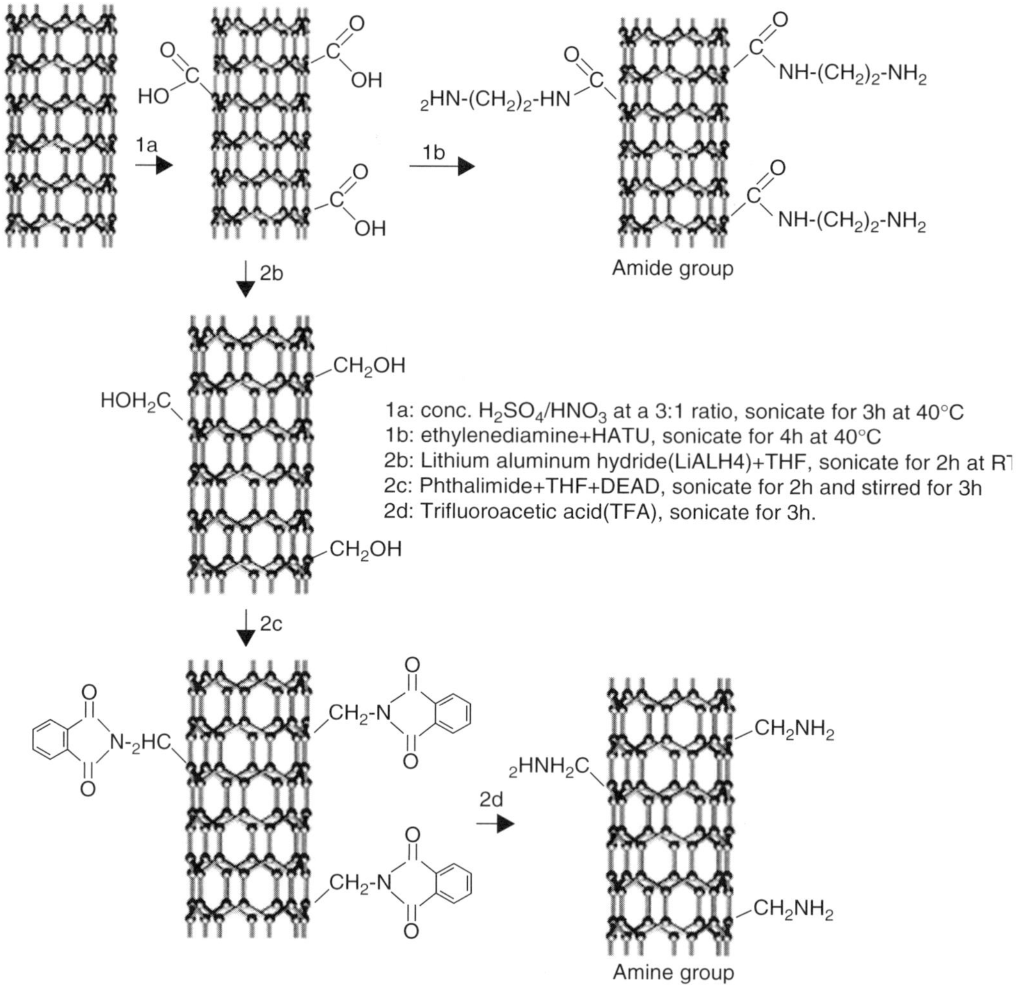

FIGURE 12.13
Schematic illustration of the reaction scheme to form SWCNT with amino functionalization: 1a and 1b shows the direct amide functionalization while 1a–2d shows the amine functionalization of the SWCNT. (From Ramanathan, T., Fisher, F. T., Ruoff, R. S., Brinson, L. C. (2005). Amino-Functionalized Carbon Nanotubes for Binding to Polymers and Biological Systems. *Chem. Mater.*, 17, 1290–1295.)

For designing sensing systems based on biofunctionlized CNTs for protein biorecognition, it is important to prevent nonspecific binding and at the same time introduce selective and specific binding probes. Surfactant-aided functionalization and poly(ethylene glycol) (PEG) [78] or polyethylene imine (PEI) [69] coating are effective and unambiguous strategies to prevent nonspecific adsorption of the desired proteins. Protein-resistant polymers such as PEG that contain ester linkages can be further used as an interface to specifically control the covalent coupling of proteins [209]. For example, specific binding of streptavidin onto SWNTs has been achieved by functionalization of CNTs with biotin and oligomeric PEG via ester-to-amide transformation reactions [68].

12.4.2 Biofunctionalization of Nanowires

In addition to CNTs, 1D nanostructures such as magnetic NWs [210,211], semiconductor [62,212,213], and conducting-polymer NWs [214] offer the prospect of high sensitivity and

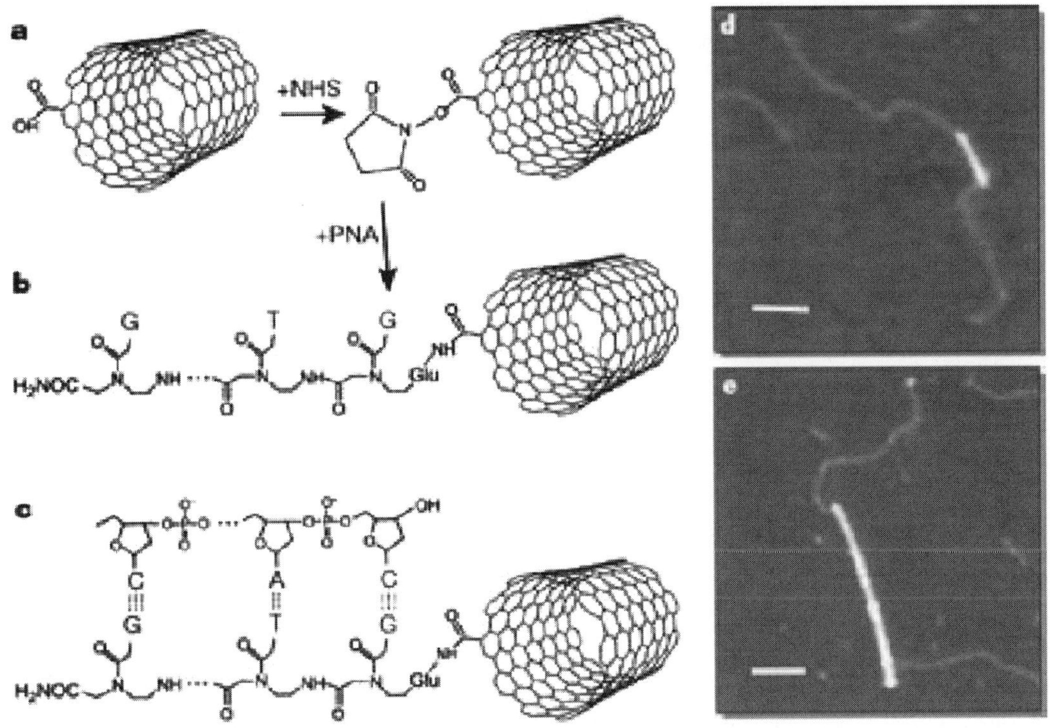

FIGURE 12.14
Attachment of DNA to carbon nanotubes. (a,b) *N*-hydroxysuccinimide (NHS) esters formed on carboxylated, single-walled carbon nanotubes (SWNTs) are displaced by PNA, forming an amide linkage. (**c**) A DNA fragment with a single-stranded, "sticky" end hybridizes by Watson–Crick base-pairing to the PNA–SWNT. (d,e) Atomic force microscope (TappingMode) images of PNA–SWNTs. SWNTs appear as bright lines; the paler strands represent bound DNA. Scale bars: 100 nm; nanotube diameters: (d) 0.9 nm; (e) 1.6 nm. (From Williams, K. A., Veenhuizen, P. T. M., De La Torre, B. G., Eritja, R., Dekker, C. (2002). Nanotechnology: Carbon Nanotubes With DNA Recognition. *Nature*, 420 (6917), 761.)

massive redundancy in nanosensor arrays, so that a large number of sensing elements can be densely packed onto a small footprint of an array device. Biological functionalization of NW surfaces is also a key point for NW-based biosensor applications [65]. The most direct approach to functionalizing NWs is to transfer the well-developed surface chemistry at planar metal interfaces to the NW geometry [124,125]. It is well known that thiols, dithiols, and thioacetates have a high affinity for gold surfaces and carboxylic acid groups strongly interact with metal oxide surfaces to form stable self-assembled monolayers [126,127]. The reactivity provides several convenient routes for functionalization of the NW surface. For example, gold nanorods can be modified by a thiol-derivatized DNA (ssDNA) [201] or a thioacetate-terminated porphyrin via an Au–S covalent bond [128]. DNA hybridization has been effectively monitored by fluorescence quenching due to the aggregation of DNA functionalized nanorods [202]. Recent advances in surface modification of silicon nanowires (SiNWs) to generate hydrophilic surface have made it possible to facilitate biofunctionalization (such as DNA, protein, and cells) of SiNWs (Figure 12.15). For example, SiNWs have been functionalized with a phosphoramidite-terminated DNA monolayer. Using the SiNWs functionalized with DNA probes, a label-free DNA analysis system was developed with a high sensitivity and selectivity [129]. The method of functionalization is based on the gas-phase method to generate a self-assembled monolayer of mercaptopropyltrimethoxysilane on silicon oxide surface. Then the functionalized silicon oxide NWs were used for covalent coupling with phosphoramidite-modified DNA.

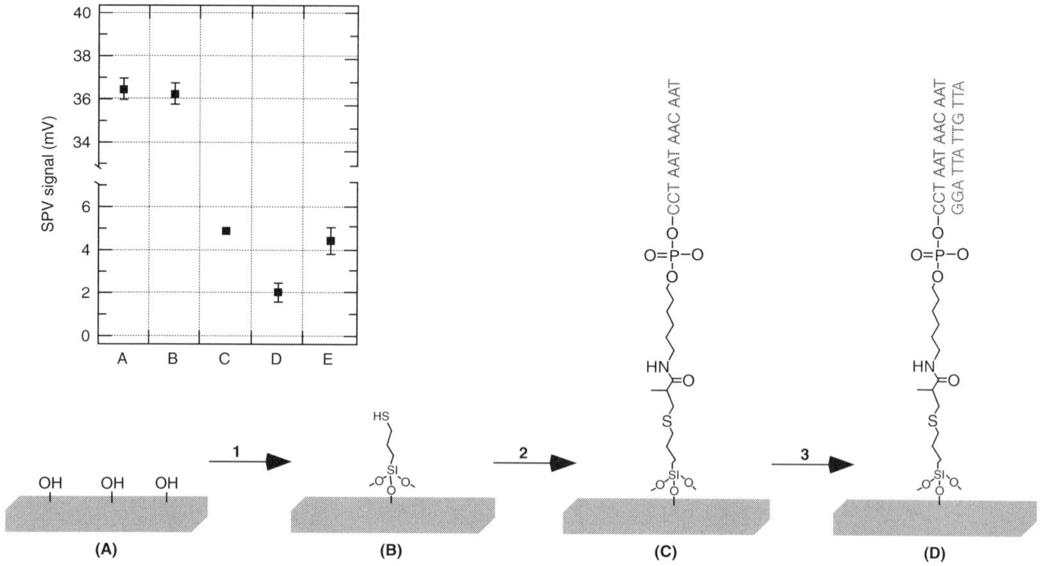

FIGURE 12.15
Modification scheme of the SiNW surface for the DNA detector: (1) self-assembly of 3-mercaptopropyltrimethoxysilane (MPTMS) by gas-phase reaction in Ar for 4 h; (2) covalent immobilization of DNA probes by exposing the previous surface to 5 μM solution of oligonucleotide CCT AAT AAC AAT modified with acrylic phosphoramidite at the 5′-end for 12 h; (3) DNA detection based on hybridization between label-free complementary DNA target GGA TTA TTG TTA and the immobilized DNA probes on the SiNW surfaces. The inset is the SPV signal on a *p*-type Si surface at different stages of the modification; A, B, and C correspond to the schematic diagrams, D is with 25 pM solution of complementary DNA target exposed to the surface C, and E is with 25 pM solution of noncomplementary DNA (GGA TCA TTG TTA) exposed to the surface C. (From Li, Z., Chen, Y., Li, X., Kamins, T. I., Nauka, K., Williams, R. S. (2004). Sequence-Specific Label-Free DNA Sensors Based on Silicon Nanowires. *Nano Lett.*, 4(2), 245–247.)

Compared with noncovalent modification, the covalent anchoring of DNA on the SiNW surface provided better stability and less nonspecific hybridization for DNA sensing.

Among the various materials of NWs, magnetic NWs are ideal candidates for applications in a magnetic cell separation system [210], which is an important step for biosensor development. Cells (such as mammalian cells and human T-cells) have an affinity for binding to the hydrophilic surface of the native nickel oxide layer or carboxylate groups on the magnetic NWs. Thus, the binding of the cells makes the magnetic NWs accessible for use in an immunomagnetic separation system based on a microfluidic device to automate and miniaturize sample preparation for biosensor application.

12.5 Design and Construction of Carbon Nanotube and Nanowire-Based Biosensors

12.5.1 Approaches to Design and Assemble Biosensors Based on Carbon Nanotubes

12.5.1.1 Carbon Nanotubes-Based Electrochemical Biosensing Platforms

12.5.1.1.1 Directly Use Carbon Nanotubes as Electrodes

Carbon nanotubes are attractive as electrodes for electrochemical sensor devices because of their high electronic conductivity, wide potential range, and the high electrochemically assessable surface area. Since the first report using CNTs as electrodes to study the oxidation of dopamine in 1996 by Britto et al. [130], CNTs have attracted many scientists in the field of analytical chemistry [66,131,132] and physics [133]. The first type of CNT electrodes were simply assembled by packing a paste of CNTs and binder materials (such as mineral oil [135–137], bromoform [75], and liquid paraffin [139]) into a glass capillary or a Perspex with a narrow cylindrical slot. The resulting CNT electrodes exhibited an improved reversibility of dopamine oxidation [137] and a good reversible voltammetric response to cytochrome C [140] and zaurin [138]. From the perspective of electrochemical biosensor application, enzyme-modified CNTs have shown superior performance by means of efficient electron transfer reactions between the underlying electrode and the redox protein, even when the redox center is embedded deep within a glycoprotein shell, such as hemoglobin, cytochrome C, zaurin, and horseradish peroxidase. The desired enzymes (such as glucose oxidase and alcohol dehydrogenase) and NAD(P)H cofactors could be incorporated within the 3D CNT/Teflon matrix. The resulting composite was then packed firmly into the electrode cavity as a sensor electrode, which performed effective low-potential amperometric biosensing of glucose and ethanol [141]. In contrast to the methods using CNTs as an aligned nanoelectrode array [142,143] or an electrocatalytic modifier that will be described below, the composite device has major advantages in terms of minimization of surface fouling and surface renewability.

12.5.1.1.2 *Carbon Nanotubes Used as Catalytic Mediators*

With the major barrier for solubilization of CNTs overcome, solubilizied CNTs have been widely used as catalytic modifiers in connection with another electrode surface. A straightforward approach to handle CNT positioning consists of proton-conducting and bioreactive polymers, such as Nafion, Teflon, or Ppy, that are used to immobilize CNTs on electrodes [144–148]. The combination of the antifouling/discriminative properties of a Nafion film with the efficient electrocatalytic activities of CNTs provides a useful avenue for improving the sensitivity of biosensors. For example, the CNT/Nafion-modified glassy carbon electrode dramatically improved the detection of catecholamine neurotransmitters by means of giving an enhanced dopamine response and discrimination against ascorbic acid [149]. Redox-polymer hydrogels provide another candidate for constructing CNT-based amperometric enzyme biosensors (Figure 12.16) [150]. The incorporation of SWCNTs modified with enzymes (such as glucose oxidase and horseradish peroxidase) into redox hydrogels resulted in a two- to threefold increase in the sensor's current output while the amount of electrochemically accessible osmium redox centers increased up to 10-fold. The attractive performance of the CNT-coated electrodes or CNT/binder composite electrodes for low potential detection of hydrogen peroxide and NADH suggests great promise for dehydrogenase- and oxidase-based amperometric biosensors, such as glucose, organophosphorus pesticides, and for nerve agents [151]. The experimental approach for organophosphorus pesticides and nerve agent analysis involves coimmobilization of CNTs and acetylcholine esterase/choline oxidase enzymes on a screen-printed electrode. The enzyme can be anchored on the CNTs with carboxylic acid groups that are created by electrochemical treatment. The large surface of the CNTs and the catalytic activity of CNTs promote the redox reaction of hydrogen peroxide produced enzymatically, which provide an ideal model for a highly sensitive analysis system based on hydrogen peroxide detection. However, it should be noted that the metal impurities present in the nanotubes could also act as the catalytic sites.

12.5.1.1.3 *The Enhanced Catalytic Performance of Nanoparticles and Carbon Nanotube Complexes*

The catalytic features of metal NPs such as gold, Pt, silver, copper, and nickel allow the CNT/metallic NP composites to act as excellent sensor materials [151–154]. We recently confirmed that combining metal particles such as Pt NPs with CNT/Nafion-modified electrodes (Figure 12.17) provides a means of further improvement in the sensitivity for

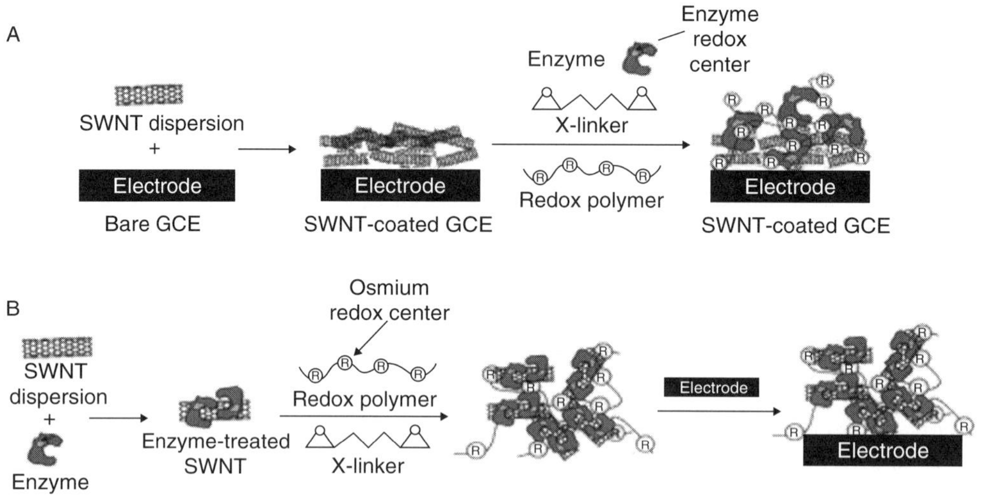

FIGURE 12.16
Schematic illustration of the construction of type A and type B sensors. (A) Fabrication of type A sensors in which a film of SWNTs was first cast onto a bare glassy carbon electrode and allowed to dry, before an aliquot of the redox hydrogel was cast on top of the SWNT-coated electrode. (B) Fabrication of type B sensors in which SWNTs were first incubated with an enzyme solution, before they were incorporated into the redox hydrogel. An aliquot of the redox hydrogel solution containing the enzyme-modified SWNTs was then cast on top of a bare glassy carbon electrode. (From Joshi, P. P., Merchant, S. A., Wang, Y., Schmidtke, D. W. (2005). Amperometric Biosensors Based on Redox Polymer–Carbon Nanotube–Enzyme Composites. *Anal. Chem.*, 77(10), 3183–3188.)

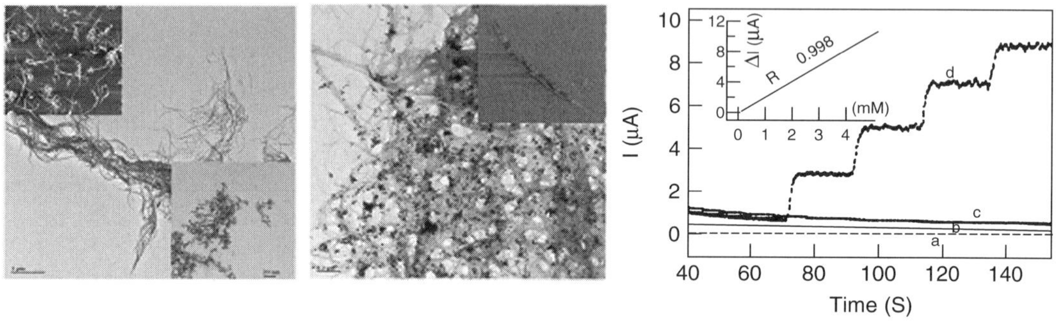

FIGURE 12.17
Left: TEM micrograph of untreated SWCNTs dissolved in 5% Nafion (diluted 10-fold with water), left insets: AFM tapping-mode height image (size, 10 μm × 10 μm; data scale, 100 nm) of SWCNTs and TEM image of Pt nanoparticles in 5% Nafion (diluted 10-fold with water); middle: TEM micrograph of SWCNT in the presence of Pt nanoparticles. Middle inset: AFM tapping-mode phase image (size, 1 μm × 1 μm; data scale, 20 nm) of one SWCNT in the presence of Pt nanoparticles; right: amperometric response for four different electrodes upon subsequent additions of 1 mM glucose solution in 50 mM phosphate buffer (pH 7.2) at 0.55 V vs. Ag/AgCl (3 M NaCl) at 25°C. (a) GC/GOx, (b) GC/CNT + GOx, (c) GC/Pt_{nano} + GOx, and (d) GC/CNT + Pt_{nano} + GOx. Inset shows calibration curve for glucose concentrations between 0.5 μM and 5 mM. (From Hrapovic, S., Liu, Y., Male, K. B., Luong, J. H. T. (2004). Electrochemical Biosensing Platforms Using Platinum Nanoparticles and Carbon Nanotubes. *Anal. Chem.*, 76(4), 1083–1088.)

glucose analysis (with a detection limit of 0.5 mM and response time of 3 s) compared with the glassy carbon electrode modified by Pt NPs or CNTs alone.

In addition, despite their catalytic properties, metal NPs can provide a full range of reactivity with different biomolecules. The biofunctionalized nanocomposite can provide the unique reorganization ability and high selectivity by attaching specific probes, which make CNT–NP composites attractive for biosensor application [155]. An electrochemical protocol for detecting DNA hybridization and even for DNA mismatch detection based on attaching thiolated DNA probes along the gold NP–CNTs hybrid has been reported recently, and such sensor systems can be extended for RNA and peptide nucleic acid (PNA) analysis [156]. This protocol (Figure 12.18) relies on the attached gold NPs with CNTs as an anchor to capture single-strand DNA probes. CNTs act as a support substrate that serves as a fast electron-transfer center and captures the catalytic oxidation current of guanine bases. In this system, a redox mediator such as $Ru(bpy)_3^{2+}$, which amplifies guanine oxidation based on an electrochemical mechanism, has often be adopted to generate larger electrochemical signals.

12.5.1.1.4 *Carbon Nanotube-Based Nanoelectrode Arrays*

The performance of electrodes with respect to response speed and sensitivity is known to scale inversely with the electrode size. It is of interest for biosensing to reduce the size of electrodes close to the size of biomolecules. Unlike employing randomly deposited CNT films as a biosensor platform, the latest advances in the fabrication of well-controlled aligned CNT arrays have attracted attention for developing ultrasensitive electrochemical biosensors [79,157–161]. Many approaches to prepare CNT arrays have been described by various workers. For example, high-temperature catalytic decomposition of hydrocarbon precursors on metal-modified substrates has yielded multiwall nanotube-based aligned array geometries [24,27].

Analogous to covalent functionalization of CNTs, carboxyl groups generated by oxidative scission of CNTs have exhibited considerable affinity for perpendicular alignment of CNTs [162,163]. For example, a dense array of DMF oxidized SWNTs has been fabricated using a metal-assisted self-assembly method via covalent bonds between the end-carboxy group of CNTs and hydroxyl group of the surface-immobilized Fe^{3+} layer [164]. Similarly, cystamine monolayer-functionalized gold substrate can also provide an excellent platform to covalently attach carboxylic acid-functionalized SWNTs for vertical alignment of CNTs. The carboxylic groups at the free ends of the standing CNTs can further form covalent bonds selectively to the biomolecular probes (such as ferrocene [158], DNA [165,166], and protein for biosensor applications [79,167]) (Figure 12.19). Although the structural alignment of the forest-like CNT arrays allows the direct contact of redox

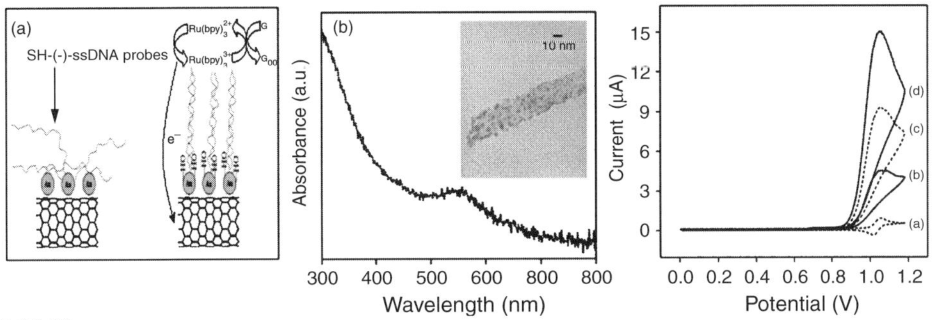

FIGURE 12.18
Left: (a) Schematic illustration of self-assembly of thiolated oligonucleotides onto Au–CNT hybrid. The use of MCH assists the erection of ssDNA and facilitates hybridization of complementary oligonucleotides, which is detected via mediator. Middle: (b) UV–Vis absorption spectrum of MWNT bound with gold nanoparticles. The inset shows the TEM image of a MWNT coated with gold nanoparticles. Right: cyclic voltammograms (CV) of $Ru(bpy)_3^{2+}$ (30 μM) in 50 mM phosphate buffer at pH 7 with 700 mM NaCl at 25 mV s^{-1}: when Au–CNT μ-electrode is modified with (a) MCH only; (b) complementary ss-oligonucleotide (2′); (c) two-mismatched ss-oligonucleotide (3′); and (d) single-mismatched ss-oligonucleotide. (From Lim, S.-H., Wei, J., Lin, J. (2004). Electrochemical Genosensing Properties of Gold Nanoparticle–Carbon Nanotube Hybrid. *Chem. Phys. Lett.*, 400, 578–582.)

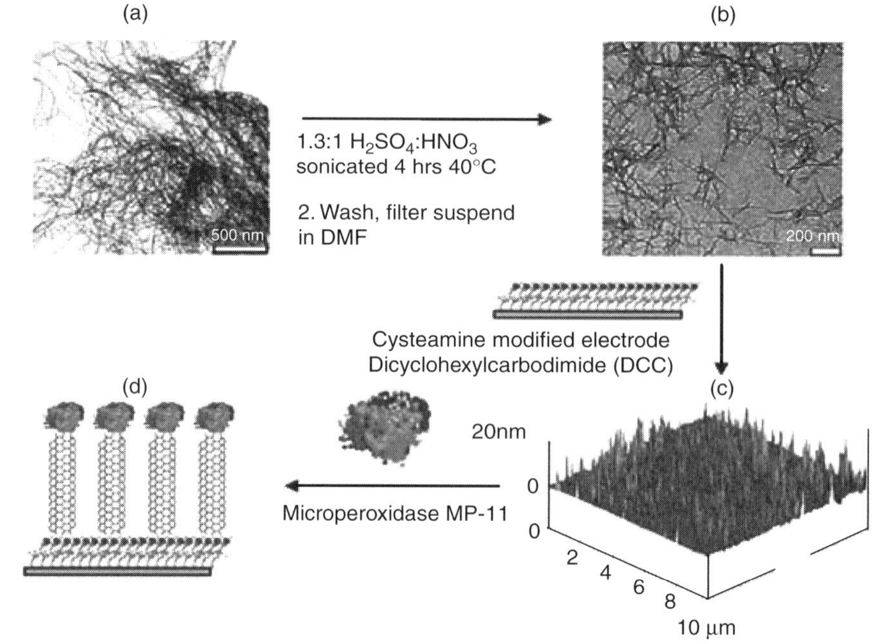

FIGURE 12.19
A schematic illustration showing the steps involved in the fabrication of aligned shortened SWNT arrays for direct electron transfer with enzymes such as microperoxidase MP-11. (From Gooding, J. J., Wibowo, R., Liu, J., Yang, W., Losic, D., Orbons, S., Mearns, F. J., Shapter, J. G., Hibbert, D. B. (2003). Protein Electrochemistry Using Aligned Carbon Nanotube Arrays. *J. Am. Chem. Soc.*, 125, 9006–9007.)

biomolecule components to CNTs, the capacitive charging–discharging current is still too high because such CNT arrays behave as a highly porous macroscopic film instead of individual nanoelectrodes.

To improve the electron transfer rate from an active open end of CNTs to the electrode and decrease background current, each CNT should individually work as a single electrode. Ideally the spacing between each CNT needs to be greater than the diameter of the nanotube to prevent the diffusion layer from overlapping. Recently, well-insulated nanoelectrode arrays based on CNTs have been fabricated by a bottom-up approach [122,168]. In contrast to the methods to fabricate forest-like CNT arrays, controlled low-site density aligned CNT arrays with a nanoelectrode-like electrochemical behavior show many advantages compared to high density-packed CNT arrays (Figure 12.20). The advantages include the increased temporal and spatial resolution, the improved signal (diffusion-controlled current) -to-noise (background or charging current) ratios, and the ability to perform analysis in solutions with high resistance such as natural water. The method for fabricating insulated CNT nanoelectrode arrays by catalytic CNT growth techniques is schematically presented in Figure 12.20, where catalyst—nickel NPs are first electrodeposited on a metal film (chromium or Pt)-covered silicon wafer. A vertically aligned MWCNT array is then grown on the surface by plasma-enhanced chemical vapor deposition using a DC-based hot-filament CVD system. The resulting substrate is dielectrically encapsulated to form a conformal SiO_2 film or is spin-coated by epoxy-based polymer [169] to fill the space between CNTs as well as to cover the substrate surface. Following the chemical mechanical polishing, a tuneable number of MWCNTs is exposed at their end positions to form nanoelectrode arrays. Thus, the MWCNTs can be well aligned vertically at the substrate surface and separated from its neighbors.

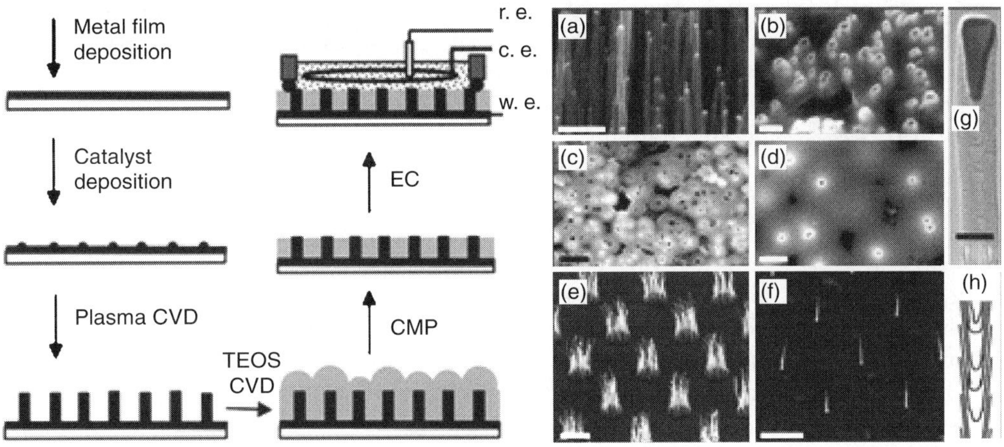

FIGURE 12.20
Left: The procedure for fabricating the MWCNT nanoelectrode array and the setup for electrochemical characterization; right: SEM images of (a) an as-grown MWCNT array, (b) the surface of a polished MWCNT array embedded in an SiO_2 matrix, (c) a high-density MWCNT nanoelectrode array (hd, $\sim 2 \times 10^9$ electrodes cm^{-2}), (d) a low-density one (ld-1, $\sim 7 \times 10^7$ electrodes cm^{-2}), (f) an as-grown MWCNT array on 2 μm Ni spots patterned with UV lithography, (g) an as-grown MWCNT array on ~100 nm diameter Ni spots patterned with e-beam lithography, and (h) a schematic of the defective bamboo-like structure. (a), (b), (f), and (g) are 45° perspective views while (c) and (d) are top views. (e) Is the TEM view of a single MWCNT. The scale bars in (a)–(g) are 500 nm, 200 nm, 500 nm, 500 nm, 50 nm, 2 μm, and 5 μm, respectively. (From Koehne, J., Li, J., Cassell, A. M., Chen, H., Ye, Q., Ng, H. T., Han, J., Meyyappan, M. (2004). The Fabrication and Electrochemical Characterization of Carbon Nanotube Nanoelectrode Arrays. *J. Mater. Chem.*, 14, 676–684.)

Since biomolecules have strong nonspecific absorption at SiO_2 surfaces, a method to localize biomolecular probes specifically at the end of CNTs is crucial. Primary amine-functionalized surfaces covered by aminopropyltriethyoxysilane monolayers on both SiO_2 and hydroxyl group-functionalized MWCNT have been used to form an ethyleneglycol moiety-terminated surface, which is known to resist nonspecific adsorption of biomolecules [121]. By a subsequent stepwise electrochemical etching treatment in NaOH, the molecular bilayer at the MWCNT ends can be removed to form a well-defined MWCNT surface terminated by hydroxyl and carboxylic groups, which allow further attachment of amine-terminated DNA [165,166] or other biomolecules [83,170] at desired sites for efficient biosensing as discussed above. The embedded-CNT array minimizes the background noise from the sidewalls and the open end of CNTs can restart growing by docking nanosized metal catalysts to construct a desirable CNT architecture with an inheritable chirality from the embedded CNTs. Using differential pulse voltammetry and CNT array electrodes modified by an amino-functionalized DNA primer, the intercalating process of daunomycin to CNT–DNA complex has been analyzed [171]. In contrast to the methods based on conventional flat electrodes for DNA analysis, the detection limit of DNA targets has been improved to be a few attomoles by using insulated CNT arrays, an improvement by several orders of magnitude in detection sensitivity [158,121].

12.5.1.2 Carbon Nanotubes as Field-Effect Transistors in Nanosensor Construction

Another promising candidate for future biosensor technology is FET-based biosensors. Since 1998, the first demonstration that an individual CNT was used as a FET [172], semiconducting SWNTs have been recognized as a very promising system for biosensors. In a typical CNT-based nanosensor, SWNT are positioned to bridge two metal electrodes, which act as the source and drain, and the conductance between them can be measured as a function of gate bias voltage. The recent bloom of nanofabrication technology and biofunctionalization methods of CNTs has stimulated significant research interest to develop FET-based nanosensors for monitoring biorecognition events and biocatalytic processes. The mechanism of the FET-based biosensor is depicted [173] in Figure 12.21, where the electrical conductance of the CNT is highly sensitive to the surrounding biological molecules and varies significantly with changes in electrostatic charges and surface adsorption of various molecules [69]. For example, by judicious immobilization of a protein-probe layer, it is then a simple matter of monitoring for a conductance change of the nanotube biosensor to detect a specific binding bioanalyte, thus obviating the need for prelabeling with fluorescent molecules. FET devices with nanotubes as the conducting channel are usually fabricated using nanotubes grown by CVD [16] on top of silicon dioxide from iron NPs with methane/hydrogen gas mixture at 800°C on doped silicon wafers [81,133,134]. To reduce the contact resistance and improve the characteristics of the carbon nanotube field-effect transistors (CNTFETs), electrical leads can be patterned by evaporation and lift-off techniques on top of the nanotubes with a small gap between the metallic source and drain, such as Ti, Co, Pt, and

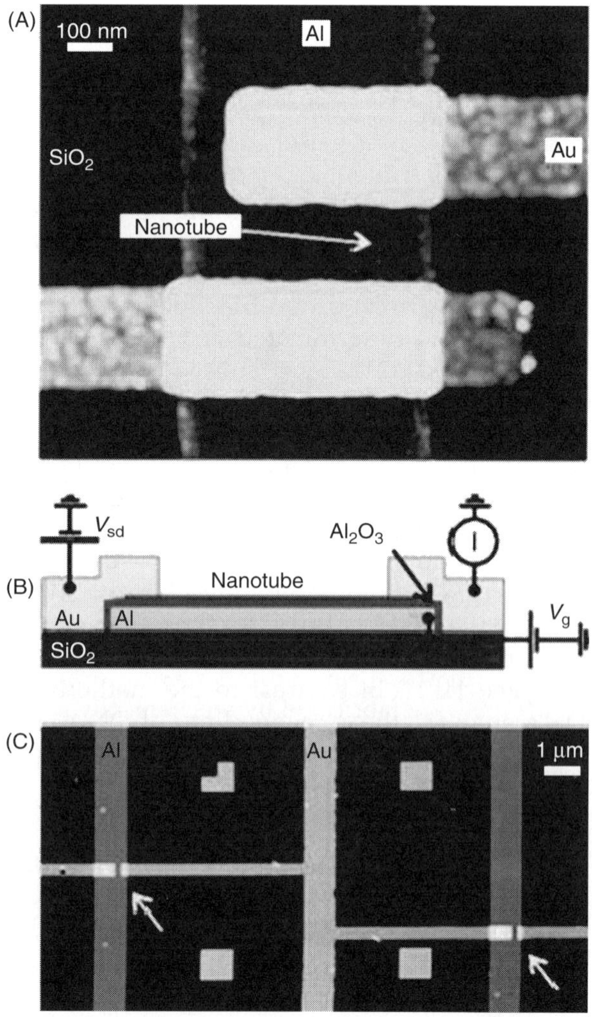

FIGURE 12.21
(See color insert) Device layout. (A) Height image of a single-nanotube transistor, acquired with an atomic force microscope. (B) Schematic side view of the device. A semiconducting nanotube is contacted by two Au electrodes. An Al wire, covered by a few-nanometers-thick oxide layer, is used as a gate. (C) Height-mode atomic force microscope image of two nanotube transistors connected by a Au interconnect wire. The arrows indicate the position of the transistors. Four alignment markers can also be seen. (From Bachtold, A., Hadley, P., Nakanishi, T., Dekker, C. (2001). Logic Circuits With Carbon Nanotube Transistors. *Science*, 294(9), 1317–1320.)

Au [174–176]. Subsequent thermal annealing leads to stronger coupling of the metal with CNTs, and the reduction of the contact resistance. Self-assembly based on molecular recognition also provides a promising approach for constructing CNTFET systems at room temperature. In fact, DNA has been used as a scaffold template to pattern SWNT at a desired address via homologous genetic recombination (such as biotin/avidin, antigen/antibody, and DNA hybridization) [177].

Nonspecific adsorption of proteins including cytochrome C, glucosidase, staphylococcal protein A, human IgG, and streptavidin onto a semiconducting CNT transistor device has been analyzed by monitoring the conductance change in the nanotubes [68,69,178]. The proposed mechanism is that the decrease in conductance is due to a reduction in the charge carriers of *p*-type SWNTs by the noncovalently binding of protein. For example, the sensitivity for the detection of cytochrome C is around 20 protein molecules per nanotube, which is in good agreement with a quartz crystal microbalance (QCM) measurement [179].

Since the nonspecific adsorption of biomolecules on the CNTFET surface is not always desirable, such binding should be avoided to improve the selectivity and specificity of the biosensing system. Polymer coating layers including PEI, PEG, polyethyleneoxide, and Tween on CNTs have been used to capture biomolecules with a high degree of control and specificity, whereas low-affinity species will be rejected [78,88]. The amino groups in irreversibly tethered polymer molecules on CNTs permit coupling of biotin or antigen probes, which still retain their antigenicity for binding their respective streptavidin or antibody with high specificity. For example, the binding of 10E3 mAbs antibody, a prototype target of the autoimmune response in patients with systemic lupus erythematosus and mixed connective tissue disease, to a recombinant human autoantigen U1A protein (a 33-kDa protein, extracted from insect cells) has been monitored in real time electronically using a CNTFETs-based sensing system [68] (Figure 12.22). For this proposal, U1A protein was immobilized on Tween-coated CNTs to capture the respective 10E3 mAbs antibody. The detection limit is lower than 1 nM, which favorably compare with the detection limit (2.3 nM) based on fluorescence array analysis.

12.5.1.3 Carbon Nanotube-Based Scanning Probe Microscopy Probe Tips for Imaging Biological Compounds and Biological Sensitive Measurements

Using CNT as atomic force microscopy (AFM) tips has great potential for structural biology, such as characterization of larger and more complex biomolecular systems. Since AFM tips modified by CNTs were first described, carbon nanotube-based tips have shown great advantages over conventional silicon microfabricated probes, including increased lateral resolution and damage resistance resulting from reversible elastic buckling to delicate organic and biological samples [181–183]. In addition to measuring topography, the probe tips can be used to probe the functionality of surface chemical groups and biological molecules by using selective chemical-functionalized CNTs [184–188]. The first CNT AFM tips were fabricated by mechanically attaching CNTs to a commercial cantilever tip by acrylic adhesive or amorphous carbon under viewing with an optical microscope or a scanning electron microscope [180]. The manual assembly method to make CNT tips is conceptually straightforward but has several important limitations, such as low reproducibility. Recently, a "bottom-up" approach in terms of directly growing nanotubes onto AFM has been reported. In this CVD-based growth approach, well-defined individual CNT tips can be fabricated by controlling reproducibly the catalyst density used for growth on the surface (Figure 12.23) [189].

Using nanotube tips, new biological structures have been investigated in the areas of amyloid-beta protein aggregation and chromatin structure [190]. The reproducibly high

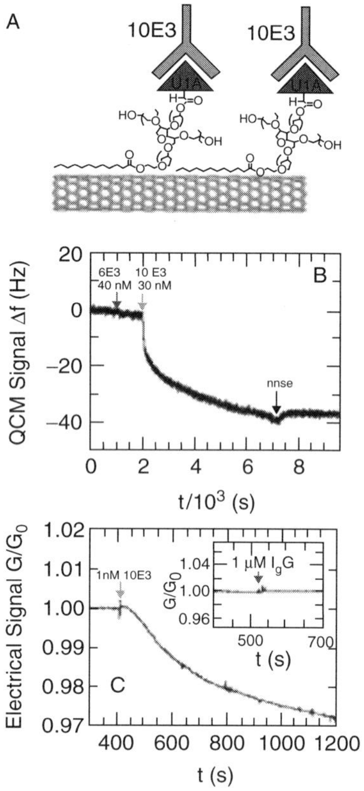

FIGURE 12.22
Specific detection of mAbs binding to a recombinant human autoantigen. (A) Scheme for specific recognition of 10E3 mAb with a nanotube device coated with a U1A antigen–Tween conjugate. (B) QCM frequency shift vs. time curve showing selective detection of 10E3 while showing rejection of the antibody 6E3, which recognizes the highly structurally related autoantigen TIAR. (C) Conductance vs. time curve of a device shows specific response to ≤1 nM 10E3 while rejecting polyclonal IgG at a much greater concentration of 1 μM (*Inset*). (From Chen, R. J., Bangsaruntip, S., Drouvalakis, K. A., Kam, N. W. S., Shim, M., Li, Y., Kim, W., Utz, P. J., Dai, H. (2003). Noncovalent Functionalization of Carbon Nanotubes for Highly Specific Electronic Biosensors. *Proc. Natl. Acad. Sci. U.S.A.*, 100, 4984–4989.)

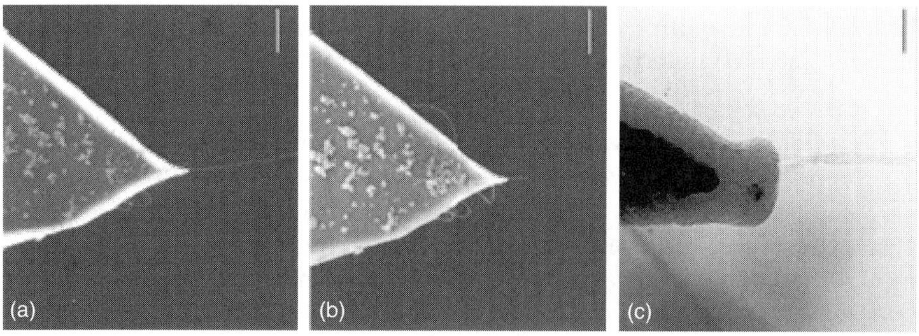

FIGURE 12.23
FE-SEM (Leo 982, 1 kV) images of a CVD nanotube tip grown from a Si cantilever/tip assembly (a) before and (b) after shortening. (c) TEM image (Philips EM420, 100 kV) of the pyramid-nanotube region of the tip shown in panel b. The Si-pyramid is in the top-center of the image, and the two nanotubes comprising the tip are in the bottom-center. Amorphous carbon on the nanotubes as deposited during FE-SEM imaging (a, b). The scale bars in panels a-c are 500, 500, and 20 nm, respectively. (From Hafner, J. H., Cheung, C. Li., Lieber, C. M. (1999). Direct Growth of Single-Walled Carbon Nanotube Scanning Probe Microscopy Tips. *J. Am. Chem. Soc.*, 121(41), 9750–9751.). (From Hafner, J. H., Cheung, C. Li., Lieber, C. M. (1999). Direct Growth of Single-Walled Carbon Nanotube Scanning Probe Microscopy Tips. *J. Am. Chem. Soc.*, 121(41), 9750–9751.)

resolution afforded by CNT probe tips also made it possible to directly read multiple polymorphic sites in DNA fragments [191,192]. The CNT-based AFM tip has been tested by identifying the spatial location of specific sequences in the M13mp18 plasmid. This demonstrates the potential of the CNT tips for single-nucleotide polymorphism detection [192].

The open-end functionalization of CNT probes allows the tips to have as few as one active molecular site localized in a relatively controlled orientation, which leads to subnanometer resolution in chemical contrast and binding site recognition even for single-molecule measurements (Figure 12.24) [181]. These functionalized CNT tips can be used to study ligand–receptor binding/unbinding with control of orientation, and to map the position of ligand–receptor binding site in proteins and on cell surfaces with nanometer or better resolution. Figure 12.24 also shows force spectra of the interaction between tips linked with biotin and immobilized streptavidin and a mica substrate. Force–displacement measurements show well-defined binding force quanta of ca. 200 pN per biotin–steptavidin pair [193].

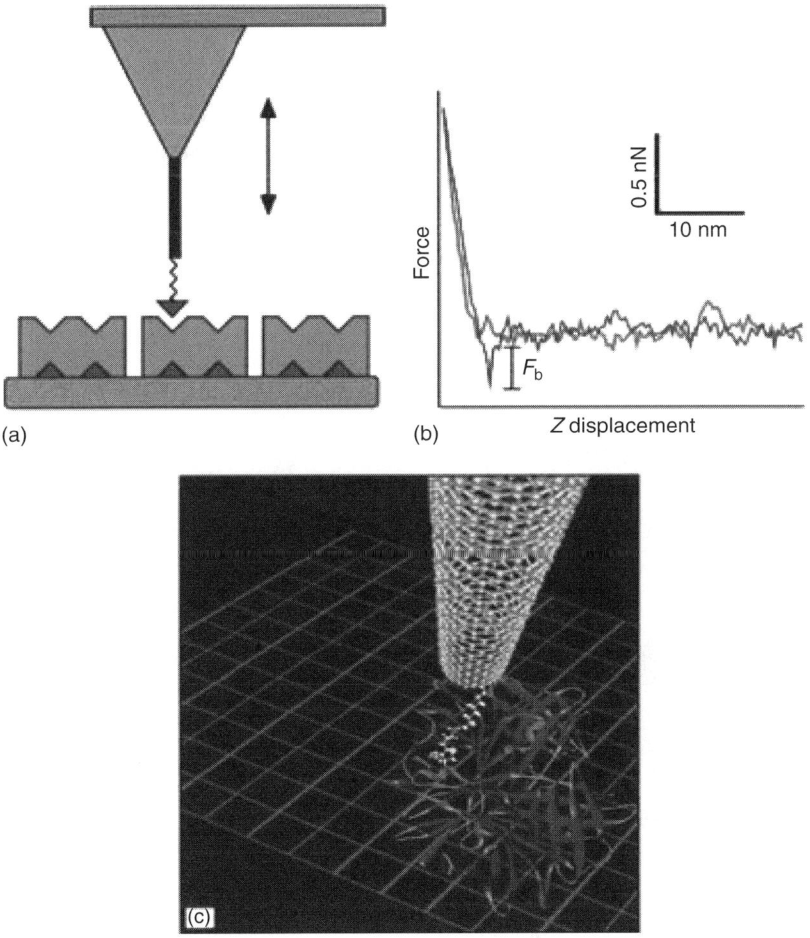

FIGURE 12.24
Nanotube AFM tips for single-molecule force spectroscopy and mapping: (a) Schematic illustration of the experiment, where the lower gray rectangle corresponds to a mica substrate, the green blocks correspond to streptavidin molecules, and the dark blue inverted triangle is biotin covalently linked to a carbon nanotube (heavy black vertical line). (b) Representative force–displacement curve recorded with a biotin-modified MWNT tip in pH 7.0 solution. The binding force, 200 pN, is consistent with unbinding of a single biotin ligand from streptavidin. (c) Model of an SWNT tip that has been modified with biotin and is interacting with a streptavidin molecule. (From Hafner, J. H., Cheung, C., Woolly, A. T., Lieber, C. M. (2001). Structural and Functional Imaging With Carbon Nanotube AFM Probes. *Prog. Biophys. Mol. Biol.*, 77, 73–110.)

12.5.2 The Use of Nanowires in Biological Detection

Similar to CNTs, semiconducting NWs such as silicon, indium, and even conducting polymer wires have attracted widespread attention because of their large surface area and novel electronic conductivity [65,129,194,213]. Using CNTs as sensor materials has limitations in terms of the controllable properties of CNTs (such as semiconducting or metallic properties and the impurity of CNTs) due to the current synthetic methods, which make systematic studies difficult. The NWs of semiconductors do not have this limitation because the dopant type and impurities of these NWs can be controlled by controlling fabrication conditions, which enables the sensitivity to be tuned in the absence of an external gate. In addition, NWs offer potential advantages of a massive body of knowledge for their surface functionalization and nanofabrication, which can be exploited directly for tailoring NWs with biological or chemical receptors. In fact, the semiconductor NWs have been first reported in 2001 to function as highly sensitive, real-time electrically based biosensors [65]. The semiconductor NWs can be functionalized with biological macromolecules [195] and incorporated into electrical circuits in this configuration; the current that flows in the NW is very sensitive to binding events of the biomolecules. For example, FET biosensors based on biotin-modified boron-doped SiNWs not only exhibited high sensitivity in terms of the detection limit of streptavidin down to the picomolar concentration range but also showed reversible antibody binding and concentration-dependent detection in real time. Combining a microfluidic system (Figure 12.25) [215] with PNA-functionalized SiNW effectively paved the way to DNA sequencing, and even to DNA mismatch detection.

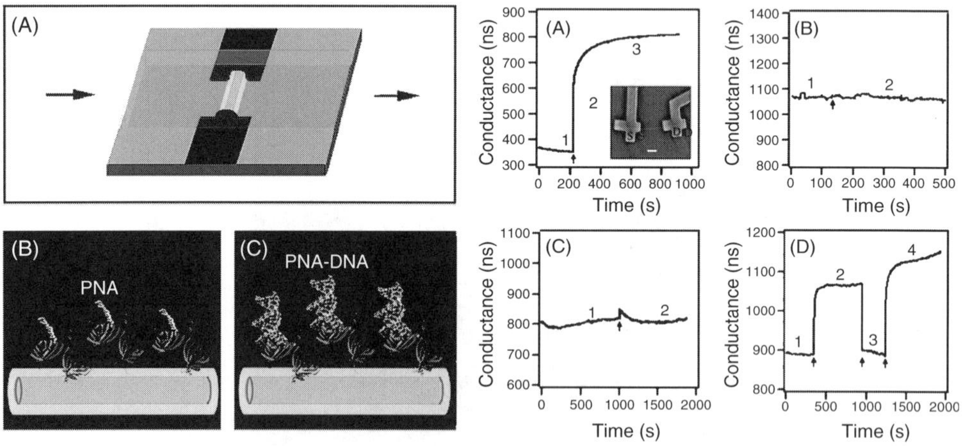

FIGURE 12.25
Left, (A) Schematic illustration of a sensor device consisting of a SiNW (yellow) and a microfluidic channel (green), where the arrows indicate the direction of sample flow. (B) The SiNW surface with PNA receptor. (C) PNA–DNA duplex formation. Right: (A) Real-time conductance response from a SiNW device functionalized with PNA receptor. The arrow marks the point in time when the 60-fM WT DNA sample was added. The inset shows a SEM image of a typical SiNW device with source (S) and a drain (D) indicated; scale bar is 1 μm. (B) Time-dependent conductance in DNA-free solution; the arrow indicates the point in time when a new solution sample was added. (C) Conductance vs. time for SiNW with linked PNA receptor in the presence of 100 fM MU DNA sample (1) and following addition (arrow) of a second 100 fM MU DNA sample. (D) Conductance vs. time for PNA-functionalized NW device during flow of DNA-free solution (1), 100 fM MU DNA (2), DNA-free solution (3), and 100 fM WT DNA (4). (From Hahm, J.-in., Lieber, C. M. (2004). Direct Ultrasensitive Electrical Detection of DNA and DNA Sequence Variations Using Nanowire Nanosensors. *Nano Lett.*, 4 (1), 51–54.)

The delta F508 mutation site in the cystic fibrosis (a common fatal genetic diseases among the population of Europe and the U.S.) transmembrane receptor gene has been distinguished using this approach. As shown in Figure 12.25, the SiNWs operating as a gate in the FET device were functionalized with PNA via intervening avidin protein layer, and a microfluidic system was used to deliver a DNA sample. The specific conductance changes due to PNA–DNA hybridization could be obtained from the time-dependent conductance recorded following introduction of wild-type and the mutant DNA samples using the same FET device. The system allowed effective and selective detection of DNA as low as ca. 10 fM without labeling and in real time. Small protein molecule inhibition of ATP binding to a tyrosine kinase (ABI) has been analyzed using a similar approach as monitoring the conductance of SiNW FET devices with ABI linked to the NW surface. The binding event and inhibition affinities (such as inhibition constants) can be directly electrically detected, which suggests that the approach could serve as a technology platform for drug discovery [196].

The latest advances in the production of well-controlled aligned NW arrays have accelerated the application of NWs in biosensor technology. Nanowire array has been performed on specific antibody-functionalized gates of FET biosensors for virus detection, and the influenza A virus can be selectively detected by measuring conductance changes (Figure 12.26) [170]. Arrays of SiNW devices were defined by using photolithography with Ni metal contacts on silicon substrates with an oxide layer. The specific antibody receptor for influenza A was covalently linked to the surfaces of SiNWs. The conductance changes due to the binding or unbinding in the presence of the virus have been observed but not the other kinds of viruses, thus confirming the selectivity and specificity of the system.

Based on the approach using NW-based FET devices, rather than metal oxide film or polycrystalline metal film, indium tin oxide NW and silver NW-based sensors have also been successfully used to detect NO_2 and NH_3 vapor with high sensitivity at room temperature [197,198].

Conducting polymers such as Ppy are more easily incorporated with biomolecules in a single step during polymer synthesis rather than the multiple steps for synthesis of surface-modified SiNWs and CNTs. The physical properties and biocompatibility have allowed conducting polymer NWs to emerge as another promising material in the development of biosensors [199,214]. Single and multiple individually addressable Ppy and polyaniline NWs of controlled dimension and location have been generated by electrodeposition within a channel between two electrodes on the surface of a silicon wafer. The source-drain resistance increased significantly with the binding of the biotin-conjugated DNA oligo with avidin-functionalized Ppy NWs. Further sensitivity enhancement can be obtained by adjusting theNW's conductivity to a value closer to the lower range of semiconductors.

In contrast to these electrical methods, NWs can also be used for optical biosensor development because of their size- and shape-dependent optical properties. Metal submicron wire (Au/Ag) barcodes have been used in conjunction with traditional florescent assays for DNA detection, where the barcode makes possible multiplexed detection [200]. A novel strategy for DNA hybridization studies based on measuring either UV or fluorescence properties of gold nanorods has been demonstrated [201,202]. DNA probe-functionalized gold nanorods show decreased fluorescence when they are mixed with complementary oligonucleotide-functionalized gold nanorods (Figure 12.27). The fluorescence properties of gold nanorods open up a new perspective for the use of gold nanorods as an alternative tag in fluorescence-based detection assays [202].

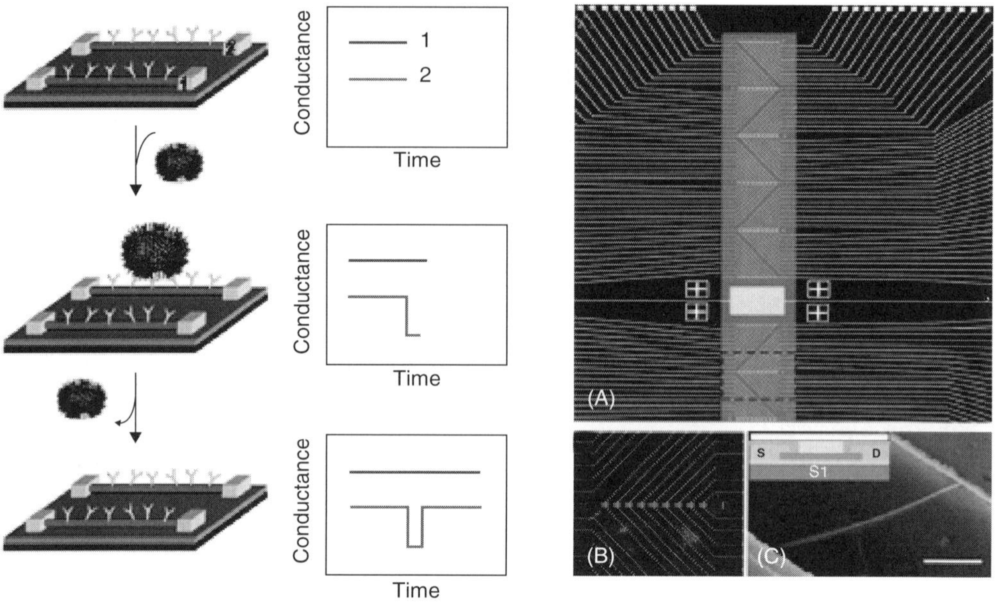

FIGURE 12.26
(See color insert) Nanowire-based detection of single viruses. Left: Schematic illustration shows two nanowire devices, 1 and 2, where the nanowires are modified with different antibody receptors. Specific binding of a single virus to the receptors on nanowire 2 produces a conductance change (middle) characteristic of the surface charge of the virus only in nanowire 2. When the virus unbinds from the surface, the conductance returns to the baseline value. Right: nanowire devices and device array. (A) Optical image of the upper portion of a device array. White lines correspond to metal electrodes that connect to individual nanowire devices. The position of the microfluidic channel used to deliver sample is highlighted in blue and has a total size of 6 mm × 500 μm (length × width). The image field is 4.4 × 3.5 mm. (B) Optical image of one row of addressable device elements from the region highlighted by the red-dashed box in A. The red arrow highlights the position of a device. The image field is 500 × 400 μm. (C) Scanning electron microscopic image of one silicon nanowire device. The electrode contacts are visible at the upper right and lower left regions of the image. (Scale bar: 500 nm.) (*Inset*) Cross-sectional schematic illustration of a single silicon nanowire device. The nanowire (orange horizontal line) is connected at its ends by source (S) and drain (D) metal electrodes (gold), and metal is insulated with a layer of silicon nitride (green). The microfluidic channel is indicated (blue). Down: selective and multiplexed single virus detection. (From Patolsky, F., Zheng, G., Hayden, O., Lakadamyali, M., Zhuang, X., Lieber, C. M. (2004). Electrical Detection of Single Viruses. *Proc. Natl. Acad. Sci. U.S.A.*, 101, 14017–14022.)

12.6 Conclusion and Future Perspective

Carbon nanotube's and various NW's high electrical conductivity, high aspect ratio, and unique catalytic properties, combined with their capability for biofunctionalization, make 1D nanostructure materials a very appealing nanoscale building block for a wide range of electronic biosensing applications. The incorporation of nanomaterials into biosensor systems has expanded rapidly, and will surely result in further impressive advances. This chapter has presented detailed information about processing techniques to synthesize CNTs and NWs, with emphasis on controlling their nanostructures. Further, we have summarized recent advances in the applications of CNTs and NWs to biosensor development during the last decade, involving CNTs and NWs functionalization, the design of CNTs-based functional electrodes, and the fabrication of FET-based biosensors. In addition, a variety of CNT- and NW-based electronic

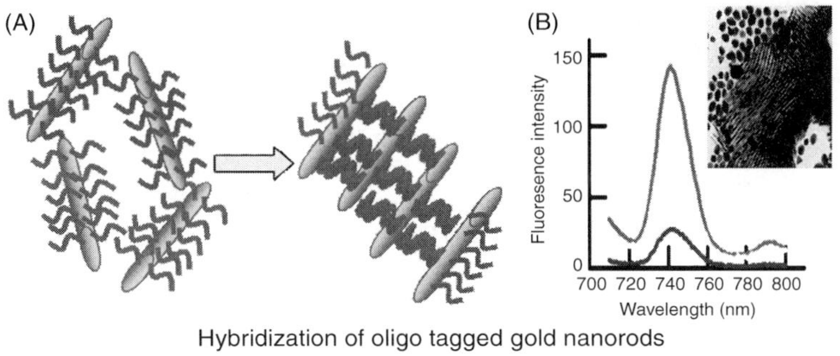

FIGURE 12.27
(See color insert) (A) Schematic illustration of gold nanorod assembly by DNA hybridization. (B) Relative fluorescence intensities of fluorescence emission spectra: oligo-functionalized gold nanorods (red line); after mixing with complementary DNA-functionalized gold nanorods (blue line). The inset is a TEM image of self-assembled gold nanorods by DNA hybridization. (From Li, C.-Z., Male, K. B., Hrapovic, S., Luong, J. H. T. (2005). Fluorescence Properties of Gold Nanorods and Their Application for DNA Biosensing. *Chem. Comm.*, 31, 3924–3926.)

biosensors including enzyme biosensors, DNA biosensors, and immunosensors have been summarized.

Although CNTs and NWs can be incorporated with biomolecules to work as a building unit in biosensors, there is still a great demand for selective coupling techniques and fabrication techniques that allow confinement of different biomolecules onto closely spaced individual NWs or CNTs to create a more powerful sensor array so that multiple analytes can be real-time monitored in parallel [203,204].

The use of biomolecules (such as protein and DNA) as a self-assembly template is anticipated to provide the potential to tailor and fabricate NW- or CNT-based devices with desirable structure and functionalities. Biomolecules linked to nanomaterials may also be used as binding elements for the specific linkage of materials to substrates in the form of addressable structures [205,206]. One example reported in *Nature* [207] suggests that enzymatic procedures combined with modern methods of bioorganic synthesis will open novel routes for the generation of well-defined biomolecule/nanomaterial components, which are useful as building blocks for the bottom-up assembly of sophisticated nanostructured architectures. Furthermore, advanced nanofabrication capabilities remain a particular challenge to fabricating a biosensor with individual addressable and reproducible sensors. The rapid development of photolithographic techniques will be beneficial for producing high-performance integrated circuits for years to come [208].

In conclusion, the combination of the unique electronic properties of CNTs and NWs and the recognition features of biomaterials will undoubtingly motivate fundamental studies and applications in connection with bioelectronic and biosensor research in interdisciplinary research involving chemistry, physics, materials science, electrical engineering, computer science, biology, and medicine. This might also open the door to catalytic and light-harvesting devices, molecular switches and transistors, biofuel cells, and other bio- and optoelectronic devices. CNT- and NW-based biosensors are expected to have a major impact upon clinical diagnostics, environmental monitoring, security surveillance, or ensuring our food safety in the future. Although the perspectives of development of nanomaterial-based biosensors is bright and fascinating, most of the fundamental results and practical applications are still ahead of us. Many within various research communities are working hard to make it real.

References

1. Daniel, M. C., Astruc, D. (2004). Gold Nanoparticles: Assembly, Supramolecular Chemistry, Quantum-Size-Related Properties, and Applications Toward Biology, Catalysis, and Nanotechnology. *Chem. Rev.*, 104, 293–346.
2. Katz, E., Willner, I. (2004). Biomolecule-Functionalized Carbon Nanotubes: Application in Nanobioelectronics. ChemPhysChem., 5(8), 1084–1104.
3. Lieber, C. M., Liu, J., Sheehan, P. E. (1996). Understanding and Manipulating Inorganic Materials Using Scanning Probe Microscopes. *Angew. Chem. Int. Ed. Engl.*, 35, 686–704.
4. Iijima, S. (1991). Helical Microtubules of Graphitic Carbon. *Nature*, 354, 56–57.
5. Beguin, F., Ehrburger, P. (Eds). (2002). Special Issue on Carbon Nanotubes. *Carbon*, 40, 1619–1842.
6. Haddon, R. C. (Ed). (2002). Special Issue on Carbon Nanotubes. *Acc. Chem. Res.*, 35, 997–1113.
7. Dresselhaus, M. S., Dai, H. (Eds). (2004). Special Issue on Carbon Nanotubes. *MRS Bull.* 29, 237–281.
8. Harris, P. J. F. (1999). *Carbon nanotubes and related structures: new materials for the 21st century.* Cambridge University Press, Cambridge, UK.
9. Dresselhaus, M. S., Dresselhaus, G., Avouris, P. (2001). *Carbon nanotubes: synthesis, structure, properties, and applications* (Topics in Applied Physics, Vol. 80). Springer-Verlag, New York.
10. Saito, R., Dresselhaus, G., Dresselhaus, M. S. (1998). *Physical properties of carbon nanotubes.* Imperial College Press, London.
11. Meyyappan, M. (2005). *Carbon nanotubes: science and applications.* CRC press, Boca Raton, FL.
12. Baughman, R. H., Zakhidov, A. A., de Heer, W. A. (2002). Carbon Nanotubes—The Route Toward Applications. *Science*, 297, 787–792.
13. Balasubramanian, K., Burghard, M. (2005). Chemically Functionalized Carbon Nanotubes. *Small*, 1, 180–192.
14. Iijima, S., Ichibashi, T. (1993). Single-Shell Carbon Nanotubes of 1-nm Diameter. *Nature*, 363, 603–605.
15. Bethune, D. S., Kiang, C. H., Devries, M. S., Gorman, G., Savoy, R., Vazquez, J., Beyers, R. (1993). Cobalt-Catalysed Growth of Carbon Nanotubes With Single-Atomic-Layer Walls. *Nature*, 363, 605–607.
16. Kong, J., Soh, H. T., Cassell, A. M., Quate, C. F., Dai, H. (1998). Synthesis of Individual Single-Walled Carbon Nanotubes on Patterned Silicon Wafers. *Nature*, 395, 878–881.
17. Cassell, A., Raymarks, J., Kong, J., Dai, H. (1999). Large Scale CVD Synthesis of Single-Walled Carbon Nanotubes. *J. Phys. Chem.*, 103, 6484–6492.
18. Moisala, A., Nasibulin, A. G., Kauppinen, E. I. (2003). The Role of Metal Nanoparticles in the Catalytic Production of Single-Walled Carbon Nnanotubes—A Review. *J. Phys. Condens. Matter.*, 15, S3011–S3035.
19. Franklin, N., Dai, H. (2000). An Enhanced CVD Approach to Extensive Nanotube Networks With Directionality. *Adv. Mater.*, 12, 890–894.
20. Su, M., Zhang, B., Liu, J. (2000). A Scalable CVD Method for the Synthesis of Single-Walled Carbon Nanotubes With High Catalyst Productivity. *Chem. Phys. Lett.*, 322, 321–326.
21. Nikolaev, P., Bronikowski, M. J., Bradley, R. K., Rohmund, F., Colbert, D. T., Smith, K. A., Smalley, R. E. (1999). Gas-Phase Catalytic Growth of Single-Walled Carbon Nanotubes From Carbon Monoxide. *Chem. Phys. Lett.*, 313, 91–97.
22. Flahaut, E., Govindaraj, A., Peigney, A., Laurent, C., Rao, C. N. (1999). Synthesis of Single-Walled Carbon Nanotubes Using Binary Fe, Co, Ni Alloy Nanoparticles Prepared In Situ by the Reduction of Oxide Solid Solutions. *Chem. Phys. Lett.*, 300, 236–242.
23. Colomer, J. F., Stephan, C., Lefrant, S., Tendeloo, G. V., Willems, I., Kanya, Z., Fonseca, A., Laurent, C., Nagy, J. B. (2000). Large-Scale Synthesis of Single-Wall Carbon Nanotubes by Catalytic Chemical Vapor Deposition CCVD Method. *Chem. Phys. Lett.*, 317, 83–89.
24. Ren, Z. F., Huang, Z. P., Xu, J. W., Wang, J. H., Bush, P., Siegal, M. P., Provencio, P. N. (1998). Synthesis of Large Arrays of Well-Aligned Carbon Nanotubes on Glass. *Science*, 282, 1105–1107.

25. Li, W. Z., Xie, S. S., Qian, L. X., Chang, B. H., Zou, B. S., Zhou, W. Y., Zhao, R. A., Wang, G., (1996). Large-Scale Synthesis of Aligned Carbon Nanotubes. *Science*, 274, 1701–1703.
26. Pan, Z. W., Xie, S. S., Chang, B. H., Wang, C. Y., Lu, L., Liu, W., Zhou, W., Li, W. Z., Qian, L. X., (1998). Very Long Carbon Nanotubes. *Nature*, 394, 631–632.
27. Fan, S., Chapline, M., Franklin, N., Tombler, T., Cassell, A., Dai, H. (1999). Self-Oriented Regular Arrays of Carbon Nanotubes and Their Field Emission Properties. *Science*, 283, 512–514.
28. Maruyama, S., Kojima, R., Miyauchia, Y., Chiashia, S., Kohno, M., (2002). Low-Temperature Synthesis of High-Purity Single-Walled Carbon Nanotubes From Alcohol. *Chem. Phys. Lett.*, 360, 229–234.
29. Hata, K., Futaba, D. N., Mizuno, K., Namai, T., Yumura, M., Iijima, S. (2005). Water-Assisted Highly Efficient Synthesis of Impurity-Free Single-Walled Carbon Nanotubes. *Science*, 306, 1362–1364.
30. Dai, H. (2002). Carbon Nanotubes: Synthesis, Integration, and Properties. *Acc. Chem. Res.*, 35, 1035–1044.
31. Chattopadhyay, D., Galeska, L., Papadimitrakopoulos, F. J. (2003). A Route for Bulk Separation of Semiconducting From Metallic Single-Wall Carbon Nanotubes. *J. Am. Chem. Soc.*, 125, 3370–3375.
32. Zheng, M., Jagota, A., Semke, E. D., Diner, B. A., McLean, R. S., Lustig, S. R., Richardson, R. E., Tassi, N. G. (2003). DNA-Assisted Dispersion and Separation of Carbon Nanotubes. *Nat. Mater.*, 2, 338–342.
33. Li, Y. M., Mann, D., Rolandi, M., Kim, W., Ural, A., Hung, S., Javey, A., Cao, J., Wang, D., Yenilmez, E., Wang, Q., Gibbons, J. F., Nishi, Y., Dai, H. (2004). Preferential Growth of Semiconducting Single-Walled Carbon Nanotubes by a Plasma Enhanced CVD Method. *Nano Lett.*, 4, 317–321.
34. Bachilo, S. M., Balzano, L., Herrera, J. E., Pompeo, F., Resasco, D. E., Weisman, R. B. (2003). Narrow (n,m)-Distribution of Single-Walled Carbon Nanotubes Grown Using a Solid Supported Catalyst. *J. Am. Chem. Soc.*, 125, 11186–11187.
35. Li, Y. M., Peng, S., Mann, D., Cao, J., Tu, R., Cho, K. J., Dai, H. (2005). On the Origin of Preferential Growth of Semiconducting Single-Walled Carbon Nanotubes. *J. Phys. Chem. B.*, 109, 6969–6971.
36. Ruland, W., Chaper, A. K., Hou, H., Greiner, A. (2003). Multi-Wall Carbon Nanotubes With Uniform Chirality: Evidence for Scroll Structures. *Carbon*, 41, 423–427.
37. Lavin, J. G., Subramoney, S., Rodney R. S., Berber, S., Tomanek, D. (2002). Scrolls and Nested Tubes in Multiwall Carbon Nanotubes. *Carbon*, 40, 1123–1130.
38. Iijima, S. (1994). Carbon Nanotubes. *MRS Bull.*, 19, 43–51.
39. Amelinckx, S., Bernaerts, D., Zhang, X. B., Tendeloo, G. V., Landuyt, J. V. (1995). A Structure Model and Growth Mechanism for Multishell Carbon Nanotubes. *Science*, 267, 1334–1338.
40. Xu, G., Feng, Z. C., Popovic, Z., Lin, J. Y., Vittal, J. J. (2001). Nanotube Structure Revealed by High-Resolution X-Ray Diffraction. *Adv. Mater.*, 13, 264–267.
41. Sun, X., Li, R., Stansfield, B., Dodelet, J. P., Ménaed, G., Désilets, S. (2006). Controlled Synthesis of Pointed Carbon Nanotubes. *Carbon*, in press.
42. Sun, X., Li, R., Villers, D., Dodelet, J. P., Désilets, S. (2003). Composite Electrodes Made of Pt Nanoparticles Deposited on Carbon Nanotubes Grown on Fuel Cell Backings. *Chem. Phys. Lett.*, 379, 99–104.
43. Bacon, R., (1960). Growth, Structure, and Properties of Graphite Whiskers. *J. Appl. Phys.*, 31, 283–290.
44. Ye, H., Naguib, N., Gogotsi, Y., Yazicioglu, A., Megaridis, C. M. (2004). Wall Structure and Surface Chemistry of Hydrothermal Carbon Nanofibres. *Nanotechnology*, 15, 232–236.
45. Samuelson, L., Thelander, C., Bjork, M. T., Borgstrom, M., Deppert, K., Dick, K.A., Hansen, A. E., Martensson, T., Panev, N., Persson, A. I., Seifert, W., Skold, N., Larsson, M. W., Wallenberg, L. R. (2004). Semiconductor Nanowires for 0D and 1D Physics and Application. *Physica E.*, 25, 313–318.
46. Hu, J. T., Odom, T. W., Libier, C. M. (1999). Chemistry and Physics in One Dimension: Synthesis and Properties of Nanowires and Nanotubes. *Acc. Chem. Res.*, 32, 435–445.
47. Xia, Y., Yang, P., Sun Y., Wu, Y., Mayers, B., Gates, B., Yin, T., Kim, F., Yan, H. (2003). One-Dimensional Nanostructures: Synthesis, Characterization, and Applications. *Adv. Mater.*, 15, 353–389.

48. Yang, P. (2003). *Chemistry of nanostructured materials.* World Scientific Pub Co Inc., Hackensack, NJ.
49. Rao, C. N. R., Muller, A., Cheetham, A. K. (2004).*The chemistry of nanomaterials: synthesis, properties and application* (Vol. 1, pp. 255–284). Wiley-VCH Verlag GmbH & Co. KGaA.
50. Cao, G. Z. (2004). *Nanostructures & nanomaterials: synthesis, properties & applications* (pp. 110–168). World Scientific Publishing Co. Pte. Ltd., Singapore.
51. Wang, Z. L. (2003). *Nanowaire and nanobelts: materials, properties and devices* (Vol. 1 & 2). Kluwer Academic Publishers, Norwell, MA.
52. Wagner, R. S., Ellis, W. C. (1964). *Appl. Phys. Lett.*, 4, 89.
53. Wu, Y., Yang, P. (2001). Direct Observation of Vapor–Liquid–Solid Nanowire Growth. *J. Am. Chem. Soc.*, 123, 3165–3166.
54. Zhang, Y., Wang, N., Gao, S., He, R., Miao, S., Liu, J., Zhu, J., Zhang, X. (2002). A Simple Method to Synthesize Nanowires. *Chem. Mater.*, 14, 3564–3568.
55. Pan, Z. W., Dai, Z. R., Wang, Z. L. (2001). Nanobelts of SemiconductingOxides. *Science*, 291, 1947–1949.
56. Jin, C., Xiang, X., Jia, C., Liu, W., Cai, W., Yao, L., Li, L. (2004). Electrochemical Fabrication of Large-Area, Ordered Bi_2Te_3 Nanowire Arrays. *J. Phys. Chem. B.*, 108, 1844–1847.
57. Takahashi, K., Limmer, S. J., Wang, Y., Cao, G. Z. (2004). Synthesis and Electrochemical Properties of Single-Crystal V_2O_5 Nanorod Arrays by Template-Based Electrodeposition. *J. Phys. Chem. B*, 108, 9795–9800.
58. Cao, G. Z. (2004). Growth of Oxide Nanorod Arrays Through sol Electrophoretic Deposition. *J. Phys. Chem. B*, 108, 19921–19931.
59. Wu, B., Heidelberg, A., Boland, J. J. (2005). Mechanical Properties of Ultrahigh-Strength Gold Nanowires. *Nat. Mater.*, 4, 525–529.
60. Huynh, W. U., Dittmer, J. J., Alivisatos, P. (2002). Hybrid Nanorod-Polymer Solar Cells. *Science*, 295, 2425–2427.
61. Favier, F., Walter, E. C., Zach, M. P., Benter, T., Penner, R. M. (2001). Hydrogen Sensors and Switches From Electrodeposited Palladium Mesowire Arrays. *Science*, 293, 2227–2231.
62. Kong, J., Franklin, N. R., Zhou, C. W., Chapline, M. G., Peng, S., Cho, K., Dai, H. J. (2000). Nanotube Molecular Wires as Chemical Sensors. *Science*, 287(5453), 622–625.
63. Someya, T., Small, J., Kim, P., Nuckolls, C., Yardley, J. T. (2003). Alcohol Vapor Sensors Based on Single-Walled Carbon Nanotube Field Effect Transistors. *Nano Lett.*, 3, 1421–1423.
64. Baughman, R. H., Zakhidow, A. A., de Heer, W. A. (2002). Carbon Nanotubes—The Route Toward Applications. *Science,* 297(2), 787–792.
65. Cui, Y., Wei, Q., Park, H., Lieber, C. M. (2001). Nanowire Nanosensors for Highly Sensitive and Selective Detection of Biological and Chemical Species. *Science,* 293(17), 1289–1292.
66. Wang, J. (2005). Carbon-Nanotube Based Electrochemical Biosensors: A Review. *Electroanalysis,* 17(1), 7–14.
67. Kohli, P., Wirtz, M., Martin, C. R. (2004). Nanotube Membrane Based Biosensors. *Electroanalysis,* 16(1–2), 9–18.
68. Chen, R. J., Bangsaruntip, S., Drouvalakis, K. A., Kam, N. W. S., Shim, M., Li, Y., Kim, W., Utz, P. J., Dai, H. (2003). Noncovalent Functionalization of Carbon Nanotubes for Highly Specific Electronic Biosensors. *Proc. Natl. Acad. Sci. U.S.A.,* 100, 4984–4989.
69. Star, A., Han, T.-R., Joshi, V., Stetter, J. R. (2004). Sensing With Nafion Coated Carbon Nanotube Field-Effect Transistors. *Electroanalysis,* 16, 1–2.
70. Burda, C., Chen, X., Narayanan, R., El-Sayed, M. A. (2005). Chemistry and Properties of Nanocrystals of Different Shapes. *Chem. Rev.*, 105, 1025–1102.
71. Lee, C.-S., Baker, S. E., Marcus, M. S., Yang, W., Eriksson, M. A., Hamers, R. J. (2004). Electrically Addressable Biomolecular Functionalization of Carbon Nanotube and Carbon Nanofiber Electrodes. *Nano Lett.,* 4(9), 1713–1716.
72. Lin, Y., Taylor, S., Li, H., Fernando, K. A. S., Qu, L., Wang, W., Gu, L., Zhou, B., Sun, Y.-P. (2004). Advances Toward Bioapplications of Carbon Nanotubes. *J. Mater. Chem.,* 14, 527–541.
73. Hirsch, A. (2002). Functionalization of Single-Walled Carbon Nanotubes. *Angew. Chem. Int. Ed.,* 11, 1853–1859.
74. Balasubramanian, K., Burghard, M. (2005). Chemically Functionalized Carbon Nanotubes. *Small,* 1(2), 180–192.

75. Davis, J. J., Coles, R. J., Hill, H. A. O. (1997). Protein Electrochemistry at Carbon Nanotube Electrodes. *J. Electroanal. Chem.*, 440, 279–282.
76. Azamian, B. R., Davis, J. J., Coleman, K. S., Bagshaw, C. B., Green, M. L. H. (2002). Bioelectrochemical Single-Walled Carbon Nanotubes. *J. Am. Chem. Soc.*, 124, 12664–12665.
77. Guo, Z., Sadler, P. J., Tsang, S. C. (1998). Immobilization and Visualization of DNA and Proteins on Carbon Nanotubes. *Adv. Mater.*, 10, 701–703.
78. Shim, M., Kam, N. W. S., Chen, R. J., Li, Y., Dai, H. J. (2002). Functionalization of Carbon Nanotubes for Biocompatibility and Biomolecular Recognition. *Nano Lett.*, 2(4), 285–288.
79. Gooding, J. J., Wibowo, R., Liu, J., Yang, W., Losic, D., Orbons, S., Mearns, F. J., Shapter, J. G., Hibbert, D. B. (2003). Protein Electrochemistry Using Aligned Carbon Nanotube Arrays. *J. Am. Chem. Soc.*, 125, 9006–9007.
80. Lin, Y., Allard, L. F., Sun, Y.-P. (2004). Protein-Affinity of Single-Walled Carbon Nanotubes in Water. *J. Phys. Chem. B*, 108, 3760–3764.
81. Besteman, K., Lee, J.-O., Wiertz, F. G. M., Heering, H. A., Dekker, C. (2003). Enzyme-Coated Carbon Nanotubes as Single-Molecule Biosensors. *Nano Lett.*, 3(6), 727–730.
82. Wang, L., Yuan, Z. (2003). Direct Electrochemistry of Glucose Oxidase at a Gold Electrode Modified With Single-Wall Carbon Nanotubes. *Sensors*, 3, 544–554.
83. Lin, Y., Lu, F., Tu, Y., Ren, Z. (2004). Glucose Biosensors Based on Carbon Nanotube Nanoelectrode Ensembles. *Nano Lett.*, 4, 191–195.
84. Zorbas, V., Ortiz-Acevedo, A., Dalton, A. B., Yoshida, M. M., Dieckmann, G. R., Draper, R. K., Baughman, R. H., Jose-Yacaman, M., Musselman, I. H. (2004). Preparation and Characterization of Individual Peptide-Wrapped Single-Walled Carbon Nanotubes. *J. Am. Chem. Soc.*, 126, 7222–7227.
85. Dieckmann, G. R., Dalton, A. B., Johnson, P. A., Razal, J., Chen, J., Giordano, G. M., Muñoz, E., Musselman, I. H., Baughman, R. H., Draper, R. K. (2003). Controlled Assembly of Carbon Nanotubes by Designed Amphiphilic Peptide Helices. *J. Am. Chem. Soc.*, 125, 1770–1777.
86. Li, H., Zhou, B., Lin, Y., Gu, L., Wang, W., Fernando, K. A. S., Kumar, S., Allard, L. F., Sun, Y.-P. (2004). Selective Interactions of Porphyrins With Semiconducting Single-Walled Carbon Nanotubes. *J. Am. Chem. Soc.*, 126, 1010–1015.
87. Murakami, H., Nomura, T., Nakashima, N. (2003). Noncovalent Porphyrin-Functionalized Single-Walled Carbon Nanotubes in Solution and the Formation of Porphyrin–Nanotube Nanocomposites. *Chem. Phys. Lett.*, 378, 481–485.
88. Chen, J., Dyer, M. J., Yu, M.-F. (2001). Cyclodextrin-Mediated Soft Cutting of Single-Walled Carbon Nanotubes. *J. Am. Chem. Soc.*, 123, 6201–6202.
89. Pompeo, F., Resasco, D. E. (2002). Water Solubilization of Single-Walled Carbon Nanotubes by Functionalization With Glucosamine. *Nano Lett.*, 2, 369–373.
90. Chen, Q., Dai, L., Gao, M., Huang, S., Mau, A. (2001). Plasma Activation of Carbon Nanotubes for Chemical Modification. *J. Phys. Chem. B*, 105, 618–622.
91. Williams, K. A., Veenhuizen, P. T. M., De La Torre, B. G., Eritja, R., Dekker, C. (2002). Nanotechnology: Carbon Nanotubes With DNA Recognition. *Nature*, 420 (6917), 761.
92. Baker, S. E., Cai, W., Lasseter, T. L., Weidkamp, K. P., Hamers, R. J. (2002). Covalently Bonded Adducts of Deoxyribonucleic Acid (DNA) Oligonucleotides with Single-Wall Carbon Nanotubes: Synthesis and Hybridization. *Nano Lett.*, 2(12), 1413–1417.
93. Hazani, M., Naaman, R., Hennrich, F., Kappes, M. M. (2003). Confocal Fluorescence Imaging of DNA-Functionalized Carbon Nanotubes. *Nano Lett.*, 3(2), 153–155.
94. Wong, S. S., Harper, J. D., Jr., P. T. L., Lieber, C. M. (1998). Carbon Nanotube Tips: High-Resolution Probes for Imaging Biological Systems. *J. Am. Chem. Soc.*, 120(3), 603–604.
95. Lee, K. M., Li, L., Dai, L. (2005). Asymmetric End-Functionalization of Multi-Walled Carbon Nanotubes. *J. Am. Chem. Soc.*, 127, 4122–4123.
96. Banerjee, I. A., Yu, L., Matsui, H. (2003). Location-Specific Biological Functionalization on Nanotubes: Attachment of Proteins at the Ends of Nanotubes Using Au Nanocrystal Masks. *Nano Lett.*, 3(3), 283–287.
97. Kovtyukhova, N. I., Mallouk, T. E., Pan, L., Dickey, E. C. (2003). Individual Single-Walled Nanotubes and Hydrogels Made by Oxidative Exfoliation of Carbon Nanotube Ropes. *J. Am. Chem. Soc.*, 125, 9761–9769.

98. Rinzler, A. G., Liu, J., Dai, H., Nikolaev, P., Huffman, C. B., Rodríguez-Macías, F. J., Boul, P. J., Lu, A. H., Heymann, D., Colbert, D. T., Lee, R. S., Fischer, J. E., Rao, A. M., Eklund, P. C., Smalley, R. E. (1998). Large-Scale Purification of Single-Wall Carbon Nanotubes: Process, Product, and Characterization. *Appl. Phys. A,* 67, 29–37.
99. Moore, V. C., Strano, M. S., Haroz, E. H., Hauge, R. H., Smalley, R. E., Schmidt, J., Talmon, Y. (2003). Individually Suspended Single-Walled Carbon Nanotubes in Various Surfactants. *Nano Lett.,* 3(10), 1379–1382.
100. Shen, K., Curran, S., Xu, H., Rogelj, S., Jiang, Y., Dewald, J., Pietrass, T. (2005). Single-Walled Carbon Nanotube Purification, Pelletization, and Surfactant-Assisted Dispersion: A Combined TEM and Resonant Micro-Raman Spectroscopy Study. *J. Phys. Chem. B,* 109, 4455–4463.
101. O'Connell, M. J., Bachilo, S. M., Huffman, C. B., Moore, V. C., Strano, M. S., Haroz, E. H., Rialon, K. L., Boul, P. J., Noon, W. H., Kittrell, C., Ma, J., Hauge, R. H., Weisman, R. B., Smalley, R. E. (2002). Band Gap Fluorescence From Individual Single-Walled Carbon Nanotubes. *Science,* 297, 593–596.
102. Richard, C., Balavoine, F., Schultz, P., Ebbesen, T. W., Mioskowski, C. (2003). Supramolecular Self-Assembly of Lipid Derivatives on Carbon Nanotubes. *Science,* 300, 775–778.
103. Zhao, W., Song, C., Pehrsson, P. E. (2002). Water-Soluble and Optically pH-Sensitive Single-Walled Carbon Nanotubes From Surface Modification. *J. Am. Chem. Soc.,* 124, 12418–12419.
104. Islam, M. F., Rojas, E., Bergey, D. M., Johnson, A. T., Yodh, A. G. (2003). High Weight Fraction Surfactant Solubilization of Single-Wall Carbon Nanotubes in Water. *Nano Lett.,* 3(2), 269–273.
105. Kamat, P. V., Thomas, K. G., Barazzouk, S., Girishkumar, G., Vinodgopal, K., Meisel, D. (2004). Self-Assembled Linear Bundles of Single Wall Carbon Nanotubes and Their Alignment and Deposition as a Film in a dc Field. *J. Am. Chem. Soc.,* 126, 10757–10762.
106. Liu, J., Rinzler, A. G., Dai, H., Hafner, J. H., Bradley, R. K., Boul, P. J., Lu, A., Iverson, T., Shelimov, K., Huffman, C. B., Rodriguez-Macias, F., Shon, Y.-S., Lee, T. R., Colbert, D. T., Smalley, R. E. (1998). Fullerene Pipes. *Science,* 280, 1253–1256.
107. Mawhinney, D. B., Naumenko, V., Kuznetsova, A., Yates, Jr., J. T., Liu, J., Smalley, R. E. (2000). Surface Defect Site Density on Single Walled Carbon Nanotubes by Titration. *Chem. Phys. Lett.,* 324, 213–216.
108. Jiang, K., Eitan, A., Schadler, L. S., Ajayan, P. M., Siegel, R. W., Grobert, N., Mayne, M., Reyes-Reyes, M., Terrones, H., Terrones, M. (2003). Selective Attachment of Gold Nanoparticles to Nitrogen-Doped Carbon Nanotubes. *Nano Lett.,* 3(3), 275–277.
109. Georgakilas, V., Tzitzios, V., Gournis, D., Petridis, D. (2005). Attachment of Magnetic Nanoparticles on Carbon Nanotubes and Their Soluble Derivatives. *Chem. Mater.,* 17, 1613–1617.
110. Chen, J., Rao, A. M., Lyuksyutov, S., Itkis, M. E., Hamon, M. A., Hu, H., Cohn, R. W., Eklund, P. C., Colbert, D. T., Smalley, R. E., Haddon, R. C. (2001). Dissolution of Full-Length Single-Walled Carbon Nanotubes. *J. Phys. Chem. B,* 105, 2525–2528.
111. Hamon, M. A., Chen, J., Hu, H., Chen, Y., Itkis, M. E., Rao, A. M., Eklund, P. C., Haddon, R. C. (1999). Dissolution of Single-Walled Carbon Nanotubes. *Adv. Mater.,* 11, 213–216.
112. Ramanathan, T., Fisher, F. T., Ruoff, R. S., Brinson, L. C. (2005). Amino-Functionalized Carbon Nanotubes for Binding to Polymers and Biological Systems. *Chem. Mater.,* 17, 1290–1295.
113. Kim, O.-K., Je, J., Baldwin, J. W., Kooi, S., Pehrsson, P. E., Buckley, L. J. (2003). Solubilization of Single-Wall Carbon Nanotubes by Supramolecular Encapsulation of Helical Amylose. *J. Am. Chem. Soc.,* 125, 4426–4427.
114. Zheng, M., Jagota, A., Strano, M. S., Santos, A. P., Barone, P., Chou, S. G., Diner, B. A., Dresselhaus, M. S., Mclean, R. S., Onoa, G. B., Samsonidze, G. G., Semke, E. D., Usrey, M., Walls, D. J. (2003). Structure-Based Carbon Nanotube Sorting by Sequence-Dependent DNA Assembly. *Science,* 302, 1545–1548.
115. Wang, S., Humphreys, E. S., Chung, S.-Y., Delduco, D. F., Lusting, S. R., Wang, H., Parker, K. N., Rizzo, N. W., Subramoney, S., Chiang, Y.-M., Jagota, A. (2003). Peptides With Selective Affinity for Carbon Nanotubes. *Nat. Mater.,* 2(3), 196–200.
116. Georgakilas, V., Kordatos, K., Prato, M., Guldi, D. M., Holzinger, M., Hirsch, A. (2002). Organic Functionalization of Carbon Nanotubes. *J. Am. Chem. Soc.,* 124, 760–761.

117. Georgakilas, V., Tagmatarchis, N., Pantarotto, D., Bianco, A., Briand, J.-P., Prato, M. (2002). Amino Acid Functionalisation of Water Soluble Carbon Nanotubes. *Chem. Commun.*, 24, 3050–3051.
118. Pantarotto, D., Partidos, C. D., Graff, R., Hoebeke, J., Briand, J. -P., Prato, M., Bianco, A. (2003). Synthesis, Structural Characterization, and Immunological Properties of Carbon Nanotubes Functionalized With Peptides. *J. Am. Chem. Soc.*, 125, 6160–6164.
119. Elkin, T., Jiang, X. P., Taylor, S., Lin, Y., Gu, L. R., Yang, H., Brown, J., Collins, S., Sun, Y.-P. (2005). Immuno-Carbon Nanotubes and Recognition of Pathogens. *ChemBioChem.*, 6, 640–643.
120. Gu, L. R., Elkin, T., Jiang, X. P., Li, H. P., Lin, Y., Qu , L. W., Tzeng, T. R. J., Joseph, R., Sun, Y.-P. (2005). Single-Walled Carbon Nanotubes Displaying Multivalent Ligands for Capturing Pathogens. *Chem. Comm.*, 7, 874–876.
121. Koehne, J., Chen, H., Li, J., Cassell, A. M., Ye, Q., Ng, H. T., Han, J., Meyyappan, M. (2003). Ultrasensitive Label-Free DNA Analysis Using an Electronic Chip Based on Carbon Nanotube Nanoelectrode Arrays. *Nanotechnology*, 12, 1239–1245.
122. Koehne, J., Li, J., Cassell, A. M., Chen, H., Ye, Q., Ng, H. T., Han, J., Meyyappan, M. (2004). The Fabrication and Electrochemical Characterization of Carbon Nanotube Nanoelectrode Arrays. *J. Mater. Chem.*, 14, 676–684.
123. Moghaddam, M. J., Taylor, S., Gao, M., Huang, S., Dai, L., McCall, M. J. (2004). Highly Efficient Binding of DNA on the Sidewalls and Tips of Carbon Nanotubes Using Photochemistry. *Nano Lett.*, 4, 89–93.
124. Seker, F., Meeker, K., Kuech, T. F., Ellis, A. B. (2000). Surface Chemistry of Prototypical Bulk II–VI and III–V Semiconductors and Implications for Chemical Sensing. *Chem. Rev.*, 100, 2505–2536.
125. Rosi, N. L., Mirkin, C. A. (2005). Nanostructures in Biodiagnostics. *Chem. Rev.*, 105, 1547–1562.
126. Love, J. C., Estroff, L. A., Kriebel, J. K., Nuzzo, R. G., Whitesides, G. W. (2005). Self-Assembled Monolayers of Thiolates on Metals as a Form of Nanotechnology. *Chem. Rev.*, 105(4), 1103–1170.
127. Bartlett, P. N., Li, C. S. (1996). Ed. Taylor, R. F., Schultz, J. S. *Handbook of chemical and biological sensors* (pp. 139–170). IOP, Philadelphia.
128. Pilloud, D. L., Chen, X., Dutton, P. L., Moser, C. C. (2000). Electrochemistry of Self-Assembled Monolayers of Iron Protoporphyrin IX Attached to Modified Gold Electrodes Through Thioether Linkage. *J. Phys. Chem. B*, 104, 2868–2877.
129. Li, Z., Chen, Y., Li, X., Kamins, T. I., Nauka, K., Williams, R. S. (2004). Sequence-Specific Label-Free DNA Sensors Based on Silicon Nanowires. *Nano Lett.*, 4(2), 245–247.
130. Britto, P. J., Santhanam, K. S. V., Ajayan, P. M. (1996). Carbon Nanotube Electrode for Oxidation of Dopamine. *Bioelectrochem. Bioenerg.*, 41, 121–125.
131. Ugo, P., Moretto, L. M., Vezza, F. (2002). Ionomer-Coated Electrodes and Nanoelectrode Ensembles as Electrochemical Environmental Sensors: Recent Advances and Prospects. *ChemPhysChem.*, 3, 917–925.
132. Zhao, Q., Gan, Z., Zhuang, Q. (2002). Electrochemical Sensors Based on Carbon Nanotubes. *Electroanalysis*, 14(23), 1609–1613.
133. Avouris, P. (2002). Molecular Electronics With Carbon Nanotubes. *Acc. Chem. Res.*, 35, 1026–1034.
134. Avouris, P., Appenzeller, J., Martel, R., Wind, S. J. (2003). Carbon Nanotube Electronics. *Proc. IEEE*, 91(11), 1772–1784.
135. Antiochia, R., Lavagnini, I., Magno, F. (2004). Amperometric Mediated Carbon Nanotube Paste Biosensor for Fructose Determination. *Anal. Lett.*, 37(8), 1657–1669.
136. Antiochia, R., Lavagnini, I., Magno, F., Valentini, F., Palleschi, G. (2004). Single-Wall Carbon Nanotube Paste Electrodes: A Comparison With Carbon Paste, Platinum and Glassy Carbon Electrodes via Cyclic Voltammetric Data. *Electroanalysis*, 16(17), 1451–1458.
137. Rubianes, M. D., Rivas, G. A. (2003). Carbon Nanotubes Paste Electrode. *Electrochem. Commun.*, 5(8), 689–694.
138. Rubianes, M. D., Rivas, G. A. (2005). Enzymatic Biosensors Based on Carbon Nanotubes Paste Electrodes. *Electroanalysis*, 17(1), 73–78.
139. Britto, P. J., Santhanam, K. S. V., Rubio, A., Alonso, J. A., Ajayan, P. M. (1999). Improved Charge Transfer at Carbon Nanotube Electrodes. *Adv. Mater.*, 11(2), 154–157.

140. Davis, J. J., Creen, M. L. H., Hill, H. A. O., Leung, Y. C., Sadler, P. J., Sloan, J., Xavier, A. V., Tsang, S. C. (1998). The Immobilization of Proteins in Carbon Nanotubes. *Inorg. Chim. Acta,* 272, 261–266.
141. Wang, J., Musameh, M. (2003). Electrodes: A Needle Microsensor for Monitoring Glucose. *Analyst,* 128, 1382–1385.
142. Heller, I., Kong, J., Heering, H. A., Williams, K. A., Lemay, S. G., Dekker, C. (2005). Individual Single-Walled Carbon Nanotubes as Nanoelectrodes for Electrochemistry. *Nano Lett.,* 5(1), 137–142.
143. Campbell, J. K., Sun, L., Crooks, R. M. (1999). Electrochemistry Using Single Carbon Nanotubes. *J. Am. Chem. Soc.,* 121, 3779–3880.
144. Wang, J., Musameh, M., Lin, Y. (2003). Solubilization of Carbon Nanotubes by Nafion Toward the Preparation of Amperometric Biosensors. *J. Am. Chem. Soc.,* 125(9), 2408–2409.
145. Gong, K. P., Dong, Y., Xiong, S. X., Chen, Y., Mao, L. Q. (2004). Novel Electrochemical Method for Sensitive Determination of Homocysteine With Carbon Nanotube- Based Electrodes. *Biosens. Bioelectron.,* 20(2), 253–259.
146. Dai, Y. Q., Shiu, K. K. (2004). Glucose Biosensor Based on Multi-Walled Carbon Nanotubes Modified Glassy Carbon Electrode. *Electroanalysis,* 16(20), 1697–1703.
147. Wang, J., Musameh, M. (2003). Carbon Nanotube/Teflon Composite Electrochemical Sensors and Biosensors. *Anal. Chem.,* 75(9), 2075–2079.
148. Gao, M., Dai, L., Wallace, G. C. (2003). Polypyrrole CNTs. *Synth. Met.,* 137, 1393–1394.
149. Jiang, L., Liu, C., Jiang, L., Peng, Z., Lu, G. (2004). A Chitosan-Multiwall Carbon Nanotube Modified Electrode for Simultaneous Detection of Dopamine and Ascorbic Acid. *Anal. Sci.,* 20, 1055–1059.
150. Joshi, P. P., Merchant, S. A., Wang, Y., Schmidtke, D. W. (2005). Amperometric Biosensors Based on Redox Polymer–Carbon Nanotube–Enzyme Composites. *Anal. Chem.,* 77(10), 3183–3188.
151. Lin, Y., Lu, F., Wang, J. (2004). Disposable Carbon Nanotube Modified Screen-Printed Biosensor for Amperometric Detection of Organophosphorus Pesticides and Nerve Agents. *Electroanalysis,* 16(1–2), 145–149.
152. Hrapovic, S., Liu, Y., Male, K. B., Luong, J. H. T. (2004). Electrochemical Biosensing Platforms Using Platinum Nanoparticles and Carbon Nanotubes. *Anal. Chem.,* 76(4), 1083–1088.
153. Deo, R. P., Lawrence, N. S., Wang, J. (2004). Electrochemical Detection of Amino Acids at Carbon Nanotube and Nickel-Carbon Nanotube Modified Electrodes. *Analyst,* 1219, 1076–1081.
154. Male, K. B., Hrapovic, S., Liu, Y., Wang, D., Luong, J. H. T. (2004). Electrochemical Detection of Carbohydrates Using Copper Nanoparticles and Carbon Nanotubes. *Anal. Chim. Acta,* 516, 35–41.
155. Zhao, Q., Nardelli, M. B., Lu, W., Bernholc, J. (2005). Carbon Nanotube-Metal Cluster Composites: A New Road to Chemical Sensors? *Nano Lett.,* 5(5), 847–851.
156. Lim, S.-H., Wei, J., Lin, J. (2004). Electrochemical Genosensing Properties of Gold Nanoparticle–Carbon Nanotube Hybrid. *Chem. Phys. Lett.,* 400, 578–582.
157. Li, J., Koehne, J. E., Cassell, A. M., Chen, H., Ng, H. T., Ye, Q., Fan, W., Han, J., Meyyappan, M. (2005). Inlaid Multi-Walled Carbon Nanotube Nanoelectrode Arrays for Electroanalysis. *Electroanalysis,* 1, 15–27.
158. Li, J., Ng, H. T., Cassell, A., Fan, W., Chen, H., Ye, Q., Koehne, J., Han, J., Meyyappan, M. (2003). Carbon Nanotube Nanoelectrode Array for Ultrasensitive DNA Detection. *Nano Lett.,* 3(5), 597–602.
159. Arrigan, D. W. M. (2004). Nanoelectrods, Nanoelectrode Arrays and Their Application. *Analyst,* 129, 1157–1165.
160. Sotiropoulou, S., Chaniotakis, N. A. (2003). Carbon Nanotube Array-Based Biosensor. *Anal. Bioanal. Chem.,* 375, 103–105.
161. Dai, L., He, P., Li, S. (2003). Functionalized Surfaces Based on Polymers and Carbon Nanotubes for Some Biomedical and Optoelectronic Applications. *Nanotechnology,* 14, 1081–1097.
162. de Heer, W. A., Châtelain, A., Ugarte, D. (1995). A Carbon Nanotube Field-Emission Electron Source. *Science,* 270, 1179–1180.

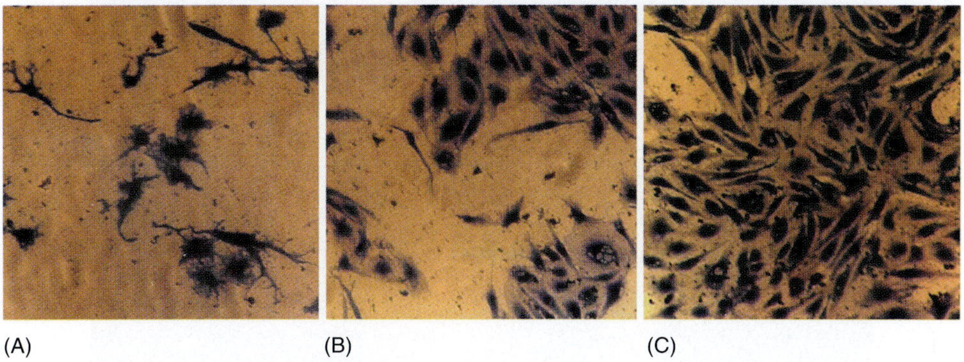

FIGURE 1.25
Light micrographs of Coomassie blue-stained ECs attached to the surface of three different electropolymerized films: (A) poly-tyrosineamide film, (B) poly-RGDY:tyrosineamide film (1:20 molar ratio), (C) poly-RGDY:tyrosineamide film (1:3 molar ratio). Reprinted from Marx, K.A., Zhou, T., McIntosh, D., Braunhut. S.J. (2004). Electropolymerization of Biomimetic Peptide-Tyrosineamide Polymer Films for Specific Cell Attachment. *Arch. Appl. Biomater. Biomol. Mater., Mater. Res. Soc.* EXS-1:169–171. With permission of the Materials Research Society.

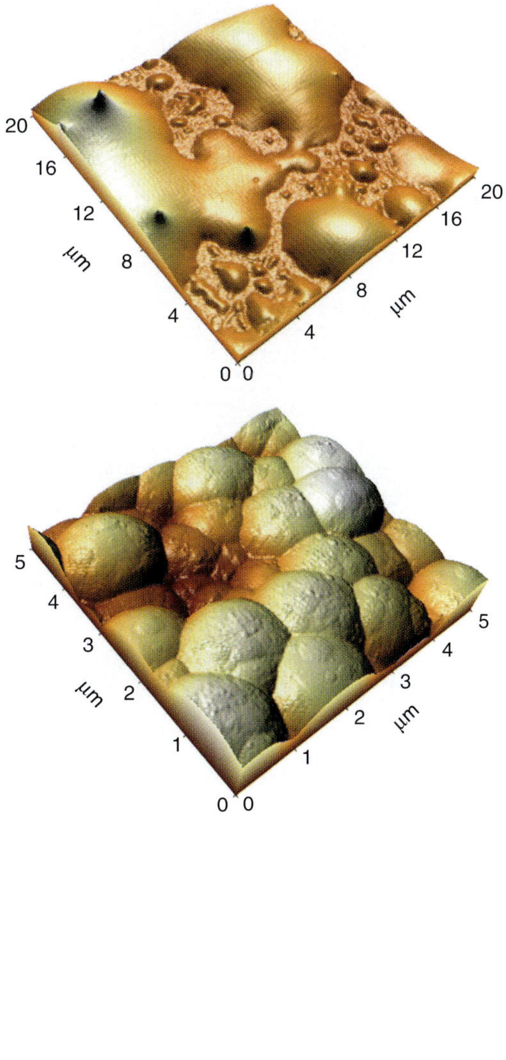

FIGURE 1.29
Noncontact AFM three-dimensional views of a 1:1 comonomer mixture of DELT (above cmc) and L-tyrosineamide prior to (upper image) and following (lower image) enzymatic polymerization with HRP and subsequent 24-h immersion of a gold-coated mica substrate to immobilize the complexes upon. Reprinted with permission from Marx, K.A., Lee, J.S., Sung, C. (2004). Enzymatic Copolymerization Alters the Structure of Unpolymerized Mixtures of the Biomimetic Monomers: The Amphiphilic Decyl Ester of L-Tyrosine and L-Tyrosineamide—An AFM Investigation of Nano- to Micrometer Scale Structure Differences. *Biomacromolecules* 5:1869–1876. Copyright (2004) American Chemical Society.

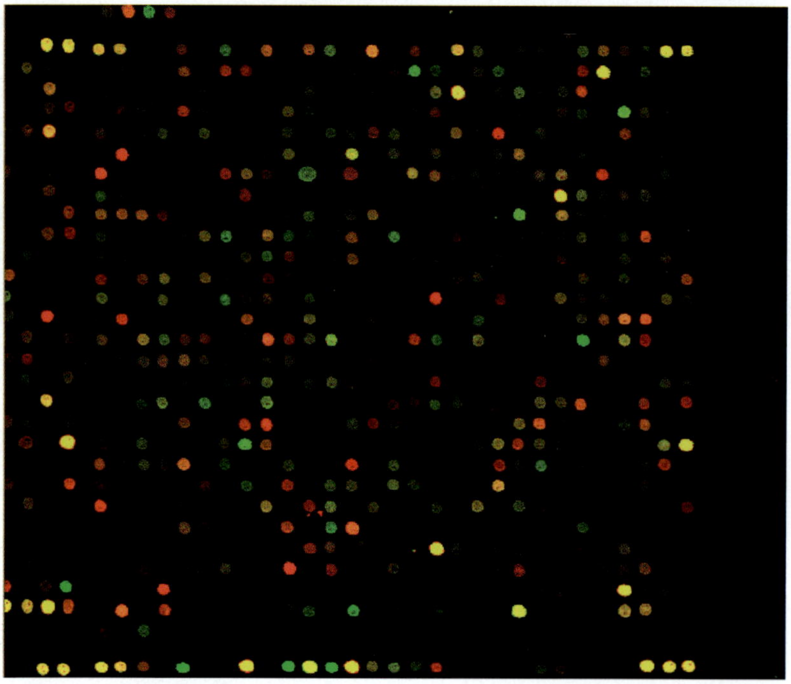

FIGURE 1.51
A low-resolution fluorescence light microscope image of the spotted array surface of a typical microarray experiment. Reprinted from website http://www.microarray.org/sfgf/jsp/home.jsp (accessed May 27, 2005). with permission of website.

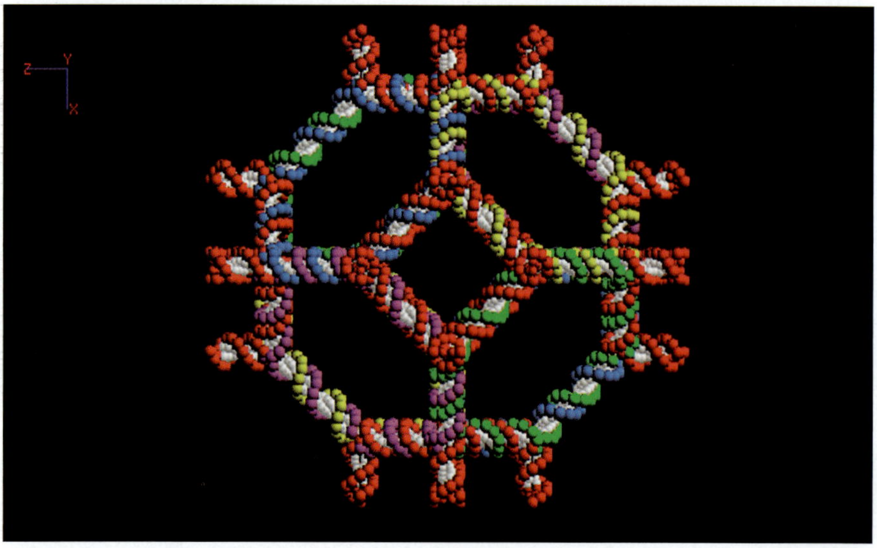

FIGURE 1.56
A schematic diagram of the top of a truncated DNA octahedron. The molecule contains 14 cyclic strands of DNA with each face of the octahedron comprising squares and hexagons, corresponding to a different cyclic strand. This view is looking down the fourfold axis of one of the squares comprising the octahedron. The approximate molecular weight of the DNA octahedron is about 790,000 Da. Reprinted with permission from Zhang, Y., Seeman, N.C. (1994). Construction of a DNA-Truncated Octahedron. *J. Am. Chem. Soc.* 116:1661–1669 and Dr. Ned Seeman. Copyright (1994) American Chemical Society.

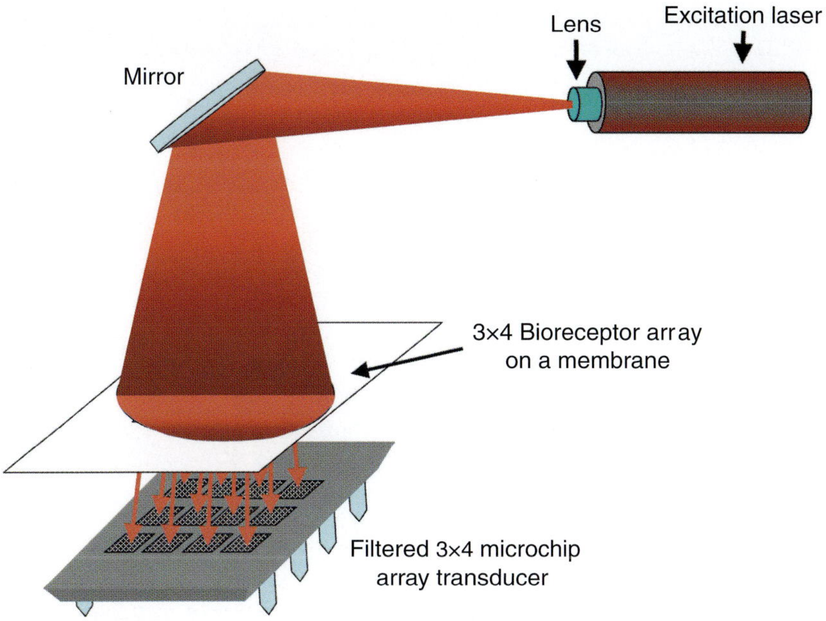

FIGURE 3.8
Schematic diagram of the basic concept of a biochip-based sensor. A light source is used to excite an array of different bioreceptors and the resulting optical emission is detected using an array detector.

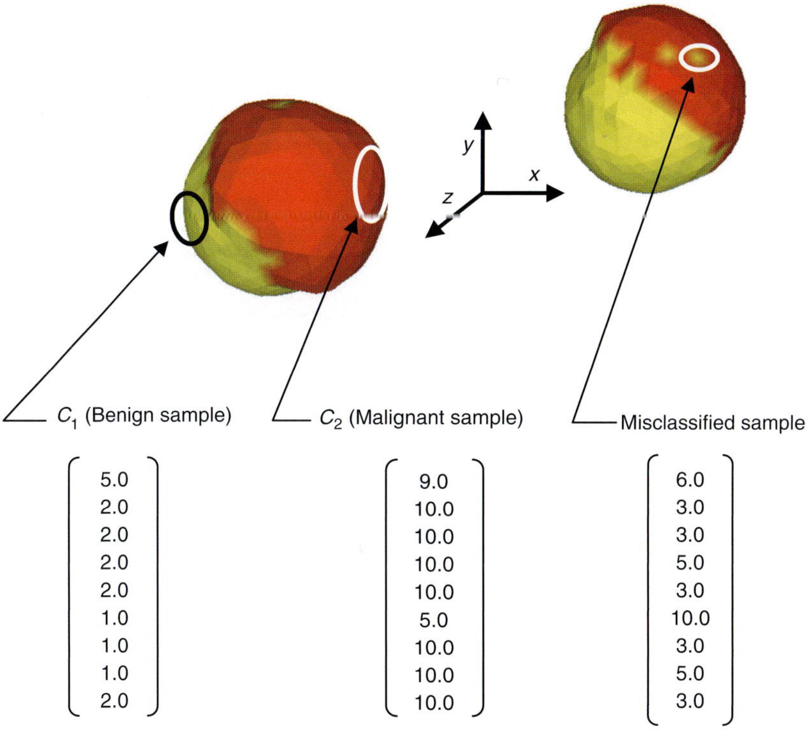

FIGURE 5.12
Two views of the same 3D colored glyph used to cluster 683 nine-dimensional data vectors into two classes (benign and malignant). Vectors assigned to the benign class (C_1) are represented by the yellow facets, while those associated with the malignant class (C_2) are shown as red. Visual identification of a misclassified data vector based on the color cue is shown in the right view.

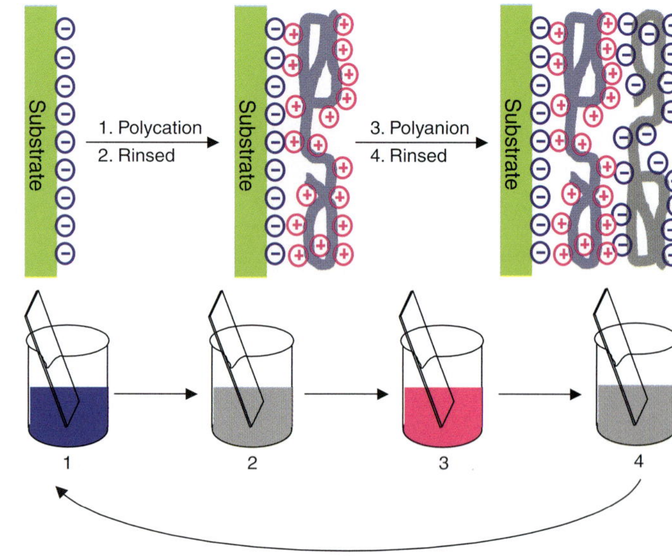

FIGURE 8.1
Multilayer assembly of oppositely charged polyelectrolytes by layer-by-layer (LBL) method.

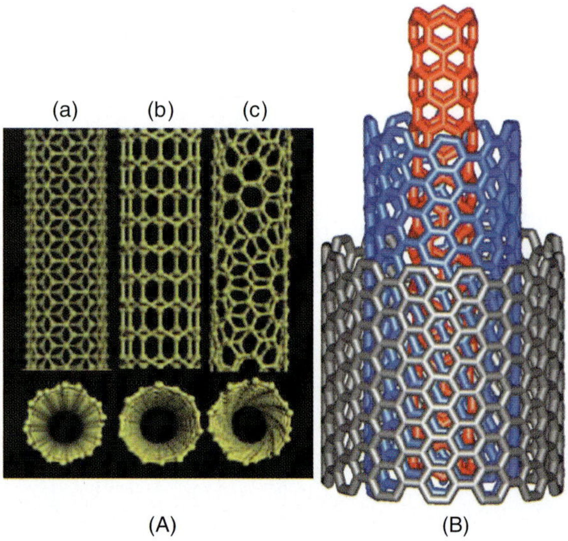

FIGURE 12.1
A. Schematic illustrations of the structures of (a) armchair, (b) zigzag, and (c) chiral SWNTs (Reproduced with permission from Baughman, R. H., Zakhidov, A. A., de Heer, W. A. (2002). Carbon Nanotubes—The Route Toward Applications. *Science*, 297, 787–792.). B. Structure of a multi-walled carbon nanotube made up of three shells of differing chirality. (Reproduced with permission from Balasubramanian, K., Burghard, M. (2005). Chemically Functionalized Carbon Nanotubes. *Small*, 1, 180–192.)

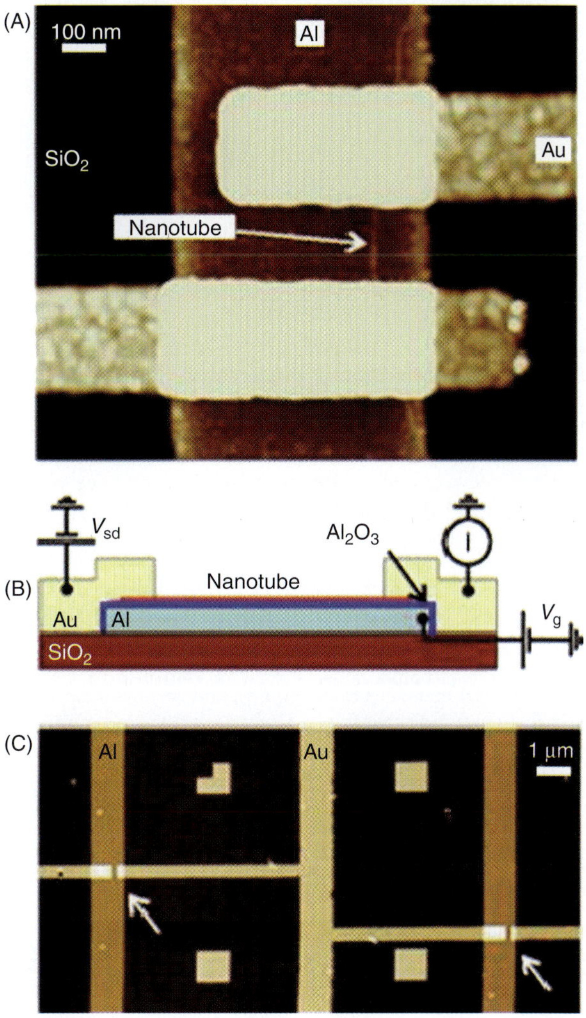

FIGURE 12.21
Device layout. (A) Height image of a single-nanotube transistor, acquired with an atomic force microscope. (B) Schematic side view of the device. A semiconducting nanotube is contacted by two Au electrodes. An Al wire, covered by a few-nanometers-thick oxide layer, is used as a gate. (C) Height-mode atomic force microscope image of two nanotube transistors connected by a Au interconnect wire. The arrows indicate the position of the transistors. Four alignment markers can also be seen. (From Bachtold, A., Hadley, P., Nakanishi, T., Dekker, C. (2001). Logic Circuits With Carbon Nanotube Transistors. *Science*, 294(9), 1317–1320.)

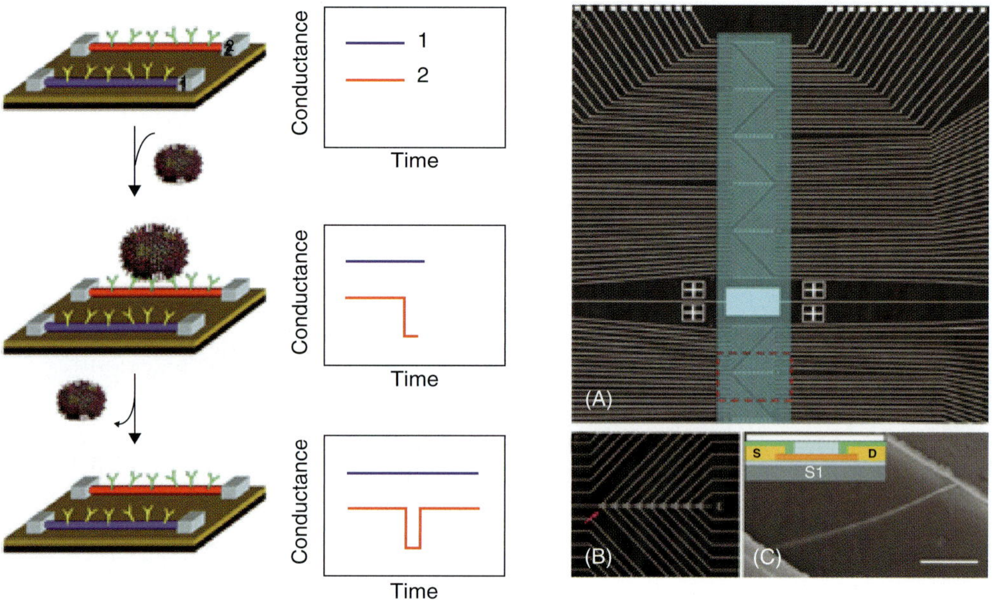

FIGURE 12.26
Nanowire-based detection of single viruses. Left: Schematic illustration shows two nanowire devices, 1 and 2, where the nanowires are modified with different antibody receptors. Specific binding of a single virus to the receptors on nanowire 2 produces a conductance change (middle) characteristic of the surface charge of the virus only in nanowire 2. When the virus unbinds from the surface, the conductance returns to the baseline value. Right: nanowire devices and device array. (A) Optical image of the upper portion of a device array. White lines correspond to metal electrodes that connect to individual nanowire devices. The position of the microfluidic channel used to deliver sample is highlighted in blue and has a total size of 6 mm × 500 μm (length × width). The image field is 4.4 × 3.5 mm. (B) Optical image of one row of addressable device elements from the region highlighted by the red-dashed box in A. The red arrow highlights the position of a device. The image field is 500 × 400 μm. (C) Scanning electron microscopic image of one silicon nanowire device. The electrode contacts are visible at the upper right and lower left regions of the image. (Scale bar: 500 nm.) (*Inset*) Cross-sectional schematic illustration of a single silicon nanowire device. The nanowire (orange horizontal line) is connected at its ends by source (S) and drain (D) metal electrodes (gold), and metal is insulated with a layer of silicon nitride (green). The microfluidic channel is indicated (blue). Down: selective and multiplexed single virus detection. (From Patolsky, F., Zheng, G., Hayden, O., Lakadamyali, M., Zhuang, X., Lieber, C. M. (2004). Electrical Detection of Single Viruses. *Proc. Natl. Acad. Sci. U.S.A.*, 101, 14017–14022.)

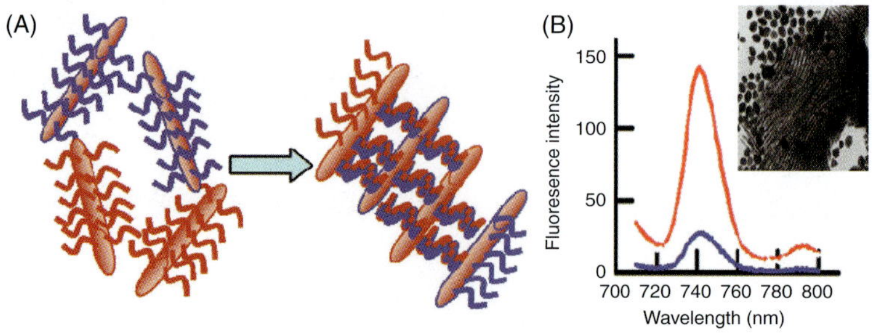

FIGURE 12.27
(A) Schematic illustration of gold nanorod assembly by DNA hybridization. (B) Relative fluorescence intensities of fluorescence emission spectra: oligo-functionalized gold nanorods (red line); after mixing with complementary DNA-functionalized gold nanorods (blue line). The inset is a TEM image of self-assembled gold nanorods by DNA hybridization. (From Li, C.-Z., Male, K. B., Hrapovic, S., Luong, J. H. T. (2005). Fluorescence Properties of Gold Nanorods and Their Application for DNA Biosensing. *Chem. Comm.*, 31, 3924–3926.)

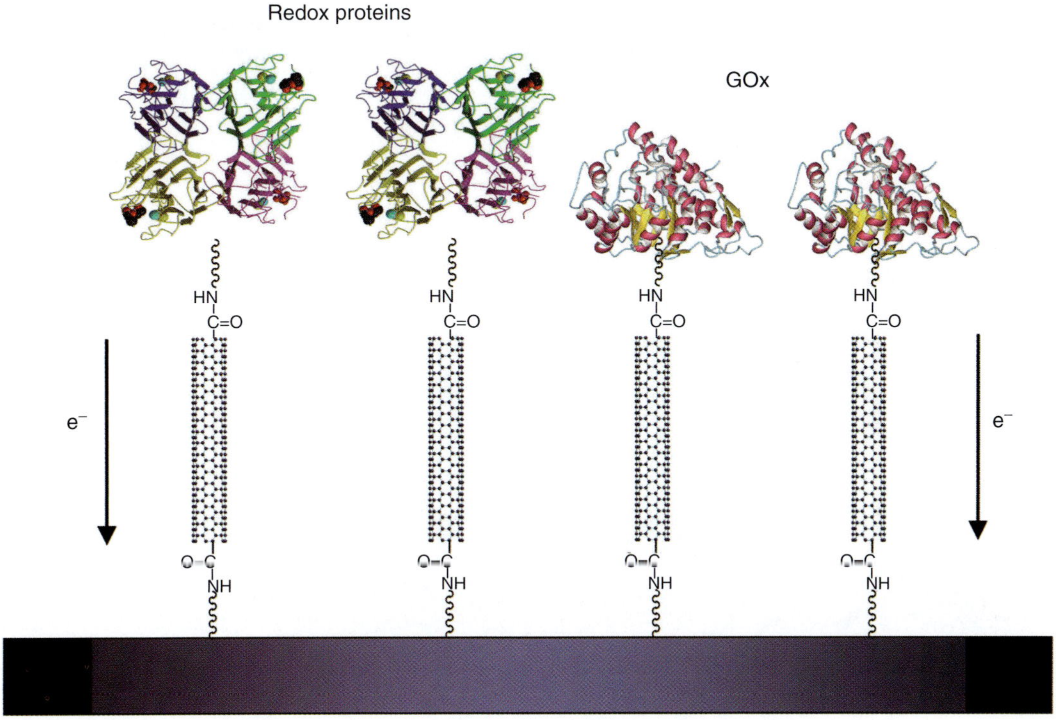

FIGURE 13.3
Direct electrochemistry of redox proteins (left) and enzymes (right).

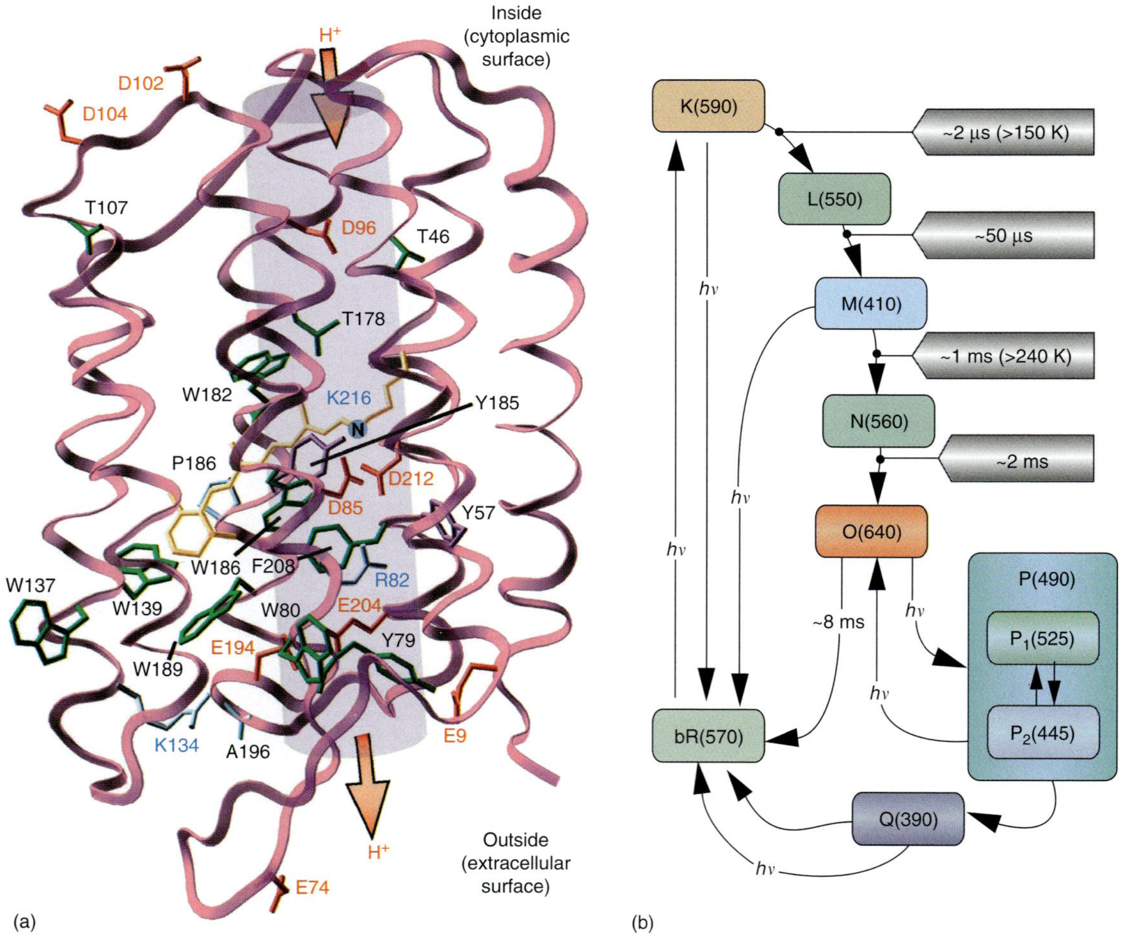

FIGURE 14.1
(a) Schematic of bacteriorhodopsin illustrating certain key amino acids and the purported path of the proton pump. The all-*trans* retinal chromophore traverses the binding pocket roughly perpendicular to the membrane normal and the alpha helices. (b) The bacteriorhodopsin photocycle, including the branched photocycle originating at the O state. Absorption maxima in nanometers are shown in parenthesis for each intermediate.

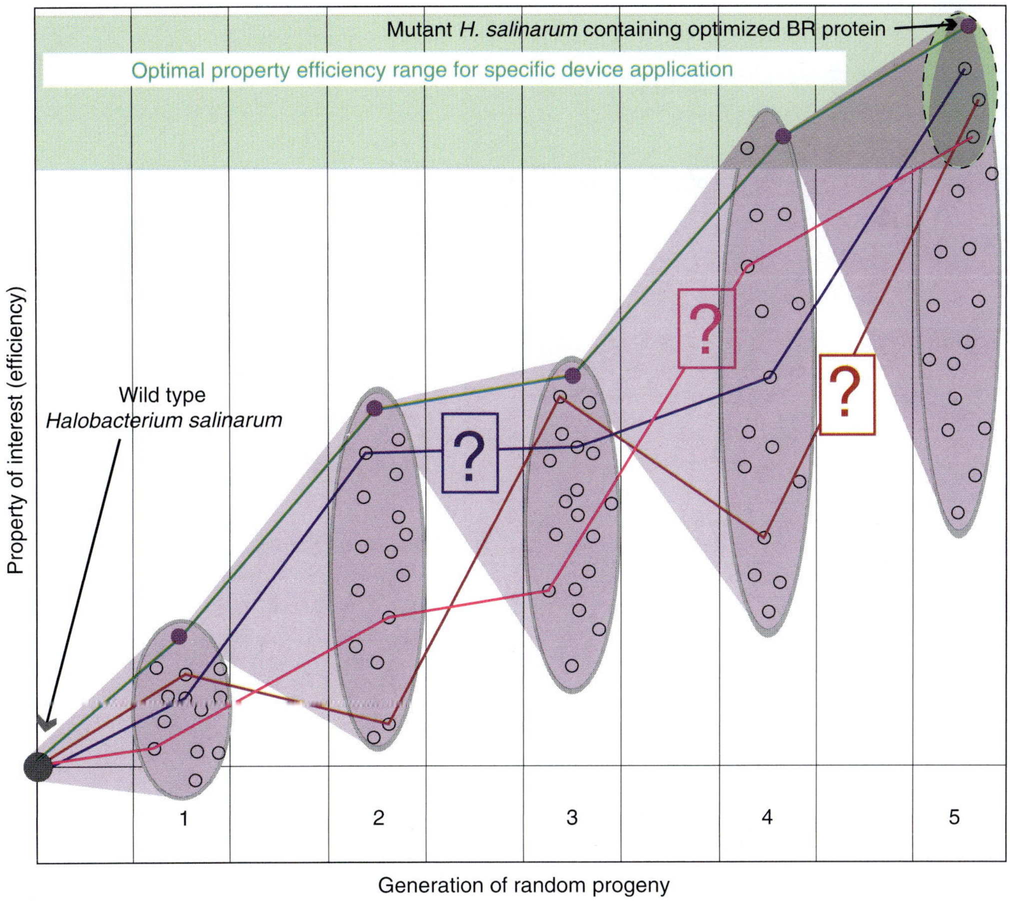

FIGURE 14.3
Schematic illustration of directed evolution of a bacterial protein toward a predefined goal. The bold green line indicates the route selected by the researcher, where, based on screening, the strain containing the most efficient mutant protein of each generation was chosen to parent the next. However, as shown by the question marks, other synergistic routes may be possible, but never discovered.

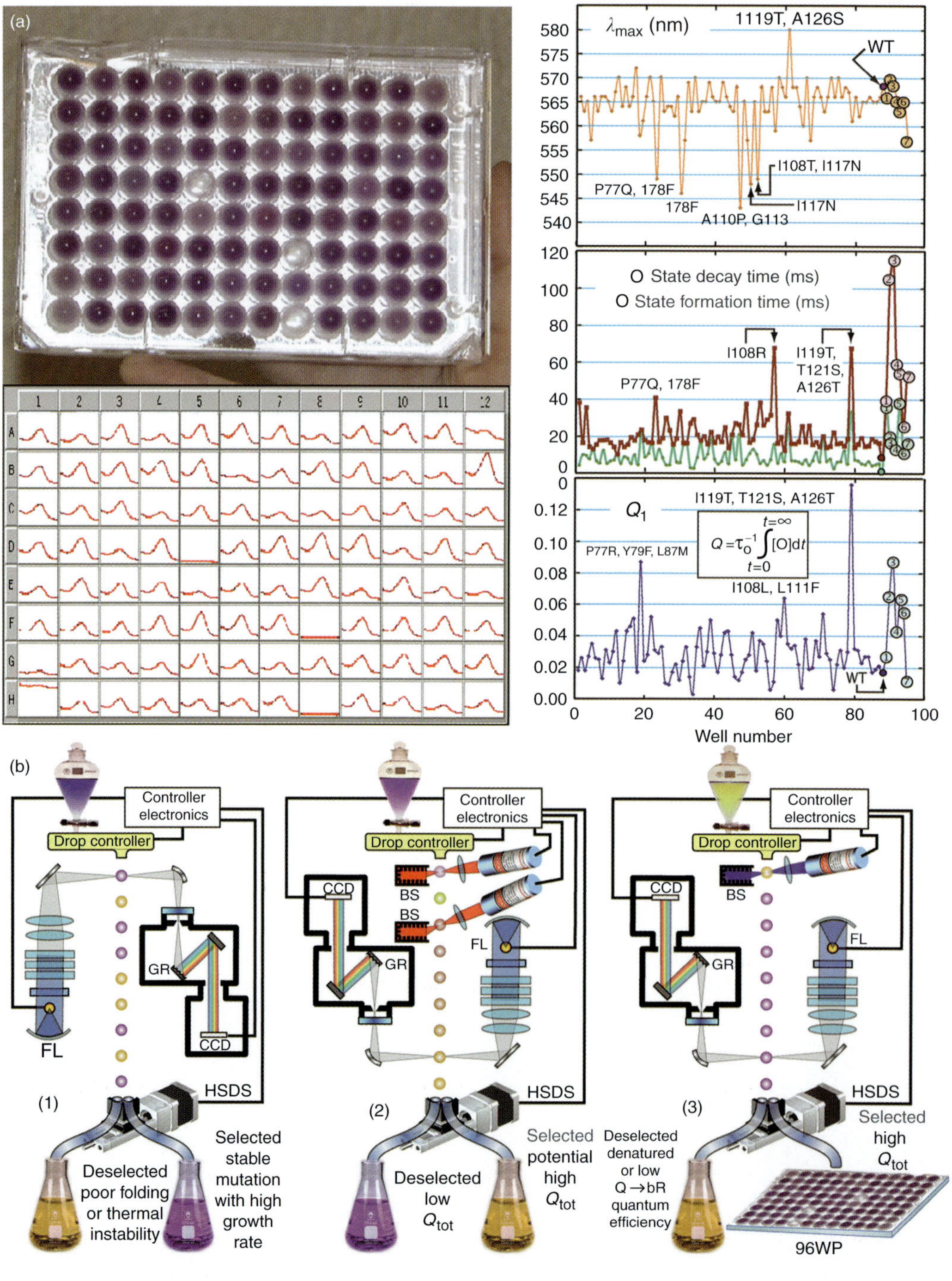
(a)
1
2
3
4
5
6
7
8
9
10
11
12
A
B
C
D
E
F
G
H
λ_{max} (nm)
580
575
570
565
560
555
550
545
540
I119T, A126S
WT
I108T, I117N
P77Q, I78F
I78F
I117N
A110P, G113
120
100
80
60
40
20
0
O State decay time (ms)
O State formation time (ms)
I108R
I119T,
T121S,
A126T
P77Q, I78F
Q_1
0.12
0.10
0.08
0.06
0.04
0.02
0.00
I119T, T121S, A126T
$Q = \tau_O^{-1} \int_{t=0}^{t=\infty} [O]dt$
P77R, Y79F, L87M
I108L, L111F
WT
0
20
40
60
80
100
Well number
(b)
Controller electronics
Drop controller
GR
FL
CCD
HSDS
(1)
Deselected poor folding or thermal instability
Selected stable mutation with high growth rate
Controller electronics
Drop controller
BS
BS
CCD
GR
FL
HSDS
(2)
Deselected low Q_{tot}
Selected potential high Q_{tot}
Controller electronics
Drop controller
BS
CCD
GR
FL
HSDS
(3)
Deselected denatured or low Q → bR quantum efficiency
Selected high Q_{tot}
96WP

FIGURE 14.5
Absorption spectra of the various bacteriorhodopsin intermediates are shown. The bR, M, and O states are of particular interest to most of the applications developed around bacteriorhodopsin's unique photochromokinetic and photovoltaic properties.

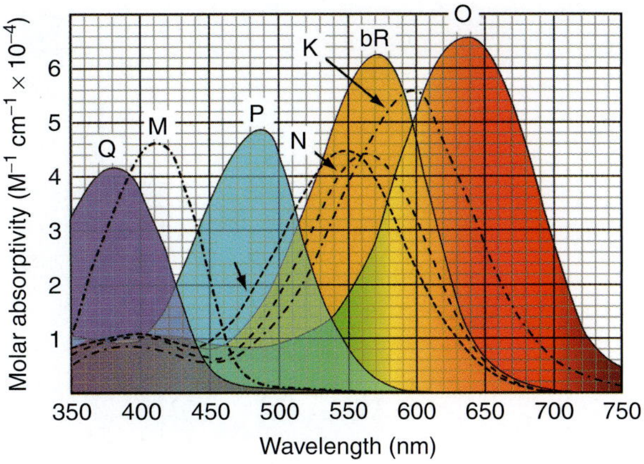

FIGURE 14.4
(a) Top left: Mutant strains of *H. salinarum* are brought up in 96-well plates and screened for production of bacteriorhodopsin; each well in the plate is analyzed by a plate scanner that generates a visible absorption spectrum for each strain, indicating whether the protein is produced (middle left). Top right: BR mutants are screened for λ_{max}, O-state formation and decay time, and *Q*, a quality factor determining how efficiently the O state is formed (maximum yield in minimum time). The cluster of mutants at the far right of each plot indicates several site-directed mutants included for comparison. The lower panel (b) illustrates a method by which flow cytometry could be used to screen for potentially valuable BR strains in vivo, cell by cell, in this case for efficient production of the branched photocycle. The three-step cell sorting system allows for in vivo high-throughput photochromokinetic screening. The first stage (1) selects for BR expression and high growth rate. The second stage (2) selects mutants with high photochemical conversion into the branched photocycle (O→P conversion). The third stage (3) selects for mutants having efficient photochemical conversion from the branched photocycle back into the bR state (Q→bR conversion). Mutants with desired photochemical characteristics are automatically deposited onto a single well of a 96-well plate (96WP). Other symbols: GR (grating), CCD (charge coupled detector), FL (flash lamp), HSDS (high-speed drop selector), BS (beam stop). See Hillebrecht et al., Optimization of protein-based volumetric optical memories and associative processors using directed evolution. *Nanobiotechnology*, 2005. **1**(2): 141–152 for more detail.

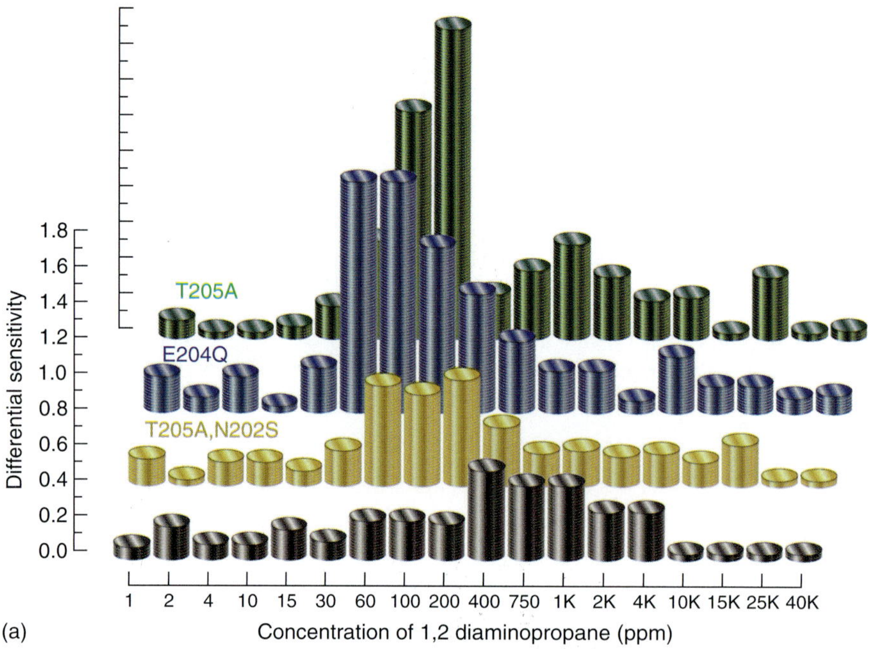

(a)

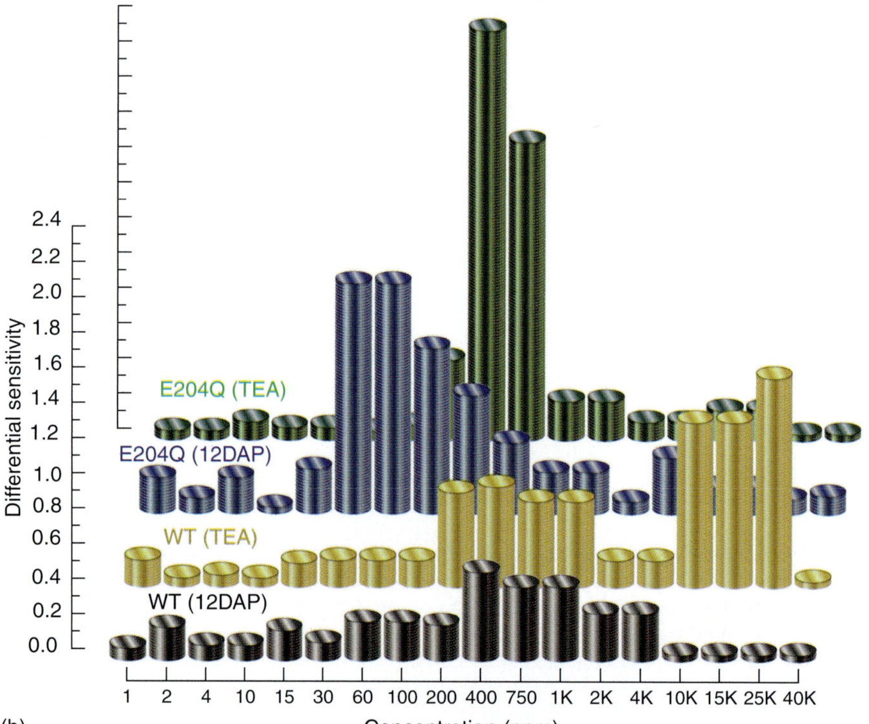

(b)

FIGURE 14.10
The sensitivity of wild-type BR and mutants in the presence of several chemical antigens is illustrated. Sensitivity is defined as the ratio of the change in the M/O decay time constant to the change in micromolar concentration of the added chemical. (a) Sensitivity to 1,2-diaminopropane (1,2-DAP) varies between the wild-type protein and BR mutants. All of the mutants demonstrated better sensitivity than wild-type BR. (b) Not only sensitivity to one chemical varies between proteins as in (a), but the same protein responds differently from one chemical to another. This is illustrated for the wild-type protein and the mutant E204Q, for 1,2-DAP and triethylamine.

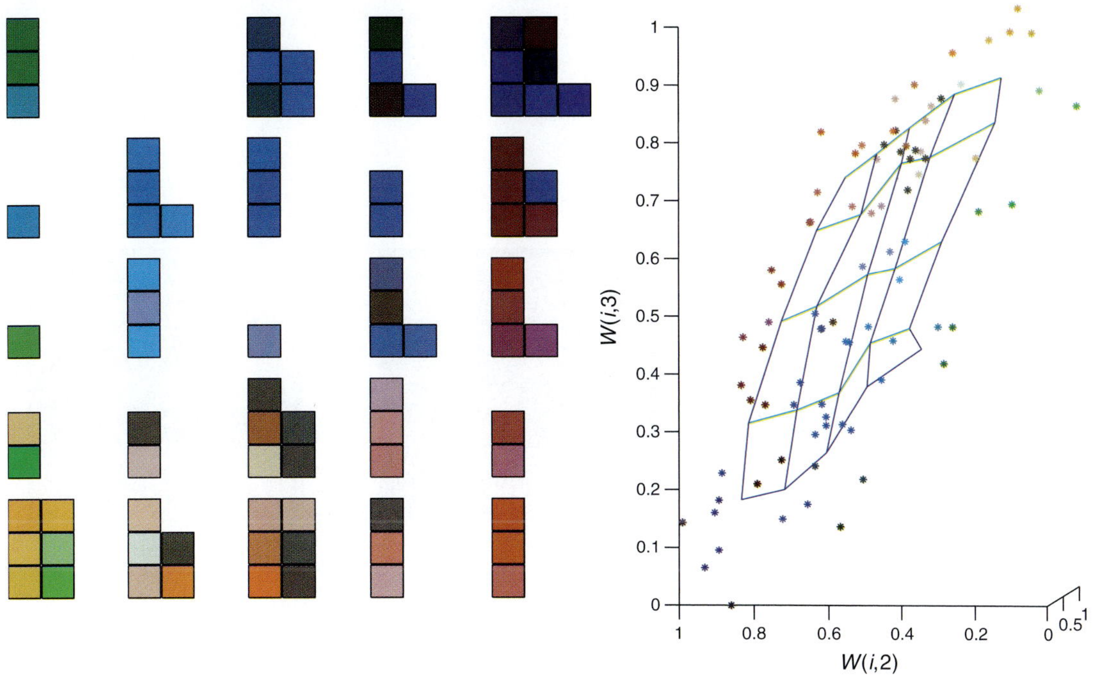

FIGURE 16.16
Visualization of the color space for BR responses by using a 5 × 5 rectangular SOM.

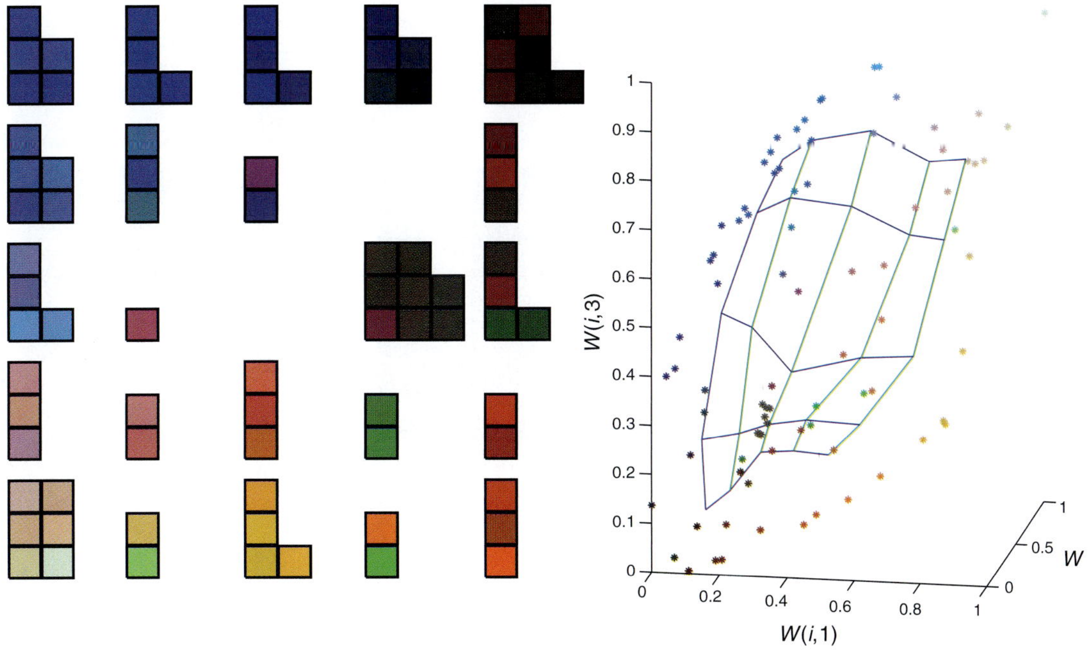

FIGURE 16.17
Visualization of the color space for simulated cone responses by using a 5 × 5 rectangular SOM.

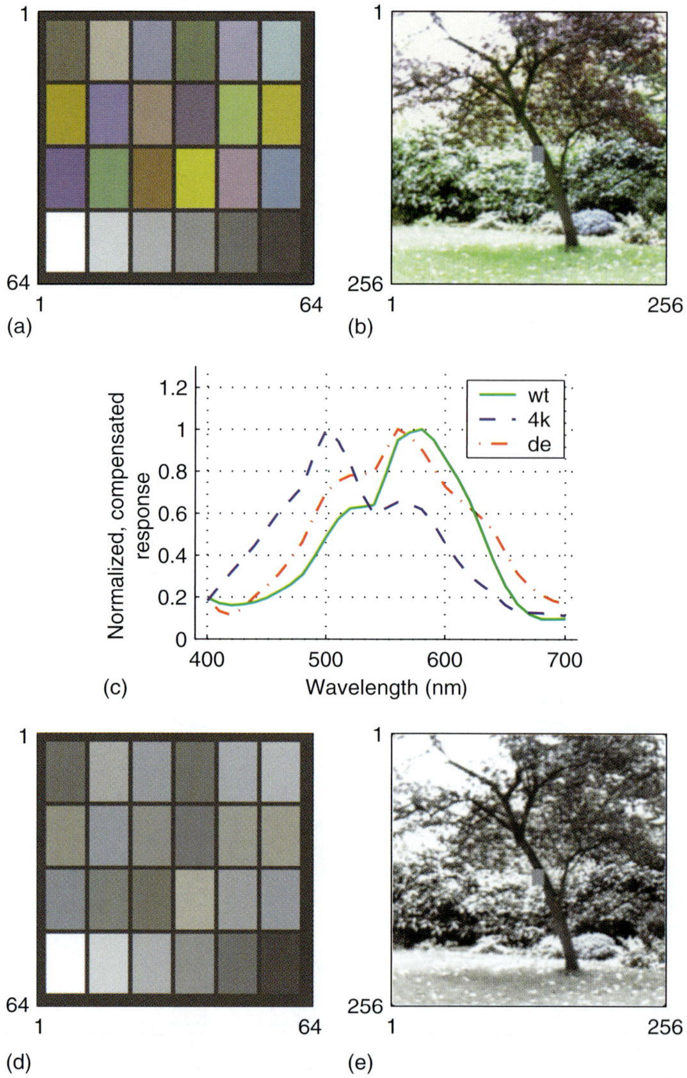

FIGURE 16.19
(a–b) Red-green-blue representations of the spectral images. (c) Measured action spectra of wild-type (wt), 4-keto (4k), and 3,4-didehydro (de) bacteriorhodopsin, and (d–e) responses of the theoretical color-sensitive imaging array to the input images.

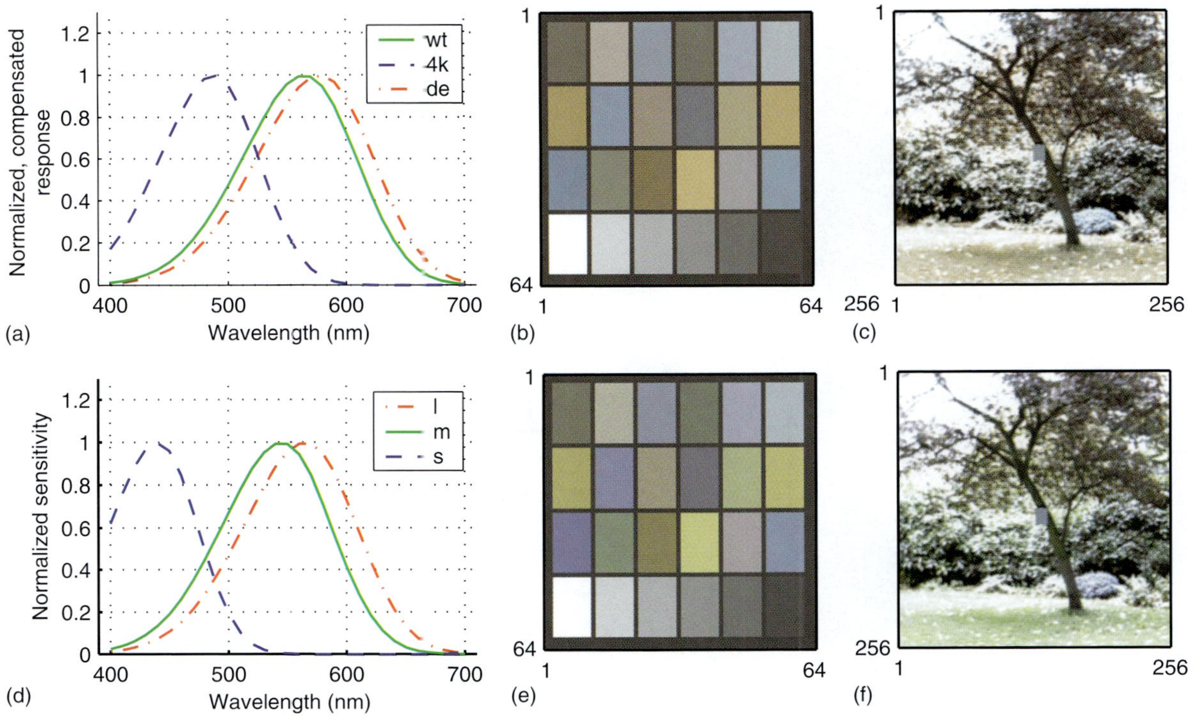

FIGURE 16.20
(a) Spectral sensitivities of visual pigment templates fitted to the absorption peaks of wild-type (wt), 4-keto (4k), and 3,4-didehydro (de) BR, (b–c) responses of the theoretical imaging array to the two input images, (d) Spectral sensitivities of *l*, *m*, and *s* cone pigments in the human retina, and (e–f) estimated response of the human eye to the two input images.

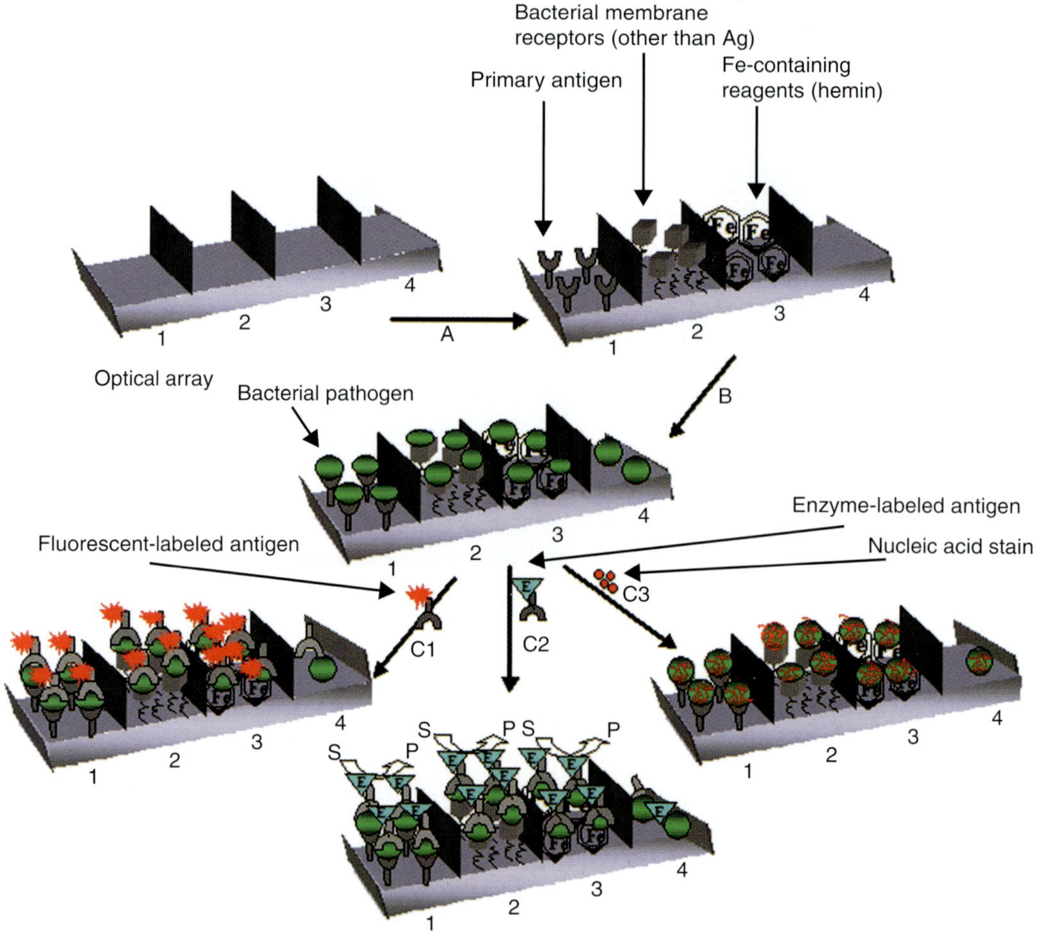

FIGURE 19.3
Possible configurations for array immunosensor.

163. Liu, Z., Shen, Z., Zhu, T., Hou, S., Ying, L., Shi, Z., Gu, Z. (2000). Organizing Single-Walled Carbon Nanotubes on Gold Using a Wet Chemical Self-Assembling Technique. *Langmuir*, 16, 3569–3573.
164. Chattopadhyay, D., Galeska, I., Papadimitrakopoulos, F. (2001). Metal-Assisted Organization of Shortened Carbon Nanotubes in Monolayer and Multilayer Forest Assemblies. *J. Am. Chem. Soc.*, 123, 9451–9452.
165. Nguyen, C. V., Delzeit, L., Cassell, A. M., Li, J., Han, J., Meyyappan, M. (2002). Preparation of Nucleic Acid Functionalized Carbon Nanotube Arrays. *Nano Lett.*, 2(10), 1079–1081.
166. Taft, B. J., Lazareck, A. D., Withey, G. D., Yin, A., Xu, J. M., Kelley, S. O. (2004). Site-Specific Assembly of DNA and Appended Cargo on Arrayed Carbon Nanotubes. *J. Am. Chem. Soc.*, *126*, 12750–12751.
167. Liu, J., Chou, A., Rahmat, W., Paddon-Row, M. N., Gooding, J. J. (2005). Achieving Direct Electrical Connection to Glucose Oxidase Using Aligned Single Walled Carbon Nanotube Arrays. *Electroanalysis*, 17, 38–46.
168. Tu, Y., Lin, Y., Ren, Z. F. (2003). Nanoelectrode Arrays Based on Low Site Density Aligned Carbon Nanotubes. *Nano Lett.*, 3(1), 107–109.
169. Tu, Y., Lin, Y., Yantasee, W., Ren, Z. (2005). Carbon Nanotubes Based Nanoelectrode Arrays: Fabrication, Evaluation, and Application in Voltammetric Analysis. *Electroanalysis*, 17(1), 79–84.
170. Patolsky, F., Zheng, G., Hayden, O., Lakadamyali, M., Zhuang, X., Lieber, C. M. (2004). Electrical Detection of Single Viruses. *Proc. Natl. Acad. Sci. U.S.A.*, 101, 14017–14022.
171. He, P., Dai, L. (2004). Aligned Carbon Nanotube–DNA Electrochemical Sensors. *Chem. Comm.*, 3, 348–349.
172. Tans, S. J., Verschueren, A. R. M., Dekker, C. (1998). Room-Temperature Transistor Based on a Single Carbon Nanotube. *Nature*, 393, 49–52.
173. Bachtold, A., Hadley, P., Nakanishi, T., Dekker, C. (2001). Logic Circuits With Carbon Nanotube Transistors. *Science*, 294(9), 1317–1320.
174. Wind, S. J., Appenzeller, J., Martel, R., Derycke, V., Avouris, Ph. (2002). Vertical Scaling of Carbon Nanotube Field-Effect Transistors Using Top Gate Electrodes. *Appl. Phys. Lett.*, 80, 3817–3819.
175. Materl, R., Derycke, V., Lavoie, C., Appenzeller, J., Chen, K., Tersoff, J., Avouris, Ph. (2001). Electrical Transport in Semiconducting Single- Wall Carbon Nanotubes. *Phys. Rev. Lett.*, 87(25), 256801–256805.
176. Materl, R., Wong, H. S. P., Chen, K., Avouris, Ph. (2001). Carbon Nanotubes Field-Effect Transistors for Logic Applications. *Proc. IEDM*, 159–161.
177. Keren, K., Berman, R. S., Buchstab, E., Sivan, U., Braun, E. (2003). DNA-Templated Carbon Nanotube Field-Effect Transistor. *Science*, 302, 1380–1382.
178. Star, A., Gabriel, J.-C. P., Bradley, K., Grüner, G. (2003). Electronic Detection of Specific Protein Binding Using Nanotube FET Devices. *Nano Lett.*, 3, 459–463.
179. Boussaad, S., Tao, N. J., Zhang, R., Hopson, T., Nagahara, L. A. (2003). In Situ Detection of Cytochrome c Adsorption With Ssingle Walled Carbon Nanotube Device. *Chem. Comm.*, 13, 1502–1503.
180. Dai, H., Hafner, J. H., Rinzler, A. G., Golbert, D. T., Smalley, R. E. (1996). Nanotubes as Nanoprobes in Scanning Probe Microscopy. *Nature*, 384, 147–150.
181. Hafner, J. H., Cheung, C., Woolly, A. T., Lieber, C. M. (2001). Structural and Functional Imaging With Carbon Nanotube AFM Probes. *Prog. Biophys. Mol. Biol.*, 77, 73–110.
182. Chen, L., Cheung, C. L., Ashby, P. D., Lieber, C. M. (2004). Single-Walled Carbon Nanotube AFM Probes: Optimal Imaging Resolution of Nanoclusters and Biomolecules in Ambient and Fluid Environments. *Nano Lett.*, 4(9), 1725–1731.
183. Burt, D. P., Wilson, N. R., Weaver, J. M. R., Dobson, P. S., Macpherson, J. V. (2005). Nanowire Probes for High Resolution Combined Scanning Electrochemical Microscopy—Atomic Force Microscopy. *Nano Lett.*, 5(4), 639–643.
184. Patil, A., Sippel, J., Martin, G. W., Rinzler, A. G. (2004). Enhanced Functionality of Nanotube Atomic Force Microscopy Tips by Polymer Coating. *Nano Lett.*, 4(2), 303–308.
185. Esplandiu, M. J., Bittner, V. G., Giapis, K. P., Collier, C. P. (2004). Nanoelectrode Scanning Probes From Fluorocarbon-Coated Single-Walled Carbon Nanotubes. *Nano Lett.*, 4(10), 1873–1879.

186. Wong, S. S., Wooley, A. T., Joselevich, E., Lieber, C. M. (1999). Functionalization of Carbon Nanotube AFM Probes Using Tip-Activated Gases. *Chem. Phys. Lett.*, 306, 219–225.
187. Shapiro, I. R., Solares, S. D., Esplandiu, M. J., Wade, L. A., Goddard, W. A., Collier, C. P. (2004). Influence of Elastic Deformation on Single-Wall Carbon Nanotube Atomic Force Microscopy Probe Resolution. *J. Phys. Chem. B,* 108, 13613–13618.
188. Nishino, T., Ito, T., Umezawa, Y. (2002). Carbon Nanotube Scanning Tunneling Microscopy Tips for Chemically Selective Imaging. *Anal. Chem.*, 74, 4275–4278.
189. Hafner, J. H., Cheung, C. Li., Lieber, C. M. (1999). Direct Growth of Single-Walled Carbon Nanotube Scanning Probe Microscopy Tips. *J. Am. Chem. Soc.*, 121(41), 9750–9751.
190. Woolley, A. T., Cheung, C. L., Hafner, J. H., Lieber, C. M. (2000). Structural Biology With Carbon Nanotube AFM Probes. *Chem. Biol.*, 7, R193–R204.
191. Shimotani, K., Shigematsu, T., Manabe, C., Watanabe, H., Shimizu, M. (2003). An Advanced Electric Probing System: Measuring DNA Derivatives. *J. Chem. Phys.*, 118(17), 8016–8022.
192. Woolley, A. T., Guillemette, C., Cheung, C. L., Housman, D. E., Lieber, C. M. (2000). Direct Haplotyping of Kilobase-Size DNA Using Carbon Nanotube Probes. *Nat. Biotech.*, 18, 760–763.
193. Wong, S. S., Joselevich, E., Woolley, A. T., Cheung, C. L., Lieber, C. M. (1998). Covalently Functionalized Nanotubes as Nanometer Sized Probes in Chemistry and Biology. *Nature,* 394, 52–55.
194. Friedman, R. S., McAlpine, M. C., Ricketts, D. S., Ham, D., Lieber, C. M. (2005). Nanowire Nanosensors. *Mater. Today*, 8, 20–28.
195. Bunimovich, Y. L., Ge, G., Beverly, K. C., Ries, R. S., Hood, L., Heath, J. R. (2004). Electrochemically Programmed, Spatially Selective Biofunctionalization of Silicon Wires. *Langmuir*, 20, 10630–10638.
196. Wang, W. U., Chen, C., Lin, K., Fang, Y., Lieber, C. M. (2005). Label-Free Detection of Small-Molecule-Protein Interactions by Using Nanowire Nanosensors. *Proc. Natl. Acad. Sci. U.S.A.*, 102, 3208–3212.
197. Murray, B. J., Walter, E. C., Penner, R. M. (2004). Amine Vapor Sensing With Silver Mesowires. *Nano Lett.*, 4(4), 665–670.
198. Li, C., Zhang, D., Lei, B., Han, S., Liu, X., Zhou, C. (2003). Surface Treatment and Doping Dependence of In_2O_3 Nanowires as Ammonia Sensors. *J. Phys. Chem. B,* 107, 12451–12455.
199. Hernández, R. M., Richter, L., Semancik, S., Stranick, S., Mallouk, T. E. (2004). Template Fabrication of Protein-Functionalized Gold-Polypyrrole-Gold Segmented Nanowires. *Chem. Mater.*, 16, 3431–3438.
200. Nicewarner-Peña, S. R., Carado, A. J., Shale, K. E., Keating, C. D. (2003). Barcoded Metal Nanowires: Optical Reflectivity and Patterned Fluorescence. *J. Phys. Chem. B,* 107, 7360–7367.
201. Dujardin, E., Hsin, L. B., Wang, C. R. C., Mann, S. (2001). DNA-Driven Self-Assembly of Gold Nanorods. *Chem. Comm.*, 14, 1264–1265.
202. Li, C.-Z., Male, K. B., Hrapovic, S., Luong, J. H. T. (2005). Fluorescence Properties of Gold Nanorods and Their Application for DNA Biosensing. *Chem. Comm.*, 31, 3924–3926.
203. Kerman, K., Kobayashi, M., Tamiya, E. (2004). Recent Trends in Electrochemical DNA Biosensor Technology. *Meas. Sci. Technol.*, 15, R1–R11.
204. Hart, J. P., Pemberton, R. M. (1999). A Review of Screen-Printed Electrochemical Sensors. Application Note 121. Perkin-Elmer Instruments, Princeton Applied Research, Oakridge, TN.
205. Stoltenberg, R. M., Woolley, A. T. (2004). DNA-Templated Nanowire Fabrication. *Biomed. Microdevices,* 6(2), 105–111.
206. Niemeyer, C. M. (2001). Nanoparticles, Proteins, and Nucleic Acids: Biotechnology Meets Materials Science. *Angew. Chem. Int. Ed.*, 40, 4128–4158.
207. Braun, E., Eichen, Y., Sivan, U., Ben-Yoseph, G. (1998). DNA-Templated Assembly and Electrode Attachment of a Conducting Silver Wire. *Nature,* 391(6669), 775–778.
208. Gates, B. D., Xu, Q., Stewart, M., Ryan, D., Willson, C. G., Whitesides, G. M. (2005). New Approaches to Nanofabrication: Molding, Printing, and Other Techniques. *Chem. Rev.*, 105(4), 1171–1196.
209. Sadana, A. (1992). Protein Adsorption and Inactivation on Surfaces. Influence of Heterogeneities. *Chem. Rev.*, 92, 1799–1818.

210. Hultgren, A., Tanase, M., Chen, C. S., Meyer, G. J., Reich, D. H. (2003). Cell Manipulation Using Magnetic Nanowires. *J. Appl. Phys.*, 93(10), 7554–7556.
211. Reich, D. H., Tanase, M., Hultgren, A., Bauer, L. A., Chen, C. S., Meyer, G. J. (2003). Biological Applications of Multifunctional Magnetic Nanowires. *J. Appl. Phys.*, 93(10), 7275–7280.
212. Qi, X., Osterloh, F. E. (2005). Chemical Sensing With $LiMo_3Se_3$ Nanowire Films. *J. Am. Chem. Soc.*, 127, 7666–7667.
213. Zhang, D., Liu, Z., Li, C., Tang, T., Liu, X., Han, S., Lei, B., Zhou, C. (2004). Detection of NO_2 Down to ppb Levels Using Individual and Multiple In_2O_3 Nanowire Devices. *Nano Lett.*, 4(10), 1919–1924.
214. Ramanathan, K., Bangar, M. A., Yun, M., Chen, W., Myung, N. V., Mulchandani, A. (2005). Bioaffinity Sensing Using Biologically Functionalized Conducting-Polymer Nanowire. *J. Am. Chem. Soc.*, 127, 496–497.
215. Hahm, J.-in., Lieber, C. M. (2004). Direct Ultrasensitive Electrical Detection of DNA and DNA Sequence Variations Using Nanowire Nanosensors. *Nano Lett.*, 4 (1), 51–54.

13

Electrochemical Biosensors Based on Carbon Nanotubes

Won-Yong Lee

CONTENTS

13.1 Introduction

Since their first discovery in 1991 [1], carbon nanotubes (CNTs) have rapidly gained importance in the fields of chemistry, physics, and material sciences [2]. Such tremendous interest stems from the fact that CNTs have unique properties such as high electrical conductivity, physical rigidity, and chemical stability [1–3]. CNTs can be made by various methods such as electric arc discharge, chemical vapor deposition, or laser evaporation [2]. The structure of a CNT is quite different from conventional carbon fiber and graphite. CNTs have tubular structure made of hexagonal honeycomb lattices built from sp^2 carbon units and are seamless structure with typically several nanometers in diameter and many microns in length. CNTs can be categorized into single-wall carbon nanotube (SWCNT) and multiwall carbon nanotube (MWCNT). SWCNT has a single cylinder formed by rolling up a single graphite sheet seamlessly into a tube as shown in Figure 13.1.

MWCNT comprises several concentric graphite tubules and the distance between graphite sheets is 3.4 Å, which is close to the interlayer spacing in graphite. Theoretical calculation predicts that CNT will behave either as metal or semiconductor depending upon its size and symmetry of the two-dimensional carbon lattice (helicity) [4].

The CNT represents a new kind of carbon-based material and is superior to other conventional carbon-based materials. The unique structural and electronic properties of CNT make them extremely attractive for the construction of chemical sensors and biosensors based on electrochemical transduction. Over the last several years, extensive efforts have been devoted into the preparation and characterization of CNT-based electrochemical sensors and biosensors. Recently, the CNT-based electrochemical sensors and biosensors have been reviewed [3, 5–6]. Recent studies demonstrate that the unique structural and electronic properties of CNT can enhance the electron-transfer reactions of hydrogen peroxide and nocotinamide adenine diclucleotide (NADH), and thus highly sensitive electrochemical biosensors based on oxidase and dehydrogenase enzymes can be fabricated with the use of CNT [7–8]. In addition to enhanced electrochemical reactivity, CNT-modified electrodes can minimize surface-fouling effects commonly involved in the NADH-oxidation process. These properties of CNT are extremely attractive for a wide range of electrochemical biosensors.

Owing to the enhanced electroactivity of the product of the enzyme label as well as guanine of target DNA, highly sensitive DNA hybridization sensors based on CNT can also be prepared. Conductive properties of CNT permit direct electron-transfer reactions between redox proteins and underlying electrodes; thus, simple reagentless biosensor can be made.

In this chapter, recent advances in the use of CNT for the construction of electrochemical biosensors are summarized. Common preparation methods for CNT-modified electrodes and their analytical characteristics are also discussed. In addition, CNT-modified amperometric biosensors based on oxidase or dehydrogenase are illustrated with examples. DNA biosensors based on the enhanced detection of the product of the enzyme label or of guanine base in the target DNA are also discussed. Finally, the field-effect transistors functionalized with biomolecules are illustrated.

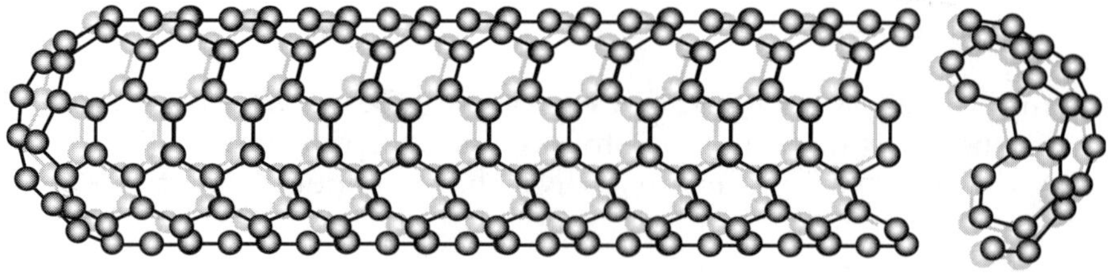

FIGURE 13.1
Structures of carbon nanotubes.

13.2 Electrochemical Behavior of CNT-Modified Electrodes

The first preliminary study of the possible applications of CNTs as electrode materials has been reported by Britto et al. [9] in 1996. They used MWCNT for the oxidation of dopamine (an important neurotransmitter). An MWCNT-modified electrode was constructed using bromoform as binder, and the oxidative behavior of dopamine was examined at this electrode. The cyclic voltammetric study in PBS (pH 7.4) at an MWCNT-modified electrode revealed a characteristic two-electron oxidation of dopamine to dopaminequinone with ideal reversibility. These results obtained in the MWCNT-modified electrode were superior to those observed at other ordinary carbon electrodes. They also studied the electrochemical oxidation of dopamine at a CNT electrode after treating the electrode with goat's brain homogenate. The cyclic voltammetric behavior of dopamine showed the same features as the untreated electrode. It demonstrated the potential applications of CNT electrodes for the determination of dopamine in vitro and in vivo.

Following the demonstration of the enhanced electroactivity of CNT-modified electrode toward dopamine oxidation, Davis et al. [10] have examined the electrochemical properties of MWCNT-modified electrodes immobilized with redox proteins. Closed CNTs, produced by arc vaporization, were opened by treatment with either nitric acid or a combination of nitric acid and potassium permanganate. The opened CNTs were packed into a glass capillary in mineral oil, deionized water, nujol, or bromoform. Redox proteins were immobilized on and within these CNTs. Well-behaved and reproducible voltammetric responses were obtained with both cychrome *c* and a "blue" copper protein, azurin. These good characteristics were mediated by hydrogen bonding interactions between protein surface lysine and negatively charged carboxylate and phenolate of the electrode. These results demonstrated the possible application of CNTs as biosensing devices.

Compared to the MWCNT, SWCNT is a well-defined system in terms of electronic properties. Individual SWCNT can be regarded as quantum wires. Luo et al. [11] investigated the electrochemical and electrocatalytic behavior of SWCNT-modified glassy carbon (GC) electrode. The carboxylic acid groups were introduced at the open ends of the SWCNT. The nitric-acid-purified SWCNT solution was cast on a GC electrode to form a CNT modified electrode. The CNT-modified electrode showed very stable cyclic voltammetric behavior and favorable electrocatalytic activity toward the oxidation of biomolecules such as dopamine and epinephrine. These results also suggested that the SWCNT-modified electrode could be used for the construction of electrochemical biosensors to study the electrochemistry of biosystems.

Toward the use of CNT for the construction of electrochemical biosensors, low-potential NADH detection at the CNT-modified electrode was first reported by Musameh et al. [7] in 2002. β-NADH is involved as a cofactor in several hundred enzymatic reactions of NAD^+/NADH-dependent dehydrogenase systems as shown in the scheme below in Equation (13.1) (e.g., alcohol dehydrogenase).

$$\text{Ethanol} + NAD^+ \xrightarrow{\text{ADH}} \text{Acetaldehyde} + \text{NADH} \tag{13.1}$$

The NADH produced from the enzymatic reaction is electrochemically detected. Problems associated with the anodic detection of NADH are the large overpotential required for its oxidation at ordinary carbon electrode and surface fouling associated with the accumulation of the reaction products. Therefore, considerable efforts are directed to the development of new electrode materials that will reduce the overpotential for the

NADH oxidation and circumvent surface passivation. Musameh et al. prepared the CNT-coated GC electrode by dispersing the SWCNT or MWCNT in a solution of concentrated sulfuric acid and casting the resulting solution of 10 μL on the GC electrode surface. The CNT-modified GC electrode exhibited a substantial (490 mV) decrease in the overpotential of the NADH-oxidation reaction compared to conventional carbon electrodes, and the oxidation started at −0.05 V (vs. Ag/AgCl). Furthermore, the CNT-modified electrode showed extremely stable responses to NADH, with 96% of the initial activity remaining after 60-min stirring in 2×10^{-4} M NADH solution compared to 20% at the bare GC-electrode surface. These results indicate that the ability of CNT to promote the oxidation of NADH along with high resistance to the surface fouling provides a useful avenue for the construction of the CNT-based dehydrogenase biosensors.

Wang et al. [8] reported on the greatly enhanced electrocatalytic activity of CNT toward the oxidation of hydrogen peroxide. They employed perfluorosulfonated inomer Nafion to solubilize the single-wall and multiwall CNT. The CNT- or Nafion-modified electrode offered a significant decrease in the overvoltage for the oxidation of hydrogen peroxide to allow convenient low-potential amperometric detection. Although a bare GC electrode exhibits no redox activity for H_2O_2 over most of the potential range, the CNT-modified electrode displays significant oxidation and reduction currents starting at around +0.20 V. This pioneering finding opened the door for the preparation of CNT-based electrochemical oxidase enzyme biosensors (e.g., glucose biosensor) as shown in the following scheme (Equations (13.2) and (13.3)).

$$\text{Glucose} + O_2 \rightarrow \text{Gluconolactone} + H_2O_2 \tag{13.2}$$

$$H_2O_2 \rightarrow O_2 + 2H^+ + 2e^- \tag{13.3}$$

The enzyme-liberated hydrogen peroxide is sensitively detected with a CNT-modified electrode. Therefore, the CNT-modified electrode coimmobilized with oxidase enzyme (e.g. glucose oxidase) can offer highly sensitive amperometric biosensors.

13.3 Enzymatic Biosensors Based on CNTs

13.3.1 Biosensors Based on the Enhanced Electrocatalytic Activity of CNT

13.3.1.1 Fabrication of CNT-Based Enzymatic Biosensors

As discussed already, CNT-based electrode permits low-potential detection as well as remarkably enhanced electrochemical activity toward the oxidation of hydrogen peroxide and NADH compared to the conventional GC electrode. Since there are over 200 dehydrogenase and 100 oxidase enzymes, CNT-based electrode offers an effective biosensing of clinically important analytes (e.g., glucose, lactate, cholesterol, and alcohol) based on the oxidase or dehydrogenate enzyme reactions. To utilize the remarkable properties of CNT in such electrochemical biosensing applications, the CNT need to be properly functionalized and also immobilized along with enzymes in electrode surface.

The purification of CNT with sulfuric acids creates the carboxylic moieties at the ends of the CNTs and also partially on sidewalls of CNT surface as defect sites. These carboxylic groups in CNTs provide sites for covalent linking of CNT to amino groups of biomolecules (or other mediators) through the carbodiimide-coupling reaction. From the perspectives of electrochemical activity, the sidewalls of CNTs are believed to have properties similar to those of the basal plane of highly oriented pyrolytic graphite (HOPG) while their open ends resemble the edge planes of HOPG.

Enzymes can be simply mixed with polymer to produce CNT-based electrochemical biosensor. Since CNT is insoluble in most solvents, a number of efforts have been made to disperse CNTs in a variety of appropriate solvents. The reported CNT-modified electrodes, which relied on casting CNTs onto a GC surface after dispersing them in bromoform [9] or dimethylformamide [11], are not compatible with the immobilization of biomolecules. Therefore, some biocompatible polymers such as Nafion, titania–nafion, Teflon, and sol–gel silica have been used for the solubilization of CNTs and the immobilization of biomolecules onto the electrode surface.

Since Nafion is known to contain both hydrophobic domain composed of fluorocarbon skeleton and hydrophilic domain composed of sulfonates, CNT can be solubilized in solutions of Nafion in phosphate buffer or alcohol [8]. The use of Nafion permits easy preparations of CNT-based electrochemical biosensors with outstanding performances because of the biocompatibility and charge discrimination ability of Nafion. For effective biosensing of glucose, glucose oxidase (GOx) was incorporated together with CNT on the surface of GC electrode.

Similar to Nafion, the composite materials composed of sol–gel-derived titania (TiO_2) and Nafion (1:1, v/v) can also solubilize both SWCNT and MWCNT [12]. The solubility of CNT in the solution of titania–Nafion composite is strongly dependent upon the Nafion content. The Nafion content in the composite ranging from 5–0.7 wt% can solubilize the single-wall and multiwall CNT. In contrast, the composite containing 0.5 wt% Nafion can solubilize single-wall CNT but cannot solubilize multiwall CNT. This behavior is exactly the same as observed in pure Nafion solution. This fact indicates that the titania–Nafion composite solution does not impair the solubility of Nafion toward CNT. Scanning electron microscopy (SEM) was employed to gain insights into the nature of the new CNT–titania–Nafion composites. Figure 13.2 shows the SEM (a) and transmission electron microscope (TEM) (b) images for MWCNT–titania–Nafion composite films. The MWCNT–titania–Nafion surface is characterized with bundles of MWCNT covering the titania–Nafion composites. The MWCNT fibers have a diameter ranging from 30–50 nm. TEM image of MWCNT dispersed in the composite indicates that CNTs are well distributed in the composite. SWCNT can also be well dispersed in the composite.

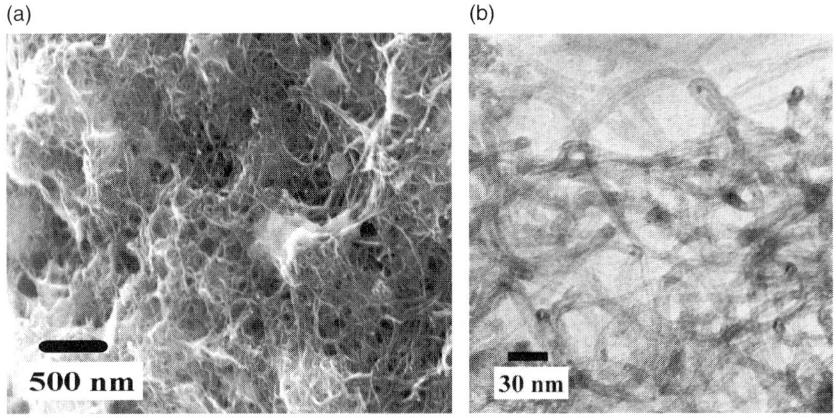

FIGURE 13.2
FE-SEM (a) and TEM (b) images of titania–Nafion (1:1, v/v) composite films containing MWCNT (1.0 mg/mL).

The titania–Nafion composite materials offer a convenient way for the preparation of chemical sensors and biosensors. The use of the titania–Nafion composite has been shown to be useful in improving the performances of the electrochemical sensors and biosensors. For example, the electrochemical sensors based on titania–Nafion composite showed extremely fast response time because the titania–Nafion composite materials have greater pore size than pure Nafion film does, thus leading to faster diffusion of analyte into the film [13]. Furthermore, the biosensors based on titania–Nafion composite films exhibited substantial improvement in long-term stability because both Nafion and titania film were very stable and biocompatible [14]. Therefore, the CNT–titania–Nafion biocomposite materials combine the major advantages of CNT with those of bulk titania–Nafion composite and thus open a new avenue for wide-range sensing application of CNT.

Another simple method for preparing effective CNT-based electrochemical biosensors involves the use of CNT–Teflon composite films [15]. The bulk of the resulting CNT–teflon electrodes play a role as a reservoir of the enzyme. The electrocatalytic properties of CNT toward hydrogen peroxide and NADH are not impaired by their association with the Teflon binder. Owing to the enhanced electron-transfer reactions of hydrogen peroxide and NADH, effective low-potential amperometric biosensors for glucose and ethanol can be fabricated with the incorporation of GOx or alcohol dehydrogenase/NAD^+ within the CNT–Teflon composite matrix. The enhanced electrochemical activity of CNT–Teflon-based biosensor was coupled with the minimization of electrode surface fouling commonly occurring in the NADH oxidation.

Gong et al. [16] reported on the sol–gel-derived ceramic–CNT nanocomposite electrodes. The ceramic–CNT electrodes prepared in the above-mentioned manner exhibited a tunable dimension ranging from conventional electrode to nanoelectrode ensembles, depending upon the amount of the MWCNT dispersed in silica sol. A high content of MWCNT resulted in the electrode in conventional dimension, while a low content produced the electrode in nanoelectrode dimension.

CNT-paste electrode was prepared by dispersing MWCNT within mineral oil [17–18]. Such composite materials combine the ability of CNTs to promote electron-transfer reactions with the attractive advantages of paste electrode materials. In particular, the substantial decrease in the hydrogen peroxide reduction potential (400 mV) with an incorporation of GOx into the composite material permits the development of a highly selective and sensitive glucose biosensor without any mediator or antiinterference membrane. Similarly, CNT-based screen-printed electrochemical sensors have been reported by Wang and Musameh [19]. Screen-printing technology is well established for the mass production of disposable electrochemical sensors. Most commonly used ink is carbon-based conducting ink incorporating graphite particles, polymer binder, and other additives. In the CNT-based screen-printed electrode, CNT was used instead of graphite for the conducting inks. Such screen-printed CNT electrodes are mechanically stable and exhibit higher electrochemical reactivity compared to conventional carbon-based screen-printed electrode. Thus, the resulting CNT-based screen-printed electrodes combine the attractive advantages of CNT and screen-printed electrodes.

13.3.1.2 Oxidase-Based Biosenors

Owing to the greatly enhanced electrocatalytic activity as well as dramatically decreased overpotential of CNT-modified electrode toward the oxidation of hydrogen peroxide, CNT-modified electrodes offer great promise for the development of selective and sensitive electrochemical biosensors based on oxidase enzymes.

Wang group [8] reported on the glucose biosensor based on the CNT–Nafion–GOx-composite on GC electrode. The accelerated electrocatalytic oxidation of hydrogen peroxide allowed glucose measurements possible at a very low potential where physiological levels of acetaminophen, uric acid, and ascorbic acid produced negligible signals. In addition, the permselectivity of Nafion offered an effective discrimination against both neutral and anionic electroactive interferents. Higher sensitivity can be obtained with the deposition of platinum nanoparticles onto the CNT–Nafion–GOx-electrode [20]. The deposited platinum nanoparticles with a diameter of 2–3 nm were in electrical contact with GC electrode through the SWCNT. Such a biosensor resulted in higher sensitivity compared with the corresponding biosensors based on CNT or Pt-nanoparticles alone and offered a detection limit of 0.5 μM glucose with good selectivity and a fast response time of 3 s.

Luo et al. [21] reported on the use of electrochemically deposited nanocomposite of chitosan and CNT for glucose biosensor. CNT dispersed in biocompatible chitosan solution was electrodeposited to form CNT–chitosan nanocomposite, where the thickness of the deposited nanocomposite film was controlled through the change of the concentration of the chitosan solution. For the construction of glucose biosensor, GOx was added into the CNT–chitosan solution before the electrodeposition. Such electrodeposition method offers simple and controllable method for the preparation of reproducible biosensors.

Davis et al. [22] reported on glucose biosensor based on SWCNT and ferrocene mediator. Prolonged incubation of SWCNT with GOx led to an effective coating of GOx onto the sidewalls of SWCNT without gross loss of enzyme activity. Although the GOx was randomly physisorbed onto the SWCNT sidewalls, the redox process of the active sites of GOx was communicated to the SWCNT π system through the diffusive mediator of ferrocene monocarboxylic acid to generate a quantifiable catalytic current. The magnitude of this catalytic current was more than one order of magnitude greater than that observed at an activated carbon electrode.

Gao et al. [23] reported on glucose biosensors based on aligned CNT coated with polypyrrole (ppy) conducting polymer. GOx was immobilized by electropolymerization of pyrrole in solution containing GOx, thus resulting in the formation of an enzyme–ppy coaxial sheath around the individually aligned CNT. Fe particles generated on the CNT during the production process led to an attractive low-potential detection of the GOx-liberated hydrogen peroxide. Such biosensor (ppy/GOx/CNT) showed the electrical contact between the enzyme and electrode through the CNTs and redox conducting polymer, and exhibited at least one order of magnitude greater signal than that obtained at the corresponding biosensor based on gold electrode (ppy/GOx/gold).

The earlier works discussed above took advantages of the bulk properties of CNTs. However, another important feature of CNT is its ultrasmall size, which is useful for the production of nanolectrode. Lin et al. [24] reported on glucose biosensors based on CNT nanoelectrode ensembles (NEEs). To make each CNT work as an individual nanoelectrode, the spacing between the CNTs should be sufficiently larger than the diameter of CNT to prevent the overlap of diffusion layer with neighboring electrodes. The NEEs consisting of millions of nanoelctrodes (each electrode being less than 100 nm in diameter) embedded in a polymer epoxy on a conductive substrate takes advantages of CNT materials over conventional macroelectrodes such as increased mass transport and the decreased influence of the solution resistance, which provides an excellent electrode in biosensing application. The GOx was immobilized on the CNT NEE using carbodiimide chemistry by forming amide bonds between amine residues of GOx and carboxylic acid groups on the CNT tips. The low-potential (−0.2V vs. saturated calomel electrode (SCE)) reductive detection of the hydrogen peroxide resulted in highly selective and sensitive amperometric measurement of the glucose, along with linearity up to 30 mM and a detection limit of 0.08 mM.

Besides glucose, several different types of electrochemical biosensors have been reported based on CNT-modified electrode coupled with oxidase enzymes. Qiaocui et al. [25]

reported on cholesterol biosensor based on the cholesterol oxidase–sol–gel film on a platinized CNT electrode. Owing to the high electrocatalytic activity of the platinized CNT electrode toward the reduction of hydrogen peroxide, the biosensor can detect as low as 1.6 μM cholesterol.

Ruianes and Rivas [26] reported on lactate and phenol biosensors based on carbon nanotube paste electrodes containing lactate oxidase or polyphenol oxidase. The promoted electron-transfer reaction of hydrogen peroxide at the CNT-based paste electrode offered a rapid low-potential (–0.10 V) detection of the substrate. CNT-based lactate biosensor can detect as low as 0.3 mM lactate; it covers not only physiological values of lactate but also pathological values. In addition, CNT-based phenol biosensor can sensitively detect phenol with a detection limit of 500 nM; this reflects that the CNT-modified electrode enhanced electrocatalytic detection of the enzymatically generated catechol-quinone.

13.3.1.3 Dehydrogenase-Based Biosensors

Dehydrogenase-based biosensor relies on the coimmobilization of dehydrogenase enzyme and their nicotinamide adenine dinucleotide (NAD^+) cofactor on an electrode. The enzymatically liberated NADH in the presence of substrates is electrochemically detected. Problems associated with the anodic detection of NADH are the large overpotential required for its oxidation at ordinary carbon electrode and surface fouling associated with the accumulation of the reaction products. Therefore, CNT-based electrode can take advantages of excellent electrocatalytic activity and minimized surface-fouling effect associated with CNTs. As discussed previously, the CNT-modified electrodes exhibit substantial decreases in the anodic peak potential as well as increases in current signals. Wang and Musameh [27] reported on a CNT-based ethanol amperometric biosensor based on the coimmobilization of ADH and its NAD^+ cofactor within the CNT–Teflon composite matrix. The marked decrease in the overvoltage for the oxidation of the liberated NADH facilitates the development of low-potential ethanol biosensor with good selectivity and sensitivity. Similar advantages are expected pertaining to the biosensing of lactate or glucose in connection with lactate dehydrogenase or glucose dehydrogenase, respectively.

Gorski group [28] reported on the MWCNT–chitosan composite for glucose biosensor based on dehydrogenase enzyme. MWCNTs were solubilized in aqueous solution of biopolymer chitosan and the resulting solution was cast onto GC electrode. The MWCNT–chitosan composite films facilitated the electrooxidation of NADH, and thus required 0.3 V less overpotential than the GC for the NADH oxidation. The glucose dehydrogenase was covalently immobilized in the MWCNT–chitosan films using glutaric dialdehyde. The resulting biosensor responded to glucose over the range 5–300 μM with a detection limit of 3 μM. Gorski group [29] further developed the MWCNT–chitosan system by covalently coimmobilizing the redox mediator of toluidine blue (TBO). Such integration of CNT and redox mediator in the dehydrogenase-based biosensor provided a remarkable synergistic augmentation of the current because of the oxidation of redox-active species.

13.3.1.4 Other Enzyme-Based Biosensors

Lin et al. [30] reported on a disposable CNT-modified screen-printed electrode for amperometric detection of organophosphorus (OP) pesticides and nerve agents. The biosensor involved the use of acetylcholinesterase (AChE) and choline oxidase (CHO) enzymes covalently attached to MWCNT-modified screen-printed electrode. The AChE works as a catalyst for the rapid hydrolysis of the acetylcholine into acetate and choline. The choline is subsequently converted by CHO into hydrogen peroxide in the presence of oxygen. The detection of OP relied on the inhibition of AChE by OP compounds. Hydrogen peroxide is then amperometrically detected with a CNT-modified electrode. Owing to the remarkable catalytic activity of CNT to promote the oxidation of hydrogen peroxide produced during the AChE–CHO enzymatic reaction with their substrate, the biosensor exhibited a high sensitivity and low detection limit of 50 nM for OP pesticide, methyl parathion. This result demonstrated the possibility of the field screening of OP pesticides and nerve agents. Joshi et al. [31] demonstrated a very simple method for the fabrication of OP biosensor. The degree of inhibition of AChE by OP pesticides was determined by measuring the oxidation current of thiocholine generated by the AChE-catalyzed hydrolysis of acetylthiocholine (ATCh). The CNT-modified screen-printed electrode exhibited remarkable electrocatalytic activity toward thiocholine and lowered oxidation overpotential to 200 mV. The biosensor can detect OP nerve agent paraoxon with a detection limit of 0.5 nM (0.145 ppb).

13.3.2 Direct Electrochemistry of Redox Proteins and Enzymes

In general, direct electron transfer of redox proteins or enzymes at unmodified electrode surface is not possible because redox-active sites of proteins or enzymes are shielded by glycoprotein shells of proteins. However, direct electron transfers are possible for some proteins and enzymes that are properly aligned to the electrode surface as shown in Figure 13.3.

Gooding et al. [32] reported on the direct protein electrochemistry using aligned SWCNT arrays. An array of SWCNTs perpendicularly aligned on a gold electrode was fabricated by covalently coupling the carboxylated SWCNT to a cystamine monolayer-modified gold

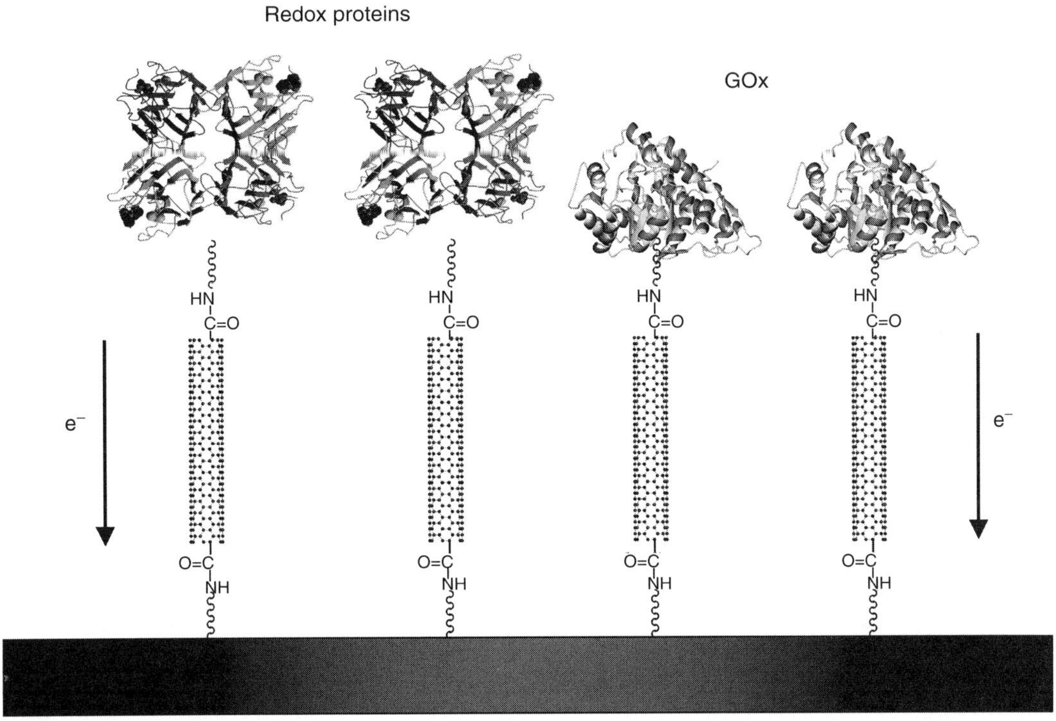

FIGURE 13.3
(See color insert following page 330) Direct electrochemistry of redox proteins (left) and enzymes (right).

electrode surface. Microperoxidase MP-11 with amino group was covalently linked to the carboxyl groups of free ends of standing SWCNTs. The MP-11 linked to the SWCNT showed a characteristic cyclic voltammogram at −390 mV vs. Ag/AgCl, indicating the MP-11 being electrically contacted with the electrode surface. The rate constant of heterogeneous electron transfer between MP-11 and the SWCNT-modified electrode was similar to that for MP-11 directly attached to cystamine-modified gold electrode. These similar rate constants indicate the efficiency of the SWCNT as molecular wires. This result allows the understanding of the redox properties of proteins and also the development of enzyme biosensors without mediators.

The redox-active site of GOx is also electrically insulated by a protein shell, and thus cannot be oxidized or reduced at an electrode. However, similar to other redox proteins, the direct electron transfer of GOx can be performed at CNT-modified electrode. For example, Guiseppi-Elie et al. [33] reported on the direct electron transfer of GOx spontaneously adsorbed onto SWCNTs cast onto the GC surface. A quasi-reversible one-electron transfer was observed possibly due to the tubular fibrils positioned within a tunneling distance of the cofactors.

Direct electron transfer of GOx and electrode by means of aligned SWCNT was reported by Willner's group [34]. An array of SWCNT perpendicularly mounted on a gold electrode was fabricated by covalently coupling the carboxylated SWCNT to a cystamine monolayer-functionalized gold-electrode surface. Flavin adenine dinucleotide (FAD) cofactor with amino groups was covalently linked to the carboxyl groups of free ends of standing SWCNTs. The FAD units linked to the SWCNT showed a quasi-reversible cyclic voltammogram, indicating the FAD unit being electrically contacted with the electrode surface. Apo-GOx was then reconstituted on the FAD units coupled to the free ends of the standing SWCNT. The biocatalytic oxidation of glucose was observed at 0.18 V vs. SCE, and the oxidation current increased as the glucose concentration increased. The electron-transfer rate showed clear dependence on the length of the SWCNTs. The GOx-CNT with an average length of 25 nm showed an about 1.5-fold enhanced electrocatalytic current compared to the GOx-CNT with an average length of 50 nm. Such enzyme reconstitution at the end of CNT represents an extremely efficient approach for the preparation of mediator-free biosensor.

Luong et al. [35] reported on the promoted electron transfer of GOx at MWCNT-modified GC electrode. 3-Aminopropylethoxysilane (APTES) was used for solubilization of CNTs as well as immobilization of GOx to fabricate a mediator-free glucose biosensor. The resulting surface acts like bundled ultramicroelectrodes that allow the access to the active FAD site and facilitate a direct electron transfer to the immobilized GOx. Cyclic voltammograms for GOx immobilized on the APTES–MWCNT-modified GC electrode showed nearly symmetrical peaks at −0.45 V vs. Ag/AgCl, which is close to the redox potential of FAD.

The determination of hydrogen peroxide is of great importance in food, pharmaceutical, and environmental analysis. The direct electrochemical detection of hydrogen peroxide requires relatively high overpotentials. Therefore, hydrogen peroxide is often detected enzymatically at low applied potential by the use of peroxidase. Yu et al. [36] reported on the direct electrochemistry of horseradish peroxidase (HRP) covalently attached onto the ends of vertically oriented SWCNT forest arrays. Quasi-reversible Fe(III)/Fe(I) voltammetry was observed for the heme group around −0.25 V (vs. SCE). This result indicates that the individual SWCNT behaves electrically similar to metal, conducting electrons from the external circuit to the redox sites of the HRP enzymes. The resulting HRP biosensor exhibited a detection limit of 100 nM.

Yamamoto et al. [37] reported on a CNT–HRP modified electrode for online monitoring of hydrogen peroxide. MWCNTs were immobilized on the surface of GC electrode by mixing with HRP. The CNT–HRP-modified electrode exhibited good sensitivity by monitoring the reduction current at −0.30 V (vs. Ag/AgCl) with a detection limit of 100 nM. This biosensor format was also applied to the determination of lactate and glucose by using CNT–HRP–lactate oxidase or CNT–HRP–glucose oxidase-modified GC electrode.

Xu et al. [38] developed a hydrogen peroxide biosensor based on attractive characteristics of the CNT. The MWCNT was used for the coimmobilization of HRP and the methylene blue (MB) mediator onto GC electrode surface. The biosensor can detect the hydrogen peroxide as low as 1 μM.

Zhang et al. [39] reported on the electrochemical behavior of monolayer film of myoglobin (Myb) strongly adsorbed on MWCNT-modified GC electrode. MWCNT in the biosensor promoted the direct electron transfer between Myb and the electrode surface. The Myb on MWCNT behaves as an enzyme-like activity toward the electrochemical reduction of nitric oxide. The unmediated MWCNT–Myb biosensor can detect as low as 80 nM nitric oxide.

Wang et al. [40] reported on the direct electrochemistry of catalase (Ct) at SWCNT-modified gold electrodes. A solution of SWCNT dispersed in DMF was cast onto the gold electrode. A well-defined redox wave corresponding to Fe(III)/Fe(II) redox center of the heme group of Ct adsorbate was observed with a reduction-peak potential at −0.41 V. The peak current linearly increased with Ct-enzyme concentration with a detection limit of 4 μM. The catalase–SWCNT modified electrode displayed a characteristic catalytic wave upon addition of hydrogen peroxide.

13.4 DNA Hybridization Biosensors Based on CNTs

The development of rapid and simple DNA hybridization biosensors is very important in clinical diagnostics for detecting genetic and infectious diseases. The DNA biosensor is composed of a nucleic acid recognition layer, which is immobilized onto a signal transducer. The role of the nucleic acid layer is to specifically hybridize with a complementary sequence of DNA. Electrochemical DNA biosensor produces meaningful electrical signal upon hybridization of a target DNA onto the DNA biosensor immobilized with a complementary single-stranded (ss-) probe DNA. Recent advances in the electrochemical DNA biosensors have been reviewed by Gooding [41]. The analytical performances of DNA biosensors can be greatly enhanced by the use of CNT. Such improvements can be attributed to the enhanced electrocatalytic activities of CNT toward the target DNA guanine base or the products of enzyme label. In addition, CNT can be used as a carrier platform for the immobilization of DNA and multiple labels because CNTs have high conductivity and extremely high surface-area-to-weight ratio.

13.4.1 DNA Hybridization Detection Based on Electroactive Label

CNT-enhanced electrochemical DNA hybridization biosensor has been reported by Cai et al. [42] based on the use of redox intercalator daunomycin. A carboxylated MWCNT-coated on a GC electrode was covalently linked to an amino-functionalized probe DNA. Complementary target DNA was then hybridized into the probe DNA-functionalized MWCNT, and the redox label daunomycin was intercalated into the double-stranded DNA formed on MWCNT as shown in Figure 13.4 (left).

The redox label daunomycin was detected by differential pulse voltammetry. This DNA biosensor allowed the detection of DNA up to 100 pM.

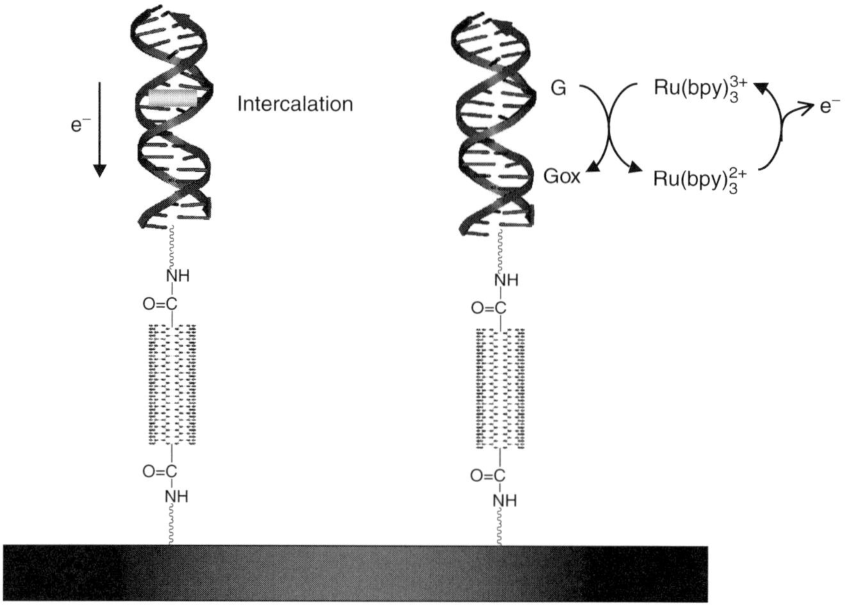

FIGURE 13.4
Carbon nanotube electrode array for the detection of DNA hybridization. Left: use of electroactive interaction compound. Right: use of Ru(bpy)$_3^{2+}$ a mediator for the oxidation of guanine base.

Similarly, SWCNT-based DNA biosensor has been reported in connection with the use of electroactive ferrocene label [43]. In the biosensor, single-stranded probe DNA was covalently linked to the carboxylated SWCNT aligned on a gold electrode. A ferrocene-labeled complementary target DNA was then hybridized with the probe DNA on an electrode to produce a reversible cyclic voltammogram of ferrocene.

13.4.2 DNA Hybridization Detection Based on Guanine Oxidation

Wang group reported on an electrochemical DNA hybridization biosensor based on the guanine oxidation. High surface area of CNT-modified electrode has been shown to permit the adsorptive accumulation of the guanine and greatly enhance the sensitivity of the electrochemical detection of DNA, which permits a label-free DNA analysis [44]. A similar enhancement of the guanine oxidation in DNA was reported by Pedano and Rivas [45]. The CNT paste electrode was suitable for adsorptive stripping potentiometric measurement of trace levels (ng/L) of nucleic acids.

Similarly, an array of vertically aligned MWCNT with Ru(bpy)$_3^{2+}$-mediated guanine oxidation has been shown useful for ultrasensitive detection of DNA hybridization [46] as shown in Figure 13.4 (right). The hybridization of subattomole DNA targets was detected by this method, showing several orders of magnitude improvement in sensitivity compared to the conventional methods where the DNA was immobilized using self-assembled monolayers.

13.4.3 DNA Hybridization Detection Based on Enzyme Label

In general, enzymes generating electroactive products are widely used as amplification labels in the assay of DNA hybridization or specific antigen–antibody interaction. Wang group [47] reported on the use of functionalized CNTs as carriers for multiple enzyme labels for electrochemical DNA hybridization detection and immunosensing. In these procedures CNT played a dual role in both the recognition and transduction events, namely as carriers for numerous enzyme tags and for accumulating the electroactive α-naphthol product of the enzyme-hydrolysis reaction. CNTs were loaded with around 9,600 alkaline phosphatase enzyme molecules per CNT by covalently coupling the enzyme to the carboxylated CNTs. The enzyme-loaded CNTs were further modified with DNA or antibody for DNA sensing or immunosensing, respectively. For example, probe DNA-functionalized magnetic particles were hybridized with the complementary target DNA leading to the DNA duplex. Free ends of target DNA were then further hybridized with the complementary DNA, which was linked to the enzyme-functionalized CNT. The enzymatic hydrolysis of α-naphthylphosphate led to the production of electrochemically detectable α-naphthol product. Such CNT-derived amplification method allowed the detection of DNA and proteins down to 1.3 and 160 zmol, respectively, in 25–50 μL samples and indicates great promise for PCR-free DNA analysis.

13.4.4 DNA Hybridization Sensor Based on Impedance Measurement

Cai et al. [48] reported on a direct electrochemical detection of DNA hybridization by AC impedance measurement. An indicator or label-free detection method was based on the probe-DNA-doped ppy-conducting film formed on MWCNT-modified GC electrode. The hybridization event with a complementary target DNA led to a decrease in impedance values as a result of the reduction of the electrode resistance. The detection limit of this label-free method was 10 nM of oligonucleotide. The sensitivity was around fivefold higher than that obtained with analogous measurements without CNT. This method offers great promise for simple and sensitive reagentless DNA hybridization analysis.

13.5 Field-Effect Transistors Based on Biomolecule-Functionalized CNTs

13.5.1 Glucose Detection

Dekker and his coworkers [49] reported on a SWCNT FET (field-effect transistor) for glucose biosensor. GOx was attached onto the sidewalls of SWCNT through a bifunctional linking molecule containing on one side a pyrene group that strongly adsorbs onto the sidewalls of CNT due to $\pi-\pi$ interactions and on the other side an active ester group for covalent binding to the lysine residues of GOx. The GOx-functionalized CNT was placed at the gate between two electrodes of source and drain, and the conductance was measured as shown in Figure 13.5 (left).

The addition of glucose to the CNTFET led to increased conductivity. This result indicates that an enzyme-activity sensor can be constructed at the single-molecule level of an individual SWCNT. The mechanism of the conductance change of the CNT as a result of the biocatalytic process is not well understood yet. Further theoretical study is necessary to explain such phenomenon.

13.5.2 Assay of Protein Bindings

Surface-protein and protein–protein binding have been studied based on the changes in the conductive properties of the single-wall CNT by Chen et al [50]. Specific antigen–antibody (associated with human autoimmune disease) interactions have been analyzed using

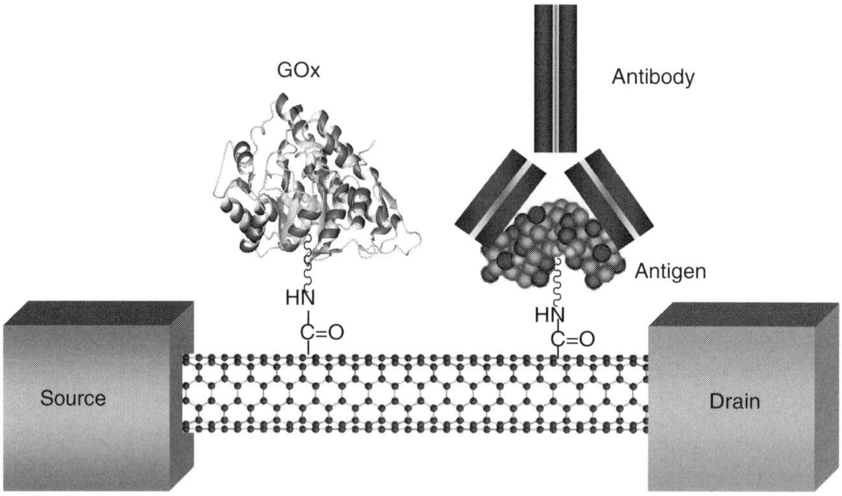

FIGURE 13.5
Field-effect transistors for the detection of glucose (left) or antigen–antibody interaction (right).

the CNTFET. The CNTFET relied on the U1A antigen (a 33-KDa protein) covalently attached to the detergent molecule, which was irreversibly adsorbed on the CNT walls. The antigen-specific 10E3 mABs antibody was then bound to the antigen-functionalized CNTs. The binding process was monitored by CNTFET electrical conductance measurements as shown in Figure 13.5 (right). The binding of antibody to antigen-functionalized CNTs resulted in a decrease in the conductance of CNTFET. The conductance measurements of CNTFET device can analyze a specific response up to 1 nM of the 10E3 antibody, whereas a much higher concentration of polyclonal IgG (1 μM) was not detected. The CNTFET device exhibited the electrical detection of the antibody–antigen interactions at a 100-fold higher sensitivity than the quartz–crystal microbalance measurements. The decrease in the conductivity of the CNT resulting from the antigen–antibody binding in CNTs is at present not fully understood theoretically.

Similarly, Boussaad et al. [51] reported on the in situ detection of a small number of protein adsorption onto individual SWCNT transistors based on the changes in the electron transport properties of the CNTFET. The adsorption of cytochrome *c* onto CNTFET led to a decrease in the conductance of SWCNT. The CNTFET device can detect as low as 20 of cytochrome *c* molecules.

13.6 Conclusions and Future Perspectives

This chapter summarizes recent advances in the rapidly developing CNT-based electrochemical biosensors. Owing to the attractive properties of CNT, recent results have demonstrated the advantages of CNT for the construction of a wide range of electrochemical enzyme-based biosensors, immunosensors, and DNA sensors. The remarkable electrocatalytic activity toward hydrogen peroxide and NADH permits the fabrication of low-potential amperometric biosensors. The electrical contacting of redox enzymes and electrode through CNT electrical wires offers great promise for the development of mediator-free biosensor. The use of CNTs offers enhanced electrocatalytic activity toward target-DNA guanine base as well as products of enzyme labels, thus providing a huge benefit for the development of PCR-free electrochemical DNA sensors.

The future perspective of CNT-based electrochemical biosensors is bright. To further envisage a variety of applications of CNT-based biosensors, better methods to control the chemical and physical properties of CNT and theoretically understand their phenomena are desired. Based on the recent advances in this field, real applications of CNT-based electrochemical biosensor can be expected in the near future.

Acknowledgment

The financial support for this work was provided by KOSEF through the Center for Bioacitve Molecular Hybrids at Yonsei University.

References

[1] Iijima, S. (1991) Helical microtubules of graphite carbon. *Nature* 354: 56–58.
[2] Ajayan, P. M. (1999) Nanotubes from carbon. *Chem. Rev.* 99: 1787–1799.
[3] Zhao, Q., Gan, Z., Zhuang, Q. (2002) Electrochemical sensors based on carbon nanotubes. *Electroanalysis* 14: 1609–1613.
[4] Klein, D. J., Seitz, W. A., Schmaltz, T. G., (1993) Symmetry of infinite tubular polymers: application to buckytubes. *J. Phys. Chem.* 97: 1231–1236.
[5] Katz, E., Willner, I. (2004) Biomolecule-functionalized carbon nanotubes: applications in nanobioelectronics. *Chem. Phys. Chem.* 5: 1084–1104.
[6] Wang. J. (2005) Carbon-nanotube based electrochemical biosensors: a review. *Electroanalysis* 17: 7–14.
[7] Musameh, M., Wang, J., Arben, M., Lin, Y. (2002) Low-potential stable NADH detection at carbon-nanotube-modified glassy carbon electrodes. *Electrochem. Commun.* 4: 743–746.
[8] Wang, J., Musameh, M., Lin, Y. (2003) Solubilization of carbon nanotubes by Nafion toward the preparation of amperometric biosensors. *J. Am. Chem. Soc.* 125: 2408–2409.
[9] Britto, P. J., Santhanam, K. S. V., Ajayan, P. M. (1996) Caron nanotube electrode for oxidation of dopamine. *Bioelectrochem. Bioenergy* 41: 121–125.
[10] Davis, J. J., Coles, R. J., Hill, H. A. O. (1997) Protein electrochemistry at carbon nanotube electrodes. *J. Electroanal. Chem.* 440: 279–282.
[11] Luo, H., Shi, Z., Li, N., Gu, Z., Zhuang, Q. (2001) Investigation of the electrochemical and electrocatalytic behavior of single-wall carbon nanotube film on a glassy carbon electrode. *Anal. Chem.* 73: 915–920.
[12] Choi, H. N. (2005). Electrochemical Sensors and Biosenors Based on Sol-Gel Metal Oxides-Nafion and Carbon Nanotube Composite Films. Ph.D. dissertation, Yonsei University, Seoul, Korea.
[13] Choi, H. N., Cho, S.-H., Lee, W.-Y. (2003) Electrogenerated chemiluminescence from tris(2,2′-bipyridyl)ruthenium(II) immobilized in titania-perfluorosulfonated ionomer composite films. *Anal. Chem.* 75: 4250–4256.
[14] Choi, H. N., Kim, M. A., Lee, W.-Y. (2005) Amperometric glucose biosensor based on sol-gel-derived metal oxide/Nafion composite films. *Anal. Chim. Acta* 537: 179–187.
[15] Wang, J., Musameh, M. (2003) Caron nanotube/Teflon composite electrochemical sensors and biosensors. *Anal. Chem.* 75: 2075–2079.
[16] Gong, K., Zhang, M., Yan, Y., Su, L., Mao, L., Xiong, S., Chen, Y. (2004) Sol-gel-derived ceramic-carbon nanotube nanocomposite electrodes: tunable electrode dimension and potential electrochemical applications. *Anal. Chem.* 76: 6500–6506.

[17] Rubianes, M. D., Rivas, G. A. (2003) Carbon nanotubes paste electrode. *Electrochem. Commun.* 5: 689–694.
[18] Valentini, F., Amine, A., Orlanducci, S., Terranova, M. L., Palleschi, G. (2003) Carbon nanotube purification: preparation and characterization of carbon nanotube paste electrodes. *Anal. Chem.* 75: 5413–5421.
[19] Wang, J., Musameh, M. (2004) Carbon nanotube screen-printed electrochemical sensors. *Analyst* 129: 1–2.
[20] Hrapovic, S., Liu, Y., Male, K. B., Luong, J. H. T. (2004) Electrochemical biosensing platforms using platinum nanoparticles and carbon nanotubes. *Anal. Chem.* 76: 1083–1088.
[21] Luo, X.-L., Xu, J.-J., Wang, J.-L., Chen, H.-Y. (2005) Electrically deposited nanocomposite of chitosan and carbon nanotubes for biosensor application. *Chem. Commun.* 2169–2171.
[22] Davis, J. J., Coleman, K. S., Azamian, B. R., Bagshaw, C. B., Green, M. L. H. (2003) Chemical and biochemical sensing with modified single walled carbon nanotubes. *Chem. Eur. J.* 9: 3732–3739.
[23] Gao, M., Dai, L., Wallace, G. G. (2003) Biosensors based on aligned carbon nanotubes coated with inherently conducting polymers. *Electroanalysis* 15: 1089–1094.
[24] Lin, Y., Lu, F., Tu, Y., Ren, Z. (2004) Glucose biosensors based on carbon natotube nanoelectrode ensembles. *Nano. Lett.* 2: 191–195.
[25] Qiaocui, S., Tuzhi, P., Yunu, Z., Yang, C. F. (2005) An electrochemical biosensor with cholesterol oxidase/sol-gel film on a nanoplatinum/carbon nanotube electrode. *Electroanalysis* 17: 857–861.
[26] Ruianes, M. D., Rivas, G. A. (2005) Enzymatic biosensors based on carbon nanotubes paste electrodes. *Electroanalysis* 17: 73–78.
[27] Wang, J., Musameh, M. (2003) A reagentless amperometric alcohol biosensor based on carbon-nanotube/Teflon composite electrodes. *Anal. Lett.* 36: 2041–2048.
[28] Zhang, M., Smith, A., Gorski, W. (2004) Carbon nanotube-chitosan system for electrochemical sensing based on dehydrogenase enzymes. *Anal. Chem.* 76: 5045–5050.
[29] Zhang, M., Gorski, W. (2005) Electrochemical sensing platform based on the carbon nanotubes/redox mediators-biopolymer system. *J. Am. Chem. Soc.* 127: 2058–2059.
[30] Lin, Y., Lu, F., Wang, J. (2004) Disposable carbon nanotube modified screen-printed biosensor for amperometric detection of organophosphorus pesticides and nerve agents. *Electroanalysis* 16: 145–149.
[31] Joshi, K. A., Tang, J., Haddon, R., Wang, J., Chen, W., Mulchandani, A. (2005) A disposable biosensor for organophosporus nerve agents based on carbon nanotubes modified thick film strip electrode. *Electroanalysis* 17: 54–58.
[32] Gooding, J. J., Wibowo, R., Liu, J. Q., Yang, W., Losic, W. D., Orbons, S., Mearns, F. J., Shapter, J. G., Hibbert, D. B. (2003) Protein electrochemistry using aligned carbon nanotube arrays. *J. Am. Chem. Soc.* 125: 9006–9007.
[33] Guiseppi-Elie, A., Lei, C., Baughman, R. (2002) Direct electron transfer of glucose oxidase on carbon nanotubes. *Nanotechnology* 13: 559–564.
[34] Patolsky, F., Weizmann, Y., Willner, I. (2004) Long-range electrical contacting of redox enzymes by SWCNT connectors. *Angew. Chem. Int. Ed.* 43: 2113–2117.
[35] Luong, J. H., Hrapovic, S., Wang, D., Bensebaa, F., Simard, B. (2004) Solubilization of multiwall carbon nanotubes by 3-aminopropyltriethoxysilane toward the fabrication of electrochemical biosensors with promoted electron transfer. *Electroanalysis* 16: 132–139.
[36] Yu, X., Chattopadhyay, D., Galeska, I., Papadimitrakopoulos, F., Rusling, J. F. (2003) Peroxidase activity of enzymes bound to the ends of single-wall carbon nanotube forest electrodes. *Electrochem. Commun.* 5: 408–411.
[37] Yamamoto, K., Shi, G., Zhou, T. S., Xu, F., Xu, J. M., Kato, T., Jin, J.Y. (2003) Study of carbon nanotube-HRP modified electrode and its application for novel on-line biosensors. *Analyst* 128: 249–254.
[38] Xu, J. –Z., Zhu, Z. Z., Wu, Q., Hu, Z., Chen, H. Y. (2003) An amperometric biosensor based on coimmobilization of horseradish peroxidase and methylene blue on a carbon nanotubes modified electrode. *Electroanalysis* 15: 219–224.
[39] Zhang, L., Zhao, G-C., Wei, X.-W., Yang, Z.-S. (2005) A nitric biosensor based on myoglobin adsorbed on multi-walled carbon nanotubes. *Electroanalysis* 17: 630–634.

[40] Wang, L., Wang, J., Zhou, F. (2004) Direct electrochemistry of catalase at a gold electrode modified with single-wall carbon nanotubes. *Electroanalysis* 16: 627–632.

[41] Gooding, J. J. (2002) Electrochemical DNA hybridization biosensors. *Electroanalysis* 14: 1149–1156.

[42] Cai, H., Xu, C., Jiang, Y., He, P., Fang, Y. (2003) Carbon-nanotube-enhanced electrochemical DNA biosensor for DNA hybridization detection. *Anal. Bioanal. Chem.* 375: 287–293.

[43] He, Pingang, Dai, L. (2004) Aligned carbon nanotube-DNA electrochemical sensors. *Chem. Commun.* 348–349.

[44] Wang, J., Kawde, A.-N., Musameh, M. (2003) Carbon-nanotube-modified glassy carbon electrode for amplified label-free electrochemical detection of DNA hybridization. *Analyst* 128: 912–916.

[45] Pedano, M., Rivas, G. A. (2004) Adsorption and electrooxidation of nucleic acids at carbon nanotubes paste electrodes. *Electrochem. Commun.* 6: 10–16.

[46] Li, J., Ng, H. T., Cassell, A., Fan, W., Chen, H., Ye, Q., Koehne, J., Meyyappan, M. (2003) Carbon nanotube nanoelectrode array for ultrasensitive DNA detection. *Nano. Lett.* 3: 597–602.

[47] Wang, J., Liu, G., Jan, M. (2004) Ultrasensitive electrical biosensing of proteins and DNA: carbon nanotube-derived amplification of the recognition and transduction events. *J. Am. Chem. Soc.* 126: 3010–3011.

[48] Cai, H., Xu, Y., He, P., Fang, Y. Z. (2003) Indicator free DNA hybridization detection by impedance measurement based on the DNA-doped conducting polymer film formed on the carbon nanotube modified electrode. *Electroanalysis* 15: 1864–1870.

[49] Besteman, K., Lee, J.-O., Wiertz, F. G., Heering, H. A., Dekker, C. (2003) Enzyme-coated carbon nanotube as single-molecule biosensors. *Nano. Lett.* 3: 727–730.

[50] Chen, R. J., Bangsaruntip, S., Drouvalakis, K. A., Kam, N., Shim, M., Li, Y., Kim, W., Utz, P. J., Dai, H. (2003) Noncovalent functionalization of carbon nanotubes fro highly specific electronic biosensors. *Proc. Natl. Acad. Sci. U S A* 100: 4984–4989.

[51] Boussaad, S., Tao, N. J., Zhang, R., Hopson, T., Nagahara, L. A. (2003) *In situ* detection of cytochrome *c* adsorption with single walled carbon nanotube device. *Chem. Commun.* 1502–1503.

Part IV

Bioelectronics

In this section, the focus is on bioelectronic devices based on bacteriorhopdopsin, a protein from the archae *Halobacter salinarium*, which has been the most studied for its photoelectric, bioelectronic, and photo-chromatic applications. In Chapter 14, Dr. Jeffrey Stuart and colleagues provide an overview of biophotonics. The first part of the chapter focuses on bacteriorhodopsin and its properties. Then bR protein-based devices and sensor applications are presented. In Chapter 15, Dr. Felix Hong first describes the principles of the photoelectric effect. Since prototype photoelectric sensors were often configured as thin films or membranes, the fundamental mechanistic aspects of light-induced charge movements in pigment-containing membranes or thin films are analyzed and discussed. Two experimental prototypes are presented to illustrate how the principles shape the interpretation of experimental data. A lipid bilayer membrane system, which contains lipid soluble magnesium porphyrins, represents perhaps the simplest light-driven electron pump. Several different reconstituted bacteriorhodopsin model membranes illustrate the next level of complexity in signal analysis. These sensors can be configured either as a photosensors, which can be modulated by small ions in the vicinity of the membrane or thin film or as a ion sensors, which can be modulated and gated by light. The pros and cons of various approaches in future smart biosensor designs are discussed. Chapter 16 by Dr. Jussi Parkkinen and his colleagues, focuses on color sensitive or photochromatic biosensors. This authors review bR and its optical properties, color vision, application of bR in color imaging. In Chapter 17, Wei Wei Wang and colleagues describe a specific application of bacteriorhodopsin incorporated on flexible substrates.

14

Bacteriorhodopsin: From Biophotonic Material to Chemical Sensor

Jeffrey A. Stuart, Robert R. Birge, Pallab Bhattacharia, Brian J. Yordy, Jeffrey Girard, William Tetley, Duane L. Marcy, Jeremy F. Koscielecki, and Jason R. Hillebrecht

CONTENTS

14.1 Introduction

Sensor technology, in its most broad definition, is the ability to use technology to detect changes in our surroundings that might adversely affect health. In this context, the term health has a fairly broad definition, in that it encompasses everything from proper mechanical and electrical operation of technology to the well-being of the human physiological condition. The ability to sense changes that potentially affect proper operation of any organized system is the goal of all sensor designs; phenomena that need to be sensed include various types of radiation (from simple light to more damaging high-energy ionizing radiation), mechanical stress (or impending failure), chemical and biological toxins, pressure, temperature, position, and sound. The human body has sensors that are capable of detecting nearly all of these factors—eyes detect position, visible light, and motion; ears detect sound; the skin contains vast numbers of pressure and temperature sensors; the nose and tongue are complicated chemical sensors; and the immune system of the body has the ability to detect and react to large numbers of pathogenic microbes. Conventional technology can detect some of these factors better than others. However, the sophisticated mechanisms provided by biology often cannot be duplicated using those same, conventional means. Detection of complex pathogens often requires equally complex—intelligent—sensor technologies, especially those capable of highly specific chemical interactions: alternative, innovative, and intelligent solutions are necessary.

The current need for sensors that detect a wide variety of chemical and biological toxins cannot be met through conventional sensor technology for a variety of reasons. The basic approach for most chemical sensors is to turn to highly specific molecular interactions, where the target molecule for detection is sensed through chemical interaction with a very specific reactant. The limitation of such detection schemes is that any given sensor can detect only a narrow range of chemicals, and similar molecules of lesser or no toxicity might result in false positives. Chemical noses offer some improvement, although they are, in essence, a large number of sensors operating in parallel, each monitoring the presence of a specific class of compounds. And such sensors are incapable of detecting the vast majority of toxic compounds, industrial pollutants, and harmful microorganisms that are deleterious to human health.

Biological molecules—specifically enzymes, proteins, and DNA—are unique, in that they have benefited from eon upon eon of natural selection and evolution, which has resulted in highly optimized properties custom-tailored for specific biological functions. These molecules have evolved to function in a wide range of environmental conditions, often with efficiencies unmatched by nonbiological or synthetic methods. Intelligent selection by the researcher can bring their novel functionalities to device applications. Often these properties offer comparative advantages over more traditional approaches to device design and construction. In fact, proteins and other biomolecules pose a unique solution to the problem of size versus function, in that they represent a class of extremely sophisticated molecules that perform specific functions under what can be a fairly broad range of conditions. From the standpoint to designing improved sensors, it follows that a better approach to detecting both biological molecules, as well as molecules that interact with biological molecules (or organisms), is to incorporate molecules of biological origin into the sensors themselves. There are many examples of naturally occurring biological interactions that can be harnessed in detection schemes; complementary DNA interactions, enzyme–substrate complexes, and antibody–antigen complexes offer just a few examples. The advantage offered by these sorts of systems is one of specificity—the tighter the interaction between the target molecule (i.e., the molecule that is the target of the sensor) and the detector, the better the sensor's response will be. DNA is perhaps the best example of such interactions, which are both strong and highly specific. As such, a wide variety of sensors for biological organisms has been developed, which are capable of detecting the presence of specific DNA or RNA sequences characteristic of disease-causing bacteria or virus. Many of these detection schemes produce fluorescence once the target sequence binds to its complement from the target organism (i.e., fluorescence resonance energy transfer, or FRET). In general, proteins are widely sensitive to the presence of many types of molecules, through both specific and nonspecific binding. This is especially true for enzymes, which participate in a large variety of specific interactions with small molecules, predominantly defined through substrate catalysis and allosterism. The primary result of these two classes of interactions is conformational modification of the protein, resulting in a change in activity. Molecules that interact with proteins through nonspecific or indirect interactions typically disrupt noncovalent bonds, which in turn alter structure. Although all proteins exhibit these types of behavior, few can be used as sensors: the challenge lies in being able to detect the changes in protein structure upon exposure to foreign molecules. The challenge is the same for sensing schemes that employ antibody–antigen complexes.

For certain classes of proteins, the problem is much simpler. The proteins either produce (or can be made to produce) an easily measurable response or have some sort of quantifiable probe of structure and function that can be detected through direct interrogation. The internal probe usually takes the form of a chromophore or a redox moiety, where light or voltage, respectively, can be used to monitor the state of the protein. Examples of the former include the rhodopsin family, photoactive yellow protein (PYP), green fluorescent protein, and phytochrome. Proteins in the latter class include cytochromes, photosynthetic reaction centers (PRCs), the various iron–sulfur cluster proteins, and chlorophyll. Photochromic proteins are better suited than most for device architectures from a number of standpoints. The internal chromophore facilitates light to chemical energy transduction, thereby providing a convenient means of interrogation, and additionally provides a constant probe of protein structure and function. A common property among many proteins in this group is the presence of multiple states that are accessed upon exposure to light, and often they can be interrogated by both optical and electronic means, which provides a direct reading of functional activity. Furthermore, the responses from many of the proteins just listed can be elicited by exposure to either light or voltage, and will respond in a reproducible manner. The remainder of this chapter will focus on this class of proteins.

The field of biophotonics is broadly defined, ranging from tissue-imaging and microscopy techniques to standard spectroscopic techniques applied to biological systems. Within the context of this chapter, the latter interpretation of the definition will be favored, in that protein-based device and sensor applications will be described that utilize spectroscopy to both excite and interrogate populations of biological molecules.

The combination of sensitivity and selectivity to chemical antigens makes biological molecules an obvious choice for active elements in sensor applications. DNA-based schemes are best suited for detection of microscopic organisms (biological warfare agents, etc.) because such organisms can be identified specifically. However, proteins offer a better solution to detection schemes for chemical species (pollutants, toxins, and chemical warfare agents). In some cases living organisms have been used in detection schemes, although such sensors have the unique requirement of maintaining a living population of cells large enough to elicit a measurable response to a given toxin [1]. This "canary in the coal mine" approach has been implemented using living organisms ranging from single-celled prokaryotes to baby bluegill fish [2,3]. Ideally, however, a sensor should employ detection schemes that exhibit long-term stability, low maintenance, high sensitivity, and selectivity. Furthermore, the signal should be easily quantified and interpreted. Proteins come close to meeting all of these requirements; their ability to be subject to multiple modes of interaction with chemical species and interrogated by multiple means makes them strong candidates for sensor applications. Furthermore, proteins can be manipulated by synthetic or genetic means, making them almost infinitely customizable. New techniques in genetic engineering are opening new avenues into protein modification, especially with the goal of introducing enhanced or new functionalities, and increased stability. Random and semirandom mutagenesis, coupled with the appropriate selection processes (i.e., directed evolution), allows the researcher to "design" proteins at the genetic level for specific applications. And advances in sol-gel encapsulation techniques are extending the protein stability from days to years [4,5].

14.2 Bacteriorhodopsin as a Biophotonic Material

Historically, bacteriorhodopsin (BR) has received more attention than any other protein with respect to device applications. Early interest was for the use of the protein as the active element in reusable holographic materials [6–16]. Later came architectures for random access memories, nonvolatile computer memories, photodiodes, and applications as active elements in microelectronic circuits [9,17–34]. A great deal of effort has been put into the integration of BR onto semiconductor surfaces for the purposes of signal transduction; upon exposure to yellow–green light (570 ± 60 nm), the protein responds in a cyclic manner, thereby modulating its visible absorption spectrum over a wide spectral range (410–640 nm) and exhibiting a well-defined voltage response in the pico- to millisecond range. The utilization of BR in microelectronics and a new class of chemical sensors will be discussed below.

14.2.1 Basic Properties

Bacteriorhodopsin is a 26-kDa integral membrane protein that is isolated from *Halobacterium salinarum*, an archaea that thrives in salt marshes where the temperature ranges between 40 and 60°C, and the average concentration of salt is approximately 4 M, roughly six times that of sea water (Figure 14.1). BR has 248 amino acids arranged into seven interconnected α-helices that span the membrane. The α-helices are arranged in a seven-helix bundle, a structural motif characterized by many G-protein-coupled receptors. An all-*trans* retinal chromophore is bound covalently to Lys-216 via a protonated Schiff base bond; this prosthetic group gives BR its characteristic purple color, and is responsible for actinic light absorption by the protein. Bacteriorhodopsin is the sole protein component of the purple membrane, a term that refers to discrete membrane patches in *H. salinarum* consisting of approximately 75% BR and 25% lipid. These patches are rugged (as compared with the nonprotein-bearing membrane component) and can be isolated by osmotically lysing the cell followed by differential centrifugation. Within the purple membrane, BR monomers are organized in a hexagonal arrangement of trimers. The retinal chromophores mirror the trimer arrangement, oriented with an angular separation between monomers of about 60°. This arrangement ensures that the purple membrane captures light of all polarizations with equal efficiency. Owing to the enhanced stability imparted to the protein by this arrangement, BR is most often utilized as the purple membrane—solubilizing the protein reduces its stability from years to days.

14.2.2 The Bacteriorhodopsin Photocycle

Absorption of light by the retinal chromophore initiates a series of events in BR that ultimately results in the translocation of a proton across the cell membrane, thereby acidifying the external medium—the protein acts as a light-driven proton pump that drives the production of ATP via a standard F_0F_1 ATPase. Proton translocation occurs as the protein goes through a cyclic response to light, typically referred to as a photocycle (Figure 14.1). Upon absorption of light, the all-*trans* retinal chromophore isomerizes to 13-*cis*, along with a concomitant shift in electron density toward the protonated Schiff base. This change in the electrostatic nature of the binding site is the driving force for a series of thermally produced intermediates, each defined by several characteristics, including absorption maximum, protonation state of key amino acids, chromophore isomerization state, lifetime, and thermal stability. Prior to light absorption, the bacteriorhodopsin resting state is referred to as bR

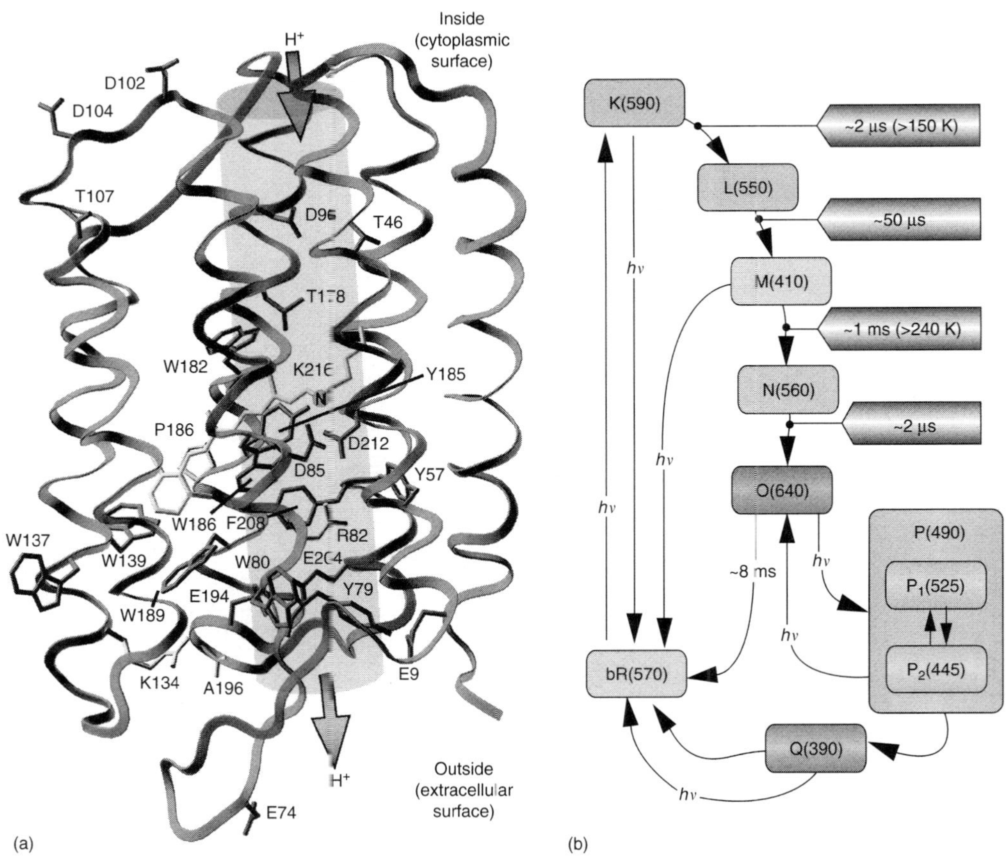

FIGURE 14.1
(See color insert following page 330) (a) Schematic of bacteriorhodopsin illustrating certain key amino acids and the purported path of the proton pump. The all-*trans* retinal chromophore traverses the binding pocket roughly perpendicular to the membrane normal and the alpha helices. (b) The bacteriorhodopsin photocycle, including the branched photocycle originating at the O state. Absorption maxima in nanometers are shown in parenthesis for each intermediate.

(as opposed to BR, which is most commonly used to represent the protein, itself). Upon light absorption, the protein is driven into the K state, which is the only photochemically generated state in the primary photocycle, and is the first intermediate after chromophore isomerization that can be trapped at low temperature (the method first used to characterize BR intermediates). The bR→K transition is often referred to as the primary event, and is characterized by a high quantum efficiency of ~65%. All subsequent intermediates are thermal, consisting of the L, M, N, and O intermediates, each being of lower free energy than the last. The protein's remarkable stability enables an extraordinarily high cyclicity, that is, the average number of times the protein can photocycle, on the order of 10^6. Throughout the photocycle, a proton is transported stepwise through a series of acidic amino acid side chains, and ultimately delivered to the extracellular medium. The mechanism by which this is accomplished is beyond the scope of this text, but many review articles are available to the interested reader [35–41].

14.2.3 Bacteriorhodopsin Photochromism

Bacteriorhodopsin is a unique protein, and has warranted a large amount of attention from the perspective of both the basic and applied sciences. It is of extreme interest for biochemical and biophysical studies because it acts as a light-driven engine that converts light into chemical energy in an extremely efficient manner. The resulting chemical energy is used to drive the formation of ATP from ADP and inorganic phosphate, and as such served as an early example of chemiosmosis [42–44]. Eventually, elaborate mechanisms were proposed to explain how protons were translocated through the protein and across the membrane. Much of the early work in genetic engineering was done on BR as a means by which the roles of individual amino acids in the proton pump could be determined. Ultimately, BR became one of the first membrane proteins to be characterized by x-ray crystallography; structures were even determined for individual photocycle intermediates [45–52].

Interest in BR by the applied sciences followed soon after the protein was initially described, and was based on the ability to manipulate several of the intermediates with light. It was discovered early on that the bR and M states could be driven back and forth using light of appropriate wavelength (570 and 410 nm, respectively). The wide spectral separation between the two states makes them an ideal pair for optical or holographic memory operations. Furthermore, chemical additives could be used to prolong the M-state lifetime, resulting in a transient bistable state (i.e., the M-state lifetime could be elongated, but not made permanent, and would decay back to the bR state); the resulting photochromism could be used as the basis for transient holograms or short-term optical computer memories [17,20,21,53,54]. The first of these efforts was made in the Soviet Union, and was denoted Project Rhodopsin; one of the goals of this military research program was to develop BR-based films with the protein suspended in a polymer matrix, which could be used in optical memory applications. Various chemical additives, mostly amines, were used to extend the M-state lifetime. The resulting product, Biochrome, was a real-time photonic and holographic material that could be written, read, and erased by application of the appropriate wavelength [11]. However, as indicated above, it was based on the bR–M binary pair and was incapable of permanent memory storage at room temperature. However, several device applications were proposed based on M-state photochromism, including low-temperature optical memories, spatial light modulators, and a two-photon three-dimensional optical memory [9,17–19,24,55,56]. A commercially viable device based on the bR–M pair has also been built and marketed by the company MIB Biochemicals in Marburg, Germany. Developed by Norbert Hampp and colleagues, Fringemaker-plus™ is a real-time holographic interferometer based on the BR mutant

D96N. Its primary application is nondestructive testing of manufactured components through microscopic deformation and interferometric analysis [57].

Despite many efforts to extend the M-state lifetime long enough to facilitate extended holographic or photonic memory applications, neither genetic nor chemical modification of the protein has managed to extend the M state to more than a few minutes. The "Holy Grail" in applied BR research has long been a truly bistable system—a two-state optical system that can be used for long-term memory applications. In 1993, Andreas Popp and colleagues [58] described a branch off of the main BR photocycle O state that produced a highly stable blue-shifted intermediate characterized by a 9-*cis* chromophore, the Q state. The so-called branched photocycle is accessed by exposure of the O state to red light—as such, it is a sequential two-photon process. First the BR photocycle is initiated with green light, followed by an exposure to red light once the O state has formed. As a result, a small percentage of the O state is driven into the branched photocycle, forming the P intermediate, a thermal precursor to the Q state. Although Q is stable for better than 10 years [58,59], the all-*trans* to 9-*cis* isomerization occurs with a very low quantum efficiency ($\sim10^{-4}$–10^{-3}) due to steric interactions within the binding site [59]. These steric interactions are introduced upon formation of the 9-*cis* chromophore and are strong enough to drive the hydrolysis of the Schiff base bond. The result is a trapped, but unbound chromophore in the binding site. Exposure of the protein to blue light reisomerizes the chromophore to all-*trans*, thereby regenerating the BR resting state. It is questionable whether the branched photocycle occurs to any extent in nature, or whether it offers the organism any competitive advantage. However, it has been suggested that the branched photocycle confers some protection to the organism during periods of high UV light flux [59].

The discovery of a permanent and reversible photochromic BR intermediate has resulted in three-dimensional architectures for both binary photonic and holographic memories based on the branched photocycle [22,60,61]. Unlike the M state, the branched photocycle is a three-state system in which the permanent states (bR and Q) are separated both energetically and temporally. This unique feature allows the branched photocycle to be utilized in nondestructive volumetric architectures [22,28–30]. The primary problem with architectures based on the branched photocycle is the low quantum efficiency with which the permanent state (Q) is accessed; this typically demands the use of either high-intensity lasers or prolonged exposure times, and both these requirements effectively preclude commercial viability for the wild-type protein. However, new molecular biological techniques are producing genetically-engineered variants custom-tailored to device applications. Some of these techniques will be described below.

14.2.4 The Bacteriorhodopsin Photoelectric Effect

The final characteristic worth mentioning with respect to BR-based device applications is the protein's photoelectric effect. Already alluded to above, the photoelectric effect is BR's concerted voltage response to light, and can be separated into three components (see Figure 14.2). The first is a picosecond-scale shift of electron density toward the Schiff base bond, concomitant with chromophore isomerization (B1). This is followed by a microsecond-scale response of opposite polarity (B2), which has recently been proposed to result from the movement of a specific amino acid (R82) toward the chromophore [62]. Finally, the B3 component originates from the action of the proton pump, and has a millisecond time-scale. The photoelectric effect has been employed as the basis for several microelectronic device applications, including wavelength-specific light-activated Fourier-effect transistors (GeAs BR-based MODFET), artificial retina prototypes, and polarization-sensitive microelectronic devices [63,64]. Introduction of polarization sensitivity is done by selectively

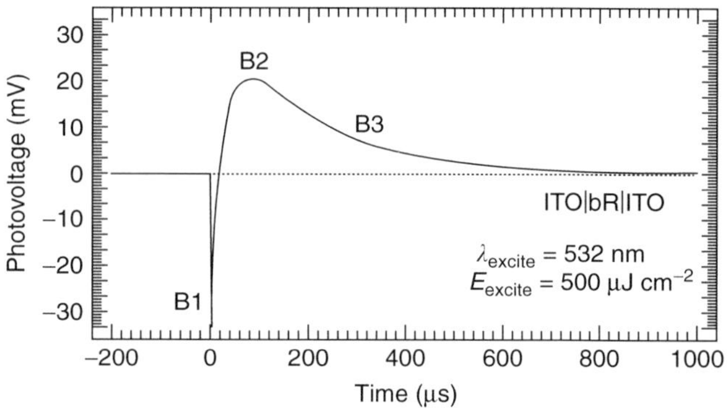

FIGURE 14.2
The bacteriorhodopsin (BR) photoelectric effect is a voltage that is produced upon absorption of light by the protein. It consists of three phases, B1, B2, and B3, which are coupled to events in the BR photocycle. See text for details.

removing the retinal chromophore from one of the three BR monomers in the trimer (photochemical bleaching using polarized light—see Li et al., 2002 [63]). To utilize the photoelectric effect for device applications, the protein must be uniformly oriented to maximize signal magnitude. Signal magnitude scales linearly with the number of oriented layers, at least for $n < 200$ (in-house data).

14.2.5 Bacteriorhodopsin Modification Through Genetic Engineering

As indicated above, BR's unique qualifications have facilitated a number of architectures for a variety of device applications. Yet after decades of effort, only one commercially viable device has been developed. Although the wild-type protein has a remarkably high quantum efficiency for the primary event, it was not designed by nature for either holography or memory storage. As a result, device applications based on the wild-type protein are inherently less efficient. For holographic applications, there is an intrinsic trade-off when considering the two photochromic approaches; the M state is produced with high efficiency, but its lifetime can only be prolonged to a few minutes. The Q state is permanent, but can only be produced with very low efficiency. The researcher needs tools by which the protein can be modified toward specific goals for specific applications. To date, two approaches have been employed: chemical modification and genetic engineering. The former technique can be done by either direct or indirect means, including synthetic modification of the retinal chromophore (e.g., 4-keto BR [65,66]) or affecting the protein's environmental milieu (e.g., pH and ionic strength). Unfortunately, neither approach is capable of altering BR's properties enough to make a difference with respect to most applications. Furthermore, chemical modification of any variety often is done at the expense of cyclicity. The latter approach, through genetic engineering, has proven successful enough to revitalize interest in bioelectronic applications of BR.

The advent of new tools in genetic engineering has resulted in new approaches to optimizing and enhancing BR's properties. Site-directed mutagenesis is the classical approach, and has proven to be a powerful technique for elucidating BR's biophysical properties, especially the proton pump mechanism. This molecular biology technique offers control of the protein at the genetic level, by enabling individual replacement of specific amino acids. The resulting mutant protein can be analyzed for new properties—such mutants provide tremendous insight into structure function studies. Although site-directed mutagenesis provided tremendous insight into BR, it became clear that this approach alone would not ultimately be successful for reengineering the protein; despite considerable understanding of protein structure and function (especially in the case of BR), even the most powerful supercomputer still does not have the ability to predict de novo the effects of most individual amino acid substitutions, especially when they are far removed from the protein's active site. And properties of multiple mutations are impossible to predict due to possible synergistic interactions. Furthermore, properties of interest to applied studies are often nonnatural (e.g., the branched photocycle). Therefore, site-directed mutagenesis is not a viable route toward a specific goal in situations that lack sufficient mechanistic understanding. Fortunately, an alternative approach exists that bypasses these limitations; random mutagenesis and directed evolution provide techniques by which a targeted goal can be achieved without specific knowledge of the changes being made to the protein.

Advances in molecular biological techniques have provided tools that offer solutions to the challenge of optimizing and enhancing protein properties toward specific functional goals. Random and semirandom mutagenesis allow the researcher to sample areas of the mutational landscape that otherwise would be ignored. It is often assumed that amino acids far removed from the active site play little or no role in protein function, other than structural, and therefore are often ignored. However, given the complexity of protein structure and function, assumptions of this nature are not always accurate, especially given the combined interactions possible in multiple mutations. Whereas site-directed mutagenesis allows the researcher to make very specific changes to proteins by design, random and semirandom methods make unpredictable changes to the genome, resulting in a library of proteins with a varied range of properties. The power in these techniques is based on the premise that mutations are made to the protein without bias—in essence, through the addition of some sort of random mutagenic factor, mutations are introduced to the genetic code with unpredictable consequences, resulting in a library of genetically altered bacterial strains. The mutant proteins ultimately isolated from those strains are screened and evaluated with respect to the properties of interest (the targeted goals); the strain producing the most efficient mutant protein can then be selected to act as parent to the next generation of random genetic progeny. This process can be repeated as many times as necessary, thereby building upon successive improvements in an incremental fashion, until the desired goal is met. When applied in this manner, the process is referred to as directed evolution [67–69]. The strength of this approach is that the researcher is not biased by a perceived understanding of protein structure–function relationships, that is, mutations are allowed that undoubtedly would be overlooked in a more systematic approach. This procedure is conceptually illustrated in Figure 14.3, and more specifically in Figure 14.4, which includes results from a typical round of random mutagenesis targeting the BR genome. Randomly mutated strains of *H. salinarum* are brought up in 96-well plates, which can be screened for BR production using a μQuant UV/VIS plate scanner made by Bio-Tek Instruments, Inc. After this initial screen, the strains in wells expressing protein are grown in quantities large enough to isolate adequate protein for characterization (Figure 14.4, top right), which in the example shown is for efficient production of the O state as determined by a quality factor (Q). Albeit effective, this process is tedious and time-consuming, and cannot be efficiently run on large numbers of mutant bacterial strains. An effective screening protocol is key to the success of both random mutagenesis and directed evolution; the ideal screening protocol can identify potential candidate proteins in vivo, thereby eliminating the need for cell culture and protein isolation procedures. As illustrated in Figure 14.4, screening for both absorption maxima (and even O-state kinetics) can be done with little difficulty in vivo. To date, all of the studies have focused on O-state optimization—the alternative approach of targeting the O to P quantum efficiency is far more complicated and poses problems with respect to an efficient screening process. The process illustrated in Figure 14.4b utilizes flow cytometry for cell selection and discrimination, based on photochemical analysis. This three-step cell-sorting system dilutes a liquid culture of genetically distinct *H. salinarum* strains and employs sequential optical illumination to screen individual cells in

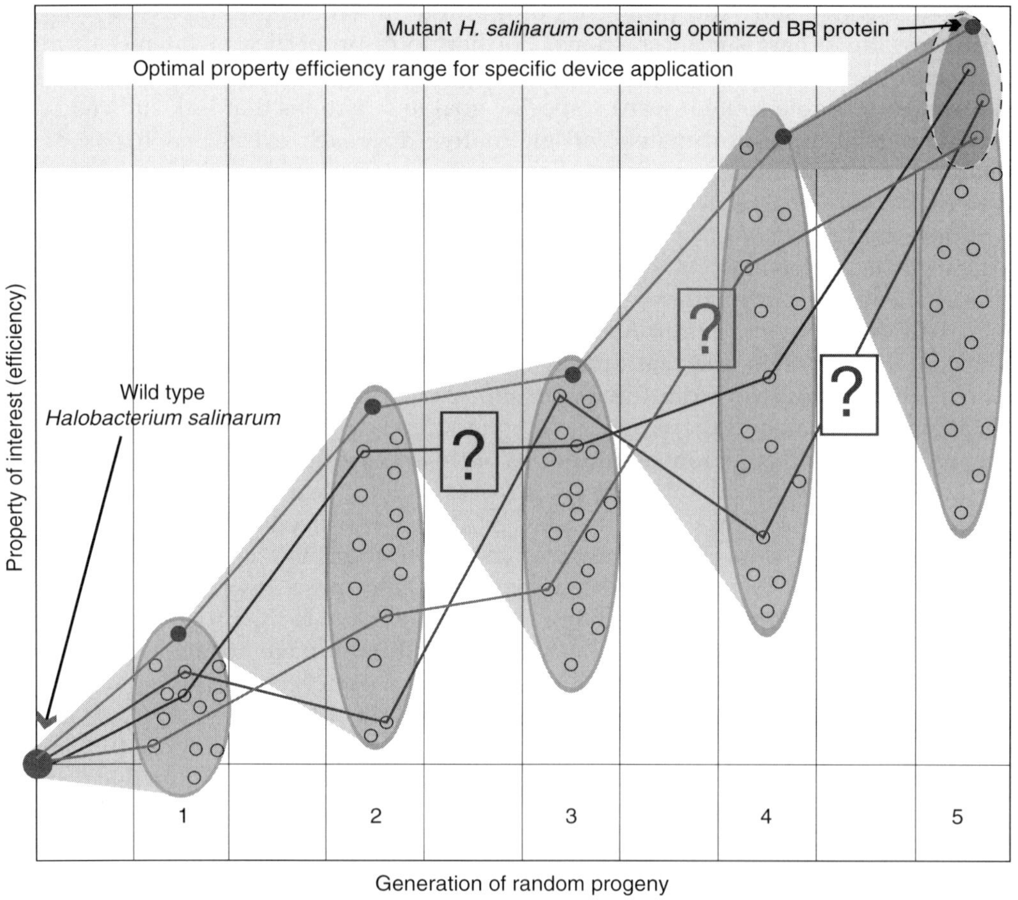

FIGURE 14.3
(See color insert) Schematic illustration of directed evolution of a bacterial protein toward a predefined goal. The bold green line indicates the route selected by the researcher, where, based on screening, the strain containing the most efficient mutant protein of each generation was chosen to parent the next. However, as shown by the question marks, other synergistic routes may be possible, but never discovered.

→

FIGURE 14.4
(See color insert) (a) Top left: Mutant strains of *H. salinarum* are brought up in 96-well plates and screened for production of bacteriorhodopsin; each well in the plate is analyzed by a plate scanner that generates a visible absorption spectrum for each strain, indicating whether the protein is produced (middle left). Top right: BR mutants are screened for λ_{max}, O-state formation and decay time, and Q, a quality factor determining how efficiently the O state is formed (maximum yield in minimum time). The cluster of mutants at the far right of each plot indicates several site-directed mutants included for comparison. The lower panel (b) illustrates a method by which flow cytometry could be used to screen for potentially valuable BR strains in vivo, cell by cell, in this case for efficient production of the branched photocycle. The three-step cell sorting system allows for in vivo high-throughput photochromokinetic screening. The first stage (1) selects for BR expression and high growth rate. The second stage (2) selects mutants with high photochemical conversion into the branched photocycle (O→P conversion). The third stage (3) selects for mutants having efficient photochemical conversion from the branched photocycle back into the bR state (Q→bR conversion). Mutants with desired photochemical characteristics are automatically deposited onto a single well of a 96-well plate (96WP). Other symbols: GR (grating), CCD (charge coupled detector), FL (flash lamp), HSDS (high-speed drop selector), BS (beam stop). See Hillebrecht et al., Optimization of protein-based volumetric optical memories and associative processors using directed evolution. *Nanobiotechnology*, 2005. **1**(2): 141–152 for more detail.

vivo. Cells are either selected or rejected on the basis of photokinetic properties of genetically altered purple membrane. Cells that are selected in the initial screen, as illustrated in Figure 14.4b, are subjected to further selection processes, including protein expression and branched photochemistry [60].

These techniques have been employed frequently for optimization of proteins and enzymes used in industrial processes. In such cases, directed evolution is used to enhance

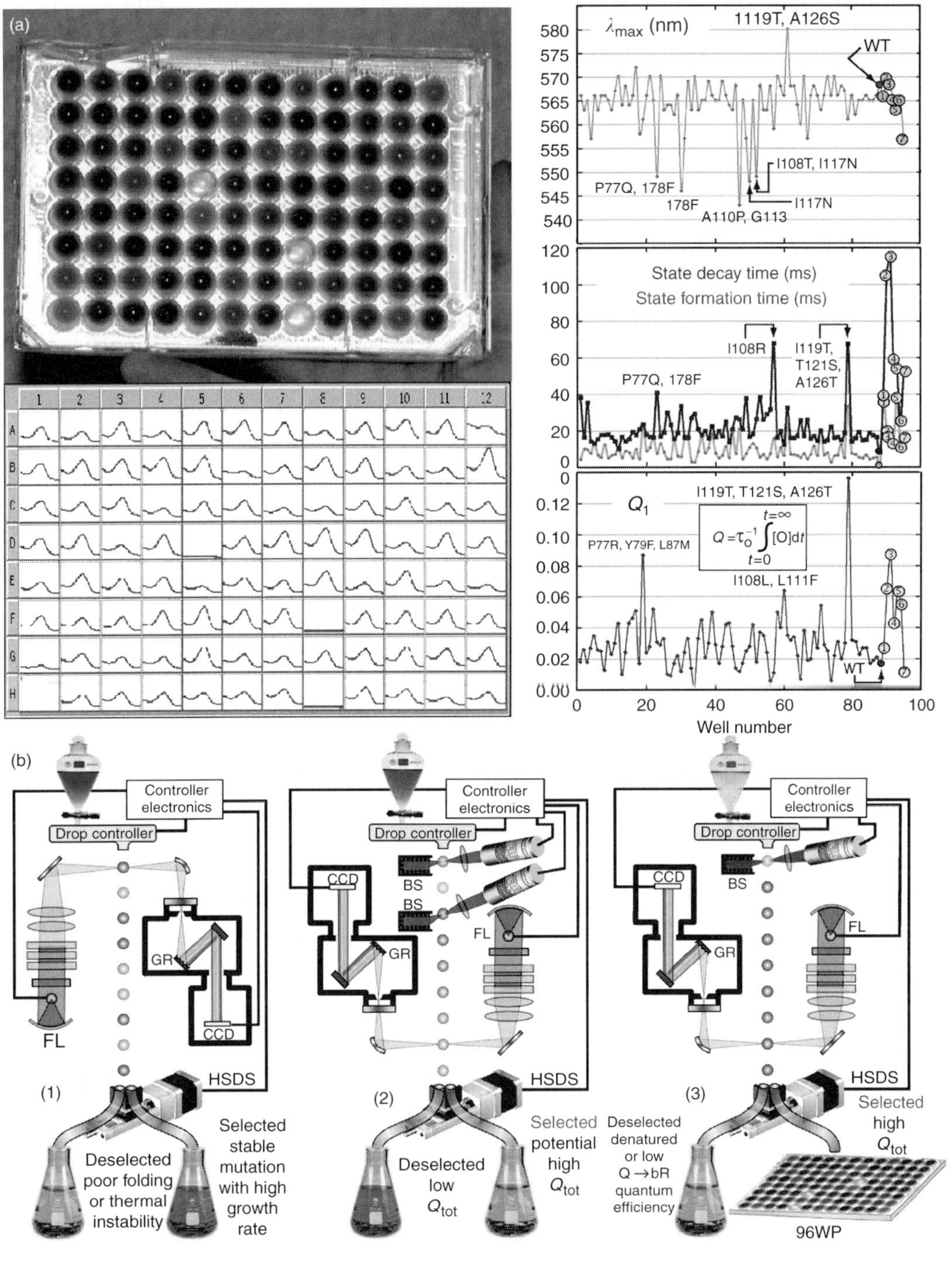

properties of industrial interest: temperature stability, catalytic function, reaction rate, and substrate specificity. All of the just-listed characteristics are variations on the given protein's natural abilities, that is, the enhancements are of preexisting naturally occurring functions. However, careful selection of the screening process, coupled with selective improvements conferred by successive generations of random progeny, makes possible the optimization of nonnatural properties custom-tailored for specific applications. Directed evolution therefore provides the applied researcher with a powerful tool for making full use of the advantages inherent to biological molecules. A full discussion of these novel techniques is beyond the scope of this chapter; many reviews exist in the literature for the interested reader [60,61,67–75].

14.3 Bacteriorhodopsin as a Sensor Element

Biological molecules can offer distinct advantages and unique functionalities to many device applications—introduction of biological molecules into inorganic environments often poses many challenges, but once accomplished, the resulting enhancements impart unique capabilities not previously possible. For sensors, these improvements take the form of better sensitivity, specificity, and often lower detection limits. The discussion below will focus on the use of genetically engineered BRs as integrated sensor elements, including the rationale for using BR for sensor technology, an overview of the architecture, technical issues that need to be resolved, and interpretation of the sensor's response.

In many ways, BR is already a sensor by natural design. It detects light with extreme efficiency, and shares a common chromophore and structural motif with the two other retinal proteins in *H. salinarum*, which have known sensory activities: sensory rhodopsins I and II. The sensory rhodopsins function as sensors that facilitate positive and negative phototaxis in the native organism [76]. Unlike these proteins, however, BR is produced by the organism in much larger quantities (~30–50:1). The key question is whether the researcher can take advantage of the protein's natural properties and apply them to new sensor architectures.

14.3.1 Bacteriorhodopsin as an Integrated Element in Microelectronics

Bacteriorhodopsin arguably has received more attention than any other protein for biomolecular electronic research, and with good reason; BR has a unique set of properties—both photochromic and photoelectronic—that are amenable to signal transduction. As expected, however, several challenges exist when integrating any biological molecule into a nonbiological environment; the two are not always compatible. This section will examine a number of these challenges, as well as some of their solutions. It should be noted that these are issues that will be critical to the success of integrating any protein into a synthetic environment.

It is worth reiterating the qualities that make BR an appealing candidate for device applications, and for sensors in particular. Many of these characteristics have been discussed above: (1) the protein is inherently rugged and robust, and is resistant to both thermal and photochemical damage; (2) the cyclicity is $>10^6$, considerably higher than most synthetic photochromic materials; (3) the protein is inherently radiation hardened and protected from free radical degradation; (4) it is inherently resistant to radiation damage; and (5) the primary photochemical event, that is, photon absorption and formation of the K state, proceeds with a very high quantum efficiency (~65%) [36]. In addition, the semicrystalline arrangement of the protein in the membrane imparts tremendous stability. Figure 14.5 illustrates the absorption spectra of the various BR intermediates, several of which are important to the protein's role in biomolecular electronics. Perhaps the most versatile aspects of BR's utility in device applications are the ability to interrogate the protein by multiple means, by examining modulation of either the optical or photovoltaic properties as a result of external stimuli, and the researchers' ability to improve BR's performance through chemical and genetic means.

To integrate BR, or any protein for that matter, into an environment designed for successful signal transduction, several conditions must be met. These include (1) maintaining the protein in a state that is functional and stable, (2) proper orientation of the protein, (3) ensuring that there is adequate coupling between the protein and the environment for efficient signal input and output, (4) proper encapsulation of the protein to hold it in place and protect it from degradation (natural or otherwise), and conversely (5) protecting the synthetic environment from any adverse affects due to the protein. For the specific case of BR, some of these issues are less important than others. Because of its native environment (the purple membrane, or PM), BR is already more rugged and stable than most proteins, and is naturally protected from oxidation and (to a large extent) microbial degradation. Loss of function due to dehydration is more of a problem. But the purple membrane has its own set of challenges, specifically due to its size (~0.5–1 μm). This size is the same order of magnitude as the visible wavelength used to excite or interrogate the protein (500–1,000 nm), and therefore will induce light scattering. Furthermore, when faced with the challenge of fabricating microelectronic devices, the average purple membrane patch size is often larger than the average lithographic feature size, making precise deposition of the protein difficult. New methodologies may have to be developed to overcome these problems. At least one methodology for patterning PM fragments has been reported in the literature; Brizzolara et al. [77–79] genetically modified BR to contain a cysteine (an amino acid containing a sulfhydral group) at the membrane surface. The sulfhydryl group can ligate to a gold surface that has been lithographically patterned previously. PM patches adhere only to the surfaces patterned with gold. However, this technique was only marginally effective, in that coverage was not complete and sulfhydryl contaminants polluted the gold surface. Large patch size may have been part of the problem, and it has been demonstrated that sonication can be used to reduce patch size (unpublished in-house data). Regardless, this ligation technique is not likely to produce a uniform PM coating on semiconductor surfaces. Other methods of depositing PM monolayers or films involve hydrophobic and hydrophilic bonding, such that the protein only bonds to hydrophilic surfaces [80]. One method relies on depositing PM over a previously patterned silicon surface that is of

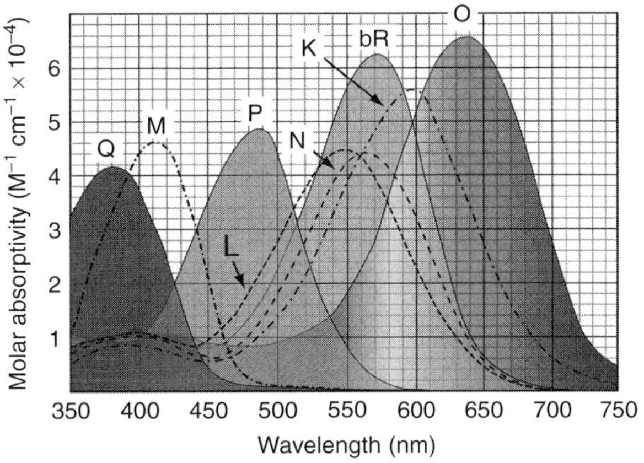

FIGURE 14.5
(See color insert) Absorption spectra of the various bacteriorhodopsin intermediates are shown. The bR, M, and O states are of particular interest to most of the applications developed around bacteriorhodopsin's unique photochromokinetic and photovoltaic properties.

varying topography. Mechanical means are then used to remove all excess proteins, except that deposited in lithographically fabricated trenches [81,82].

The challenge of maintaining BR in a functional and stable state is only partly satisfied by the purple membrane. Although it does impart some level of protection from weathering, it is unable to maintain the appropriate moisture level for proper function. Encapsulation of the protein in a suitable polymer matrix can often fix the moisture at a functional level, and also ensure enhanced stability. The matrix must allow efficient interrogation of the protein (i.e., it must be optically transparent), as well as adequate coupling to the semiconductor surface. Selection of a matrix can be difficult because the protein prefers an aqueous environment and water can often be fatal to semiconductors.

The ability to couple the protein signal output to the microelectronic substrate will vary for different proteins, and will depend upon the nature of the output signal: optical, chemical, or electrical. Specific orientation of the protein may be required. For BR, orientation is facilitated by a net charge difference (–3) across the purple membrane. Application of the appropriate voltage will orient the purple membrane and deposit it on the substrate surface [83]. This is particularly important for applications that rely on BR's photoelectric effect, which will be cancelled out in a random orientation of PM fragments. Effective orientation will maximize the voltage signal resulting from photoexcitation, thereby ensuring effective coupling for signal transduction. Assuming that the signal is strong enough, optical coupling is only important if the detection scheme is wavelength dependent. Several examples of devices utilizing BR (in purple membrane form) as a light to voltage transducer exist in the literature, and a few will be discussed herein, as they are relevant to BR's sensor capabilities.

14.3.1.1 Microelectronic Devices Employing Bacteriorhodopsin for Enhanced Function

There have been several devices proposed and in some cases constructed, which utilize BR to enhance function by integration of the protein into a semiconductor device. This latter challenge—integration and stabilization of BR in a semiconductor environment—must be met to make BR-based sensors a commercially viable reality. Work along these lines has progressed over the last few years, resulting in several microelectronic devices that utilize BR to modulate circuit response through the photoelectric effect. Several such devices have been fabricated by Bhattacharya and coworkers [32], including a BR field-effect transistor (FET) photoreceiver and a MODFET with BR as a light-activated gate. In the latter case, purple membrane was oriented and selectively deposited at the gate of a GaAs-based modulation-doped FET; the gate was activated by the photovoltage that developed across the purple membrane in response to light, thereby allowing current flow (Figure 14.6). The photo-induced current was transient with a picosecond-scale rise time and a slower decay—the voltage modulation was produced only with changes in the actinic light intensity, indicating that the resulting photoreceiver cannot be used with continuous illumination. However, the transient voltage response produced by the protein facilitates any application that requires detection of dynamic changes in light intensity, that is, any situation where the object of interest is characterized by modulated light levels. This sort of sensitivity is referred to as differential responsivity, and examples include edge detection and motion sensitivity [30,32,84].

Artificial vision and construction of an artificial retina based on BR has been a long-term goal for a number of researchers [84,85]; BR is a strong candidate for such architectures due not only to its characteristic ability of differential responsivity, but also its obvious similarity to the visual rhodopsins. Differential responsivity is a necessary behavior for any optical system designed to mimic human vision. From a physiological perspective, the

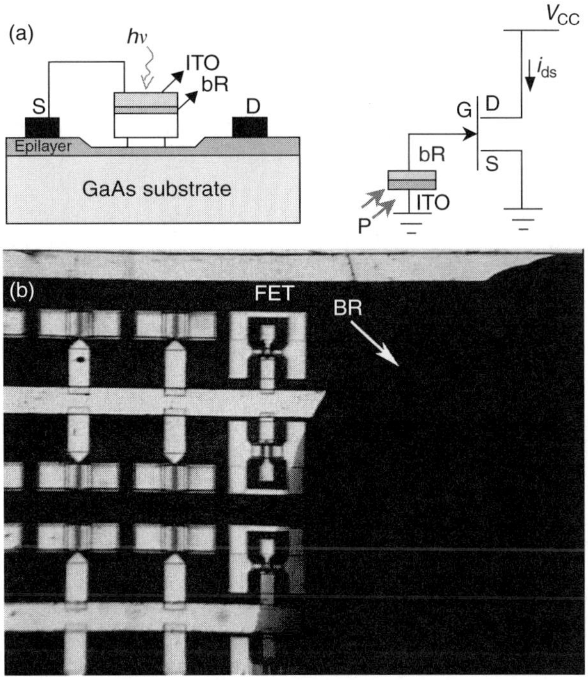

FIGURE 14.6
(a) Schematic (left) and equivalent circuit (right) of a monolithically integrated BR/GaAs MODFET photoreceiver. (b) Photomicrography of a fabricated chip illustrating a series of MODFET devices and the selectively deposited and oriented dried BR film.

term refers to the human retina's ability to preprocess image information prior to interpretation by the brain, especially with respect to motion detection and edge discrimination. The two closely related effects are common in the human visual field, which is frequently characterized by changing levels of intensity, brought about either by a changing physical environment, or facilitated by actual head and eye movement. Implementation of these properties into a conventional semiconductor device is not easily accomplished.

The first attempt to employ BR in an artificial retina was made by Miyasaka et al. in 1992, who fabricated an 8 × 8 patterned array (64 pixels) of transparent SnO_2 electrodes on a glass substrate. Purple membrane was applied to the array in an oriented Langmuir–Blodgett film, which was coated with an electrolyte gel. The array was capped with a gold-coated glass plate, and each individual pixel was connected to an external amplifier-detector circuit. When the array was illuminated, a current flowed through the external circuit connecting the transparent electrodes and the system exhibited differential responsivity behavior, that is, the array did not generate a signal under constant illumination, but only in response to changes in illumination [85].

An artificial retina architecture capable of higher resolution was built and tested by Chen and Birge in 1993 [84]. In this case, a 256 × 256 array of charge-sensitive semiconductor elements (CID array) was used, covered with a thin poly(vinyl alcohol)-based BR film (Figure 14.7). An x–y addressable CID imaging array is capable of producing an electrical output proportional to the light incident upon each addressed pixel location—the need for individually addressing each pixel is eliminated, making the design much simpler. The oriented BR-based film deposits charge on the array surface upon illumination, and the CID spatially samples the protein-induced photocurrent at discrete sites by applying the appropriate x–y address. Although the prototype worked as expected, cation contamination of the semiconductor surface made the device unreliable. The problem was ultimately resolved by utilizing chemically modified BR, in which the inorganic cations [86] were replaced by organic analogs [87].

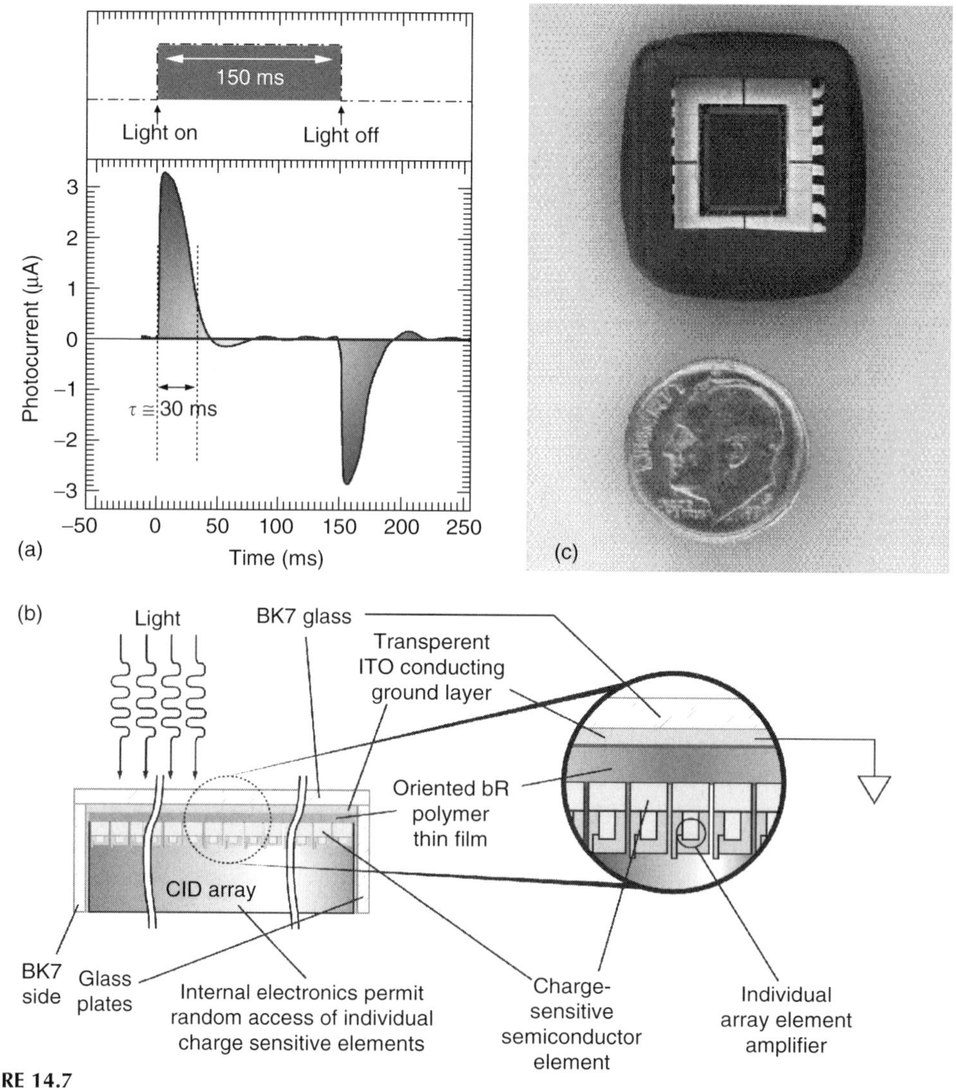

FIGURE 14.7
A hybrid semiconductor bacteriorhodopsin-based artificial retina is fast and inexpensive, and mimics the motion sensitivity of the human retina. By utilizing differential responsivity (a), the artificial retina is capable of both edge enhancement and motion detection. (b) Device schematic illustrating the oriented bacteriorhodopsin layer and the CID array. The protein is encapsulated in a polymer for stability and longevity. The array is modified so that the charge-sensitive semiconductor elements (SCCE) are in direct contact with the protein film. A transparent indium tin oxide (ITO) conductive coating on BK7 glass is placed over the top, sealing the protein from the environment along with two other glass plates. Each SCCE has an individual amplifier that provides 100 photon per pixel per second sensitivity (photon pixel-1s-1). The differential responsivity of the protein photocurrent generates a final image which is selective of those elements that have changed within the last 3–100 ms. Thus this artificial retinal simulates the motion-sensitivity characteristics of its biological counterpart. (c) Photograph of actual device.

14.3.2 Bacteriorhodopsin as a Sensor of its Chemical Environment

The complex photophysical properties that characterize BR make it a natural candidate for architectures that rely on modulations of either light or voltage. In this regard, all previous applications have revolved around BR's ability to respond to light of the appropriate wavelength, thereby eliciting a photochromic response. In essence, the protein senses the actinic light and responds by entering the photocycle. In the applications discussed earlier, the desired result was the production of either a specific intermediate (M or Q), or a photovoltage. However, the protein is capable of providing much more information to the researcher, specifically in the form of the photokinetics of both responses. These three responses, photochromism, photovoltage, and photokinetics, can all be used to facilitate sensor design.

14.3.2.1 Bacteriorhodopsin and Chemical Sensitivity

Bacteriorhodopsin has been long known to display an extreme sensitivity to its environment, responding to changes in pH, relative humidity, and ionic strength. The results of such environmental changes are manifested as modulations of the photocycle kinetics or the photoelectric effect. Furthermore, a large number of chemicals have well-defined effects on the protein. One such class of chemical modulators is organic amines, which prolong the M intermediate. The M state decays as a function of the deprotonation of amino acid D96, and environmental conditions that favor reduced proton mobility will prolong this state (e.g., high pH or dehydration). The O state, also characterized by proton transfer, is similarly sensitive to the presence of environmental additives that alter pH or reduce proton mobility. Furthermore, Bryl and Yoshihara [88] demonstrated in 2000 that the BR photoelectric effect can be modulated as a function of the chemical environment experienced by the protein.

Other classes of chemicals that modulate BR's properties include alcohols (methanol, ethanol, propanol, and butanol) [88], anesthetics [89–102], and azides [103–107]. The protein's sensitivity to alcohols is manifested as modulations of the light-induced photocurrent, and can be enhanced by chemical modification of the chromophore or genetic manipulation of the protein—the resulting protein variants exhibit enhanced responses to various alcohols as compared to the wild type [88]. One plausible mechanism for the alcohol-induced sensitivity is local dehydration effects that mimic reduced relative humidity. Numerous studies have demonstrated the effect of anesthetics on BR, which exhibit several general modes of action. Anesthetics modify ion transport at membrane channels and are of interest in connection with the nerve system. The concentration-induced effects consist of a slight initial blue shift (569–567 nm) to a state with an accelerated M state. Higher anesthetic concentrations result in a larger blue shift (480 nm), a prolonged M state, and a loss of both proton-pumping function and the purple membrane lattice structure. Some anesthetics will produce a 380-nm-absorbing species. At lower anesthetic concentrations, most effects were found to be reversible. X-ray studies indicate that anesthetics bind at the lipid protein interface in the center of a BR trimer [91,96]. Other studies have posited distinct binding sites [93,99,108]. Generally speaking, therefore, chemical agents have several modes by which they can modulate protein response, including direct interaction with the protein, or indirect interaction with the lipids proximal to the protein.

Various chemicals have been used in the past to modulate BR's response to light through modulation of the M and O states. Although these attempts were targeted at photonic devices (e.g., optical and holographic memories and spatial light modulators), they are indicative of the BR's ability to respond to and register the state of its environment. Furthermore, as detailed above, a wide variety of other chemical species have proven to modulate the protein's response, indicating the potential role BR might play as a chemical sensor in hybrid protein-semiconductor architectures. Given the ability to interrogate with light, signals can be extracted that are a reflection of the protein's environment.

Bacteriorhodopsin's potential in hybrid protein-semiconductor devices stems from its sensitivity to external molecular stimuli, which registers as either a modulation of the photocycle kinetics or the photovoltaic effect, or a combination of both [89–102,109]. Thus, it is possible to use either photocycle kinetics [93,95,96] or photovoltaic responsivity

[89,100,109] to monitor the concentration of volatile species. Most of the work, thus far, has concentrated on anesthetics, but these and other studies indicate that many organic compounds will directly alter BR's photophysical properties [91,99,109]. The complexity of both the photochromic and photovoltaic effects, and the sensitivity of these signals to the external environment, combine to make BR a strong candidate for the design of hybrid semiconductor sensors.

14.3.2.2 Bacteriorhodopsin-Based Chemical Sensor Architecture

The ability to utilize BR as a chemical sensor depends on two assertions. The first is that BR has proven to be sensitive to a remarkable range of chemical antigens, and that sensitivity is manifested as measurable changes in the protein's photochromic and photoelectric properties. Sensitivity to its chemical environment, combined with the researcher's ability to integrate it into device architectures, its rugged stability, high cyclicity, and the ability to interrogate the protein by multiple means, makes BR a strong candidate for sensor architectures. However, it is doubtful that wild-type BR will be useful as a sensor element on its own. Even though the wild-type protein can respond to a wide variety of chemical antigens, it is unlikely that wild-type BR alone will be able to discriminate between more than a few chemical species—it simply cannot elicit a wide-enough variety of responses. Fortunately, this limitation can be easily overcome by turning to genetically engineered variants for the detection scheme. The second assertion, therefore, is that the utilization of BR mutants will introduce differential sensitivity and chemical antigen specificity to the sensor architecture. Directed evolution can be used to custom-design proteins that will respond specifically and selectively to individual chemical agents. The ultimate approach would be to employ several BR mutants, each highly optimized to detect the presence of a unique chemical species. The problem with this approach is that it involves several rounds of directed evolution, consisting of random mutagenesis and screening thousands of mutant strains for desirable proteins. Utilization of advanced BR mutations therefore represents a long-term approach to sensor development.

A far simpler approach than employing directed evolution is possible, however, by taking advantage of the differential sensitivities exhibited by a library of mutant BR proteins: as long as the protein is functional, it will respond in some way to external chemical stimuli. Looking for an optimized response is somewhat secondary to the ability to characterize a preexisting response; in other words, as engineering a specific change in the BR photocycle is no longer the goal, screening need be done only for functional proteins, not for a specific quality. Chemical agents are certain to affect the photokinetic and photovoltaic responses of mutant BR proteins, and as long as their responses are uniquely different from the native protein, they will be successful in chemical agent detection schemes. Although individually, neither the wild type nor any specific mutant protein is capable of accurately determining the identity of an unknown chemical agent, a collection of such mutants with varied responses to the same toxin will enable a far more accurate determination, thereby reducing the possibility of false negatives. The result is a numeric matrix describing the responses of each mutant protein that acts as a fingerprint for each chemical agent. Furthermore, access to a library of BR mutants with characterized responses to external chemical stimuli has definitive implications for highly specific detection schemes.

A sensor based on a collection of BR mutants has the potential of displaying both sensitivity and selectivity. The approach to implementing such a sensor is to simultaneously monitor the response of a large number of mutant proteins to a given chemical antigen. As detailed above, each mutant will exhibit a specific response, different from each and every other mutant. By examining the simultaneous responses of a large set of mutant proteins to a given chemical antigen, we now have the possibility of generating a numerical fingerprint characteristic of that antigen. Such a mutant BR matrix sensor would operate with great sensitivity and selectivity. In this way, the need for highly optimized BR mutant proteins is eliminated by the use of multiple mutants with unique and varied responses, enabling a far more accurate determination. Furthermore, the "mutant matrix sensor" would be capable of differential identification of various chemical agents, giving a different numeric fingerprint for each agent to which the sensor is exposed—this detection scheme can be thought of as a biochemical "nose," capable of differentiating between various external chemical stimuli. Increasing the number of mutants used in the detection matrix will result in a higher confidence level for the identification. The introduction of highly optimized BR mutants produced through directed evolution will only enhance confidence levels further.

The mutant matrix would be designed and operate as illustrated in Figure 14.8. An array of individual sensor elements, each containing a different BR mutant, is exposed to a chemical antigen. Upon exposure, the sensor elements are exposed to light and monitored individually for their photokinetic response—in this case the kinetics of the M and O intermediates are recorded. Each intermediate can produce as many as three numeric matrices, including rise time, decay time, and photoelectric response, all specific to chemical species interacting with the sensor. The resulting numeric fingerprints can be used to identify the specific chemical toxin through database comparison. For situations in which the sensor is exposed to multiple toxins, signal deconvolution techniques, such as signature analysis, will be used to parse out the signals for identification. The M and O matrices will be fit to exponential formation and decay functions to provide a set of four matrices, two for each intermediate—these sets represent a numerical signature that will be different for each chemical antigen. Using signature analysis, the datasets can be manipulated mathematically to provide a numerical correlation coefficient with reference

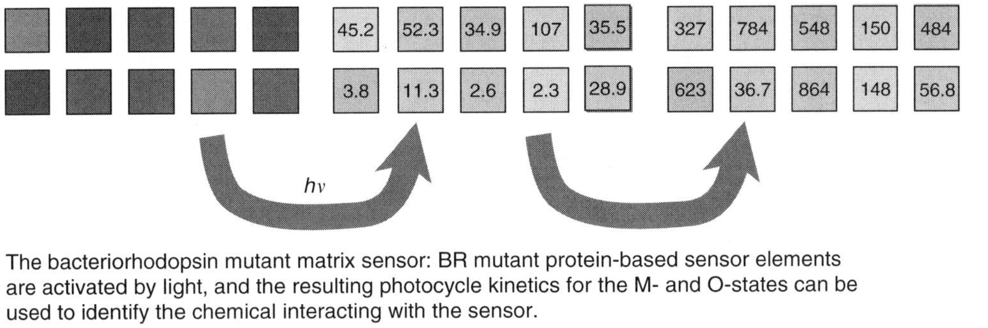

FIGURE 14.8
The sensor elements contain a random variety of bacteriorhodopsin mutants that are characterized by individual and specific responses to any given chemical species interacting with the sensor matrix. The matrix is illuminated upon exposure to a chemical (e.g., industrial pollutants, chemical warfare agents, and toxins), and the BR mutants' photocycle into the M state. Upon the decay of M, the O state is formed. Kinetic parameters (M- and O-state rise and decay time constants) are determined and the resulting matrices constitute a numeric signature specific to the chemical interacting with the matrix.

to stored matrices (signatures) for known targets (chemicals). The mathematical methods for carrying out signature analysis are well known [110–113], and in cases where each measurement is given equal weight, can be represented numerically as a least-squares sum over the deviations of the observed parameters from the values measured for each specific target. Signature analysis gains comparative advantage by allowing weighting of components based on iterative optimization to enhance both sensitivity and accuracy.

Implementation of the sensor architecture described above requires a platform that can simultaneously measure both the protein's photochromokinetic response and photoelectric effect. Design of such an architecture poses a challenge, in that the photoelectric effect can be much more difficult to detect than the protein's photokinetics. To register both responses, use of a concentrated sample deposited on a conducting, and perhaps transparent, substrate will be necessary, with a second conducting lead in contact with the sample surface. The photokinetic response is relatively easy to monitor, given the appropriate optics and circuitry; a fairly basic flash photolysis set-up monitoring the protein's response at two wavelengths (M and O) is all that is needed. However the photoelectric effect, whether measuring light-induced current or voltage, requires that the sample essentially be sandwiched in a capacitor that can be used to measure light-induced displacement currents. Measuring the photocurrent component may require a hydrated sample to allow the protein to pump protons. However, the studies of Xu et al. [64] and Bryl et al. [88] detected photo-induced current in dried samples, indicating that ambient humidity provided enough moisture to elicit a signal. There are several examples in the literature where dried biological preparations (including BR) have proven to be stable and maintain high levels of activity. Greenbaum et al. [114], demonstrated the use of dried tissues for detection of chemical antagonists; samples consisted of immobilized photosynthetic algae deposited on filter paper disks, and utilized a fluorescence detection scheme. BR is nonliving and extremely stable and robust when dried into films (as required for the photoelectric effect). Such films can be deposited in a variety of ways, including electrophoresis, simple evaporation, or chemical ligation to modified surfaces. As discussed above, additional stability can be achieved by polymer or sol gel encapsulation, which maintains protein immobilization and stability, while providing a porous structure to facilitate diffusion of volatile agents. Ultimately, sensor miniaturization onto a semiconductor platform will be required. Issues with integration of the protein-based component of the sensor have been covered above, and a discussion of overall miniaturized sensor design is beyond the scope of this chapter.

14.3.2.2.1 Preliminary Results

Currently, work is progressing along the lines of characterizing the response of the wild-type protein and a selection of genetically engineered BR variants to a variety of chemicals known to elicit a photochromokinetic response [115]. Preliminary results from those trials are presented here. A bench-scale flash-photolysis prototype has been developed in-house that is capable of measuring the rise and decay kinetics of the M and O states. The prototype was designed to demonstrate proof-of-principle, and does not fully represent the technology ultimately needed for an in-field sensor platform. The actinic beam is provided by a 532 nm and 20 mW laser that initiates the BR photocycle. Blue (450 nm) and red (650 nm) light emitting diodes are used to monitor the photokinetics of the M and O states, respectively. A pair of photo detectors records the resulting data traces, which are fit to exponential curves to determine the M- or O-state lifetimes and decay rates. The current prototype lacks the capability to measure the photoelectric effect. Aqueous samples of the wild-type protein, as well as a number of randomly selected genetically engineered variants, were prepared carefully with a series of chemicals known to modulate photocycle kinetics [115]. Included were 1,2-diaminopropane (1,2-DAP) and triethylamine (TEA), among others, in concentrations ranging from 1 to 50,000 ppm. A control was set up for each sample, consisting of an equivalent volume of water added to the protein solution as a substitute for the chemical of interest. Figure 14.9 shows typical M- and O-state datasets derived from the wild-type protein, as a function of increasing concentration of a variety of amine compounds. The raw O-state data shown in Figure 14.9 was collected at 650 nm by the flash-photolysis prototype, and is typical for wild-type BR. Data collected for the M state appears very similar, except the protein is monitored at 450 nm and the time constants are generally shorter. Trends typical for the M-state lifetime as a function of several chemical additives are also shown. The ability of the chemical to modulate the protein's photokinetic response is clearly illustrated, and similar trends were seen for each of the

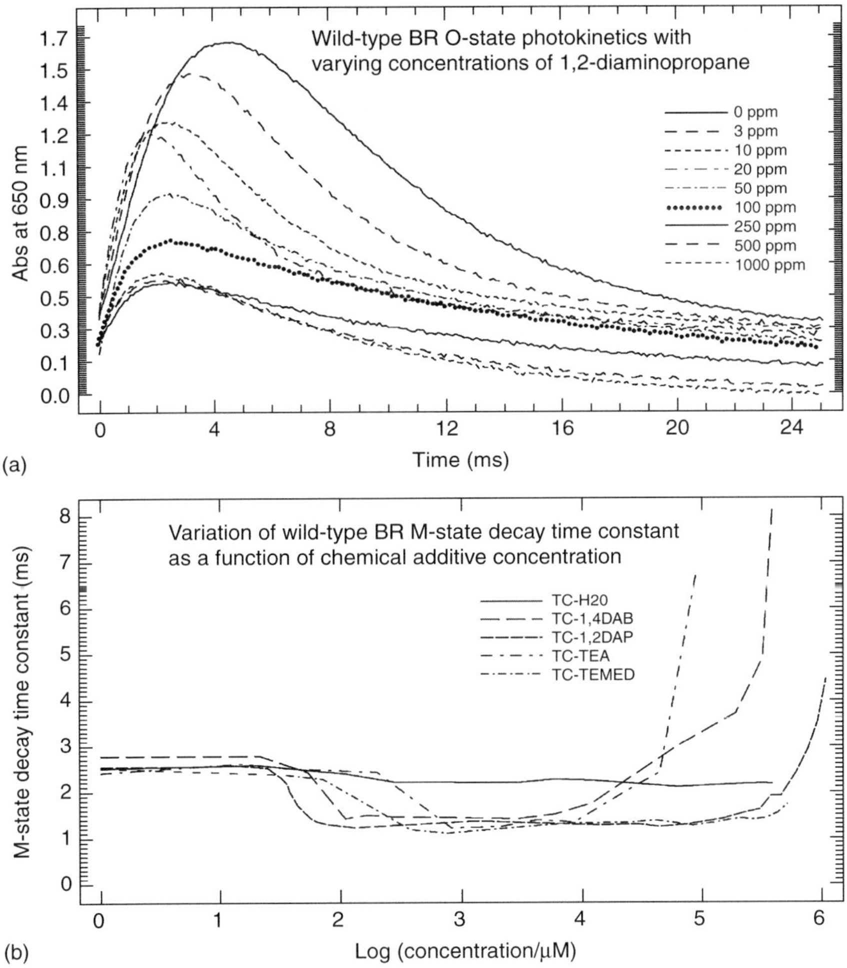

FIGURE 14.9
(a) Wild-type BR photokinetics as a function of the concentration of 1,2-diaminopropane (1,2-DAP). A series of concentrations of 1,2-DAP were prepared in a cuvette containing wild-type BR (~1.5 OD), and flash-photolysis experiments were performed on the resulting samples to monitor the photokinetic response of the O state. Similar photokinetic traces were collected for the M state. The rise and decay constants were determined for both states at each concentration of 1,2-DAP. (b) The data shown in the lower panel were compiled in a plot as for the wild-type M state, where the M decay time constant is plotted as a function of concentration for a variety of chemicals: 1,4-diaminobutane (1,4-DAB), 1,2-DAP, triethylamine (TEA), and *N,N,N′,N′*-tetramethylethylenediamine (TEMED). Water was used as a control (solid line).

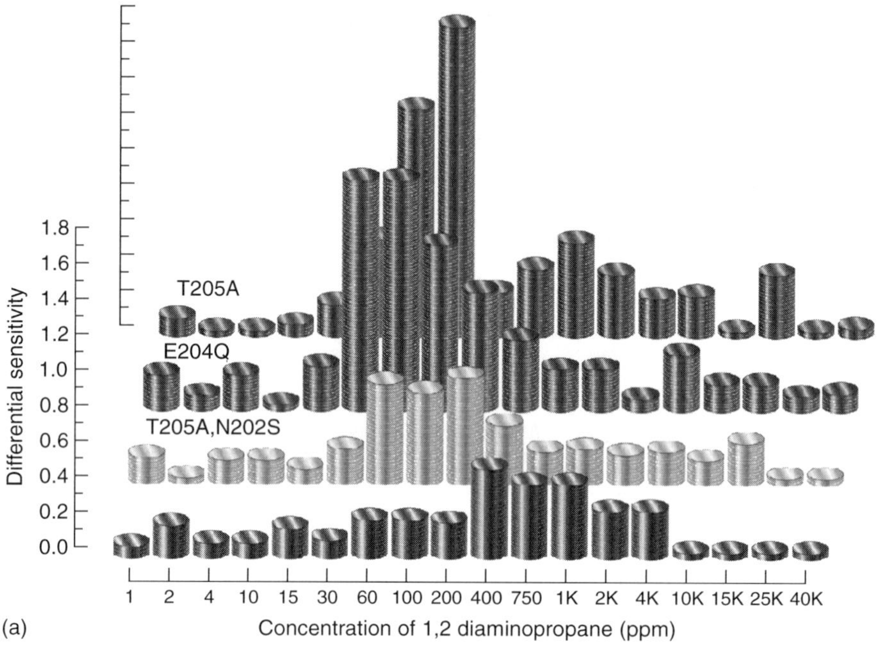

(a)

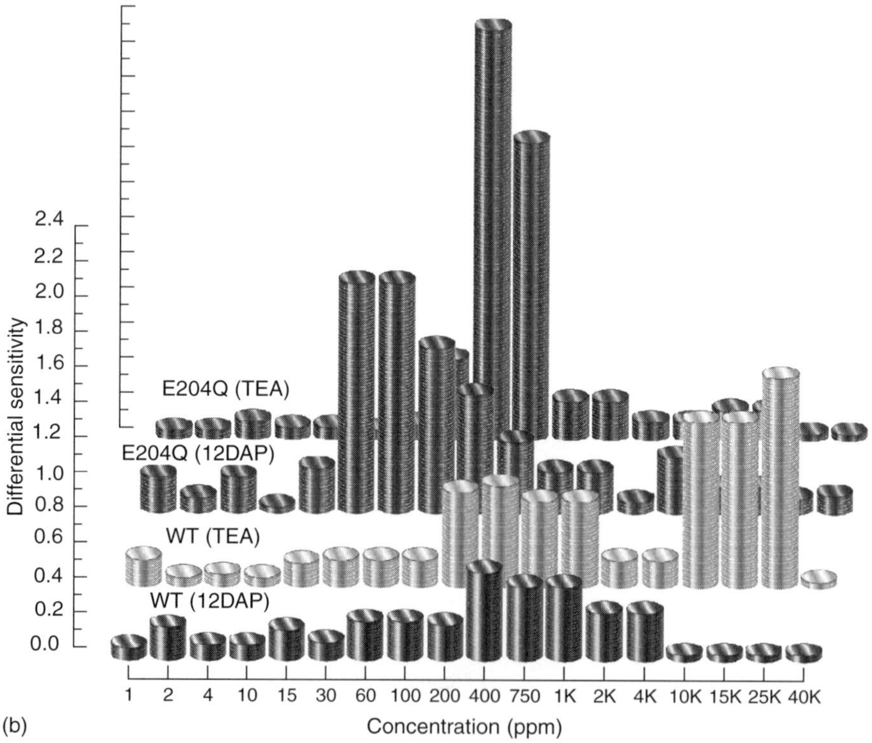

(b)

FIGURE 14.10
(See color insert) The sensitivity of wild-type BR and mutants in the presence of several chemical antigens is illustrated. Sensitivity is defined as the ratio of the change in the M/O decay time constant to the change in micromolar concentration of the added chemical. (a) Sensitivity to 1,2-diaminopropane (1,2-DAP) varies between the wild-type protein and BR mutants. All of the mutants demonstrated better sensitivity than wild-type BR. (b) Not only sensitivity to one chemical varies between proteins as in (a), but the same protein responds differently from one chemical to another. This is illustrated for the wild-type protein and the mutant E204Q, for 1,2-DAP and triethylamine.

proteins tested. Despite the fact that several of the chemicals elicit similar trends in the wild-type protein as their concentrations are increased, introduction of mutant proteins enabled a wider variety of responses. As evident in Figure 14.10, all proteins examined exhibited distinct and specific responses to a given chemical, and the protein's individual responses varied strongly from chemical to chemical. Furthermore, the responses exhibited by genetically engineered BR variants varied from wild type, and from mutant to mutant. The sensitivity is defined in Figure 14.10 as the ratio of percent change in the time constant to that of the change in the micromolar concentration of the added chemical. Defining sensitivity in this way provides a unit-less, quantitative method that facilitates comparisons. The best overall sensitivities were found when combining the responses of both the M- and O-state data to the given chemical. At this point the detection limit seems to be approximately the same as the concentration of the protein.

14.3.2.2.2 Other Photoactive Proteins

Given that a sensor based on BR has proven potential for detection of wide variety chemicals, it stands to reason that other proteins might be capable of the similar behavior and activity. Most examples of protein-based sensors in the literature refer to proteins that play a biological sensing role by detecting the levels of various cellular constituents (sugars, ions, and phosphates). However, there are a large number of light-activated proteins that may have use in architectures similar to those described above, including the recently described proteorhodopsins (PRs), PRCs, phytochromes, and PYP. Proteorhodopsin refers to a wide variety of related proteins originating from marine organisms (phytoplankton) worldwide. This group was discovered only recently and shares many attributes with BR, including a common structural motif, an all-*trans* retinal chromophore, and proton-pump activity. Unlike BR, however, there are a large number of naturally occurring variants, and the native organisms have yet to be identified (the first protein was produced after searching an RNA library isolated from the marine waters of Monterey Bay in California; see the references by Beja et al. [116,117]). As such, all PR reported thus far has been the result of heterologous expression in *E. coli*, and no information exists about the protein's physiological arrangement in the native cell membrane. The PR expressed in *E. coli* has some other significant differences from BR, but perhaps the most significant from the standpoint of device design is that the solubilized protein is remarkably stable. Solubilized BR has a lifetime of only a few days and needs to be used in purple membrane form for device applications. The purple membrane imposes a number of limitations on device architectures, especially for optical memory and holography. PR's enhanced stability in the solubilized state may give it an advantage over BR in several aspects, especially with respect to integrating it into a semiconductor environment. PR's potential in sensor architectures (optical, chemical, or otherwise) remains to be evaluated.

Photosynthetic reaction centers (PRCs) are redox proteins that transport electrons as part of Photosystem I; the physiological role of this widespread group of proteins is energy transduction in photosynthesis. This class of proteins is isolated from plant tissues with relative ease, and remains stable for prolonged periods in dried preparations (e.g., see Lee et al. [118]). Several chemical sensor platforms have been proposed in the literature, some utilizing the isolated reaction center [119,120] while others immobilized tissues containing PRCs [1,114]. However, while this group of proteins has the potential for highly sensitive detection schemes, it remains to be seen whether PRCs can be used in chemical sensors capable of both specificity and selectivity.

Phytochrome and PYP are further examples of photoactive proteins that produce quasi-stable thermal intermediates in photocycles analogous to BR and PR. Phytochromes are photo-regulatory proteins found in plants that are responsible for a host of signal transduction and regulatory pathway mechanisms; they respond to modulations in solar energy, including light quality and quantity, duration, and direction—in essence, this group of proteins enables plants to respond to varying light conditions (for a recent review, see Montgomery and Lagarias [121]). They also facilitate light-mediated control at a genetic or transcriptional level, and have complex cofactor and prosthetic group associations. Two optically interconvertible forms exist, Pr (red-light absorbing) and Pfr (far red-light absorbing)—conversion between the two forms is accompanied by isomerization of a bilin chromophore and the concomitant structural rearrangement of the protein, which modulates phytochrome's biochemical activity. The bilin chromophore is a derivatized tetrapyrrole, and the light-initiated isomerization causes the fourth pyrrole ring to rotate 180°. The exceedingly complex nature of both the protein's structure and its light-mediated response give phytochrome potential in sensor architectures; the large number of interactions provides many avenues for chemical antigens to intercalate and disrupt. However, that same complexity may prove to be a liability, in that stabilizing the protein might be problematic, and its response might be difficult to interpret. Nonetheless, phytochrome's potential needs to be investigated, if only due to its already powerful physiological role as a biosensor. Photoactive yellow protein, in contrast, is a much simpler sensor protein that mediates negative phototaxis for its host organism, *Ectothiorhodospira halophila*. Just as for the rhodopsins and phytochrome, PYP's biological activity is initiated by a blue-light-induced chromophore isomerization, which in this case is *para*-hydroxy cinnamic acid. A photocycle follows light absorption, which like BR, encompasses a time scale ranging from picoseconds to milliseconds (for a review, see Hellingwerf, 2000 [122]). The extent to which the PYP's biophysical properties can be exploited in sensor architectures remains to be determined.

14.4 Future Directions

Biological molecules hold great promise for introducing novel function to conventional technologies, and as the researcher's ability to integrate these molecules becomes more advanced, the potential for more sophisticated sensor platforms can be realized. Bacteriorhodopsin is one of a broad class of light- and or signal-transducing proteins that has potential in biosensor architectures. Sensor platforms for light modulation, voltage, and chemical detection and identification are conceivably possible. More advanced applications include conferring advanced sensing capabilities to preexisting devices (microelectronic or otherwise), and construction of an artificial retina. Central to these efforts is the ability to modify the protein to better suit the needs of a given application—the techniques of site-directed and semirandom mutagenesis provide the researcher with the ability to introduce novel attributes at the genetic level, enabling the production of custom-designed proteins for specific device applications. Given this ability, the potential exists for an entire new class of sensors that brings new functionalities to conventional designs.

The data presented above clearly demonstrate proof of principle for the BR-based chemical sensor. A manuscript containing a more detailed account of the latter research effort is currently being prepared [115]. The range of experiments must be expanded to encompass a wider variety of both BR variants and toxic chemicals that are of potential hazard to human health, including industrial pollutants and chemical warfare agents. Characterizing and integrating a larger number of mutants into the sensor platform will result in a higher probability of positively identifying the toxin. Furthermore, as demonstrated by Bryl et al. [88], the BR photoelectric effect is also a valuable tool for distinguishing between toxins, and work to that effect is ongoing at the University of Connecticut. Designs for a miniaturized sensor platform implementing the mutant matrix with both photochromokinetic and photoelectric capabilities are also being developed.

Acknowledgments

The authors gratefully acknowledge the following funding organizations for their generous support of the research efforts described herein: The New York State Center for Excellence in Infotonics UC-004-0007, Nation Science Foundation grants #NSF-0432151, including REU support (JAS, RRB) and # NSF-0412387 (PB, RRB), and the Nation Institutes of Health, NIH-GM34548 (RRB).

References

1. Rodriguez, M., Jr., C.A. Sanders, and E. Greenbaum, Biosensors for rapid monitoring of primary-source drinking water using naturally occurring photosynthesis. *Biosens. Bioelectron.*, 2002. **17**(10): 843–849.
2. van der Schalie, W.H., T.R. Shedd, M.W. Widder, and L.M. Brennan, Response characteristics of an aquatic biomonitor used for rapid toxicity detection. *J. Appl. Toxicol.*, 2004. **24**(5): 387–394.
3. Fleming-Michael, K., Aquatic sentries keep watch on water quality. 2003. US Army Environmental Center: Environmental Update.
4. Wei, Y., J. Xu, Q. Feng, M. Lin, H. Dong, W.J. Zhang, and C. Wang, A novel method for enzyme immobilization: direct encapsulation of acid phosphatase in nanoporous silica host materials. *J. Nanosci. Nanotechnol.*, 2001. **1**(1): 83–93.
5. Wei, Y., H. Dong, J. Xu, and Q. Feng, Simultaneous immobilization of horseradish peroxidase and glucose oxidase in mesoporous sol-gel host materials. *ChemPhysChem*, 2002. 3(9): 802–808.
6. Birge, R.R., P.A. Fleitz, R.B. Gross, J.C. Izgi, A.F. Lawrence, J.A. Stuart, and J.R. Tallent, Spatial light modulators and optical associative memories based on bacteriorhodopsin. *Proc. IEEE EMBS*, 1990. **12**: 1788–1789.
7. Birge, R.R., K.C. Izgi, J.A. Stuart, and J.R. Tallent, Wavelength dependence of the photorefractive and photodiffractive properties of holographic thin films based on bacteriorhodopsin. *Proc. Natl. Res. Soc.*, 1991. **218**: 131–140.
8. Hampp, N., A. Popp, C. Bräuchle, and D. Oesterhelt, Diffraction efficiency of bacteriorhodopsin films for holography containing bacteriorhodopsin wildtype BR_{WT} and its variants BR_{D85E} and BR_{D96N}. *J. Phys. Chem.*, 1992. **96**(11): 4679–4685.
9. Birge, B., P. Fleitz, R. Gross, J. Izgi, A. Lawrence, J. Stuart, and J. Tallent. *Spatial light modulators and optical associative memories based on bacteriorhodopsin*. 1990, Materials Research Society: Boston, MA.
10. Vsevolodov, N.N., A.B. Druzhko, and T.V. Djukova, Actual possibilities of bacteriorhodopsin application in optoelectronics, in *Molecular Electronics: Biosensors and Biocomputers*, F.T. Hong, Editor. 1989, Plenum Press: New York. pp. 381–384.
11. Vsevolodov, N.N. and V.A. Poltoratskii, Holograms in biochrome, a biological photochromic material. *Sov. Phys. Tech. Phys.*, 1985. **30**: 1235.
12. Hampp, N., C. Bräuchle, and D. Oesterhelt, Bacteriorhodopsin wildtype and variant aspartate-96 Æ asparagine as reversible holographic media. *Biophys. J.*, 1990. **58**: 83–93.
13. Hampp, N., C. Bräuchle, and D. Oesterhelt, Mutated bacteriorhodopsins: competitive materials for optical information processing? *MRS Bull.*, 1992. **XVII**(11): 56–60.

14. Hampp, N., C. Bräuchle, and D. Oesterhelt. Mutated bacteriorhodopsins: new materials for optical storage and information processing, in *4th International Symposium on Bioelectronic and Molecular Electronic Devices*. 1992, R&D Association for Future Electron Devices: Miyazaki, Japan.
15. Oesterhelt, D., C. Bräuchle, and N. Hampp, Bacteriorhodopsin: A biological material for information processing. *Quart. Rev. Biophys.*, 1991. **24**: 425–478.
16. Druzhko, A.B. and S.K. Zharmukhamedov, Biochrome film based on some analogues of bacteriorhodopsin, in *Photosensitive Biological Complexes and Optical Recording of Information*, G.R. Ivanitskiy and N.N. Vsevolodov, Editors. 1985, USSR Academy of Sciences: Biological Research Center, Institute of Biological Physics: Pushchino, Moscow, Russia. pp. 119–125.
17. Birge, R.R., Protein based optical computing and memories. *IEEE Comp.*, 1992. **25**(November): 56–67.
18. Birge, R.R., Three-dimensional optical memories. *Am. Sci.*, 1994. **82**: 349–355.
19. Birge, R.R., Protein-based three-dimensional memory. *Am. Sci.*, 1994. **82**(July–August): 348–355.
20. Birge, R.R., Protein-based computers. *Sci. Am.*, 1995. **272**(3): 90–95.
21. Birge, R.R., Z. Chen, R.B. Gross, S.B. Hom, K.C. Izgi, J.A. Stuart, J.R. Tallent, and B.W. Vought, Biomolecular photonics based on bacteriorhodopsin, in *CRC Handbook of Organic Photochemistry and Photobiology*, P.-S.S. W. M. Horspool, Editor. 1995, CRC Press: New York. pp. 1568–1585.
22. Birge, R.R., N.B. Gillespie, E.W. Izaguirre, A. Kusnetzow, A.F. Lawrence, D. Singh, Q.W. Song, E. Schmidt, J.A. Stuart, S. Seetharaman, and K.J. Wise, Biomolecular electronics: protein-based associative processors and volumetric memories. *J. Phys. Chem.*, 1999. **103**: 10746–10766.
23. Birge, R.R. and R.B. Gross, Biomolecular optoelectronics, in *Introduction to Molecular Electronics*, M.C. Petty, M.R. Bryce, and D. Bloor, Editors. 1995, Edward Arnold: London. pp. 315–344.
24. Birge, R.R., D.S.K. Govender, R.B. Gross, A.F. Lawrence, J.A. Stuart, J.R. Tallent, E. Tan, and B.W. Vought, Bioelectronics, three-dimensional memories and hybrid computers. *IEEE IEDM Tech. Digest*, 1994. **94**: 3–6.
25. Birge, R.R., B. Parsons, Q.W. Song, and J.R. Tallent, Protein-based three-dimensional memories and associative processors, in *Molecular Electronics*, M.A. Ratner and J. Jortner, Editors. 1997, Blackwell Science Ltd.: Oxford. pp. 439–471.
26. Birge, R.R., C. Zhang, and A.F. Lawrence, Optical random access memory based on bacteriorhodopsin, in *Molecular Electronics: Biosensors and Biocomputers*, F.T. Hong, Editor. 1989, Plenum Press: New York. pp. 369–379.
27. Stuart, J.A., D.L. Marcy, and R.R. Birge. Photonic and optoelectronic applications of bacteriorhodopsin, in *Bioelectronic Applications of Photochromic Pigments*. 2000a. IOS Press: Szeged, Hungary.
28. Stuart, J.A., D.L. Marcy, and R.R. Birge. Bacteriorhodopsin-based three-dimensional optical memory, in *Bioelectronic Applications of Photochromic Pigments*. 2000b. IOS Press: Szeged, Hungary.
29. Stuart, J.A., D.L. Marcy, K.J. Wise, and R.R. Birge, Volumetric optical memory based on bacteriorhodopsin. *Syn. Metals*, 2002. **127**: 3–15.
30. Stuart, J.A., D.L. Marcy, K.J. Wise, and R.R. Birge, Biomolecular electronic device applications of bacteriorhodopsin, in *Molecular Electronics: Bio-sensors and Bio-computers*, L.e.a. Barasanti, Editor. 2003, Kluwer Academic Publishers: Norwell, MA. pp. 265–299.
31. Stuart, J.A., J.R. Tallent, E.H.L. Tan, and R.R. Birge, Protein-based volumetric memory. *Proc. IEEE Nonvol. Mem. Tech. (INVMTC)*, 1996. **6**: 45–51.
32. Bhattacharya, P., J. Xu, G. Váró, D.L. Marcy, and R.R. Birge, Monolithically integrated bacteriorhodopsin-GaAs field-effect transistor photoreciever. *Opt. Lett.*, 2002. **27**(10): 839–841.
33. Xu, J., P. Bhattacharya, and G. Váró, Photo-induced anisotropic photoelectric response in oriented bacteriorhodopsin films. *Opt. Mater.*, 2003. **22**: 321–326.
34. Rayfield, G., Ultra high speed bacteriorhodopsin photodetectors, in *Molecular Electronics*, F.T. Hong, Editor. 1989, Plenum: New York. pp. 361–368.
35. Ebrey, T.G., Light energy transduction in bacteriorhodopsin, in *Thermodynamics of Membrane Receptors and Channels*, M.B. Jackson, Editor. 1993, CRC Press: Boca Raton, FL. pp. 353–387.

36. Stuart, J.A. and R.R. Birge, Characterization of the primary photochemical events in bacteriorhodopsin and rhodopsin, in *Biomembranes*, A.G. Lee, Editor. 1996, JAI Press: London. pp. 33–140.
37. Birge, R.R., Nature of the primary photochemical events in rhodopsin and bacteriorhodopsin. *Biochim. Biophys. Acta*, 1990. **1016**: 293–327.
38. Lanyi, J.K., Bacteriorhodopsin. *Int. Rev. Cytol.*, 1999. **187**: 161–202.
39. Lanyi, J.K. and H. Luecke, Bacteriorhodopsin. *Curr. Opin. Struct. Biol.*, 2001. **11**(4): 415–419.
40. Lanyi, J.K. and G. Váró, The photocycles of bacteriorhodopsin. *Isr. J. Chem.*, 1995. **35**: 365–385.
41. Lanyi, J., Bacteriorhodopsin: a paradigm for proton pumps? *Biophys. Chem.*, 1995. **56**: 143–151.
42. Danon, A. and W. Stoeckenius, Photophosphorylation in Halobacterium halobium. *Proc. Natl. Acad. Sci. U.S.A.*, 1974. **71**(No. 4): 1234–1238.
43. Stoeckenius, W., Purple membrane of Halobacteria: a new light-energy converter. *Acc. Chem. Res.*, 1980. **13**: 337–344.
44. Oesterhelt, D. and W. Stoeckenius, Functions of a new photoreceptor membrane. *Proc. Natl. Acad. Sci. U.S.A.*, 1973. **70**(No. 10): 2853–2857.
45. Luecke, H., B. Schobert, H.-T. Richter, J.-P. Cartailler, and J.K. Lanyi, Structure of bacteriorhodopsin at 1.55 Å resolution. *J. Mol. Biol.*, 1999. **291**: 899–911.
46. Luecke, H., B. Schobert, H.-T. Richter, J.-P. Cartailler, and J.K. Lanyi, Structural changes in bacteriorhodopsin during ion transport at 2 angstrom resolution. *Science*, 1999. **286**: 255–260.
47. Luecke, H., B. Schobert, J.-P. Cartailler, H.-T. Richter, A. Rosengarth, R. Needleman, and J.K. Lanyi, Coupling photoisomerization of retinal to directional transport in bacteriorhodopsin. *J. Mol. Biol.*, 2000. **300**: 1237–1255.
48. Luecke, H., H.T. Richter, and J.K. Lanyi, Proton transfer pathways in bacteriorhodopsin at 2.3 angstrom resolution. *Science*, 1998. **280**(5371): 1934–1937.
49. Lanyi, J.K. and B. Schobert, Mechanism of proton transport in bacteriorhodopsin from crystallographic structures of the K, L, M1, M2, and M2′ intermediates of the photocycle. *J. Mol. Biol.*, 2003. **328**(2): 439–450.
50. Lanyi, J. and B. Schobert, Crystallographic structure of the retinal and the protein after deprotonation of the Schiff base: the switch in the bacteriorhodopsin photocycle. *J. Mol. Biol.*, 2002. **321**(4): 727–737.
51. Schobert, B., L.S. Brown, and J.K. Lanyi, Crystallographic structures of the M and N intermediates of bacterorhodopsin: assembly of a hydrogen-bonded chain of water molecules between Asp-96 and the retinal Schiff base. *J. Mol. Biol.*, 2003. **330**: 553–570.
52. Schobert, B., J. Cupp-Vickery, V. Hornak, S.O. Smith, and J.K. Lanyi, Crystallographic structure of the K intermediate of bacteriorhodopsin: conservation of free energy after photoisomerization of retinal. *J. Mol. Biol.*, 2002. **321**: 715–726.
53. Birge, R.R., Photophysics and molecular electronic applications of the rhodopsins. *Annu. Rev. Phys. Chem.*, 1990. **41**: 683–733.
54. Gross, R.B., K.C. Izgi, and R.R. Birge, Holographic thin films, spatial light modulators and optical associative memories based on bacteriorhodopsin. *Proc. SPIE*, 1992. **1662**: 186–196.
55. Birge, R.R., Introduction to molecular and biomolecular electronics. *Adv. Chem.*, 1994. **240**: 1–14.
56. Birge, R.R. and D.S.K. Govender, Three-dimensional optical memory. 1993, Syracuse University: Syracuse, NY.
57. Hampp, N. and T. Juchem. Fringemaker—the first technical system based on bacteriorhodopsin, in *Bioelectronic Applications of Photochromic Pigments*, 2000, IOS Press: Szeged, Hungary.
58. Popp, A., M. Wolperdinger, N. Hampp, C. Bräuchle, and D. Oesterhelt, Photochemical conversion of the *O*-intermediate to 9-*cis*-retinal-containing products in bacteriorhodopsin films. *Biophys. J.*, 1993. **65**(October): 1449–1459.
59. Gillespie, N.B., K.J. Wise, L. Ren, J.A. Stuart, D.L. Marcy, J. Hillebrecht, Q. Li, L. Ramos, K. Jordan, S. Fyvie, and R.R. Birge, Characterization of the branched-photocycle intermediates P and Q of bacteriorhodopsin. *J. Phys. Chem.*, 2002. **106**: 13352–13361.
60. Hillebrecht, J.R., J.F. Koscielecki, K.J. Wise, D.L. Marcy, W. Tetley, R. Rangarajan, J. Sullivan, M. Brideau, M.P. Krebs, J.A. Stuart, and R.R. Birge, Optimization of protein-based volumetric optical memories and associative processors using directed evolution. *Nanobiotechnology*, 2005. **1**(2): 141–152.

61. Hillebrecht, J.R., K.J. Wise, J.F. Koscielecki, and R.R. Birge, Directed evolution of bacteriorhodopsin for device applications. *Methods Enzymol.*, 2004. **388**: 333–347.
62. Lee, I., E. Greenbaum, S. Budy, J.R. Hillebrecht, R.R. Birge, and J.A. Stuart, Photoinduced surface potential change of bacteriorhodopsin mutant D96N measured by scanning surface potential microscopy. *J. Phys. Chem. B*, 2006. **110**(22): 10982–10990.
63. Li, Q., J.A. Stuart, R.R. Birge, J. Xu, A. Stickrath, and P. Bhattacharya, Photoelectric response of polarization sensitive bacteriorhodopsin films. *Biosens. Bioelectron.*, 2002. **19**: 869–874.
64. Xu, J., P. Bhattacharya, D.L. Marcy, J.A. Stuart, and R.R. Birge, Photoconduction in bacteriorhodopsin/GaAs heterostructures. *Electr. Lett.*, 2001. **37**(10): 1–2.
65. Boehm, M.F., M.A. Gawinowicz, A. Foucault, F. Derguini, and K. Nakanishi, Photoaffinity labeling studies of bacteriorhodopsin with [15-^{3}H]-3-diazo-4-keto-all-*trans*-retinal. *J. Am. Chem. Soc.*, 1990. **112**(21): 7779–7782.
66. Druzhko, A.B. and H.H. Weetall, Photoinduced transformation of wild-type and D96N-mutant 4-keto-bacteriorhodopsin gelatin films. *Thin Solid Films*, 1997. **293**: 281–284.
67. Arnold, F. and J.C. Moore, Optimizing industrial enzymes by directed evolution. *Adv. Biochem. Eng.*, 1997. **58**: 1–14.
68. Kuchner, O. and F. Arnold, Directed evolution of enzyme catalysts. *Trends Biotech.*, 1997. **15**: 523–530.
69. Arnold, F.H., Design by directed evolution. *Acc. Chem. Res.*, 1998. **31**: 125–131.
70. Arnold, F.H. and A.A. Volkov, Directed evolution of biocatalysts. *Curr. Opin. Chem. Biol.*, 1999. **3**(1): 54–59.
71. Arnold, F., P.L. Wintrode, K. Miyazaki, and A. Gershenson, How enzymes adapt: lessons from directed evolution. *Trends Biochem. Sci.*, 2001. **26**(2): 100–106.
72. Crameri, A., S. Raillart, E. Bermudez, and W.P.C. Stemmer, DNA shuffling of a family of genes from diverse species accelerates directed evolution. *Nature*, 1998. **391**: 288–291.
73. Miyazaki, K., P.L. Wintrode, R.A. Grayling, D.N. Rubingh, and F.H. Arnold, Directed evolution study of temperature adaptation in a psychrophilic enzyme. *J. Mol. Biol.*, 2000. **297**(4): 1015–1026.
74. Stemmer, W.P.C., DNA shuffling by random fragmentation and reassembly: in vitro recombination for molecular evolution. *Proc. Natl. Acad. Sci. U.S.A.*, 1994. **91**: 10747–10751.
75. Tobin, M.B., C. Gustafsson, and G.W. Huisman, Directed evolution: the 'rational' basis for 'irrational' design. *Curr. Opin. Struct. Biol.*, 2000. **10**: 421–427.
76. Oesterhelt, D., The structure and mechanism of the family of retinal proteins from halophilic archaea. *Curr. Opin. Struct. Biol.*, 1998. **8**: 489–500.
77. Brizzolara, R.A., J.L. Boyd, and A.E. Tate, Evidence for covalent attachment of purple membrane to a gold surface via genetic modification of bacteriorhodopsin. *J. Vac. Sci. Technol.*, 1997. **15**(3): 773–778.
78. Brizzolara, R.A., A method for patterning purple membrane using self-assembled monolayers. *BioSystems*, 1995. **35**: 137–140.
79. Brizzolara, R.A. and B.C. Beard, Control of purple membrane adsorption to a glass surface using self-assembled monolayers. *J. Vac. Sci. Technol.*, 1994. **12**(5): 2981–2987.
80. Takamatsu, S., K. Hoshino, K. Matsumoto, and I. Shimoyama. Biomolecular image sensor of bacteriorhodopsin patterned by electrodeposition, in *18th IEEE International Conference on Micro Electro Mechanical Systems*. 2005. Miami, Florida.
81. Libertino, S., M. Fichera, G. D'Arrigo, A. La Mantia, and D. Ricceri, Characterization and patterning of bacteriorhodopsin films on Si-based materials. *Syn. Metals*, 2003. **138**: 71–74.
82. Libertino, S., M. Fichera, A. La Mantia, and D. Ricceri, Optical and structural characterization of bacteriorhodopsin films on Si-based materials. *Syn. Metals*, 2003. **138**: 141–144.
83. Keszthelyi, L., Orientation of membrane fragments by electric field. *Biochim. Biophys. Acta*, 1980. **598**: 429–436.
84. Chen, Z. and R.R. Birge, Protein based artificial retinas. *Trends Biotech.*, 1993. **11**: 292–300.
85. Miyasaka, T., K. Koyama, and I. Itoh, Quantum conversion and image detection by a bacteriorhodopsin-based artificial photoreceptor. *Science*, 1992. **255**(17 January 1992): 342–344.
86. Tan, E.H.L., D.S.K. Govender, and R.R. Birge, Large organic cations can replace Mg^{2+} and Ca^{2+} ions in bacteriorhodopsin and maintain proton pumping ability. *J. Am. Chem. Soc.*, 1996. **118**(11): 2752–2753.

87. Martin, C.H., Z.P. Chen, and R.R. Birge, Towards a bacteriorhodopsin-silicon neuromorphic photosensor. In: *Pac. Symp. Biocomput.*, R.B. Altman, et al. Editors 1997, World Scientific: Maui. 268–279.
88. Bryl, K. and K. Yoshihara, Can purple membrane play the role of biosensor under extreme conditions, in *Bioelectric Applications of Photonic Pigments*, A. Der and L. Keszthelyi, Editors. 2000, IOS Press: Amsterdam. pp. 211–224.
89. Boucher, F., S.G. Taneva, S. Elouatik, M. Dery, S. Messaoudi, E. Harvey-Girard, and N. Beaudoin, Reversible inhibition of proton release activity and the anesthetic-induced acid–base equilibrium between the 480 and 570 nm forms of bacteriorhodopsin. *Biophys. J.*, 1996. **70**(2): 948–961.
90. Gao, M.M. and F. Boucher, The uncoupling of bacteriorhodopsin by high temperature and anaesthetics. *Toxicol. Lett.*, 1998. **100–101**: 393–396.
91. Hamanaka, T., T. Nakagawa, Y. Kito, S. Nishimura, I. Uchida, and T. Mashimo, Binding of volatile anesthetics to purple membranes studied by X-ray diffraction. *Toxicol. Lett.*, 1998. **100–101**: 397–403.
92. Lee, K.H., A.R. McIntosh, and F. Boucher, The interaction between halogenated anaesthetics and bacteriorhodopsin in purple membranes as examined by intrinsic ultraviolet fluorescence. *Biochem. Cell Biol.*, 1991. **69**(2–3): 178–184.
93. Lin, C.T., Y.G. Chyan, G.C. Kresheck, H.C. Bitting, Jr., and M.A. el-Sayed, Interaction of dibucaine.HCl local anesthetics with bacteriorhodopsin in purple membrane: a spectroscopic study. *Photochem. Photobiol.*, 1989. **49**(5): 641–648.
94. Lin, C.T., C.J. Mertz, H.C. Bitting, and M.A. El-Sayed, Fluorescence anisotropy studies of dibucaine.HCl in micelles and bacteriorhodopsin. *J. Photochem. Photobiol.*, 1992. **13**(2): 169–185.
95. Nakagawa, T., T. Hamanaka, S. Nishimura, I. Uchida, T. Mashimo, and Y. Kito, The quantitative analysis of three action modes of volatile anesthetics on purple membrane. *Biochim. Biophys. Acta*, 2000. **1468**(1–2): 139–149.
96. Nakagawa, T., T. Hamanaka, S. Nishimura, T. Uruga, and Y. Kito, The specific binding site of the volatile anesthetic diiodomethane to purple membrane by X-ray diffraction. *J. Mol. Biol.*, 1994. **238**(3): 297–301.
97. Nishimura, S., T. Mashimo, K. Hiraki, T. Hamanaka, Y. Kito, and I. Yoshiya, Volatile anesthetics cause conformational changes of bacteriorhodopsin in purple membrane. *Biochim. Biophys. Acta*, 1985. **818**(3): 421–424.
98. Perozo, E. and W.L. Hubbell, Voltage activation of reconstituted sodium channels: use of bacteriorhodopsin as a light driven current source. *Biochemistry*, 1993. **32**(39): 10471–10478.
99. Renthal, R., L. Brogley, and J. Vila, Altered protein–chromophore interaction in dicyclohexylcarbodiimide-modified purple membrane sheets. *Biochim. Biophys. Acta*, 1988. **935**(2): 109–114.
100. Taneva, S.G., J.M. Caaveiro, I.B. Petkanchin, and F.M. Goni, Electrokinetic charge of the anesthetic-induced bR480 and bR380 spectral forms of bacteriorhodopsin. *Biochim. Biophys. Acta*, 1995. **1236**(2): 331–337.
101. Uchida, I., T. Mashimo, and I. Yoshiya, Pressure–anesthetic interaction on the molecular conformation of bacteriorhodopsin. *Ann. N.Y. Acad. Sci.*, 1991. **625** (*Mol. Cell. Mech. Alcohol Anesth.*): 767–769.
102. Uruga, T., T. Hamanaka, Y. Kito, I. Uchida, S. Nishimura, and T. Mashimo, Effects of volatile anesthetics on bacteriorhodopsin in purple membrane, *Halobacterium halobium* cells and reconstituted vesicles. *Biophys. Chem.*, 1991. **41**(2): 157–168.
103. Dyukova, T.V. and N.N. Vsevolodov, *Photochromic compositions and materials containing bacteriorhodopsin*. U. S. Patent No. 5,518,858, May 21, 1996.
104. Misra, S., R. Govindjee, T.G. Ebrey, N. Chen, J.X. Ma, and R.K. Crouch, Proton uptake and release are rate-limiting steps in the photocycle of the bacteriorhodopsin mutant E204Q. *Biochemistry*, 1997. **36**(16): 4875–4883.
105. Radionov, A. and A. Kaulen, Two bacteriorhodopsin M intermediates differing in accessibility of the Schiff base for azide. *FEBS Lett.*, 1996. **387**: 122–126.
106. Steinhoff, H., M. Pfeiffer, T. Rink, O. Burlon, M. Kurz, J. Riesle, E. Heuberger, K. Gerwert, and D. Oesterhelt, Azide reduces the hydrophobic barrier of the bacteriorhodopsin proton channel. *Biophys. J.*, 1999. **76**: 2702–2710.

107. Takao, K.-i., T. Kikukawa, T. Araiso, and N. Kamo, Azide accelerates the decay of M-intermediate of pharaonis phoborhodopsin. *Biophys. Chem.*, 1998. **73**(1–2): 145–153.
108. Lo, K.M., S.S. Jones, N.R. Hackett, and H.G. Khorana, Specific amino acid substitutions in bacterioopsin: Replacement of a restriction fragment in the structural gene by synthetic DNA fragments containing altered codons. *Proc. Natl. Acad. Sci. U.S.A.*, 1984. **81**: 2285–2289.
109. Hong, F.T., Molecular sensors based on the photoelectric effect of bacteriorhodopsin: origin of differential responsivity. *Mater. Sci. Eng.*, 1997. **C4**: 267–385.
110. Ahmad, A., N.K. Nanda, and K. Garg, Are primitive polynomials always best in signature analysis? *IEEE Des. Test Comput.*, 1990. **7**: 36–38.
111. Braun, S., *Mechanical Signature Analysis*. 1986, Academic Press: London.
112. David, R., Signature analysis for multiple-output circuits. *IEEE Trans. Comput.*, 1986. **35**: 830–837.
113. Dekker, A.G., T.J. Malthus, M.M. Wijnen, and E. Seyhan, The effect of spectral band width and positioning on the spectral signature analysis of inland waters. *Remote Sens. Environ.*, 1992. **41**: 211–226.
114. Sanders, C.A., M. Rodriguez, Jr., and E. Greenbaum, Stand-off tissue-based biosensors for the detection of chemical warfare agents using photosynthetic fluorescence induction. *Biosens. Bioelectron.*, 2001. **16**(7–8): 439–446.
115. Yordy, B., J. Girard, J.F. Koscielecki, J.R. Hillebrecht, W. Tetley, D.L. Marcy, and J.A. Stuart, Utilization of bacteriorhodopsin mutants in chemical sensor architectures. Manuscript in preparation, 2006.
116. Beja, O., L. Aravind, E.V. Koonin, M.T. Suzuki, A. Hadd, L.P. Nguyen, S.B. Jovanovich, C.M. Gates, R.A. Feldman, J.L. Spudich, E.N. Spudich, and E.F. DeLong, Bacterial rhodopsin: evidence for a new type of phototrophy in the sea. *Science*, 2000. **289**(5486): 1902–1906.
117. Beja, O., E.N. Spudich, J.L. Spudich, M. Leclerc, and E.F. DeLong, Proteorhodopsin phototrophy in the ocean. *Nature*, 2001. **411**(6839): 786–789.
118. Lee, I., J.W. Lee, A. Stubna, and E. Greenbaum, Measurement of electrostatic potentials above oriented single photosynthetic reaction centers. *J. Phys. Chem. B*, 2000. **104**(11): 2439–2443.
119. Nakamura, C., M. Hasegawa, N. Nakamura, and J. Miyake, Rapid and specific detection of herbicides using a self-assembled photosynthetic reaction center from purple bacterium on an SPR chip. *Biosens. Bioelectron.*, 2003. **18**(5–6): 599–603.
120. Koblizek, M., J. Maly, J. Masojidek, J. Komenda, T. Kucera, M.T. Giardi, A.K. Mattoo, and R. Pilloton, A biosensor for the detection of triazine and phenylurea herbicides designed using Photosystem II coupled to a screen-printed electrode. *Biotechnol. Bioeng.*, 2002. **78**(1): 110–116.
121. Montgomery, B.L. and J.C. Lagarias, Phytochrome ancestry: sensors of bilins and light. *Trends Plant Sci.*, 2002. **7**(8): 357–366.
122. Hellingwerf, K.J., Key issues in the photochemistry and signalling-state formation of photosensor proteins. *J. Photochem. Photobiol.*, 2000. **54**(2–3): 94–102.

15

Photoelectric Biosensors: Fundamentals and Innovative Designs

Felix T. Hong

CONTENTS

15.1 Introduction

The molecular machines for vision and for photosynthesis are Nature's optoelectronic devices. Both devices comprise a membrane with two planar layers of phospholipid molecules (lipid bilayers) in which oriented light-sensitive molecules are embedded. The photosynthetic membrane and the visual membrane serve two radically different biological functions: the former converts light energy into electrochemical energy for the purpose of sustaining life (photon energy converter) whereas the latter detects light signals for the purpose of imaging the external world (photon energy sensor). However, the primary events in both types of devices are similar: light-induced charge separation. Because of the asymmetric orientation of the photopigments in these membranes, light-induced charge separation results in a unidirectional charge movement. This vectorial charge movement can be detected as a photovoltage or a photocurrent via a pair of electrodes placed across the membrane. The structure of photosynthetic and visual membranes has inspired useful structural configurations of molecular optoelectronic devices. Prototype molecular and molecular electronic devices were often constructed as layered thin films. Using the Langmuir–Blodgett (LB) technique to configure thin organic films with embedded organic dyes and other functional molecules on a solid support (substrate), Kuhn and coworkers [1–3] pioneered the study of supramolecular photochemistry and have been able to demonstrate fundamental processes mimicking photosynthetic membranes.

The electrical phenomena associated with light-induced vectorial charge movement in membranes or thin films are collectively referred to as the *photoelectric effect*. If brief light pulses are used to excite the membrane, the response is known as the fast photovoltage or the fast photoelectric effect. This particular effect is relevant for biosensor designs. The fast photoelectric effect was first discovered by Brown and Murakami [4], in monkey retina, about four decades ago; the signal was named the *early receptor potential* (ERP). Rigorous analysis of the ERP was not feasible prior to the advent of the method of reconstituting artificial membranes. Visual membranes with complex geometry contribute to shaping (and distortion) of the signal waveforms. Too many extraneous electric parameters made it difficult to construct mathematical models with satisfactory precision.

In 1962, Mueller et al. [5–7] discovered a method to form planar bilayer lipid membranes (BLM) in vitro. Rigorous mathematical modeling became a serious possibility. However, the necessity of using organic solvents in the original method hampered the incorporating of proteins and other biomolecules into the artificial membrane. Subsequently, Takagi and coworkers [8] succeeded in reconstituting a protein—the visual pigment rhodopsin—into a BLM free of organic solvent. Tien [9] succeeded in recording a fast photovoltage in a BLM reconstituted from a chloroplast extract. Using a modified method similar to that of Takagi et al., Montal and coworkers [10,11] observed an ERP-like fast photosignal in both a reconstituted rhodopsin membrane and a reconstituted bacteriorhodopsin (bR) membrane. bR is a protein pigment similar to rhodopsin in chemical structure, but it is found in the purple membrane of *Halobacterium salinarum* (formerly *Halobacterium halobium*) [12–19]. bR has a chemical structure similar to that of rhodopsin but is considerably more stable than rhodopsin. Vsevolodov and coworkers [20,21] pioneered the use of bR as an advanced functional material ("Biochrom-BR"). Subsequently, a variety of prototype devices were reported by a number of laboratories (for reviews, see [22–26]).

The present chapter focuses on the fundamentals of the photoelectric effect and illustrates smart sensor designs with several representative prototype examples reported in the literature. An equivalent circuit simulating the photoelectric effect is developed from first principles, based on realistic molecular models. Experimental photoelectric data are interpreted in mechanistic and molecular terms. The pitfalls that have led to misinterpretation and confusion in the earlier literature are also pointed out. We shall first consider a minimalist experimental prototype in which the underlying chemistry reflects a single-reaction step—an artificial light-driven electron pump. We shall then consider several reconstituted bR model membranes, which exhibit the next higher level of complexity. The technical issues related to molecular sensor designs will be discussed, using several prototypes reported in the literature. Anticipated features of future smart sensors are speculated. On the basis of our new insight into human creativity, speculation and suggestions about future biosensor design are also presented.

15.2 AC and DC Photoelectric Effects

Direct transplantation of classical electrophysiological methodology and methodology used in solution-phase photochemistry has generated considerable controversy and confusion in the early literature regarding the interpretation of measured photoelectric signals. Additional consideration must be given to the presence of photoelectromotive force (photoemf) as well as the interaction of signal-generating element, bR, and properties of inert membrane structures, resistance and capacitance. In addition, a novel capacitance appears because the current source is either deep inside the membrane or at the membrane–water interface(s). This is to be contrasted with classical electrophysiological studies, in which all current sources are outside of the membrane.

Substantial clarification can be made by taking into account the presence of consecutive electron (or proton) transfer reactions and the almost ubiquitous presence of reverse reactions. The most common charge transfer reactions are redox (electron-transfer) reactions and acid–base (proton-transfer) reactions. Consecutive electron-transfer reactions are present in both photosynthetic membranes and in the inner membrane of mitochondria, thus forming the electron-transport chain. Similarly, consecutive proton-transfer reactions are present in the purple membrane of *Halobacteria*. A detailed description of coupled consecutive charge transfer reactions will be presented in Section 15.5. When these reactions take place in a highly organized and anisotropic medium, such as layered membranes or thin films, the net-forward charge transfer results can be observed externally via a pair of electrodes as DC electric currents whereas reversible rapid charge movements accompanying reverse reactions can be detected as AC electric currents. This way of classifying photosignals greatly simplifies the interpretation and facilitates designing photoelectric biosensors.

As will be demonstrated later, the transient photosignal generated by a brief light pulse contains mainly a transient *capacitative* component, which reflects reversible charge transfers with no net charge transport across the membrane. In contrast, the steady-state photosignal reflects the net unidirectional charge transport (vectorial charge transport). Under continuous illumination, the capacitative component sometimes appears only briefly at the onset and again at the cessation of illumination. These transients may not be detected under certain experimental conditions. It can further be shown analytically that this descriptive classification can be replaced by a technically accurate classification: the DC and the AC photoelectric signals [27–30].

The fast photovoltage mentioned earlier is therefore an AC photoelectric signal. That the ERP is a capacitative signal was established by an experiment reported by Hagins and McGaughy [31]. These investigators found that the time-integral of the measured current associated with the ERP approaches zero after the signal decays, indicating no net charge transport across the membrane (see Section 15.3 and Section 15.4).

It was widely recognized that the ERP is generated by rapid charge displacement rather than by ionic diffusion [32,33]. The ERP are therefore different from other bioelectric signals commonly encountered in biology, for example, action potentials and synaptic potentials. The AC and the DC photoelectric signals are related to the underlying photochemistry and are manifestations of interactions of the intrinsic photochemical processes and the inert support structure (substrate). The latter feature is absent in solution-phase photochemistry but these factors must be taken into consideration in smart sensor designs.

15.3 Electron as a Charge Carrier: An Artificial Light-Driven Electron Pump

We shall begin the analysis with a minimalist approach by devising a simple model membrane system with well-defined photochemical reactions that exhibits both the AC and the DC photoelectric effects [34–36]. The choice of such a simple system is necessary to avoid ambiguity in molecular and mechanistic interpretation. This requirement is fulfilled by redox reactions of lipid-soluble magnesium porphyrins. Magnesium octaethylporphyrin or long-chain alkyl esters of magnesium mesoporphyrin IX were used as the membrane-bound light-sensitive redox component, whereas potassium ferricyanide, $K_3Fe(CN)_6$, and ferrocyanide, $K_4Fe(CN)_6$, were used as the water-soluble electron acceptor and donor, respectively. The redox reaction in the dark is a reversible one-electron-transfer reaction [37,38]:

$$P + Fe(CN)_6^{3-} \rightleftarrows P^+ + Fe(CN)_6^{4-},$$

where P and P^+ are the ground-state neutral magnesium porphyrin and the corresponding oxidized monocation, respectively. The oxidized pigment monocation is thermodynamically stable. The photochemistry is also simple and devoid of side reactions:

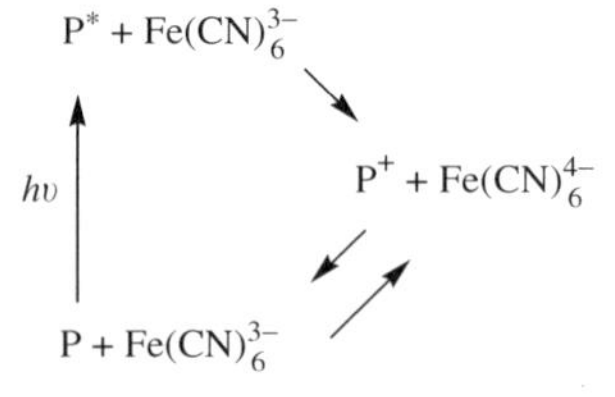

where P^* is neutral magnesium porphyrin in the excited state.

In the model membrane, the magnesium porphyrin molecules are confined to the membrane phase because of hydrophobicity conferred by esterifying magnesium porphyrins with long alkyl chain alcohols (up to 20 carbons). The electron-transfer (redox) reactions between the membrane-bound magnesium porphyrin and the aqueous-borne ferricyanide/ferrocyanide can only take place across a membrane–water interface (interfacial electron-transfer reactions). Thus, there are two interfacial redox (electron-transfer) reactions, one at each interface. Since magnesium porphyrin molecules are mobile in the membrane phase, vectorial electron transport is made possible by imposing an aqueous redox gradient across the membrane. The transmembrane diffusion of P and P^+ allows the two interfacial electron-transfer reactions to exchange reactants and products, that is, the two interfacial reactions are coupled by transmembrane diffusion of P and P^+.

This model system constitutes a rudimentary photosynthetic system; light pumps electrons unidirectionally across the membrane. Although this simple system is not as efficient as natural photosynthetic systems, it captures the essence of a minimalist scheme for photosynthesis and serves to conceptually link chlorophyll-based photosynthetic membranes to the purple membrane of *Halobacteria*.

In a typical experiment, one of the two aqueous phases (called the oxidant side) contains predominantly ferricyanide ions (e.g., 100 mM potassium ferricyanide with or without a much smaller amount of potassium ferrocyanide ranging from 0.5 to 3 mM) whereas the other aqueous phase (the reductant side) contains predominantly potassium ferrocyanide (with or without a trace amount of ferricyanide) (Figure 15.1). The asymmetry in the redox gradient across the membrane allows more P^+ to be produced at the oxidant side than at

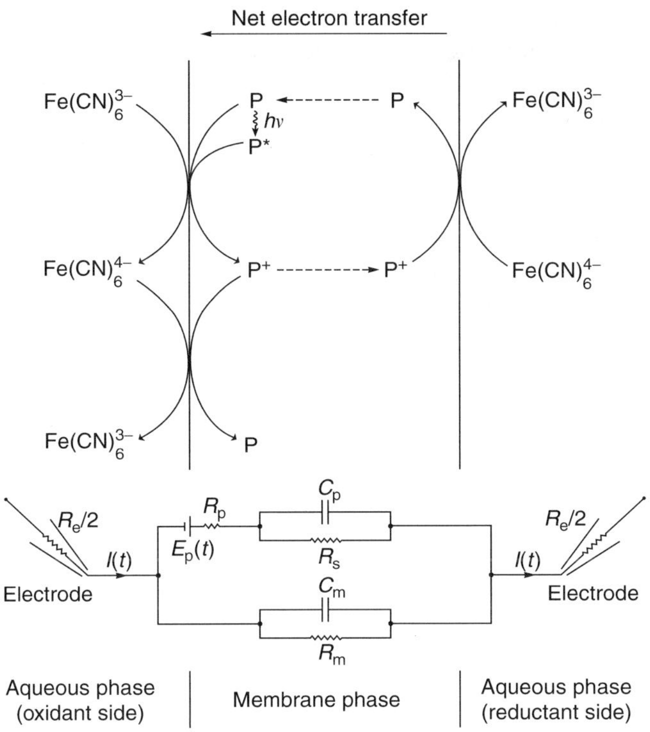

FIGURE 15.1
A simple light-driven electron pump and the equivalent circuit. The bilayer lipid membrane (BLM) contains a lipid-soluble magnesium porphyrin and separates two aqueous phases with asymmetric concentrations of potassium ferricyanide (electron acceptor) and potassium ferrocyanide (electron donor). The symbol *P* stands for the ground-state pigment, P^* for the excited state pigment, and P^+ for the corresponding monocation. Interfacial electron-transfer reactions between the membrane-bound pigment and the aqueous-borne electron donor/acceptor are coupled by diffusion of P and P^+. Electrons are preferentially transported from the reductant side (where the donor concentration is much greater than the acceptor concentration) to the oxidant side (where the acceptor concentration is much greater than the donor concentration). The reverse reaction at the oxidant interface is also shown. The equivalent circuit is shown at the bottom of the diagram. The photoelectromotive force (photoemf) $E_p(t)$ with its internal resistance, R_p, is connected in series with the chemical capacitance, C_p, and the transmembrane DC resistance, R_s. The ordinary membrane (ionic) resistance, R_m, and capacitance, C_m, are placed in parallel with the photoemf. The access impedance (resistance), R_e, includes the amplifier input impedance, the electrode impedance, and the impedance of the intervening electrolyte solution. See text for further explanation. (From Hong, F. T., Mauzerall, D. (1974). Interfacial photoreactions and chemical capacitance in lipid bilayers. *Proc. Natl. Acad. Sci. USA* 71:1564–1568.)

the reductant side, and, conversely, causes less neutral P to appear at the oxidant side than at the reductant side. As a consequence, the net transmembrane diffusion of P^+ generates an electric current that flows from the oxidant side to the reductant side. Net diffusion of the neutral ground-state P in the opposite direction is electrically silent. Thus, in the membrane phase, the pigment molecules undergo cyclic oxidation and reduction—oxidation at the oxidant side and reduction at the reductant side. The "counter-current" transmembrane flows of P and P^+ thus give rise to a sustained DC photoelectric current (from the oxidant side to the reductant side). The two interfacial redox reactions are coupled by diffusion of P and P^+ in opposite directions (coupled interfacial electron-transfer reactions). The hydrophobic porphyrin molecules thus serve as the electron shuttles.

Under illumination with continuous light, a steady-state (sustained) photocurrent or photovoltage is observed; the polarity is consistent with electron movement from the reductant side to the oxidant side and is independent of direction of illumination (from the reductant side to the oxidant side or the other way around). Similar observations have been reported using other types of redox reactions (reviewed by Tien [39]).

Upon illumination with a brief (microsecond or shorter) light pulse, a transient photocurrent initially flows in the direction expected by vectorial electron transfers but

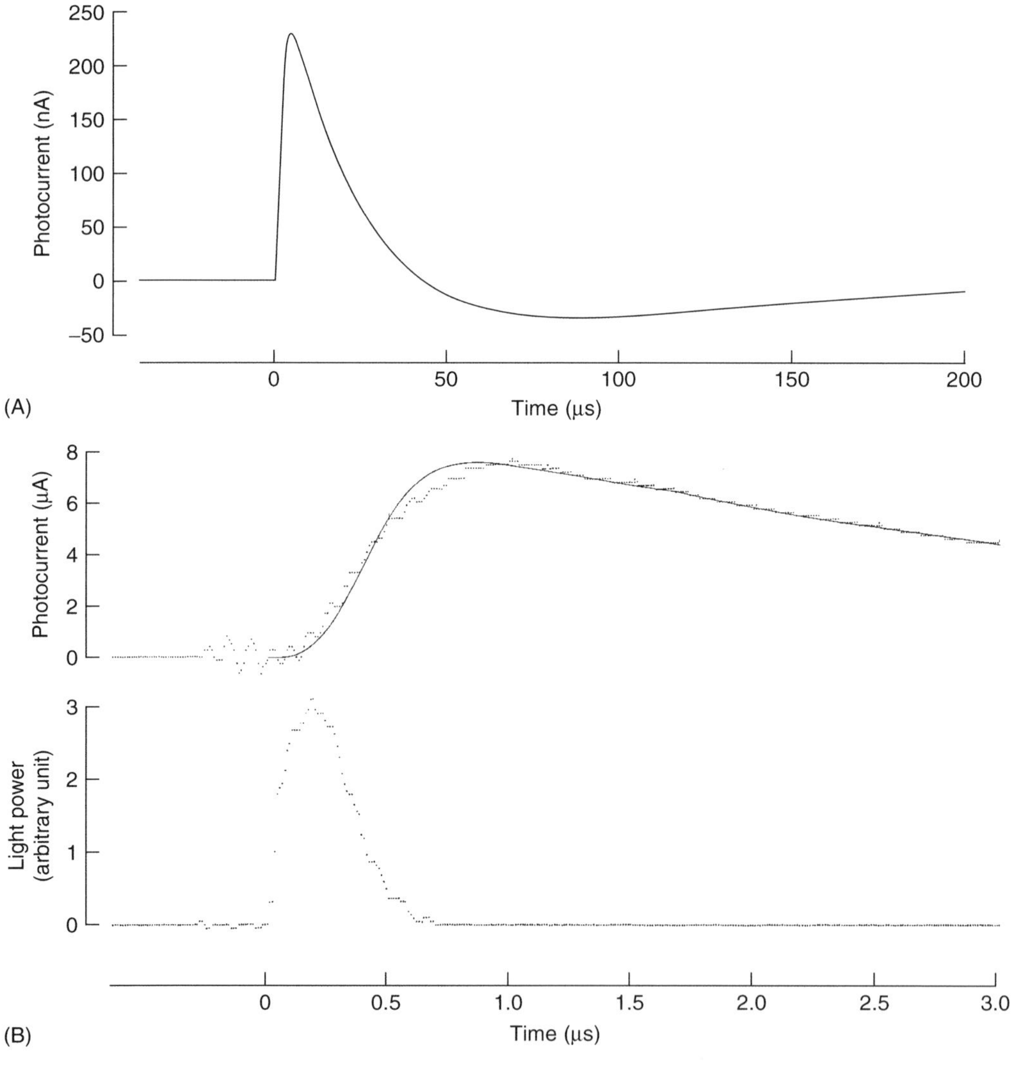

FIGURE 15.2
Comparison of experimentally measured with predicted photoelectric responses. The BLM contained about 5 mM magnesium mesoporphyrin IX di-*n*-amyl ester. Both aqueous phases contained 1 M NaCl and 10 mM phosphate buffer at pH 7.2. In addition, the oxidant side contained 20 mM potassium ferricyanide and 0.5 mM potassium ferrocyanide whereas the reductant side contained 20 mM potassium ferrocyanide (the omission of potassium ferrocyanide at the reductant side has no major effect). The access impedance was 5.1 kΩ in Record A and 380 Ω in Record B. The instrumental time constant was 1.5 μs in Record A and 150 ns in Record B. The photoemf has the time course of the power (intensity) of the dye laser pulse (output 590 nm, shown as the bottom trace in Record B). See text for further explanation. (From Hong, F. T., Mauzerall, D. (1974). Interfacial photoreactions and chemical capacitance in lipid bilayers. *Proc. Natl. Acad. Sci. USA* 71:1564–1568. (A) and Hong, F. T., Mauzerall, D. (1976). Tunable voltage clamp method: application to photoelectric effects in pigmented bilayer lipid membranes. *J. Electrochem. Soc.* 123:1317–1324. (B))

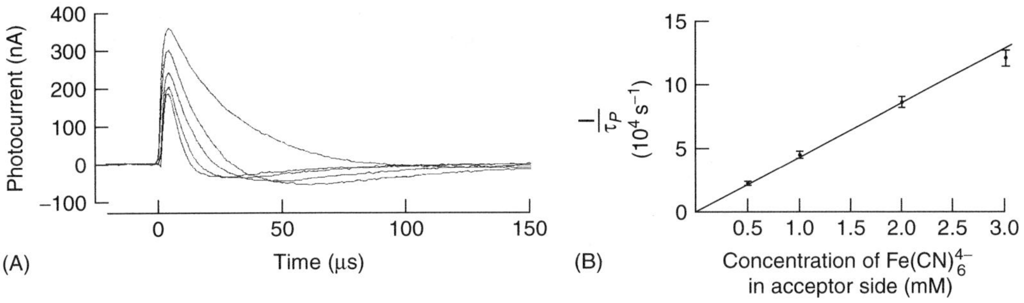

FIGURE 15.3
The effect of varying the aqueous redox compositions on the time course of the AC photosignal from the Mg porphyrin BLM. (A) The reductant side was free of electron acceptors and donors. The oxidant side initially contained 10 mM of potassium ferricyanide but no ferrocyanide. The concentration of potassium ferrocyanide was then increased from 0 to 0.5, 1.0, 2.0, and finally to 3.0 mM. The relaxation of the photocurrent became progressively faster and faster. (B) Deconvolution yields the pseudofirst-order rate constant of the interfacial charge recombination, $1/\tau_p$, which is plotted against the concentration of $Fe(CN)_6^{-4}$. This plot yields a straight line passing through the origin, as expected from a pseudofirst-order regime. The slope is the second-order rate constant of the reverse electron transfer at the oxidant interface $(4.3 \pm 0.5) \times 10^7\ M^{-1}\ s^{-1}$. (From Hong, F. T. (1976). Charge transfer across pigmented bilayer lipid membrane and its interfaces. *Photochem. Photobiol.* 24:155–189.)

subsequently the current reverses its polarity (Figure 15.2). If the concentration of ferrocyanide at the oxidant side is increased, the decay rate of the photocurrent (i.e., the relaxation of the photoelectric process) also increases (Figure 15.3A). This observation suggests that increasing ferrocyanide (electron donor) concentration at the oxidant side enhances the rate of the reverse (back) electron transfer. Furthermore, the relaxation-time course of the photocurrent can be fit with the following equations [35]:

$$I(t) = A_s \cdot \exp\left(-\frac{t}{\tau_s}\right) - A_1 \cdot \exp\left(-\frac{t}{\tau_1}\right) \tag{15.1}$$

where the amplitudes and the time constants are inversely related:

$$\frac{A_s}{A_1} = \frac{\tau_1}{\tau_s} \tag{15.2}$$

Equations (15.1) and (15.2) guarantee that the photocurrent obeys the following condition:

$$\int_0^\infty I(t)\,dt = 0 \tag{15.3}$$

which happens to be the *zero time-integral* condition discovered by Hagins and McGaughy [31]. Together with the ferrocyanide dependence of the photosignal relaxation, the zero time-integral condition suggests that the transient photocurrent in the Mg porphyrin membrane is a manifestation of forward and reverse electron transfers at the oxidant interface. It is not caused by the reversal of transmembrane diffusion of P and P^+ because the latter is much too slow. Nor is it caused by interfacial electron transfer at the reductant interface because the AC photosignal is insensitive to the redox composition there (the AC signal generated at the reductant side is much smaller; see [35,40]). In other words, the two interfacial redox reactions are *decoupled chemically* during the brief transient but are coupled in the steady state. It is observed that the amplitude of transient photocurrent exceeds that of the steady-state photocurrent by several orders of magnitude. This means that the transient photocurrent is preferentially amplified. The reason for the preferential amplification can be understood in terms of AC-coupling of the photocurrent source as indicated by the zero time-integral condition above. AC-coupling can be implemented with a high-pass *RC* filter. Thus, the simplest electrical equivalent circuit that can account for all these observations is shown in the bottom half of Figure 15.1. Since the transient photoelectric signals were DC-coupled to the measuring circuitry and there were no external high-pass filters, the phenomenological high-pass RC filter might be intrinsic.

In the equivalent circuit, the photocurrent (or photovoltage) source is light-dependent (and therefore time-dependent) if a brief light pulse, instead of continuous light, is used to illuminate the membrane. This time-dependent electromotive force is designated by $E_p(t)$ in Figure 15.1. The combination of a capacitance C_p and a resistance R_p constitutes the high-pass filter mentioned above. A transmembrane resistance R_s is added to provide the pathway for the steady-state (DC) photocurrent. These circuit elements represent the process responsible for the generation of a light-induced transient signal (AC current driven by E_p through R_p and C_p) and a steady-state signal (DC current driven by E_p through R_p and R_s). It is however necessary to connect this *RC* circuit to another circuit formed by R_m (transmembrane ionic resistance) and C_m (ordinary membrane capacitance). Here C_p and C_m are regarded as two separate elements. Whether C_p is indeed a distinct physical entity has been a topic of much debate. Contrasting views using alternative approaches can be found in several old articles [41–51]. Detailed experimental and theoretical basis of chemical capacitance has been published [35,52,53]. The equivalent circuit in Figure 15.1 is equally applicable to other types of pigment-containing membranes. In brief, the molecular origin of C_p is light-induced charge separation and subsequent recombination. C_p has thus been named *chemical capacitance* because of its physicochemical origin [34]. Contrary to a recent claim [51], it can be shown analytically and experimentally that chemical capacitance is an intrinsic property of pigment-containing membranes with built-in asymmetry.

The equivalent circuit shown in Figure 15.1 predicts that the time course of the observed photoelectric signal is strongly dependent on the access impedance, R_e, which usually comprised the input impedance of the instrument, the electrode resistance (or impedance), and the resistance of the intervening electrolyte solution (between the tip of electrodes and the membrane surfaces). Short-circuit conditions are achieved by rendering R_e much smaller than the source impedance of the membrane. The entire photocurrent thus proceeds to the measuring device, which is usually a current amplifier. Upon pulsed light illumination, the photocurrent first rises as a sharp spike, and then reverses its polarity and decays in a single exponential time constant of τ_p, which is a function of R_p, C_p, and R_s (Figure 15.4). To a good approximation, τ_p is equal to R_pC_p (see below). Under continuous illumination, the photocurrents exhibit transient spikes when the light source is turned on ("on" overshoot) and when it is turned off ("off" undershoot). The steady-state signal between the two transients during continuous illumination represents the DC photocurrent (Figure 15.4). The two spikes, which decay with the time constant τ_p, reflect the AC photoelectric effect.

If, however, R_e is rendered much greater than the source impedance, almost the entire photocurrent proceeds to charge C_m and only a trace fraction reaches the measuring device—usually an electrometer (open-circuit condition). The photosignal relaxation will then be dominated by the *RC* relaxation with a decay constant of R_mC_m (τ_1 in Figure 15.4 is approximately equal to R_mC_m).

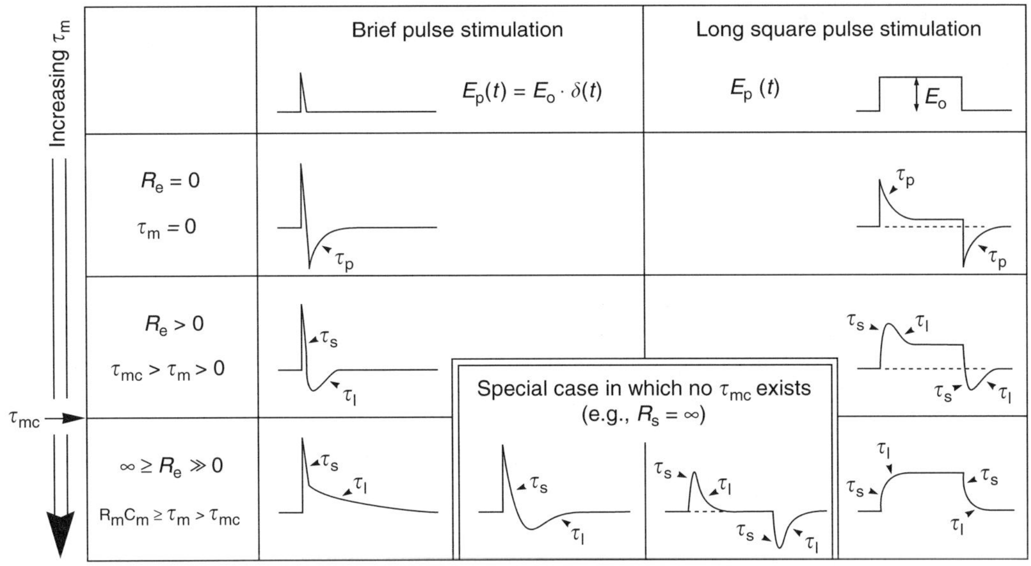

FIGURE 15.4
Relaxation-time course of AC photoelectric responses under different measurement conditions and in response to different kinds of light sources. The parameter τ_m is the characteristic *RC* relaxation time as defined by Eq. (15.7). The parameter τ_{mc} is the critical value at which the time course is changed from monotonic to biphasic and *vice versa*. See text for further explanation. (From Hong, F. T. (1980). Displacements photocurrents in pigment-containing biomembranes: artificial and natural systems. In: Blank, M. (Ed.). *Bioelectrochemistry: Ions, Surfaces, Membranes* (Advances in Chemistry Ser. No 188). Washington, DC: American Chemical Society, pp. 211–237.)

When the access impedance is neither very large nor very small compared with the source impedance (e.g., when matched to the source impedance), the photocurrent splits into two branches, one branch to charge C_m and the other to reach the external measuring device. The photosignal relaxation then reflects the interaction of the two *RC* networks and decays with two exponential time constants, which are neither R_pC_p nor R_mC_m. In such a brief time interval, the access impedance R_e remains constant, and an exact expression for the photocurrent can be obtained [34]:

$$I(t) = \frac{1/\tau_s - 1/R_sC_p}{R_eC_m(1/\tau_s - 1/\tau_l)}\int_0^t \frac{E_p(u)}{R_p}\exp\left(\frac{u-t}{\tau_s}\right)du - \frac{1/\tau_l - 1/R_sC_p}{R_eC_m(1/\tau_s - 1/\tau_l)}\int_0^t \frac{E_p(u)}{R_p}\exp\left(\frac{u-t}{\tau_l}\right)du \tag{15.4}$$

where τ_s and τ_l are defined by the following equations:

$$\frac{1}{\tau_s} = \frac{1}{2}\left[\left(\frac{1}{R_pC_m} + \frac{1}{\tau_p} + \frac{1}{\tau_m}\right) + \sqrt{\left(\frac{1}{R_pC_m} + \frac{1}{\tau_p} + \frac{1}{\tau_m}\right)^2 - 4\left(\frac{1}{R_pC_mR_sC_p} + \frac{1}{\tau_p\tau_m}\right)}\right] \tag{15.5}$$

$$\frac{1}{\tau_l} = \frac{1}{2}\left[\left(\frac{1}{R_pC_m} + \frac{1}{\tau_p} + \frac{1}{\tau_m}\right) - \sqrt{\left(\frac{1}{R_pC_m} + \frac{1}{\tau_p} + \frac{1}{\tau_m}\right)^2 - 4\left(\frac{1}{R_pC_mR_sC_p} + \frac{1}{\tau_p\tau_m}\right)}\right] \tag{15.6}$$

The parameters τ_s and τ_l are defined by the following two equations:

$$\frac{1}{\tau_m} = \frac{1}{R_e C_m} + \frac{1}{R_m C_m} \tag{15.7}$$

$$\frac{1}{\tau_p} = \frac{1}{R_p C_p} + \frac{1}{R_s C_p} \tag{15.8}$$

Thus, the time constant τ_p cannot be directly obtained by exponential decomposition of the photosignal but it can be recovered by means of deconvolution, using only input parameters that can be determined experimentally (τ_s, τ_l, R_e, R_m, C_m, R_s). Contrary to the common practice of curve fitting using adjustable parameters, the present equivalent circuit model has no freely adjustable parameters. We observed that when R_e is comparable to the source impedance of the membrane, the data obtained by means of deconvolution are more accurate than at other R_e values. In fact, individual values of R_p and C_p can be so determined. In contrast, a short-circuit measurement can only determine the value of the product, R_pC_p, but not the individual values of R_p and C_p. We routinely tuned the value of R_e to optimize the measurements of R_p and C_p. For this reason, this method is named the *tunable voltage clamp* (TVC) method [36].

The source impedance of the membrane is much reduced in the high-frequency range because the chemical capacitance is connected in series to the photoemf. In other words, the built-in high-pass filter consisting of R_p and C_p selectively suppresses signal components representing slower electrical processes and preferentially lets signal components representing faster processes pass through without attenuation. This is why the DC photocurrent is usually below the noise level when the AC photocurrent is measured by means of the TVC method. This is also why an intended short-circuit measurement with a commercial picoammeter (input impedance about 100 kΩ) can actually be closer to open-circuit than to short-circuit conditions because the source impedance at high frequency can be easily reduced to the order of 10 kΩ (that of R_p) [54] (see also discussion in [53] and Section 11 of [29]).

A typical comparison of the prediction of the equivalent circuit with the experimental measurement is shown in Figure 15.2. The time constant τ_p, as obtained by means of deconvolution, is 44 μs. A remarkable picture emerges if we identify the reciprocal of this time constant as the rate constant of the pseudofirst-order reverse electron transfer. In Figure 15.3B, the reciprocal of this time constant is plotted against the concentration of ferrocyanide at the *oxidant side*. The data fit a straight line that intercepts both axes at the origin. This result is consistent with the fact that the reverse reaction is a bimolecular second-order reaction. The slope gives the second-order rate constant of $(4.3 \pm 0.5) \times 10^7\ M^{-1}\ s^{-1}$. The agreement between the predicted and the measured time course of the AC signal motivated us to investigate the molecular origin of the chemical capacitance and to prove the assumption that $1/\tau_p$ is actually the pseudofirst-order rate constant of the reverse electron transfer [35].

15.4 Proton as a Charge Carrier: Bacteriorhodopsin Membranes

The investigation of the purple membrane of *H. salinarum* is a natural extension of the study of the Mg porphyrin membrane. The purple membrane contains a single membrane-bound protein, bR. It is essentially a light-driven proton pump. It exhibits a DC photoelectric signal. However, the unidirectionality of the DC photocurrent is not a consequence of an asymmetric chemical gradient across the membrane. Rather, it arises from the asymmetrical orientation of the photopigment in the membrane.

Both bR and rhodopsin contain a covalently bound vitamin A aldehyde as the chromophore, and both consist of seven segments of transmembrane α-helices with the *C*-terminus located on the cytoplasmic side and the *N*-terminus on the opposite side (see [15]). Both pigments exhibit similar AC photoelectric signals in their native membranes. We therefore use analogous terminology for the signal components. The ERP consists of a faster R_1 component that is temperature-insensitive and a slower R_2 component that can be reversibly inhibited by low temperature (0°C) [33]. The analogous AC signal components elicited by a brief light pulse from a reconstituted purple membrane are named B_1 and B_2, which are differentiated on the basis of temperature sensitivity as in the ERP [55].

15.4.1 Mechanistic Aspects of Signal Generation

In the literature, the ERP has been interpreted as the electrical manifestation of light-induced *intramolecular* charge displacement in rhodopsin and the molecular mechanism is usually referred to as the dipole mechanism, based on the following reasoning. Rhodopsin contains several charged amino-acid residues and maintains a fixed orientation in the visual membrane. Photoactivation of rhodopsin leads to a sequence of conformational changes and concurrent changes of absorption spectra known as the photobleaching sequence. The concurrent charge shifts can, in principle, generate a transient electric dipole in the direction perpendicular to the membrane surface. Such charge separation may or may not span the entire membrane thickness and may or may not lead to a transmembrane net charge transport. There is no evidence that photoactivation of rhodopsin leads to a net charge transport (see Section 15.5). Presumably, all separated charges subsequently recombine upon recovery of rhodopsin from the photoexcitation. Such a process is depicted as the *oriented dipole* (OD) mechanism shown in Figure 15.5 [52]. Thus, the AC photoelectric signal is a manifestation of charge separation during the formation of a transient array of electric dipoles and the subsequent charge recombination following its recovery. This process is tantamount to the charging and discharging of a capacitance, C_p. By applying the Gouy–Chapman diffuse double layer theory and chemical kinetic analysis to this simple model, an equivalent circuit can be derived for the process of charge separation and recombination [35,52].

The OD mechanism is however not the only possible mechanism to generate an AC photosignal. In the aforementioned Mg porphyrin membrane, there is no electric dipole formation during photoactivation because Mg porphyrin molecules are free to diffuse (translationally and rotationally) and, therefore, they cannot maintain a fixed orientation in the membrane. Yet the AC photoelectric signal so generated exhibits all major characteristics of the ERP [35]. Indeed, an interfacial electron (or proton) transfer constitutes another kind of charge separation (across a membrane–water interface). In the case of the Mg porphyrin membrane, when an electron is transferred from the membrane-bound porphyrin molecule to the aqueous acceptor ferrocyanide, the membrane-bound porphyrin molecule becomes a monocation and a pair of charges becomes separated across the interface. In the case of rhodopsin, there is evidence suggesting the presence of an interfacial proton-transfer reaction; so far there is no evidence suggesting the presence of an interfacial electron-transfer reaction. The R_2 component of the ERP has been known to appear at the time when the photointermediate metarhodopsin I is converted to metarhodopsin II [33]. This latter reaction is an acid–base reaction and an aqueous proton is transferred to the membrane-bound pigment, that is, the proton donor is a hydronium ion, which is present in the adjacent aqueous phase. Thus, when a proton is transferred from the aqueous phase to the membrane phase, a "counterion," which takes the form of molecular water, must be left behind in the adjacent aqueous phase, leading to charge separation across the membrane–water interface. The subsequent charge recombination results in a reverse transfer of protons from the membrane phase back to the aqueous phase. We named this model the *interfacial charge transfer* (ICT) mechanism [52]. A similar analysis of the latter mechanism leads to a slightly different equivalent circuit (Figure 15.5). From the point of view of an external observer who conducts an electrical measurement, these two equivalent circuits are equivalent to the same irreducible circuit shown at the bottom of Figure 15.5. This irreducible circuit is part of the circuit shown in Figure 15.1. The differentiation of these two mechanisms may be made by chemical means. For example, since the AC signals shown in Figure 15.3A is generated by interfacial electron transfers at the oxidant interface, the pseudofirst-order relaxation-time constants are expected to be proportional to the ferricyanide concentration at the oxidant side. Qualitatively, the

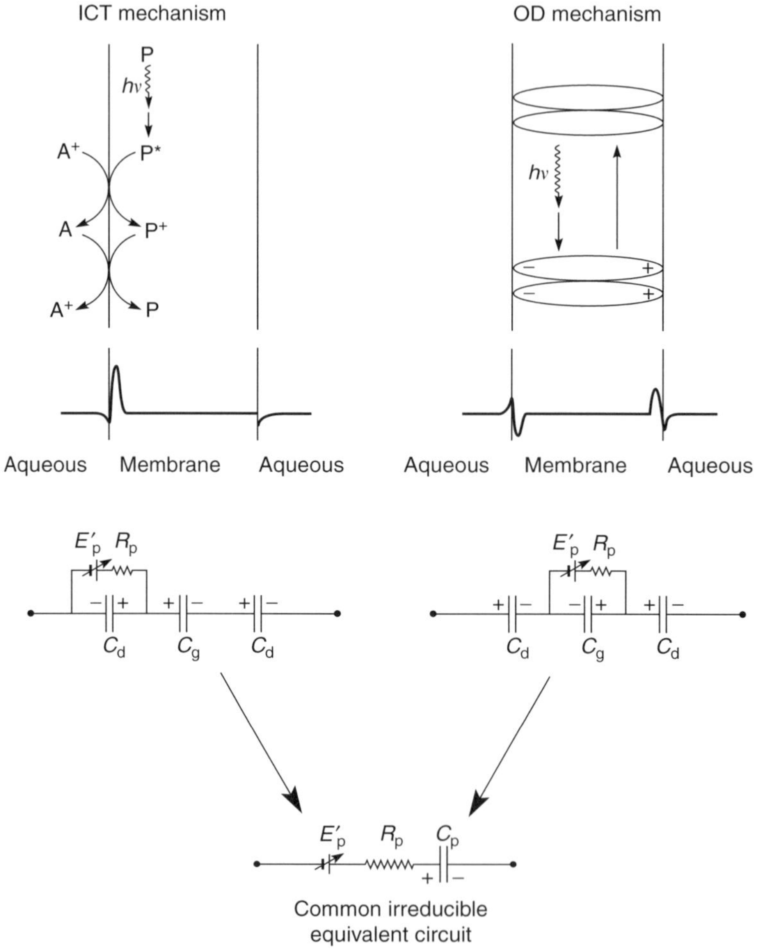

FIGURE 15.5
Interfacial charge transfer (ICT) mechanism and oriented dipole (OD) mechanism. The symbols P, P^*, and P^+ have the same meanings as in Figure 15.1. In the ICT mechanism, the symbol A stands for either proton acceptor or electron donor. The symbol A^+ stands for either proton donor or electron acceptor. Symbols A and A^+ can be replaced by B^- and B, if necessary. The reverse reaction is also shown. The OD mechanism is self-explanatory. The space charge profiles are shown just above the two microscopic equivalent circuits. In the two models, the photoemf, (with an internal resistance, R_p) is located at different places with respect to the geometric capacitance, C_m, and the two double-layer capacitances, C_d. The irreducible equivalent circuit is shown below them. (From Hong, F. T. (1978). Mechanisms of generation of the early receptor potential revisited. *Bioelectrochem. Bioenerg.* 5:425–455.)

shortening of the two exponential time constants, τ_s and τ_l, correlates well with increasing ferrocyanide concentrations. Figure 15.3B shows a quantitative test, in which $1/\tau_P$ (instead of $1/\tau_s$ or $1/\tau_l$) is plotted against the ferrocyanide concentration. Likewise, if an AC signal is generated by the interfacial proton-transfer mechanism, the law of mass action predicts that the signal is sensitive to pH because H^+ (or, rather, hydronium ion) is one of the reactants. Since the R_2 signal is generated during the metarhodopsin I to metarhodopsin II reaction, the R_2 signal is expected to be pH-sensitive. Yet most, if not all, previous investigations did not reveal this expected pH dependence [54]. This constitutes an apparent paradox that needs an explanation.

Since bR is a proton pump, interfacial proton transfer at the two surfaces of the purple membrane is an obligatory process. However, the ERP-like photosignal from bR was also found to be pH-insensitive [10,56]. The apparent paradox for both rhodopsin and bR can be explained by the equivalent circuit analysis presented above. As mentioned earlier, the relaxation of an open-circuit photovoltage reflects mainly the membrane *RC* relaxation and is therefore a poor indicator of the intrinsic photokinetic relaxation (Figure 15.4). Since most previous measurements of the ERP and the ERP-like photosignals in reconstituted bR membranes were carried out under open-circuit conditions, the apparent lack of pH dependence is thus more expected than surprising [27].

We subsequently measured the AC photoelectric signal of a reconstituted bR membrane under near-short-circuit conditions. It was then found that the B_2 component is highly sensitive to pH changes: it is reversibly inhibited by low pH [55]. Comparison of the near-short-circuit data with the corresponding open-circuit data reveals two important differences. The B_1 component is smaller than the B_2 component under open-circuit conditions but the opposite is true under near-short-circuit conditions. Furthermore, the apparent relaxation of the two components is considerably faster under near-short-circuit conditions than under open-circuit conditions. That is, the time course of the AC photosignal varies with the change of measurement conditions. Conventional approaches offer no satisfactory explanation of these kinetic features. In contrast, the equivalent circuit analysis presented above predicts these features [27]. Similar observations in a reconstituted rhodopsin membrane were reported by Ostrovsky and coworkers [57,58], when the membrane was short-circuited by shunting. The latter observation suggests that similar photoelectric phenomena appear in reconstituted bR and rhodopsin membranes.

15.4.2 Component Analysis

Semiquantitative investigations of the AC photosignal in reconstituted bR membranes indicate that it is a composite signal comprising at least two separate chemical processes: one for B_1 and another for B_2 [55]. The test of the prediction of the equivalent circuit as applied to bR data requires an unequivocal decomposition of the AC signal into several (pure) components for the following reasons. The equivalent circuit was derived for a single chemical relaxation process, such as the case of Mg porphyrin membrane. For a single relaxation chemical process, there are no freely adjustable parameters in the equivalent circuit because all input parameters for the computation can be determined experimentally. If, however, there are two or more chemical processes, two or more separate sets of parameters must be used, one set for each process. These parameters can no longer be uniquely determined by direct experimental measurements. Investigators often turned to curve fitting by trial and error to get the best fit. This procedure appeared to be less satisfactory than for the case of single relaxation because it is tantamount to the introduction of one or more adjustable parameters, which determine how these signal components are to be decomposed. An agreement between the equivalent circuit and the measured(composite) signal would be less compelling evidence for its validity. In contrast, a conventional approach commonly found in the literature decomposes the photosignal into as many exponential terms as possible. This is still curve fitting by trial and error, even though the arbitrariness becomes hidden in the "black box" of computer algorithm. The experimental data thus impose *no constraints* on the individual parameters—the conventional approach enlists twice as many free parameters as the exponential terms: one for the decay time constant and another for the amplitude of each exponential term (e.g., see [50]). The conventional approach thus guarantees an agreement between the model and the data under *any* circumstance. This pitfall was seldom appreciated in the literature of photoelectric effects. But the conventional approach appeared to be problematic because it often generated inconsistencies in the same membrane system under different conditions and in similar systems under similar conditions in different laboratories (see discussion in [59]). The resulting discrepancy cannot be explained without taking into account the subtle effect of the presence of a series capacitance—chemical capacitance. Furthermore, since the time-integral of a single exponential decay function is nonzero $\left(\int_0^\infty \exp(-t/\tau)dt \neq 0\right)$, the conventional practice of exponential decomposition is a direct violation of the zero time-integral condition (cf. Figure 15.10 in Chapter 5 of [48]). Therefore, the individual exponential components cannot possibly represent distinct molecular relaxation processes.

While the conventional approach essentially decomposes the signal by curve fitting with multiple exponential terms, we sought to decompose the signal by *physical means*. We used three different reconstitution methods to dissect and decompose the experimentally measured photosignal into several individual components. The decomposition is thus independent of the curve fitting process and is therefore model-independent. The detailed experimental evidence in support of our approach can be found in the literature [59–66].

If bR is incorporated by means of the BLM technique into a genuine lipid bilayer membrane, all three AC components, B_1, B_2, and B_2', as well as the DC component, can be detected. If a method, which was originally developed by Trissl and Montal [10], is used reconstitute bR, only the B_1 and the B_2 components can be detected. In the latter method, oriented purple membrane sheets are deposited on a transparent Teflon film with the extracellular surface attached right next to the Teflon film. The Teflon film prevents the extracellular surface of the purple membrane sheets from a direct contact with water, and, therefore, the B_2' component, which reflects interfacial proton transfers at the extracellular surface, is missing. The DC component of the photosignal is also missing because Teflon is an excellent insulator. We shall refer to this method as *Trissl–Montal* (TM) method.

Okajima and Hong [59] modified the TM method by depositing multiple layers of oriented purple membrane sheets on the Teflon film and by allowing the multilayered (ML) purple membrane-Teflon film assembly to dry in air for about 4 days or longer. When such a film assembly is mounted in a chamber with two compartments and rehydrated with electrolyte solution, the photosignal exhibits a dramatically different waveform as compared with the signal generated by means of the TM method. We shall refer to the latter method of reconstitution as the ML thin film method.

Typical photoresponses from a TM film and from an ML film are shown in Figure 15.6. All photoelectric signals have a biphasic waveform, exhibiting a positive peak and a negative peak. The signals from a TM film are highly sensitive to both pH and temperature (Figure 15.6A and Figure15.6B). Lowering the pH (or the temperature) causes an increase of the positive peak but a concurrent decrease of negative peak. In contrast, the signals elicited from an ML film exhibit no detectable pH sensitivity (Figure 15.6C) and are only slightly sensitive to temperature (Figure 15.6D). When the photosignals elicited from an ML film are subject to equivalent circuit analysis, the measured signals are in good agreement with the calculated response based on the equivalent circuit, again using only experimentally measurable input parameters for the computation. The agreement holds under various conditions including

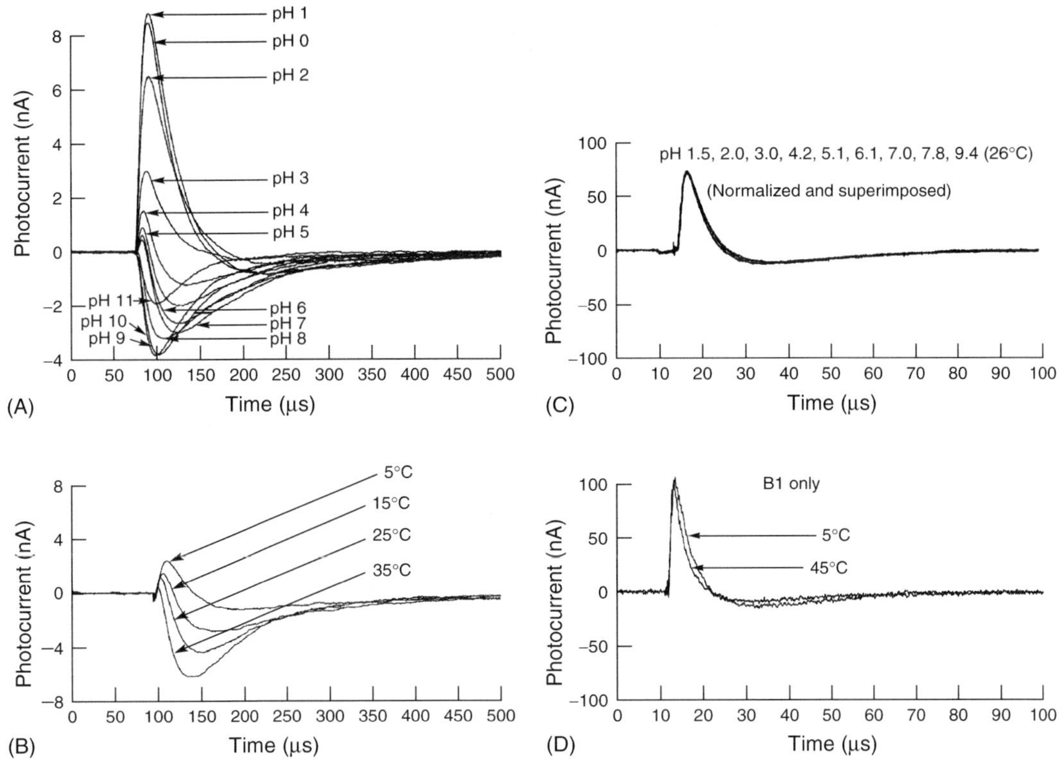

FIGURE 15.6
Effect of pH and temperature on B_1 and B_2. (A) The data, taken from a TM film at 25°C, illustrate the effect of varying pH on a composite signal consisting of both B_1 and B_2. (B) The data, taken from a TM film at pH 7, show the temperature dependence of the composite signal (B_1 and B_2). (C) The data, taken from an multilayered (ML) film at 26°C, illustrate the total lack of pH dependence of an isolated B_1 signal. (D) The data, taken from an ML film at pH 2, shows a small temperature dependence of an isolated B_1 signal. The temperature and/or pH effect is fully reversible. All data were measured at R_e = 40 kΩ. For detailed experimental conditions, consult the cited sources. (From Okajima, T. L., Hong, F. T. (1986). Kinetic analysis of displacement photocurrents elicited in two types of bacteriorhodopsin model membranes. *Biophys. J.* 50:901–912. (A), Hong, F. T., Okajima, T. L. (1987). Rapid light-induced charge displacements in bacteriorhodopsin membranes: an electrochemical and electrophysiological study. In: Ebrey, T. G., Frauenfelder, H., Honig, B. Nakanishi, K. (Eds.). *Biophysical Studies of Retinal Proteins.* Urbana-Champaign, IL: University of Illinois Press, pp. 188–198. (B, D), and Michaile, S., Hong, F. T. (1994). Component analysis of the fast photoelectric signal from model bacteriorhodopsin membranes: Part I. Effect of multilayer stacking and prolonged drying. *Bioelectrochem. Bioenerg.* 33:135–142. (C))

changes of temperature, pH, and electrolyte concentrations [59,60]. The inclusion of the access impedance in the equivalent circuit is absolutely essential. Furthermore, as the access impedance is varied, the apparent relaxation-time course varies in accordance with the prediction of the equivalent circuit. In other words, regardless of the value of the access impedance, the intrinsic photochemical relaxation-time constant (τ_p) extracted by means of deconvolution of the measured apparent relaxation-time course remains the same: 12.3 ± 0.7 μs. We also found that the apparent relaxation-time course of the B_1 component varies with the thickness of the Teflon substrate, but then, again, τ_p remains unchanged [66]. B_1 exhibits a slight temperature dependence, which corresponds to an activation energy of 2.54 ± 0.24 kcal mole^{-1} [59].

The dramatic pH dependence exhibited by a TM film (Figure 15.6A) is to be contrasted with the absence of pH dependence, which was widely reported in the literature. As explained in Section 15.3, the commonly used open-circuit method lacks the sensitivity to detect the pH-induced change of the intrinsic photochemical relaxation. The insensitivity is further aggravated by the use of the peak amplitude as an indicator of the intensity of the underlying photoelectric process. This latter pitfall can be made apparent by referring to a schematic diagram shown in Figure 15.7. Both B_1 and B_2 are shown to have a biexponential decay curve, as predicted by the equivalent circuit. The composite signal ($B_1 + B_2$), which can be observed at neutral pH and room temperature, is the algebraic sum of B_1 and B_2. Because of the partial overlap of the decay of B_1 and the rise of B_2, inhibition of B_2 alone diminishes the negative peak but increases the positive peak, giving rise to the apparent effect of a concurrent enhancement of B_1. In data reported by Drachev et al. [56], inhibition of B_2 by low pH had been partially attributed to an apparent enhancement of B_1, which did not actually happen. This misinterpretation further conceals the already inconspicuous pH dependence of B_2 under open-circuit conditions. A similar problem also appears in the ERP elicited from a reconstituted rhodopsin membrane and can be similarly explained [54].

What then is the molecular mechanism for B_1 and for B_2? B_1 and B_2 respond very differently to the manipulations of the aqueous phases. B_2 is highly sensitive to pH [59], to electrolyte composition [60,65], to proton–deuterium exchange [59], to chemical modification of the membrane surface [62], and to site-directed mutagenesis [63]. In contrast, B_1 is insensitive to all these treatments. These differences suggest that B_1 is generated by molecular processes occurring deep inside the hydrophobic portion of bR, whereas B_2 is generated at the hydrophilic domain at the membrane–water interface. In other words, B_1 is generated by the OD mechanism whereas B_2 is generated by the ICT mechanism, most likely interfacial proton transfer. This assignment is further supported by an experiment to be described next. In view of the drastically different effects of temperature, pH, proton–deuterium exchange, chemical modification, and amino-acid substitution, on the two components, the physical methods of decomposition must be a clean one. Ad hoc models found in the literature usually could not achieve such a sweeping consistency.

A lingering question remains: why is B_2 absent in an ML film? The answer to this question is provided by a "Q-tip" experiment carried out by Michaile and Hong [61]. In the "Q-tip" experiment, an ML film can be converted into a TM film by stripping the multiple layers with a cotton swab (Q-tips; registered trademark of Unilever HPC-NA). The working hypothesis is depicted in Figure 15.8, which shows a TM film with a single oriented layer of purple membrane and an ML film with five oriented layers of purple membranes (in an actual experiment, it ranges from eight to ten layers). The B_1 component, being

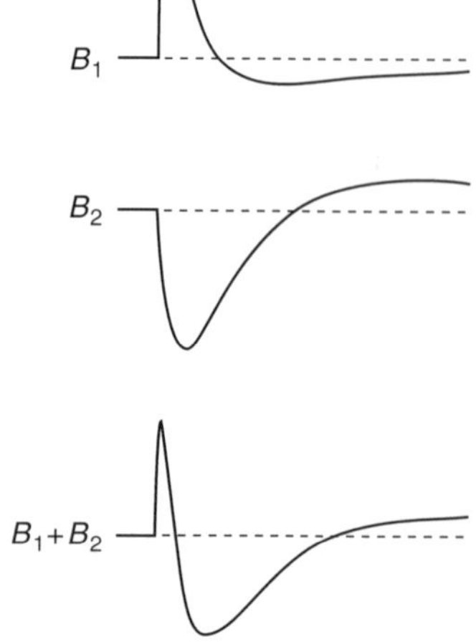

FIGURE 15.7
Schematic diagram showing the decomposition of the composite signal ($B_1 + B_2$) when the access impedance is not zero. Each component decays in two exponential terms as dictated by the equivalent circuit and the zero time-integral condition. Note that inhibition of the B_2 component results in a decrease of the negative peak but an increase of the positive peak. An inhibition of B_2 alone could be misinterpreted as a concurrent enhancement of B_1. (From Hong, F. T., Montal, M. (1979). Bacteriorhodopsin in model membranes: a new component of the displacement photocurrent in the microsecond time scale. *Biophys. J.* 25:465–472.)

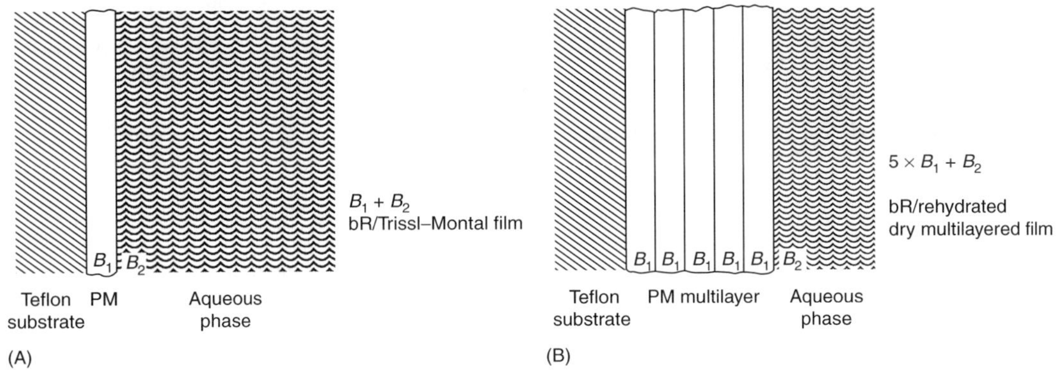

FIGURE 15.8
Working hypothesis for the "Q-tip" experiment. The symbol PM stands for the purple membrane. See text for further explanation. (From Hong, F. T. (1994). Electrochemical processes in membranes that contain bacteriorhodopsin. In: Blank, M., Vodyanoy, I. (Eds.). *Biomembrane Electrochemistry* (Advances in Chemistry Ser. No 235). Washington, DC: American Chemical Society, pp. 531–560.)

generated by intramolecular charge separation (OD mechanism), contributes one unit of B_1 in the TM film but five units of B_1 in the ML film because of the signal additivity. In contrast, there is only one unit of B_2 signal in both types of films because only the top layer that has access to aqueous protons or other ions can generate B_2 (ICT mechanism).

As compared with the photoelectric signal from a typical TM film (Figure 15.9A), that of a freshly prepared ML film (Figure 15.9B) exhibits increased amplitude but diminished pH sensitivity in accordance with the expected contribution of B_2 to the overall signal. Stripping the ML film with a cotton swab reduces the amplitude of the photosignal but does not completely eliminate it, presumably because the bottom layer adheres tightly to the Teflon surface (Figure 15.9C). However, the remaining signal shows a pattern of pH and temperature dependence, which is almost indistinguishable from that of a TM film (compare Figure 15.9A with 15.9C).

If, however, the ML film is allowed to dry for 4 days or longer, the photosignal from the ML film, though still enhanced seven to ten times in amplitude, exhibits no trace of pH dependence (Figure 15.9D). Subsequent stripping reduces the amplitude to sizes comparable to that of a TM film, but the signal does not regain its characteristic pH and temperature dependence (Figure 15.9E). In fact, the diminished signal, after normalization, superimposes with the corresponding signal before stripping (Figure 15.9F). This observation is consistent with the complete elimination of the B_2 component and the persistence of a pure component, B_1. Apparently, prolonged drying has permanently damaged the B_2 generating mechanism but it has left the B_1 generating mechanism intact. This interpretation is also consistent with the assignment of the OD mechanism to B_1 and the ICT mechanism to B_2 (Section 15.4.1).

Can the B_2 component be eliminated by other means? Lowering the temperature to 0°C reversibly but incompletely suppresses B_2. We have also attempted to suppress the B_2 component by means of chemical modification [62]. We found that prior treatment of the purple membrane with fluorescamine partially (and irreversibly) inhibits B_2. However, such a treatment has no effect on B_1. We also found that B_2 is naturally absent (at pH > 3) in a mutant D212N but is present at low pH (<2) with a reversed polarity, that is, B_2 at low pH has the same polarity as B_1 [63]. This latter finding suggests that it may be possible to titrate the pH so that B_2 becomes completely suppressed in the wild-type bR TM film [64] (Figure 15.10). The end point of pH titration is the point where B_2 reverses its polarity. Figure 15.10A shows the monotonic and continuous variation of the time course of the

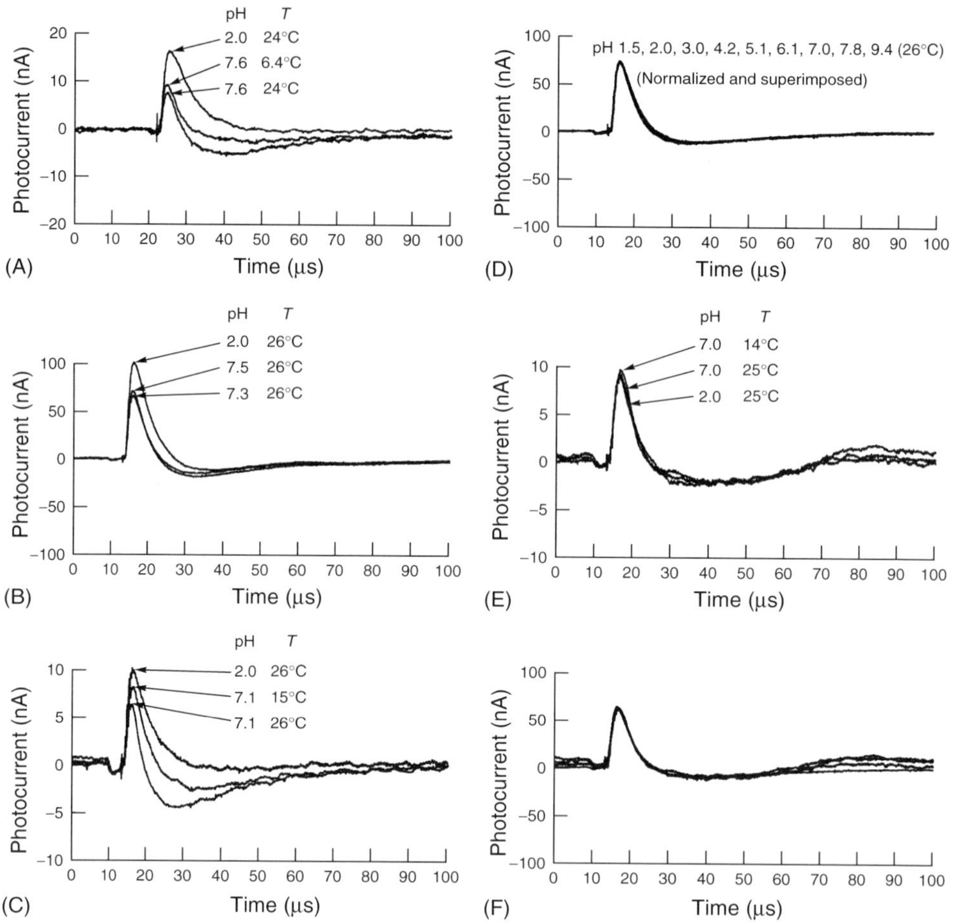

FIGURE 15.9
"Q-tip" experiments. Photosignals from bacteriorhodopsin (bR) thin films, measured at various pH values and/or temperatures, are shown for: (A) typical TM film. (B) Fresh ML film before stripping with a cotton swab. (C) Fresh ML film after stripping. (D) aged ML film before stripping. (E) Aged ML film after stripping. (F) In an aged ML film, photosignals after stripping (from Record E) is shown to superimpose with photosignal before stripping (pH 6.1 in Record D), if they were normalized to the positive peak of the latter signal. See the cited source for detailed experimental conditions. (From Michaile, S., Hong, F. T. (1994). Component analysis of the fast photoelectric signal from model bacteriorhodopsin membranes: Part I. Effect of multilayer stacking and prolonged drying. Bioelectrochem. Bioenerg. 33:135–142.)

composite signal ($B_1 + B_2$) as the pH varies from 6.0 to 1.7. Each signal trace is then compared with a standard B_1 signal taken from Figure 15.9D after normalization of the peak amplitude. Figure 15.10B shows the superposition of a B_2 signal at pH 2.7 (from Figure 15.10A) with the standard B_1 after normalization of the peak amplitude. Comparison of Figure 15.10B with Figure 15.10C reveals that the end point is sharply defined, as the end point must depend on a reasonable superposition of the entire two curves rather than a single point. This titration experiment permitted us to isolate the pure B_2 component by subtracting the normalized B_1 signal from each composite signal ($B_1 + B_2$) shown in Figure 15.10A (Figure 15.10D). The resulting manipulation shows that B_2 indeed reverses its polarity at pH 2.7. Furthermore, the B_2 component itself appears to be a composite signal. We have tentatively decomposed it into a B_2-c subcomponent with negative polarity and a B_2-a subcomponent with positive polarity. We further found that the B_2-c component is modulated by divalent cations such as Ca^{2+} and Mg^{2+} whereas the B_2-a component is modulated by halide ions instead [65].

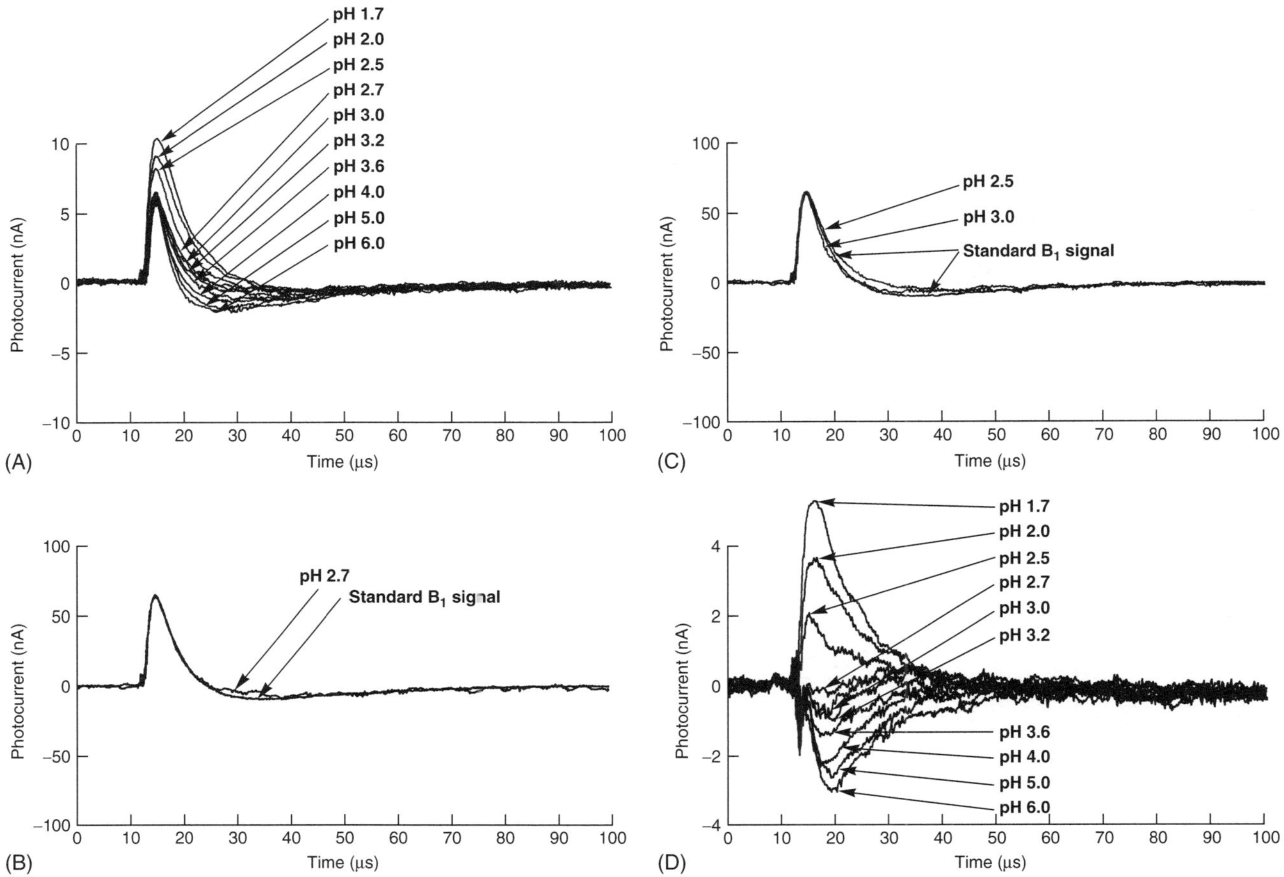

FIGURE 15.10
Isolation of pure B_2 signals. (A) The composite signals ($B_1 + B_2$) were measured from a TM film at pH ranging from 1.7 to 6.0. The temperature was 26°C. The signals are plotted without normalization. (B) The composite signal measured at pH 2.7 is normalized and superimposed on a standard B_1 signal (26°C, pH 6.1, taken from an ML thin film, Figure 15.9D). (C) The composite signals taken at pH 2.5 and 3.0 are similarly superimposed to the standard B_1 signal. (D) Isolated B_2 signals at various pH are obtained by subtracting the normalized B_1 signal from each of the composite signals shown in (A). The reversal of the B_2 polarity at pH 2.7 is apparent. See the cited source for detailed experimental conditions. (From Hong, F. H., Hong, F. T. (1995). Component analysis of the fast photoelectric signal from model bacteriorhodopsin membranes: Part 4. A method for isolating the B2 component and the evidence for its polarity reversal at low pH. *Bioelectrochem. Bioenerg.* 37:91–99.)

The B_2' component is conveniently suppressed by adsorbing purple membranes to the surface of a Teflon film. If a genuine BLM is used, both B_2 and B_2' are expected to be observable but their differentiation may be problematic for the following reasons. Both B_2 and B_2' represent charge movement in the same direction, and therefore both will have the same polarity. Since both components are related to interfacial proton transfer, both are expected to be pH-dependent. However, the differentiation between B_2 and B_2' can be achieved by the following experiment [40,60]. It was pointed out in Section 15.4.2 that the transmembrane diffusion of P and P^+ in the Mg porphyrin membrane is slow compared with interfacial electron transfer—the two interfacial reactions are *chemically decoupled* on the microsecond time scale. We suspected that this might also be true for transmembrane proton movement in the purple membrane. If so, the interfacial proton-transfer reaction at either membrane surface will depend solely on the pH of the adjacent aqueous phase but not on the pH of the opposite aqueous phase (concept of *local reaction conditions*) [40].

An experiment based on this strategy is shown in Figure 15.11 in which purple membranes are reconstituted into a collodion membrane according to a method developed by Drachev et al. [67]. When both sides of the membrane face neutral pH, both B_1 and B_2 are observable. B_2 can be reversibly inhibited by lowering the pH of the cytoplasmic side as expected. When the pH of both the cytoplasmic side and the extracellular side are kept low, a negative component appears. This negative component is identified as the hypothetical B_2' component, because it has a different relaxation-time course and because its pH dependence is opposite to that of the B_2 component; B_2' is enhanced by low pH.

15.4.3 DC Photoelectric Effect: The Null-Current Method

The methodology described above for the analysis of AC photoelectric signals is not suitable for the detection of the DC photoelectric signal for several reasons. The use of a Teflon film, which has a high resistance, does not permit a DC current to pass through. A genuine

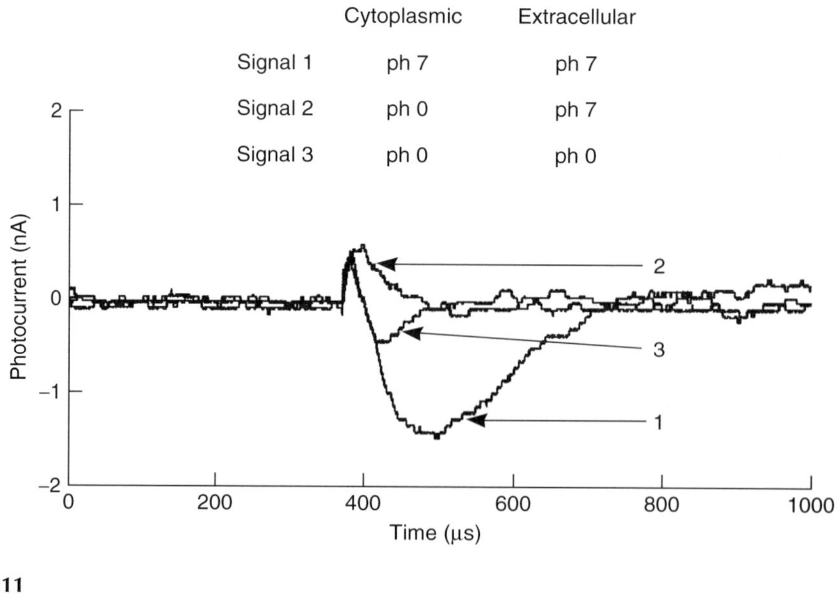

FIGURE 15.11
"Differential experiment" showing the separation of the B_2 and B_2' component. See text for further detail. (From Hong, F. T., Okajima, T. L. (1987). Rapid light-induced charge displacements in bacteriorhodopsin membranes: an electrochemical and electrophysiological study. In: Ebrey, T. G., Frauenfelder, H., Honig, B. Nakanishi, K. (Eds.). *Biophysical Studies of Retinal Proteins*. Urbana-Champaign, IL: University of Illinois Press, pp. 188–198.)

BLM is required to reconstitute bR. But even if a genuine BLM is used to reconstitute bR, a high-bandwidth current amplifier will allow the AC photosignal to become so prominent as to overwhelm the concurrent DC signal for reasons presented earlier. Detection of DC photocurrents requires substantial amplification of the signal and the use of a low-pass filter to suppress the concurrent AC component. Illumination with continuous light allows the DC current to emerge after the initial AC transient, if any, subsides (Figure 15.4). Instead of the TVC method, we used a null-current method [68,69] to analyze the DC photoelectric effect.

There are two important attributes of the DC photoelectric effect: the DC *photoconductive* effect and the DC *photovoltaic* effect. The photovoltaic effect is the manifestation of photoemf. Photoconductivity is the manifestation of a specific current pathway activated by light. In bR, the photoconductive pathway is thought to be a series of protonation sites extending from one membrane surface to the other [18,19]. This pathway is different from an ionic conduction pathway provided by ion channels or by leakage through the membrane. Several pertinent questions can be raised here. Does the specific proton pathway stay open in the dark or only during illumination? Is it possible to reverse the proton current by applying an opposing transmembrane potential? If the answer to the latter question is yes, then it ought to be possible to cancel the light-induced DC current by concurrently applying a transmembrane voltage with an appropriate magnitude in the opposite direction. This reasoning forms the basis of the null-current method, which is based on the potentiometric principle in electrochemistry: the photoemf is equal but opposite to the applied voltage that cancels the DC photocurrent. Detailed discussion explaining why the photoemf must be measured with a potentiometric method and how photoconductivity can exist without a concurrent photovoltaic effect can be found in [69]. It suffices to say that measurements under conditions other than null diminishes the measured emf values by virtue of a voltage drop caused by internal resistance.

The equivalent circuit in Figure 15.1 can be used for the DC photoelectric with proper identification of terminology as follows. The resistance encountered by the DC photocurrent is equal to $(R_s + R_p)$, or approximately R_s since R_p is much smaller than R_s. The chemical capacitance C_p can be ignored in the DC photoelectric effect for obvious reason. Here, we define the photoconductance G_p as the reciprocal of $(R_s + R_p)$. Note that G_p is actually closer to $1/R_s$ than to $1/R_p$ even though the notation suggests that the erroneous relation: $G_p = 1/R_p$. Also we define G_m as the reciprocal of R_m.

We used a conventional voltage clamp amplifier to measure the DC photocurrent. The procedure of a null-current measurement is illustrated by the schematic diagram in Figure 15.12A and by an actual example in Figure 15.12B. For the sake of simplicity, let us assume that the conductance G_p is much greater than G_m so that the shunting effect of the parallel current path through G_m can be ignored. We further assume that the conductance G_p has the same value in the dark as in the light. These restrictions will be subsequently removed after an intuitive picture is presented.

At the beginning of a null-current measurement, the clamping voltage is set to 0 and the illuminating light source is also turned off (current level 1). While the light source remains off, the clamping voltage is then set to a preselected value V_c, and the current responds by settling at a new value (level 2) after a brief capacitative transient (shown in Figure 15.12B but not in Figure 15.12A). While the clamping voltage is maintained at V_c, the light is then turned on. The current responds, again after a brief AC photoelectric transient, by settling at yet another new value (level 3). The difference, level 3 minus level 2, gives the DC photocurrent, I_p (corresponding to the clamping voltage V_c). While the illumination continues at a fixed intensity (power), the clamping voltage is then adjusted to a new value so that the measured current is "tuned" back to level 2. That is, the end point of tuning (V_c') is the new voltage that is required to cancel ("nullify") the DC photocurrent. In other words, by

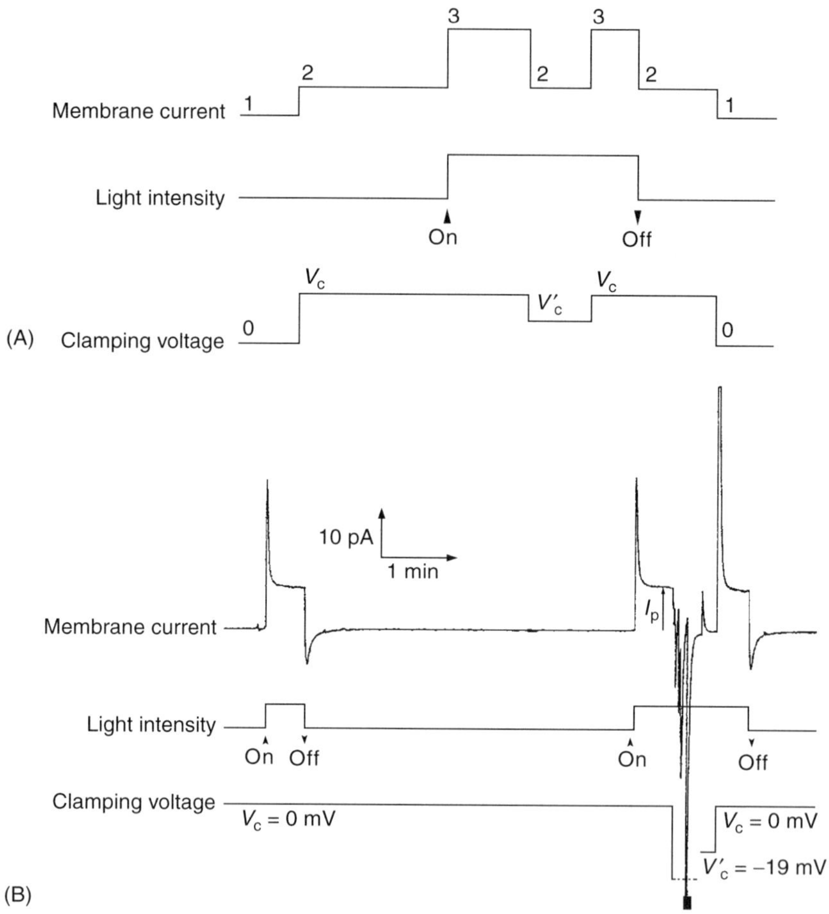

FIGURE 15.12
Null-current method (A, schematic diagram; B, actual experimental record taken from a reconstituted bR-containing BLM). The experimental conditions are as follows: The pH was 6.9, and the temperature was 27.5°C. The instrumental time constant was 1 s. The first measurement was a "light on" and "light off" operation, which allowed I_p to be measured (corresponding to the clamping potential 0 mV). In the second measurement, the clamping potential was adjusted to −19 mV so as to obtain the null-current condition. See the cited source for detailed experimental conditions. (From Fuller, B. E., Okajima, T. L., Hong, F. T. (1995). Analysis of the d.c. photoelectric signal from model bacteriorhodopsin membranes: d.c. photoconductivity determination by means of the null current method and the effect of proton ionophores. *Bioelectrochem. Bioenerg*. 37:109–124.)

applying an offset voltage, $-E_0=(V_c'-V_c)$, it is sufficient to abolish the DC photocurrent and move the measured current from level 3 back to level 2 (the preillumination level).

By invoking the principle of potentiometry, the offset voltage can be taken as equal to the photoemf at the clamping voltage V_c, but with an opposite polarity. Thus, the photoemf is

$$E_p = -(V_c' - V_c) = E_0 \tag{15.9}$$

Subsequently, the clamping voltage is then brought back to V_c, so as to ascertain that the measured current actually returns to level 3, which is the level attained before applying the offset voltage. When the light is finally turned off, the measured current settles back to level 2, again after a brief AC photoelectric current transient. Finally, with the light remaining off, the clamping voltage is returned to zero and the measured current is brought back to level 1, again after a capacitative transient. This completes the cycle of measurements in which two acts of measurement have been carried out: the measurement of the photocurrent I_p (level 3 minus level 2), and the measurement of the photoemf E_0. The DC photoconductance, G_p, is then obtained by dividing the DC photocurrent with the photoemf by virtue of Ohm's law.

$$G_p = \frac{I_p}{E_0} \tag{15.10}$$

By varying the value V_c in a systematic way, both the DC photocurrent and the photoemf can be measured as a function of the transmembrane (clamping) voltage, V_c. The DC photoconductance G_p can be determined as a function of the transmembrane voltage by repeatedly using Eq. (15.10). Note that the background ionic conductance in the absence of illumination, G_m, can also be obtained from the measured currents shown in Figure 15.12 by dividing the current (level 2 minus level 1) with the imposed voltage V_c, again by virtue of Ohm's law.

We shall refer to the above-determined photoemf and the photoconductance as the *apparent photoemf* and the *apparent photoconductance*, respectively, since additional corrections of these directly measured values are required when the restrictions imposed earlier are relaxed. First, when G_m is comparable to G_p, the shunting effect of G_m must be taken into account. Shunting will reduce the measured value of the apparent photoemf by a factor of $G_p/(G_p + G_m)$, as is intuitively evident from the consideration of the effect of a voltage divider in elementary electronics. Again by intuitive reasoning, the apparent photoconductance so measured actually contains a contribution from G_m, that is, the apparent photoconductance represents the combined value of $(G_p + G_m)$.

Another correction is required for the following reason. As measured experimentally, G_p turns out to be zero in the dark and is activated by illumination to a fixed nonzero value. We shall refer to this feature as *"step function" photoswitching* of the proton conductance channel. In other words, the proton channel is either completely on or completely off, and does not take on any intermediate values. In this case, when G_p suddenly turns nonzero upon illumination, both V_c and E_p can each drive a current through G_p. Both currents will be treated as photocurrents by the definition given above. Thus, the photocurrent must be decomposed into two fractions: one driven by light and the other by the transmembrane potential. Likewise, the apparent photoemf can be decomposed into two parts: the applied voltage and the true photoemf. This correction would not be necessary if G_p had the same value during illumination as in the dark because the V_c-driven current has already been included as part of the dark current (e.g., see the DC photoelectric effect of the Mg porphyrin membrane [35,68,30]).

Thus, to correct the above-mentioned effects, Eqs. (9) and (10) must be replaced by

$$E_p = E_0 \cdot \frac{G_p + G_m}{G_p} - V_c \tag{15.11}$$

and

$$G_p = \frac{I_p}{E_0} - G_m \tag{15.12}$$

respectively. In other words, the measured photoemf by means of the null-current method as described by Eq. (15.9) is the apparent photoemf, the effectiveness of which is reduced by shunting, whereas the measured conductance during illumination actually contains a contribution from the dark ionic conductance. The photoconductance as determined by Eq. (15.10) is more appropriately referred to as the *combined conductance during illumination* ($G_p + G_m$), or simply *apparent photoconductance* for brevity. Since G_m can be independently measured, the true photoemf E_p and the true photoconductance G_p can be calculated by virtue of Eqs. (11) and (12). The rigorous mathematical derivation of Eqs. (11) and (12) has been published elsewhere [69].

The result of a typical null-current measurement is shown in Figure 15.13. We confirmed a widely reported observation [70]: the photocurrent and the apparent photoemf are linearly dependent on the applied voltage, and the photocurrent reverses its polarity at a certain applied voltage (−70 mV in Figure 15.13A). In contrast, the true photoemf and the photoconductance are voltage independent (Figure 15.13B and Figure 15.13C). Our interpretation of the linear voltage dependence of the photocurrent and its polarity reversal is therefore quite different from what were offered by other investigators (discussed in [69]). Our experimental result indicates that the fraction of photocurrent driven by light is voltage independent (Figure 15.13B). The apparent linear dependence of the photocurrent is the consequence of the current driven by the applied voltage through the *same* proton conductance channel activated by light. Therefore, the apparent linear dependence reflects the linear dependence of the voltage-driven current on the applied voltage by virtue of Ohm's law. Furthermore, the polarity reversal is a consequence of applying an increasingly opposing voltage, which eventually overcomes the true photoemf. Thus, the simple result here actually justifies the use of a potentiometric method—the null-current method. A simple explanation of the linear voltage-dependence of the photocurrent was thus made possible.

So far, we have ignored the "spike" transients that appear at the onset and at the cessation of illumination (Figure 15.12B). Elementary analysis indicates that a linear high-pass *RC* filter exhibits a similar waveform (cf. Figure 15.4). Thus, the transient spikes are a manifestation of chemical capacitance. However, the nature of chemical capacitance (a built-in high-pass filter) alone does not explain the asymmetry of the two peaks. In Figure 15.12B, the positive spike is more prominent and decays faster than the negative spike. This asymmetry can be explained on the basis of "step-function" photoswitching: R_s (or $1/G_p$) assumes two different values, being much greater (virtually infinity) in the dark than in the light (a finite value). Therefore, the symmetry of the two spikes predicted by the equivalent circuit analysis on the assumption that R_s is a constant (or infinity, as assumed in the computation of the time course of the AC photoelectric signal) and does not hold any longer. The decay of the transient will be faster in the light than in the dark because of the effect of light upon R_s (cf. Eq. (15.8)). Again the condition of the zero time-integral requires that the area enclosed by the two transient decay curves be equal. This is possible only if the "on" peak is more prominent than the "off" peak.

Our experimental result indicates that the proton conduction pathway is *not rectified*: the current driven by the applied voltage encounters the same resistance in either direction during illumination, that is, bR is not a photodiode in the strict sense (Figure 15.14). The notion that the photoemf and the applied transmembrane voltage are interchangeable for driving a transmembrane proton current is however consistent with the basic principle of bioenergetics. According to the chemiosmotic theory of Mitchell [71] (see also a similar theory by Williams [72]), the converted energy by bR photoreaction is stored as a transmembrane *electrochemical* gradient of protons. Furthermore, both the electrical component (ΔV) and the chemical component (ΔpH) of this gradient are equivalent and are both available for ATP synthesis. The latter condition holds because of the common pathway for light-driven and voltage-driven proton currents.

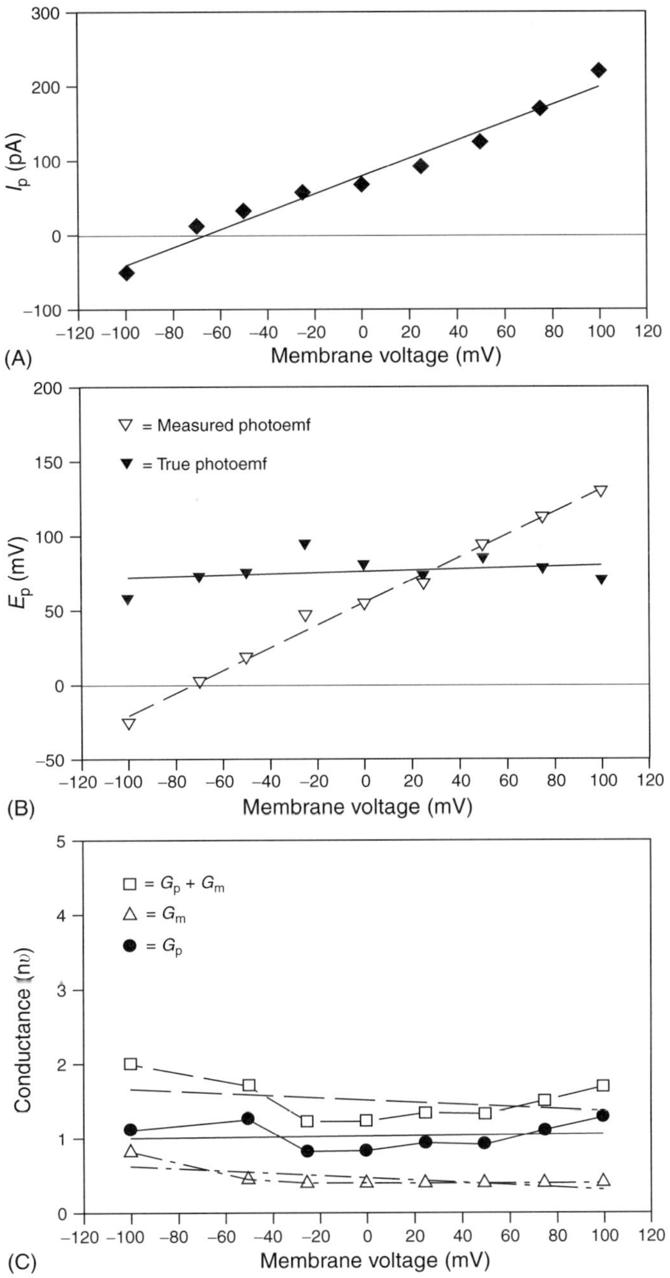

FIGURE 15.13
DC photoelectric response of a bR-containing BLM. The pH was 6.9, and the temperature was 24°C. (A) The photocurrent (I_p) is plotted as a function of membrane voltage. (B) The photoemf (E_p) is plotted as a function of membrane voltage. The apparent photoemf is shown as open inverted triangles whereas the true photoemf, derived according to Eq. (15.11), is shown as filled inverted triangles. (C) The apparent photoconductance (G_p+G_m) (open squares, calculated by means of Eq. (15.10)), the measured dark conductance (G_m) (open triangles), and the true photoconductance (G_p) (filled circles, obtained by subtracting G_m from the apparent photoconductance) are plotted as functions of membrane voltage. The conductance data are connected with line segments for the sake of easy data reading. Each set of data is also fitted with a straight line by the least square method. All the data were taken from the same membrane. (From Fuller, B. E., Okajima, T. L., Hong, F. T. (1995). Analysis of the d.c. photoelectric signal from model bacteriorhodopsin membranes: d.c. photoconductivity determination by means of the null current method and the effect of proton ionophores. *Bioelectrochem. Bioenerg.* 37:109–124.)

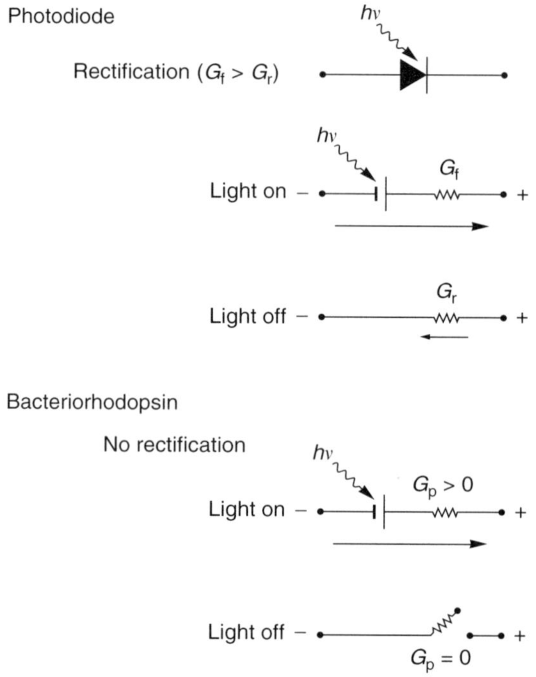

FIGURE 15.14
Schematic diagram comparing rectification in a conventional photodiode and the "step-function" photoswitching in bR. G_f is the forward conductance, and G_r is the reverse conductance of the photodiode. Photoactivation causes a photocurrent to flow in the forward direction as indicated. As a consequence, the photodiode becomes back-biased. Back-biasing drives a current in the reverse direction during illumination, but the reverse current encounters a much greater resistance because of rectification ($G_r \ll G_f$). In the dark, charge recombination is minimized because the conductance G_r is much lower than G_f. In bR, there is no rectification: the forward photocurrent and the reverse current (driven by back-biasing) encounter the same resistance ($1/G_p$). Thus, the observed photocurrent is actually the difference of these two currents. The constancy of the measured true photoemf (Figure 15.13B) indicates that illuminated bR's ability to maintain a net "driving force" is not affected by the extent of back-biasing. In other words, the performance of bR during illumination has not been seriously compromised by the lack of rectification but it would be seriously compromised if the proton channel remained open in the dark. "Step-function" photoswitching averts this potential problem: $G_p = 0$ in the dark. The plus and the minus signs indicate the direction of biasing. In the case of bR, the signs also indicate the polarity of the photovoltage. (From Hong F. T. (1999). Interfacial photochemistry of retinal proteins. *Prog. Surf. Sci.* 62:1–237.)

As for enhancing the efficiency of photoconversion, bR apparently relies on a different strategy than rectification. We argue here that "step-function" photoswitching allows bR to prevent wasteful dissipation of converted energy *in the dark*. The generation of a transmembrane proton gradient is the intermediate step between photon energy conversion and bioenergetic synthesis of ATP (the universal energy currency in living cells). The formation of the transmembrane electrical gradient (i.e., the photovoltage) constitutes a situation similar to back-biasing a photodiode. Back-biasing provides the driving force to drive protons back into the cell (strictly speaking, the proton backflow is driven by both the electrical component and the chemical component of the electrochemical potential). In principle, there are several ways the electrochemical potential can drive protons back to the cytoplasm: (a) protons reentering the cell via a reverse proton flow through the *same* proton channel in bR, (b) protons or other small ions reentering the cell via leakage through the phospholipid portion of the purple membrane, and thus dissipating the electrical portion of converted energy, (c) protons reentering the cell via the proton channel of ATP synthase (residing in the red membrane), thus synthesizing ATP, and (d) protons reentering the cell via the flagella motor apparatus, and thus powering the motion of the flagellas. The first two possibilities would lead to wasteful dissipation of the converted energy. They are prevented (or minimized) in the purple membrane by "step-function" photoswitching (in the dark) *and* by having a relatively small area of phospholipid portion in the two-dimensional crystalline structure of the purple membrane [13,15]. In the reconstituted bR BLM, G_p in the dark is considerably smaller than G_m. This result means that the insulation against a proton backflow via the proton channel of bR *in the dark* is much more effective than the insulation of phospholipid portion of the BLM. Quantitative comparison of G_p with G_m in a native purple membrane is not available. But we think that the ratio G_p/G_m should be much greater in the native purple membrane than in our experimental system because reconstitution in our experimental system could not have achieved the same degree of high-density, crystalline packing of bR as in the native purple membrane.

From a mechanistic point of view, rectification in a photodiode is more important in the dark than during illumination. While light-induced charge separation encounters the forward resistance ($1/G_f$ in Figure 15.14), charge recombination during illumination and *in the dark* encounters a much higher resistance ($1/G_r$ in Figure 15.14). What transpires in the purple membrane indicates that photon energy conversion is still possible without rectification. The lack of rectification in the purple membrane may be the consequence of the universal presence of reverse reactions in the scheme of coupled consecutive charge transfer reactions (see later, Section 15.5). How *net*-forward charge transfers are still possible in spite of the lack of rectification is discussed in greater detail in Section 18 of [29]. Thus, by means of "step-function" photoswitching, bR achieves the same goal (in the dark, at least) as does a silicon photodiode by means of rectification.

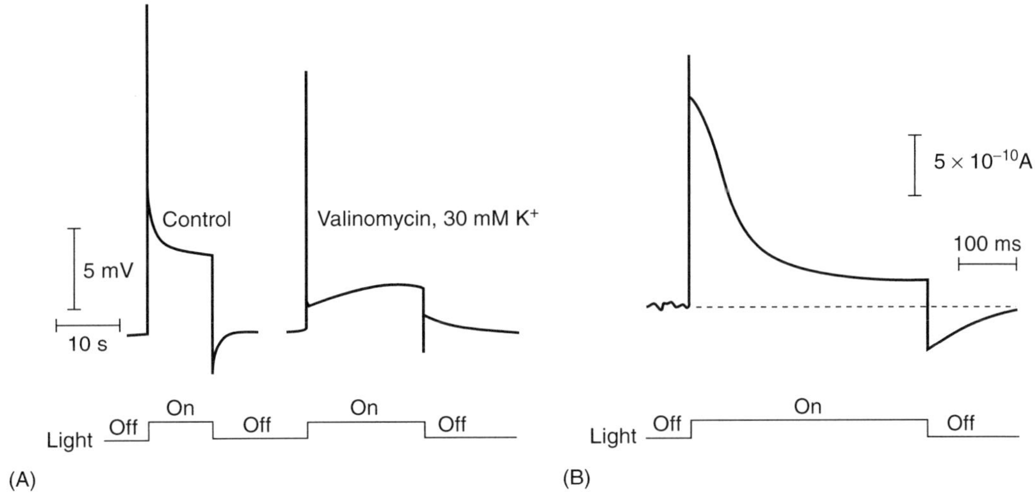

FIGURE 15.15
Photoelectric signals from photosynthetic membranes in response to a long square-wave light pulse. (A) The open-circuit photovoltage was recorded intracellularly from an intact chloroplast of *Peperomia metallica*. The control signal, taken before the addition of valinomycin, is shown at left, and the effect of adding 1 μM valinomycin and 30 mM potassium ions in the external medium is shown at right. (B) The photosynthetic reaction center of *Rhodobacter sphaeroides* (8 μM), supplemented with 100 μM of ubiquinone-10, was reconstituted into BLM. One of the two aqueous phases contained 25 μM reduced cytochrome *c*, and the other contained 1 mM ferricyanide. See text for discussion. (From Vredenberg, W. J., Bulychev, A. A. (1976). Changes in the electrical potential across the thylakoid membranes of illuminated intact chloroplasts in the presence of membrane-modifying agents, *Plant Sci. Lett.* 7:101–107. (A) and Packham, N. K., Mueller, P., Dutton, P. L. (1988). Photoelectric currents across planar bilayer membranes containing bacterial reaction centers: the response under conditions of multiple reaction-center turnovers. *Biochim. Biophys. Acta* 933:70–84. (B))

We further suspect that the "step-function" photoswitching mechanism of a light-dependent proton channel may also be present in chlorophyll-based photosynthetic membranes. The asymmetrical transient current spikes that appear upon the onset and the cessation of illumination of a giant chloroplast from *Peperomia metallica* [73] suggest that the charge conduction pathway in the photosynthetic reaction center may also be activated only by light (Figure 15.15A). The same waveform persisted in a BLM reconstituted with the bacterial reaction center of *Rhosobacter spheroides* [74] (Figure 15.15B), thus reflecting an intrinsic property of the membrane-bound pigment. Evidence from direct electrical measurement of G_p (as compared with G_m) is presently lacking. But if this interpretation is correct, then it appears that Nature has implemented the same design principle by using completely different molecular constructs (see discussion in [75–77]).

15.5 Generalization to Other Photoelectric Systems

The interpretation of the asymmetric capacitative transients in chloroplasts presented in the previous section tacitly assumes that similar AC photoelectric events take place in chlorophyll-based photosynthetic membranes. The OD and the ICT models are general in the sense that they capture most, if not all, of the *mechanistic* features of light-induced charge movements in various photobiological membranes. The analysis of charge separation and recombination of either OD or ICT type also suggests that the charged species being moved in the membrane need not be limited to electrons or protons and may include small ions other than protons. The inherent generality of the analysis thus allows us to make additional predictions. For example, the ICT model predicts that a halorhodopsin (hR) membrane, which transports chloride ions across the membrane upon illumination, will exhibit a photoelectric signal by means of interfacial Cl^- transfer (hR is another retinal-containing protein found in the red membrane fraction of *H. salinarum*, and is a light-driven Cl^-pump). This component should be sensitive to the aqueous Cl^-concentration rather than the proton concentration. This prediction was confirmed experimentally: the photoelectric signal has a B_1-like component (H_1) and a B_2-like component (H_2). The H_2 component, but not H_1, is sensitive to Cl^-concentration [78] (Figure 15.16).

Can the scheme of analysis for bR be extended to cover more complex light-driven proton pumps (photosynthetic reaction centers)? Superficially, the structure of photosynthetic reaction centers, such as that of *Rhodopseudomonas viridis* [79], appears to be considerably more complex than that of bR. At a mechanistic level, they are rather similar. Both systems can be characterized as systems of *coupled ICT reactions* [80,81]. The two interfacial proton-transfer reactions are coupled by transmembrane charge (electrons or protons) transport. Such transmembrane charge transport is accomplished by breaking up the charge movement into several small steps of consecutive charge transfers with charge donors and acceptors arranged in a chainlike spatial organization (electron transport chain). The reactants of a given reaction are the products of the preceding reaction whereas its products become the reactants of the subsequent reaction. Schematically, these coupled consecutive charge transfer reactions are shown in Figure 15.17A, for bacterial photosynthetic reaction center, and in Figure 15.17B, for bR. The mechanistic similarity is apparent.

Although not shown in Figure 15.17A, reverse charge transfer reactions are quite prevalent in photosynthetic membranes. For each forward charge transfer reaction (charge separation) there is a reverse charge transfer (charge recombination) including the two interfacial reactions. The imbalance between the forward charge transfer and reverse

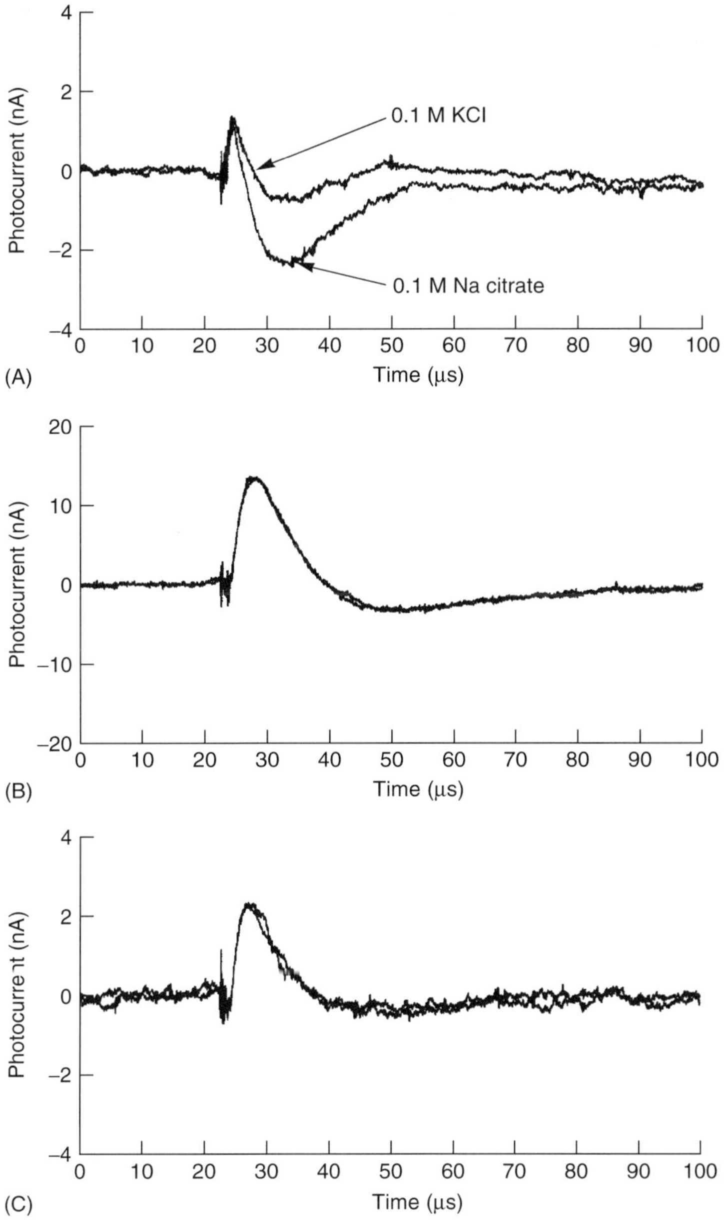

FIGURE 15.16
Photoelectric signals from reconstituted halorhodopsin (hR) thin films. (A) The change of the photosignal, measured from a TM film, was reversible, when the aqueous solution was changed from KCl (pH 6) to sodium citrate (pH 6) and vice versa. H_1 and H_2 components have opposite polarities: positive and negative, respectively. (B) Photosignals from an ML film (H_1 component only) in KCl (pH 7) and in sodium citrate (pH 6), before stripping with a cotton swab, are shown to be superimposable. (C) Corresponding photosignals, after stripping with a cotton swab, are also shown to be superimposable. All measurements were made at 25°C. (From Michaile, S., Duschl, A., Lanyi, J. K., Hong, F. T. (1990). Chloride ion modulation of the fast photoelectric signal in halorhodopsin thin films. In: *Proceedings 12th Annual International Conference of IEEE Engineering in Medicine and Biology Society*. Philadelphia, PA, November 1–4, pp. 1721–1723.)

charge transfer contributes to the net transmembrane charge transport (DC photoelectric effect). The universal presence of reverse reactions in photobiological membranes also implies the universal presence of either the OD or the ICT mechanism. It is true that the OD-type charge separation–recombination in the reaction center does not span the entire

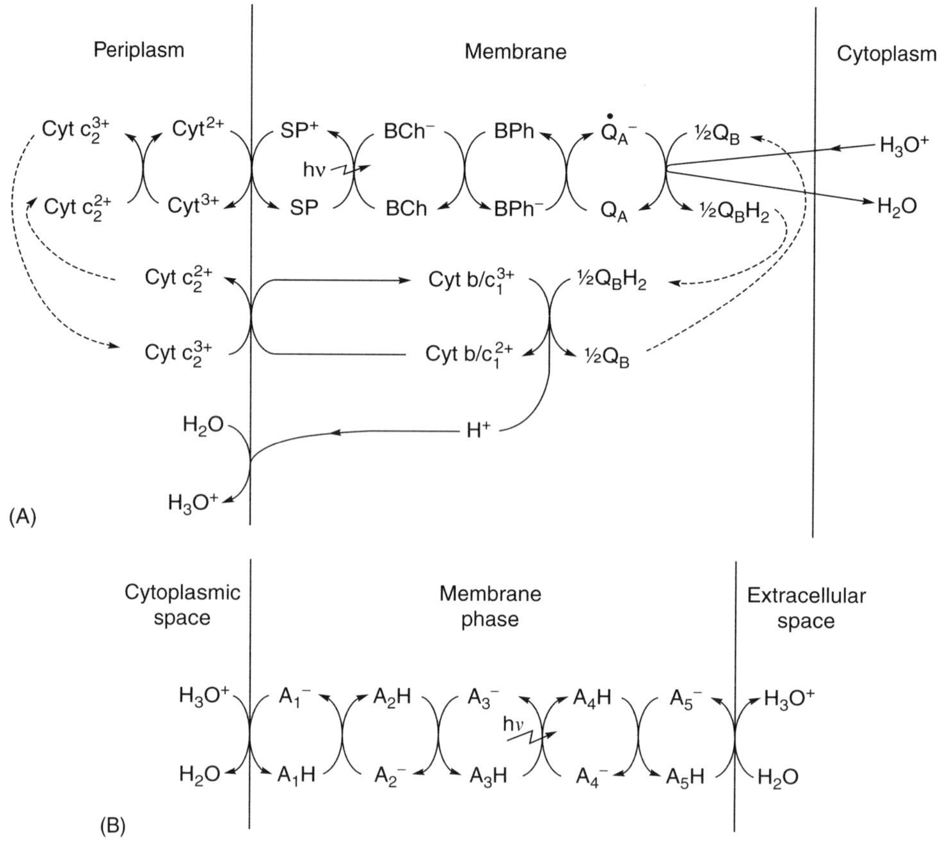

FIGURE 15.17
Coupled consecutive charge transfer reactions. (A) The scheme shows coupled consecutive electron-transfer reactions with reverse electron–proton cotransport in the photosynthetic reaction center of *Rhodopseudomonas viridis*. SP is the "special pair", BChl is voyeur bacteriochlorophyll, BPhe is bacteriopheophytin, Q_A is the tightly bound quinone, and Q_B is the loosely bound quinone. The electron-transfer reactions in the cytochrome b–c_1 complex are simplified to become a single step, so are those in the four bound cytochromes on the periplasmic surface. Cyt c_2 is the peripheral protein cytochrome c_2. The dotted arrows indicate the diffusion of Q_B and Cyt c_2. The reverse reactions are not shown in the schematic. (B) The scheme shows coupled consecutive proton-transfer reactions across the purple membrane of *Halobacterium salinarum*. The actual number of binding sites is not known. For simplicity, only five binding sites (A_1 through A_5) are shown. Site A_3 is the proton-binding site at the Schiff's base linkage. Its two adjacent sites, A_2 and A_4, are the residue 96 aspartate and the residue 85 aspartate, respectively. It is possible that additional sites exist between site A_1 and site A_2, as well as between site A_4 and A_5. It is understood that the Schiff's base is neutral when unprotonated and is positively charged when protonated. The reverse reactions are not shown. (From Hong, F. T. (1994). Retinal proteins in photovoltaic devices. In: Birge, R. R. (Ed.). *Molecular and Biomolecular Electronics* (Advances in Chemistry Series No. 240). Washington, DC: American Chemical Society, pp. 527–559. (A) and Hong, F. T. (1990). Bacteriorhodopsin as an intelligent material. In: Wang, C.-Y., Chen, C.-T., Cheng, C.-K., Huang, Y.-Y., Lin, F.-H. (Eds.). *Biomedical Engineering in the 21st Century*. Taipei, Taiwan: National Taiwan University Center for Biomedical Engineering. pp. 85–95. (B))

membrane thickness, but the OD model is readily modified to accommodate this variation (Figure 15.8 in [53]). Evidence for the presence of the AC photoelectric signal in the reaction center can be found in the data of Hara et al. [82] and those of Govorunova et al. [83] (Figure 15.18). The reported signals elicited from immobilized bacterial reaction center of *R. viridis* and those elicited from that of *Chlamydomonas rheinhardtii* exhibit a biphasic waveform and satisfies the zero time-integral condition, indicating that it is indeed a capacitative photocurrent, most likely generated by an OD mechanism.

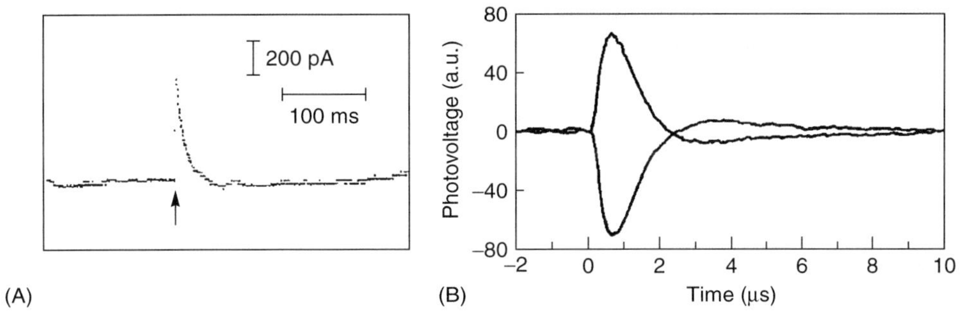

FIGURE 15.18
Pulsed-light-induced photoelectric responses in reconstituted photosynthetic membranes. (A) Purified photosynthetic reaction centers of *Rhodopseudomonas viridis* were immobilized, by means of avidin–biotin coupling technique, on transparent metal electrode. (B) Purified photosynthetic reaction centers of *Chlamydomonas reinhardtii* were oriented and immobilized in gel. (From Hara, M., Majima, T., Miyake, J., Ajiki, S., Sugino, H., Toyotama, H., Kawamura, S. (1990). Oriented immobilization of bacterial photosynthetic membrane. *Appl. Microbiol. Biotechnol.* 32:544–549. (A) and Govorunova, E., Dér, A., Tóth-Boconádi, R., Keszthelyi, L. (1995). Photosynthetic charge separation in oriented membrane fragments immobilized in gel. *Bioelectrochem. Bioenerg.* 38:53–56. (B))

The striking structural and chemical similarities between bR and rhodopsin suggest that there may be something in common between the two pigments [77]. Functionally speaking, the two systems serve diametrically opposite purposes. Efficiency of photon energy conversion is crucial to photosynthetic membranes such as the purple membrane and the chloroplast membrane, which store the converted energy as a transmembrane proton gradient. In contrast, rhodopsin is a photon sensor of which the efficiency of photon energy conversion is not crucial, but the sensitivity and the dynamic range of the sensory response are.

It is well known that under optimal conditions, a single photon, if absorbed by a single molecule of rhodopsin, can be detected psychophysically [84]. What transpires during visual transduction is the amplification of the initial photon energy by 100,000 fold. This amplification is achieved by means of a biochemical scheme known as the cyclic GMP cascade [85]. Apparently, a transmembrane proton gradient serves no useful purpose in vision. Ostrovsky and coworkers [86,58] demonstrated the presence of interfacial proton uptakes at the cytoplasmic surface of visual membranes and the absence of interfacial proton releases at the opposite side. The absence of interfacial proton releases implies that there is no transmembrane proton transport in visual membranes. The B_2 and the R_2 components are manifestation of interfacial proton uptake (and rerelease) at the cytoplasmic surface of the membrane. In contrast, B_2' reflects extracellular proton release in bR but its counterpart R_2' is not expected to exist in rhodopsin, in view of the finding of Ostrovsky and coworkers.

The presence of similar AC photoelectric signals suggests a common scheme for various types of retinal proteins. The fast signals B_1, R_1, and H_1 are characterized by ultrafast risetime, which is estimated to be shorter than 5 ps [87,88]. Such fast signals are unlikely to be caused by charge motion accompanying conformational changes or proton motion during the pumping process. Rather they are most likely the manifestation of rapid electron movements that accompany photoisomerization of the chromophore. Apparently, photon energy is temporarily stored as an electric dipole, which subsequently vanishes, and the stored energy is transferred to the apoprotein moiety of rhodopsin or bR. In other words, the initial energy is stored in a "spring"-like mechanism, which subsequently drives the protein conformational change when the "spring" recoils. This mechanism is similar to a "slingshot" model previously proposed by Keszthelyi [41] for bR. But there is an important difference. Keszthelyi interpreted the B_1 component as deprotonation of the Schiff's base that links the chromophore to the apo-protein moiety of bR, whereas we interpret B_1 as electron–hole pair production, which is similar to what happens in a chlorophyll-based photosynthetic reaction center. The relaxation of the electric dipole residing in the chromophore then provides the energy to deprotonate the Schiff's base. The detail arguments leading to our alternative interpretation can be found in [80]. Thus, the molecular processes underlying B_1, R_1, and H_1 may constitute the common pathway for photon energy conversion in these retinal proteins. The subsequent conformational changes then diverge in their functional purposes, leading to visual transduction (rhodopsin), proton and chloride ion transport (bR and hR, respectively). For bR and rhodopsin, the common pathway may be further extended to the step of the generation of B_2 and R_2, but not the B_2' generation. What then is the possible role of the R_2 signal in visual transduction?

Historically, the ERP was regarded as an epiphenomenon—an evolutionary vestige serving no known physiological function. However, just because no known function is known does not mean a heretofore-unsuspected function does not exist. While direct evidence is lacking, circumstantial evidence abounds, which points to a possibility of the R_2 component being a mechanistic trigger for visual transduction [75,89,90]. Our interpretation of the mechanistic origin of R_2 predicted that the molecular process responsible for the R_2 signal also generate a surface potential [52]. The prediction was subsequently confirmed by Cafiso and Hubbel [91], using a hydrophobic spin-label to monitor surface potential changes. Such a surface potential might serve as an electrostatic trigger to initiate the cyclic GMP cascade. How a light-induced surface potential works as a mechanistic trigger is best illustrated with an experiment performed by Drain et al. [92] (Figure 15.19).

These investigators studied the voltage-driven transport of a lipophilic anion, tetraphenylborate, across an artificial BLM. At the same time, they configured the BLM such that light illumination would generate a symmetric positive surface potential on both surfaces of the membrane. These surface potentials are generated by a light-induced electron transfer from the membrane-bound magnesium octaethylporphyrin molecule to an aqueous electron acceptor with equal concentrations in the two aqueous phases (Figure 15.19A). No macroscopic photovoltaic electric signal can be detected because the photocurrents generated by interfacial electron transfers at both surfaces are equal in magnitude but opposite in polarity. However, the positive surface potential induced by light at the two membrane surfaces increases the surface concentration of the negatively charged tetraphenylborate ions suddenly. As a consequence, the ionic current carried by tetraphenylborate ions is also increased (Figure 15.19B). Thus, illumination triggers an increase of ionic current.

If, however, tetraphenylborate ions are replaced by the lipophilic cation tetraphenylphosphonium ions, illumination then causes a sudden decrease of the ionic current (Figure 15.19C). This surface potential-induced change of ionic concentration would have been even more pronounced if the ionic species being transported were multivalent. This prototype device illustrates that light-induced surface potentials can be configured to implement a field-effect transistor (FET) or phototransistor. Thus it is not inconceivable that visual transduction process may depend on a surface potential-based FET to couple the photochemical reaction of rhodopsin to the biochemical processes of the cyclic GMP cascade.

The actual mechanistic event during visual transduction is of course more complex. Although electrostatic interactions feature prominently, other more subtle interactions are also present. In the crucial step of initiation of the cyclic GMP cascade, the binding surface of transducin (a G-protein) appears to have a patch of negative charge that can conceivably steer towards rhodopsin's exposed cytoplasmic surface via electrostatic attraction. There is no quantitative estimate of how much the R_2-generated surface potential contributes to the binding process. It is nevertheless quite certain that transducin binding is not a random process of diffusion and collision. It is also reasonably certain that binding

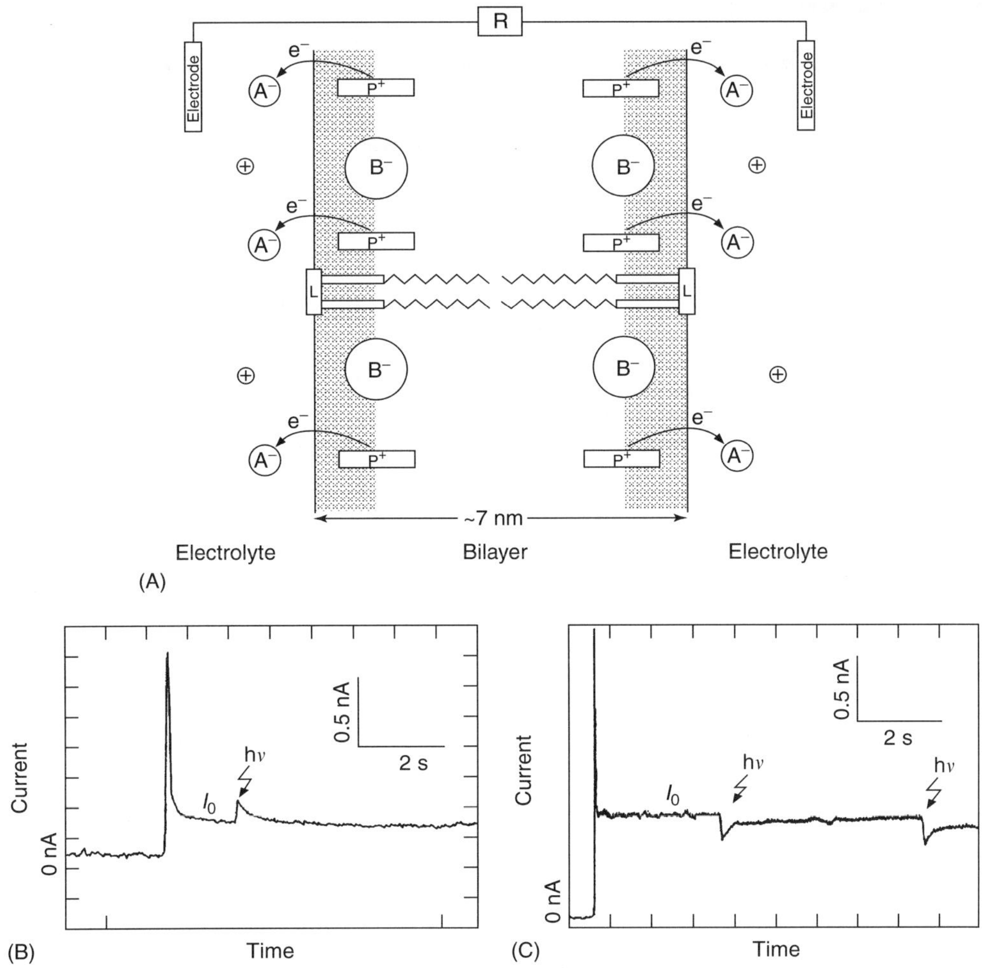

FIGURE 15.19
An experimental prototype showing switching of ionic currents based on light-induced surface potentials. (A) The BLM contained magnesium octaethylporphyrin, which is lipid-soluble (3.6 mM). The aqueous phases contained the electron acceptor methyl viologen in equal concentrations on both sides (20 mM). Tetraphenylborate ions, B^-, are partitioned into the membrane at the region of polar head groups of lipid. Photoactivation of magnesium octaethylporphyrin generates two symmetrical positive surface potentials, which increase the surface concentration of B^- but decrease the concentration of tetraphenylphosphonium ions (positively charged). (B) The ionic current of 1 μM tetraphenylborate ions increases 50% upon illumination with a 1 μs laser pulse. (C) The ionic current of 5 mM of tetraphenylphosphonium ions decreases 25% upon illumination. In both (B) and (C), the spike on the left is the capacitative transient upon imposition of an applied voltage of +50 mV. Light pulses are indicated by arrows. (From Drain, C. M., Christensen, B., Mauzerall, D. (1989). Photogating of ionic currents across a lipid bilayer. *Proc. Natl. Acad. Sci. USA* 86:6959–6962.)

of arrestin to photophosphorylated rhodopsin is facilitated by the cluster of multiple negative charges, which appears as a consequence of phosphorylation of multiple serine or threonine residues. However, allosteric effect as a result of conformational changes also play a role (see [90] for more detail).

In contrast, the work of Drain et al. demonstrated that the electrostatic switching mechanism inspired by partially reverse-engineering the visual membrane could be implemented in the laboratory in terms of completely different materials. This is the essence of biomimetic science: using the same principle but different materials to construct a device.

15.6 Prototype Molecular Sensors Based on Photoelectric Effects

15.6.1 Two Ways of Configuring Photoelectric Sensors

The term "photoelectric sensors" suggests two ways to configure sensors. From the above discussion, it is evident that bR is a *bifunctional* electronic material: it is sensitive to light as well as to small ions, such as H^+, Cl^-, Ca^{2+}, and Mg^{2+}. The amplitude of a photoelectric signal depends on the stimulating light intensity. The dependence is linear until it approaches saturation [93]. The photoelectric effects can be exploited to design light sensors. In addition, by exploiting the ion sensitivity of the photoelectric signals, sensors can be configured to monitor specific ion concentrations. The variety of ions to be detected depends on the type of pigment and its photochemistry. The design principle can be generalized to pigments other than rhodopsins or porphyrins. In view of the vast repertoire of photochemical reactions accumulated in the literature, the variety and specificity of ion-sensing photoelectric sensors can be rather impressive.

Here, I deliberately use the term "photoelectric effects" in the plural form instead of the singular "photoelectric effect." The intent is to remind us that there are two ways to deliver light stimuli: pulsed light and continuous light, and there are two effects: AC and DC photoelectric effects. Both types of illumination have been used in sensor designs (see below). Note that photoelectric effects can also be exploited to design artificial solar energy converters. But the design considerations are quite different. The essence of artificial solar energy conversion is the DC photoelectric effect. In this regard, the accompanying AC photoelectric effect is an unwanted side effect because charge recombination internally short-circuits the device and dissipates the partially converted photon energy as heat. The designers of artificial solar energy converters are constrained to utilize continuous light as the energy source and to focus on improving and maximizing the DC photoelectric effect. As far as sensor designs are concerned, there is no such constraint. In addition, the AC photoelectric effect is more advantageous than the DC photoelectric effect. Pulsed light is more versatile than continuous light in eliciting a response from photoelectric sensors.

Earlier, Trissl [94] and Rayfield [95] have entertained the idea of using a bR thin film as a photodiode. Although, strictly speaking, bR is not a photodiode, bR thin films remain one of the prototype biological thin films capable of delivering photon-switched electric signals with an ultrafast risetime. Two specific examples will be described and discussed in detail here. One was configured as a photon sensor whereas the other was configured as a specific ion sensor.

15.6.2 A Light Sensor Based on the AC Photoelectric Effect

Miyasaka et al. [93,96] successfully constructed a motion detector using wild-type bR (Figure 15.20). The detector was constructed as a square array of 8 by 8 patches of oriented bR thin films deposited on transparent metal electrodes, using photomicrolithography (Figure 15.20A and Figure 15.20B). Each bR electrode is connected by a pair of metal circuit lines, which allow the sensor array to be interfaced to conventional microelectronic circuitry for further signal processing. A typical photocurrent response from a patch of bR electrode is shown in Figure 15.20C. A transient photocurrent develops upon the onset of illumination and another transient photocurrent of opposite polarity develops upon the cessation of illumination. The photocurrent subsides during steady illumination and during the dark after the initial transient. Such a waveform is the "signature"

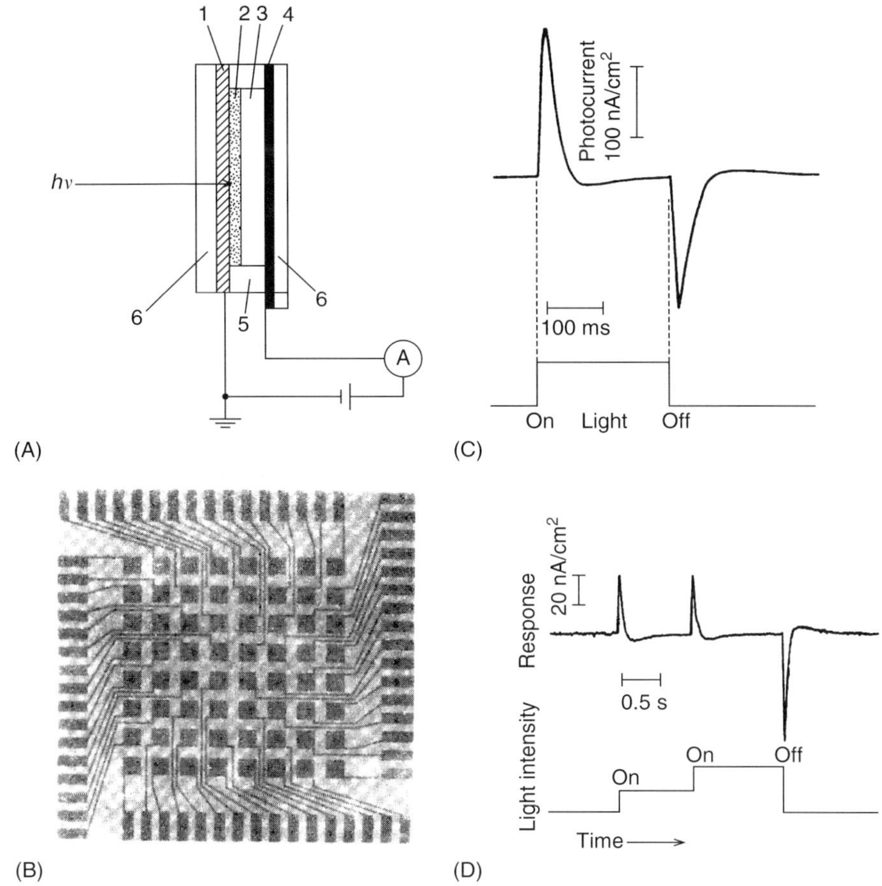

FIGURE 15.20
A motion-sensitive sensor array based on the AC photoelectric effect of bR. (A) The cross-section of a photocell, formed by immobilizing bR on a transparent electrode, is shown: 1, SnO_2 transparent conductive layer; 2, purple membrane LB film (typically six to ten layers) 3, aqueous electrolyte gel layer (200 μm thick); 4, Au layer (<1000 Å) as counter-electrode; 5, Teflon ring space; 6, glass substrate. (B) ITO (indium tin oxide) electrode was patterned with 64 pixels for image-sensing. Pixels of ITO (2.5 mm by 2.5 mm, 1000 Å thick transparent layer) form a two-dimensional array on a glass plate. Each pixel has a separate wire leading to the four edge terminals along the sides for interfacing with an amplifier circuit. (C) The photocurrent is generated in each pixel by a square light pulse of about 200-ms duration. The positive spike ("on" response) and the negative spike ("off" response) are main characteristics of the response of a linear high-pass *RC* filter to a long square pulse of applied current (see also Figure 15.4). The linear high-pass filter is formed by the irreducible equivalent circuit shown in Figure 15.5. (D) Photoresponses to step illumination shows the differential responsivity, which is also a characteristic of a linear high-pass *RC* filter: the amplitude of the photosignal transient is proportional to the magnitude of step illumination. (From Miyasaka, T., Koyama, K., Itoh, I. (1992). Quantum conversion and image detection by a bacteriorhodopsin-based artificial photoreceptor. *Science* 255:342–344. (A, B, C), and Miyasaka, T., Koyama, K. (1993). Image sensing and processing by a bacteriorhodopsin-based artificial photoreceptor. *Appl. Optics* 32:6371–6379. (D))

of a linear high-pass *RC* filter. The photocurrent response is therefore the manifestation of the AC photoelectric effect of bR (see Figure 15.4). The expected linearity of the high-pass filter response is demonstrated in Figure 15.20D, which shows that the amplitude of the AC transients are proportional to the change of the level of input illumination rather than the level per se (see also Wang et al. [97]). The capability of this photosensor to detect motion relies on differential responsivity as well as the linearity of responses.

Additional prototypes, which were based on the AC photoelectric effect include an artificial retina developed by Birge and Chen [98], an artificial neural network developed by

Haronian and Lewis [99], a color-discriminator developed by Boyer et al. [100], a color-sensitive artificial retina developed by Frydrych et al. [101] (Chapter 16 by Lensu et al. in this volume), a photodetector by Xu et al. [102], and by Wang et al. [103] (Chapter 17 by Wang et al. in this volume).

15.6.3 Specific Ion Sensors Based on the DC Photoelectric Effect

Seki et al. [104] developed an experimental prototype in which the gate region of an ion-sensitive field effect transistor (ISFET) was coated with a hR layer (Figure 15.21). The design of this sensor was similar to an earlier one developed by Tanabe et al. [105], who used bR instead as the light-sensing element. Seki et al. covered the gate region of the ISFET with a polyvinylbutyral (PVB) resin layer. They then immobilized hR-containing membrane vesicles in the matrix of PVB resin layer. They measured the gate voltage of hR-ISFET directly with a silver–silver chloride reference electrode. The gate voltage in response to a long square-wave pulse of light (duration about 2 min) is shown in Figure 15.21A.

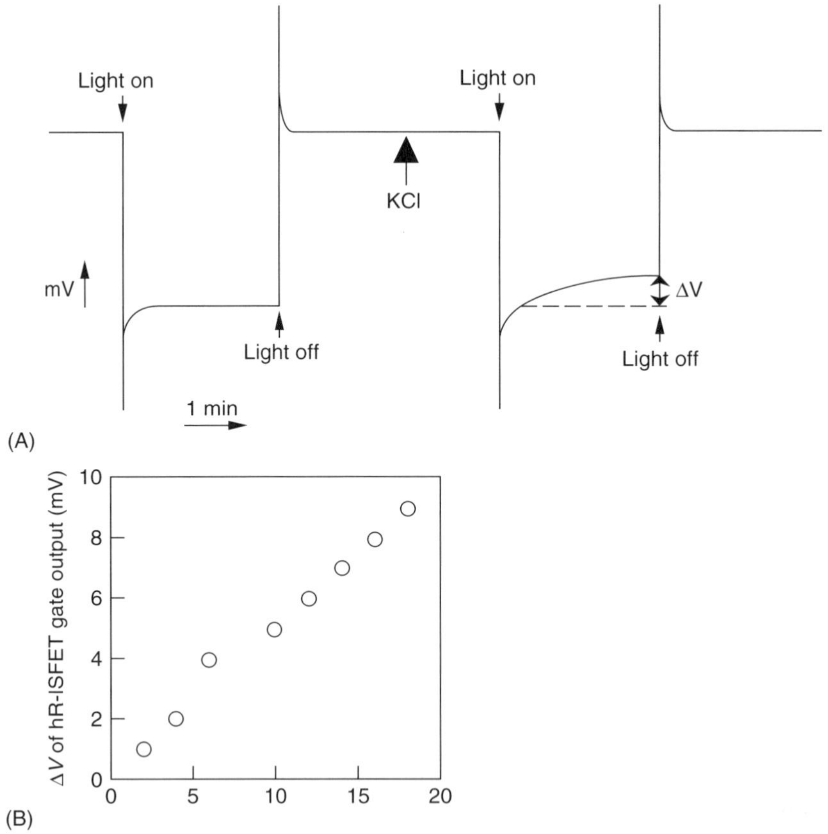

FIGURE 15.21
An ion-sensitive field effect transistor (ISFET) sensor based on the DC photoelectric effect of hR. hR-containing vesicles were immobilized on an ISFET. (A) The photovoltage in response to a 3-min square-wave light pulse was recorded before and after the addition of KCl. (B) The difference ΔV shown in (A) was plotted against the Cl^- concentration. (From Seki, A., Kubo, I. Sasabe, H., Tomioka, H. (1994). A new anion-sensitive biosensor using an ion-sensitive field effect transistor and a light-driven chloride pump, halorhodopsin. *Appl. Biochem. Biotechnol.* 48:205–211.)

The gate voltage responded with an initial spike upon illumination and then reaches a steady-state level. Likewise, another spike of opposite polarity appeared upon cessation of illumination. Seki et al. attributed the spike response to the intrinsic property of the ISFET itself. However, we suspect that residual interfacial ion transfers of hR layer, even in the absence of Cl^-, may have an additional contribution. Seki et al. found a decline of the steady-state photosignal upon the addition of Cl^-. They attributed this additional change of gate voltage to Cl^- pumping of hR because it was not observed in the ISFET alone or the PVB-coated ISFET. Nor the change could be observed by Cl^- addition in the dark. They further demonstrated that this gate voltage change is proportional to the Cl^- concentration (Figure 15.21B).

On the basis of the foregoing discussion, the gate voltage changes were most likely because of the interfacial Cl^- transfers rather than intramolecular charge displacement of hR. If this interpretation is correct, the performance of this hR-ISFET can be vastly improved by using pulsed light to elicit the Cl^--dependent signal and to monitor the AC photoelectric signal. There are several advantages for using pulsed light stimulation (cf. Figure 15.16).

First, the sensitivity of the device can be vastly enhanced. The AC photoelectric effect of bR has a picosecond rise time, and the signal amplitude is at least two orders of magnitude greater than that of the DC photosignal. We suspect that the corresponding signal from hR may have a similarly fast risetime. Second, the short duration of measurements increases the speed of measurement. Third, the use of repeated light pulses allows for modulation for the purpose of excluding spurious signals: the signals are measured only when the sensor is gated by a light pulse (a technique known as gating).

Such a design can be modified to construct a light-addressable multisensor array, which was first proposed by Haneman et al. [106]. The increased speed of measurement of AC photoelectric signals and the use of sharply collimated or focused laser pulses make such a design particularly attractive. Each sensor cell can then be interrogated by a scanning pulsed laser beam in a rapid sequential fashion, thus giving an illusion of parallel processing, much like the common technique of multiplexing in conventional microelectronics. A two-sensor array based on mixed patches of hR and bR thin films illustrates such a design (Figure 15.22). The spatially discrete patches require rapid steering of the interrogating laser beam. If, however, the various pigments used in configuring a multisensor array have nonoverlapping absorption maxima, a single sensor patch made of a mixture of various pigments can be constructed and a rapid steering mechanism of interrogating beam becomes unnecessary. Instead, a composite laser beam made of different wavelengths must be used to interrogate the sensor patch. Even if the absorption maxima of constituent pigments partially overlap, the composite output signal can be further analyzed and processed by means of artificial neural network technology (see Section 15.8). In this way, investigators have a vast repertoire of known photochemistry and an equally large variety of pigments or dyes to exploit for design purposes.

15.7 Bioelectronic Interfacing

In the examples presented in Section 15.6, the photoelectric signal is capacitance-coupled (AC-coupled) to the metal electrode. The devices pose no interfacing problem in hooking up with conventional microelectronic circuitry for further signal processing. However, a number of investigators have found that no DC photoelectric current can be observed in bR-coated metal-electrode sensor systems. This was most likely due to the absence of appropriate electrodic reactions in the system to convert a current carried by protons in

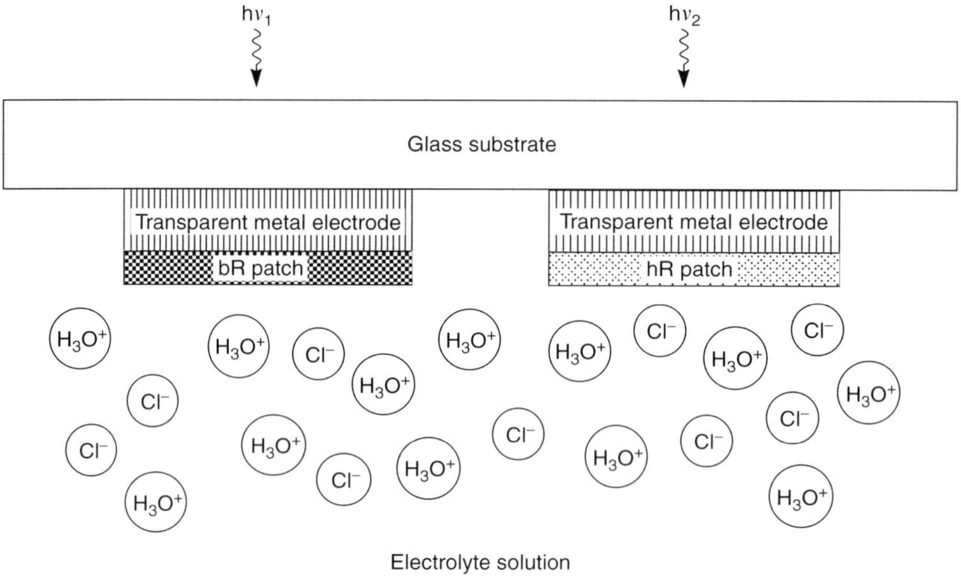

FIGURE 15.22
Multisensor array. Electrode patches similar to what is shown in Figure 15.20 are made of different pigment thin films, such as bR and hR. Each patch is interrogated by focused laser beams to elicit photoelectric responses, which report local concentrations of H^+ and Cl^-, respectively.

the purple membrane into a current carried by electrons in the conduction band of the metal electrode. With the exception of some redox proteins, this interfacing problem will, in general, arise in reconstituted protein systems that generate DC photocurrents. Even with redox proteins, interfacing that permits the DC current to go through the electrode requires special bridging molecules that link the protein to the metal electrode.

Greenbaum [107,108] studied the photoelectric signal in isolated thylakoid (chloroplast) membranes containing the photosystem I $NADP^+$-ferredoxin-oxidoreductase. The photosystem I transfers electrons from plastocyanine at the thylakoid space (i.e., the space enclosed by a thylakoid) across the membrane to the enzyme $NADP^+$-ferredoxin-oxidoreductase, located at the stromal surface. Greenbaum deposited colloidal platinum on the reducing end (stromal surface). A DC photocurrent generated in the system was thus made observable. Electrons are intercepted by the colloidal platinum before they are processed by $NADP^+$-ferredoxin-oxidoreductase.

Aizawa [109] studied an enzyme (metal) electrode based on immobilized glucose oxidase, which is a redox enzyme. Interfacing in this type of biosensor requires facilitation of electron transfers from the active site of the enzyme protein to the metal electrode. Various methods have been used, such as electron mediators, electron promoters, and molecular wires, to accomplish this purpose. For example, a conducting polymer, polypyrrole, deposited on the electrode surface, was used to form electron wires that facilitate electron transfers to and from the metal electrode.

Interfacing technology developed for redox proteins is unsuitable for detecting the DC photocurrent generated by the purple membrane primarily because bR is not a redox protein. A solution for this problem could be proposed on the basis of reverse engineering of the photoreactions in the photosynthetic reaction centers of green plants [75]. Let us take a closer look at how the above-mentioned Photosystem I operates in vivo (see [110] for a concise review of green plant photosynthesis). In chlorophyll-based photosynthetic systems (e.g., Photosystem I and Photosystem II of green plants and the reaction center of most photosynthetic bacteria), the primary reaction is charge separation involving the production of electrons and holes. Electron transfers dominate the process of energy conversion along the so-called electron transport chain, except at the membrane–water interfaces where the moving part is switched from electrons to protons. The net result is a transmembrane proton transport in the opposite direction. This conversion is implemented with the intervention of a special class of mediators known as quinoid compounds. They form the cofactors for some redox enzymes: NAD^+ (nicotinamide adenine dinucleotide) in the mitochondria inner membrane and $NADP^+$ (nicotinamide adenine dinucleotide phosphate) in the thylakoid membrane of Photosystem I. These enzymes operate in the aqueous phase adjacent to the membrane surface. Two additional quinoid compounds, plastoquinone and ubiquinone, serve as membrane-bound mobile electron carriers in the thylakoid and the mitochondrial inner membrane, respectively. They are both redox and acid/base compounds. The prototype reaction of quinoid compounds is shown in the benzoquinone–benzohydroquine reaction:

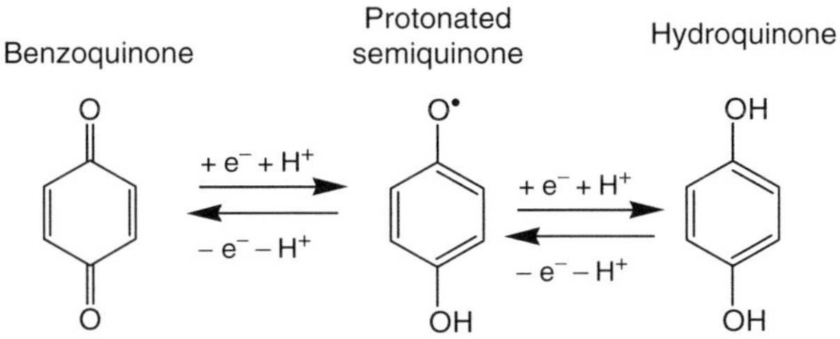

The structure of quinoid compounds mentioned above contains the benzoquinone moiety. Electrons transferred to enzyme $NADP^+$-ferredoxin-oxidoreductase are converted into mobile protons:

$$2\,Fd_{red} + 2e^- + NADP^+ + 2H^+ \rightleftarrows 2\,Fd_{ox} + NaDPH + H^+$$

where Fd_{red} and Fd_{ox} are the reduced and oxidized form of $NADP^+$-ferredoxin-oxidoreductase, respectively. Therefore, a solution for DC-current interfacing of an oriented purple membrane thin film to a metal electrode can be found by choosing a suitable quinoid compound to make the required conversion. It may not be necessary to search for such a compound in natural sources because it may be obtained by means of molecular engineering, using the recombinant DNA technology.

15.8 Discussions and Concluding Remarks

The high-technology revolution during the latter half of the 20th century, which made personal computers standard office equipment and home appliance, was mainly fueled by the astonishing advances in microelectronics. Miniaturization allowed more and more circuit elements to be packed into a small integrated circuit (IC). As indicated by the well-known Moore's law, the number of the device components contained in a single IC grew exponentially with the passage of time. Will this trend continue so that device size eventually reaches the atomic scale? According to experts, the trend is still going strong but it certainly cannot continue beyond bounds [111,112]. This consideration has ushered in a multidisciplinary research effort in molecular electronics [113–117]. bR played a prominent role in earlier efforts of prototype device research, presumably because of its exceptional thermal and chemical stability. bR exhibits two properties that could be exploited for sensor constructions: photochromic and photoelectric effects. This chapter focuses on the photoelectric side of bR-based sensor designs.

Electrical events are common control processes in living systems. Many such events take place in the solution phase; these events are only indirectly observable because the direction of charge movements is somewhat random. A particular class of electrical phenomena involves biological membranes and has been the subject of electrophysiological investigations. Biological membranes provide a spatially anisotropic medium for electrical events to take place; the direction of charge movements is, therefore, largely perpendicular to the membrane (vectorial charge movement), thus greatly facilitating electrical measurements and permitting unequivocal interpretations. Bioelectric events such as action potentials and synaptic potentials have their origin in passive electrodiffusion of small inorganic ions through water-filled transmembrane ion channels. The photo-induced phenomena in biomembranes call for somewhat different analytical methodology because these phenomena, just like those caused by electrogenic pumps, require movement of electrons or ions *against* a transmembrane electrochemical gradient. These charge movements reflect an active process of charge transport fueled either by metabolic energy (in the case of mitochondria) or by absorbed photons (in the case of photosynthesis and purple membranes). The presence of additional electric parameters (such as photoemf) renders the application of Ohm's law alone insufficient to analyze the electric events. Visual photoreceptor membranes do not perform active charge transport but are rather equipped with mechanisms of tranducing absorbed photon signals into electric signals suitable for processing by the nervous system. However, visual membranes and photosynthetic membranes share a common feature due to similarity in the primary photophysical event upon illumination. Such primary events are reflected in changes of absorption spectra and in charge movements far too rapid to be accounted for in terms of electrodiffusion. Rather, they reflect rapid charge displacements either at the membrane surfaces or in the interior of functional membrane-bound proteins. Such events occur at a considerable faster time scale than electrodiffusion. This fast photoelectric effect is highly suitable for biosensor designs. However, some prototype photoelectric sensors reported in the literature utilized the steady-state photoelectric effect rather than the fast photoelectric effect. We have shown that these two types of effects correspond to the DC photoelectric effect and the AC photoelectric effect, respectively. Direct transplantation of electrophysiological methodology appeared to be inadequate. Many years ago, we introduced the method of null-current measurement and TVC measurement, for handling the DC and the AC photoelectric effects, respectively. We demonstrated how a combined electrophysiological and electrochemical approach could provide a coherent description and analysis of the molecular and mechanistic events associated with the fast photoelectric effect. This approach was initially adopted to better our understanding of Nature's general scheme for light-sensitive membranes, and to gain a deeper insight into the process of electric signal generation. However, we got more than what we had bargained for, because of its potential impact on bR-based sensor applications.

While the DC photoelectric effect is highly relevant for the construction of artificial solar cells, rapid light-induced responses to the chemical environment offer a venue for sensor construction. Mining of sensor properties needs not be an exercise of trial and error because understanding of molecular mechanisms provides a road map. Most sensor applications were based on "differential responsivity." However, elucidation of the underlying mechanisms has never been straightforward. Even to date, alternative rival interpretations thrive among different persuasions of bR researchers. This concern brings us to the topic of mathematical modeling. Recently, Gauch [118] complained about biomedical researchers' complacency in the lack of pursuing mathematical modeling, while pointing out that modeling often leads to deeper understanding. However, there was no lack of effort in mathematical modeling regarding the fast photoelectric effect. Rather, the problem was a plethora of ad hoc models generated by means of what is known as exponential analysis, that is, using computer-based curve fitting to generate a mathematical formula to account for the kinetics of measured electrical responses. As was discussed in detail elsewhere (e.g., Section 20 of [29]), exponential analysis alone imposes no constraint on the mathematical model so derived, and is, therefore, not experimentally *falsifiable*. As explained lucidly by science philosopher Popper [119], a scientific theory or model must be experimentally falsifiable. This puts models or theories generated by means of exponential analysis in the same league with the "theory of intelligent design," which was proposed by its proponents to be a rival theory of evolution for the consumption of public school children in the United States. The work of Yao et al. [50] is a case in point. Nevertheless, in a subsequent publication, Yao et al. [51] did make a scientific statement: differential responsivity is not an intrinsic property of bR, but rather the consequence of AC-coupling of the measuring circuit. Evidence to be invoked to falsify the latter statement abounds: No AC-coupling was involved in the measurement of Figure 15.12B, Figure 15.15A, Figure 15.15B, and Figure 15.21. Additional evidence can also be found in [120]. From the point of view of photoelectric sensor designs, exponential analysis offers no design aid because it cannot predict data other than the very set, on which the analysis has been conducted. In Gauch's words [121,122], models based on the exponential analysis offer only *postdictions* but no *predictions*.

We have placed a great deal of emphasis on the materials science aspect of biosensor technology. However, the search for suitable biomaterials for sensor construction is not the only reason to launch a biological approach. I believe reverse-engineering Nature is a viable approach. In spite of misgivings about reverse engineering, raised by investigators from time to time in the past, imitating Nature at the level of fundamental principles, rather than superficial and "verbatim" imitation, could still yield handsome dividends [123]. There are obvious advantages in utilizing biomaterials for sensing purposes. The mere incorporation of biomaterials in the sensor construction vastly increases the variety and the specificity of sensors, since we have the resource of a vast repertoire of biomolecular recognition at our command—the myriad of signal transduction processes involves molecular recognition. However, advances made in digital computer technology have already ushered in an unprecedented trend of information explosion. Merely increasing the variety and the degree of specificity of sensors would probably not make human users more informed. Rather, these improvements may further exacerbate information explosion. Human users have already become overwhelmed with a huge amount of undigested raw information, as the U.S. intelligence community had learned the hard way in the aftermath of the 9/11 attacks. A popular technique of electronic eavesdropping depends largely on matching of keywords. Contrary to popular belief, keyword matching, in which digital computers excel, does not always lead to understanding of what raw information can provide, as astute Internet users know how to paraphrase common keywords so as to evade cyber-surveillance. Even the so-called semantic web was of limited help [124]. A reverse-engineering approach to understand how the human brain comprehends raw information is therefore desirable. This brings us to the field of biocomputing research (for reviews, see [125–128]).

A digital computer is often compared with the human brain, yet there are significant differences in terms of information processing [129]. Computer scientist Conrad [125] previously pointed out that the predominant feature of biological information processing is "shape-based" computing. In contrast, digital computing is switch-based, since the latter recognizes only "0" and "1." Conrad further surmised that the "computer" architecture of the human brain is hierarchical, and it involves two levels of information processing: a microscopic level for intracellular information processing and a macroscopic level of interneuronal communications. Following the lead of Conrad's ideas, the author subsequently conducted a detailed survey of biocomputing schemes and found that both "shape-based" and "switch-based" information processing are utilized alternatingly [130]. Furthermore, biocomputing is neither absolutely deterministic, as in digital computing, nor completely random, as suggested by molecular diffusion and collisions in solution-phase biochemistry. At all levels, extending from submolecular protein folding all the way to intermolecular communications, information process exhibits the combination of a "top-down" constraint and a "bottom-up" exploration—apparently a compromise between absolute determinism and extreme randomness. It is the hierarchical separation that prevents randomness from propagating beyond bound and keeps randomness at bay. This insight was extended to investigate creative problem-solving. Again, human cognition at the system-level mirrors events at lower hierarchical levels. Creative problem-solvers do not resort to a highly random process of trial and error. Nor do they follow predictable dogmatic and algorithmic processes of finding the answers to a novel problem. Our new insights into geniuses' thinking process have only recently appeared in print [131,132].

What we did was treat creative problem-solving as a process of *pattern recognition*: recognizing potential solutions to match a given problem. The idea of treating creative problem-solving as pattern recognition could be traced back to 1988 Peano Lecture presented by economist and computer scientist Simon [133]. Recently, high-tech entrepreneur Hawkins [134] also invoked pattern recognition to interpret his personal experience in design innovations. Here, we consider two kinds of pattern recognition: analog and digital pattern recognition. To make a long story short, these two terms correspond to visual thinking and verbal thinking, respectively, in the psychology literature. Visual thinking is nothing new [135,136]. It was frequently advocated by scientific geniuses, such as Albert Einstein [137], Richard Feynman, [138], and Stephen Hawking [139]. Inventor Tesla [140] had a vivid account of how he had been tormented by persistent visual imagery and claimed that he could visualize how an electric motor turned without actually carrying out an experimental test. However, mainstream psychologists have dismissed these introspective reports as mere anecdotes. In contrast, verbal thinking, which is synonymous as logical thinking, is regarded as being objective and reliable by conventional wisdom, and is often heavily reinforced by higher education. In fact, dogmatism, which was characterized by thinking along a predictable, well-trodden path, is a manifestation of verbal thinking, whereas keyword matching, which was often invoked by students to answer a standardized test, is an extreme form of verbal thinking.

Obviously, digital pattern recognition requires strict matching of criteria, which are often expressed in terms of explicit symbolic or verbal rules. Analog pattern recognition requires holistic judgment, which is often based on ill-defined criteria and gut feeling; the matches are often imprecise and loose rather than rigorous and strict. Obviously, digital pattern recognition is more reliable than analog pattern recognition, and it is also far more objective than analog one. However, there is a serious drawback of digital pattern recognition. Because of the use of strict criteria, potential solutions tend to be excluded prematurely. In contrast, analog pattern recognition is known to be highly subjective and error-prone. What then is the merit of visual thinking that had lured so many geniuses of the past?

Most models of creative problem-solving reported in the literature include a *solution-generating phase* and a *solution-verifying phase* (so-called double-checking). Scientific geniuses used intuition to generate solution while invoked strict logic for solution verification, as mathematician Henri Poincaré [135] aptly proclaimed, "It is by logic that we prove, but by intuition that we discover." Here, we identify intuition with visual thinking (see [131,132] for detailed evidence). The saving grace for the peril of visual thinking is that its fallibility can be rescued and rectified by subsequent verification based on strict logic and explicit rules, that is, verbal reasoning. In brief, visual thinkers invoke visual reasoning during the solution-generating phase, while invoking verbal reasoning during the solution-verifying phase.

The above scheme was based on Poincaré's hint, as cited above. However, he gave only half of the story, presumably because he had no direct knowledge of the thinking process of people who were severely handicapped in creativity. It is possible to invoke verbal, logical reasoning during the solution-generating phase so as to discover solutions to certain types of problems. High-school students, who learn Euclidean geometry, know that some theorems can be found as logical consequences of other more fundamental theorems or the combination thereof. These less fundamental theorems are often referred to as corollaries or lemmas. But important discoveries as such have long been exhausted. Modern investigators must rely on intuition to make major discoveries. As Poincaré once said, "Pure logic could never lead us to anything but tautologies; it could create nothing new; not from it alone can any science issue" [141].

It is apparent that the practice of visual reasoning in the solution-generating phase ensures highly *inclusive* gathering of *potential* solutions, whereas the practice of verbal reasoning in the solution-verifying phase ensures highly *exclusive* gathering of *accepted* solutions. In other words, visual reasoning during the first phase avoids exclusion of false-negative solutions, whereas verbal reasoning during the second phase avoids inclusion of false-positive solutions. It is therefore obvious that this combined approach offers a better chance of solving difficult and novel problems; failure in either or both phases diminishes the chance of success, except by dumb luck. Obviously, those who stick to verbal thinking during the entire problem-solving session are at a significant disadvantage. It is particularly precarious, if the problem-solver combines the solution-generating phase and the solution-verifying phase into a singe step of verbal thinking. It is even faster than the two-phase routine, but it lacks the benefit of double-checking. Needless to say, the chance of solving a given problem by luck alone diminishes exponentially with increasing complexity of the problem—a well-known consequence of combinatorial explosion. Visual reasoning is a top-down constraint to limit the search for solutions to a most likely range (i.e., free associations by pictures rather than by keywords), while allowing sufficient exploration by means of loose matching. It is an approach known as heuristic searching in the literature of artificial intelligence and operations research [133,142].

This general principle of heuristic searching is demonstrated by the agent technology in problem-solving computer programs [143]. Software agents, of which "cookies" are better known examples, are empowered with limited freedom. Here, the computer programmer provides a general strategy in the main program but does not micromanage the detailed task. The general strategy prevents the computational resource from spreading out too thin, whereas limited freedom allows the software agent to explore a wider range of options than the programmer could have conceived ahead of time.

Our interpretation of human creativity also has the merit of explaining difficult concepts such as intuition, the "aha" phenomenon, and the unpredictability of creativity. The key concepts to be invoked are parallel and sequential processing. In brief, visual reasoning is a parallel process, whereas verbal reasoning is a sequential process. This division of work is reflected in the well-known cerebral lateralization: the dominant (usually left) hemisphere is specialized for verbal reasoning, whereas the nondominant hemisphere is equipped to perform visual reasoning. Mainstream psychologists usually dismissed any correlation between creativity and the use of the right brain. But I believe it was a premature—i.e., false-negative—conclusion based on misinterpretation of experimental data (see Section 4.15 of [132]).

The implication of creativity research to biosensor technology is obvious. Future smart sensors ought to be empowered with the capability of analog pattern recognition and, therefore, the ability of holistic judgment. Clearly, this capability cannot be implemented in terms of sequential algorithm, as Conrad repeatedly pointed out. However, there is reason for optimism in the future. An effective approach is again imitating Nature at the principle level rather than outright imitation.

Decision-making in digital computers depends on logical operations of numerous discrete flip-flop circuits, while decision-making in the human brain depends on a holistic process of evaluating the interactions of numerous synapses at the neuronal level and arbitration of conflicting factors at the cognitive level. The idea led to the conception of artificial neural network computing, implemented in the environment of digital computers, that is, it is a virtual machine within a digital machine. Artificial neural network research is among the fastest growing science and technological endeavors, since Hopfield [144] published his seminal article in 1982. Equally explosive is the growth of the literature of artificial neural network; it defies a serious review here (see a highly readable review by Churchland [145]). It suffices to cite a single example of its application to sensor technology.

Among all five human (special) senses, primary detection in the visual sense is relatively straightforward. The interpretation of color in humans is based on the relative degrees of activation of red, blue, and green (cone-shaped) photoreceptors (essentially Helmholz's trichromatic theory). The process is somewhat discrete and algorithmic. That is not to say human's recognition of visual forms and motion is also straightforward. In contrast, the chemical sense provided by the olfactory (smell) and the gustatory (taste) systems is somewhat difficult to articulate in words and to simulate with computer algorithm because differentiation of different smells or tastes depends more heavily on holistic judgment than color discrimination does. While it was feasible to construct different chemical sensors with a wide variety of specificities, criteria of differentiating a vast number of volatile organic chemicals cannot be readily specified with a set of explicit rules or sequential computer algorithm.

Olfaction provides two categories of physiological functions. One is general odor sensing, whereas the other has important behavioral consequence through sensing of pheromones [146]. It turned out that general odor sensing and pheromone sensing are segregated in different neural pathways. Even in general odor sensing, chemical structures offer few clues. A slight change in chemical structures sometimes leads to a dramatic shift in odor. Even a change of concentration can turn a pleasant odor into an offensive one and vice versa (smelling of a skunk's secretion is a common experience).

The advent of modern molecular biology allowed a detailed dissection of olfactory receptor molecules. The olfactory receptor molecules are encoded with multiple genes [147]. While these receptor molecules are discrete, the sheer large number of them transforms discrete sensing, as in color discrimination, into a *virtual continuum*, thus making holistic judgment (or analog pattern recognition) possible. Nature's approach parallels the transformation of discrete flip-flop operations into complex synaptic interactions. Of course, comprehension of this vast continuum of information must be implemented at the architecture level, as advocated by Conrad [125]. (So do visual comprehension of forms, motion, and other high-level cognition [148]). Sometime ago, Pearce [149,150] relied on an artificial neural network program to accomplish this feat (Artificial Electronic Nose). Essentially, information is stored, in a distributed sense, among numerous "synapses" of the network, much like what transpires in human's nervous system. Given a sufficient number of "synapses," discrete properties of individual synapses also coalesce into a virtual continuum, thus converting a basically digital process into an equivalent analog process. Again, this is a testimonial of the power of analog pattern recognition.

Another exciting development in nontraditional computing was what is now known as DNA computing. Adleman [151] used standard tools of molecular biology to solve a case of the directed Hamiltonian path problem—the so-called "traveling salesman" problem.

This problem was known, for a long time, to be refractory to solutions based on sequential algorithm. The success of DNA computing offers another reason of optimism in future sensor designs. This approach allows mixing of parallel and sequential processes in computation. Exactly how this insight can be translated into novel sensor designs depends on ingenuity of future investigators/inventors. It is not something I can explicitly describe ahead of (its) time.

Acknowledgments

The author acknowledges the contribution of the following individuals whose experimental work on bR was cited: Man Chang, Albert Duschl, Brian Fuller, Filbert Hong, Sherie Michaile, Baofu Ni, Ting Okajima, Michelle Petrak, and Wita Wojtkowski. The author wishes to thank his collaborators: Janos Lanyi, Lowell McCoy, Mauricio Montal, and Richard Needleman. The author is indebted to Professor David Mauzerall of The Rockefeller University, under whose supervision the work on magnesium porphyrins-containing BLM was performed. By setting personal examples, Professor Mauzerall proved to be instrumental in the author's subsequent pursuit of the problem of human creativity. The author is also indebted to the late Professor Michael Conrad of Wayne State University for his profound influence on the author's work on biocomputing.

References

1. Kuhn, H. (1979). Synthetic molecular organizates. *J. Photochemistry* 10:111–132.
2. Kuhn. H. (1983). Functionalized monolayer assembly manipulation. *Thin Solid Films* 99:1–16.
3. Kuhn, H., Möbius, D. (1993). Monolayer assemblies. In: Rossiter, B. W., Baetzold, R. C. (Eds.). *Investigations of Surfaces and Interfaces, Part B.* Physical Methods of Chemistry Series. Vol. IXB. 2nd edn. New York: John Wiley and Sons, pp. 375–542.
4. Brown. K. T., Murakami, M. (1964). A new receptor potential of the monkey retina with no detectable latency. *Nature* 201:626–628.
5. Mueller, P., Rudin, D. O., Tien, H. T., Wescott, W. C. (1962). Reconstitution of cell membrane structure *in vitro* and its transformation into an excitable system. *Nature* 194:979–980.
6. Tien, H. T., Ottova-Leitmannova, A. (2000). *Membrane Biophysics As Viewed From Experimental Bilayer Lipid Membranes (Planar Lipid Bilayers and Spherical Liposomes).* Amsterdam, Lausanne, New York, Oxford, Shannon, Singapore and Tokyo: Elsevier.
7. Ottova A., Tien, H. T. (2005). The lipid bilayer principle: a historic perspective and some highlights. In: Tien, H. T., Ottova-Leitmannova, A. (Eds.). *Advances in Planar Lipid Bilayers and Liposomes, Volume 1.* Amsterdam, Boston, Heidelberg, London, New York, Oxford, Paris, San Diego, San Francisco, Singapore, Sydney and Tokyo: Elsevier Academic Press, pp. 1–76.
8. Takagi, M., Azuma, K., Kishimoto, U. (1965). A new method for the formation of bilayer membranes in aqueous solution. *Ann. Report Biol. Works Fac. Sci. Osaka Univ.* 13:107–110.
9. Tien, H. T. (1968). Light-induced phenomena in black lipid membranes constituted from photosynthetic pigments. *Nature* 219:272–274.
10. Trissl, H.-W. and Montal, M. (1977). Electrical demonstration of rapid light-induced conformational changes in bacteriorhodopsin. *Nature* 266:655–657.
11. Darszon, A., Philipp, M., Zarco, J., Montal, M. (1978). Rhodopsin-phospholipid complexes in apolar solvents: formation and properties. *J. Membr. Biol.* 43:71–90.
12. Oesterhelt, D., Stoeckenius, W. (1971). Rhodopsin-like protein from the purple membrane of *Halobacterium halobium*. *Nature* 233:149–152.

13. Blaurock, A. E., Stoeckenius, W. (1971). Structure of the purple membrane. *Nature* 233:152–155.
14. Stoeckenius, W., Bogomolni, R. A. (1982). Bacteriorhodopsin and related pigments of Halobacteria. *Ann. Rev. Biochem.* 51:587–616.
15. Henderson, R., Baldwin, J. M., Ceska, T. A., Zemlin, F., Beckmann, E., Downing, K. H. (1990). Model for the structure of bacteriorhodopsin based on high-resolution electron cryomicroscopy. *J. Mol. Biol.* 213:899–929.
16. Birge, R. R. (1990). Nature of the primary photochemical events in rhodopsin and bacteriorhodopsin, *Biochim. Biophys. Acta* 1016:293–327.
17. Lanyi, J. K., Luecke, H. (2001). Bacteriorhodopsin. *Curr. Opin. Struct. Biol.* 11(4):415–419.
18. Luecke, H., Lanyi, J. K. (2003). Structural clues to the mechanism of ion pumping in bacteriorhodopsin. *Adv. Protein Chem.* 63:111–30.
19. Lanyi, J. K. (2004). Bacteriorhodopsin. *Ann. Rev. Physiol.* 66:665–688.
20. Bunkin, F. V., Vsevolodov, N. N., Druzhko, A. B., Mitsner, B. I., Prokhorov, A. M., Savranskii, V. V., Tkachenko, N. V., Shevchenko, T. B. (1981). Diffraction efficiency of bacteriorhodopsin and its analogs. *Sov. Tech. Phys. Lett.* 7(12):630–631.
21. Ivanitsky, G. R., Vsevolodov, N. N. (Eds.). (1985). *Collected Articles of the Scientific Biological Research Center of the Academy of Sciences USSR*, Pushchino, pp. 98–163. (in Russian).
22. Vsevolodov, N. N., Dyukova, T. V. (1994). Retinal-protein complexes as optoelectronic components. *Trend. Biotechnol.* 12:81–103.
23. Hong, F. T. (1996). Biomolecular electronics, In: Taylor, R. F., Schultz, J. S. (Eds.). *Handbook of Chemical and Biological Sensors*. Bristol: Institute of Physics Publishing Ltd., pp. 257–286.
24. Vsevolodov, N. N. (1998). *Biomolecular Electronics: An Introduction via Photosensitive Proteins*. Boston, Basel and Berlin: Birkhäuser.
25. Hampp, N. (2000). Bacteriorhodopsin as a photochromic retinal protein for optical memories. *Chem. Rev.* 100:1755–1776.
26. Wise, K. J., Gillespie, N. B., Stuart, J. A., Krebs, M. P., Birge, R. R. (2002). Optimization of bacteriorhodopsin for bioelectronic devices. *Trends Biotech.* 20:287–394.
27. Hong, F. T. (1980). Displacements photocurrents in pigment-containing biomembranes: artificial and natural systems. In: Blank, M. (Ed.). *Bioelectrochemistry: Ions, Surfaces, Membranes* (Advances in Chemistry Ser. No 188). Washington, DC: American Chemical Society, pp. 211–237.
28. Hong, F. T. (1994). Electrochemical processes in membranes that contain bacteriorhodopsin. In: Blank, M., Vodyanoy, I. (Eds.). *Biomembrane Electrochemistry* (Advances in Chemistry Ser. No 235). Washington, DC: American Chemical Society, pp. 531–560.
29. Hong F. T. (1999). Interfacial photochemistry of retinal proteins. *Prog. Surf. Sci.* 62:1–237.
30. Hong, F. T. (2003). Interfacial photochemistry of biomembranes. In: Nalwa, H. S. (Ed.) *Handbook of Photochemistry and Photobiology*. Vol. 4. Stevenson Ranch, CA: American Scientific Publishers, pp. 383–430.
31. Hagins, W. A., McGaughy, R. E. (1967). Molecular and thermal origins of fast photoelectric effects in the squid retina. *Science* 157:813–816.
32. Cone, R. A. (1967). Early receptor potential: photoreversible charge displacement in rhodopsin. *Science* 155:1128–1131.
33. Cone R. A., Pak, W. L. (1971). The early receptor potential. In: Loewenstein, W. R., (Ed.). *Handbook of Sensory Physiology*. Vol. 1. Berlin: Springer-Verlag, pp. 345–365.
34. Hong, F. T., Mauzerall, D. (1974). Interfacial photoreactions and chemical capacitance in lipid bilayers. *Proc. Natl. Acad. Sci. USA* 71:1564–1568.
35. Hong, F. T. (1976). Charge transfer across pigmented bilayer lipid membrane and its interfaces. *Photochem. Photobiol.* 24:155–189.
36. Hong, F. T., Mauzerall, D. (1976). Tunable voltage clamp method: application to photoelectric effects in pigmented bilayer lipid membranes. *J. Electrochem. Soc.* 123:1317–1324.
37. Fuhrhop, J.-H., Mauzerall, D. (1968). The one-electron oxidation of octaethylporphin. *J. Amer. Chem. Soc.* 90:3875–3876.
38. Fuhrhop, J.-H., Mauzerall, D. (1969). The one-electron oxidation of metalloporphyrins. *J. Am. Chem. Soc.* 91, 4174–4181.
39. Tien, H. T. (1989). Membrane photobiophysics and photochemistry. *Prog. Surf. Sci.* 30:1–199.

40. Hong, F. T. (1987). Effect of local reaction conditions on heterogeneous reactions in bacteriorhodopsin membrane: an electrochemical view. *J. Electrochem. Soc.* 134:3044–3052.
41. Keszthelyi, L. (1984). Intramolecular charge shifts during the photoreaction cycle of bacteriorhodopsin. In: Bolis, C. L., Helmreich, E. J. M., Passow, H. (Eds.). *Information and Energy Transduction in Biological Membranes*. New York: Alan R. Liss, Inc. pp. 51–71.
42. Bamberg, E., Fahr, A., Szabó, G. (1984). Photoelectric properties of the light-driven proton pump bacteriorhodopsin. In: Blaustein, M. P., Lieberman, M. (Eds.) *Electrogenic Transport—Fundamental Principles and Physiological Implications* (Soc. Gen. Physiol. Ser. Vol. 38), New York: Raven Press, pp. 381–394.
43. Skulachev, V. P. (1988). *Membrane Bioenergetics*. Berlin, Heidelberg, New York, London, Paris and Tokyo: Springer-Verlag.
44. Holz, M., Laudau, M., Heyn, M. P. (1988). Distributed kinetics of the charge movements in bacteriorhodopsin: evidence for conformational substates. *Biophys. J.* 53:623–633.
45. Liu, S. Y., Ebrey, T. G. (1988). Photocurrent measurements of the purple membrane oriented in a polyacrylamide gel. *Biophys. J.* 54:321–329.
46. Keszthelyi, L., Ormos, P. (1989). Protein electric response signals from dielectrically polarize systems. *J. Membr. Biol.* 109:193–200.
47. Trissl, H.-W. (1990). Photoelectric measurements of purple membranes. *Photochem. Photobiol.* 51:793–818.
48. Läuger, P. (1991). *Electrogenic Ion Pumps*. Sunderland, MA: Sinauer Associates, Inc.
49. Ludmann, K., Gergely, C. Dér, A., Váró, Gy. (1998). Electric signals during the bacteriorhodopsin photocycle, determined over a wide pH range. *Biophys. J.* 75:3120–3126.
50. Yao, B. L., Xu, D. L., Hou, X., Hu, K. S., Wang, A. J. (2001). Analyses and proofs of multi-exponential process of bacteriorhodopsin photoelectric response. *J. Appl. Phys.* 89: 795–797.
51. Yao, B., Wang, Y., Lei, M., Zheng, Y. (2003). Characteristics and mechanisms of the two types of photoelectric differential response of bacteriorhodopsin-based photocell. *Biosens. Bioelectron.* 19:283–287.
52. Hong, F. T. (1978). Mechanisms of generation of the early receptor potential revisited. *Bioelectrochem. Bioenerg.* 5:425–455.
53. Hong, F. T., Okajima, T. L. (1986). Electrical double layers in pigment-containing biomembranes. In: Blank, M. (Ed.) *Electrical Double Layers in Biology*. New York: Plenum Press, pp. 129–147.
54. Trissl, H.-W. (1979). Light-induced conformational changes in cattle rhodopsin as probed by measurements of the interfacial potential. *Photochem. Photobiol.* 29:579–588.
55. Hong, F. T., Montal, M. (1979). Bacteriorhodopsin in model membranes: a new component of the displacement photocurrent in the microsecond time scale. *Biophys. J.* 25:465–472.
56. Drachev, L. A., Kaulen, A. D., Khitrina, L. V., Skulachev, V. P. (1981). Fast stages of photoelectric processes in biological membranes. I. Bacteriorhodopsin. *Eur. J. Biochem.* 117:461–470.
57. Drachev, L. A., Kalamkarov, G. R., Kaulen, A. D., Ostrovsky, M. A., Skulachev, V. P. (1981). Fast stages of photoelectric processes in biological membranes. I. Visual rhodopsin. *Eur. J. Biochem.* 117:471–481.
58. Ostrovsky, M. A. (1989). Animal rhodopsin as a photoelectric generator. In: Hong, F. T. (Ed.) *Molecular Electronics—Biosensors and Biocomputers*. New York: Plenum Press, pp. 187–201.
59. Okajima, T. L., Hong, F. T. (1986). Kinetic analysis of displacement photocurrents elicited in two types of bacteriorhodopsin model membranes. *Biophys. J.* 50:901–912.
60. Hong, F. T., Okajima, T. L. (1987). Rapid light-induced charge displacements in bacteriorhodopsin membranes: an electrochemical and electrophysiological study. In: Ebrey, T. G., Frauenfelder, H., Honig, B. Nakanishi, K. (Eds.). *Biophysical Studies of Retinal Proteins*. Urbana-Champaign, IL: University of Illinois Press, pp. 188–198.
61. Michaile, S., Hong, F. T. (1994). Component analysis of the fast photoelectric signal from model bacteriorhodopsin membranes: Part I. Effect of multilayer stacking and prolonged drying. *Bioelectrochem. Bioenerg.* 33:135–142.
62. Okajima, T. L., Michaile, S., McCoy, L. E., Hong, F. T. (1994). Component analysis of the fast photoelectric signal from model bacteriorhodopsin membranes: Part II. Effect of fluorescamine treatment. *Bioelectrochem. Bioenerg.* 33:143–149.

63. Hong, F. H., Chang, M., Ni, B., Needleman, R.B., Hong, F. T. (1994). Component analysis of the fast photoelectric signal from model bacteriorhodopsin membranes. Part III. Effect of the point mutation aspartate 212 asparagines 212. *Bioelectrochem. Bioenerg.* 33:151–158.
64. Hong, F. H., Hong, F. T. (1995). Component analysis of the fast photoelectric signal from model bacteriorhodopsin membranes: Part 4. A method for isolating the B2 component and the evidence for its polarity reversal at low pH. *Bioelectrochem. Bioenerg.* 37:91–99.
65. Petrak, M. R., Hong, F. T. (1998). Component analysis of the fast photoelectric signal from model bacteriorhodopsin membranes: Part V. Effects of chloride ion transport blockers and cation chelators. *Bioelectrochem. Bioenerg.* 45:193–201.
66. Hong, F. T. (1986). The bacteriorhodopsin model membrane system as a prototype molecular computing element. *BioSystems* 19:223–236.
67. Drachev, L. A., Kaulen, A. D., Skulachev, V. P. (1978). Time resolution of the intermediate steps in the bacteriorhodopsin-linked electrogenesis. *FEBS Lett.* 87:161–167.
68. Hong, F. T., Mauzerall, D. (1972). The separation of voltage-dependent photoemfs and conductances in Rudin-Mueller membranes containing magnesium porphyrins. *Biochim. Biophys. Acta* 275:479–484.
69. Fuller, B. E., Okajima, T. L., Hong, F. T. (1995). Analysis of the d.c. photoelectric signal from model bacteriorhodopsin membranes: d.c. photoconductivity determination by means of the null current method and the effect of proton ionophores. *Bioelectrochem. Bioenerg.* 37:109–124.
70. Drachev, L. A., Jasaitis, A. A., Kaulen, A. D., Kondrashin, A. A., Liberman, E. A., Nemecek, I. B., Ostroumouv, S. A., Semenov, A. Yu., Skulachev, V. P. (1974). Direct measurement of electric current generation by cytochrome oxidase, H-ATPase and bacteriorhodopsin. *Nature* 249:321–324.
71. Mitchell, P. (1966). Chemiosmotic coupling in oxidative and photosynthetic phosphorylation. *Biol. Rev.* 41:445–502.
72. Williams, R. J. P. (1978). The multifarious couplings of energy transduction. *Biochim. Biophys. Acta* 505:1–44.
73. Vredenberg, W. J., Bulychev, A. A. (1976). Changes in the electrical potential across the thylakoid membranes of illuminated intact chloroplasts in the presence of membrane-modifying agents, *Plant Sci. Lett.* 7:101–107.
74. Packham, N. K., Mueller, P., Dutton, P. L. (1988). Photoelectric currents across planar bilayer membranes containing bacterial reaction centers: the response under conditions of multiple reaction-center turnovers. *Biochim. Biophys. Acta* 933:70–84.
75. Hong, F. T. (1989). Relevance of light-induced charge displacements in molecular electronics: design principles at the supramoleuclar level. *J. Molec. Electron.* 5:163–185.
76. Hong, F. T. (1992). Intelligent materials and intelligent microstructures in photobiology. *Nanobiology* 1:39–60.
77. Hong, F. T. (1991). Does nature utilize a common design for photoactive transport and sensor proteins? In: Lazarev, P. I. (Ed.). *Molecular Electronics—Materials and Methods*. Dordrecht, The Netherlands: Kluwer Academic Publishers, pp. 291–310.
78. Michaile, S., Duschl, A., Lanyi, J. K., Hong, F. T. (1990). Chloride ion modulation of the fast photoelectric signal in halorhodopsin thin films. In: *Proceedings 12th Annual International Conference of IEEE Engineering in Medicine and Biology Society*. Philadelphia, PA, November 1–4, pp. 1721–1723.
79. Deisenhofer, J., Michel, H. (1989). The photosynthetic reaction center from the purple bacterium *Rhodopseudomonas viridis*. *Science* 245:1463–1473.
80. Hong, F. T. (1994). Retinal proteins in photovoltaic devices. In: Birge, R. R. (Ed.). *Molecular and Biomolecular Electronics* (Advances in Chemistry Series No. 240). Washington, DC: American Chemical Society, pp. 527–559.
81. Hong, F. T. (1990). Bacteriorhodopsin as an intelligent material. In: Wang, C.-Y., Chen, C.-T., Cheng, C.-K., Huang, Y.-Y., Lin, F.-H. (Eds.). *Biomedical Engineering in the 21st Century*. Taipei, Taiwan: National Taiwan University Center for Biomedical Engineering. pp. 85–95.
82. Hara, M., Majima, T., Miyake, J., Ajiki, S., Sugino, H., Toyotama, H., Kawamura, S. (1990). Oriented immobilization of bacterial photosynthetic membrane. *Appl. Microbiol. Biotechnol.* 32:544–549.

83. Govorunova, E., Dér, A., Tóth-Boconádi, R., Keszthelyi, L. (1995). Photosynthetic charge separation in oriented membrane fragments immobilized in gel. *Bioelectrochem. Bioenerg.* 38:53–56.
84. Hecht, S., Shlaer, S., Pirenne, M. H. (1942). Energy, quanta, and vision. *J. Gen. Physiol.* 25:819–840.
85. Stryer, L. (1986). Cyclic GMP cascade of vision. *Annu. Rev. Neurosci.* 9:87–119.
86. Shevchenko, T. F., Kalamkarov, G. R., Ostrovsky, M. A. (1987). The lack of H^+ transfer across the photoreceptor membrane during rhodopsin photolysis. *Sensory Systems (USSR Acad. Sci.)* 1:117–126. (in Russian)
87. Groma, G. I., Szabó, G., Váró, Gy. (1984). Direct measurement of picosecond charge separation in bacteriorhodopsin. *Nature* 308:557–558.
88. Simmeth, R., Rayfield, G. W. (1990). Evidence that the photoelectric response of bacteriorhodopsin occurs in less than 5 picoseconds. *Biophys. J.* 57:1099–1101.
89. Hong, F. T. (2004). The early receptor potential and its analog in bacteriorhodopsin membranes. In: Horspool, W., Lenci, F. (Eds.). *CRC Handbook of Organic Photochemistry and Photobiology*. Boca Raton, FL: CRC Press, pp. 128–1 to 128–25.
90. Hong, F. T. (2004). Molecular electronics switches in photobiology. In: Horspool, W., Lenci, F. (Eds.). *CRC Handbook of Organic Photochemistry and Photobiology*. Boca Raton, FL: CRC Press, pp. 134–1 to 134–26.
91. Cafiso, D. S., Hubbell, W. L. (1980). Light-induced interfacial potentials in photoreceptor membrane. *Biophys. J.* 30:243–263.
92. Drain, C. M., Christensen, B., Mauzerall, D. (1989). Photogating of ionic currents across a lipid bilayer. *Proc. Natl. Acad. Sci. USA* 86:6959–6962.
93. Miyasaka, T., Koyama, K., Itoh, I. (1992). Quantum conversion and image detection by a bacteriorhodopsin-based artificial photoreceptor. *Science* 255:342–344.
94. Trissl, H.-W. (1987). Eine biologische photodiode mit hochster zeitauflosung. *Optoelecktron. Mag.* 3:105–107.
95. Rayfield, G. W. (1994). Photodiodes based on bacteriorhodopsin. In: Birge, R. R. (Ed.). *Molecular and Biomolecular Electronics* (Advances in Chemistry Series No. 240). Washington, DC: American Chemical Society, pp. 561–575.
96. Miyasaka, T., Koyama, K. (1993). Image sensing and processing by a bacteriorhodopsin-based artificial photoreceptor. *Appl. Optics* 32:6371–6379.
97. Wang, J. P., Li, J. R., Tao, P. D., Li, X. C., Jiang, L. (1994). Photoswitch based on bacteriorhodopsin Langmuir-Blodgett films. *Adv. Mater. Optics Electron.* 4:219–224.
98. Chen, Z., Birge, R. R. (1993). Protein-based artificial retinas. *Trend. Biotech.* 11:292–300.
99. Haronian, D., Lewis, A. (1991). Elements of a unique bacteriorhodopsin neural network architecture. *Appl. Optics* 30:597–608.
100. Boyer, A., Déry, M., Selles, P., Arbour, C., Boucher, F. (1995). Colour discrimination by forward and reverse photocurrents in bacteriorhodopsin-based photosensor. *Biosens. Bioelectron.* 10:415–422.
101. Frydrych, M. Silfsten, P. Parkkinen, S. Parkkinen, J. Jaaskelainen, T. (2000). Color sensitive retina based on bacteriorhodopsin. *Biosystems* 54(3):131–140.
102. Xu, J., Bhattacharya, P., Váró, G. (2004). Monolithically integrated bacteriorhodopsin/semiconductor opto-electronic integrated circuit for a bio-photoreceiver. *Biosens. Bioelectron.* 19:885–892.
103. Wang, W. W., Knopf, G. K., Bassi, A. S. (2006). Photoelectric properties of a detector based on dried bacteriorhodopsin film. *Biosens. Bioelectron.* 21:1309–1319.
104. Seki, A., Kubo, I. Sasabe, H., Tomioka, H. (1994). A new anion-sensitive biosensor using an ion-sensitive field effect transistor and a light-driven chloride pump, halorhodopsin. *Appl. Biochem. Biotechnol.* 48:205–211.
105. Tanabe, K., Hikuma, M., SooMi, L., Iwasaki, Y., Tamiya, E., Karube, I. (1989). Photoresponse of a reconstituted membrane containing bacteriorhodopsin observed by using an ion-selective field effect transistor. *J. Biotechnol.* 10:127–134.
106. Hafeman, Parce, J. W., McConnel, H. M. (1988). Light-addressable potentiometric sensor for biochemical systems. *Science* 240:1182–1185.

107. Greenbaum, E. (1985). Platinized chloroplasts: a novel photocatalytic material. *Science* 230:1373–1375.
108. Greenbaum, E. (1989). Biomolecular electronics: observation of oriented photocurrents by entrapped platinized chloroplasts. *Bioelectrochem. Bioenerg.* 21:171–177.
109. Aizawa, M. (1994). Molecular interfacing for protein molecular devices and neurodevices. *IEEE Eng. Med. Biol.* 13(1): 94–102.
110. Barber, J. (1983). Photosynthetic electron transport in relation to thylakoid membrane composition and organization. *Plant Cell Environ.* 6:311–322.
111. Keyes, R. W. (1975). Physical limits in digital electronics. *Proc. IEEE* 63:740–767.
112. Chiabrera, A., Di Zitti, E., Costa, F., Bisio, G. M. (1989). Physical limits of integration and information processing in molecular systems. *J. Phys. D: Appl. Phys.* 22:1571–1579.
113. Carter, F. L. (Ed.). (1982). *Molecular Electronic Devices*. New York and Basel: Marcel Dekker.
114. Aviram, A. (Ed.). (1992). *Molecular Electronics—Science and Technology*. New York: American Institute of Physics.
115. Sienicki, K. (Ed.). (1993). *Molecular Electronics and Molecular Electronic Devices*. Boca Raton, FL; Ann Arbor, MI; London; Tokyo: CRC Press.
116. Birge, R. R. (Ed.). (1994). *Molecular and Biomolecular Electronics* (Advances in Chemistry Series No. 240). Washington, DC: American Chemical Society.
117. Barsanti, L., Evangelista, V., Gualtieri, P., Passarelli, V., Vestri, S. (Eds.). (2003). *Molecular Electronics: Bio-sensors and Bio-computers* (NATO Science Series II, Vol. 96). Dordrecht, Boston, and London: Kluwer Academic Publishers.
118. Gauch, H. G., Jr. (2006). Winning the accuracy game. *American Scientist (Sigma Xi)* 94:133–141.
119. Popper, K. R. (1968). *The Logic of Scientific Discovery*. Revised edition. London: Hutchinson, London. Original German version: *Logik der Forschung* (Vienna, 1934). Reprinted (London and New York: Routledge, 1992).
120. Hong, F. T. (1997). Molecular sensors based on the photoelectric effect of bacterirhodopsin: origin of differential responsivity. *Mater. Sci. Engg. C* 4:267–285. Erratum *Mater. Sci. Engg. C* 5:61–79.
121. Gauch, H. G. Jr. (1993). Prediction, parsimony and noise. *Am. Sci. (Sigma Xi)* 81:468–478.
122. Gauch, H. G., Jr. (2003). *Scientific Method in Practice*. Cambridge: Cambridge University Press.
123. Hong, F. T. (1992). A reflection on the first International Conference on Intelligent Materials. *Intell. Mater. Newslett. (Intelligent Materials Forum, Tokyo)* 2(2):15–18.
124. Berners-Lee, T., Hendler, J., Lassila, O. (2001). The semantic web. *Sci. Am.* 284(5):34–43.
125. Conrad, M. (1990). Molecular computing, In: Yovits, M. C. (Ed.). *Advances in Computers*. Vol. 31. Boston, San Diego, New York, London, Sydney, Tokyo and Toronto: Academic Press, pp. 235–324.
126. Rambidi, N. G. (1997). Biomolecular computer: roots and promises. *Biosystems* 44:1–15.
127. Rambidi N. (2002). Roots and promises of chemical-based computing. *Biosystems* 64:169–178.
128. Zauner, K. P. (2005). Molecular Information Technology. *Crit. Rev. Solid State Mater. Sci.* 30:33–69.
129. Rambidi, N. G. (1994). Biomolecular computing: from the brain-machine disanalogy to the brain-machine analogy. *BioSystems* 33:45–54.
130. Hong, F. T. (2005). A multi-disciplinary survey of biocomputing: 1. molecular and cellular levels. In: Bajic, V. B., Tan, T. W. (Eds.). *Information Processing and Living Systems*. (Series on Advances in Bioinformatics and Computational Biology, Vol. 2). London: Imperial College Press, pp. 1–137.
131. Hong, F. T. (2003). The enigma of creative problem solving: a biocomputing perspective. In: Barsanti, L., Evangelista, V., Gualtieri, P., Passarelli, V., Vestri, S. (Eds.). *Molecular Electronics: Bio-sensors and Bio-computers*. (NATO Science Series II, Vol. 96). Dordrecht, Boston and London: Kluwer Academic Publishers, pp. 457–542.
132. Hong, F. T. (2005). A multi-disciplinary survey of biocomputing: 2. Systems and evolutionary levels, and technological applications. In: Bajic, V. B., Tan, T. W. (Eds.). *Information Processing and Living Systems*. (Series on Advances in Bioinformatics and Computational Biology, Vol. 2). London: Imperial College Press, pp. 141–573.
133. Simon, H. A. (1992). Scientific discovery as problem solving. In: Egidi, M., Marris, R. (Eds.). *Economics, Bounded Rationality and the Cognitive Revolution*. Hants, UK, and Brookfield, VT: Edward Elgar Publishing, pp. 102–119.

134. Hawkins, J. (with Blakeslee, S.) (2004). *On Intelligence*. New York: Owl Books, pp. 177–204.
135. Arnheim, R. (1969). *Visual Thinking*. Berkeley, Los Angeles and London: University of California Press, p. 274.
136. West, T. G. (1997). *In the Mind's Eye*. Updated edition. New York: Prometheus Books.
137. Hadamard, J. (1945). *The Psychology of Invention in the Mathematical Field*. Princeton, NJ: Princeton University Press, pp. 142–143. Reprinted: *The Mathematician's Mind: The Psychology of Invention in the Mathematical Field*, 1996.
138. Gleick, J. (1993). *Genius: The Life and Science of Richard Feynman*. New York: Vintage Books, p. 131.
139. Hawking, S. (1993). *Black Holes and Baby Universes and Other Essays*. New York, Toronto, London, Sydney and Auckland: Bantam Books, p. 35.
140. Tesla, N. (1977). *Moji Pronalasci/My Inventions*, in both Serbian and English. Zagreb: Školska Knjiga. Reprinted English version: Belgrade: Nikola Tesla Museum, 2003.
141. Poincaré, H. (1913). *The Foundations of Science: Science and Hypothesis, The Value of Science, Science and Method*. Translated by G. B. Halsted. New York and Garrison, NY: Science Press, pp. 214–215.
142. Simon, H. A., Newell, A. (1958). Heuristic problem solving: the next advance in operations research. *Oper. Res.* 6:1–10.
143. Jennings, N. R., Wooldridge, M. J. (Eds.). (1998). *Agent Technology: Foundations, Applications, and Markets*. Berlin, Heidelberg and New York: Springer-Verlag.
144. Hopfield, J. J. (1982). Neural networks and physical systems with emergent collective computational abilities. *Proc. Natl. Acad. Sci. USA* 79:2554–2558.
145. Churchland, P. M. (1990). Cognitive activity in artificial neural networks, In: Osherson, D. N., Smith, E. E. (Eds.). *An Invitation to Cognitive Science, Volume 3: Thinking*. Cambridge, MA, and London: MIT Press, pp. 199–227.
146. Buck, L. B. (2000). The molecular architecture of odor and pheromone sensing in mammals. *Cell* 100:611–618.
147. Malnic, B., Hirono, J., Sato, T., Buck, L. B. (1999). Combinatorial receptor codes for odors. *Cell* 96:713–723.
148. Marr, D. (1982). *VISION, A Computational Investigation into the Human Representation and Processing of Visual Information*. New York: Freeman.
149. Pearce, T. C. (1997). Computational parallels between the biological olfactory pathway and its analogue 'The Electronic Nose': Part I. Biological olfaction. *BioSystems* 41: 43–67.
150. Pearce, T. C. (1997). Computational parallels between the biological olfactory pathway and its analogue 'The Electronic Nose': Part II. Sensor-based machine olfaction. *BioSystems* 41:69–90.
151. Adleman, L. M. (1994). Molecular computation of solutions to combinatorial problems. *Science* 266:1021–1024.

16

Color-Sensitive Biosensors for Imaging Applications

Lasse Lensu, Michael Frydrych, Jussi Parkkinen, Sinikka Parkkinen, and Timo Jaaskelainen

CONTENTS

16.1 Introduction

Computers are being increasingly used to process visual information in the digital form. The amount of data processing has rapidly increased due to the introduction of digital cameras and scanning devices. In contrast, future computers are expected to have more intuitive user interfaces, and there are complex problems that cannot be solved even using state-of-the-art computers. All of these factors generate a constant need for more powerful information-processing devices.

Several alternatives to current technologies used in microelectronics are being investigated to continue the performance trend of computing devices after the limits of current technology have been met. The alternatives are based on mechanics, molecular electronics, photonics, quantum mechanics, and chemical reactions (1). Biomolecules are highly optimized in their functions in living organisms. All living organisms are composed of biomolecules that have been extensively studied to gain a better understanding of life and the means of survival. The evolution in nature has, in many cases, optimized the properties of biomolecules according to their changing and sometimes harsh living conditions (2). Therefore, numerous things can be learned from the evolution by studying the biomolecules and their interactions. These findings can be applied even to information processing. Most of the studies concerning information processing using biomolecules have been concentrated in DNA and photoactive biomolecules, for example rhodopsins, chloroplasts, photosynthetic reaction centers and light-harvesting complexes, and retinal proteins including bacteriorhodopsin (BR) (3).

BR as the light-sensitive protein of an archaean group participates in its energy-balancing mechanism. The photoelectric properties of BR have been widely studied in various setups since 1974 (4), and also mechanically stable films where BR has been immobilized in a water-soluble polymer have raised interest (5,6). Several applications in optics and information processing have been suggested for BR (7). For example, it has been used to implement a prototype for a three-dimensional optical memory (8), a real-time interferometer (9), and an imaging device (10–13). The highest spatial resolutions of implemented imaging devices are 8×8 for a gray-scale device (12), and 3×2 for a three-component color-sensitive device (11,13). Even preprocessing of visual information has been proposed for such devices (14).

The objective of this chapter is to discuss the design factors for the application of BR to color-sensitive imaging. The motivation to use biomolecules in this case arises from the following factors: no filters are needed to achieve color sensitivity; thin films and coatings offer numerous possibilities for novel array geometries; and optically active molecules offer the possibility of preprocessing visual data. The practical goal of our research project was to develop photosensors and imaging devices based on BR as a case study of a material proposed even to molecular electronics and nanoscale applications. The designed characteristics for the imaging devices included a spatial resolution higher than 3×2 together with color sensitivity, and capability to capture video. In this chapter, we discuss the requirements and modeling of color-vision systems and present the design principles and photoelectric properties of thick-film photosensors and imaging devices based on BR. As a necessary background, the following are described: the design aspects of artificial color-vision systems, the characteristics of BR and BR-in-polyvinylalcohol films, applications of BR to photosensing and imaging, and theoretical aspects of a photosensor based on BR.

16.2 Bacteriorhodopsin in Imaging Applications

To use BR as a material for an imaging device has been proposed in a few studies (10–17). Changes in incident light intensity cause relatively synchronous proton displacements in BR. These displacements can be sensed with suitable design for the sensor. An imaging array based on BR fulfills the purpose of a transducer in a general imaging system. The proposed applications of BR demonstrate that BR can be used to process optical information.

Disregarding some special applications, an imaging system is expected to capture the spatial distribution of collected radiation. This is possible using a combination of one sensor and an appropriate scanning system (18), but most imaging devices rely on one-dimensional or two-dimensional arrays of sensors. This is because the devices are usually expected to capture the information at high speed. The geometry of an individual sensor element can be square, rectangle, or hexagon. Sensitivity to color is achieved by using spectral color filters, and the spatial or time resolution will be diminished if only a single array of sensors is used for all colors in a multicomponent color system.

In the proposed system, the imaging array elements have been implemented by enclosing BR solution between two electrodes, or drying a BR film on top of an array of electrodes, or charge-sensitive devices. The element dimensions of the imaging arrays in the actually implemented imaging arrays are from 20×20 mm^2 down to 2.5×2.5 mm^2 but even sizes of 15×15 μm^2 have been proposed. In these proposals, the film thickness varies from 30 nm to about 150 μm, and the spatial resolutions are from 3×2 elements to 8×8 elements. Even an artificial retina based on BR that mimics the functionality of X-type retinal ganglion cells with motion sensitivity has been proposed (10,17).

In the color-sensitive imaging devices, color sensitivity is based on a combination of BR and flavin (15), or three types of BR (11,13). The spatial resolution is 3×2 elements (11,13) or 3×3 elements (15). In the former case, the resolution must be divided by the number of color components, whereas the latter contains a layered structure to produce a two-component color response of nine elements. In addition, a model for the color-vision system based on BR has been devised (11,19), and it consists of photosensitive and preprocessing layers.

16.3 What is Color?

Color is among the most salient features of visual environment detected by the human visual system. Perception of color makes it easier and faster to detect, recognize, discriminate, and classify objects. Moreover, it provides an esthetic component to our visual experiences. Color also gained attention in artificial vision systems and it is central to a number of image analysis and computer-vision methods.

Color, as commonly understood in hue terms, such as red, green, yellow, etc., does not have a direct counterpart in a physical world; experience of color is only a product of our visual perception. The sensation of color originates from responses of three types of light-sensitive cells in the retina, or more precisely cones, to incoming light. The cone responses afterwards undergo complex neural processing in the brain, which ultimately results in a single percept that we call color. Using a physical approach, complete characterization of color requires a specification of the spectral energy of light originating from the observed object and approaching the eye (20). Therefore, color is a continuous function of wavelength.

Color is usually represented and communicated using three-dimensional color coordinate systems—tristimulus color spaces, like CIE XYZ, CIELAB, and CIELUV color spaces. These spaces are closely related to human visual system, in which there are three different classes of photoreceptors in the retina. A color space can be thought of as a subspace in infinite-dimensional space of light spectra. The color coordinates associated with the spectra are projections, which are results of inner products between the spectra and basis functions. In the CIE XYZ coordinate system, these functions are color-matching functions. Let $E(\lambda)$ be the spectral energy of light originated from the object and $\bar{x}_i$ be a set of basis. The color coordinates X_i, $i = 1, \ldots, n$ of the light are calculated as

$$X_i = \int E(\lambda)\bar{x}_i(\lambda)\mathrm{d}\lambda \tag{16.1}$$

The basis functions can be also considered as color filters through which the light is observed. The projections, or coordinates, express the total energy passed through these filters.

16.3.1 Origin of Color Vision

We might ask how many different classes of photoreceptors are needed to sufficiently discriminate between the given color spectra, and we also want to evaluate the discrimination accuracy. In human visual system, the infinite dimensionality of spectral representation is greatly reduced to just three by the three types of cones. Once photons are absorbed by a visual pigment in photoreceptors, the only remaining information is the photon count; information about the wavelength of each individual photon is lost. This property is called the principle of univariance (Rushton, 1972), (21). The color of the objects is assigned by the visual system by comparing the signals from different cones.

From the principle of univariance, it is followed that the sensor with only one class of photoreceptor will not discriminate among wavelengths. To gain some color vision, we need at least two classes of photoreceptors. Each of these photoreceptors contains a visual pigment with different absorption spectrum. Each given monochromatic stimuli would activate the two receptors differently from any other monochromatic stimuli.

In Figure 16.1 are shown two examples of ideal photosensitivity functions, which correspond to the basis functions of Eq. 16.1. In Figure 16.1a, it is shown that the problem of equal stimuli for different colors may also exist in two sensitivity cases. Two monochromatic stimuli of 450 nm and 650 nm give the same response to both sensors. This same response to both sensors is also given by one 550-nm stimulus. It means that for this color-vision system the light with two monochromatic light sources (450 and 650 nm) would look exactly the same as with one monochromatic light source (550 nm). In the human visual system, there are three light-sensitive cells (Figure 16.1b). In this system, the former light would appear purple and the latter one green.

16.3.2 Evaluation of Accuracy and Acuteness

The discrimination accuracy of color vision shows actual ability to discriminate between color stimuli. This ability can be characterized in several ways (22). The simplest characterization evaluates the ability to discriminate between monochromatic stimuli. Hallikainen (23) plotted inverse of the normalized Euclidean distance between the lateral geniculate nucleus (LGN) responses to adjacent pairs of equally spaced equal-energy monochromatic stimuli to receive an estimate of wavelength discrimination of the human visual system. Figure 16.2 presents a wavelength discrimination curve calculated from the XYZ color coordinates of equal-energy monochromatic stimuli in the visible spectrum interval spaced in 1-nm steps. Naturally, the best discrimination is where the slope of color matching functions x, y, and z is steep. It is convenient to describe wavelength discrimination ability by a single number. The number can be defined as either the Euclidean distance between the vector of normalized distances and the unity vector of the same dimension, or the angle between these two vectors. In either case, smaller number will mean better overall discrimination.

Accuracy of human color vision across the whole color space can be evaluated by recording minimum perceivable color difference, so-called just-noticeable difference (JND) (21,24). An observer is presented with two color stimuli; one fixed and the other alterable. The observer is then asked to adjust the alterable stimuli so that the difference between visual appearances of the two stimuli would be just noticeable. By repeating this scheme for various colors, MacAdam (21,25) has evaluated a change of JND across the

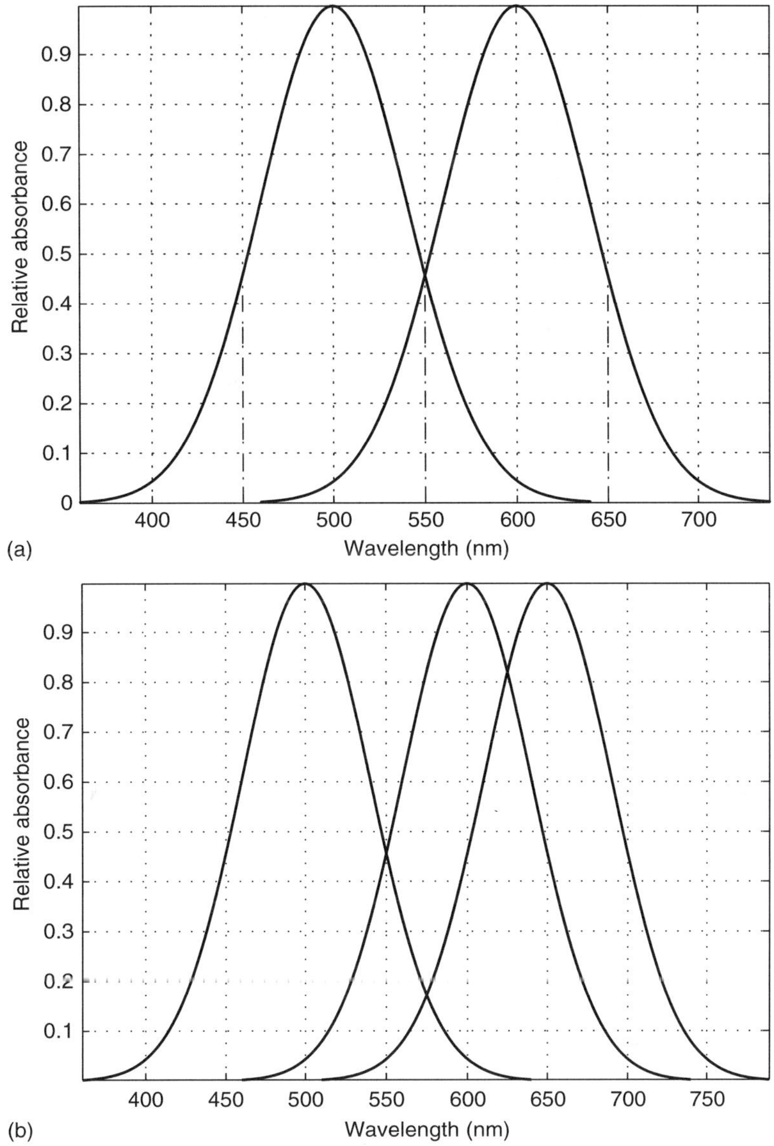

FIGURE 16.1
Relative absorption for (a) two (b) three hypothetical pigments. The absorbance functions are constructed as Gaussians with corresponding means.

possible chromaticities. When the JNDs are plotted to a chromaticity diagram, they appear as ellipses of different sizes. This clearly shows the nonuniform nature of the color space with respect to human vision.

16.4 Model of the Color-Vision System

Organization of visual systems of living species inspired inclusion of preprocessing ability to artificial vision sensors—vision chips. These are special-purpose silicon circuits that

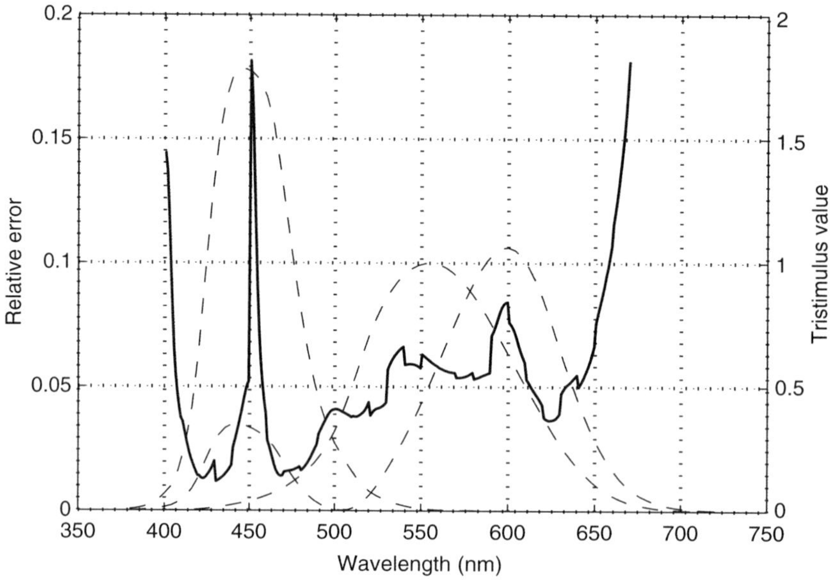

FIGURE 16.2
Wavelength discrimination capability of CIEXYZ tristimulus values. Lower value means better discriminability. The CIE 2° color matching functions x, y, and z are plotted for comparison as dashed lines on the background.

integrate both photosensors and parallel-processing elements on a single chip (26). The vision chips provide many advantages over conventional vision-processing systems consisting of a camera and a digital processor. Among the advantages are speed and size. The speed of many vision chips exceeds that of the conventional systems mainly due to the tight coupling of image sensing and processing in a single chip, which allows for a high degree of parallelism. Also, the data transfer bottleneck between sensor and processor is alleviated to a significant extent.

The preprocessing capability is also included in our model of a color-vision system outlined in Figure 16.3. The system consists of two layers: a photosensitive layer and a preprocessing layer. The photosensitive layer converts the image of the scene into electrical signals, which are passed on to the preprocessing layer. The output of the sensor is communicated to a processor. The preprocessing can include some image processing, for example, edge detection or color-coordinate transformations and color grouping.

16.4.1 Photosensitive Layer

The photosensitive layer consists of one or more arrays of photosensitive elements. In general, the elements can be charge-coupled device (CCD) elements, cones in the retina, patches of BR film, or any other suitable photosensitive material. The geometrical arrangement of the elements is not restricted. The output of the preprocessing layer is an array of n-tuples.

Each photosensitive element is characterized by its relative spectral response function $r(\lambda)$, which expresses the relative strength of response to equal-energy monochromatic stimuli. The response of an element to a stimulus with spectral energy $P(\lambda)$ is

$$R = \int P(\lambda) r(\lambda) d\lambda \qquad (16.2)$$

This can be seen as an inner product of $P(\lambda)$ and $r(\lambda)$.

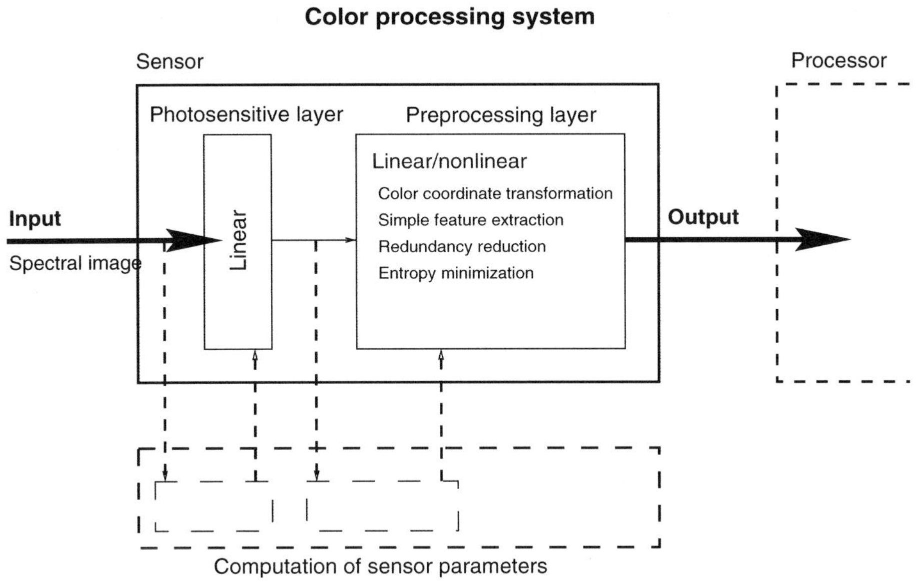

FIGURE 16.3
Model of the color-processing system.

16.4.2 Preprocessing Layer

We are now interested in color processing. The preprocessing layer is designed to learn color space for the responses of photosensitive layer. The learning is done by training self-organizing feature map (SOM), which yields a grouping of similar colors.

The preprocessing layer may operate in two modes: off- and online mode, respectively. The purpose of the offline or learning mode is to adjust the internal parameters of the preprocessing layer. Training the SOM is the offline mode of operation of the preprocessing layer. In the online mode, the sensor is presented with arbitrary colors and it responds with their color coordinates in learned color space.

16.5 Imaging with Bacteriorhodopsin

To show the potential of BR in photosensing, we have prepared single-pixel photosensors based on wild-type BR and its analogs (27,28). We selected polyvinylalcohol (PVA) as the polymer matrix to prepare the thick films, which is supported by the following facts:

- The preparation process is straightforward and relatively simple (29).
- The PVA matrix forms a stable environment for the purple membrane (PM) fragments containing BR (30).
- The PVA matrix permits measurements concerning the alternating-current photoelectric response (PER) and optical properties of the film (6).
- The films are mechanically stable.

The preparation method has a disadvantage. If direct control of the homogeneity of the films or the orientation of the PM fragments in the films is desired, the preparation process becomes much more difficult.

16.5.1 Gray-Scale Imaging Array

The properties of BR make it an attractive material to imaging purposes. Wild-type BR offers high quantum yield and repeatability (8). The chemical environment of the BR molecule defines its absorption characteristics and sensitivity to photon energies (31). The wavelength of maximum absorbance for the wild-type BR is 412 nm (dark-adapted BR (32)) or 568 nm (light-adapted BR (32)). Therefore, the absorption spectrum of BR is close to the one for the *l*-cones in the human retina (20).

The gray-scale BR imaging array consists of (i) a two-dimensional array of electrodes on the printed circuit board (see Figure 16.4a), (ii) continuous BR film on top of the electrode array, and (iii) continuous layer of gold. The concentration of the wild-type BR after extraction and concentration was 29.2 mg/ml. The final concentration of BR after mixing it with the solution of PVA and phosphate buffer was 8.32 mg/ml. The solution was spread on top of the printed circuit board, and the PM fragments were immobilized by the gelation of PVA. A thin layer of gold was sputtered on top of the BR-PVA film to form a counter electrode. The complete gray-scale BR imaging array is shown in Figure 16.4b.

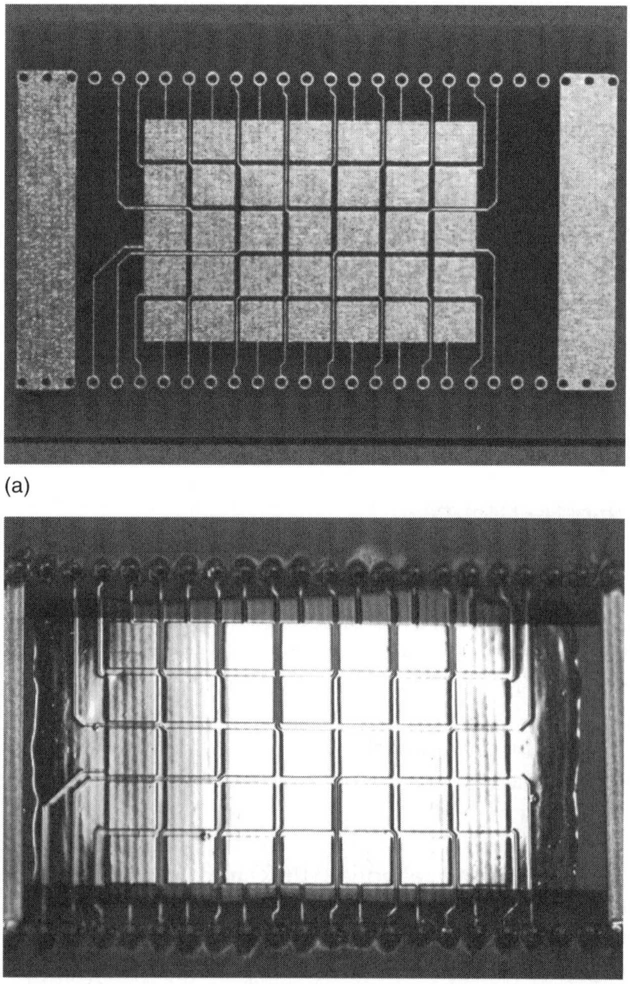

(a)

(b)

FIGURE 16.4
(a) The printed circuit board for the imaging array contains 7 × 5 electrodes of size 5 × 5 mm. (b) The wild-type BR-PVA thick film and a thin layer of gold are processed on top of the board to form the gray-scale imaging array.

The wild-type BR-PVA elements in the gray-scale array are 144 μm thick on an average (standard deviation 26 μm).

According to the displacement current model (33,34), the PER linearly depends on the number of BR molecules per unit distance in the direction perpendicular to the electrodes. This states that the variation in the film thickness is the dominant factor of quality, assuming that the BR suspension is homogeneous.

16.5.2 Color-Sensitive Imaging Array

Color sensitivity can be achieved in an imaging device by capturing the radiant energy from a color scene by at least two sensor elements having different spectral sensitivities. Three color components are sufficient for general use if the image is viewed by a human being. However, if accuracy is wanted, a three-component color representation is not enough for all purposes (35). The spectral properties of the sensor elements are important, since discrimination of colors is expected. Depending on the application, a sound color-sensitive imaging device should capture the information of the energy spectrum, and offer a signal-to-noise ratio (SNR) good enough for discrimination of different colors.

To modify the spectral properties of BR, retinal substitution was used. The absorption and action spectra of the single-pixel photosensors based on BR are shown in Figures 16.5 and 16.6. All figures show that there is solid support to achieve color sensitivity.

The color-sensitive BR imaging array was prepared in the same procedure as the gray-scale array, except that individual array elements were prepared one by one using the three types of BR. Layout of the array elements and the complete color-sensitive BR imaging array are shown in Figure 16.7. Average thickness of the elements in the color-sensitive array with different types of BR is 101–107 μm (standard deviation 16–20 μm).

Other geometries for the array elements, for example hexagons, could be employed with ease due to the flexible preparation process. This would enable the use of interpolation methods to enhance the spatial resolution when captured data is converted into a rectangular grid. It would be possible to prepare the film on top of very diverse substrates too, as long as the gravity coating method can be used.

16.5.3 Electronics

The electronics for an element in the imaging array consist of an instrumentation amplifier, signal amplifier, peak-hold circuit, and buffer circuit. The peak photovoltages from each element are digitized and processed by a microcontroller (36). The electronics based on the microcontroller was used for demonstration purposes but the photoelectric measurements were performed using the electronics developed for the single-pixel photosensors.

16.5.3.1 Photosensor Electronics

The relatively large size of the photosensor causes it to have small source capacitance (31). This needs to be compensated by proper impedance matching between the sensor and signal-conditioning electronics. Therefore, the photosensor has been connected as closely as possible to the voltage follower shown in Figure 16.8a. According to the manufacturer of the operational amplifier (TL082MJG), the input resistance is 10^{12} Ω, rise time is 50 ns, and unity gain bandwidth is 3 MHz.

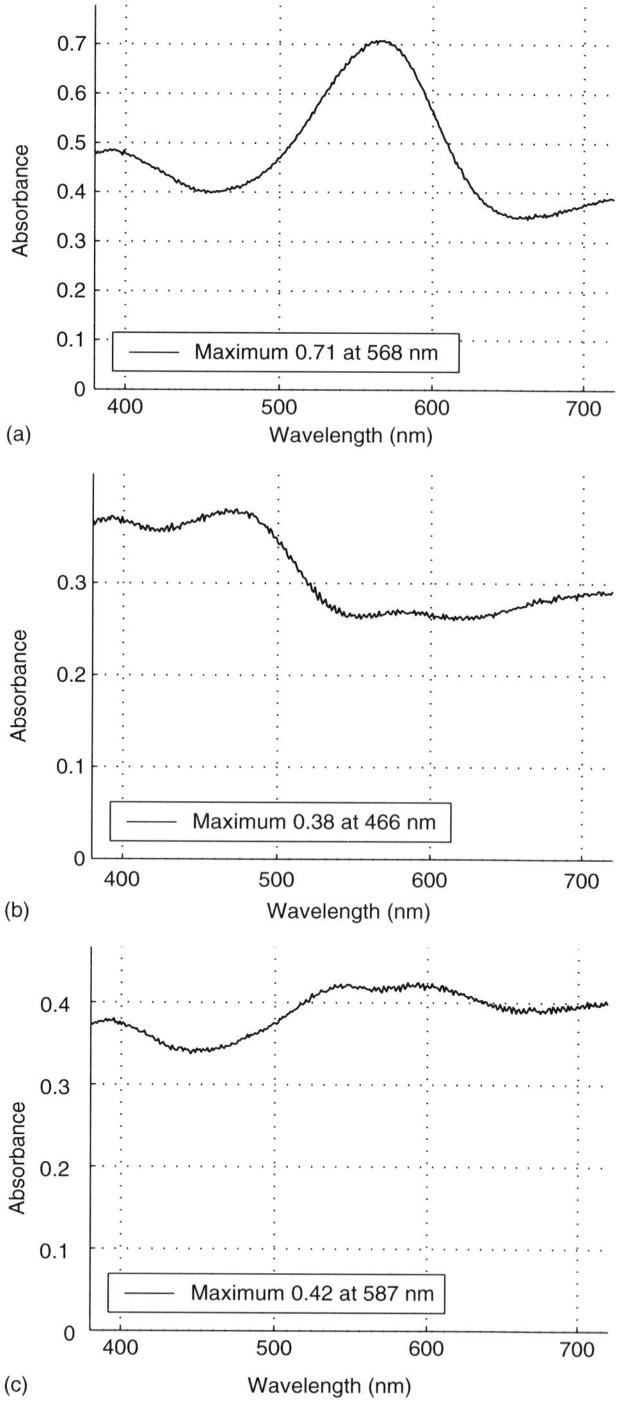

FIGURE 16.5
Absorption spectra of the (a) wild-type, (b) 4-keto, and (c) 3,4-didehydro BR-PVA photosensors.

Signal conditioning is also needed because of electromagnetic interference from the environment. The signal-conditioning stage shown in Figure 16.8 consists of three active low-pass filters of second order (−3 dB cutoff frequency 5 kHz), and a passive high-pass filter of first order (−3 dB cutoff frequency 7 Hz). Amplification of the signal amplifier is 56. The measured frequency response of the signal-conditioning stage is shown in Figure 16.9.

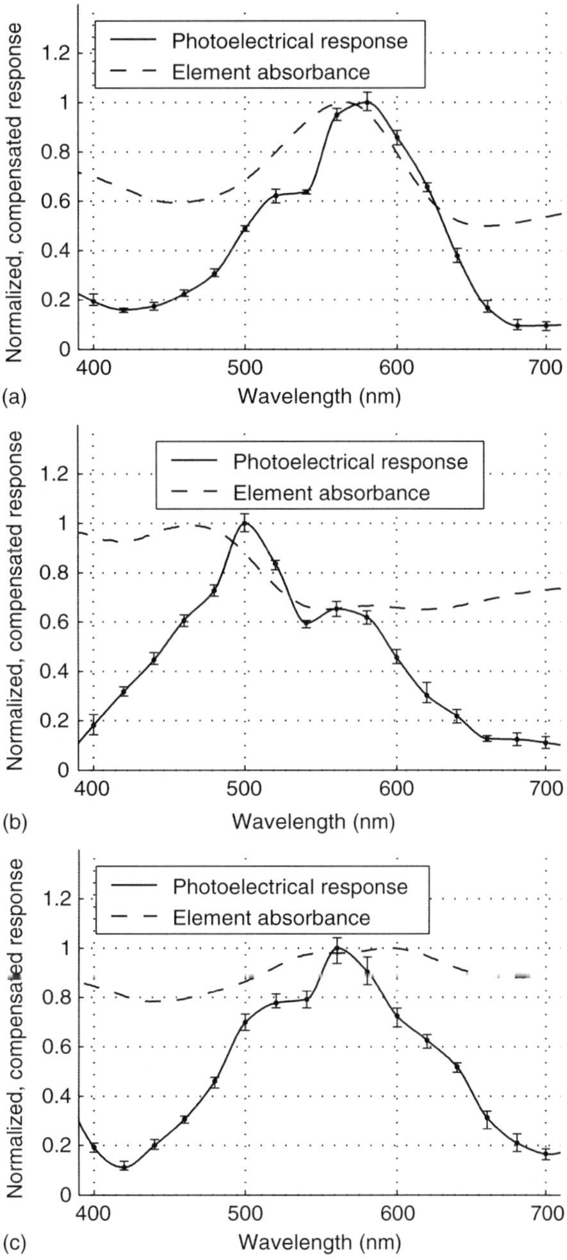

FIGURE 16.6
Action spectra for (a) wild-type, (b) 4-keto, and (c) 3,4-didehydro BR-PVA photosensor based on the amplitudes of the PER. The dashed lines denote the uncorrected absorption spectra of the photosensors.

16.5.3.2 Imaging Array Electronics

The design principles are the same as for the photosensors, although the amplification is increased to 980 because of expected degradation of the PER. The response is amplified with an instrumentation amplifier (INA126), and the peak photovoltage is stored in a small capacitor as shown in Figure 16.10. A microcontroller (PIC17C43) is used to digitize the stored charges and transfer them to a computer for postprocessing (36).

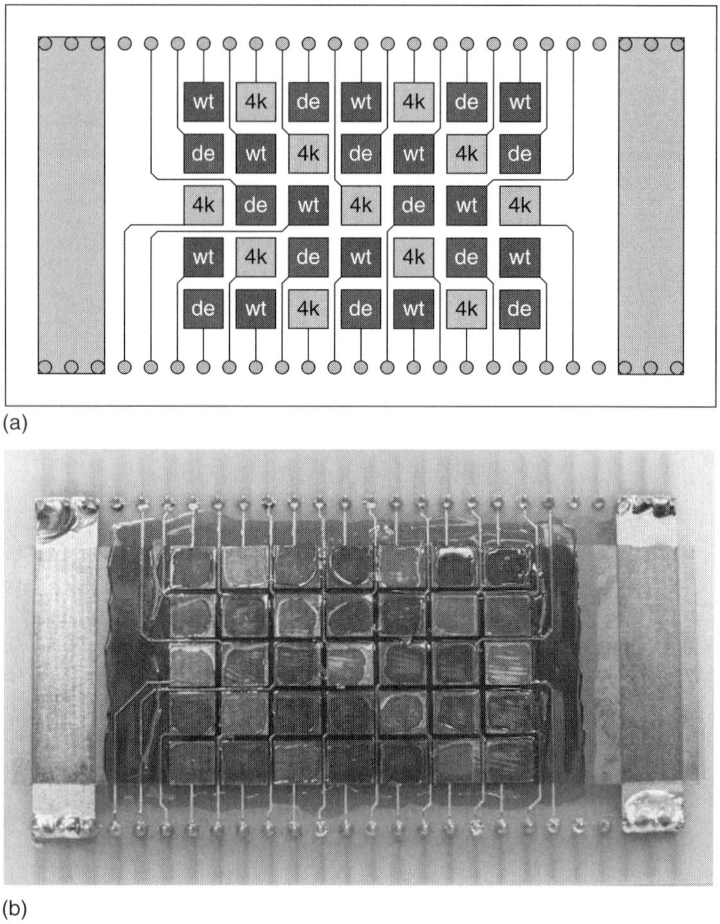

FIGURE 16.7
(a) Schematic for the color-sensitive imaging array containing three types of BR: wild type (wt), 4-keto (4k), and 3,4-didehydro (de). (b) The different absorption properties of the three types are clearly visible in the finished color-sensitive imaging array.

16.5.4 Measurements

The measurement of photoelectric properties requires a light source and an instrument to register the electric response. In most studies, the PER is induced by flash excitation, and the time course of either the photocurrent (closed circuit) or photovoltage (open circuit) is measured (31). In our studies, a pulsed xenon light source and a regular photographic camera flash were used for the excitation, and neutral density, interference and wide-band color filters to change the spectrum of light. We registered the photovoltage because the sensitivity is approximately invariant of time (31).

The AC PERs from individual elements in the gray-scale imaging device were measured by using the voltage follower and signal conditioning designed for the photosensors. The PER of an element using constant discharge energy of 160 mJ and pulsing frequency of 2.0 Hz for the pulsed light source is shown in Figure 16.11. Based on the figure, it is apparent that there are at least two components with opposite polarities in the response. The response contains more noise from the environment at 50 Hz but the amplitude was comparable to the photosensor with proper focusing of light. The slow negative component is visible in Figure 16.11a. Note that signal conditioning containing the high-pass filter affects the slow part of the response, so no time constants related to the components were determined from the response.

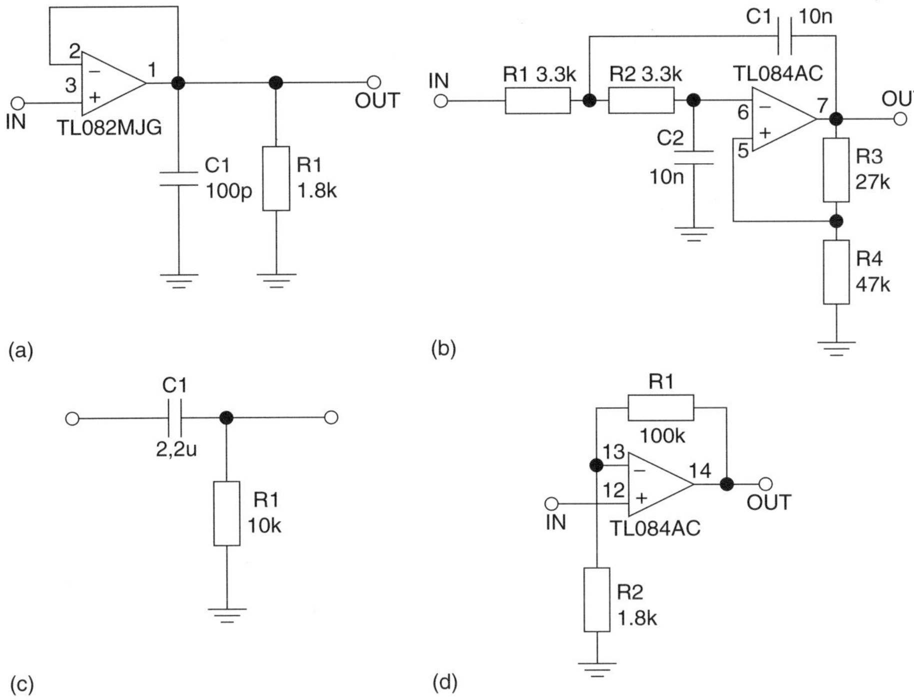

FIGURE 16.8
Electronics for the photosensors: (a) voltage follower, (b) low-pass filter, (c) high-pass filter, and (d) signal amplifier.

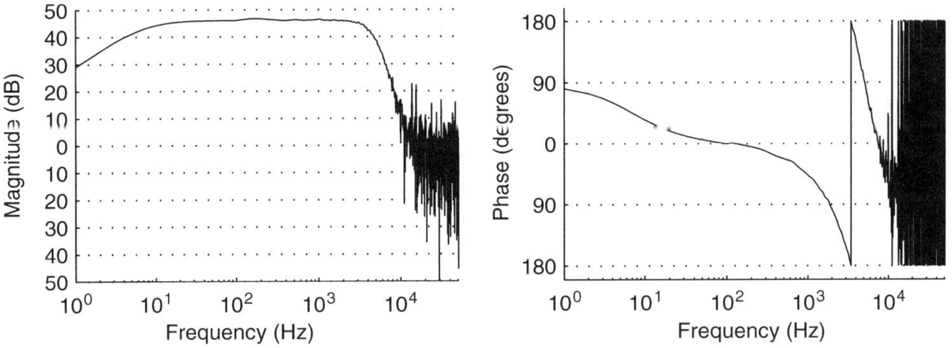

FIGURE 16.9
Frequency response of signal conditioning (low-pass filter, high-pass filter, and amplifier) including (a) magnitude and (b) phase.

The PERs of each BR type were measured for the color-sensitive array. The results presented in Figure 16.12 show that the response amplitudes were considerably lower than the ones generated by the gray-scale array elements. This holds even for the wild-type elements. The explanation to the degraded responses was low conductance of the gold layer, which was confirmed by conductivity measurements.

The quality of individual elements in the imaging arrays can be determined by measuring the PER for each element. Uniformity of the PER was determined by measuring one array element at a time. The average amplitudes for sets of 20 measurements per

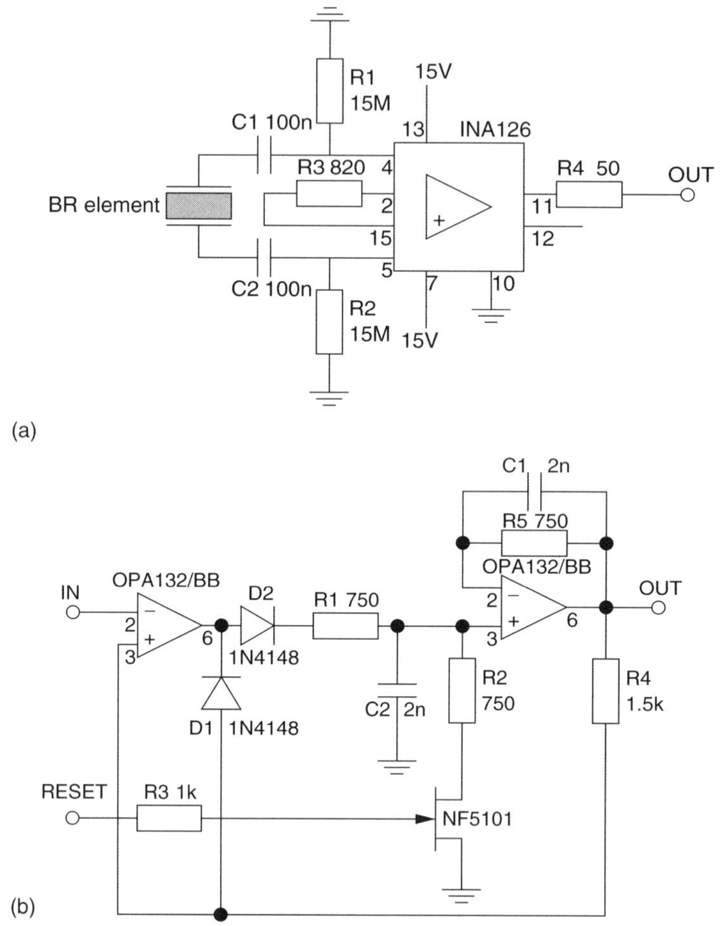

FIGURE 16.10
Electronics for the imaging arrays: (a) the voltage follower and (b) signal amplifier and peak-hold circuitry.

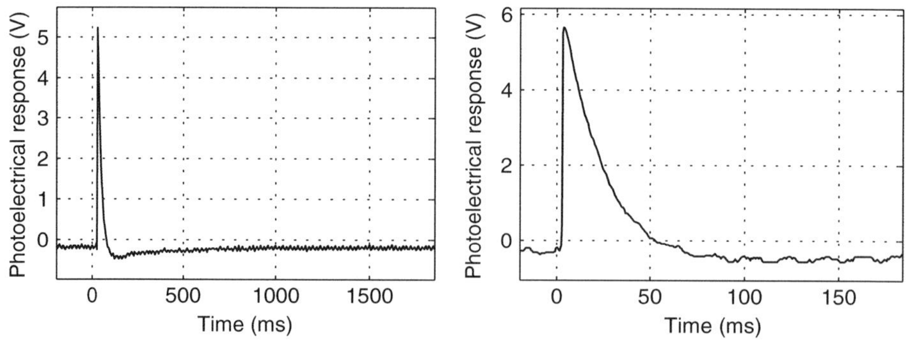

FIGURE 16.11
Time course of the PER for an element of the gray-scale imaging array. Two time resolutions are used to illustrate the components in the response.

array element are shown in Figure 16.13. Based on the measurement, the imaging array and the preparation process cannot be conceived as of industrial quality. However, the preparation process is straightforward and robust. All prepared photosensors and array elements generated a PER, and differences between the responses can be compensated using calibration.

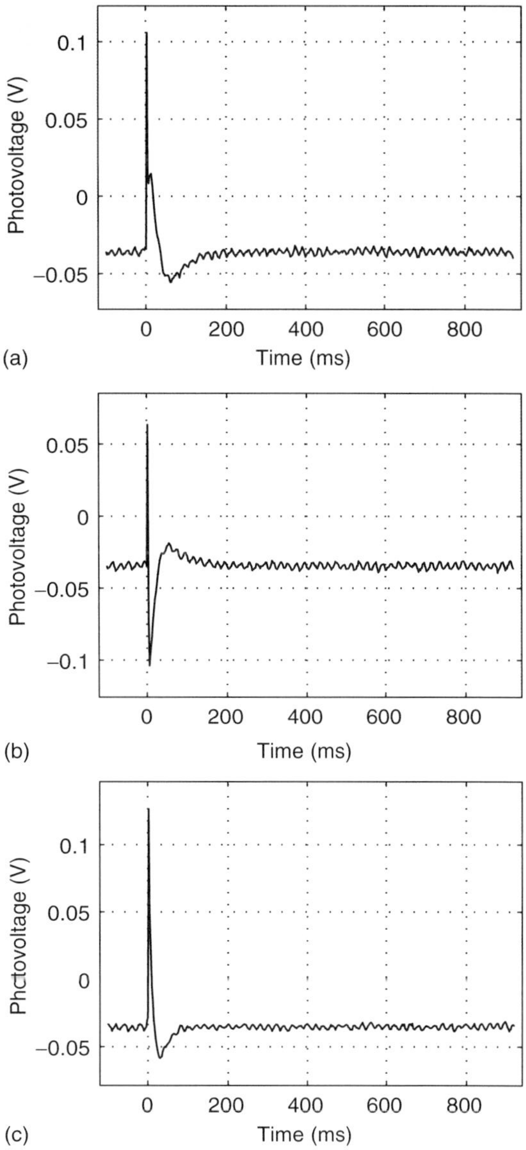

FIGURE 16.12
Time courses of the PERs of (a) wild-type, (b) 4-keto, and (c) 3,4-didehydro BR elements of the color-sensitive imaging array.

To use the imaging arrays for capturing standard video, the device must be able to handle pulsing frequencies of 25 Hz. Responsivity of an element in the gray-scale imaging array to pulsing was tested with frequencies higher than 1 Hz. The average photovoltage generated by at least 20 pulses at frequencies ranging from 1 to 100 Hz was measured by using the pulsed light source with constant discharge energy. The result including the photovoltage amplitude span for each measured frequency is shown in Figure 16.14. The average photovoltage was about 25% less at 100 Hz than at 1–5 Hz. Response variance was generally increased at higher frequencies, which was at least partly due to the light source. The result shows that the gray-scale imaging array was capable of handling high pulsing frequencies when the level of optical power was low. No saturation effects were noticed even when 200 pulses were registered at 100 Hz. Considering the relaxation times in the BR photocycle and the time course of the

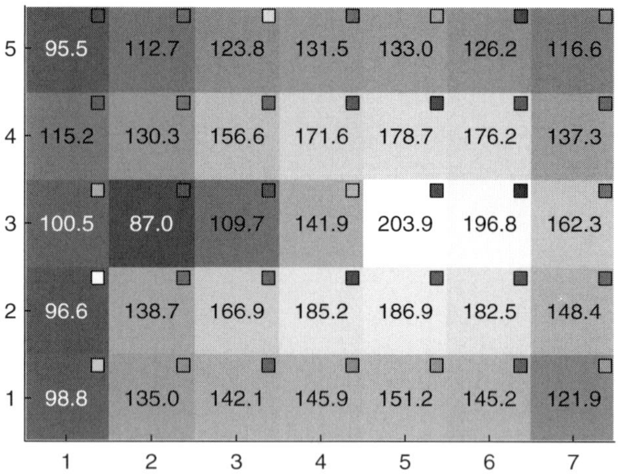

FIGURE 16.13
Uniformity of the PER for the gray-scale imaging array. Densities of the gray patches correspond to the response amplitudes shown in millivolts. Densities of the smaller patches represent the response variances relative to the amplitudes.

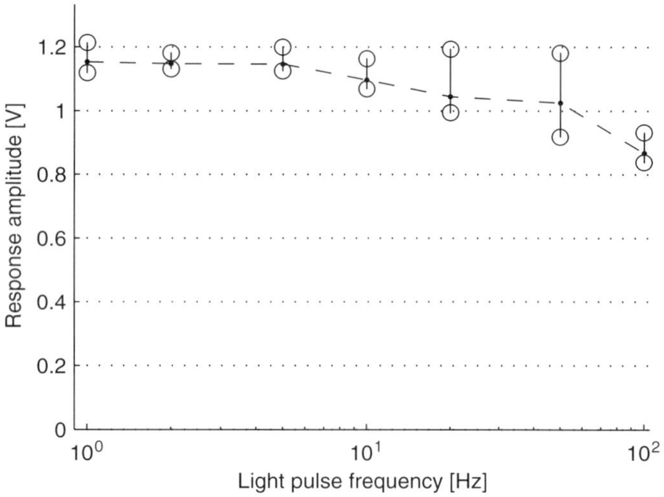

FIGURE 16.14
Pulsing frequency dependence of the PER for an element in the gray-scale imaging array when pulsing at low intensity of light. The vertical lines represent the range of at least 20 responses for each frequency.

photovoltage, it is clear that the arrays were not able to handle higher pulsing frequencies at high intensities for a long time. When the level of optical power was increased close to or above the saturation limit, the PER dropped to zero (see Figure 16.15). If the array was used in an application requiring high frame rates, the populations at each intermediate state could be estimated by models developed for analyzing the photocycle (37). If necessary, photostimulation could be used to return the BR molecules to the ground state. Frequency sensitivity of the other two types of BR was not measured but it would give interesting information where the different photocycles are concerned.

16.5.5 Results

The spatial resolution of the imaging arrays is limited but it is possible to downscale the element size with suitable design. The PERs generated by the gray-scale imaging array

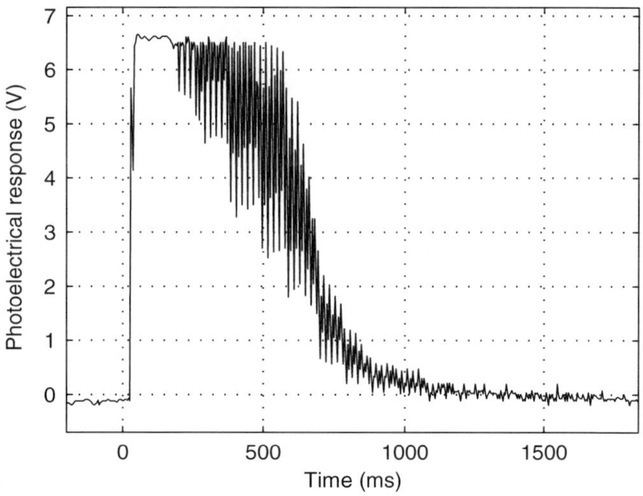

FIGURE 16.15
Time course of the PER for an element in the gray-scale imaging array when pulsing at 100 Hz with high intensity of light.

are similar to the responses from the single-pixel photosensors. Uniformity of the responses has been shown to be moderate but the preparation procedure for the BR-thick films is robust. The differences in the responses could be easily compensated by calibrating the sensor in the application. The elements in the color-sensitive imaging array generate detectable responses but the amplitudes are small. This is because of the concentrations of BR in the elements, and low conductivity of the gold layer. The sensitivity of the array elements to radiant energy and spectrum of light have not been thoroughly measured, since elementary experiments have showed them to function like the photosensors. The pulsing experiment has not been performed to 4-keto and 3,4-didehydro BR because of the limited responses from the color-sensitive array. The experiment with the gray-scale array shows that wild-type BR could be pulsed even at 100 Hz when the intensity of light is limited. This indicated that at least the gray-scale array can be used in principle for video capturing. The characteristics of the imaging arrays based on BR are summarized in Table 16.1.

16.5.6 SOM

The color space formed by training SOM with responses of BR elements to the collection of 84-color lights can be visualized. Figure 16.16 shows learned color space. Each group of color patches are the colors assigned to one neuron in the map after the training is completed. Figure 16.17 shows in three-dimensional space the self-organizing map embedded into the training data set; the weights corresponding to the neurons are the vertices of the grid. It can be seen from Figure 16.16 that the colors have been sensibly organized; similar colors have been clustered to occupy close regions of the color space. Analogically, Figure 16.17 shows the color space formed by training the SOM with the simulated responses of human cones to the same collection of 84-color lights.

The sensible organization of colors may also be observed by looking at how individual neurons in the trained map respond to monochromatic stimuli. Each neuron in the map has a distinct response curve to colors throughout the color space. Figure 16.18 shows the response curves of neurons in the map and the wavelength regions for which each particular

TABLE 16.1
Summarized Properties of BR-PVA Imaging Arrays

Characteristic and Notes	Gray-Scale Array	Color-Sensitive Array
Structure	BR-PVA film on gold	
Concentration (mg/ml)	8.32	n.a.
Active area (mm^2)	25	18–22
Thickness (μm)	144	101–107
Peak-to-peak photovoltage (40 μJ, with electronics)	6.10 V	84–186 mV
Response full width at 50% (ms) (40 μJ, with electronics)	18	23
Response sensitivity to BR concentration	Linear	n.a.
Response sensitivity to illuminated area	Linear	n.a.
Response sensitivity to radiant energy	Linear	Linear
Response resolution (bins) (2–40 μJ, with electronics)	53	36
Wavelength of peak response (nm)	580 ± 10	500–580
Spectral width of response peak (nm) (50% responsivity)	129 ± 10	129–158
Maximum pulsing frequency (Hz) (90% of the maximum response)	50	n.a.

Notes: n.a. = not available.

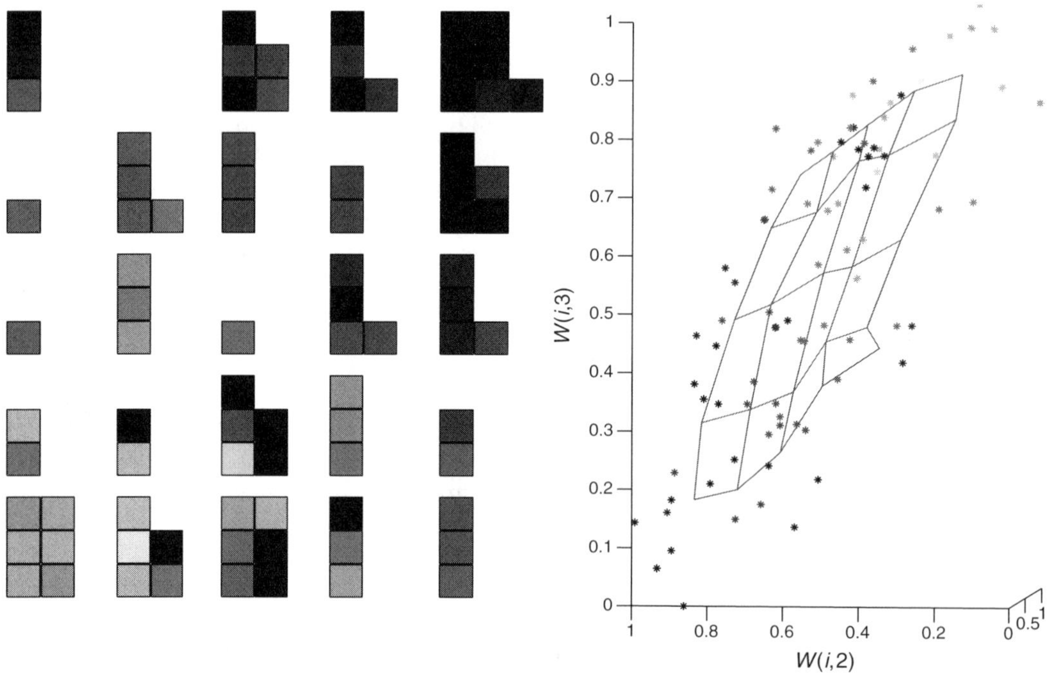

FIGURE 16.16
(See color insert following page 330) Visualization of the color space for BR responses by using a 5 × 5 rectangular SOM.

neuron gives the maximum response among all neurons. The response curves express the strength of a neuron's response to stimuli at different wavelengths. The regions of maximum response are indicated by filled areas under the curves. We may observe that the neurons responding maximally to adjacent wavelength regions are also neighbors in the map.

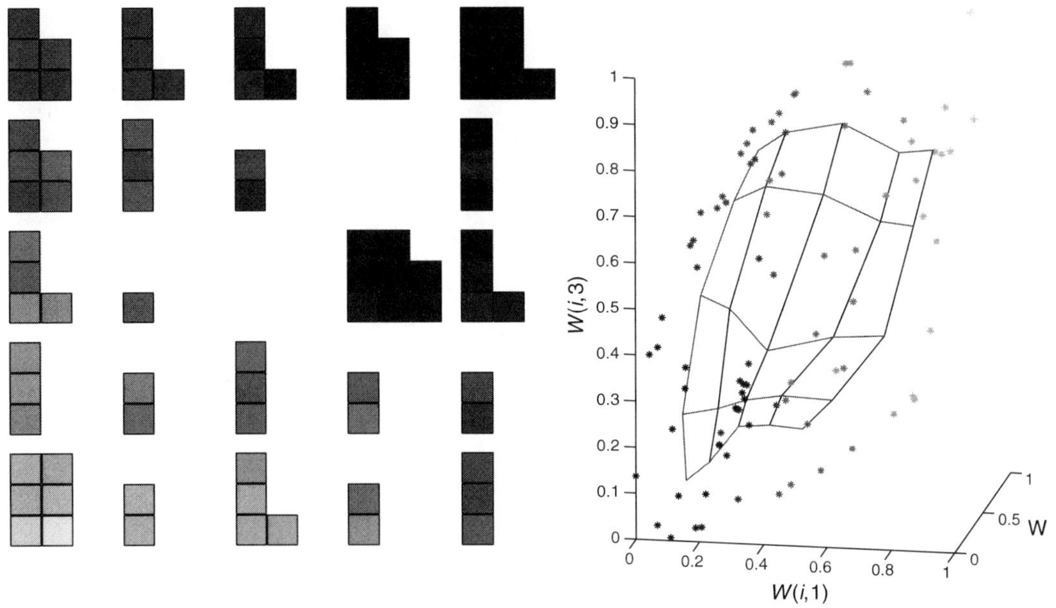

FIGURE 16.17
(See color insert) Visualization of the color space for simulated cone responses by using a 5 × 5 rectangular SOM.

16.5.7 Imaging Spectral Scenes

The PERs and action spectra of the imaging arrays have been measured and their sensitivities have been calculated. This information enables functional simulation of a theoretical high-resolution imaging device based on BR. The dynamic range of such a device would be limited by the number of molecules in the BR elements. Depending on the number of BR molecules in each array element, shutter mechanism, SNR of the electronics, and resolution of the analog-to-digital conversion, the dynamic range and number of detectable intensity levels would be sufficient for general use. To visualize the capabilities of such an imaging device, two test images were used. The first one was a spectral reflectance image of a GretagMacBeth™ color-rendition chart (see Figure 16.19a), and the red1 spectral image from the spectral data set of natural scenes measured at University of Bristol (see Figure 16.19b). The resolutions of the images were 64 × 64 and 256 × 256 in the spatial dimension, whereas the numbers of spectral bands were 151 and 31 (400–700 nm; 2-nm and 10-nm intervals). The response images in Figure 16.19 show that the imprecision in the measured action spectra caused limited color discrimination capability. When templates developed for visual pigments were fitted to the absorption peaks of the three BR types, the result better represented the true capabilities of the theoretical device. The response together with an estimated response of cone pigments in the human retina is shown in Figure 16.20. It is evident that for good color separation, further processing of sensor information is needed.

16.6 Conclusions

Several factors affect and limit the future development of computers based on technologies in current use. Biomolecules offer an alternative approach to computing, and some of

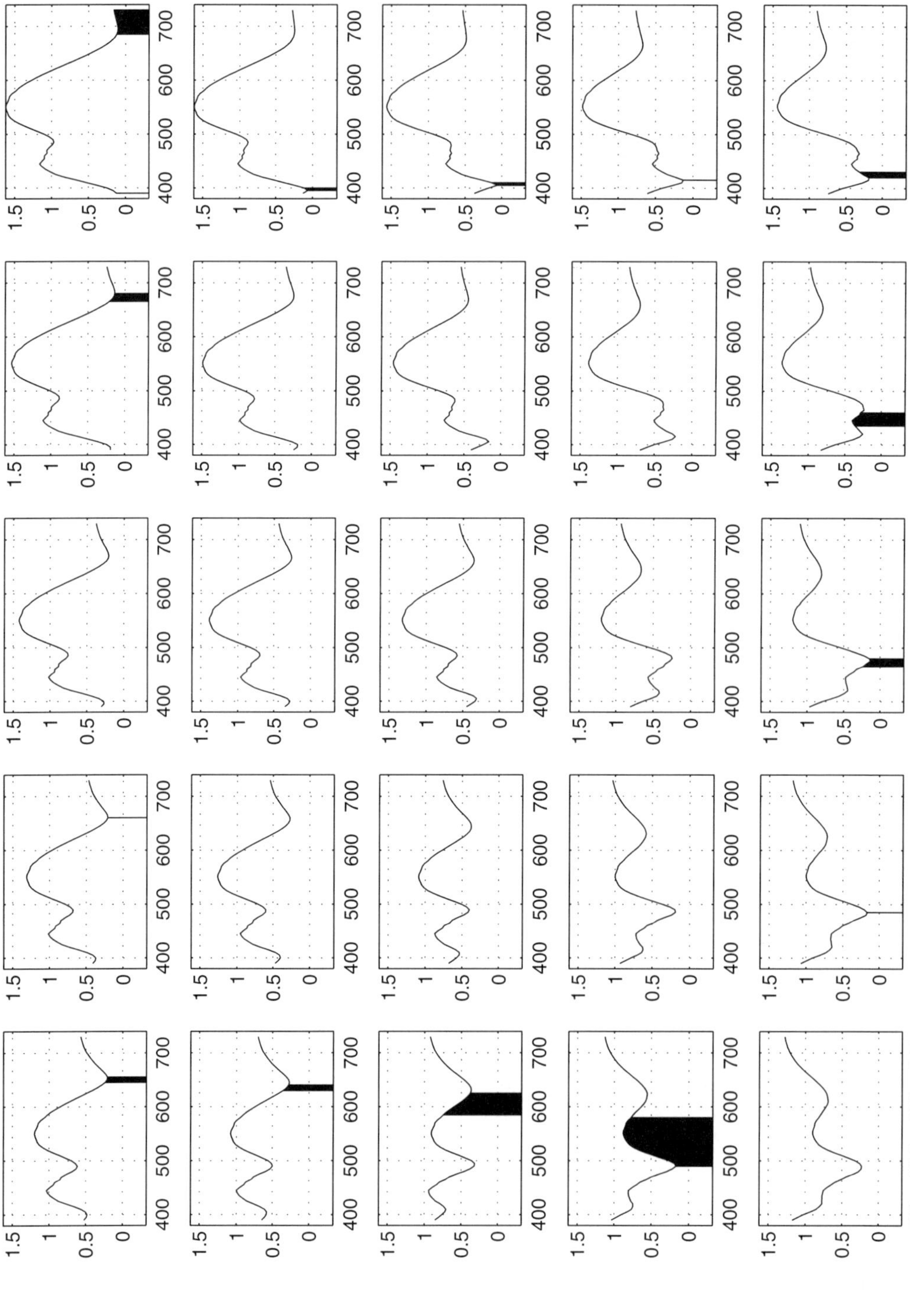

FIGURE 16.18
Responses of the neurons in the SOM to monochromatic stimuli.

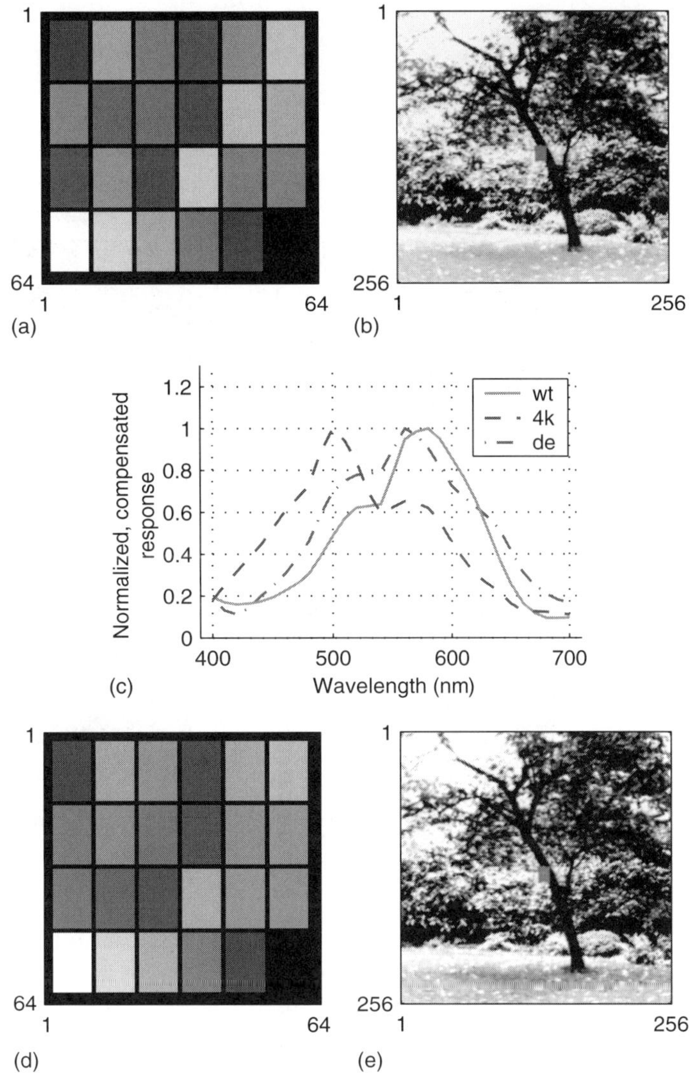

FIGURE 16.19
(See color insert) (a–b) Red-green-blue representations of the spectral images. (c) Measured action spectra of wild-type (wt), 4-keto (4k), and 3,4-didehydro (de) bacteriorhodopsin, and (d–e) responses of the theoretical color-sensitive imaging array to the input images.

them are well suited for processing of visual information. Among these molecules, BR is one of the most studied one.

The objective of this chapter was to discuss the nature of color and the design factors for the application of BR to imaging. Since color plays an important role in visual perception, it should be properly considered in every design of color-sensitive or color-reproduction systems in which human beings have some part. The practical goal of our research has been to develop photosensors and imaging devices based on BR as a case study of a material proposed even to molecular electronics and nanoscale applications. The experiments with the imaging arrays based on different types of BR showed reliable photoelectric responsivity and usability of the arrays to color-sensitive and even video applications.

Based on the results, it is apparent that these biomolecules have characteristics distinct from the current technologies for imaging and visual information processing: no filters are needed to achieve color sensitivity, thin films and coatings offer numerous possibilities for novel array geometries, and optically active molecules enable preprocessing of visual data.

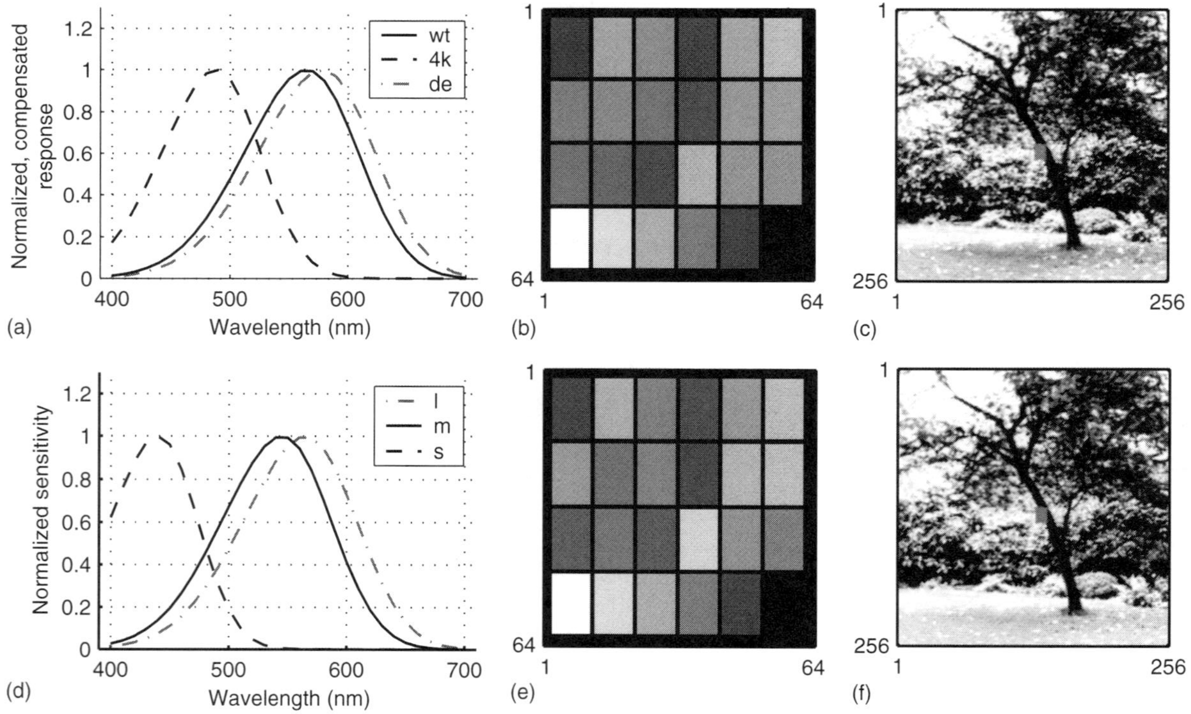

FIGURE 16.20
(See color insert) (a) Spectral sensitivities of visual pigment templates fitted to the absorption peaks of wild-type (wt), 4-keto (4k), and 3,4-didehydro (de) BR, (b–c) responses of the theoretical imaging array to the two input images, (d) Spectral sensitivities of *l*, *m*, and *s* cone pigments in the human retina, and (e–f) estimated response of the human eye to the two input images.

References

1. D. Goldhaber-Gordon, M.S. Montemerlo, J.C. Love, G.J. Opiteck, J.C. Ellenbogen (1997). Technologies and designs for electronic nanocomputers. MITRE Corporation. http://www.mitre.org/tech/nanotech/IEEE_article.html (accessed March 10, 2006).
2. J. Hecht (1994). "Rare" bug dominates the oceans. *Science*, 284:21.
3. R. Companó, ed. (2001). *Technology Roadmap for Nanoelectronics*. European Communities, Luxembourg, 2nd ed.
4. L.A. Drachev, A.D. Kaulen, S.A. Ostroumov, V.P. Skulachev (1974). Electrogenesis by bacteriorhodopsin incorporated in a planar phospholipid membrane. *FEBS Lett.*, 39(1):43–45.
5. A. Dér, P. Hargittai, J. Simon (1985). Time-resolved photoelectric and absorption signals from oriented purple membranes immobilized in gel. *J. Biochem. Biophys. Methods*, 10:295–300.
6. K. Bryl, G. Váró, R. Drabent (1991). The photocycle of bacteriorhodopsin immobilized in poly(vinylalcohol) film. *FEBS Lett.*, 285(1):66–70.
7. N. Hampp (2000). Bacteriorhodopsin as a photochromic retinal protein for optical memories. *Chem. Rev.*, 100:1755–1776.
8. R.R. Birge (1992). Protein-based optical computing and memories. *Computer*, 25(11):56–67.
9. T. Renner, N. Hampp (1993). Bacteriorhodopsin-films for dynamic time average interferometry. *Optics Commun.*, 96:142–149.
10. Z. Chen, R.R. Birge (1993). Protein-based artificial retinas. *TIBTECH*, 11:292–300.
11. M. Frydrych, P. Silfsten, S. Parkkinen, J. Parkkinen, T. Jaaskelainen (2000). Color sensitive retina based on bacteriorhodopsin. *Biosystems*, 54(3):131–140.
12. T. Miyasaka, K. Koyama, I. Itoh (1992). Quantum conversion and image detection by a bacteriorhodopsin-based artificial photoreceptor. *Science*, 255:342–344.
13. P. Silfsten, S. Parkkinen, J. Luostarinen, A. Khodonov, T. Jaaskelainen, J. Parkkinen (1996). Color sensitive biosensors for imaging. In the Proceedings of the 13th International Conference on Pattern Recognition, ICPR'96, Vol. 3, pp. 331–335.
14. H. Takei, A. Lewis, Z. Chen, I. Nebenzahl (1991). Implementing receptive fields with excitatory and inhibitory optoelectrical responses of bacteriorhodopsin films. *Appl. Optics*, 30:500–509.
15. H.-G. Choi, W.-C. Jung, J. Min, W.H. Lee, J.-W. Choi (2001). Color image detection by biomolecular photoreceptor using bacteriorhodopsin-based complex LB films. *Biosens. Bioelectron.*, 16:925–935.
16. J. Min, H.-G. Choi, B.-K. Oh, W.H. Lee, S.-H. Paek, J.-W. Choi (2001). Visual information processing using bacteriorhodopsin-based complex LB films. *Biosens. Bioelectron.*, 16:917–923.
17. J. Yang, G. Wang (1998). Image edge detecting by using the bacteriorhodopsin-based artificial ganglion cell receptive field. *Thin Solid Films*, 324:281–284.
18. B. Jähne (1999). Introduction. In B. Jähne, H. Haussecker, P. Geissler, ed., *Handbook of Computer Vision and Applications, Volume 1. Sensors and Imaging*. Academic Press, San Diego, CA.
19. M. Frydrych (1999). Color vision system based on bacteriorhodopsin. Ph.D. dissertation, Lappeenranta University of Technology, Finland.
20. B.A. Wandell (1995). *Foundations of Vision*. Sinauer Associates, Sunderland, Massachusetts.
21. G. Wyszecki, W.S. Stiles (1982). *Color Science*. Wiley, London, 2nd ed.
22. P.K. Kaiser, R.M. Boynton (1996). *Human Color Vision*. Washington, DC: Optical Society of America, 2nd ed.
23. J. Hallikainen (1991). Acousto-optic color signal processing. Ph.D. dissertation, University of Kuopio, Finland.
24. R. Robertson (1977). The CIE 1976 color-difference formulae. *Color. Res. Appl.*, 2(1):7–11.
25. D.L. MacAdam (1942). Visual sensitivities to color differences in daylight. *J. Opt. Soc. Am.*, 32(5):247–273.
26. A. Moini (1997). Vision chips or seeing silicon. Technical report, Centre for High Performance Integrated Technologies and Systems, The University of Adelaide.
27. L. Lensu, J. Parkkinen, S. Parkkinen, M. Frydrych, T. Jaaskelainen (2003). Photoelectrical properties of protein-based optoelectronic sensor. *Opt. Mater.*, 21:783–788.

28. L. Lensu, M. Frydrych, J. Parkkinen, S. Parkkinen, T. Jaaskelainen (2004). Photoelectric properties of bacteriorhodopsin analogs for color-sensitive optoelectronic devices. *Opt. Mater.*, 27:57–62.
29. L. Lensu (2002). Photoelectric properties of bacteriorhodopsin films for photosensing and information processing. Ph.D. dissertation, Lappeenranta University of Technology.
30. L.R. Lindvold, P.S. Ramanujam (1995). The use of bacteriorhodopsin in optical processing: a review. *J. Scientific Industrial Res.*, 54:55–66.
31. H.-W. Trissl (1990). Yearly review: photoelectric measurements of purple membranes. *Photochem. Photobiol.*, 51(6):793–818.
32. A.A. Khodonov, O.V. Demina, L.V. Khitrina, A.D. Kaulen, P. Silfsten, S. Parkkinen, J. Parkkinen, T. Jaaskelainen (1997). Modified bacteriorhodopsins as a basis for new optical devices. *Sens. Actuators, B Chem.*, 38–39:218–221.
33. L. Keszthelyi, P. Ormos (1983). Displacement current on purple membrane fragments oriented in a suspension. *Biophys. Chem.*, 18:397–405.
34. L. Keszthelyi, P. Ormos (1989). Protein electric response signals from dielectrically polarized systems. *J. Membr. Biol.*, 109:193–200.
35. M. Hauta-Kasari, K. Miyazawa, S. Toyooka, J. Parkkinen (1999). A prototype of the spectral vision system. In Proceedings of the 11th Scandinavian Conference on Image Analysis, pp. 79–86.
36. M. Palviainen (1998). Matrix camera based on bacteriorhodopsin (in Finnish). Master's thesis, Lappeenranta University of Technology.
37. J. Vanhanen, V.-P. Leppanen, T. Jaaskelainen, S. Parkkinen, J.P.S. Parkkinen (1999). Nonlinear transmittance of the 4-keto bacteriorhodopsin. *Opt. Mater.*, 12:473–480.

17

Protein-Based Photoreceptor Array on Flexible Plastic Substrates

Wei Wei Wang, George K. Knopf, and Amarjeet S. Bassi

CONTENTS

17.1 Introduction

17.1.1 Bioelectronics

Bioelectronics, first defined by visionary biologist Albert Szent-Gyorgyi in 1968, has emerged as one of the most rapidly expanding interdisciplinary research frontiers (1). As silicon-based devices continue to reduce in scale, their limitations become evermore apparent. Future developments indicate that electronic systems will migrate to systems that are based on organic or biological materials. Evolution has optimized biomaterials to exhibit highly dynamic characteristics at the molecular scale in ways that cannot be attained by current technologies. First, biological systems can store and process large amounts of information in extremely compact structures. Second, biological devices are assembled from individual molecules that are arranged spatially with respect to one another. Third, biological systems are extremely sensitive to their surroundings and are able to detect and discriminate changes in the environment. The performance of biological materials, united with their small size, energy efficiency, and inexpensive fabrication techniques, make them a promising alternative for designing various sensing devices.

Photosensitive proteins that can convert light directly into an electrical signal are becoming widely investigated biomaterials. Their optical and photoelectric features permit them to behave as "smart materials" in various signal processing and image sensing applications. Over the past three decades, biochemists and biophysicists have studied the structures and functions of such proteins in detail. Among them, bacteriorhodopsin (bR) is the most notable example. It exhibits light-sensitive characteristics similar to that of *rhodopsin*, a protein found in human eyes. Compared with other protein photoreceptors, bR is highly resistant to thermal and photochemical degradation, thereby exhibiting excellent long-term stability. As bR shows application potential in low-light detection, artificial vision, parallel associative processors and memories, it is quickly becoming an advanced material for constructing bioelectronic devices.

17.1.2 Bacteriorhodopsin: Structure and Biological Function

Bacteriorhodopsin is the light-harvesting protein present in the plasma membrane of *Halobacterium salinarium* (2). Under anaerobic conditions the cell membrane grows purple membrane (PM) patches in the form of a hexagonal two-dimensional (2D) crystalline lattice of uniformly oriented bR trimers. These patches roughly consist of a 10:1 molar ratio of lipids to bR (3). They are irregular in shape and have a constant thickness of 5 nm with lateral dimensions up to 5 μm. PM is stable against sunlight exposure in the presence of oxygen for several years. In water, its stability holds for temperatures over 80°C, pH values ranging from 0 to 12, and in the presence of high ionic concentrations (up to 3 M NaCl) (4). PM preserves its color and photochemical activity under dry conditions and can withstand temperatures up to 140°C (5).

The crystalline structure is the root of bR's chemical and thermal stability. The bR molecules convert sunlight into chemical energy by transporting protons from the cytoplasm to the cell exterior. Each bR molecule consists of 248 amino acid residues that are arranged in seven α-helices (6) (Figure 17.1). A retinal residue is bound to lysine-216 (K216) via a protonated Schiff-base linkage (8). All bR molecules within the patch are uniformly oriented, where their carboxy termini are located in the cytoplasm and amino termini are located in the periplasm. The seven transmembrane helices enclose an interhelical region that contains the retinylidene residue, key amino-acid residues, and water molecules (9). The amino acids in this region influence absorption properties of the chromophore and control the proton transfer pathway.

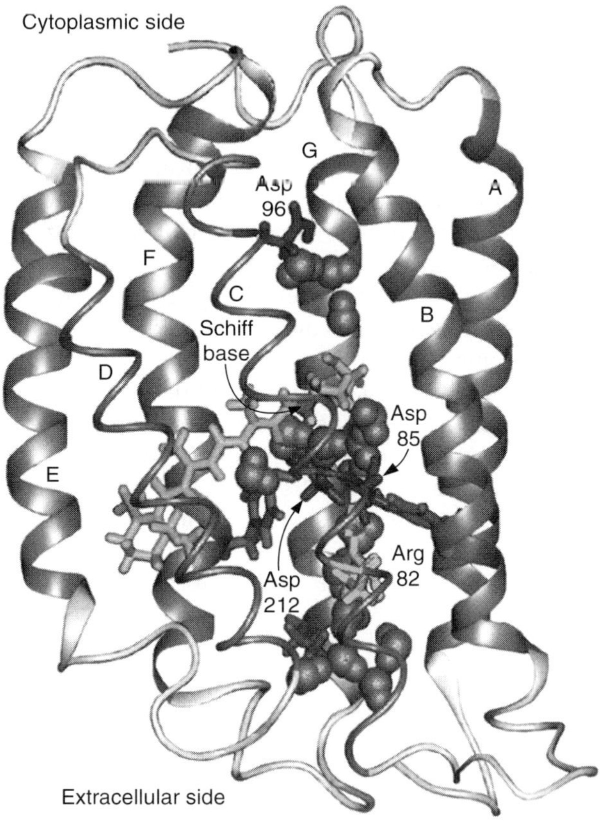

FIGURE 17.1
Structural model of bacteriorhodopsin. The seven transmembrane helices are labeled A to G. The retinal Schiff base and the amino acids most relevant for proton transport, Asp96, Asp85, Asp212, and Arg82, are shown explicitly. (From http:// www.ks. uiuc. edu/ Research/newbr/br_fig.html (accessed November 10, 2005.)

The retinal chromophore of bR naturally exists in one of two states: all-*trans* and 13-*cis* (10). In the dark, a bR population contains a mixture of these two retinal states and is called "dark-adapted". Under continuous illumination, the 13-*cis* retinal is converted into the all-*trans* retinal and all chromophores are in the all-*trans* form, which is referred to as "light-adapted bR". The absorption of a photon by light-adapted bR initiates isomerization of the retinal from an all-*trans* to a 13-*cis* conformation. This is followed by proton transport from the cytoplasm of the cell to the exterior, thus converting light energy into chemical energy needed for cell metabolism (11). Proton transport is accompanied by a complex photocycle, which involves several intermediate states with distinct spectral absorption maxima. The Schiff-base protonation and the retinal configuration cause the shift of the absorption bands. Wild-type bR molecules need approximately 10 ms to complete one photocycle and they do not have a refractory period after completing a photocycle under light-saturated conditions (12).

17.1.3 Proton Transfer Mechanism in Reconstituted Bacteriorhodopsin Films

A major challenge for scientists and engineers is to find ways of incorporating biomaterials into practical devices. This requires appropriate immobilization techniques that can orient the protein molecules and prevent them from denaturing. Thin films are considered to be the most practical architectures for heterostructural systems containing photosensitive materials because they improve photoelectric functionality compared with the bulk state. Furthermore, thin films provide a basic arrangement that allows biological components to be interfaced with microelectronic systems.

Methods recently used for assembling bR thin films onto solid substrates include electrophoretic sedimentation (EPS), Langmuir–Blodgett deposition (LB), self-assembly, electrostatic layer-by-layer adsorption, antibody-mediated monolayer, and immobilization within polymer gels (13–18). Success of these methods is largely attributed to bR's highly stable crystalline structure. The photoelectric conversion efficiency of a bR film is dependent on the uniformity of molecule orientation within the film. Most of the aforementioned assembly methods are able to pattern bR in manners suitable for various applications. However, it must be noted that different fabrication methods alter the biomolecule's environment from that of its natural state, resulting in different phototransducing mechanisms.

After light-adapted bR is excited by incident light, the absorbed photons electronically excite the chromophore and drive retinal isomerization around the $C_{13}=C_{14}$ double bond, transforming the retinal from an all-*trans* into a 13-*cis* configuration (19). Photoisomerization triggers proton translocation and initiates a series of thermally driven reaction steps that take place within a millisecond time frame. The protein structure that surrounds the retinal has crucial impact on the photocycle and the proton transfer mechanism. In the aqueous phase, proton transport starts with the release of a proton to the extracellular side during the $L \rightarrow M$ transition and terminates with a proton uptake from the cytoplasmic side during the following $M \rightarrow N$ transition (20) (Figure 17.2a). The aspartate residues, Asp85 and Asp96, act as proton acceptor and donor for reversible protonation and deprotonation of the Schiff-base linkage, respectively. When the pH $\geq$ 7.0, proton release precedes proton uptake. The sequence is reversed at pH $<$ 7.0, whereby proton uptake is followed by proton release (21). In contrast, absorption of a photon by bR in its dried form initiates twisting of the retinal. However, this process does not cause proton pumping due to the lack of water molecules in the extracellular channel. The corresponding photocycle stops at the *M* state before the Schiff base can be reprotonated (Figure 17.2b) (22). Ultra fast photoinduced charge separation is followed by a slow charge recombination as the protein returns to its ground state.

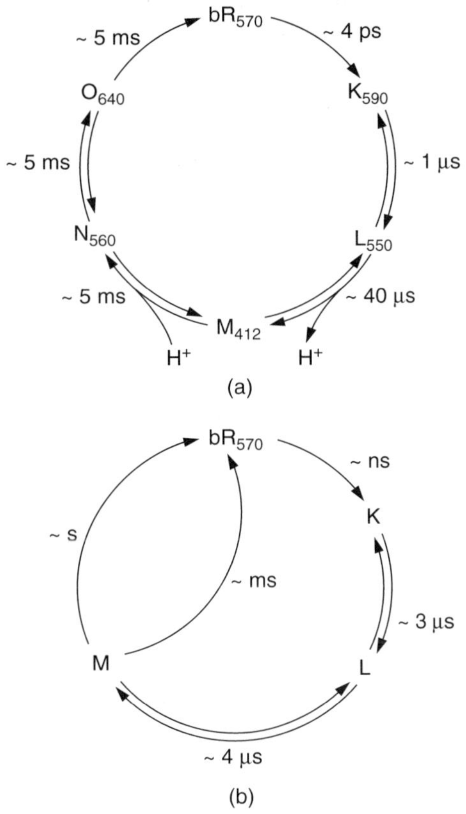

FIGURE 17.2
Photochemical cycles of the bR molecule: (a) in aqueous phase, where the proton transfer begins with the release of a proton to the extracellular side during the *L*→*M* transition and ends with proton uptake from the cytoplasmic side during the *M*→*N* transition (17). (b) In dried form, only the *K*, *L*, and *M* intermediates are involved in the photochemical cycle; the protein returns to its ground state through several paths, each with a different lifetime (18).

17.1.4 Bacteriorhodopsin Applications

Its unusual stability and high quantum efficiency, coupled with unique photochromic and photoelectronic properties, make bR a promising biological material for technical applications. Various device designs based on bR have been explored since its discovery. Most early works have applied its photochromic properties to optical applications. Recently, emerging interests lie in the exploration of its photoelectric properties, which can be realized within hybrid bioelectronic architectures.

Photochromic applications of bR range from all-optical logic gates and holographic interferometry to optical volumetric memories and parallel associative processors (23–26). These applications are based on bR's photochromic properties, where absorption of a photon leads to reversible photochemical and thermal cycles. bR can be reversibly converted between different intermediates of the photocycle by exploiting photoexcitation at different wavelengths. Genetically engineered bR can provide significant spectral shifts and long lifetimes within certain intermediate states, showing additional potential in such optical applications (27). Simple alignment of PM patches is required; therefore, embedding PM in an appropriate matrix is sufficient to meet application needs.

Photosensitive properties of bR provide a basis for its sensing and imaging applications. Immobilization of bR molecules to charge-sensitive substrates is required to utilize such properties. The generated photoelectric response from immobilized bR on various substrates has been investigated. Widely adopted substrates include silica glass coated by conductive oxides such as tin oxide or indium tin oxide (ITO), and metal electrodes such as gold and platinum (28; 29). Some new substrates have emerged recently toward micro and nano applications, involving the gate terminal of a GaAs-based MOSFET and nano-black lipid membranes (30; 31). Research groups worldwide have been exploring the effective methods to pattern bR in 2D arrays that are well suited for the development of membrane-based sensing devices. An 8 × 8 pixel array of bR photocells was fabricated by Miyasaka et al. (32) for the purpose of image detection. The bR LB film is uniformly deposited on a glass substrate patterned with an ITO coating. The resultant bR film is then covered with an electrolyte and a gold electrode to complete the sandwich-type structure. Haronian and Lewis (33) investigated methods that allow microfabrication of bR without losing its photoelectronic properties. A bR array with 50 × 50 μm pixels is created by depositing bR onto an ITO electrode and then ablating both the film and substrate using an argon fluoride excimer laser. A novel protein–silicon hybrid photodetector array was proposed by Birge and his group (34). To construct the high-resolution bR-silicon photosensor, the detection grid of a charge injection device is utilized as the substrate for a thin bR-polymer film. This architecture also incorporates processing circuitry to monitor the photoresponse of the bR film. Researchers in Italy recently demonstrated a compact bR microarray, which is fabricated by patterning dehydrated bR films on microstructured Si and SiO_2 surfaces without orientation (35).

Design and construction of bR-based bioelectronic devices to mimic various image processing features of biological vision systems have been reported since the last decade. At the material level, bR has the inherent ability to process simple visual information. Two fundamental visual functions, feature extraction and pattern recognition, have captured the attention of many researchers. In general, feature extraction involves detection and location of spatial and temporal variation in image intensity, which is necessary for edge enhancement and motion detection applications. In biological vision systems, the ganglion cell-receptive field forms a basic edge detection unit, which is sometimes referred to as a zero-crossing filter. Takei et al. and Yang et al. (36; 37) presented two types of bR-based artificial receptive field. The polarity of bR photoelectric response is used to emulate excitation and inhibition within neural computation. The use of bR for parallel processing, as based on fundamental neurobiological principles, is further explored by Haronian and Lewis. They present a rapidly reprogrammable neural network architecture with the possibility of employing a large synapse matrix (38).

Differential responsivity that is exhibited by reconstituted bR films provides the essential function in motion detection. It compresses visual information by removing DC levels and amplifying interesting features. A bR-based photoreceptor array that is capable of detecting motion and edge information in real time was developed by Miyasaka et al. (39). Their detector also responds to static images when incident light is modulated within a given frequency range. Fukuzawa (40) later presented a position-sensitive motion sensor using the image-recording capability of bR at high pH. This sensor is capable of detecting an object's displacement at certain times and positions.

Visual pattern recognition consists of high-level information processing in vision systems. Valuable research toward realizing pattern recognition function on a molecular level has been conducted by Min et al. (41). Their bR-based photoreceptors were constructed in different configurations, where some degree image extraction and pattern recognition were achieved by applying an initial learning process to film configurations. A similar visual information processing method is also suitable for color recognition. Two types of color-sensitive sensor arrays have been realized thus far: Frydrych et al. (42) use bR with its variants, whereas Choi et al. (43) integrate bR with flavins. Using artificial neural networks trained for color space recognition allows such systems to learn the observed colors and adapt to their environment.

17.1.5 Flexible Electronics Technology

Microelectronics are commonplace in nearly all aspects of daily life, but interests continue to grow toward creating high-performance, low-cost, and large-area photoelectric devices that are difficult to achieve with the conventional technologies. Examples include the large flat panel displays, solar cell arrays, and x-ray sensors large enough to image the whole body. Spreading electronics over complex surfaces to monitor or control the surface characteristics has attracted both scientific and commercial interest. Flexible electronics technology presents a new approach that may provide solutions to such expanding needs (44).

Extending electronics to flexible substrates encompasses many potential advantages. Flexible electronics can reduce the space, weight, complexity, and cost of a system by integrating electronics into the physical structure. Plastic substrates are particularly attractive because they enable construction of lightweight and durable devices that can potentially be rolled up or folded when not in operation (45).

Fabricating devices on flexible plastic substrates provides diverse possibilities. Using the surface areas of large structures within an intelligent sensor network for structural health monitoring illustrates a promising application area (46). Local signal processing of sensor data, such as amplification, routing, and switching, can also be employed by incorporating transistor-based electronics onto flexible substrates (47). Compared with standard printed circuit board solutions, such integration can save space and reduce weight. While current interests lie in replacing existing technologies, there are many opportunities for developing new applications and capabilities. High-scale integration of flexible electronics can be used to create intelligent skin surfaces for air and land vehicles (48). New classes of photosensor arrays can also be developed for visual navigation or medical imaging.

17.1.6 Motivation of This Work

Within the past several decades, scientists and engineers strive to uncover the mysteries of biological vision and attempt to design artificial vision systems to mimic natural functions (49–51). Biological vision has evolved over several millions of years into very efficient sensors that provide some unique features. Biological photoreceptors can be extremely sensitive; for example, the retina of human eyes and that of several other vertebrates can respond to individual photons (52). These photoreceptors can also respond to signals over a remarkably wide range of intensities. Human eyes, for example, exhibit a dynamic range that is ten orders in magnitude (53). Typically, sensing elements arranged in high spatial densities yield high-resolution visual information. Transmitting this wealth of information to the brain requires a large degree of data compression. Spatial and temporal compression is accomplished by the preprocessing capabilities of front-end sensing elements and their highly parallel and interconnected neural pathways. In addition, other preprocessing capabilities include edge detection and contrast enhancement. Unlike planar silicon-based image arrays, nearly all biological vision systems exhibit curved geometry. Insect eyes, for example, typically have their photoreceptors situated on the convex side of the surface. Alternatively, all vertebrate have their photoreceptors arranged within the inner concave surface of their eyes. These geometries can provide various advantages, such as increased field of view or enhanced focusing and tracking capabilities.

As stated earlier, bR exhibits photoresponse characteristics similar to that of *rhodopsin*, including high sensitivity, large dynamic range, high spatial resolution, and preprocessing capabilities. Most recently-proposed bR sensor designs are composed of planar 2D arrays that are fabricated on solid substrates. Flexible electronics offers a great opportunity to develop bR-based vision devices with curved or free-form geometries. In this work, complementary features of bioelectronics and flexible electronics produce the primary motivation for integrating these two technologies. This combination can grant artificial system with the benefits of biological vision and its unique geometries. However, there are numerous engineering challenges that must be overcome to transform this material into reliable technology. Current efforts are directed toward refining material properties and fabrication technology to produce desired thin-film structures. Design of high-performance processing circuitry for bR-based photoreceptor array is also a critical task. Applying bR to high-level vision tasks is an emerging area of research. Most proposed devices exploiting the photoelectric properties of bR only exploit amplification to process the detected signals. Successful incorporation of bR onto flexible substrates is also dependent on whether the desired visual task can be realized.

17.2 Flexible Photoreceptor Array Based on Bacteriorhodopsin Film

17.2.1 Purple Membrane Preparation

Purple membrane patches containing bR are extracted from wild-type *H. salinarium* by using a standard procedure (54). To remove sucrose and salt ions, the PM suspension is purified by diluting with bidistilled water and then isolated by centrifugation. The ultracentrifuge operates at 22,400 rpm at 4°C for 30 min with a 50-Ti rotor. After the rinsing cycle, the supernatant is carefully removed and the PM pellet is resuspended in bidistilled water. This centrifugation process is repeated two more times and the final PM pellet is suspended in bidistilled water to achieve a desired concentration of 16 mg/ml. This sample is then softly sonicated for 10 s to break up particle aggregation. Upon the completion of three rinse cycles, the pH of the PM suspension is roughly 6.7, which eliminates the need for acidification or alkalization of the sample.

Studies have shown that the salt concentration within the PM suspension influences the electric signal generated by the reconstituted bR films (55). A film that contains a high salt concentration will have a lower resistance, resulting in decreased signal-to-noise ratio (SNR) during signal measurement. Therefore, the low salt concentration is necessary for acceptable signal measurement. The salt concentration of a PM suspension is monitored by measuring its ionic strength in term of conductivity. The final conductivity value is maintained less than 10 μS. If the conductivity is still too high, additional centrifugation is required. Note that extra and unnecessary deionization cycles will produce undesirable results. More specifically, cation loss occurs in the photochromic group of the bR molecule and the PM patches become blue in color. Upon absorption of light, the retinals of the blue membrane will isomerize from all-*trans* to both 13-*cis* and 9-*cis*, where the 9-*cis* retinal is a long-lived photoproduct (56). Such a membrane exhibits lower quantum efficiency and is thus not suitable for photoelectric applications.

17.2.2 Flexible Polyethylene Terephthalate Film With Patterned Indium Tin Oxide Coating

17.2.2.1 Overview of Materials and Techniques

Emerging applications such as foldable display, large-area solar cells, wearable monitoring systems, and conformal sensor sheets have motivated electronic devices to be grown on flexible substrates. Plastics continue to replace traditional substrate materials due to their ease of manufacture, chemical resistance, and wide availability. Therefore, it would be advantageous to utilize plastic substrates for electronic devices. Plastics can form thin films to cover large areas, while only requiring low-temperature manufacturing techniques. Some commonly used plastic substrates include polyethylene terephthalate (PET), polyethylene naphthalate, polycarbonate, and polyimide (57; 58).

Optoelectronic devices require a transparent conductive layer with high electrical conductivity and low visible light absorption. This layer can be prepared from a wide variety of materials, including doped semiconducting oxides of tin, indium, zinc, or cadmium and titanium nitride ceramics, as well as metals such as silver and gold (59). Among them, the most widely adopted transparent conducting oxides (TCO) are that of tin (SnO_2:F), zinc (ZnO_2:F), and indium (In_2O_3:Sn, ITO).

For flexible device applications, TCO layers must be deposited at low temperatures because plastic substrates are thermally sensitive. Fluorine-doped tin dioxide (SnO_2:F) is most common due to its low manufacturing cost; however, it requires high deposition temperatures and is very difficult to etch; hence, it is now widely used for glass substrates. Fluorine-doped zinc oxide (ZnO_2:F) can be deposited at lower temperatures and has the greatest light transparency. Despite these qualities, it suffers from rather low conductivity, precluding its use from sensitive electronics like optical sensors and flat-panel displays. Although indium is a costly raw material and its tin-doped oxide (ITO) is somewhat more difficult to etch than zinc oxide, it allows very low-deposition temperatures ($\leq$200°C) and high conductivity, making it ideal for deposition on thin flexible plastic substrates (60).

The properties of a TCO depend not only on its chemical composition but also on the method used for its preparation. Numerous methods are available for TCO deposition. Some of the most common methods used for flexible substrate coating include chemical vapor deposition, magnetron sputtering, and pulsed-laser deposition (61–63). Chemical vapor deposition is suitable for large-scale applications and can provide very high deposition rates; however, many chemicals tend to be toxic or unstable. Magnetron sputtering provides high deposition rates at low temperatures and can provide uniform coatings over large areas. Pulsed laser deposition is a relatively new process and well suited for research laboratories. Some industrial implementations are able to pattern the electrode directly and eliminate the etching process, but this method is costly and limited to small areas.

17.2.2.2 Fabrication of Indium Tin Oxide Patterned Polyethylene Terephthalate Film

In this work, the flexible substrate is made of a 175-μm-thick PET film. This material is commonly used for flat panel displays, touch screens, and flexible electronic circuits because it is inexpensive, mechanically durable, and provides high light transmission (typically >86%). The ITO electrode pattern is designed and then transferred to an image, which is then used to develop a photomask. A conductive ITO layer is deposited onto the substrate via pulsed laser deposition (Sheldahl Inc.). The ITO-coated film has surface resistance of 35 Ω/sq. The flexible ITO electrode is imprinted as a 4 × 4 pixel array pattern. Each pixel is 2 × 2 mm and is separated from the neighboring pixels by 1 mm. An independent ITO-coated wire with a width of 300 μm is connected to each pixel. These wires lead to two sets of terminals along the sides. Each terminal is 1 × 3 mm, separated by 1 mm from each other. The overall area of a patterned ITO electrode is 15 by 23 mm.

17.2.3 Device Fabrication

17.2.3.1 Principles of Electrophoretic Sedimentation Fabrication

Purple membrane patches, like many other cell membrane materials, maintain a small potential across the membrane. This small potential arises from differences in charge densities of each side. The C- and N-terminals of the polypeptide chain are located on opposite sides of the membrane. The cytoplasmic side of the membrane is negative with respect to the extracellular side at pH values between 5 and 7; the polarity reverses when the pH falls below

5 (64). This charge asymmetry results in a permanent dipole moment across the membrane, directed from the cytoplasmic side to the extracellular side. EPS occurs when the PM suspension is inserted between two parallel electrodes while an electric field is applied. The PM patches are oriented and transported onto the anode due to the more negatively charged cytoplasmic side. This provides a simple and effective means of achieving high degrees of bR orientation during film formation. Note that either the cytoplasmic side or extracellular side of the PM can be attached to the electrode by selecting the appropriate suspension pH. The EPS method produces thick PM films where the film thickness is a function of the applied DC electric field, its duration, and PM concentration.

17.2.3.2 Fabrication Conditions

Fabrication conditions, such as the electric field intensity, exposure time, and humidity, affect the photoelectric response and topology of bR films. A large aggregation of bR molecules is observed at low electric field intensities. These aggregates cannot be oriented due to their hydrophobic behavior and thus form poorly oriented films, resulting in reduced photoelectric responses. High electric field intensities reduce aggregation and create more uniform bR films. However, signals measured from bR films formed at high electric field strengths are found to be relatively small. This may be caused by protein degradation (65). The long exposure time under an electric field can also cause decreased photoresponse because of degradation. Therefore, an intermediate electric field produces the best results. It has been reported that larger photocurrents are generated by the bR film when dried under higher humidity; therefore, humidity must be carefully controlled during the drying process. Relative humidity in the range of 50 ~ 60% typically yields the best results (66). Constant humidity can be achieved by drying the film in the presence of vapor generated by a saturated salt solution.

17.2.3.3 Fabrication Process

Figure 17.3 illustrates the fabrication process used to create the flexible 16-pixel array. One hundred microliters of PM suspension is ejected from a pipette evenly over the patterned surface. A 1.0-mm plastic spacer (not shown) is placed around the array to separate the two electrodes. An external power supply is connected such that the bottom electrode becomes the anode and the top becomes the cathode. An electric field of 40 v/cm is applied for 5 min. The top electrode is then lifted and the remaining bulk water is carefully removed with a pipette. The film is allowed to dry by placing it in a humidity-regulated chamber for 12 h. When dry, a PET film with a continuous ITO coating is then aligned with the dried bR pixels and sealed carefully along the edge by using fast-curing epoxy resin. A photograph of the bR photoreceptor array on flexible PET substrate is shown in Figure 17.4. Two flexible ribbon cables are connected to the array via two flat connectors to the readout electronics. Such a connection maintains device flexibility while keeping good alignment.

17.2.3.4 Absorption Spectrum of the Flexible Bacteriorhodopsin Film

To verify deposition quality, the absorption spectrum of the dried bR film is compared against the absorption spectra of the PM suspension and ITO-coated substrate, respectively. An ultraviolet-visible spectrophotometer (Varian, Cary 50Bio) is used to record the absorption spectra. The absorption peak of the PM suspension in the visible light range was found at 570 nm (Figure 17.5a), whereas no peaks appeared in the visible light range for the ITO-coated PET film (Figure 17.5b). The spectrum of bR-deposited flexible film was obtained by subtracting the absorption spectrum of the ITO-PET substrate from the

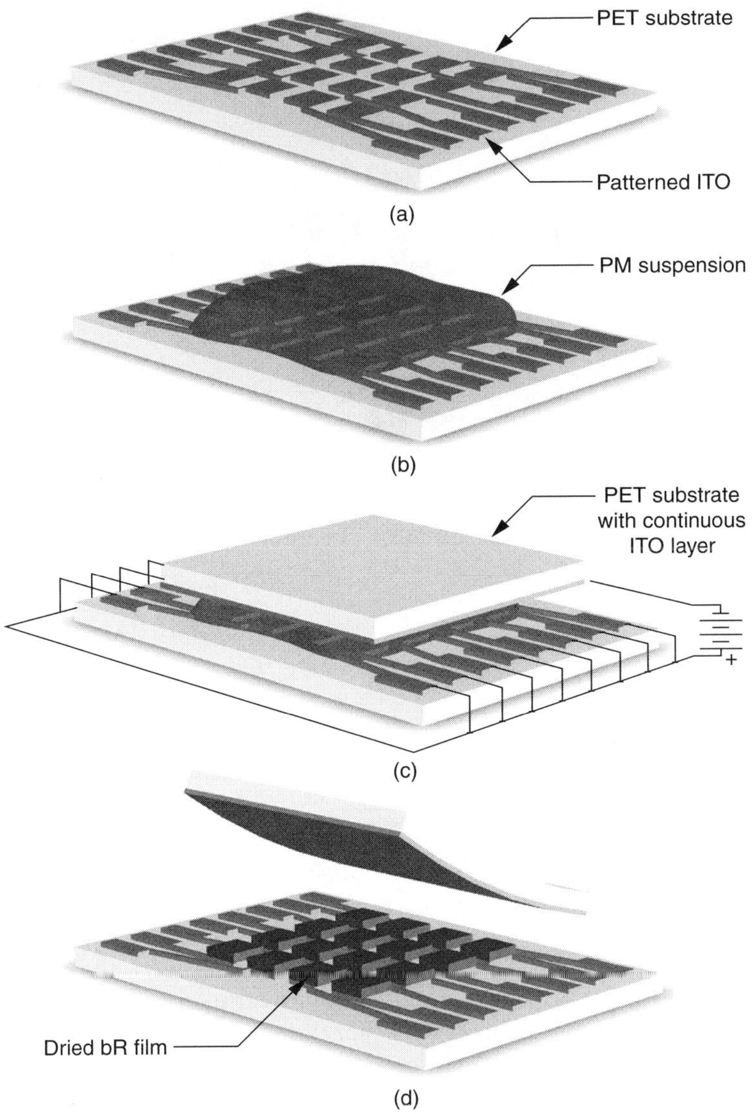

FIGURE 17.3
Fabrication process: (a) Patterned electrode surface is cleaned of all residues, (b) 100 μL of PM suspension is spread evenly over the patterned surface, (c) A 1.0-mm plastic spacer (not shown) is placed around the array to separate the two electrodes, where the bottom ITO electrode is patterned and the top ITO electrode is continuous, and they are connected to the positive and negative terminals of the power supply, respectively. An electric field of 40 V/cm is then applied. (d) After drying, the top electrode is attached and sealed carefully along the edge by using fast-curing epoxy resin.

bR-ITO-PET film spectrum (Figure 17.5c). A broad peak, from 550 to 580 nm, was observed for the dried bR film. Successful deposition shows that the 570-nm absorption peak found in the PM suspension is also preserved in the dried film. The peak is broader than the PM suspension due to influences of dehydration on the Schiff base of bR retinal chromophore. The dried film exhibits an optical density of approximately 0.5 due to high bR concentration in the film sample, which is greater than the bulk value of 0.3.

FIGURE 17.4
Photograph of the 16-pixel bR photoreceptor array on flexible PET substrate.

17.3 Array Circuit Design

17.3.1 Equivalent Circuit Model of the Individual Pixel

During photoisomerization, charge separation inside the bR film causes changes in its surface charge density (67). This alters the surface alignment of bR molecules in contact with the solid substrate; therefore, the physical model of dried-bR film must take into account the interactions between the surface polarization and the surface charge changes to accurately describe the physical process. Rather than employing the permanent dipole moment to describe time-dependent surface charge changes, the transition dipole moment is more appropriate in this instance. The reason lies in the fact that the former is caused by the surface charge asymmetry in PM, and the latter is induced by temporal charge translocation in bR. Increased dipole moment relates directly to retinal structural changes and likely drives the accompanying torsional motion (68).

The electric dipole moment provides a useful means of measuring the effects of charge separation between atoms and molecules; however, the distances between charges are typically too small to be easily measured. In electrostatic theory, a dipole is defined as two opposite point charges that are separated by a distance. No current flows; only static charge is present. However, the effect of time-dependent dipole moment changes must be considered in the model because it is invalid to assume that charge changes instantaneously. Thus, the current dipole concept is introduced; it is defined by two current-carrying points separated by a distance. This is a fundamental model for representing endogenous signal sources in biology (69). The potential field generated by a current dipole in a continuous uniform medium is given as

$$\Phi_d = \frac{I}{4\pi\sigma}\left(\frac{1}{r_1} - \frac{1}{r_2}\right) \tag{17.1}$$

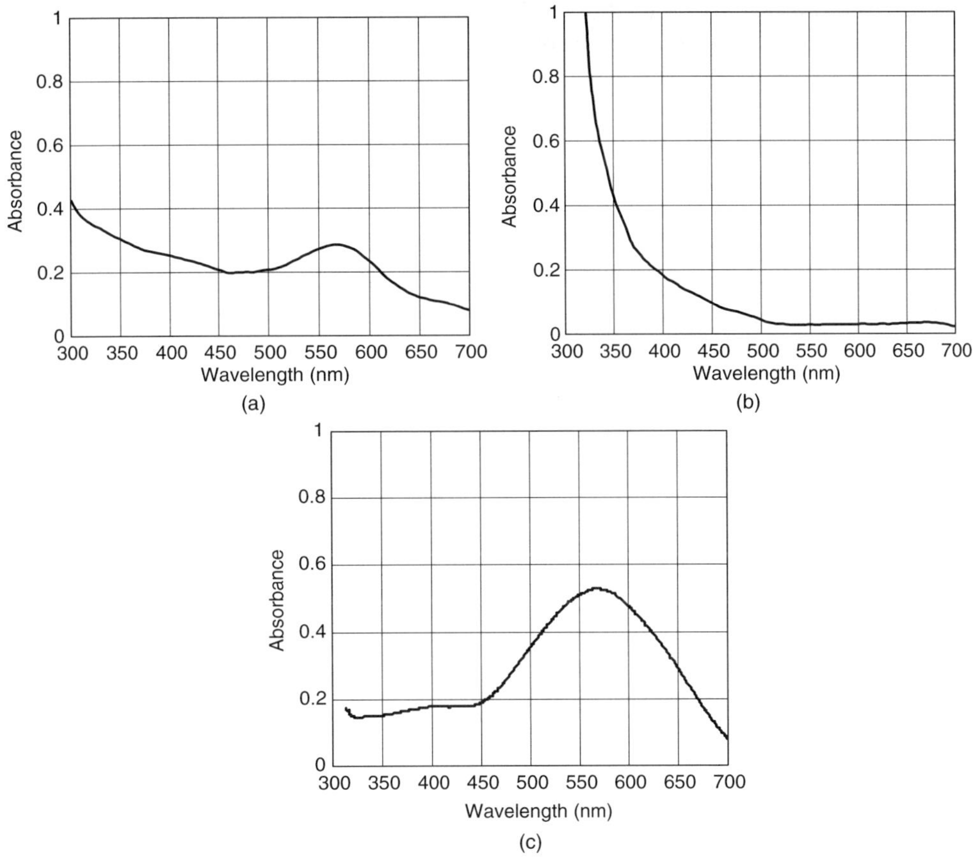

FIGURE 17.5
Absorption spectra of (a) PM suspension, (b) ITO-coated PET film, and (c) dried bR film deposited on ITO–PET substrate.

where σ represents conductivity of the medium, r_1 and r_2 are the distances from each pole to the position Φ_d, and the displacement current is I. This equation is valid at all distances from the dipole.

If the distance between the dipole and the measured point is large in comparison to the dipole length, the current dipole potential can be approximated by

$$\Phi_d = \frac{I\vec{d} \cdot \vec{r}}{4\pi\sigma r^3} \tag{17.2}$$

The vector product $\vec{d} \cdot \vec{r}$ indicates directional sensitivity, so the potential intensity falls as a function of $1/r^2$ in all directions. The only fundamental difference between the dipole moment and the current dipole is that charge is replaced by charge per second in the latter. Current dipole values for each excited state of bR photocycle can be derived by dividing the measured dipole moment by the lifetime of that state.

To determine the total induced dipole moments of the bR film, the population distribution of all excited states within the bR film is required. Referring to the photocycle introduced in Section 17.1, the time-dependent population of each excited state can be obtained by applying the rate equations of the photochemical cycle of bR. The concentrations of different intermediate states can be described as:

$$dC_i(t)\big/_{dt} = k_{(i-1)\to i}C_{(i-1)}(t) - k_{i\to(i+1)}C_i(t) \tag{17.3}$$

where the parameters $k_{(i-1)\to i}$ and $k_{i\to(i+1)}$ ($i = L, M,$ and M_1) represent the rate constants of thermal relaxation from intermediate $(i-1)$ to i and from intermediate i to $i + 1$, respectively. The parameter k_{BK} represents the rate constant of the photochemical reactions from the ground state B to state K initiated by incident light. This can be written as

$$k_{BK}(t,I) = (1-\eta)\sigma_B\Phi_{BK}F(t,I) \tag{17.4a}$$

where

$$F(t,I) = I(t)\big/_{h\upsilon} = I(t)\lambda_i\big/_{hc} \tag{17.4b}$$

Herein, η is the refractive index of the substrate; σ_B is the absorption cross-section of B state; Φ_{BK} is the quantum yields of photoreaction $B\to K$; and $F(t,I)$ is the photon density flux of the incident light. The frequency of the incident light and the speed of light in vacuum are represented as υ and c, respectively, where h is Planck's constant.

Modeling a bR-based photoreceptor is an essential step in integrating it with signal processing circuit. An equivalent circuit model of a bR photoreceptor is shown in Figure 17.6a. Each component in the equivalent circuit model represents an analog found in the bR photoreceptor's physical model (Figure 17.6b). The overall behavior of the bR molecule is that of a current generator, which is given by the dipole current, $I(t)$. Considering the finite resistance of protein medium, R_s, the dipole current source can also be described by an equivalent electric potential source, $E_{ph}(t)$. Since the lipids within the PM exhibit a high dielectric constant, they can be modeled as the combination of a series capacitor C_l and a series resistor R_l. The parallel combination of resistor R_m and capacitor C_m, where the current source describes the behavior in nonilluminated bR molecules in conjunction with lipids that are perpendicular to the PM orientation direction. The series resistance R_c originates from ohmic contacts between film surfaces and electrodes. Capacitor C_j is the junction capacitance connecting the external circuit and the bR film. Load resistance R_f is the measuring instrument's input impedance. Contact resistance R_c and junction capacitance C_j can be ignored for analysis purpose because their values are insignificant in comparison to the other components.

17.3.2 Front-End Circuit Design

17.3.2.1 Three Common Front-End Circuit Designs

Photoreceptor performance depends primarily on its front-end circuit design. Generally, front-end circuits should demonstrate low noise and provide sufficiently wide bandwidth to ensure accurate representation of input light signals. They should also provide any necessary amplification and buffering. The choice of front-end circuits is determined by the photoreceptor's intrinsic characteristics and the intended application. Conventional circuit designs may not be appropriate for extremely small-current

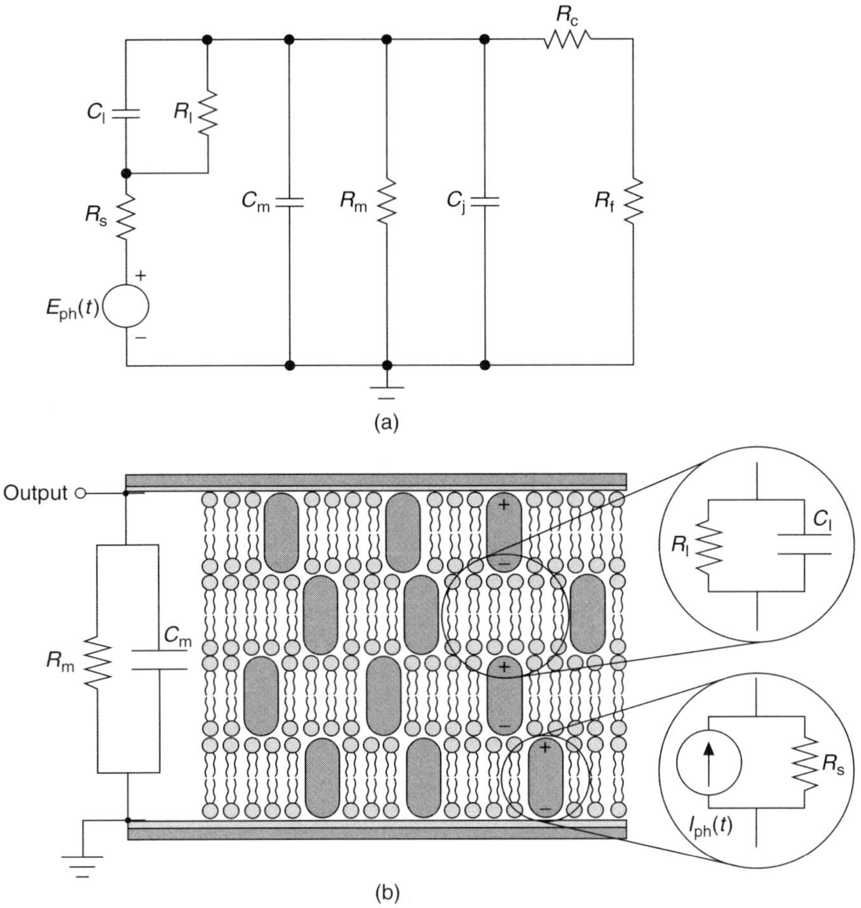

FIGURE 17.6
(a) Equivalent RC circuit of one pixel of bR-based photoreceptor array. The bR molecules are modeled as a current generator and represented by an equivalent electric potential source $E_{ph}(t)$ in series with the source resistance R_s. (b) Physical model of one bR pixel. The dipole moment of each bR molecule crosses the membrane in the same direction. The bR molecules are surrounded by the lipid bilayers.

signals that are generated by detectors with virtually infinite resistance, as is demonstrated by a bR-based photoreceptor. Therefore, highly optimized circuit designs are required, where high-input impedance and low-noise amplifiers should be applied. Figure 17.7 illustrates the simplified equivalent bR circuit model in combination with three common front-end designs: a voltage follower, a transimpedance amplifier, and a switched integrator.

A voltage follower is simply an operational amplifier with a voltage gain of unity where the high-input impedance is limited by its input bias current (Figure 17.7a). This is commonly referred to as a voltage buffer and is used here to isolate the high photoreceptor load from the remaining processing circuitry. A dried bR photoreceptor exhibits particularly high resistance, which significantly limits the bandwidth of this amplifier design.

Rather than measuring photovoltage, photocurrent generated by the bR photoreceptor can be amplified directly by using a transimpedance amplifier (Figure 17.7b). In this configuration, the load resistor R_f is connected in the feedback loop of the amplifier. The inverting input of the amplifier is connected to the photoreceptor output, where the noninverting input is grounded. This technique effectively shorts out the membrane

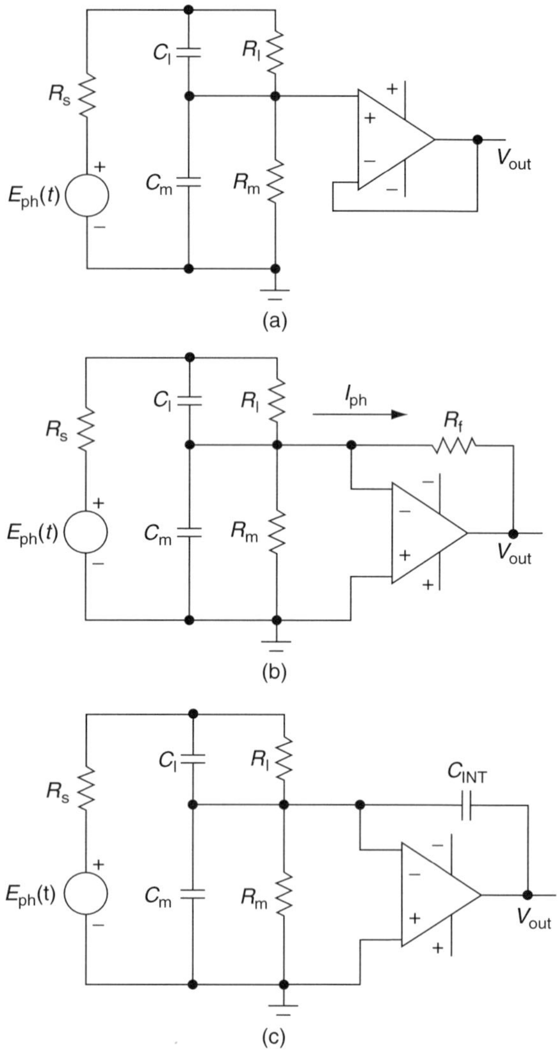

FIGURE 17.7
Schematic representation of simplified equivalent bR circuit model in combination with three front-end circuit designs: (a) voltage follower, (b) transimpedance amplifier, and (c) switched integrator.

capacitance C_m and resistance R_m, thus significantly improving bandwidth. The output voltage of the transimpedance amplifier is described by

$$V_{out} = -I_{ph} R_f \tag{17.5}$$

The transimpedance gain is determined by R_f. To detect very low signals, a large feedback resistor is required (on the order of 1 GΩ); unfortunately, this approach introduces system instability and thermal noise.

The switched integrator provides similar functionality to that of the transimpedance amplifier (Figure 17.7c). It converts an input current into an output voltage, but a capacitor replaces the large resistor in the feedback loop. The input current flows through the feedback capacitor C_{INT} charging it at a rate proportional to the input current. With an input current I_{INT}, the voltage at the output is given by

$$V_{out} = -\frac{1}{C_{INT}} \int I_{INT}(t) dt \tag{17.6}$$

The output voltage V_{out} is proportional to the integration time T_{INT} and inversely proportional to the feedback capacitor C_{INT}. Therefore, the effective transimpedance gain is T_{INT}/C_{INT}, where very high gain can be achieved by reducing the integration capacitor values and increasing the integration time. The integrating behavior of this design reduces noise by averaging noise generated by the photoreceptor, amplifier, and external sources. The switched integrator design is well matched to the bR photoreceptor because it provides adjustable gain-bandwidth, low noise, and low power consumption.

17.3.2.2 Analysis of the Switched Integrator

Figure 17.8 shows a more detailed circuit diagram in which a precision switched integrator, IVC102, combines an FET op amp, integrating capacitors, and low-leakage FET switches onto a single chip. The pixel output connects to the inverting input via a sample switch S_1, where the noninverting input is grounded. The reset switch S_2 is in parallel with the integrating capacitors to provide a discharge path. Amplifier output is connected to a holding capacitor via a readout switch S_3. The overall circuit operates as follows: (1) integration of the input current begins when S_1 is closed and S_2 and S_3 are opened. Referring to the timing diagram in Figure 17.9, signal integration occurs over time *a*, where the input

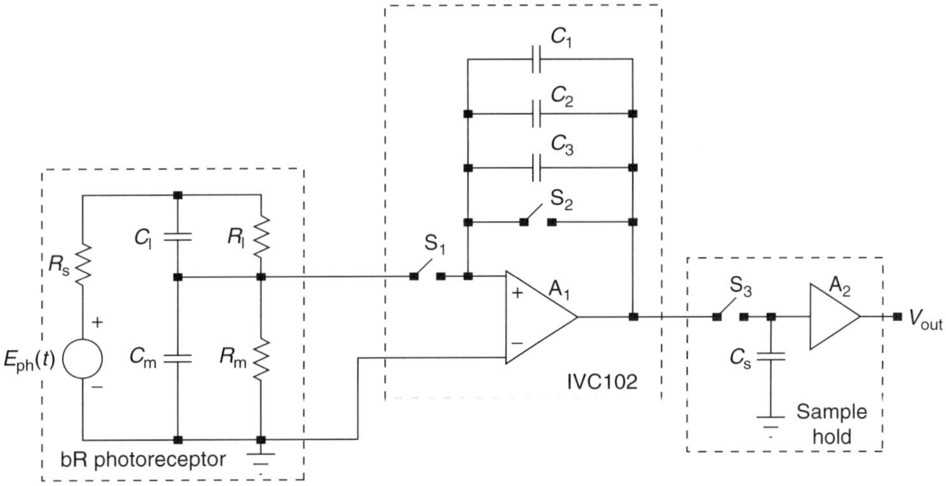

FIGURE 17.8
Schematic representation of a complete front-end configuration for one bR pixel. Three dashed boxes include the simplified bR model, IVC102 switched integrator, and sample/hold circuitry.

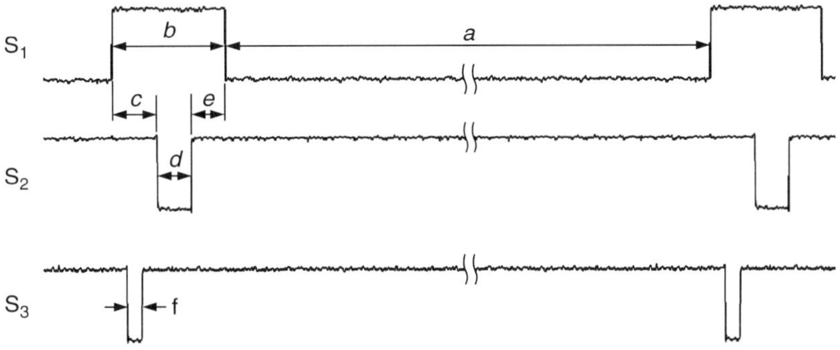

FIGURE 17.9
Timing diagrams of switches S_1, S_2, and S_3 recorded from the microcontroller output.

current causes amplifier A_1 to charge capacitors C_1-C_3 at a proportional rate. (2) The sample switch S_1 is then opened, allowing the output voltage of IVC102 to stabilize. After approximately 5 μs, the readout switch S_3 is closed to transfer the charge from C_1-C_3 to the sample hold capacitor C_S. (3) Following successful signal readout, S_3 is reopened and S_2 is toggled for 10 μs to completely discharge C_1-C_3 and allow A_1 to stabilize. (4) A preintegration hold period *e* is required to ensure that S_2 is fully open before S_1 is closed, thus preventing any input signal loss. Repeating this process provides continuous sampling, generating a stepwise approximation of the true signal. Provided that the integrator does not saturate, sampled voltages are proportional to the integrated intensity over the integration period. By adjusting the timing parameters or feedback capacitance, measurement sensitivity can be adjusted over a wide range.

The bR photoreceptor has a very high source resistance, which is in the range of $10^{10}-10^{12}\ \Omega$. The source capacitance value is around 50 pF, which plays a key role in operation of the switched integrator. This front-end design provides continuous integration of the input signal. During the short period *b* that S_1 is opened, any signal current produced by the photoreceptor charges the source capacitance C_m and is transferred to C_1-C_3 immediately after S_1 is closed. As a result, no charge produced by the sensor is lost, and the effective integration time is equivalent to the sampling period $(a + b)$.

The digital timing inputs to S_1-S_3 are compatible with standard complementary metal-oxide-semiconductor (CMOS) or TTL logic signals. A PIC12F675 microcontroller is used to control the digital timing functions. Special attention must be paid when these logic signals are routed to their respective input pins on the circuit board, since significant noise can be introduced by capacitive coupling between the logic signal traces and the sensitive input pins of IVC102.

17.3.3 Overall Circuit Architecture

17.3.3.1 Overview of Array Circuit Architecture

In most imaging applications, photosensors are arranged in either 1D or 2D arrays. The application may require the photosensors to be combined with processing electronics to achieve the desired functionality. For visible and near infrared detection, the charge couple device (CCD) imager and the CMOS imager are two of the most common silicon arrays (70). The general circuit architecture of these arrays can be divided into two stages: the pixel-level and the array-level processes. At the pixel level, two elementary configurations include passive pixel sensors (PPS) and active pixel sensors (APS) (71). For PPS, each pixel contains a photosensor and a sampling switch to synchronize output. To improve the image quality, more processing components, such as active buffers and preamplifiers, are incorporated into each pixel, which is thus referred to as APS. An essential task for array-level design is to develop an efficient way to transfer information from the array elements to the outside system. This is typically dictated by the system requirements. For arrays that generate 2D outputs, the common methods of transferring data from arrays include scanning using shift registers, scanning using decoders, and multichannel readout (72; 73).

Each pixel of a CCD imager is composed of a capacitor that is charged by incident light; thus, no signal processing occurs. The most widely used method of transferring data from a CCD is progressively scanning the array using shift registers (74). A row of a sensor array is selected, followed by the sequential readout of each pixel in the column. The collected charges are then converted into voltage signals using amplifiers and digitized using analog to digital converters. This method reduces the number of wires required to connect each pixel. It also maximizes the pixel fill-factor, which reduces measurement noise. However, CCD imagers tend to be slow and consume significant power.

The fundamental difference between CCD and CMOS is that pixel level signal processing is now possible. The CMOS sensing element consists of two key components: the photodiode that produces a small photocurrent proportional to the incident light and an amplifier to boost the signal. Early sensors suffer from significant noise, but steady advances in signal processing have dramatically improved image quality. Although pixel readout can be accomplished using multiple channels, sequential scanning is still much more effective. Instead of using shift registers to read pixels in a sequential fashion, two decoders in the X and Y axes can be used to provide random-access readout from individual pixels (73). Other supporting circuitry can potentially be integrated into a single chip, such as image stabilization and image compression.

Conventional photoreceptor arrays only capture images, and processing is accomplished by general-purpose computers.In contrast the image capture layers of biological vision systems are in close interaction with subsequent processing stages in the neural pathways. Vision chips are developed to mimic this interaction by integrating both stages on the same chip and performing signal processing at the pixel level (51). Data compression is also possible at the pixel level, increasing the effective bandwidth during transmission. The sum of these capabilities makes vision chips much faster than conventional CMOS systems.

17.3.3.2 Circuit Architecture for Bacteriorhodopsin-Based Photoreceptor Array

A hybrid architecture is applied to the flexible bR photoreceptor array design. The signal processing electronics are separated from the bR sensor array due to differing fabrication methods. Active circuitry is applied to each individual pixel, where it has its own front-end processing stage. As discussed in Section 17.3.2, the switched integrator is used as a preamplifier for a single pixel and it converts the weak photoinduced current into a measurable voltage. In addition to amplification, other functions like spatial and temporal signal processing are required at the pixel level to benefit high-level imaging applications. In vision chips, such functions have been achieved by integrating filtering components within each pixel. However, by applying bR-based photoreceptor, similar functions can be accomplished at the material level without the need for extra circuitry. When exposed to light, a bR photoreceptor readily demonstrates differential temporal filtering inherent in its photoresponse. This feature can be exploited to assist high-level vision tasks, such as motion detection, edge detection, and contrast enhancement.

For this work, a multichannel readout approach is used to transfer the information processed by each pixel to the processing circuitry. This greatly simplifies array fabrication, since no shift registers need to be considered at this time. It has been shown that bR photoreceptors can be manufactured into 256-pixel arrays on glass substrates, using a multichannel readout scheme (39). The multichannel design reduces the information to be transferred and increases the transfer bandwidth. Depending on the intended application, this approach also allows additional per-pixel and array-level signal processing prior to transmitting the image data to the external application.

In this prototype design, the bR photosensor and its processing circuitry are physically separated. The sensor array is deposited on an ITO-patterned PET film. The processing circuitry is fabricated on a standard PCB circuit board using discrete surface-mount components. Even though the design in this manner has limitations (i.e., moderate noises and large physical size), from a system design perspective, the hybrid architecture itself does have a great potential for imaging applications. Several advantages of applying such architecture include increased density, higher fill factor, and the ability to independently optimize the sensor array and processing circuits. As the processing circuitry within each pixel become increasingly large and complex, limitations of a planar 2D architecture become evident. Either the photosensitive area must be reduced, resulting in increased noise, or the pixel itself must become larger, which decreases resolution. These conflicts drive photosensor array development toward 3D architectures. The same multilayered approach is evident in nearly all biological vision systems. A 3D-sensor array has been proposed, where three circuit layers with different functions are stacked vertically into one chip (75). The first layer only consists of a photosensor array, the second layer contains analog circuits to amplify the light-induced signal, and the third layer consists of circuits to digitize the signal and read out the data. Each pixel in the sensor array is directly connected with its processing circuits in parallel via the vertical interconnection. As the signals from individual pixels are processed and transferred in parallel, real-time and high-performance image processing can be realized. Even though multichannel designs are limited to relative small arrays at the current stage, rapidly developing thin-film transistor technology can allow improved readout capabilities for future bR photoreceptor arrays. This technology will allow circuits to be constructed in many thin-3D layers, while retaining its flexible characteristics.

17.4 Performance Analysis

The performance of the proposed bR photoreceptor array is evaluated at both the pixel level and overall array level. The key characteristics concerning individual pixels involve SNR, linearity and dynamic range, spectral response, and response time. Pixel uniformity and overall responsivity under mechanical bending are applied for array performance testing.

17.4.1 Noise and Signal-to-Noise Analysis

17.4.1.1 An Overview of Noise Sources

The smallest signal that a photodetector can detect is often limited by the amount of noise that is present. Typically, lower noise levels offer better sensitivity. Hence, noise characterization plays an important role in quantifying detector performance.

Shot noise and thermal noise are the two predominant noise sources in all photodetection systems (76). Shot noise is caused by random fluctuations in the arrival of discrete particles, such as individual electrons. In semiconductors, shot noise is present when charge carriers are generated and recombined at randomly varying rates. The power of shot noise, in terms of current, is given by the Schottky formula

$$i^2 = 2qI_{ph}\,\Delta f \tag{17.7}$$

where q is the charge of one electron (1.6×10^{-19} C), I_{ph} is the constant photocurrent, and Δf is the measurement bandwidth.

Thermal noise is dominant when photocurrents are small and it is generated by thermal fluctuations of moving electrons as they collide with atoms and other electrons in a conductive material. The power of thermal noise, given in terms of current, is described by the Johnson formula, also known as the Nyquist formula

$$i_n^2 = \frac{4KT\,\Delta f}{R} \tag{17.8}$$

where K is Boltzmann's constant, T is the absolute temperature, R is the conductor resistance, and Δf is the measurement bandwidth. The internal photodetector resistance and load resistance are the major sources of thermal noise.

Applying noise analysis to biological systems tends to be more difficult due to their inherent structural complexity, dynamic nature, and inhomogeneous properties. It is well accepted that thermal noise, present in all physical systems, is the limiting factor in detection of weak signals. In biomaterials, thermal noise is a key contributor to the overall system noise. When a proton channel within a cell membrane is in thermal equilibrium, it is considered equivalent to a simple resistor (77). Although it has been reported that shot noise also exists in biological systems, the mechanisms of generating such noise in biosystems are far different from that of silicon-based devices. Shot noise in biological processes is modeled as fluctuations of a variable (i.e., ion current), due to its discrete nature. When significant energy flows into or out of a system (i.e., in cases where it is not in thermal equilibrium with its environment), motion of the charged particle introduces shot noise (78).

It has been widely observed that the electrical responses from both vertebrate and invertebrate photoreceptors are accompanied by significant noise (79; 80). Signals carried by light are noisy because there are variations in the rate at which photons are emitted. The resultant fluctuations within biological phototransduction processes are recognized as noise sources. There are three major noise categories under consideration herein (80; 81): (1) photon-induced fluctuation. The observed signal, under the conditions of steady illumination, is the superposition of all signals produced by every incident photon. Variation in the rate of incoming photons and in their rate of absorption will produce noise. (2) Discrete noise in darkness. In complete darkness, there is a low rate of events leading to thermal isomerization. These events set the limits of performance in the detection of dim flashes. (3) Continuous noise in darkness. It appears that thermal isomerizations of rhodopsin and fluctuations in biochemical intermediates are the dominant fluctuations in photoreceptors that have been exposed to darkness for prolonged periods of time (i.e., fully dark adapted). In the presence of moderate illumination levels, the dominant noise source arises from photon fluctuations.

17.4.1.2 Signal-to-Noise Analysis in Bacteriorhodopsin Photoreceptor

In terms of the component parameters that comprise the photoreceptor, an analysis of the SNR is important to optimize its performance. High SNR can be achieved by improving photoreceptor fabrication, as well as front-end amplifier design. A general noise equivalent circuit of a bR photoreceptor is shown in Figure 17.10. Five noise sources contribute to the output signal: photoreceptor shot noise V_{iph}, photoreceptor thermal noise V_{Rm}, amplifier current noise V_{in}, amplifier voltage noise V_{en}, and feedback thermal noise V_{Rf}.

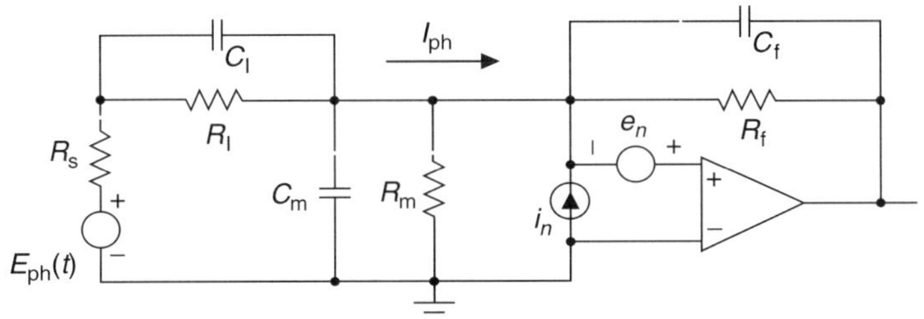

FIGURE 17.10
Schematic representation of the simplified equivalent bR circuit used for noise analysis. This equivalent RC circuit shows one pixel with its amplifier and includes the different noise sources.

The signal current I_{ph} produces shot noise when it flows through the feedback impedance Z_f. Its output voltage is given by

$$V_{iph} = \sqrt{2qI_{ph}\,\Delta f} \cdot Z_f \tag{17.9}$$

where the feedback impedance Z_f is

$$Z_f = \frac{R_f Z_{Cf}}{R_f + Z_{Cf}} \tag{17.10}$$

and $Z_{Cf}=1/sC_f$. Eq. (17.9) indicates that higher photocurrent will increase shot noise. However, higher photocurrent also produces a much larger signal. The resulting SNR improves proportionally with the square root of the photocurrent increase.

Photoreceptor thermal noise caused by thermal isomerization is apparent inside the bR film at any temperature above absolute zero. Photoreceptor thermal noise output voltage is given as

$$V_{Rm} = \sqrt{\frac{4KT}{R_m}\Delta f} \cdot Z_f \tag{17.11}$$

Normally, front-end amplifiers introduce two frequency-dependent noise sources: a current noise V_{in} and a voltage noise V_{en}. The current noise is actually shot noise produced by the bias current in the amplifier input stages; its output voltage is in the form of

$$V_{in} = i_n \cdot Z_f \tag{17.12}$$

Amplifier voltage noise is attributed to thermal noise in resistive components of input stage transistors and is represented as

$$V_{en} = e_n \cdot \left(1 + \frac{Z_f}{Z_m}\right) \tag{17.13}$$

where the photoreceptor impedance Z_m is given as

$$Z_m = \frac{R_m Z_{Cm}}{R_m + Z_{Cm}} \tag{17.14}$$

and $Z_{Cm}=1/sC_m$ is the equivalent impedance of the bR film capacitance. Eq. (17.13) indicates that the high bR film resistance can help reduce amplifier voltage noise.

The feedback resistor also causes thermal current noise; however, this noise is avoided by employing a switched integrator design, where only capacitors are located within the feedback loop.

To obtain the output signal, the transfer function for the overall photoreceptor is required and it can be described as

$$T(s) = -\frac{Z_f}{Z_1} = -\frac{1}{R_s C_f}\frac{(s + 1/\tau_3)}{(s + 1/\tau_1)(s + 1/\tau_2)} \tag{17.15}$$

where

$$Z_1 = R_s + \frac{R_1 Z_{C1}}{R_1 + Z_{C1}} \tag{17.16}$$

$Z_{C1} = 1/sC_1$, $\tau_1 = (R_s \,//R_1)\cdot C_1$, $\tau_2 = R_f \cdot C_f$, and $\tau_3 = R_1 \cdot C_1$.

Note that uncorrelated noise contributions must be added as sums of squares. Therefore, SNR can be written as

$$S/N = \frac{E_{ph} \cdot T}{(V_{in}^2 + V_{en}^2 + V_{iph}^2 + V_{Rm}^2)^{1/2}} \tag{17.17}$$

17.4.1.3 Noise Measurement

To investigate noise contribution, the responses of a bR photoreceptor under the conditions of complete darkness and constant illumination are compared with the open-input noise of the front-end circuit. Figure 17.11 gives time-domain responses that have been recorded under each of the three test cases. The figures show that there are no obvious differences among them. The Matlab quantile–quantile plot is used to determine whether these datasets are normally distributed. As shown in Figure 17.12, the nearly straight line relationship indicates that a normal distribution can be assumed for the three samples. The normal distribution can be parameterized by a mean value and standard deviation. Slight variations are found when comparing the three standard deviation values. The noise recorded by the bR photoreceptor in darkness has a higher standard deviation value than that of the open-input circuit noise. This may be caused by the thermal noise of bR in the darkness. In addition, the open-input circuit configuration may decrease contributions from amplifier voltage noise. As R_m in Eq. (17.14) is equal to infinity for an open input, Z_m increases, resulting in the reduction in amplifier voltage noise in Eq. (17.13). The photoreceptor response under steady illumination demonstrates the highest standard deviation, indicating that photon-induced fluctuation still occur while bR photoreaction is at the steady state. Since the standard deviation is equivalent to the root-mean-square (RMS) value for a signal with a zero mean, the standard deviation can be used to quantify the noise as an RMS voltage. The RMS noise voltages for open-input circuit, no light, and constant light are 3.1, 3.6, and 4.2 mV, respectively.

Signal analysis in the frequency domain provides advantages over those in the time domain. Time-domain signals are likely to be obscured by noises that normally have wide bandwidth. Using the fast Fourier transform (FFT) to obtain a power spectrum is a common method to represent a signal in the frequency domain. The power spectra of noises generated by the open-input circuit configuration and constant illumination configuration are plotted as shown in Figure 17.13. Both plots illustrate that a wave pattern with decreased amplitude superimposes on a nearly flat noise floor. Both shot noise and thermal noise can be well approximated by white noise, implying that the noise power spectrum is constant at all frequencies. The observed patterns are actually caused by the switching noise of the front-end amplifier. The switched integrator greatly reduces input noise by averaging the noise present in the incident light, photoreceptor, and amplifier. However, it introduces switching noise at the output as a consequence of high gain and discrete design. This type of noise is caused by the injection of charges across the parasitic gate-to-source, gate-to-drain, and source-to-drain capacitances of the FET switches (82). It can be eliminated by using correlated double sampling technique.

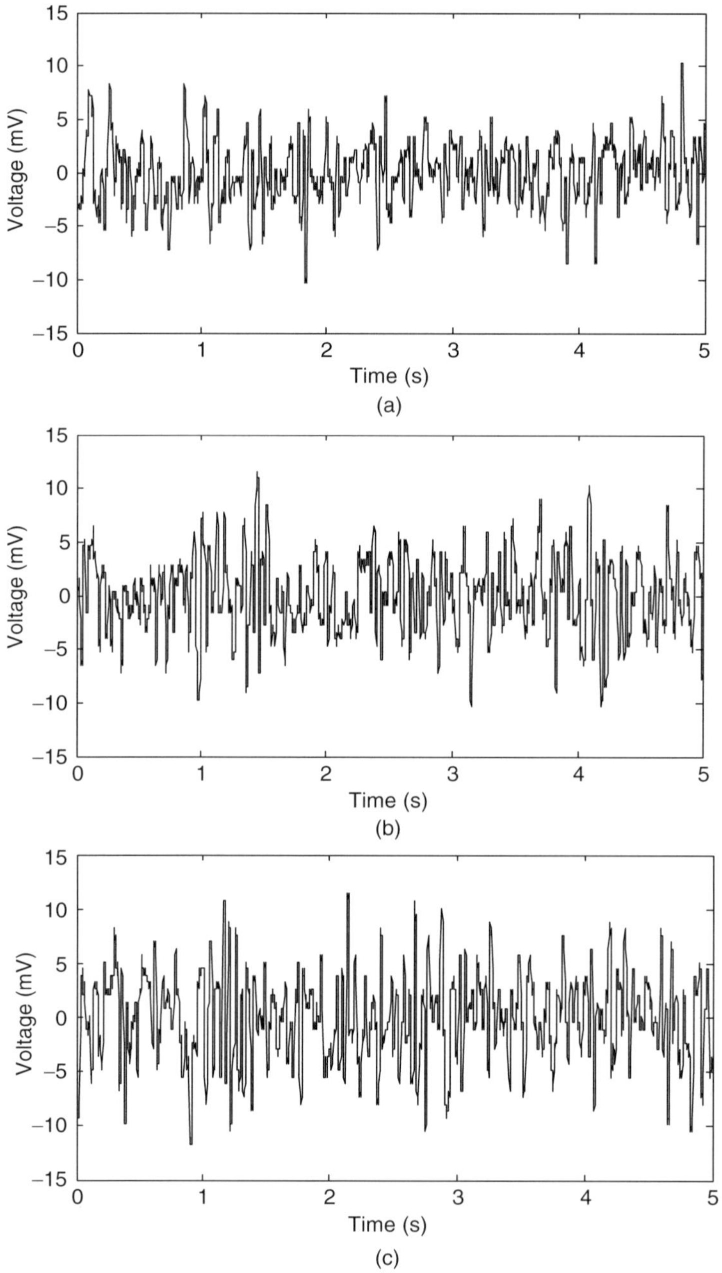

FIGURE 17.11
Typical time domain noises recorded by oscilloscope under the following conditions: (a) open-input front-end circuit (no bR photoreceptor is attached), (b) bR photoreceptor in the dark, and (c) bR photoreceptor under constant illumination of a laser beam with a wavelength of 568 nm and illumination power of 0.75 mW.

To measure the SNR, the input light to the bR photoreceptor is modulated, which allows the bR photoreceptor to respond to light intensity changes. A laser beam, with a wavelength of 568 nm, is used as the light source and is modulated at a frequency of 10 Hz by means of an optical chopper that is placed in the beam path. An oscilloscope is used to record the time-domain signal. The data acquisition rate is 400 Hz, since the integration time of the switched integrator is set to be 20 ms (i.e., 50 Hz), thus preventing the aliasing

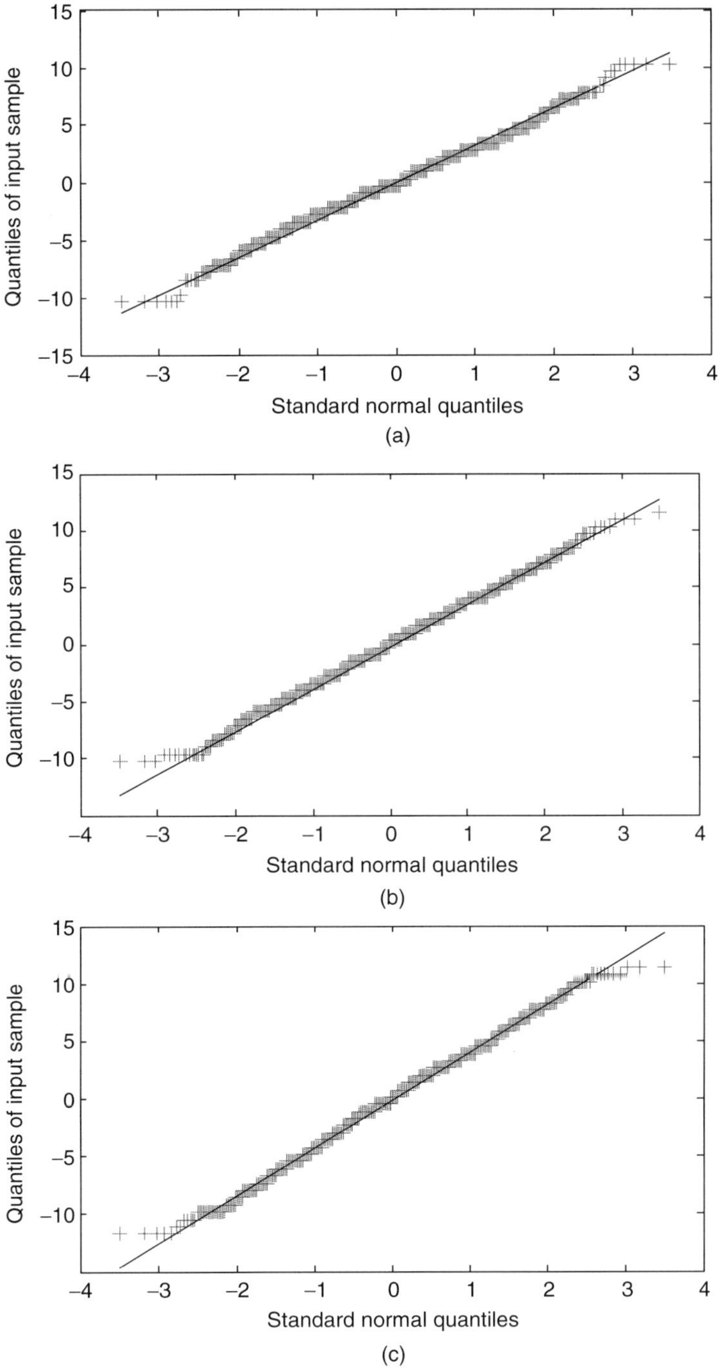

FIGURE 17.12
Normal quantile plots, as calculated by QQplot function in Matlab. (a) Open-input frond-end circuit, (b) bR photoreceptor in darkness, and (c) bR photoreceptor under the constant illumination of a laser beam with a wavelength of 568 nm and illumination power of 0.75 mW.

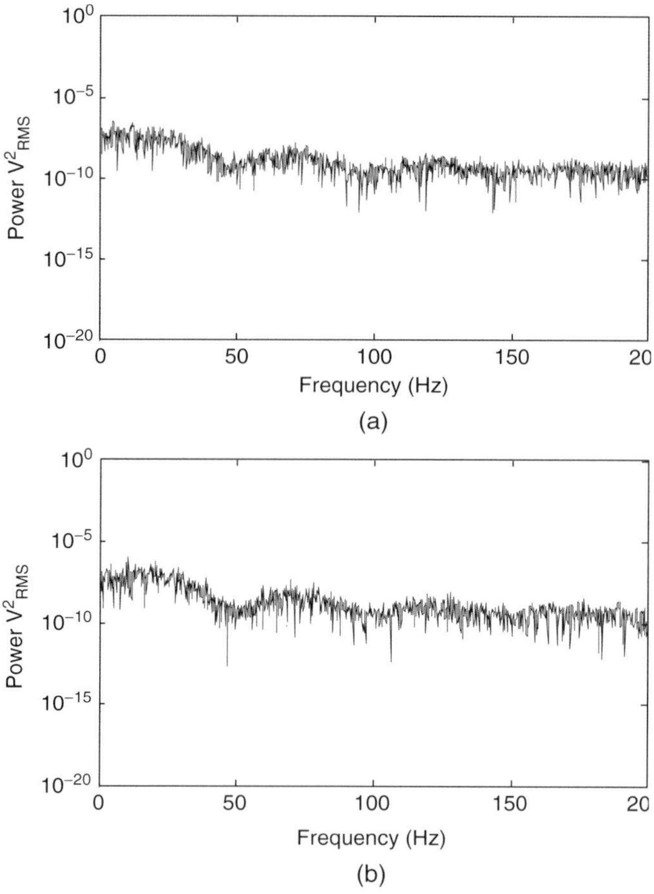

FIGURE 17.13
Power spectra of the output signals corresponding to (a) an open-input frond-end circuit and (b) bR under the constant illumination of a laser beam with a wavelength of 568 nm and illumination power of 0.75 mW.

of frequency above the Nyquist frequency (i.e., 200 Hz). The SNR is typically shown on a logarithmic scale, in the unit of decibels (dB) as defined by

$$\mathrm{SNR} = 20\log_{10}\left(\frac{V_\mathrm{s}}{V_\mathrm{n}}\right)\mathrm{dB} \tag{17.18}$$

where V_s and V_n are the RMS amplitudes of the signal and noise, respectively. The amplitude spectrum in RMS volts is obtained by taking the FFT and plotting it as shown in Figure 17.14. The SNR is approximated as the ratio of the peak height at the light modulation frequency to the noise level in a 200-Hz measurement band. Note that the generated ratio is equal to $(V_\mathrm{s} + V_\mathrm{n})/V_\mathrm{n}$, since the peak at the modulation frequency is offset by the baseline noise. Nonetheless, this gives a close approximation to the true SNR.

As discussed in the previous two sections, the thermal noise is dominant at low photocurrent levels. It no longer plays a significant role in the SNR as the light power level increases; instead, shot noise tends to dominate. Under intense illumination, the noise power is a function of illumination power; therefore, this relation is investigated further by plotting the bR photoreceptor SNR against illumination power. Figure 17.15 shows that the SNR is logarithmically dependent with the incident light power. The highest SNR value, 35.37 dB, is obtained from an illumination power of 22 mW.

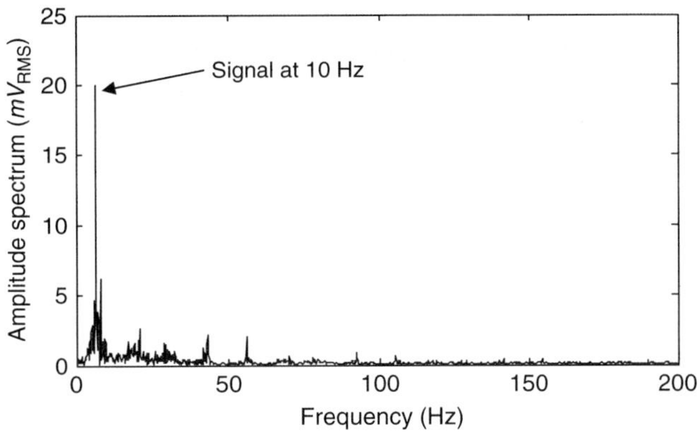

FIGURE 17.14
Amplitude spectrum of the signal generated by the bR photoreceptor in response to a 10-Hz input signal. This amplitude spectrum in volts RMS is calculated by using FFT function in Matlab.

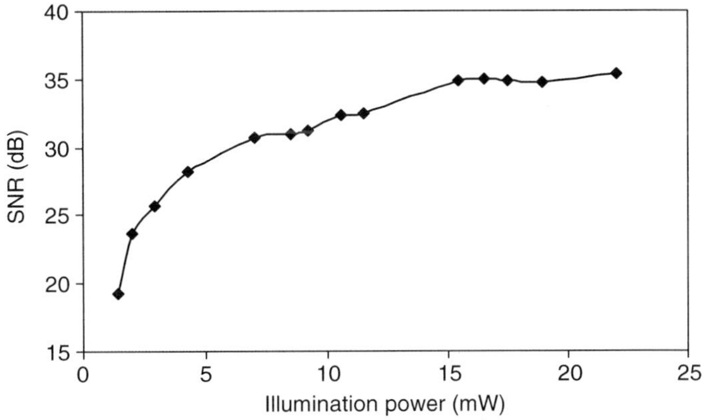

FIGURE 17.15
Signal-to-noise ratio as a function of light power for a single bR photoreceptor. A laser light source operating at a wavelength of 568 nm, with an output power ranging between 1.45 and 22 mW, is used to acquire the data for this experiment.

17.4.2 Linearity and Dynamic Range

Bacteriorhodopsin photoreceptors exhibit differential response when illuminated by a step light signal. Analysis of the equivalent circuit developed in Section 17.3.1 reveals the origin of this differential photosensitivity. The photoreceptor model consists of a serially connected resistor and capacitor, as well as a parallel connected resistor and capacitor. When a switched integrator is used as the front-end amplifier, the membrane capacitance C_m and resistance R_m are effectively shorted out, allowing the bR photoreceptor to function as a high-pass filter. Polarity of the photoinduced current changes at the rising and falling edges of the input-step signal, which corresponds to the charge displacement inside the bR film.

Photodetection systems can be characterized by their linearity over a wide dynamic range so that the data can be reproduced with accuracy. A photodetector is considered linear if the generated photocurrent increases proportionally to the incident light power. The maximum signal level that can be detected by a photodetector is its saturation level. The lowest level of the detectable light is determined by noise in the system.

The peak of the differential response produced by a bR photodetector varies directly with changes in light intensity (provided that saturation does not occur). It has been reported that a bR detector can maintain its linearity even when exposed directly to sunlight (4). This material also responds to very small change of light intensity, demonstrating high sensitivity. This property is investigated experimentally, where a tunable argon/krypton laser system with nine selectable wavelengths ranging from 476 nm (blue) to 676 nm (red) (Melles Griot, 35KAP431-220) is used as the light source. Compared with other sources of light, laser can produce an intense beam of light with an extremely pure wavelength (monochromatic), a fixed phase relationship (coherence), and a very low rate of expansion (highly collimated).

These properties make laser a good tool for high-precision measurements. Linearity of a single bR photoreceptor pixel is tested for three wavelengths (488, 568, and 647 nm, respectively), where the laser beam is adjusted using a beam expander to cover the entire pixel area. The illumination power level is plotted against the pixel photoresponse as shown in Figure 17.16. It shows that the peak signal amplitudes for three wavelengths keep a good linearity over the light power range between 200 μW and 12 mW.

17.4.3 Spectral Response

The magnitude of the photocurrent that is generated by a bR photoreceptor is dependent upon the wavelength of the incident light. bR responds mostly to visible light, where its responsivity peak is located around 570 nm. Wavelength-dependent response is measured using the same tunable argon/krypton laser system, and seven wavelengths covering the whole visible range from blue, green, yellow to red are selected. The output beam power for each wavelength is also tunable over the range between 200 μW and 12 mW. Experimental data are plotted in a 3D, where the photoresponse is a function of both illumination power and wavelength (Figure 17.17). It can be seen that the photoreceptor responds with higher sensitivity to some wavelengths than to others. Therefore, the spectral response is also expressed in a plot of responsivity versus wavelength (Figure 17.18). Compared with the absorption spectrum, Figure 17.5c, the maximum photoelectric response occurs in the vicinity of the optical absorption peak of approximately 568 nm, indicating good agreement between the two experiments.

17.4.4 Response Time

Another beneficial feature of the bR-based photoreceptor is its fast response time. Unlike many bioelectric signals that are generated by slow ionic diffusion process, the fast response of the bR photoreceptor is caused by light-induced charge displacement. Fast signal rise time, on the order of picoseconds, can be obtained using ultrafast laser pulses (83). The overall response time of a bR photoreceptor is also dependent on the amplification

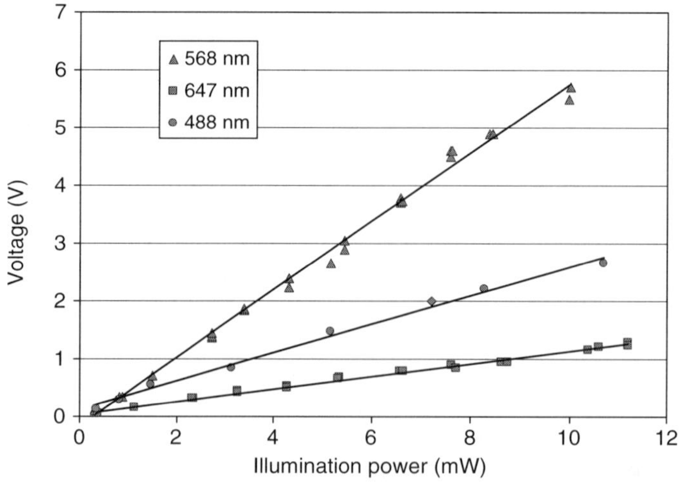

FIGURE 17.16
Linearity of photoelectric signals generated by a bR photoreceptor in response to the laser beam at wavelengths of 568, 647, and 488 nm, respectively. Each data point corresponds to the peak photoreceptor response to a step signal, showing that response is proportional to the input light power.

circuitry. To illustrate such effects, the response of bR photoreceptor to step light signals is tested using durations of 2 s and 33 ms. A fast camera shutter is used to modulate a 568-nm laser beam, and is capable of producing rise times that are less than 2 ms. In Figure 17.19a, the top trace represents the output from a silicon photodetector operating in photovoltaic mode and the bottoms trace is the output of a bR photoreceptor. It can be seen that the bR photoreceptor reaches maximum response in approximately 40 ms. A step light signal with 33-ms time duration is chosen to better observe the response lag produced by the bR photoreceptor. The comparison of the two traces in Figure 17.19b shows that the delayed response is not an intrinsic characteristic of the bR protein, but a limitation of the front-end

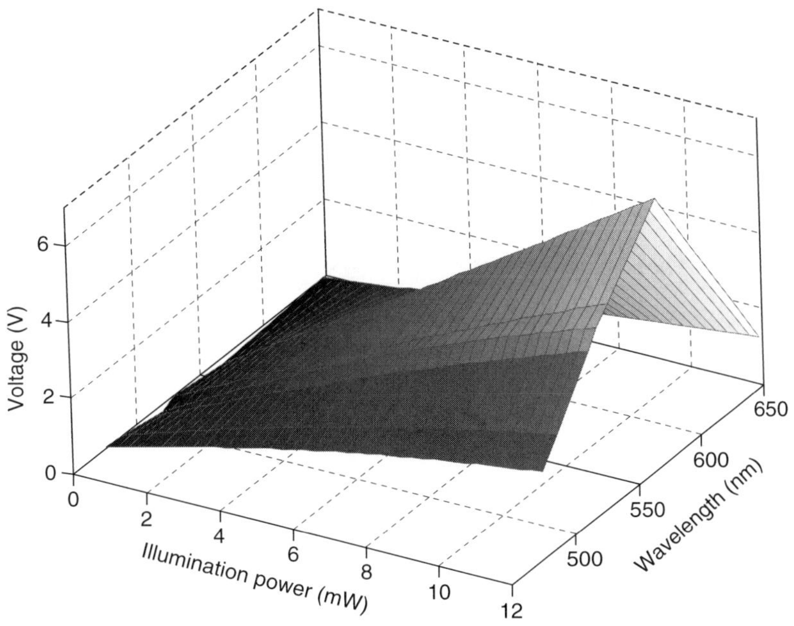

FIGURE 17.17
Measured response of a bR photoreceptor over varying wavelength and light power.

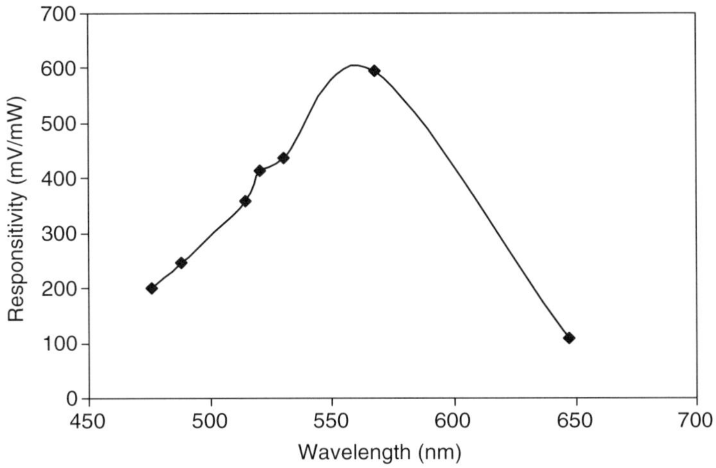

FIGURE 17.18
Responsivity of a bR photoreceptor as a function of incident light wavelength.

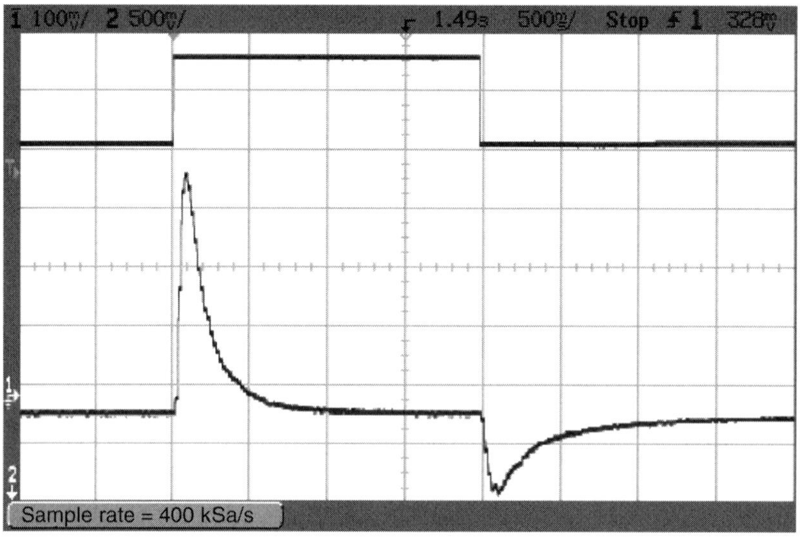

(a)

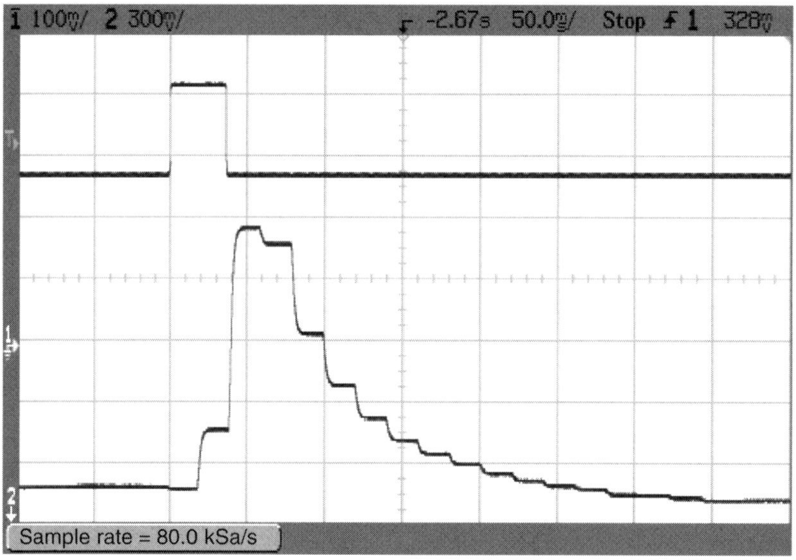

(b)

FIGURE 17.19
The differential photoelectric response of a bR-based photoreceptor to a step light signal having a duration of (a) 2 s and (b) 33 ms.

amplifier. Here, the photoreceptor response time is on the order of 20 ms, but this is caused by the integration time of the switched integrator that is set to 20 ms in the measurement. To reduce the response time, the integration time must be decreased; however, lower sensitivity is the consequence.

17.4.5 Pixel Uniformity

Variations in light sensitivity may occur between individual pixels within the bR photoreceptor array due to variations in local fabrication conditions and lead resistances of

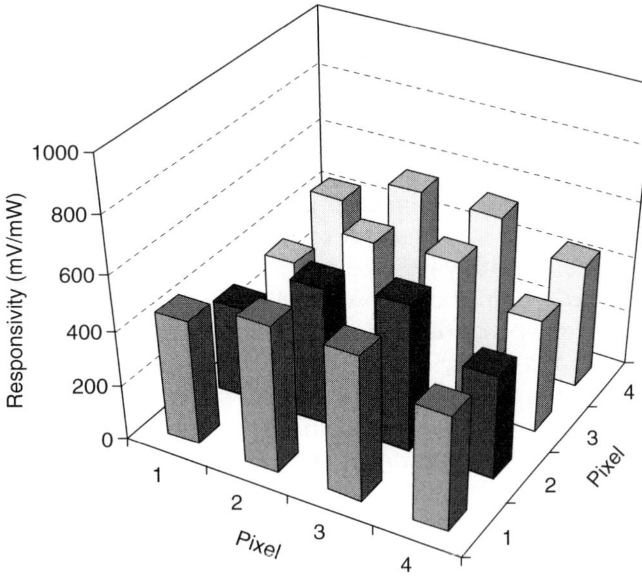

FIGURE 17.20
Relative uniformity of bR photoreceptor responses among all pixels in the 4 × 4 array.

individual pixels. Uniformity is investigated by measuring the photoresponse of each pixel under identical illumination conditions. Figure 17.20 shows the pixel-to-pixel uniformity of responsivity within the bR photoreceptor array that is illuminated by a laser beam with a single wavelength of 568 nm and power of 2.5 mW. It is found that uniformity of photoresponsivity between pixels across the array is more than 71% of the average value, 465.25 mV/mW.

17.4.6 Array Performance Under Mechanical Bending

A bR photoreceptor array with a plastic substrate provides the advantage of mechanical flexibility. However, bending may give rise to some additional problems. Deposition of a thick bR film using EPS may produce cracks in its surface under excessive bending conditions. Delamination between electrodes and loss of contact with the bR film are also possible. A simple experiment is conducted to demonstrate the effect of mechanical bending on the signal. The photoreceptor array is wrapped around rigid tubes of different diameters. It is found that the proposed array keeps good photoresponse characteristics without any cracks and exfoliation, even at the bend radius of 10 mm.

17.5 Application—Motion Detection

17.5.1 Overview of Motion Detection

Motion detection plays an important role in many applications, including autonomous robot navigation and remote sensing. Conventional hardware implementation employs progressive-scanned CCD or CMOS sensors for image capture and standard computers to process the image sequences. It is often difficult or costly to achieve real-time performance using such a serial architecture because it involves transmitting and processing large quantities of raw data. Parallel processing techniques that combine image sensing and motion computation into a single stage are becoming an increasingly attractive alternative. A variety of motion sensors based on analog very-large-scale-integration (AVLSI) technology have been developed for motion detection purposes.

Most reported AVLSI motion sensors incorporate silicon photoreceptors and parallel processing circuits on a single chip (84; 85). Some problems that have been encountered include limited dynamic range and low spatial resolution. Simple photoreceptors in biological vision systems can respond to stimuli over extremely wide ranges. In addition, their visual pathways are highly parallel-connected, allowing simple processing elements situated within the pathways. It is difficult for silicon-based AVLSI systems to achieve such high spatial resolution within a size comparable to a biological system. Furthermore, the use of a planar imaging surface and perspective projection optics often leads to undesirable image distortion or bulky vision systems. Instead, using curved or spherically shaped retinas can provide uniform sampling across the entire field of view. For these reasons, biological materials have received much attention from artificial vision researchers because they can perform many complex functions at the molecular level.

Experimental results reveal that the photoelectric signals generated by bR photoreceptors exhibit differential photosensitivity. Both wavelength and intensity of incident light influence the signal amplitude. The photoresponse peak is linearly proportional to the illumination intensity over a wide range. These unique photoelectric properties make bR a viable material for real-time image-processing applications. Most proposed bR-based devices that exploit these photoelectric properties are simply light detectors or image sensors that have not been efficiently incorporated into vision systems (49–51). Image data acquired from biological vision systems is highly compressed to eliminate unnecessary information, thereby improving transmission and processing speed. In this section, a motion detection sensor that uses a bR photoreceptor array is described as a sample application. It is capable of detecting both speed and direction of a moving light source.

17.5.2 Motion Detection Algorithm

Algorithms for motion detection can be divided into two major categories: intensity-based algorithms and feature-based algorithms (51). Intensity-based methods use image irradiance directly to estimate optical flow. The two most popular types of intensity-based motion algorithms are those based on gradients or on correlation. Feature-based methods are commonly used in computer vision to extract unique features from an image, such as edges, corners, or patterns. Motion is estimated by tracking these features over time. Most motion detection algorithms tend to be computationally expensive and difficult to realize in hardware.In contrast, motion sensors based on biologically inspired vision models provide a simpler architecture and are more suitable for hardware implementations.

The Reichardt motion model is a model to describe an insect's visual response; it is one of the most common motion detection models implemented in hardware (86). As shown in Figure 17.20, this model employs a bilocal delayed-coincidence detector that compares light intensity at two locations separated by a time delay. It can resolve both direction and speed of the observed motion. The signal from one photoreceptor is correlated with the delayed signal from its adjacent photoreceptor. Spatiotemporal filters extract spatial and temporal edges from the input signals, while removing the DC components. When the delayed and nondelayed signals arrive simultaneously at the multiplier, the resulting signal indicates a particular direction of motion. Subtraction between the two half-detector signals results in an output, which is positive for rightward movement, negative for leftward movement, and zero for no motion or flickering stimuli. The delayed time between adjacent photoreceptors contains the movement velocity information.

17.5.3 Implementation

The delay-and-correlate algorithm is an improvement upon the Reichardt model and it is adopted in the proposed motion detector design. The schematic block diagram including two bR-pixels is shown in Figure 17.21. It uses temporal edges as the image features and correlates them by using binary pulses (87). This binary pulse correlation significantly increases motion detection robustness. Implementation can be divided into three separate stages: feature extraction, pulse shaping, and direction/speed detection. To simplify prototype development the bR photoreceptor array and its front-end circuitry are combined with a microcontroller to realize the above three processing stages. The later two stages are simulated by the microcontroller.

Spatiotemporal features of a moving object are typically extracted by applying edge detectors to the input image. The differential responsivity of a bR photoreceptor allows a simpler design to be realized. Use of unreliable temporal differentiation circuitry can be avoided by combining each bR pixel with a switched integrator. The former transduces light into electric signals and the latter transforms these signals into current pulses containing edge information. Current pulses that are generated are then converted into voltage pulse using voltage comparators. An appropriate output trigger level is set according to the minimum expected light power. Consequently, this threshold value determines the contrast sensitivity of the motion detector. Generally, lower thresholds yield higher contrast sensitivity. When the bR pixel generates a sufficiently high signal, the comparator will generate a corresponding pulse that is compatible with 5-V TTL circuitry.

The time delay within the motion detection algorithm is usually realized by pulse-shaping circuitry. A delayed pulse is generated at the rising edge that is incoming from the previous stage. The duration of the delay is set long enough to allow the object to travel between two neighboring pixels. A 4 MHz PIC16F684 microcontroller is used to simulate the pulse shaping and direction/speed detection stages for four bR pixels. Four digital input pins are used to sample the TTL signals from four pixel channels. Reading digital

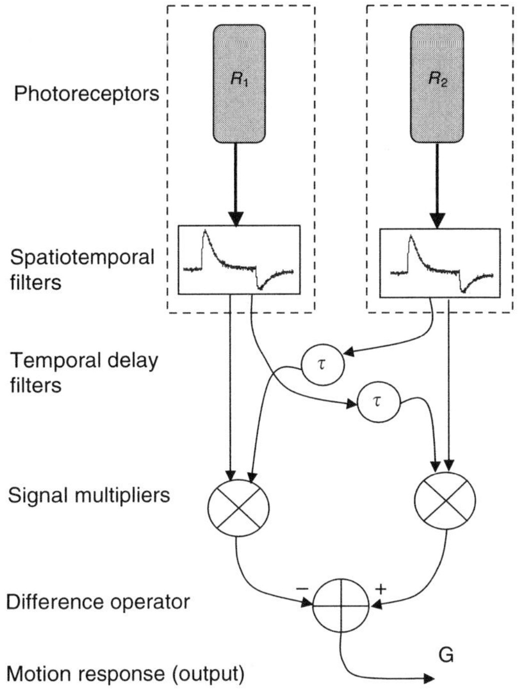

FIGURE 17.21
Block diagram of Reichardt's correlation model for motion detection. The dashed boxes indicate the functions that are accomplished by bR photoreceptors.

signals directly simplifies programming and eliminates the need for time-consuming analog to digital conversion. As shown in Figure 17.22, the microcontroller reads four input channels in parallel. Based on the current implementation, the four input pins are sampled at a rate of 500 HZ. Each new input triggers an internal counter on that channel, emulating the pulse delay. The microcontroller continues to scan input channels for additional digital signals on other input channels. Direction and speed can be determined between two neighboring pixels whenever their delays are overlapping in time. If no other digital signals are detected within a pixels' delay period, the time delay simply terminates until a new signal arrives.

Motion is determined by correlating the delay digital input from one pixel with the appearance of a digital signal from a neighboring pixel. Pixel positions (input channels) are used to determine the motion direction, and a binary local direction signal is set for each pixel pair. Speed is extracted by measuring the time interval between two pixels. The outputs of every two neighboring pixels are integrated to produce a globe motion measure, which is proportional to the average motion speed. Depending on the intended application, the global motion information can be transmitted to the external system in various formats. Figure 17.22 shows a DC motor connected to the microcontroller through a PWM driver. Two output channels send direction and speed information to the DC motor.

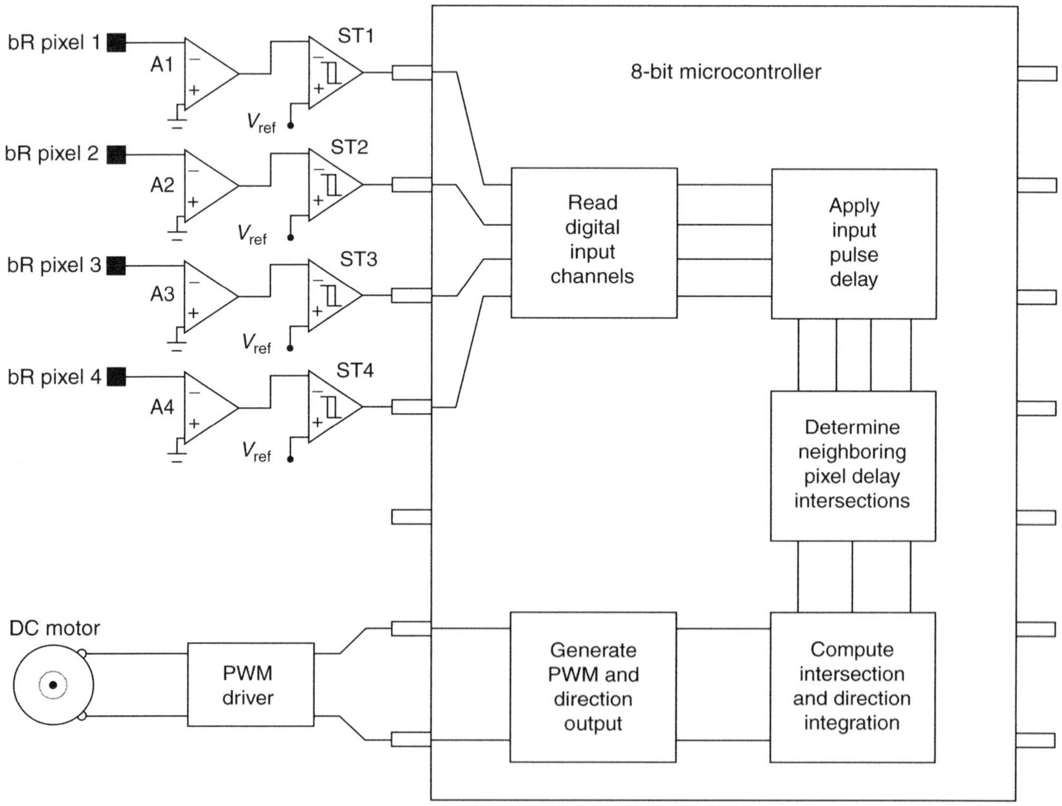

FIGURE 17.22
Architecture block diagram illustrating the implementation of bR photoreceptor motion detector. Several functions are simulated within a PIC microcontroller to simplify prototyping and experimentation.

17.5.4 Experimental Setup and Results

The photoreceptor array is tested experimentally by detecting a moving light spot generated by a 2D laser scanning system, as illustrated in Figure 17.23. A tunable argon/krypton laser system is used as the light source, and a two-axis acousto-optic deflector (AA.DTS.XY-250) scans the laser beam across the array surface. A two-channel function generator is connected to the deflector driver and controls the deflector's RF frequency. The beam deflection angle is adjusted by varying the amplitude of the signal being fed into the deflector driver, whereas the signal frequency controls the sweeping speed directly. The photoreceptor array and detection circuitry are located inside a shielded enclosure with an opening to expose the array. Output data are observed with a four-channel oscilloscope.

Figure 17.24 shows signals that correspond to the different motion detection stages as a laser beam scans a pair of bR pixels from left to right. The top two traces indicate the output signals produced by the left and right bR photoreceptor amplifiers. Traces three and four are produced by two voltage comparators. Each comparator is set to trigger at a 0.5 V threshold and produces a 5 V binary signal as its output. These are then read by the microcontroller's digital input pins. The input signals are then delayed internally for a period of 200 ms, as indicated by traces five and six. The overlap between the two delayed pulses produces a region that is proportional to the light spot velocity. This region is integrated to produce a ramp-like profile over time, as indicated by the final trace. However, the integrated value must be scaled to an appropriate PWM duty cycle to drive an external motor. Thus, the effective power delivered to the motor will be proportional to the observed light spot velocity. Direction of motor rotation is determined by the sign of the integrated value.

17.6 Conclusions

17.6.1 Overview

This chapter presents a novel photoreceptor array that is fabricated by immobilizing bR onto a flexible plastic substrate with a patterned conductive coating. As a retinal protein, bR shows great potential in many optical and photoelectrical applications due to its high sensitivity, large dynamic range, fast response time, small spatial size, and long-term stability. Extending electronics to flexible substrates introduces a new approach that has promising future in lightweight and durable sensing devices with curved or free-form geometries. The possibility of combining bioelectronics and flexible electronics is investigated in this work; the ultimate goal is to develop an artificial retina that can mimic functions and geometries inherent in biological vision systems.

In this work, PM patches, obtained from wild-type bR, are deposited onto a PET substrate coated with a patterned ITO layer using the EPS technique. A flexible 4×4 pixel array is carefully assembled and packaged. Positive spectral absorption results and the stable photoelectric response generated from each individual pixel prove EPS to be a viable fabrication technique.

Designing suitable signal processing circuitry for the proposed array is critical to its success. An equivalent circuit model for a bR photoreceptor is presented. Modeling the dried-bR film in this study is a vital step in integrating it with signal processing circuitry. The underlying physical mechanisms of photoelectric transduction are studied to ensure

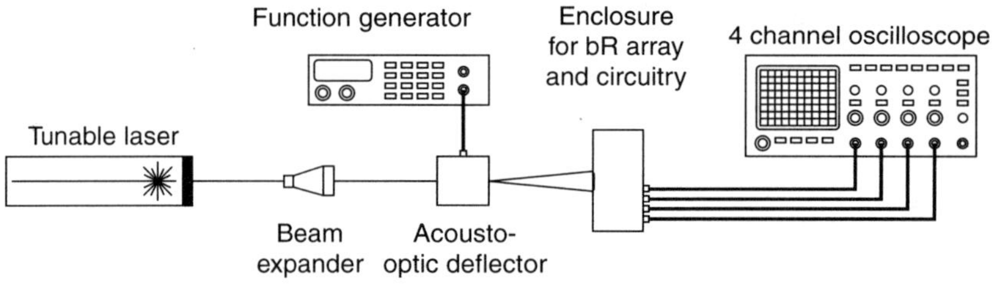

FIGURE 17.23
Experimental setup for motion detection measurement. An acousto-optic deflector is used to sweep the laser beam across the bR array so that the light spot velocity can be estimated by the motion detection circuitry.

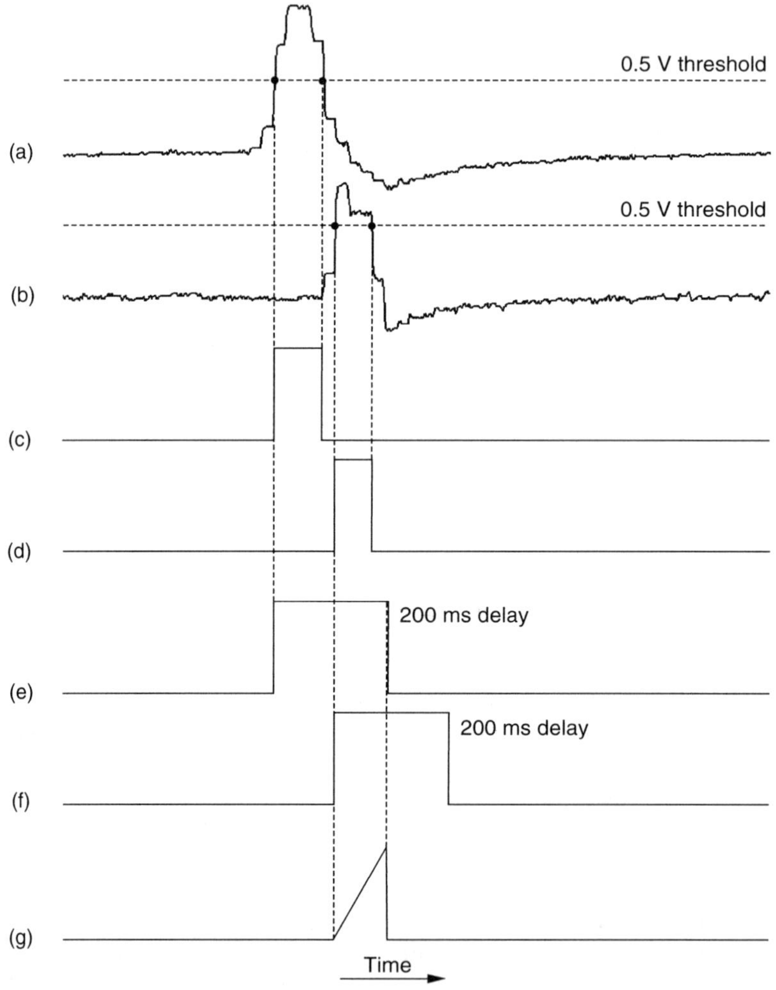

FIGURE 17.24
Waveforms corresponding to the different motion detection stages for two neighboring pixels as a light spot moves from left to right: (a, b) are the measured amplifier outputs for each pixel; (c, d) are the 5-V output signals generated by the voltage comparators, which are triggered at a 0.5-V threshold. (e, f) Represent the sample delay that is applied to the input signals. The delay overlap between the two channels is integrated to produce the output (g).

model correctness. The switched integrator design is selected as the most appropriate front-end circuitry design over two other common designs. The integrating behavior of the chosen design significantly reduces output noise by averaging the input noise generated by the light source, photoreceptor, and amplifier. Furthermore, the switched integrator design provides adjustable gain-bandwidth and low power consumption, making it a practical solution for creating highly integrated bR photoreceptor array.

The photoelectric response of the proposed bR photoreceptor array is characterized experimentally to provide a means of performance evaluation. The following characteristics are addressed: noise, linearity and dynamic range, spectral response, response time, pixel uniformity, and responsivity under mechanical bending. The SNR is logarithmically dependent with the light power. The highest SNR value, 35.37 dB, is obtained from an illumination power of 22 mW at the wavelength of 568 nm. The photoelectric response of an individual pixel is approximately linear over the light power range of 200 μW–12 mW. bR photoreceptors respond mostly to visible light, where the spectral response peak is located at 568 nm. Depending on the selected integrating time, signal response times within the millisecond range can be achieved. The overall array is also evaluated in terms of response uniformity of each pixel and responsivity under mechanical bending. Uniformity of the photoresponse among all pixels is greater than 71% of the average value 465.25. Good photoresponse characteristics, without any cracking or delamination, are achieved even when the bending radius is 10 mm.

Inherent image processing capabilities of the proposed bR array are demonstrated in a motion detection application. Reichardt's delay-and-correlate algorithm is implemented in hardware to detect both the speed and direction of a moving light spot. Binary pulse correlation identifies direction of motion, where the pulse overlapping area determines the speed. Such a motion detection scheme can be exploited by real-time machine vision systems. Differential photosensitivity inherent in bR films provides the principal mechanism for motion detection. This unique property simplifies the signal processing stage and effectively reduces the need for differential circuitry used by conventional spatiotemporal detection systems.

17.6.2 Limitations and Recommendations

From a system design perspective, protein-based photoreceptor arrays on the flexible substrates have great advantages over their conventional silicon-based counterparts. High-resolution and high-speed vision is possible, although many challenges and limitations must be overcome. Several limitations have to be addressed before bR photoreceptor array can become commercially viable. The reminder of this section discusses a number of recommendations.

The primary challenge in developing a bR photoreceptor array is to develop effective fabrication and packaging techniques. In this work, PM patches are immobilized onto a flexible substrate by EPS. This process gives rise to some unavoidable problems, such as film fragility and nonuniform thickness. Thus, it is difficult to control the interface quality at the film–electrode boundary. LB deposition is a method widely used to form highly ordered and uniform thin films of organic molecules. However, bR film created in this manner is also fragile under bending conditions. Encapsulating oriented PM patches within a polymer gel and covering the bR film with an aqueous electrolyte gel provide two viable alternatives to the above methods. Nonetheless, these two methods have critical requirements for packaging as simple shelter cannot maintain device lifetime. Moreover, additional materials may influence the optical and electrical properties of the hybrid bR devices.

Dried-bR photoreceptors are inherently susceptible to electrical noise because of their extremely high film resistance. Moreover, photoinduced currents flowing through the bR photoreceptor are very small. Measuring such small signals requires circuitry that is highly insensitive to noise. Fabricating switched integrator circuitry on a standard PCB board using discrete surface-mount components is impractical for precision low-noise applications. The most significant problem is switching noise, which is caused by charge injection across the parasitic capacitances of FET switches. To reduce the overall noise, it is recommended that the switched integrator with correlated double sampling circuitry be combined into a single-chip design.

The 4 × 4 photoreceptor prototype is provided as a proof of concept. Designing it in this manner imposes some limitations, such as limited spatial resolution and excessive physical size. Creating high-resolution motion detectors requires reducing the pixel scale and improving the readout architecture. Unfortunately, as photosensitive areas decrease in size and complex processing circuitry consume larger areas, the limitations of 2D architectures become evermore apparent. Extending the array architectures to 3D provides a promising solution to this conflict. By arranging the photoreceptor, amplifier, and motion circuitry in layers, maximum utilization of sensor area can be achieved. Continuing innovation in thin-film transistor technology may provide a means to realize practical and flexible bR photoreceptor imaging devices.

Acknowledgments

This work has been supported by research grants to Professors A.S. Bassi and G.K. Knopf from the Natural Sciences and Engineering Research Council of Canada (NSERC), the Canada Foundation for Innovation (CFI), and Photonics Research Ontario (PRO). The authors would like to thank Nicholas Jankovic, Qing Zhang, Gerrit Aartsen, and Eugen Porter for their constant assistance and valuable discussions.

References

1. Bone, S., Zaba, B. (1992). *Bioelectronics.* Chichester: John Wiley & Sons Ltd.
2. Oesterhelt, D., Stoeckenius, W. (1971). Rhodopsin-like protein from the purple membrane of *Halbacterium. Nature* 233(39): 149–152.
3. Blaurock, A. E., Stoeckenius, W. (1971). Structure of the purple membrane. *Nat. New. Biol.* 233: 152–154.
4. Hampp, N. (2000). Bacteriorhodopsin as a photochromic retinal protein for optical memories. *Chem. Rev.* 100(5): 1755–1776.
5. Shen, Y., Safinya, C. R., Liang, K., Ruppert, A., Rothschild, K. (1993). Stabilization of the membrane-protein bacteriorhodopsin to 140°C in 2-dimensional films. *Nature* 366(39): 48–50.
6. Henderson, R., Baldwin, J. M., Ceska, T. A., Zemlin, F., Beckmann, E., Downing, K. H. (1990). Model for the structure of bacteriorhodopsin based on high-resolution electron cryomicroscopy. *J. Mol. Biol.* 213: 899–929.
7. http://www.ks.uiuc.edu/Research/newbr/br_fig.html (accessed November 10th, 2005).
8. Lemke, H.-D., Oesterhelt, D. (1981). Lysine 216 is a binding site of the retinyl moiety in bacteriorhodopsin. *FEBS Lett.* 128(2): 255–260.
9. Lanyi, J. K. (1998). Understanding structure and function in the light-driven proton pump bacteriorhodopsin. *J. Struct. Biol.* 124: 164–178.

10. Maeda, A., Iwasa, T., Yoshizawa, T. (1977). Isomeric composition of retinal chromophore in dark-adapted bacteriorhodopsin. *J. Biochem. (Tokyo)*. 82(6): 1599–1604.
11. Lanyi, J. K. (1993). Proton translocation mechanism and energetics in the light-driven pump bacteriorhodopsin. *Biochim. Biophys. Acta*. 1183: 241–261.
12. Dancshazy, Z., Groma, G. I., Oesterhelt, D., Tittor, J. (1986). The photochemical cycle of bacteriorhodopsin has no refractory period. *FEBS Lett.* 196(2): 198–202.
13. Váró, G. (1981). Dried oriented purple membrane samples. *Acta. Biol. Acad. Sci. Hung*. 32(3–4): 301–310.
14. Pepe, I. M., Nicolini, C. (1996). Invited review: Langmuir–Blodgett films of photosensitive proteins. *J. Photochem. Photobiol. B: Biol*. 33: 191–200.
15. Choi, H.-C., Min, J., Lee, W. H., Choi, J.-W. (2002). Adsorption behavior and photoelectric response characteristics of bacteriorhodopsin thin films fabricated by self-assembly technique. *Colloid. Surf. B: Biointerf.* 23(4): 327–337.
16. He, J.-A., Samuelson, L., Li, L., Kumar, J., Tripathy, S. K. (1998). Oriented bacteriorhodopsin/polycation multilayers by electrostatic layer-by-layer assembly. *Langmuir* 14: 1674–1679.
17. Koychi, K., Yamaguchi, N., Miyasaka, T. (1994). Antibody-mediated bacteriorhodopsin orientation for molecular devices architectures. *Science* 265: 762–765.
18. Dér, A., Hargittai, P., Simon, J. (1985). Time-resolved photoelectric and absorption signals from oriented purple membranes immobilized in gel. *J. Biochem. Biophys. Methods*. 10(5–6): 259–300.
19. Luecke, H., Richter, H.-T., Lanyi, J. K. (1998). Proton transfer pathways in bacteriorhodopsin at 2.3 angstrom resolution. *Science* 280: 1934–1937.
20. Edman, K., Nollert, P., Royant, A., Belrhall, H., Pebay-Peyoula, E., Hajdu, J., Neutze, R., Landau, E. M. (1999). High-resolution X-ray structure of an early intermediate in the bacteriorhodopsin photocycle. *Nature* 401: 822–826.
21. Ludmann, K., Gergely, C., Váró, G. (1992). Kinetic and thermodynamic study of the bacteriorhodopsin photocycle over a wide pH range. *Biophys. J.* 75: 3110–3119.
22. Ganea, C., Gergely, C., Ludmann, K., Váró, G. (1997). The role of water in the extracellular half channel of bacteriorhodopsin. *Biophys. J.* 73: 2718–2725.
23. Sharmar, P., Roy, S. (2004). All-optical biomolecular parallel logic gates with bacteriorhodopsin. *IEEE Trans. Nanobiosci*. 3(2): 129–136.
24. Renner, T. Hampp, N. (1993). Bacteriorhodopsin-films for dynamic time average interferometry. *Opt. Commun.* 96: 142–149.
25. Cullin, D. W., Vsevolodov, N. N., Dyukova, T. V. (1995). Holographic properties of triton X-100-treated bacteriorhodopsin embedded in gelatin films. *BioSystems*. 35(2–3): 141–144.
26. Birge, R. R., Gillespie, N. B., Izaguirre, E. W., Kusnetzow, A., Lawrence, A. F., Singh, D., Song, Q. W., Schmidt, E., Stuart, J. A., Seetharaman, S., Wise, K. J. (1999). Biomolecular electronics: Protein-based associative processors and volumetric memories. *J. Phys. Chem. B*. 103: 10746–10766.
27. Bräuchle, C., Hampp, N., Oesterhelt, D. (1991). Optical applications of bacteriorhodopsin and its mutated variants. *Adv. Mater*. 3(9): 420–428.
28. Wang, W. W., Knopf, G. K., Bassi, A. S. (2005). Photoelectric properties of a detector based on dried bacteriorhodopsin film. *Biosens. Bioelectr*. 21: 1309–1319.
29. Saga, Y., Watanabe, T., Koyama, K., Miyasaka, T. (1999). Mechanism of photocurrent generation from bacteriorhodopsin on gold electrodes. *J. Phys. Chem. B*. 103: 234–238.
30. Xu, J., Bhattacharya, P., Váró, G. (2004). Monolithically integrated bacteriorhodopsin/semiconductor opto-electronic integrated circuit for a bio-photoreceiver. *Biosens. Bioelectr.* 19: 885–892.
31. Horn, C., Steinem, C. (2005). Photocurrents generated by bacteriorhodopsin adsorbed on nano-black lipid membranes. *Biophys. J.* 89: 1046–1054.
32. Miyasaka, T., Koyama, K., Itoh, I. (1992). Quantum conversion and image detection by a bacteriorhodopsin-based artificial photoreceptor. *Science* 255: 342–344.
33. Haronian, D., Lewis, A. (1992). Microfabricating bacteriohodopsin films for imaging and computing. *Appl. Phys. Lett.* 61(18): 2237–2239.

34. Martin, C. H., Chen, Z. P., Birge, R. R. (1997). Towards a bacteriorhodopsin-silicon neuromorphic photosensor. *Proc. Pacific Symp. Biocomput*. 268–279.
35. Libertino, S., Fichera, M., Arrigo, G. D., Mantia, A. La, Ricceri, D. (2003). Characterization and patterning of bacteriorhodopsin films on Si-based materials. *Syn. Metals*. 138: 71–74.
36. Takei, H., Lewis, A., Chen, G. P., Nebenzahl, I. (1991). Implementing receptive fields with excitatory and inhibitory optoelectrical responses of bacteriorhodopsin films. *Appl. Opt*. 30(4): 500–509.
37. Yang, J., Wang. G. (1998) Image edge detecting by using the bacteriorhodopsin-based artificial ganglion cell receptive field. *Thin Solid Films* 324: 281–284.
38. Haronian, D., Lewis, A. (1991). Elements of unique bacteriohodopsin neural network architecture. *Appl. Opt*. 30(5): 597–608.
39. Miyasaka, T., Koyama, K. (1993). Image sensing and processing by a bacteriorhodopsin-based artificial photoreceptor. *Appl. Opt*. 32(31): 6371–6379.
40. Fukuzawa, K. (1994). Motion-sensitive position sensor using bacteriorhodopsin. *Appl. Opt*. 33(31): 7489–7495.
41. Min, J., Choi, H.-G., Oh, B.-K., Lee, W. H., Paek, S.-H., Choi, J.-W. (2001). Visual information processing using bacteriorhodopsin-based complex LB films. *Biosens. Bioelectr*. 16: 917–923.
42. Frydrych, M., Silfsten, P., Parkkinen, S., Parkkinen, J., Jaaskelainen, T. (2000). Color sensitive retina based on bacteriorhodopsin. *BioSystems* 54(3): 131–140.
43. Choi, H.-G., Jung, W.-C., Min, J., Lee, W. H., Choi, J.-W. (2001). Color image detection by biomolecular photoreceptor using bacteriorhodopsin-based complex LB films. *Biosens. Bioelectr*. 16: 925–935.
44. Reuss, R. H., Chalamala, B. R., Moussessian, A., Kane, M., Kumar, A., Zhang, D.C., Rogers, J.A., Hatalis, M., Temple, D., Moddel, G., Eliasson, B.J., Estes, M.J., Kunze, J., Handy, E.S., Harmon, E.S., Salzman, D.B., Woodall, J.M., Alam, M. A., Murthy, J.Y., Jacobsen, S.C., Olivier, M., Markus, D., Campbell, P., Snow, E. (2005). Macroelectronics: Perspective on technology and applications. *Proc. IEEE*. 93(7): 1239–1256.
45. Xu, J. M. (2000). Plastic electronics and future trends in microelectronics. *Synth. Met*. 115: 1–3.
46. Qing, X., Kumar, A., Zhang, C., Gonzalez, I. F., Guo, G., Chang, F. (2005). A hybrid piezoelectric/fiber optic diagnostic system for structural health monitoring. *Smart Mater. Struct*. 14: S98–S103.
47. Dimitrakopoulos, C. D., Mascaro, D. J. (2001). Organic thin-film transistors: A review of recent advances. *IBM J. Res. Dev*. 45(1): 11–27.
48. Lumelsky, V., Shur, M., Wagner, S. (2001). Sensitive skin. *IEEE Sens. J*. 1(1): 41–51.
49. Harrison, R., Koch, C. (1998). An analog VLSI model of the fly elementary motion detector. In: Jordan, M. I., Kearns, M. J., Solla, S. A., Eds. *Advances in Neural Information Processing Systems*. Cambridge, MA: MIT Press, Vol. 10. pp. 880–886.
50. Liu, S. C. (2000). A neuromorphic aVLSI model of global motion processing in the fly. *IEEE Trans. Circuits Syst. II: Analog Digital Signal Process*. 47(12): 1458–1467.
51. Moini, A. (2000). *Vision Chips*. Norwell, MA: Kluwer Academic Publishers.
52. Baylor, D. A., Lamb, T. D., Yau, K. W. (1979). Response of retinal rods to single photons. *J. Physiol*. 288: 613–634.
53. Barlow, H. B., Mollon, J. D. (1982). *The Senses*. Cambridge: Cambridge University Press.
54. Oesterhelt, D., Stoeckenius, W. (1974). Isolation of the cell membrane of *Halobacterium halobium* and its fractionation into red and purple membrane. *Methods. Enzymol*. 31: 667–678.
55. Ludmann, K., Gergely, C., Dér, A., Váró, G. (1998). Electric signals during the bacteriorhodopsin photocycle, determined over a wide pH range. *Biophys. J*. 75(6): 3120–3126.
56. Liu, S. Y., Ebrey, T. G. (1987). The quantum efficiency for the interphotoconversion of the blue and pink forms of purple membrane. *Photochem. Photobiol*. 46: 263–267.
57. Logothetidis, S. (2005). Polymeric substrates and encapsulation for flexible electronics: Bonding structure, surface modification and functional nanolayer growth. *Rev. Adv. Mater. Sci*. 10: 387–397.
58. Sessier, G. M., Hahn, B., Yoon, D. Y. (1986). Electrical conduction in polyimide films. *J. Appl. Phys*. 60(1): 318–326.
59. Ginley, D. S., Bright, C. (2000). Transparent conducting oxides. *MRS. Bull*. 25(8): 15–18.
60. Gordon, R. G. (2000). Criteria for choosing transparent conductors. *MRS. Bull*. 25(8): 52–57.

61. Deshmukh, S. C., Aydil, E. S. (1994). Low-temperature plasma enhanced chemical vapor deposition of SiO_2. *Appl. Phys. Lett.* 65: 3185–3187.
62. Lewis, B. G., Paine, D. C. (2000). Applications and processing of transparent conducting oxides. *MRS. Bull.* 25(8): 22–27.
63. Ginley, D., Roy, B., Couutts, T., Readey, D., Hosono, H., Perkins, J. (2003). Non-vacuum and PLD growth of next generation TCO materials. *Thin Solid Films* 445: 193–198.
64. Alexiev, U., Marti, T., Heyn, M. P., Khorana, H. G., Scherrer, P. (1994). Surface charge of bacteriorhodopsin detected with covalently bound pH-indicators at selected extracellular and cytoplasmic sites. *Biochemistry* 33: 298–306.
65. Min, J., Choi, H.-G., Choi, J.-W., Lee, W. H. (1998). Optimal fabrication condition of bacteriorhodopsin films by electrophoretic sedimentation technique. *Supramolec. Sci.* 5: 687–690.
66. Kononenko, A. A., Lukashev, E. P., Chamorovsky, S. K., Maximychev, A. V., Timashev, S. F., Chekulaeva, L. N., Rubin, A. B., Paschenko, V. Z. (1986). Orientated purple-membrane films as a probe for studies of the mechanism of bacteriorhodopsin functioning. I. The vectorial character of the external electric-field effect on the dark state and the photocycle of bacteriorhodopsin. *Biochim. Biophys. Acta.* 892: 162–169.
67. Alexiev, U., Scherrer, P., Marti, T., Khorana, H. G., Heyn, M. P. (1995). Time-resolved surface charge change on the cytoplasmic side of bacteriorhodopsin. *FEBS Lett.* 373: 81–84.
68. Schenkl, S., Mourik, F. V., Zwan, G. V. D., Haacke, S., Chergui, M. (2005). Probing the ultrafast charge translocation of photoexcited retinal in bacteriorhodopsin. *Science* 309(5): 917–920.
69. Grimnes, S., Martinsen, Ø. G. (2000). *Bioimpedance & Bioelectricity Basics.* London: Academic Press Inc. (London) Ltd.
70. Donati, S. (2000). *Photodetectors: Devices, Circuits, and Applications.* Upper Saddle River, NJ: Prentice Hall PRT.
71. Fossum, E. R. (1997). CMOS image sensors: Electronic camera-on-a chip. *IEEE Trans. Electron Devices* 44: 1689–1698.
72. Nakamura, J. (2005). *Image Sensors and Signal Processing for Digital Still Cameras.* Boca Raton, FL: Taylor & Francis.
73. Yadid-Pecht, O., Fossum, E. R. (1997). Readout schemes to increase dynamic range of image sensors. *NASA Tech. Briefs.* 21: 32–33.
74. Theuwissen, A. J. P. (1995). *Solid-State Imaging with Charged-Coupled Devices.* Dordrecht, Boston, MA: Kluwer Academic Publishers.
75. Kurino, H., Nakagawa, M., Lee, K. W., Nakamura, T., Yamada, Y., Park, K. T., Koyanagi, M. (2000). Smart vision chip fabricated using three dimensional integration technology. *Neural Inform. Process. Syst.* 13: 720–726.
76. Johnson, M. (2003). *Photodetection and Measurement: Maximizing Performance in Optical Systems.* New York; London: McGraw-Hill.
77. Abshire, P. A., Andreou, A. G. (2001). A communication channel model for information transmission in the blowfly photoreceptor. *BioSystems* 62: 113–133.
78. Lacoste, D. Lau, A. W. C. (2005). Dynamics of active membranes with internal noise. *Europhys. Lett.* 70(3): 418–424.
79. Niven, J. E., Vahasoyrinki, M., Juusola, M., French, A. S. (2004). Interactions between light-induced currents, voltage-gated currents, and input signal properties in drosophila photoreceptors. *J. Neurophysiol.* 91: 2696–2706.
80. Lamb, T. D. (1987). Sources of noise in photoreceptor transduction. *J. Opt. Soc. Am. A.* 4(12): 2295–2300.
81. Holcman, D., Korenbrot, J. I. (2005). The limit of photoreceptor sensitivity: Molecular mechanisms of dark noise in retinal cones. *J. Gen. Physiol.* 125: 641–660.
82. Becker, B. (1994). Comparison of noise performance between a FET transimpedance amplifier and a switched integrator. *Burr-Brown Appl. Bull.* AB-57A: 1–7.
83. Xu, J., Stickrath, A. B., Bhattacharya, P., Nees, J., Váró, G., Hillebrecht, J. R., Ren, L., Birge, R. R. (2003). Direct measurement of the photoelectric response time of bacteriorhodopsin via electro-optic sampling. *Biophys. J.* 85: 1128–1134.
84. Koch, C., Li, H., Eds. (1995). *Vision Chips: Implementing Vision Algorithms with Analog VLSI Circuits.* Piscataway, NJ: IEEE Computer Press.

85. Etienne-Cummings, R., Van der Spiegel, J., Mueller, P., Zhang, M. Z. (2000). A foveated silicon retina for two-dimensional tracking. *IEEE Trans. Circuits Syst. II.* 47(6): 504–517.
86. Reichadt, W. (1961). Autocorrelation: A principle for the evaluation sensory information by the central nervous system. In: Rosenblith, W. A., Ed. *Sensory Communication.* New York: Wiley, pp. 303–317.
87. Kramer, J., Sarpeshkar, R., Koch, D. (1997). Pulse-based analog VLSI velocity sensors. *IEEE Trans. Circuits Syst. II.* 44: 86–101.

Part V

Applications in Detection and Monitoring

In this section, the focus is on applications of smart biosensors in environmental and medical applications. In Chapter 18, Tao Geng and Professor Arun Bhunia describe the development of optical biosensors for the monitoring of foodborne pathogens. Conventional bacterial testing methods are very time-consuming and require several days. The proposed techniques provide "real-time" and user-friendly approach. In Chapter 19, Professor Omowunmi Sadik and colleagues discuss the challenges and opportunities in designing multiarray biosensors for monitoring of toxins and bacterial pathogens. Examples of biosensor devices being developed for monitoring chemical toxicants and bacterial pathogens are presented with special emphasis on their use for toxicity screening. In Chapter 20, Professor Anthony Turner and his colleagues have described approaches to allergy detection using apta-sensors.

Chapters 21 and 22 by Professor Wilkins and his colleagues focus on biosensors for the detection of viruses. Viral diseases are the primary cause of serious disorders that do not require hospitalization among people who reside in developed countries, according to epidemiologic studies. Among infants and children, they exact a heavy toll in mortality and permanent disability. Emerging viral diseases, such as those brought about by HIV, Ebola virus, and Hantavirus, appear regularly. In addition, while antibiotics effectively combat most bacterial-based infections, viral infections are not so readily controlled; in comparison, they pose a greater threat to people's health. In Chapter 23, some thoughts of the coeditors of this book on future perspectives and directions of smart biosensor technologies are presented.

18

Optical Biosensors in Foodborne Pathogen Detection

Tao Geng and Arun K. Bhunia

CONTENTS

18.1 Introduction

Food safety is a major concern to consumers as well as to the producers and processors alike. Increasing illnesses and fatalities due to foodborne pathogen outbreaks demand rapid analysis of products for contaminants. Contamination of food also incites product recalls bringing inevitable economic disasters to food manufacturers. In recent years, however, global political instability has also raised serious concerns about potential food bioterrorism. Detection tools are an essential component of food safety and control of food bioterrorism. Government and industrial agencies recognize the need for sensitive and low-cost detection tools. Biosensor technology offers the best solutions because of its ultrasensitivity and potential for automation and online use; however, the specificity and cost must be addressed as we continue to develop sensors for specific applications. Often sensor developments are dictated by the relative significance of the agent, numbers of fatalities, or associated economic losses. Broadly, sensors can be grouped as electrochemical (impedance-based, amperometric), optical (fiber optic, surface plasmon resonance (SPR)), thermometric (thermister, pyroelectric), and mass-based (piezoelectric, surface acoustic). Optical sensors appear to be the most appealing because of their sensitivity, available instrumentation, and relative ease of data interpretation. Applications of some optical sensors for detection of foodborne pathogens including *Salmonella*, *Escherichia coli*

O157:H7, *Listeria monocytogenes*, *Campylobacter*, staphylococcal enterotoxin B, *Clostridium botulinum* toxin, and others are discussed in this review.

18.2 Foodborne Pathogens

Foodborne pathogens cause either food poisoning or foodborne infections due to consumption of preformed toxins or viable microorganisms, respectively. Most infections are life threatening to immunologically challenged individuals. Children, the elderly people, persons with acquired immunodeficiency syndrome (AIDS), persons with organ transplants, persons receiving chemotherapy to treat malignancy, and pregnant women are susceptible to many of the foodborne pathogens. Foodborne pathogens may cause gastroenteritis, fatal hemolytic uremic syndrome, abortion, meningitis, atherosclerosis, and neurological malady. Centers for Disease Control and Prevention (CDC) estimates that there are about 76,000,000 illnesses due to foodborne pathogens, 325,000 hospitalizations and 5,000 deaths each year in the United States (1). Economic losses for foodborne diseases are estimated to be 6.9 billion dollars per year, and the losses are due to product recalls, medical intervention, loss of work hours, and death. In the recent years the threat of food bioterrorism unfolds a new perspective to the control and management of foodborne infections. The threat of intentional introduction of food pathogens into products means that those products that would not naturally be associated with a particular pathogen may now need to be screened for several pathogens or toxins capable of surviving in the product. A single event of food bioterrorism can have tremendous impact on our economy and lifestyle, since food and water are the most crucial entities for life support.

Foodborne pathogens comprise bacteria, viruses, protozoa, molds, and nematodes. Among the bacterial pathogens, some produce potent exotoxins like botulinum toxin produced by *C. botulinum*, staphylococcal enterotoxin (SE) by *Staphylococcus aureus*, epsilon toxin by *Clostridium perfringens*, and diarrheagenic or emetic enterotoxins by *Bacillus cereus*. Exogenous administration or production of toxin in nanogram quantities in the food could lead to serious consequences. Alternatively, when viable pathogenic bacterial cells are ingested, they can cause infection. Organisms belonging to this group are *L. monocytogenes*, *C. perfringens*, *E. coli*, *Salmonella enterica*, *Vibrio*, *Shigella*, *Campylobacter*, and *Yersinia* species. In addition, spores also could present serious problems with foodborne infections. Spores are resistant to most food-processing treatments and heat may trigger their germination into vegetative cells, which can then produce toxins. If spores are introduced into food (naturally or intentionally), they can germinate in the intestine and initiate infection. For example, *Bacillus anthracis* spores, if consumed with food, could germinate in the gut into vegetative cells, which can pass through intestinal epithelial cell linings inducing severe inflammation and gastroenteritis, referred to as intestinal anthrax. *B. anthracis* spores are most deadly when inhaled (causing inhalation anthrax). In contrast, cutaneous anthrax occurs when spores lodge on the skin causing milder disease. Examples of viral foodborne diseases are Norwalk and Hepatitis A. An infective protein like prion, a causative agent for mad cow disease or bovine spongiform encephalopathy (BSE), can also be transmitted through meat and cause fatal neurologic disease (variant Creutzfeldt-Jacob disease—vCJD) in humans. Important foodborne protozoan diseases are caused by *Giardia lamblia*, *Entamoeba histolytica*, *Toxoplasma gondi*, *Cryptosporidium parvum*, *Cyclospora cayetanensis*, and *Isospora belli*. Mycotoxins produced by various mold species are also problematic when contained in cereal foods and meats, since they are carcinogenic, heptotoxic, or nephrotoxic.

Analysis of foods for the presence of both pathogenic and spoilage bacteria is a standard practice for ensuring food safety and quality. Conventional bacterial testing methods rely on specific microbiological media to isolate and enumerate viable bacterial cells in foods. Generally, it consists of five steps involving preenrichment, selective enrichment, selective plating, biochemical tests, and serological tests. These methods are sensitive, and can give both qualitative and quantitative information on the number and the nature of the microorganisms present in a food sample. However, this method is time consuming, requiring 5–7 days, and relies on the bacteria's ability to multiply and form visible colonies. Moreover, culture-medium preparation, inoculation of plates, colony counting, and biochemical characterization make these methods very labor intensive. Especially in the food industry, there is a need for more rapid and user-friendly methods to provide adequate information on the possible presence of pathogens in raw materials and ready-to-eat food products, for manufacturing process control, and for the monitoring of cleaning and hygienic practices.

18.3 Detection of Foodborne Pathogens Using Optical Biosensors

Biosensors use a combination of biological receptors and physical or chemical transducers, which represent a new and unique technology with great potential to meet the need for rapid detection of low levels of biomolecules (2–5). Optical sensors measure a parameter of the reaction between a receptor and an analyte as a quantifiable optical signal. Optical biosensors offer the advantages of noninvasive, nondestructive, continuous, and simultaneous multianalyte detection (6). Optical systems do not interfere with metabolism, and thus in vivo measurements are possible. In this chapter, some applications of the following optical sensing principles in foodborne pathogen detection are discussed: (i) SPR, (ii) resonance mirror (RM), (iii) fiber-optic biosensor, (iv) array biosensor, (v) Raman spectroscopy, and (vi) light-addressable potentiometric sensor (LAPS).

18.3.1 Surface Plasmon Resonance

SPR is the phenomenon that occurs as a result of total internal reflection (TIR) of light between a dielectric and a metal interface. SPR propagates along the surface and presents itself as an electromagnetic field. Energy from the incident light is absorbed by the metal, resulting in a decrease in reflected-light intensity. This reflectance minimum appears in the reflected light at an acutely defined incident angle (resonance angle), which is dependent on the refractive index of the medium close to the metal-film surface. Changes in refractive index within the evanescent field result in a shift of resonance angle, defined as an SPR response. When biomolecules are adsorbed or they interact with already immobilized molecules within a probed volume, defined by the size of the illuminated area and the evanescent wave (EW) depth, an increase in surface concentration occurs and the resonance angle shifts to greater values (7–9). When used to detect biomolecules, three types of transducers are typically used in SPR (Figure 18.1). These transducers are (i) prisms coated with a thin film of metal, usually gold or silver (~55-nm thick), typically known as the Kretschmann configuration; (ii) waveguide coupled with a thin film of metal; and (iii) metallized diffraction gratings where the metal thickness can be much greater (up to 150-nm thick) (10). Several manufacturers now offer SPR sensors under the above configuration including Biacore (Biacore International SA, Switzerland), Affinity Sensors

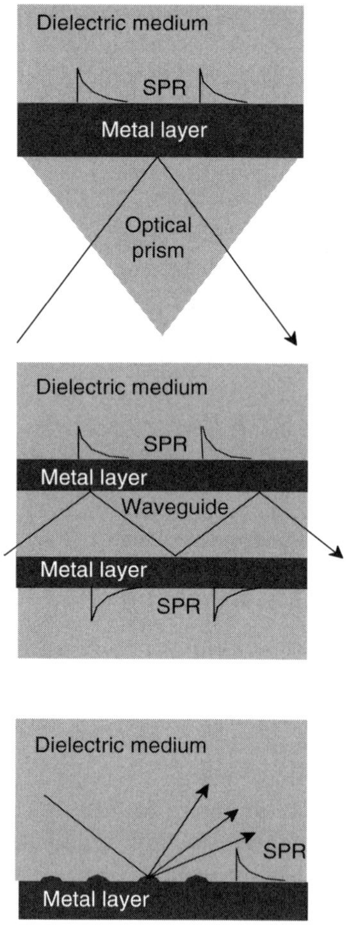

FIGURE 18.1
Excitation of surface plasmon resonance (SPR): (a) by prism coupling: a light wave passes through a high-refractive-index prism and is totally reflected at the prism−metal layer; (b) by optical waveguide: the light wave is guided by an optical waveguide and, when entering the region with a thin metal layer, it evanescently penetrates through the metal layer exciting an SPR at its outer boundary; and (c) by light diffraction on a diffraction grating: the component of the wave vector of the diffracted waves parallel to the interface is diffraction-increased by an amount that is inversely proportional to the period of the grating and can be matched to that of an SPR.

(Cambridgeshire, UK), Nippon Laser and Electronic Lab (Nagoya, Japan), and Texas Instrument (Dallas, TX) (9).

Foodborne pathogens or toxins are detected using specific antibodies on the gold-SPR surface allowing a label-free detection of analytes. SPR has been employed to detect *E. coli* O157:H7 at 10^7 cfu/ml (11), at 10^2 cfu/ml (12), or at 10^5–10^8 cfu/ml (13). *Salmonella* was detected at 10^2–10^9 cfu/ml (14–17), and *L. monocytogenes* cells were detected at 10^5–10^7 cfu/ml (15,18,19). The sensitivity of a majority of the SPR-based assays is equivalent to standard ELISA-based assays. In many of these applications, specificity of the assay was not adequately addressed; however, most sensor applications were focused on demonstrating proof of concept. SPR was also used for its potential to detect soluble toxin preparations. Nedelkov and coworkers (20,21) detected and identified staphylococcal enterotoxin B (SEB) at 1 ng/ml in milk or mushroom samples using the Biacore SPR. This assay employed biomolecular interaction analysis and mass spectrometry (BIA-MS) together to improve analysis. Antibodies immobilized on the sensor chip captured the molecules, and identity was analyzed by MALDI-TOF-MS (matrix assisted laser desorption or ionization time-of-flight mass spectrometry). A miniature integrated two-channel SPR sensor based on Spreeta™ (Texas Instrument) was also employed to detect SEB at femtomolar levels (22,23). Homola et al. (24) also developed a two-channel SPR device and were able to detect SEB at 0.5–5 ng/ml in buffer and milk samples.

18.3.2 Resonant Mirror

The resonant mirror (RM) sensor has been extensively used in the field of chemical and biological sample analyses (25). The RM device consists of a high-index substrate, a thin low-index spacer and a very thin monomode waveguiding layer (Figure 18.2). These dielectric films are very durable, unlike the thin gold films used in SPR sensors, enabling the devices to be cleaned and reused.

Light incident above the critical angle on the substrate–spacer interface is coupled into the waveguiding layer via the evanescent field in the spacer layer when the propagation constants in the substrate and the waveguide match. For monochromatic light, this occurs over a very narrow range of angles, typically spanning less than one degree. Alternatively, at a fixed input angle the coupling occurs over a narrow range of wavelengths. This device has been termed the RM because it contains a resonant cavity (the waveguide) that acts as an almost perfect reflector for light incident above the critical angle. However, if an optically absorbing species is present in the evanescent field above the waveguide layer, the reflectivity of light overlapping the absorbance band will decrease. Similar modulations of the reflected mode are also encountered on variations of the refractive index of the guiding medium. Both these factors can be exploited to produce optical sensor devices. An RM biosensor is commercially available as the IAsys immunosensor (affinity sensors). Dmitreiv et al. (25,26) conducted a detailed kinetic-analysis study determining antibody–antigen (purified protein) interactions using IAsys sensor. Rasooly and Rasooly (27) employed IAsys affinity sensor to detect staphylococcal enterotoxin A in various food matrices including hotdog, potato salad, milk, and mushrooms at a concentration of 10–100 ng/g of product in less than 4 min. They used a sandwich-assay format to improve sensitivity. Lathrop et al. (28) detected crude surface protein preparations from *L. monocytogenes* in IAsys at concentrations of 200 μg/ml in 20–30 min using a specific antibody. This sensor gave several-fold higher signal for *L. monocytogenes* over other *Listeria* species. The assay also gave positive signals with *L. innocua*, a nonpathogenic species, because of the cross-reactivity of the antibody with this species (29,30). RM is most suitable for the detection of soluble toxins or proteins and was unable to detect whole bacterial cells because of the limited depth (~300 nm) of the sensing layer (28).

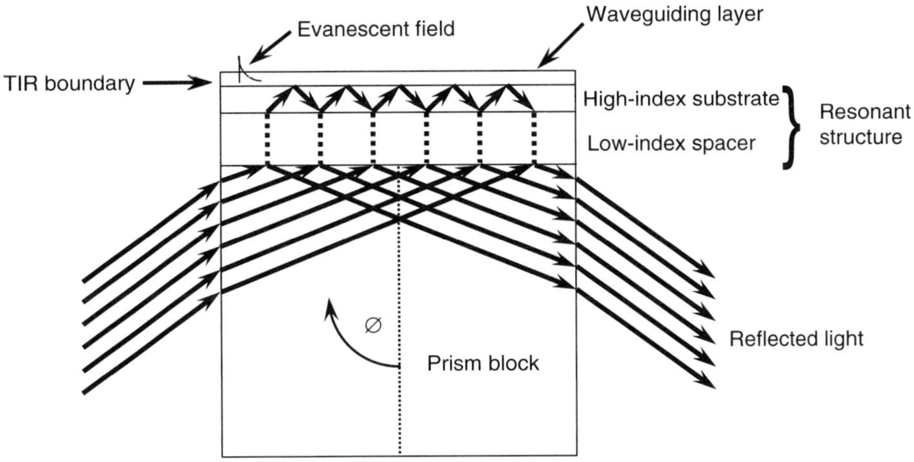

FIGURE 18.2
Configuration of resonant mirror (RM) biosensor. (Adapted from IAsys Manual, Affinity Sensors, Cambridgeshire, UK.)

18.3.3 Fiber-Optic Biosensor

The fiber-optic biosensor is a powerful tool in biological and chemical detection (31,32), and much of its success is due to the revolutionary developments in the telecommunication industry. There are four main characteristics that differentiate fiber-optic biosensors from other types of biosensors: The very high bandwidth of optical fibers allows them to carry a large amount of signal through a single fiber because optical fiber is a dielectric and it is not subject to interference from electronic waves that might be present in the sensing environment; and fiber-optic biosensors can function under physiological environment with high concentrations of electrolytes and wide ranges of pH values that could interfere with the measurement and erode metals at a rapid rate. In addition, fiber-optic biosensors are intrinsically safe in explosive environments (no sparks), lightweight, compact, robust, and potentially inexpensive. Indeed, fiber-optic biosensors can perform the functions of virtually any conventional sensor—often faster and with greater sensitivity—and they can also perform measurement tasks that cannot be achieved with conventional biosensors.

The field of optical biosensors has experienced rapid growth in the past few years (33). This is partly due to the ever-improving optoelectronics designed for telecommunications, and advances in material sciences, which have led to better fabrication of materials and improved methods of signal generation and measurement (34). Fiber-optic biosensors can be described as either direct or indirect, just like the other immunoassay platforms. The direct systems rely solely on antigen–antibody binding to modulate the signal being measured, while indirect sensors depend on the use of labels or fluorophores to sense the binding event. The advantage of the direct type is that the assay is essentially "reagentless" in that no additional substances are needed for detection. The disadvantages are that sensitivity may be limited by nonspecific binding and the limited number of analytes that could be detected using this format (35,36). The indirect format has the advantage of improved sensitivity and selectivity due to the label, as well as a reduced amount of nonspecific binding effects. Normal-assay formats can still be used with indirect sensors, since the labels will serve to amplify the optical sensor response even to small molecules bound to the transducer. The main disadvantage of the indirect sensors is the need for labeled reagents.

Optical biosensors based on the EW use the technique of attenuated total reflection (ATR) spectroscopy and SPR to measure real-time interaction between biomolecules. The basis of ATR is the reflection of light inside the core of a waveguide when the angle of incidence is less than the critical angle. Waveguides can be slab guides, planar integrated optics, or optical fibers. Light waves are propagated along fibers by the law of TIR. Even though the light is totally internally reflected, the intensity does not abruptly fall to zero at the interface. The intensity exponentially decays with distance, starting at the interface and extending into the medium of lower refractive index. The EW is the electromagnetic field created in the second medium. It is characterized by the penetration depth defined as the distance from the interface at which it decays to $1e^{-1}$ of its value at the interface (36). The wavelength of light, ratio of the refractive indices, and angle of the light at the interface determine the penetration depth (37). Penetration depths are typically 50–1000 nm; thus, the EW is able to interact with many monolayers at the surface of the probe. Reactions occurring very close to the interface perturb the evanescent field and the change in signal can be related to the amount of binding between the target and immobilized ligand at the interface.

Fluorescent measurements can also be used to monitor the binding events occurring on the surface of optical biosensors. When light is traveling through the optical waveguide, it excites fluorophores within the evanescent field, and the fluorescent signal is propagated back up the fiber and is detected by a fluorometer (Figure 18.3). By exploiting the detection of fluorescence-emitting labels, specific antibody–antigen complexes can be monitored. Hirschfeld and coworkers (38–40) demonstrated that EW sensing excites fluorophores that

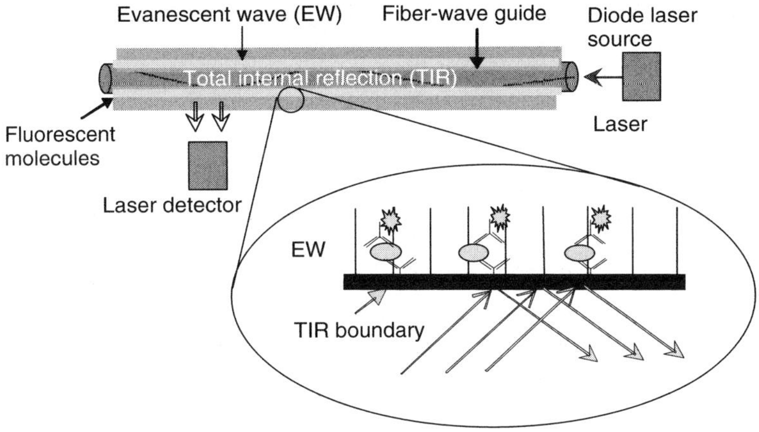

FIGURE 18.3
Configuration and detection principle of fiber-optic biosensor.

are primarily bound to the fiber as opposed to those in the bulk solution. Fluorescent radiation propagates back through the fiber in high-order modes.

Overall, EW-based fiber-optic biosensors exploit the measurement of fluorescent light excited by an EW generated at the interface of an optical waveguide in which a laser light undergoes TIR, to quantitatively detect biomolecules immobilized on the fiber surface (6). A commercially available fiber-optic biosensor (Analyte 2000) has been developed using the above principle by Research International (Monroe, WA). Now a portable integrated version of this system called RAPTOR (Research International) is commercially available for the detection of multiple foodborne and bioterrorism agents (41,42). The assay principle is based on a sandwich immunoassay, where a capture antibody is immobilized onto the optical fibers, and a Cyanine 5 fluorescent dye (Cy5) or Alexa-flour 647-labeled antibody is used for detection of bimolecular attachment to the fiber. This assay has been used to detect staphylococcal enterotoxin B (43–45), *C. botulinum* toxin (46) at nanogram concentrations, and *S. enterica* serovar Typhimurium at 10^4 cfu/ml (47). RAPTOR system was also used to detect *S.* Typhimurium from spent sprout-irrigation water after 67 h of initial inoculation of seeds with 50 cfu/g (48). Possible use of this sensor for online monitoring of this pathogen in irrigation water used during sprouting was proposed. Fiber-optic sensor was also developed for *E. coli* O157:H7 and was detected at 3–30 cfu/ml in spiked ground-beef samples (49). This sensor was also developed to detect PCR products of *Listeria* species by allowing hybridization of complementary sequence coated on the fiber surface (50). Later, this sensor was used to detect *L. monocytogenes* cells at 1×10^8 cfu/ml (51). Most recently, an even more sensitive fiber-optic assay for *L. monocytogenes* was developed using a combination of polyclonal and monoclonal antibodies in a sandwich format (52). Rabbit polyclonal antibody was used as a capture antibody on the fiber waveguide and subsequent detection was accomplished by using a monoclonal antibody C11E9 (29). This combination improved the sensitivity of the fiber-optic biosensor to a range of 4.3×10^3 to 4×10^4 cfu/ml of *L. monocytogenes* after 2.5 h of sampling even in the presence of common food contaminants or stress conditions. This sensor was able to detect *L. monocytogenes* from hotdogs or bologna that was naturally contaminated or artificially inoculated with 10^1–10^3 cfu/g after enrichment in buffered Listeria enrichment broth in less than 24 h. This biosensor was specific for *L. monocytogenes* and showed significantly higher signal than other *Listeria* species or other microorganisms and was an important advancement for detection of *L. monocytogenes* in ready-to-eat foods. Varshney et al. (53) used immunomagnetic separation method to capture *S.* Typhimurium on beads, and chemiluminescence signals emitted from the bacteria were measured by using a fiber-optic light guide.

18.3.4 Array Biosensor

Array biosensors can interrogate multiple samples at discrete regions on the sensing surface and often it is facilitated by applying evanescence wave technology that can excite multiple fluorophores. Taitt et al. (54) used a patterned array of antibodies against multiple analytes immobilized on the surface of planar waveguide (microscope slides) to capture antigen. Fluorescent-labeled tracer antibodies were allowed to bind to the antigen and subsequent excitation with a diode laser launched at the edge of the glass slide emitted a fluorescent signal, which was captured by a charge-coupled device (CCD) camera. Positive-signal-emitting antigens were identified by using an image-analysis software (54). This system was developed primarily to address biothreat agents and simultaneously detected multiple pathogens including *B. anthracis*, *S. aureus* enterotoxin B, cholera toxin, ricin, *Franciscella tularensis* and *Brucella abortus* (55). Bacterial cells were detected in the range of 10^3–10^6 cfu/ml, and toxins were in the ng/ml range. With appropriate modifications, this system could be applied for detection of foodborne pathogens and was employed for rapid detection of *S.* Typhimurium with a detection limit of 8×10^4 cfu/ml from different food matrices such as cantelope, chicken washings, sprouts, and liquid egg (56). This system is now made as a portable device that contains a power supply, peristaltic pumps, interface circuit boards, diode laser, and CCD camera for onsite use (57).

An array biosensor was also configured to detect a single agent from different matrices. Shriver-Lake et al. (58) detected SEB from six different types of food samples, including beverages, homogenates of fruit and meat, and carcass washings, which could be completed in less than 20 min without sample preconcentration. However, the fluorescence intensity varied between samples and the detection limit was 0.5 ng/ml. Later, the multi-analyte array biosensor (MAAB) was developed. The MAAB can detect and identify multiple target agents in complex samples with minimal user manipulation. Upon binding of a fluorescent analyte or fluorescent immunocomplex, the patterns of fluorescent spots were detected using a CCD camera. The location of the spot and the mean fluorescence intensity were used to determine the toxin or microorganism identity and concentration. *S.* Typhimurium and *L. monocytogenes* were detected at 10^3-10^6 cells. Staphylococcal enterotoxin B, ricin, cholera toxin, botulinum toxoids, trinitrotoluene, and the fumonisin, a mycotoxin, were detected at levels as low as 0.5 ng/ml (59,60).

Moreno-Bondi et al. (61) described another array sensor platform for multiple analytes on a metal-oxide silicon biochip. Essentially, this system contains four-by-four microarrays of antibodies. Binding of analytes to the capture antibody could be detected by using a second antibody labeled with Cy-5 and by acquisition of signals using a diffractive optical element and CCD camera.

18.3.5 Raman Spectroscopy

Raman spectroscopy is based on the phenomena of the shifted wavelength scattering of molecules excited with monochromatic light due to inelastic collisions of photons with molecules. When the incident light is applied to bacterial cells, photons can be directly absorbed or scattered. The scattered photons have the same energy (wavelength) as the incident photons. However, a small fraction of light ($1/10^8$) is scattered at optical frequencies different from the frequency of the incident photons. The process leading to this inelastic scattering is called the Raman effect.

Raman spectroscopy is a noninvasive, nondestructive, and water-insensitive tool that provides information on the structure and intramolecular and intermolecular interaction of pathogenic bacteria or toxins (62). The recording of a Raman spectrum often requires small sample amounts and least sample preparation, and unlike IR spectroscopy, water can be readily used as a solvent. However, biosystems are prone to auto-fluorescence, which may deteriorate or completely mask spectra. What's more, the conversion efficiency of the Raman effect is fairly poor. Only a small amount, $10^{-6}-10^{-8}$, of the laser photon is converted into Raman photons, limiting the sensitivity of the detection. Thus, surface-enhanced Raman scattering (SERS) was developed to overcome this problem.

The effect of SERS from molecules adsorbed on an electrochemically roughened surface with silver colloid was first discovered by Fleischmann et al. (63). This enhancement is due to electromagnetic and chemical enhancement factors. When the incident light is applied to a metal surface, molecules adsorbed or in close proximity to the surface create an exceptionally large electromagnetic field emitting an SPR. The chemical enhancement is due to a charge transfer interaction between the metal and adsorbed molecules. Taking advantage of these two factors, Raman signal can be increased by up to six orders of magnitude or more.

Raman spectroscopy has been employed to identify clinically relevant microcolonies of *S. aureus*, *Staphylococcus epidermidis*, *E. coli*, *Enterococcus faecium*, and *Candida albicans* (64–67). This technique was also employed to detect and differentiate foodborne pathogens including *E. coli* O157:H7, *B. cereus*, *Saccharomyces cerevisiae*, and *Aspergillus niger* on the surface of apple (68). Micro-Raman spectroscopy was able to identify bacterial colonies of *Micrococcus luteus*, *Bacillus subtilis*, *Bacillus sphaericus*, *Rhodotorula mucilaginosa*, and *Pseudomonas fluorescence* on agar plates (69,70). The micro-Raman technique appeared to be suitable for identification of pigment-producing microorganisms, but for detection of colorless colonies, confocal micro-Raman is ideal (69). Recently, the micro-Raman technique was employed for possible use in a clean room setting for food processing or pharmaceutical manufacturing for online identification of bacterial cells (70).

Grow et al. (71) described a μSERS technology for fingerprinting of individual bacterial cells captured selectively on a biochip by antibodies. This technology utilizes a light microscope to localize a single microorganism on a chip, and is hence called μSERS. Antibodies to target analytes were immobilized on a SERS-active biochip surface and allowed to capture antigens. Surface-enhanced Raman-spectroscopy fingerprints were collected and compared with a database for identification. This system was able to differentiate viable from nonviable cells and could detect *Listeria* species, *Legionella*, and *Cryptosporidium* oocysts at subspecies and strain levels. This system was also able to detect aflatoxin. The μSERS gave overall weaker signals when antibody-captured cells were analyzed compared to directly deposited cells on SERS surfaces.

Scanning electron microscopy was also used to localize regions in cells with SERS substrate (sliver colloid) for improved vibrational Stokes spectra to discriminate bacterial strains (72,73). This strategy was used to discriminate cells of *B. subtilis* and *E. coli*. Zeiri et al. (74) was able to probe the presence of flavin components associated with cell wall (membrane) of *E. coli*, *Pseudomonas*, *Acinetobacter*, and *Bacillus* using SERS.

18.3.6 Light-Addressable Potentiometric Sensor

Potentiometric sensors work by catalyzing substrates with enzymes, and the resulting ionic species cause changes in pH, which can be detected by an ion-selective electrode with sensitivity several order magnitude higher than the standard enzyme immunoassays (75). In most applications, such as in LAPS, antigen is captured on a solid surface such as

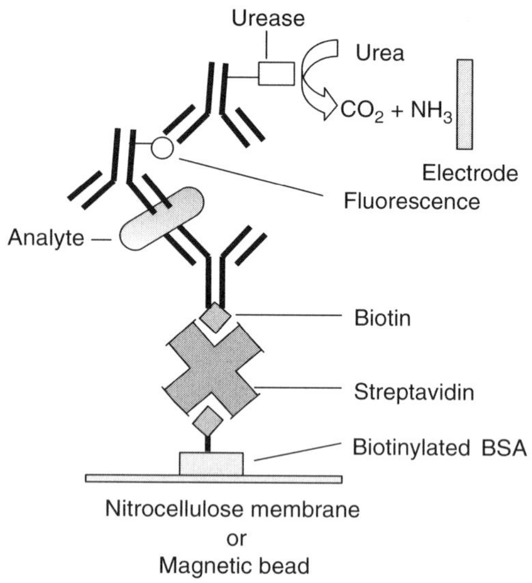

FIGURE 18.4
Schematics of light-addressable potentiometric sensor (LAPS). (Adapted from Tu et al. (1999) *J. Rapid Methods Automat. Microbiol.* 7: 69-79.)

nitrocellulose membrane or paramagnetic beads by a capture antibody (Figure 18.4). A fluorescent-labeled antibody is allowed to react with the analyte. Antifluorescent antibody conjugated to urease is then allowed to bind to the fluorescent-labeled antibody, and subsequent exposure to urea (substrate) will result in the production of CO_2 and NH_3, which could be detected by an *n*-type silicon semiconductor-based sensor coated with a pH-sensitive insulator (76,77). The magnitude of current generated depends on the potential across the sensor. Gehring et al. (77) demonstrated that *E. coli* O157:H7 cells could be detected at 10^4-10^5 cfu/ml of buffer using LAPS. Later, Tu et al. (78,79) used immunomagnetic beads to capture and detect *E. coli* O157:H7 at one cfu/g of ground beef after a 6-h enrichment using LAPS. Ercole et al. (80,81) used LAPS to detect *E. coli* from vegetables; lettuce, sliced carrots, and rucola at 10 cells/ml in 1.5 h in liquid suspension of vegetables acted as a bioindicator for possible contamination with pathogens.

A LAPS coupled with a flow-through immunofiltration-enzyme assay system was developed to provide an indication for the presence or absence of a biothreat agent like *Bacillus* spores rapidly (in 15 min) (82). This has been used in the Biological Integrated Detection System (BIDS) introduced by the U.S. Department of Defense (DOD) for detection of biological agents to protect the military from airborne agents.

18.4 Concluding Remarks

We stand at the dawn of the nano biotechnology era. Many biosensor tools show great promise for the detection of foodborne pathogens and in some cases proof of concept has been demonstrated. The ruggedness and robustness of these sensors for their ability to detect pathogens from complex samples like food matrices, environmental swabs, floors, drains, and water samples must still be examined. In reality, attractiveness and mass appeal are possible only if they are user-friendly, inexpensive, and made of disposable materials. Furthermore, the looming threat of bioterrorism appears as a blessing in disguise for the field of biosensor technologies that continue to generate promise and fascination leading to more support through grants. The progress in this new frontier must be sustained to continue revolutionizing not only the foodborne pathogen detection in food safety and food biosecurity applications but also the biomedical applications for early detection of many life-threatening diseases such as cancer and heart disease.

Acknowledgments

Sincere thanks to Prof. Mark T. Morgan for critical reading of the manuscript. Biosensor-related research projects in the authors' laboratory is supported through a cooperative agreement with the Agricultural Research Service of the U.S. Department of Agriculture project number 1935-42000-035 and the Center for Food Safety and Engineering at Purdue University.

References

1. Mead, P. S., Slutsker, L., Dietz, V., McCaig, L. F., Bresee, J. S., Shapiro, C., Griffin, P. M., Tauxe, R. V. (1999). Food-related illness and death in the United States. *Emerg. Infect. Dis.* 5: 607–625.
2. Baeumner, A. J. (2003). Biosensors for environmental pollutants and food contaminants. *Anal. Bioanal. Chem.* 377: 434–445.
3. Bhunia, A. K., Lathrop, A. (2003). Pathogen detection, food-borne. *McGraw-Hill Yearbook of Science and Technology*. New York, McGraw-Hill, pp. 320–323.
4. Ivnitski, D., Abdel-Hamid, I., Atanasov, P., Wilkins, E. (1999). Biosensors for detection of pathogenic bacteria. *Biosens. Bioelectron.* 14: 599–624.
5. Vo-Dinh, T., Cullum, B. M., Stokes, D. L. (2001). Nanosensors and biochips: frontiers in biomolecular diagnostics. *Sens. Actuators, B Chem.* 74: 2–11.
6. Taitt, C. R., Anderson, G. P., Ligler, F. S. (2005). Evanescent wave fluorescence biosensors. *Biosens. Bioelectron.* 20: 2470–2487.
7. Jonsson, U., Fagerstam, L., Ivarsson, B., Johnsson, B., Karlsson, R., Lundh, K., Lofas, S., Persson, B., Roos, H., Ronnberg, I., Sjolander, S., Stenberg, E., Stahlberg, R., Urbaniczky, C., Ostlin, H., Malmqvist, M. (1991). Real-time biospecific interaction analysis using surface plasmon resonance and a sensor chip technology. *Biotechniques* 11: 620–627.
8. Malmqvist, M. (1993). Surface plasmon resonance for detection and measurement of antibody antigen affinity and kinetics. *Curr. Opin. Immunol.* 5: 282–286.
9. Rich, R. L., Myszka, D. G. (2000). Advances in surface plasmon resonance biosensor analysis. *Curr. Opin. Biotechnol.* 11: 54–61.
10. Robinson, A. K., Marsh, C. D., Bussmann, U., Kilner, J. A., Li, Y., Vanhellemont, J., Reeson, K. J., Hemment, P. L. F., Booker, G. R. (1991). Formation of thin silicon films using low energy oxygen ion implantation. *Nuclear Instruments & Methods in Physics Research Section B-Beam Interactions with Materials and Atoms* 55: 555–560.
11. Fratamico, P. M., Strobaugh, T. P., Medina, M. B., Gehring, A. G. (1998). Detection of *Escherichia coli* O157: H7 using a surface plasmon resonance biosensor. *Biotechnol. Tech.* 12: 571–576.
12. Oh, B. K., Lee, W., Lee, W. H., Choi, J. W. (2003). Nano-scale probe fabrication using self-assembly technique and application to detection of *Escherichia coli* O157: H7. *Biotechnol. Bioprocess Eng.* 8: 227–232.
13. Su, X. L., Li, Y. (2005). Surface plasmon resonance and quartz crystal microbalance immunosensors for detection of *Escherichia coli* O157: H7. *Transact. ASAE* 48: 405–413.
14. Bokken, G. C. A. M., Corbee, R. J., van Knapen, F., Bergwerff, A. A. (2003). Immunochemical detection of *Salmonella* group B, D and E using an optical surface plasmon resonance biosensor. *FEMS Microbiol. Lett.* 222: 75–82.

15. Koubova, V., Brynda, E., Karasova, L., Skvor, J., Homola, J., Dostalek, J., Tobiska, P., Rosicky, J. (2001). Detection of foodborne pathogens using surface plasmon resonance biosensors. *Sens. Actuat., B Chem.* 74: 100–105.
16. Medina, M. B., VanHouten, L., Cooke, P. H., Tu, S. I. (1997). Real-time analysis of antibody binding interactions with immobilized *E coli* O157:H7 cells using the BIAcore. *Biotechnol. Tech.* 11: 173–176.
17. Oh, B.-K., Kim, Y.-K., Park, K. W., Lee, W. H., Choi, J.-W. (2004). Surface plasmon resonance immunosensor for the detection of *Salmonella typhimurium. Biosens. Bioelectron.* 19: 1497–1504.
18. Leonard, P., Hearty, S., Quinn, J., O'Kennedy, R. (2004). A generic approach for the detection of whole *Listeria monocytogenes* cells in contaminated samples using surface plasmon resonance. *Biosens. Bioelectron.* 19: 1331–1335.
19. Leonard, P., Hearty, S., Wyatt, G., Quinn, J., O'Kennedy, R. (2005) Development of a surface plasmon resonance-based immunoassay for *Listeria monocytogenes. J. Food Prot.* 68: 728–735.
20. Nedelkov, D., Rasooly, A., Nelson, R. W. (2000). Multitoxin biosensor-mass spectrometry analysis: a new approach for rapid, real-time, sensitive analysis of staphylococcal toxins in food. *Int. J. Food Microbiol.* 60: 1–13.
21. Nedelkov, D., Nelson, R. W. (2003). Detection of staphylococcal enterotoxin B via biomolecular interaction analysis mass spectrometry. *Appl. Environ. Microbiol.* 69: 5212–5215.
22. Naimushin, A. N., Soelberg, S. D., Nguyen, D. K., Dunlap, L., Bartholomew, D., Elkind, J., Melendez, J., Furlong, C. E. (2002). Detection of *Staphylococcus aureus* enterotoxin B at femtomolar levels with a miniature integrated two-channel surface plasmon resonance (SPR) sensor. *Biosens. Bioelectron.* 17: 573–584.
23. Naimushin, A. N., Spinelli, C. B., Soelberg, S. D., Mann, T., Stevens, R. C., Chinowsky, T., Kauffman, P., Yee, S., Furlong, C. E. (2005). Airborne analyte detection with an aircraft-adapted surface plasmon resonance sensor system. *Sens. Actuat., B Chem.* 104: 237.
24. Homola, J., Dostalek, J., Chen, S., Rasooly, A., Jiang, S., Yee, S. S. (2002). Spectral surface plasmon resonance biosensor for detection of staphylococcal enterotoxin B in milk. *Int. J. Food Microbiol.* 75: 61–69.
25. Dmitriev, D. A., Massino, Y. S., Segal, O. L., Smirnova, M. B., Pavlova, E. V., Gurevich, K. G., Gnedenko, O. V., Ivanov, Y. D., Kolyaskina, G. I., Archakov, A. I., Osipov, A. P., Dmitriev, A. D., Egorov, A. M. (2002). Analysis of the binding of bispecific monoclonal antibodies with immobilized antigens (human IgG and horseradish peroxidase) using a resonant mirror biosensor. *J. Immunol. Methods* 261: 103–118.
26. Dmitriev, D. A., Massino, Y. S., Segal, O. L. (2003). Kinetic analysis of interactions between bispecific monoclonal antibodies and immobilized antigens using a resonant mirror biosensor. *J. Immunol. Methods* 280: 183–202.
27. Rasooly, L., Rasooly, A. (1999). Real time biosensor analysis of Staphylococcal enterotoxin A in food. *Int. J. Food Microbiol.* 49: 119–127.
28. Lathrop, A. A., Jaradat, Z. W., Haley, T., Bhunia, A. K. (2003). Characterization and application of a *Listeria monocytogenes* reactive monoclonal antibody C11E9 in a resonant mirror biosensor. *J. Immunol. Methods* 281: 119–128.
29. Bhunia, A. K., Ball, P. H., Fuad, A. T., Kurz, B. W., Emerson, J. W., Johnson, M. G. (1991). Development and characterization of a monoclonal antibody specific for *Listeria monocytogenes* and *Listeria innocua. Infect. Immun.* 59: 3176–3184.
30. Bhunia, A. K., Geng, T., Lathrop, A. A., Valadez, A., Morgan, M. T. (2004). Optical immunosensors for detection of *Listeria monocytogenes* and *Salmonella Enteritidis* from food. *Proc. SPIE* 5271: 1–6.
31. Marazuela, M. D., Moreno-Bondi, M. C. (2002). Fiber-optic biosensors—an overview. *Anal. Bioanal. Chem.* 372: 664–682.
32. Monk, D. J., Walt, D. R. (2004). Optical fiber-based biosensors. *Anal. Bioanal. Chem.* 379: 931–945.
33. Mehrvar, M., Bis, C., Scharer, J. M., Moo-Young, M., Luong, J. H. (2000). Fiber-optic biosensors—Trends and advances. *Anal. Sci.* 16: 677–692.
34. Aizawa, M. (1994). Immunosensors for clinical analysis. *Adv. Clin. Chem.* 31: 247–275.
35. Paddle, B. M. (1996). Biosensors for chemical and biological agents of defence interest. *Biosens. Bioelectron.* 11: 1079–1113.

36. Squillante, E. (1998). Applications of fiber-optic evanescent wave spectroscopy. *Drug Develop. Indust. Pharmacy* 24: 1163–1175.
37. Anderson, G. P., Golden, J. P., Ligler, F. S. (1993). A fiber optic biosensor—combination tapered fibers designed for improved signal acquisition. *Biosens. Bioelectron.* 8: 249–256.
38. Glass, T. R., Lackie, S., Hirschfeld, T. (1987). Effect of numerical aperture on signal level in cylindrical wave-guide evanescent fluorosensors. *Appl. Optics* 26: 2181–2187.
39. Hirschfeld, T., Callis, J. B., Kowalski, B. R. (1984). Chemical sensing in process analysis. *Science* 226: 312–318.
40. Hirschfeld, T. (1985). Instantaneous source wavelength identification with a minimal number of detectors. *Appl. Optics* 24: 2484–2485.
41. Anderson, G. P., Nerurkar, N. L. (2002). Improved fluoroimmunoassays using the dye Alexa Fluor 647 with the RAPTOR, a fiber optic biosensor. *J. Immunol. Methods* 271: 17–24.
42. Jung, C. C., Saaski, E. W., McCrae, D. A., Lingerfelt, B. M., Anderson, G. P. (2003). RAPTOR: A fluoroimmunoassay-based fiber optic sensor for rapid detection of biological threats. *IEEE Sens. J.* 3: 352–360.
43. Slavik, R., Homola, J., Brynda, E. (2002). A miniature fiber optic surface plasmon resonance sensor for fast detection of staphylococcal enterotoxin B. *Biosens. Bioelectron.* 17: 591–595.
44. Strachan, N. J. C., John, P. G., Millar, I. G. (1997). Application of a rapid automated immunosensor for the detection of *Staphylococcus aureus* enterotoxin B in cream. *Int. J. Food Microbiol.* 35: 293–297.
45. Tempelman, L. A., King, K. D., Anderson, G. P., Ligler, F. S. (1996). Quantitating staphylococcal enterotoxin B in diverse media using a portable fiber-optic biosensor. *Anal. Biochem.* 233: 50–57.
46. Ogert, R. A., Brown, J. E., Singh, B. R., Shriver-Lake, L. C., Ligler, F. S. (1992). Detection of *Clostridium botulinum* toxin A using a fiber optic-based biosensor. *Anal. Biochem.* 205: 306–312.
47. Zhou, C. H., Pivarnik, P., Auger, S., Rand, A., Letcher, S. (1997). A compact fiber-optic immunosensor for *Salmonella* based on evanescent wave excitation. *Sens. Actuat., B Chem.* 42: 169–175.
48. Kramer, M. F., Lim, D. V. (2004). A rapid and automated fiber optic-based biosensor assay for the detection of salmonella in spent irrigation water used in the sprouting of sprout seeds. *J. Food Prot.* 67: 46–52.
49. DeMarco, D. R., Saaski, E. W., McCrae, D. A., Lim, D. V. (1999). Rapid detection of *Escherichia coli* O157:H7 in ground beef using a fiber-optic biosensor. *J. Food Prot.* 62: 711–716.
50. Strachan, N. J. C., Gray, D. I. (1995). A rapid general method for the identification of PCR products using a fiberoptic biosensor and its application to the detection of *Listeria. Lett. Appl. Microbiol.* 21: 5–9.
51. Bhunia, A. K., Jaradat, Z. W., Naschansky, K., Shroyer, M., Morgan, M. T., Gomez, R., Bashir, R., Ladisch. M. (2001). Impedance spectroscopy and biochip sensor for detection of *Listeria monocytogenes Proc. SPIE* 4206: 32–39.
52. Geng, T., Morgan, M. T., Bhunia, A. K. (2004). Detection of low levels of *Listeria monocytogenes* cells by using a fiber-optic immunosensor. *Appl. Environ. Microbiol.* 70: 6138–6146.
53. Varshney, M., Li, Y. B., Nanapanneni, R., Johnson, M. G., Griffis, C. L. (2003). A chemiluminescence biosensor coupled with immunomagnetic separation for rapid detection of *Salmonella typhimurium. J. Rapid Methods Automat. Microbiol.* 11: 111–131.
54. Taitt, C. A. R., Golden, J. P., Feldstein, M. J., Cras, J. J., Hoffman, K. E., Ligler, F. S. (2000). Array biosensor for detection of biohazards. *Biosens. Bioelectron.* 14: 785–794.
55. Taitt, C. A. R., Hazzard, J. W., Hoffman, K. E., Cras, J. J., Golden, J. P., Ligler, F. S. (2000). Simultaneous detection of six biohazardous agents using a planar waveguide array biosensor. *Biosens. Bioelectron.* 15: 579–589.
56. Taitt, C. R., Shubin, Y. S., Angel, R., Ligler, F. S. (2004). Detection of *Salmonella enterica* serovar Typhimurium by using a rapid, array-based immunosensor. *Appl. Environ. Microbiol.* 70: 152–158.
57. Golden, J. P., Taitt, C. R., Shriver-Lake, L. C., Shubin, Y. S., Ligler, F. S. (2005). A portable automated multianalyte bisensor. *Talanta* 65: 1078–1085.
58. Shriver-Lake, L. C., Shubin, Y. S., Ligler, F. S. (2003). Detection of staphylococcal enterotoxin B in spiked food samples. *J. Food Prot.* 66: 1851–1856.

59. Ligler, F. S., Taitt, C. R., Shriver-Lake, L. C., Sapsford, K. E., Shubin, Y. S., Golden, J. P. (2003). Array biosensor for detection of toxins. *Anal. Bioanal. Chem.* 377: 469−477.
60. Taitt, C. R., Golden, J. P., Shubin, Y. S., Shriver-Lake, L. C., Sapsford, K. E., Rasooly, A., Ligler, F. S. (2004). A portable array biosensor for detecting multiple analytes in complex samples. *Microb. Ecol.* 47: 175−185.
61. Moreno-Bondi, M. C., Alarie, J. P., Vo-Dinh, T. (2003). Multi-analyte analysis system using an antibody-based biochip. *Anal. Bioanal. Chem.* 375: 120−124.
62. Petry, R., Schmitt, M., Popp, J. (2003). Raman Spectroscopy—A prospective tool in the life sciences. *Chemphyschem* 4: 14−30.
63. Fleischmann, M., Hendra, P. J., McQuilla, A. J. (1974). Raman spectra of pyridine adsorbed at a silver electrode. *Chem. Phys. Lett.* 26: 163−166.
64. Choo-Smith, L. P., Maquelin, K., van Vreeswijk, T., Bruining, H. A., Puppels, G. J., Thi, N. A. G., Kirschner, C., Naumann, D., Ami, D., Villa, A. M., Orsini, F., Doglia, S. M., Lamfarraj, H., Sockalingum, G. D., Manfait, M., Allouch, P., Endtz, H. P. (2001). Investigating microbial (micro)colony heterogeneity by vibrational spectroscopy. *Appl. Environ. Microbiol.* 67: 1461−1469.
65. Kirschner, C., Maquelin, K., Pina, P., Thi, N. A. N., Choo-Smith, L. P., Sockalingum, G. D., Sandt, C., Ami, D., Orsini, F., Doglia, S. M., Allouch, P., Mainfait, M., Puppels, G. J., Naumann, D. (2001). Classification and identification of enterococci: a comparative phenotypic, genotypic, and vibrational spectroscopic study. *J. Clin. Microbiol.* 39: 1763−1770.
66. Maquelin, K., Choo-Smith, L. P., van Vreeswijk, T., Endtz, H. P., Smith, B., Bennett, R., Bruining, H. A., Puppels, G. J. (2000). Raman spectroscopic method for identification of clinically relevant microorganisms growing on solid culture medium. *Anal. Chem.* 72: 12−19.
67. Maquelin, K., Choo-Smith, L. P., Endtz, H. P., Bruining, H. A., Puppels, G. J. (2002). Rapid identification of Candida species by confocal Raman micro spectroscopy. *J. Clin. Microbiol.* 40: 594−600.
68. Yang, H., Irudayaraj, J. (2003). Rapid detection of foodborne microorganisms on food surface using Fourier transform Raman spectroscopy. *J. Mol. Struct.* 646: 35.
69. Rosch, P., Schmitt, M., Kiefer, W., Popp, J. (2003). The identification of microorganisms by micro-Raman spectroscopy. *J. Mol. Struct.* 661: 363−369.
70. Rosch, P., Harz, M., Schmitt, M., Peschke, K. D., Ronneberger, O., Burkhardt, H., Motzkus, H. W., Lankers, M., Hofer, S., Thiele, H., Popp, J. (2005). Chemotaxonomic identification of single bacteria by micro-Raman spectroscopy: Application to clean-room-relevant biological contaminations. *Appl. Environ. Microbiol.* 71: 1626−1637.
71. Grow, A. E., Wood, L. L., Claycomb, J. L., Thompson, P. A. (2003). New biochip technology for label-free detection of pathogens and their toxins. *J. Microbiol. Methods* 53: 221−233.
72. Jarvis, R. M., Brooker, A., Goodacre, R. (2004). Surface-enhanced Raman spectroscopy for bacterial discrimination utilizing a scanning electron microscope with a Raman spectroscopy interface. *Anal. Chem.* 76: 5198−5202.
73. Jarvis, R. M., Goodacre, R. (2004). Discrimination of bacteria using surface-enhanced Raman spectroscopy. *Anal. Chem.* 76: 40−47.
74. Zeiri, L., Bronk, B. V., Shabtai, Y., Eichler, J., Efrima, S. (2004). Surface-enhanced Raman spectroscopy as a tool for probing specific biochemical components in bacteria. *Appl. Spectroscopy* 58: 33−40.
75. Warsinke, A., Benkert, A., Scheller, F. W. (2000). Electrochemical immunoassays. *Fresenius J. Anal. Chem.* 366: 622−634.
76. Dill, K., Song, J. H., Blomdahl, J. A., D. Olson, J. (1997). Rapid, sensitive and specific detection of whole cells and spores using the light-addressable potentiometric sensor. *J. Biochem. Biophys. Methods* 34: 161−166.
77. Gehring, A. G., Patterson, D. L., Tu, S. I. (1998). Use of a light-addressable potentiometric sensor for the detection of *Escherichia coli* O157: H7. *Anal. Biochem.* 258: 293−298.
78. Tu, S. I., Uknalis, J., Gehring, A. (1999). Detection of immunomagnetic bead captured *Escherichia coli* O157: H7 by light addressable potentiometric sensor. *J. Rapid Methods Automat. Microbiol.* 7: 69−79.

79. Tu, S. I., Uknalis, J., Gore, M., Irwin, P., Feder, I. (2003). Factors affecting the bacterial capture efficiency of immuno beads: A comparison between beads with different size and density. *J. Rapid Methods Automat. Microbiol.* 11: 35−46.
80. Ercole, C., Gallo, M. D., Pantalone, M., Santucci, S., Mosiello, L., Laconi, C., Lepidi, A. (2002). A biosensor for *Escherichia coli* based on a potentiometric alternating biosensing (PAB) transducer. *Sens. Actuat., B Chem.* 83: 48−52.
81. Ercole, C., Del Gallo, M., Mosiello, L., Baccella, S., Lepidi, A. (2003). *Escherichia coli* detection in vegetable food by a potentiometric biosensor. *Sens. Actuat., B Chem.* 91: 163−168.
82. Uithoven, K. A., Schmidt, J. C., Ballman, M. E. (2000). Rapid identification of biological warfare agents using an instrument employing a light addressable potentiometric sensor and a flow-through immunofiltration-enzyme assay system. *Biosens. Bioelectron.* 14: 761−770.

19

Multiarray Biosensors for Toxicity Monitoring and Bacterial Pathogens

Silvana Andreescu, Jason Karasinski, and Omowunmi A. Sadik

CONTENTS

19.1 Introduction

During the last two decades, biosensor technology has rapidly developed for sensitive and selective determination of analytes of interest in many practical applications (1). One of the most promising applications of biosensors is for monitoring toxic compounds, ranging from simple chemicals to complex microorganisms or bacterial pathogens. In this direction, the main idea is to design a biosensor that will operate as an "alarm system" able to provide an early warning signal about the presence and toxicity of an analyte, with minimal reagents and sample preparation (2,3). Ideally such system should be portable, robust, and easy to use.

The major problem associated with toxicity monitoring is the large number of samples and analytes that must be tested and the fact that traditional methods require excessive sample preparation steps, tend to be expensive and time-consuming and are generally restricted to a single analyte or class of analytes. Thus, the availability of a screening device able to rapidly provide qualitative information about the presence of a toxin in multiple samples before more complex tests could be performed would significantly reduce the cost and time of analysis (3). Such screening methods could be used to provide a simple "yes or no" answer (4) regarding the toxicity of the sample and complement the traditional techniques. Owing to their unique characteristics, biosensors are able to provide rapid qualitative information about the toxicity of an analyte prior to extensive laboratory analysis (1). However, most biosensors developed to date are designed in a single sensor format and the analysis is usually restricted to a single component in a specific sample. In this configuration, samples are analyzed one at a time and, in some cases, additional polishing, reequilibration, and calibrations are required. Advances have been reported with the development of disposable single-use sensors that can be easily mass-produced by photolithography or chemical vacuum deposition (5–7). While most of these sensors are often used for single analysis, their reproducibility and stability must be strictly ensured. To increase the sample throughput and respond to current requirements for multianalysis and multidetection in toxicity monitoring, recently, there has been a great interest in designing multiarray biosensors.

In this chapter the challenges and opportunities in designing multiarray biosensors for toxicity monitoring, using example from our laboratory and existing published literature data, are described. This includes the description of existing fabrication techniques, transduction and amplification of sensor signal, and data quantification. Applications of such devices for monitoring chemical toxicants and bacterial pathogens are presented with special emphasis on their use for toxicity screening.

19.2 Multiarray Biosensors: Concept, Design, and Opportunities for Toxicity Monitoring

Currently, a significant amount of scientific research has been devoted to developing biosensor systems for monitoring toxic chemicals. Almost all types of biosensors (enzyme-, DNA-, cell-, and tissue-based sensors with electrochemical, optical, thermal, impedimetric, or piezoelectric, detection) have been used. It is now generally recognized that such systems provide unique solutions to many problems encountered with traditional technologies for toxicity monitoring and screening (e.g., time-consuming, in situ analysis, selectivity and sensitivity, and price) (1–3). One of the advantages of biosensors is their size and the possibility of assembling several sensors to form arrays, thus increasing the sample throughput, which is almost impossible using conventional methods.

In a typical arrangement, a biosensor has two main components: a (bio)receptor that provides specificity and selectivity of the sensor and a physicochemical transductor that is responsible for conversion and amplification of the biomolecular recognition process into a quantifiable analytical signal (1–3). In a multiarray sensor, this conventional design is completely changed. For instance, the system may consist of multiple transductors with similar or different functions, onto which one or more biological receptors are deposited. Owing to a large number of variables that have to be considered, fabrication of multiarray biosensors is clearly more complex and involves additional design and optimization steps specific to multichannel devices and also depends on the type and number of sensing elements as well as the final application of the system. This includes: (i) multiarray fabrication and assembly; (ii) availability of specific instrumentation for signal transduction, processing, and recording data generated on multiple channels; and (iii) availability of adequate software for data collection and quantification of the results. Using such devices, the amount of data is considerably higher, and therefore designated software and special computational methods such as complex neuronal networks or support vector machines (SVMs) (4) have to be used to collect and interpret the results. In addition, the performance criteria requires validation that are specific to multichannel systems and have to be clearly defined for the entire set of sensors composing the arrays. These include sensitivity, specificity, reproducibility, noise, stability, and response time and have to be considered as a combination of the contributions of individual sensors to the final biosensing device. Moreover, many multiarray biosensors are fabricated using microfabrication techniques that may affect the activity of biological systems.

Another important point that has to be considered when working with multiarray sensors is "cross-reactivity," important when multicomponent mixtures or complex matrices have to be analyzed. This problem can be solved by carrying out differential measurements using sensors specifically designed for each analyte and subtracting the nonspecific responses due to the interfering compounds by using appropriate control samples (8). One of the advantages of using multiarray sensors is the large number of electrodes available that makes possible designing all the necessary configurations to respond to a diversity of crossreactive species, while the measurements are performed simultaneously for all the electrodes under the same experimental variables. This is almost impossible to achieve with a simple sensor configuration.

Multiarray biosensors could be designed as a combination of multiple sensors that are able to (i) provide real-time information on the toxicity of multiple samples simultaneously and (ii) detect multiple components in a given sample using arrays of sensors specifically designed to sense a particular analyte. The performance of such devices is largely dependent on the configuration of the sensor array and the detection method used (8–10). When used in arrays, the reproducibility, selectivity, and sensitivity of individual sensor are of fundamental importance for obtaining optimum overall system performance. The configuration of the multiarray sensor is selected in accordance with the transduction system used to convert the chemical and biological information on multiple channels and is strongly dependent upon the number of channels, detection method, and the analytes of interest. Finally, due to the large amount of variables, computational methods and chemometric analysis have to be used for data processing and quantification (8,10,11). The use of computational chemistry in combination with multiarray sensors introduces additional advantages to the final sensing system (e.g., ability to classify and superior data processing). By using appropriate control samples and optimized experimental conditions, a multiarray biosensor could provide accurate identification, differentiation, and classification of analytes. In the context of toxicity monitoring, these systems could provide analytical information (type and amount) related to the presence of a specific toxicant in the sample or to a total toxicity level induced by a class of toxicants. Depending on the system used and its performance, this information can be quantitative, semiquantitative, or simply qualitative by providing a "yes or no" answer (3). A schematic representation of a multiarray (bio)sensor including detection, data processing, and quantification is shown in Figure 19.1.

The most preferred arrangement for a multiarray sensor is a 96-well electrode system, but sensors composed of 4, 16, or 32 electrodes have also been reported. These sensing systems can be designed as a combination of multiple sensors deposited on a single-sensor substrate to form an array or as an array of individual sensors arranged in a multiarray format. Figure 19.2 shows examples of three sensor arrays used in our laboratory. The first one (A) is a laboratory-made sensor array comprising six sensors deposited onto the same chip. In such a configuration, each sensor can be specifically tailored toward a particular analyte so that the final device will allow the detection of multiple analytes in a particular sample. The second (B) and the third (C) are produced by Daikin Corporation, Japan and

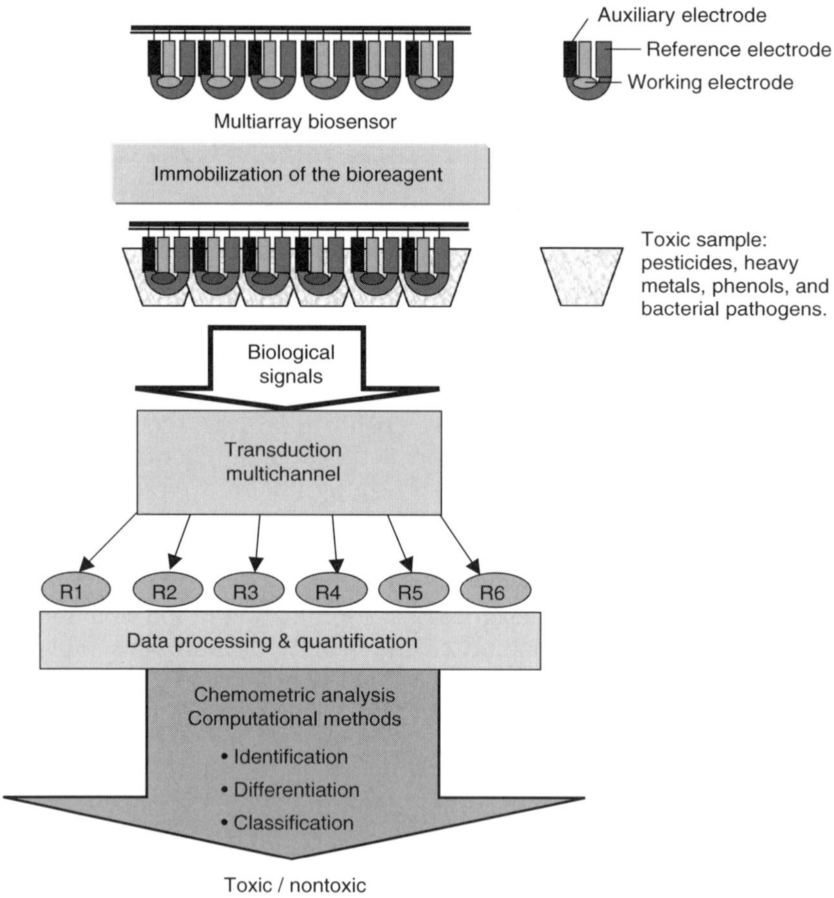

FIGURE 19.1
Schematic diagram of multiarray (bio)sensor system.

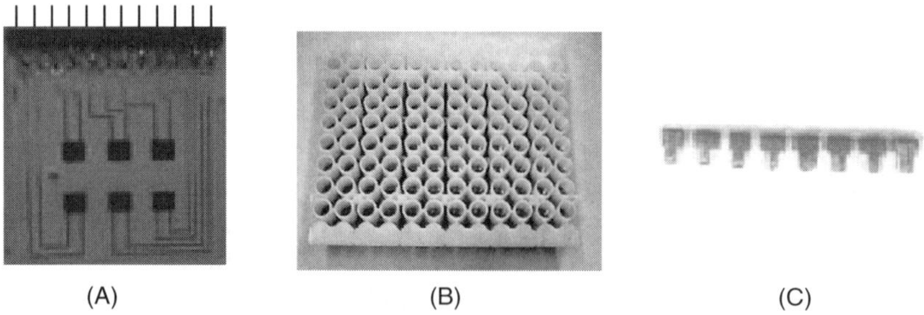

FIGURE 19.2
Examples of sensor arrays used in our laboratory.

are arrays of individual sensors assembled together to form a 96-electrode array. These have been designed to monitor multiple samples simultaneously and, also, each sensor could be modified to detect one or multiple analytes. In the last two configurations (B and C), the sensors are assembled in a format that is compatible with a conventional 96-well plate. In our laboratory, we have used these sensors for monitoring cell-specific metabolic activity via electrochemical detection of oxygen consumed by the cells over time and for cytotoxicity monitoring (12). Table 19.1 is a summary of multiarray sensors described in

TABLE 19.1
Example of Multiarray (Bio)Sensors Described in Literature for Toxicity Monitoring

Multiarray (Bio)Sensors	Array/ Fabrication	Bioreagent/ Immobilization	Detection and Data Processing	Analyte/ Application	References
Optical fiber waveguide biosensor	Glass microscope slides, 12-channel PDMS	Sandwich immunoassay using a cocktail of fluorescent antibodies/ avidin–biotin coupling	Immunofluorescence/ custom-made acquisition software	Simultaneous detection of bacteria and protein toxins	[32]
Electronic nose	32 Conducting polymer sensor array/ electropolymerization PPy	—	Electronic nose coupled with support vector machines	Detection and classification of organophosphate nerve agents	[4]
DOX oxygen multisensor	96-well gold-plated electrode array	—	Oxygen detection/ amperometry/DOX software and Excel®	Bacterial pathogens	[12]
Screen-printed biosensor array	Screen-printing electrode array (16 or 32 electrodes) /screen printing technology	Lectins that binds to cell-surface lipopolysaccharides /porous-activated membranes (ImmunodyneABC®)	Electrochemistry (reduction of ferricyanide) in conjunction with actor analysis	Identification and discrimination of four *E. coli* subspecies: JM10, B, HB101, and neotype	[53]
4-Electrode thick-film sensors	Screen-printing electrodes	Native and recombinant mutants of AChEs from *Dm, Ee,* and *Be*/ cross-linking with glutaraldehyde and physical entrapment in a PVA-SbQ membrane	Enzyme inhibition/ amperometry coupled with artificial neural networks	Simultaneous detection and discrimination of organophosphate and carbamate pesticides	[6,7]
Multibiosensor	Pairs of gold interdigitated electrodes/vacuum deposition	AChE, BuChE, urease/ cross-linking with glutaraldehyde and BSA	Enzyme inhibition/dual potentiometric pH-field effect transistors and conductometric thin-films interdigitated electrodes/ multivariate correspondence analysis	Simultaneous detection of pesticides (trichlorfon, carbofuran) and heavy metals (Hg and Ag)	[5]
96-well	96-well plate	*E. coli* and *Daphnia magna* strains immobilized in agar	Bioluminescence	Toxicity classification of phenol, H_2O_2, and mitomycin	[27,28]

Note: PPy: polypyrrole; Dm: Drosophila melanogaster; Ee: Electric eel; Be: bovine erythrocytes; PDMS: poly(dimethylsiloxane); AChE: acetylcholinesterase; BuChE: butyrylcholinesterase; PVA-SbQ: photocrosslinkable polyvinyl alchohol; BSA: bovine serum albumin.

the literature for monitoring toxic compounds, including the electrodes and the method used for signal transduction and quantification of the results. Most of the multiarray biosensors reported in literature are optical sensors, while relatively few use electrochemical detection. These biosensors have a large number of application including environmental monitoring, homeland security, agriculture, and toxicology.

A typical example of an array-sensor configuration is the electronic nose technology consisting of an array of chemical sensors coupled with appropriate transducers and pattern-recognition techniques. This technique is commonly used to detect analytes by mimicking the human sense of smell. In the context of toxicity monitoring, this technology has been successfully used for identifying and correctly classifying organophosphate (OP) nerve agents and bacterial pathogens and will be discussed in detail later on in this chapter. Another class of multiarray biosensors that has received considerable attention for toxicity monitoring is the immunosensors, considered as one of the most promising methods for identifying biological species (13). Although immunological sensors have been the most successful, sensors based on DNA hybridization reactions have also been employed (14–16). Compared with other techniques, these sensors have provided enhanced sensitivities toward toxicants and by adapting the well-known DNA microarrays it may be possible to construct new multiarray sensors with enhanced performance (17). At present, DNA sensors have been mainly reported in a single sensor configuration, while many of the multiarray formats are simple DNA probes. More innovative methods have reported the development of a microarray containing over 1,000 electrodes per square centimeter using conventional integrated circuitry in conjunction with sandwich-immunoassay protocols and electrochemical-based enzyme amplification detection (18).

In the following section, specific examples of multiarray biosensors for monitoring toxic chemicals and bacterial pathogens, including details regarding their fabrication, detection limits, and applications, are discussed.

19.3 Multiarray Biosensors for Monitoring Toxic Chemicals

The most studied toxic chemicals are pesticides, polychlorinated biphenyls, phenols, and heavy metals (2). Recent reports from our group presented the most recent advances and trends in analytical technologies (3) including biological sensors for environmental monitoring and public safety (2,3). In this work, we focused on several examples of multiarray biosensors designed for multidetection and multianalysis of chemical toxicants.

19.3.1 Electrochemical Multisensor Array

Among sensors with electrochemical detection, the most successful are the enzyme sensors mainly based on enzyme inhibition. An important line of development is directed toward disposable biosensors for semiquantitative analysis and direct on field simultaneous screening of multiple samples. For this purpose, electrochemical multichannel instrumentation is now available while other custom-made portable electrochemical multisensors are currently being developed in many biosensor laboratories. Most of these sensors are designed for amperometric or chronoamperometric measurements at constant applied potential.

Recently, Solna et al. (19) reported an electrochemical array sensor for simultaneous determination of pesticides (carbaryl, heptenophos, and fenitrothion) and phenols (*p*-cresol, catechol, and phenol). The array consists of four disposable screen-printed electrodes with immobilized or coimmobilized enzymes belonging to two different classes: oxidases (peroxidase and tyrosinase) and hydrolases (acetylcholinesterase (AChE) and butyrylcholinesterase (BuChE)). The results showed that when used in array, the biosensor system could provide complementary information about the composition of the sample, although the electrochemical signals of one enzyme electrode might be affected by the substrate of the other enzyme used in the same array. Schmid et al. (6,7) reported a similar screen-printed electrodes array combined with artificial neural networks (ANNs) for simultaneous and selective determination and discrimination of organophosphorus and carbamate pesticides in mixtures. The sensor is based on the inhibition of AChE by pesticides. While both organophosphorus and carbamate pesticides inhibit the AChEs, discrimination was possible by using recombinant enzymes, specifically modified to possess different sensitivities toward a particular pesticide. To construct the sensor array, four types of native and genetically modified AChEs from different sources have been deposited onto four screen-printed working electrode surfaces. The array was then connected to a commercial four-channel potentiostat, which applies a constant potential to all four electrodes and the resulting currents were recorded on a multichannel recorder for further processing, evaluation, and crossvalidation with the ANNs method. The optimized sensor was tested using solutions of pesticides separately or in mixture, while sample discrimination in real environmental matrices was carried out on spiked samples using several sets of modified electrodes.

Some other multibiosensor systems have been reported as a combination of amperometric, potentiometric, and conductometric sensors coupled with multichannel transducers. For instance, Arkhypova et al. (5) reported a three-enzyme (urease, AChE, and BuChE)-based multibiosensor coupled with potentiometric pH-sensitive transistors and conductometric thin-film interdigitated electrodes for the detection of two different classes of toxic analytes: pesticides and heavy metals. Silber et al. (20) developed a seven-channel enzyme and ion-selective multibiosensor containing 14 working and counter electrodes fabricated by screen-printing technique and a reference Ag/AgCl electrode for dual amperometric–potentiometric measurements of ions (K^+, Li^+), glucose, urea, and lactate simultaneously (20).

19.3.2 Electronic Nose Technology

In the electronic nose technology, cross-reactive chemical-sensing arrays are coupled with an appropriate sample collection, transduction system, and special computational programs and function as an effective analytical tool for detection, quantification, and classification of volatile or semivolatile organics based on their odors (21). These less conventional sensing systems are referred to as electronic noses. A general overview of this technology, the chemistry and fabrication of the sensor arrays, and a description of their characteristics and applications have been extensively discussed in numerous reviews (8–10,22). Most of these sensors are based on quartz crystal piezoelectric (23), metal oxide semiconductor, and conducting polymers technologies (9,24).

The principle of a gas sensor is based on the changes in the electrical resistance when a volatile chemical is absorbed onto its surface. This change in electrical resistance is characteristic to each analyte and can be monitored and used to create a distinct pattern. The samples are collected using a sample collector system, equilibrated, and exposed to the sensor array under controlled temperature, pH, and pressure. In addition to the analyte samples, reference air samples are passed through the sensor array to produce a baseline signal. Finally, the changes in resistance versus time are monitored and recorded on each channel. We have recently used this technology for the detection and classification of OP nerve agents: paraoxon, parathion, malathion, dichlorvos, trichlorfon, and diazinon (4). In this application, an array of 32 conducting polymer sensors was used as a detector for an AS32/8S commercial gas-sensing labstation. The sensor array was fabricated electrochemically by electrodeposition of polypyrrole onto a metallic substrate. The deposition was carried out by cyclic voltammetry using a three-electrode system, in which the metallic substrate was the working electrode. The sensor responses are analyzed using ANN and compared with a new type of classifier, namely SVMs.

Electronic nose systems have been applied for the detection of a variety of toxic analytes such as organophosphorus nerve agent insecticides (24) and volatile organic compounds such as *n*-caproic acid, isoamyl acetate, and ethyl caproate (23). For instance, an electronic nose employing conducting polymer sensor array and pattern recognition techniques was employed in the analysis of chlorinated phenols and PAHs (21). The method was successfully used for the identification of unknown analytes with a recognition rates ranging from 98.8 to 100% for the tested compounds. Using modular sensor system MOSES II electronic nose, Baby et al. (25) detected two water contaminants: lindane and nitrobenzene at concentrations ranging from 1 to 500 ppm. The sensor was also useful for the discrimination of these contaminants based on their odors. In another study, an electronic nose was applied for the detection of different microorganisms (*Enterobacter aerogenes, Echerichia coli,* and *Pseudomonas aeruginosa*) alone and in the presence of different heavy metals (As, Cd, Pb, and Zn) (26).

19.3.3 High-Throughput Multiarray Biosensors for Toxicity Screening

Multiarray biosensors could be integrated to construct high-throughput sensing devices for screening multiple samples. These assays should be fast and suitable for in-field monitoring and could be designed to sense either a single analyte or a family of analytes belonging to the same class. The information provided could be a simple qualitative YES or NO response and could serve to select the samples possessing a certain degree of toxicity (toxicity of risk). These samples will be further exposed to a more advanced quantitative analysis in a specialized analytical laboratory. Depending on the configuration used, the detection method, and the final application of the device, the arrays can contain one or more immobilized bioreagents or can function as a simple chemical sensor. For instance, depending on the type, number, and nature of the analytes of interest, specific bioreagents can be immobilized or coimmobilized onto the surface of individual electrodes. Thus, using a combination of various biosensing approaches in the same array it may be possible to obtain complete information on the composition and the toxicity of the sample.

Today, a variety of high-throughput analytical assays are currently being developed, most of these applied in drug discovery and toxicology. For these specific applications, a typical experimental set-up involves the use of cells or bacteria plated in a 6-, 24-, or 96-well plate and exposed to chemical toxins. The multiarray system is then used to monitor the changes in their activity upon treatment with toxicants. Typically, the detection is carried out using traditional methodology such as spectrophotometric or fluorescence methods and the final system is a probe rather than a sensor. The challenge in adapting this type of analytical set-up for biosensing concept involves (i) designing electrodes in a multiarray arrangement that will correspond to well plate pattern and (ii) developing a transduction system able to convert, analyze, and process information on multiple channels and allow automatization and miniaturization.

In our laboratory, we have used an autonomous high-throughput DOX (dissolved oxygen) multiarray biosensor system for a variety of applications: (i) monitoring bacterial pathogens, (ii) studying the effect of antibiotics on bacterial pathogens, and (iii) cytotoxicity of polyphenols isoflavonoids on cancerous cells (12). On-going works with this system focus on (iv) monitoring enzymatic reactions for the detection of enzyme substrates and inhibitors and (v) immunoassays for the detection of pathogens. The system is fabricated by Daikin Corp., Japan and uses 96 electrodes disposed in a conventional 96-well plate, connected to a multipotentiostat controlled by a computer. It was designed as an oxygen sensor but it can be adapted for any type of amperometric and chronoamperometric measurements involving an oxido-reduction process at constant applied potential. This system is portable, is suitable for monitoring 96 samples simultaneously and is ideal for in-field toxicity screening.

Kim and Gu (27,28) designed a 96-well high-throughput system for toxicity classification and screening of chemicals based on the bioluminescent properties of bacteria. Evidence suggests that in the presence of toxicants, the bioluminescence is affected in a dose-dependent manner and this change can be used as a measure of their toxicity (27,28). Natural bioluminescence is affected by toxicants, but without providing information about the type or the nature of these chemicals. Therefore, changes in bioluminescence can be assigned to a global toxicity of the sample. To increase the selectivity of the device toward a group or a particular toxicant, recombinant bioluminescent bacteria have been used. For instance, Gu et al. utilized four recombinant bioluminescent *E. coli* strains (DPD2511, DPD2540, DPD2794, and TV1061) that respond differently to various toxicants (phenols, hydrogen peroxide, and mitomycin) so that the information could be used to differentiate and classify between various analytes. In this case, the experimental set-up involved the immobilization of cells in a 96-well plate, followed by addition and incubation with toxicants for a specific time. Finally, the bioluminescence was quantified using a 96-well plate luminometer.

19.4 Multiarray Biosensors for Pathogen Detection

The selective and sensitive detection and identification of harmful cells, spores, or viruses in air, water, and food samples are areas of biosensor research with unique challenges. One of the major problems with pathogen detection is long incubation time required. Traditional methods for identifying a particular species of bacteria typically begin by culturing the sample on a selective medium and collimate with identifying specific DNA sequences from the genome of the sample using the polymerase chain reaction (29). While accurate, these methods may take several days to complete. Considering that many of these pathogenic species have generation times of less than 30 min once inside the human body, rapid identification systems are needed to prevent serious illness or even death (30). Currently, it may be necessary to administer treatment to all individuals exposed to a potentially dangerous pathogen before it has been identified.

The most common platform for developing a sensor for early identification of pathogens is the immunosensor (31). Immunosensors in a single or multiarray format have been designed for detection and classification of biological agents ranging from simple bacterial pathogens (e.g., *E. coli*) to more complex biological warfare agents (e.g., *Bacillus anthracis*). These devices are utilized to capture antibodies, which are commonly deposited onto the spots of a single-patterned sensor array surface. The antibody captures its specific target in a multicomposite mixture and the binding is then converted into a signal and measured by a suitable transducer. This transducer may be optical (32), electrochemical (33), or gravimetric (34). There are several factors that must be considered with immunosensor design for pathogen detection: (i) as with any immunoassay, there is always a concern about false positives due to nonspecific binding of other particles in the sample being analyzed; (ii) additionally, the antibodies may lose specificity due to the immobilization process and labeling with fluorescent dyes (35); and (iii) in most cases, an immunoassay cannot differentiate between live and dead cells. For example, in food and water samples, where nonviable cells may not pose a problem, a standard immunoassay may not provide enough information; and (iv) an immunosensor is designed to detect a specific target because the antibody for that target is immobilized within the sensor. With this in mind, it is obvious that the investigator must know what he or she is looking for in the sample. This of course requires some previous knowledge about the sample, and more importantly, other potentially harmful pathogens may go undetected. The multiarray sensors reviewed herein were specifically designed to solve or partially solve one or more of the limitations of existing methods discussed above.

It is pertinent to note that most research laboratories use "simulants" when testing prototype systems for pathogen detection. For example, *Bacillus globigii* is a nonpathogenic, sporulating bacterium, and is a commonly used simulant for the detection of *Bacillus anthracis* (anthrax). Owing to its widespread use, *B. globigii* is being cited in this review as a model to compare the sensitivities of different systems.

19.4.1 Optical Multiarray Sensors for Pathogen Detection

Portability is an important consideration in pathogen detection; consequently, fluorescence-based sensors are the most common type of optical systems reported for this purpose. There is a large variety of commercially available, low-power-consuming diode lasers, photodiode detectors as well as many different bright fluorophores covering a wide spectral range. Fluorescent immunoassays are typically performed in the sandwich format where a capture antibody is immobilized on a surface. After the target antigen is captured, it is tagged with a second "tracer" antibody that is labeled with a fluorescent molecule or conjugated to an enzyme that will catalyze a reaction producing a fluorescent product.

The two most obvious advantages of a multiarray sensor are as follows: (i) individual analytes and controls can be simultaneously assayed multiple times; and (ii) capture surfaces (e.g., antibodies) for different targets can be integrated onto the same sensor surface permitting multiple sample screening for more than one analyte. Fluorescent immunosensors designed in this manner will then require an antibody "cocktail" made up of the required fluorescent-labeled antibodies for detection. This concept has been applied to a 6×6 array patterned onto a glass slide (36). The immunosensor was designed to simultaneously detect three different viral and bacterial analytes (37). Anti-*B. globigii*, SEB, and MS2 bacteriophage antibodies were immobilized onto discrete areas of a glass slide. Detection was accomplished using fluorescent-labeled antibodies and a charge-coupled array for detection. The system correctly detected and identified the analytes in 126 blind samples in approximately 14 min with detection limits similar to ELISA methods. This system was further developed for the detection of nine analytes and produced a limit of detection of 10^5 cfu/ml for *B. globigii* (38). The feasibility of using the fluorescent sandwich immunoassay format on glass slides for actual food samples was examined (39). The system was able to identify the intestinal disease-causing microorganisms, *Campylobacter* and *Shigella*, in a variety of food samples with detection limits as low as 4.9×10^4 cfu/ml. The planar array immunoassay format was coupled to microarray printing technology to deposit antibodies for different analytes onto the glass slides (40). The microarray technology has the advantage of automated production and detection as well as small sample volume requirement. The system used a prototype flow module containing five channels to introduce samples and fluorescent-labeled antibody cocktails onto the microarray slide. Each channel could contain up to 88 capture antibody spots. Using a sandwich immunoassay format with fluorescent-labeled detection antibodies, the system produced a limit of detection for *B. globigii* spores of 6.2×10^4 cfu/ml in 15 min under flow conditions.

Fluorescent sandwich immunoassays can be performed using the wave-guiding properties of optical fibers (41). In this setup, the capture antibody is immobilized on the outside of the optical element. The excitation source is guided through the core of the element producing an evanescent field on the outside of the fiber. The field can excite fluorescent-labeled probe antibodies and guide the excitation signal to the detector. The RAPTOR fiber optic sensor uses a polystyrene optical fiber arrays and can simultaneously test for four different analytes (42). The instrument is completely portable and automated using a diode laser for excitation and photodiodes for detection. Using antibody cocktails for the detection of multiple analytes and a 10-min assay time, the RAPTOR was used to detect 98% of the target analytes with detection limits of 5×10^5 and 5×10^4 cfu/ml for *Francisella tularensis* and *B. globigii*, respectively.

A 4×4 array sensor utilizing a fluorescent ELISA format has been developed (43). Again, the sandwich immunoassay was employed. In contrast to other sensors that use fluorescent-labeled antibodies for detection, this system utilizes signal amplification from alkaline phosphatase-labeled secondary antibodies. Enzymatic cleavage of the substrate produces a product that can be excited by a low-power diode laser. The system uses a multifunctional CMOS biochip that contains a photodiode array and signal processing integrated into the chip (44). The limit of detection for *B. globigii* was 100 spores.

The autonomous pathogen detection system is a multiplexed, liquid array system that utilizes polystyrene beads and a flow cytometer (45). The 5.5-μm commercially available polystyrene beads are imbedded with red and infrared dyes in precise ratios giving each bead a unique spectral address. Antibodies specific for each target were then immobilized onto one type of spectrally encoded bead. After the target is captured, it is sandwiched with a green fluorescent-labeled secondary antibody. When the bead passes through the flow cytometer, the red laser identifies the bead (and therefore, identifies the captured analyte) and the green laser quantifies the amount of target captured. In theory, this system can analyze a sample for up to 100 different analytes simultaneously. Using a 65-min assay, the system was tested for four different analytes and produced limits of detection of 1.54×10^6 and 5×10^6 cfu/ml for *B. globigii* and *Erwinia herbicola*, respectively, with minimal cross-reactivity.

A multiarray sensor that relies on the intrinsic fluorescence of cellular components has been designed (46). This portable system addresses several important issues: the fluorescence is detected from excited metabolites that are present in the cell only during cell respiration. Therefore, only viable cells should be detected. The intrinsic fluorescence detection concept also eliminates the need for indirect detection with a secondary molecule such as fluorescent-labeled antibodies. This detection system is sensitive enough to detect as few as 20 cells/ml. The system uses a 4×8 array patterned on a glass slide. As with the detection system, antibodies are not used for cell capture. Instead, small molecules such as the iron-containing porphyrin, hemin, are patterned onto the slide (47). A large variety of pathogens that require iron from their hosts will attach to hemin. This strategy can be applied for screening food or water supplies for general contamination. For specific detection of organisms such as *Staphylococcus aureus*, small peptides identified from phage display libraries were also immobilized on the chip. Peptides identified in this manor have been shown to be very specific for cells and spores, although the exact binding site is not known (48,49). The above strategy illustrates the concept of multiligand capture of one target. Detection schemes such as this can provide more confident results compared with systems that exclusively use a single antibody for pathogen capture.

Figure 19.3 illustrates an example of an "optimal" sensor array based on some of the technologies discussed above: (A) the sensing surface comprised multiple ligands (e.g., antibodies, peptides, and hemin) specific for selected-target bacteria. A fourth site,

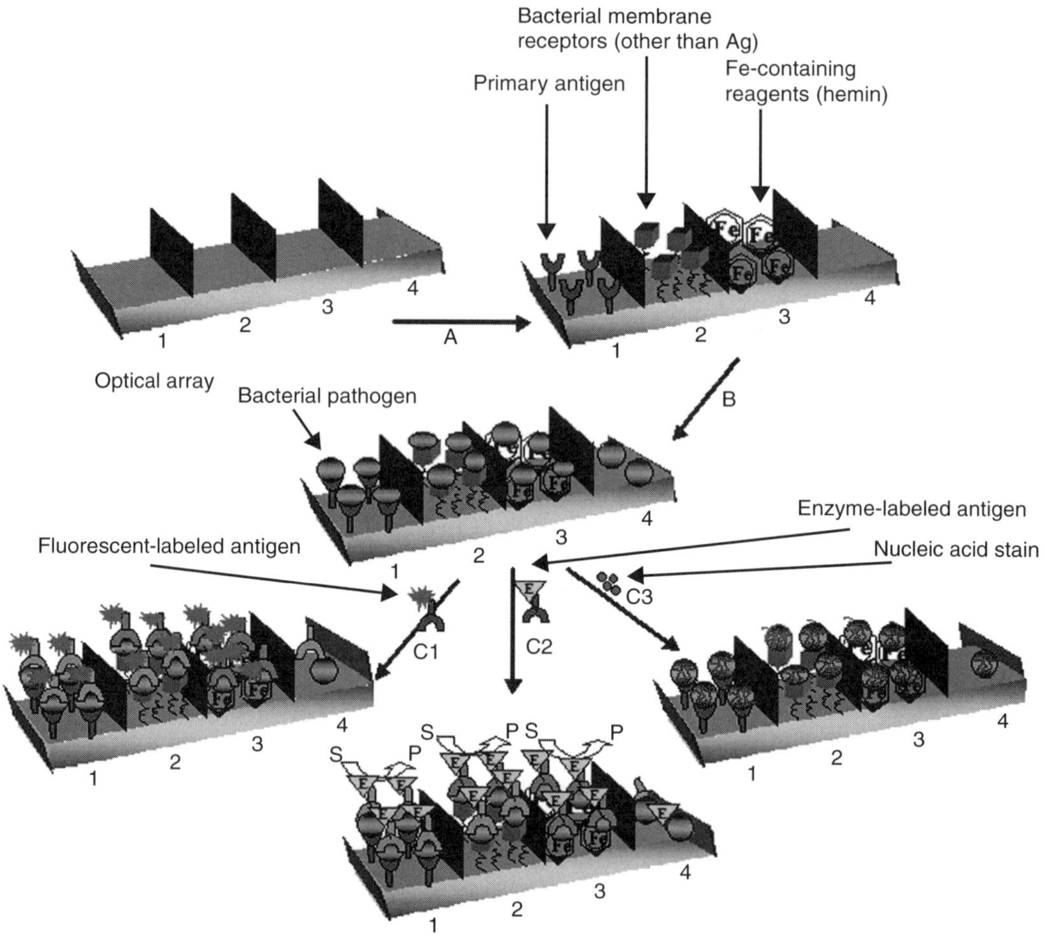

FIGURE 19.3
(See color insert following page 330) Possible configurations for array immunosensor.

with no immobilized molecule, is shown as a control. (B) Binding to all three sites would provide a more confident result as compared with a sensor with a single type of molecule comprising the capture surface. (C1) The captured cells can then be tagged with a fluorescent-labeled antibody for detection. (C2) Alternatively, signal amplification can be employed using an antibody–enzyme conjugate and an appropriate substrate (ELISA). (C3) Cells can also be labeled with nucleic acid stains that are selective for live cells only or all cells (viable cells and those with compromised membranes).

19.4.2 Electrochemical Multiarray Sensors for Pathogen Detection

Electrochemical sensors for pathogen detection are an attractive area of research because of the simple instrumental requirements that are amenable to miniaturization. Unlike optical systems, electrochemical sensors do not require excitation sources, filters, and lenses. Additionally, they are not susceptible to interference from turbid samples. Electrochemical sensors, including immunosensors, have been reviewed elsewhere (50).

A microelectrode immunosensor array for the detection of *E. coli* 0157:H7 was developed (51). Each sensor chip had a total of 1,700 electrodes in a total area of 9.6 mm^2.

Antibodies were immobilized onto oxide surface of the chip between the gold electrode fingers. The change in impedance from 100 Hz to 10 MHz at an applied potential of 50 mV upon cell binding was measured. The sensor showed a linear response to changes in cell concentration, an analysis time of 10 min and a detection limit of 10^4 cfu/ml, and the ability to detect *E. coli* from real food samples. The system also performed well in the presence of other microbes, giving no false positives.

A multianalyte immunosensor was developed using a commercially available microelectrode array chip (52). The chip contains 1000 individually addressable electrodes in a 1-cm^2 area. Short-oligonucleotide sequences are immobilized onto the electrodes allowing for attachment of biomolecules conjugated to the complementary sequence. Using this strategy a sandwich immunoassay was developed. Antibodies for five different targets, including *B. globigii* spores, were conjugated to DNA sequences and immobilized onto selected electrodes. Captured targets were then tagged with a secondary antibody conjugated to biotin. Finally to a horseradish peroxidase enzyme–avidin conjugate was added. The product of the enzymatic reaction was then reduced at the platinum electrode surface at 0.3 V. The system was successfully tested for the simultaneous detection of all five different analytes using a checkerboard pattern on the electrodes. The limits of detection for most analytes were in the attomolar range with an assay volume of 50 μl. Assays could be completed in 12 min, with maximum sensitivity occurring in 60 min.

A screen-printed array based on coulometry was developed to identify different strains of *E. coli* (53). In this strategy, lectins, proteins that bind oligosaccharides with high affinity and specificity, were immobilized onto porous membranes. It was shown that different strains of *E. coli* would bind different lectins with specific patterns. Although ten different lectins were used, only five were required to differentiate between the four strains of *E. coli* tested. The membranes were incubated in the sample containing the cells as well as the respiratory substrates succinate and formate and the redox mediator ferricyanide. The membranes were then placed over the electrode arrays. Viable cells reduce the ferricyanide, which is then oxidized at the electrode surface and measured coulometrically. Computer software was used to analyze the lectin-binding pattern and correlate it to patterns of known standards. This system shows the capability to capture a wide variety of bacteria because antibodies are not used for capture. Additionally, only viable cells are detected at the electrode making this an attractive method for food quality control.

An electronic tongue was used to recognize six microbial species, including bacteria, mold, and yeast commonly found as food contaminates (54). The system is based on voltammetry where a pulse is applied over an array of metal electrodes and the resulting current is measured. Redox-active species, commonly found as metabolites of respiration, are measured. Each species in the sample will have characteristic metabolites present at different phases of the growth cycle. Sample data is analyzed using multivariate methods and compared with stored data patterns obtained from standards of pure microbial strains. The method addresses two important problems in food quality control: first, because the measured analytes are products of metabolism, only viable microbes are detected. Second, the system itself does not require a biological component, such as antibodies for capture or detection. The work demonstrated that the method could identify and differentiate between microorganisms by electrochemical detection of metabolites but statistical measurements such as the limit of detection were not made.

Similarly, the electronic nose has been shown to be capable of identifying microbial species based on the volatile compounds excreted from bacteria and yeast growing on nutrient agar plates (55). The setup consisted of a 16-sensor array of electroconductive polymers. The system correctly classified 228 out of 244 samples (93.4%) containing 12 different types of bacteria and 1 type of yeast. For seven species samples, all were classified correctly. Limits of detection were not established. The electronic nose was also used to sample the headspace of microbes in potable water (56). Two different electronic nose systems were tested, both using electroconducting polymer arrays. Purified water samples were inoculated with 10^4 cfu/ml of three bacterial species. Additionally, the samples contained three fungal species. After inoculation, the samples were incubated at 25°C for 24 h. Both systems correctly identified the six species when compared with purify water samples. When the inoculation concentration of the bacteria was 10^2 cfu/ml, the discrimination between species was not clear, although the samples were still distinctly different than the controls. Another important observation from this work is that the volatile compound profiles of bacterial samples in the presence of low concentrations of heavy metals were different than those obtained in pure water samples. This is an important consideration in systems such as the electronic nose when samples are identified from pattern recognition of stored "digital fingerprints."

19.4.3 Mass-Sensitive Sensors for Pathogen Detection

Mass-sensitive sensors, such as the quartz crystal microbalance (QCM), have been successfully applied for the detection of antibody–antigen binding (57). Small mass changes on the surface of the crystal are detected by changes in the crystals oscillation frequency. While sensitive, sensors such as the QCM are usually not adaptable for multiarray production. A four-element magnetoelastic sensor array has been developed that works in a similar manner as the QCM (58). The sensor itself is made of a gold-coated alloy, ribbon-like microcantilevers that oscillate at a characteristic frequency in the presence of a magnetic field. Antibodies for *E. coli* were immobilized onto the cantilevers. After cell binding, an alkaline phosphatase enzyme-linked tracer secondary antibody was added. The product of catalysis formed a precipitate which mass loaded the cantilever changing its oscillation frequency. The sandwich immunoassay ELISA format produced a limit of detection for *E. coli* cells of 10^2 cfu/ml.

19.5 The Role of Computational Techniques in Processing Multiarray Biosensor Data

When working with multiarray biosensors, computational techniques are necessary to interpret, process, and analyze data generated on multiple channels, especially when a precise and accurate classification and differentiation of analytes is required. Experiments with multiarray biosensors can generate enormous amounts of data in a very short time that may be difficult to interpret using commercially available programs. In addition, due to the large amount of data generated, data reduction software may also be necessary. Therefore, appropriate software connected to the sensing system that can rapidly search and identify a specific toxicant in a database of well-characterized toxic compounds could significantly improve the overall performances of a multiarray sensor and speed up the investigation process. With the recent advances in microarray and high-throughput systems, chemometrics and computational methods become more and more important. These methods can be used to create a model system able to extract and analyze relevant multivariate data from chemical information and provide a better and more useful interpretation and understanding of the process being investigated (11). In a recent review, Lavine and Workman discussed the role of chemometric analysis in sensors and microarrays and presented an overview of the existing methods and their applications in instrumental analysis. Jurs et al. (10) presented a comprehensive review on computational methods for the analysis of chemical sensor array data covering the basis of each approach, its suitability and applications, and a critical discussion of each method.

Data processing methods include chemometric techniques, ANNs, pattern recognition, or a combination of these (8). The selection of an appropriate method for data analysis is strongly dependent upon a number of experimental variables as well as the nature and the type of information under consideration (10). Depending on the final application and required performance of the system, computational methods in conjunction with multiarray sensors can provide qualitative or quantitative analysis or both. Typically, the data processed using computational methods are graphically presented. However, when the amount of data is very high, graphical presentation may be difficult, and therefore additional methods such as principal component analysis are needed to reduce the dimensionality of datasets (10). In some cases, two or more computational methods are used and compared so as to achieve a higher effectiveness. For instance, in a recent study, we have compared the performance of two computational methods when investigating detection and classification of OP nerve agents: ANNs and a novel learning scheme, SVMs. For this specific application, three SVM kernels were tested: the Scholkopf kernel, the polynomial, and the Gaussian radial basis function. The use of SVM methods is a recent development, almost unknown from analytical chemistry perspective. Experimental results showed that SVMs provide a more accurate classification of OP than the traditional ANN as machine learning and recognition programs (4). The sensor design utilized arrays of conducting polymers from a commercial electronic nose instrument to generate molecular patterns for the six different OPs: parathion, malathion, dichlorvos, trichlorfon, paraoxon, and diazinon. In terms of classification, dichlorvos, trichlorfon, diazinon, parathion, and paraoxon were classified perfectly. This is remarkable considering the resemblances in the chemical structures of paraoxon and parathion with only a single atom difference. Malathion was detected at 95%. When the ANN was used, the positive predictions were inconsistent for the range of chemicals tested.

19.6 Conclusion and Future Trends

The current trends in the development of biosensors for toxicity monitoring and screening are conducted through small and fast multiarray sensors that will allow detection of multiple samples simultaneously, with substantial decrease in costs per sample throughput. These devices are expected to function as alarm systems by sensing the presence of toxicants up to a fixed threshold value. Ideally, multiarray biosensors should generate a pattern of known analytical signals with distinct characteristics for samples or analytes that could be subsequently used to correctly classify and differentiate between unknown samples possessing various degrees of toxicity. The information could be qualitative or semiquantitative and extremely useful for an initial screening of toxicants in the field, before more sophisticated investigations could be performed in a specialized analytical laboratory. The real success in the development of a reliable multiarray biosensor depends on several parameters including the type and the number of electrodes composing the array, the sophistication of the transductor system, which may facilitate efficient signal conversion on multiple channels, as well as the availability of a powerful data processing and quantification program. This chapter provided a general overview of current multiarray biosensors including details regarding their design, signal transduction data treatment, and applications for the toxicity monitoring and bacterial pathogens.

In the future, the development of multiarray sensors for toxicity monitoring could find major use as "*screening devices*" of field samples capable of recognizing and classifying a chemical or biological toxicant. Advances in the field of bioelectronics could facilitate the design of novel transduction systems able to collect and process more efficient analytical signals generated on multiple channels. Current trends are directed toward miniaturization and integration of multiarray sensors with computational techniques for data processing and analysis that could simplify and extend toxicity monitoring outside a central laboratory.

References

1. Thevenot, D.R., Toth, K., Durst, R.A., Wilson, G.S. (2001). Electrochemical biosensors: recommended definitions and classification. *Biosens. Bioelectron.* 16:21–131; (b) Patel, P.D. (2002). (Bio)sensors for measurement of analytes implicated in food safety: a review. *Trends Anal. Chem.* 21(2):96–115.
2. Andreescu, S., Sadik, O.A. (2004). Trends and challenges in biochemical sensors for clinical and environmental monitoring. *Pure Appl. Chem.* 76(4):861–878.
3. Sadik, O.A., Wanekaya, A.K., Andreescu, S. (2004). Advanced in analytical technologies for environmental protection and public safety. *J. Environ. Monitor.* 6:413–522; (b) Sadik, O.A., Witt, D.M. (1999). Monitoring endocrine-disrupting chemicals. *Environ. Sci. Technol.* 1:368–374.
4. Sadik, O.A., Land, W.H., Wanekaya, A.K., Uematsu, M., Embrechts, M.J., Wong, L., Leibensperger, D., Volykin, A. (2004). Detection and classification of organophosphate nerve agents simulants using support vector machines with multiarray sensors. *J. Chem. Inf. Comput. Sci.* 44:499–507.
5. Arkhypova, V.N., Dzyadevych, S.V., Soldatkin, A.P., El'skaya, A.V., Jaffrezic-Renault, N., Jaffrezic, H., Martelet, C. (2001). Multibiosensor based on enzyme inhibition analysis for determination of different toxic substances. *Talanta* 55:919–927.
6. Bachmann, T.T., Leca, B., Vilatte, F., Marty, J.L., Fournier, D., Schmid, R.D. (2000). Improved multianalyte detection of organophosphates and carbamate with disposable multielectrode biosensors using recombinant mutants of *Drosophila* acetylcholinesterase and artificial neural networks. *Biosens. Bioelectron.* 15:193–201.
7. Bachmann, T.T., Schmid, R.D. (1999). A disposable multielectrode biosensor for rapid simultaneous detection of the insecticides paraoxon and carbofuran at high resolution. *Anal. Chim. Acta.* 401:95–103.
8. Albert, K.J., Lewis, N.S., Schauer, C.L., Sotzing, G.A., Stitzel, S.E., Vaid, T.P., Walt, D.R. (2000). Cross-reactive chemical sensor arrays. *Chem. Rev.* 100:2595–2626.
9. Albert, K.J., Walt, D.R. (2003). Information coding in artificial olfaction multisensor arrays. *Anal. Chem.* 75:4161–4167.
10. Jurs, P.C., Bakken, G.A., McClelland, H.E. (2000). Computational methods for the analysis of chemical sensor array data from volatile analytes. *Chem. Rev.* 100:2649–2678.
11. Lavine, B., Workman, J. (2004). Chemometrics. *Anal. Chem.* 76:3365–3372.
12. Andreescu, S., Sadik, O.A., McGee, D.W., Suye, S. (2004). Autonomous multielectrode system for monitoring the interaction of isoflavonoids with lung cancer cells. *Anal. Chem.* 76(8):2321–2878.
13. Iqbal, S.S., Mayo, M.W., Bruno, J.G., Bronk, B.V., Batt, C.A., Chambers, J.P. (2000). A review of molecular recognition technologies for detection of biological threat agents. *Biosens. Bioelectron.* 15:549–578.
14. Yan, F., Sadik, O.A. (2001). Enzyme-modulated cleavage of dsDNA for studying interfacial biomolecular interactions. *J. Am. Chem. Soc.* 123:11335–111340.
15. K'owino, I.O., Agarwal, R., Sadik, O.A. (2003). Novel electrochemical detection scheme for DNA binding interactions using monodispersed reactivity of silver ions. *Langmuir* 19(10):4344–4350.

16. Fojita, M. (2002). Electrochemical sensors for DNA interactions and damage. *Electroanalysis* 14(21):1449–1463.
17. Vo-Dinh, T., Isola, N., Alarie, J.P., Landris, D., Griffin, G.D., Allison, S. (1998). Development of a multiarray biosensor for DNA diagnostics. *Instrument. Sci. Technol.* 26(5):503–514.
18. Dill, K., Montgomery, D.D., Ghindilis, A.L., Schwarzkopf, K.R., Ragsdale, S.R., Oleinikov, A.V. (2004). Immunoassays based on electrochemical detection using microelectrode arrays. *Biosens. Bioelectron.* 20: 736–742.
19. Solna, R., Sapelnikova, S., Skladal, P., Winter-Nielsen, M., Carlsson, C., Emneus, J., Ruzgas, T. (2005). Multienzyme electrochemical array sensor for determination of phenols and pesticides. *Talanta* 65:349–357.
20. Silber, A., Bisenberger, M., Brauchle, C., Hampp, N. (1996). Thick-film multichannel biosensors for simultaneous amperometric and potentiometric measurements. *Sens. Actuat.* 30:127–132.
21. Masila, M., Sadik, O.A. (2000). *Chemical and Biological Sensors for Environmental Monitoring* (A. Mulchandani and O.A. Sadik, Eds.), p. 37–59. ACS Symposium Series 762, Oxford University Press, Washington D.C.
22. Ampuero, S., Bosset, J.O. (2003). The electronic nose applied to dairy products: a review. *Sens. Actuat.* 94:1–12.
23. Wu, T.Z. (1999). A piezoelectric biosensor as an olfactory receptor for odour detection: electronic nose. *Biosens. Bioelectron.* 14:9–18.
24. Grate, J.W., Rose-Pehrsson, S.L., Venezky, D.L., Klusty, M., Wohltjen, H. (1993). Smart sensor system for trace organophosphorus and organosulfur vapour detection employing a tem perature-controlled array of surface acoustic wave sensors, automated sample preconcentration and pattern recognition. *Anal. Chem.* 65:1868–1881.
25. Baby, R.E., Cabezas, M., Walsoe de Reca, E.N. (2000). Electronic nose: a useful tool for monitoring environmental contamination. *Sens. Actuat.* 69:214–218.
26. Canhoto, O.F., Magan, N. (2003). Potential for detection of microorganisms and heavy metals in potable water using electronic nose technology. *Biosens. Bioelectron.* 18:751–754.
27. Kim, B.C., Gu, M.B. (2003). A bioluminescent sensor for high throughput toxicity classification. *Biosens. Bioelectron.* 18:1015–1021; (b) Kim, B.C., Park, K.S., Kim, S.D., Gu, M.B. (2003). Evaluation of a high throughput toxicity biosensor and comparison with a *Daphnia magna* bioassay. *Biosens. Bioelectron.* 18:821–826.
28. Kim, J.H., Gu, M.B. (2005). An integrated minibiosensor system for continuous water toxicity monitoring. *Biosens. Bioelectron.* 20(9):1744–1749
29. Wilson, W.J., Erler, A.M., Nasarabadi, S.L., Skowronski, E.W., Imbro, P.M. (2004). A multiplexed PCR-coupled liquid bead array for the simultaneous detection of four biothreat agents. *Molec. Cell. Probes* 19(2):137–144.Prescott, L.M., Harley, J.P., Klein, D.A. (1999). *Microbiology.* 4th ed. WCB/Mcgraw Hill, Boston.
30. Prescott, L.M., Harley, J.P., Klein, D.A. (1999). *Microbiology.* 4th ed. Boston: WCB/Mcgraw Hill.
31. Luppa, P.B., Sokoll, L.J., Chan, D.W. (2001). Immunosensors—principles and applications to clinical chemistry. *Clin. Chim. Acta* 314:1–26.
32. Ligler, F.S., Breimer, M., Golden, J.P., Nivens, D.A., Dodson, J.P., Green, T.M., Haders, D.P., Sadik, O.A. (2002). Integrating waveguide biosensor. *Anal. Chem.* 74:713–719.
33. Ruan, C., Yang, L., Li, Y. (2002). Immunobiosensor chips for detection of *Escherichia coli* 0157:H7 using electrochemical impedance spectroscopy. *Anal. Chem.* 74:4814–4820.
34. Cavic, B.A., Hayward, G.L., Thompson, M. (1999). Acoustic waves and the study of biochemical macromolecules and cells at the sensor-liquid interface. *Analyst* 124:1405–1420.
35. Hoyer-Hansen, G., Hamers, M.J.A.G., Pedersen, A.N., Nielsen, H.J., Brunner, N., Dano, K., Stephens, R.W. (2000). Loss of ELISA specificity due to biotinylation of monoclonal antibodies. *J. Immunol. Meth.* 235:91–99.
36. Rowe, C.A., Scruggs, S.B., Feldstein, M.J., Golden, J.P., Ligler, F.S. (1999). An array immunosensor for simultaneous detection of clinical analytes. *Anal. Chem.* 71:433–439.
37. Rowe, C.A., Tender, L.M., Feldstein, M.J., Golden, J.P., Scruggs, S.B., MacCraith, B.D., Cras, J.J., Ligler, F.S. (1999). Array biosensor for simultaneous identification of bacterial, viral and protein analytes. *Anal. Chem.* 71:3846–3852.

38. Rowe Taitt, C., Anderson, G.P., Lingerfelt, B.M., Feldstein, M.J., Ligler, F.S. (2002). Nine analyte detection using an array based biosensor. *Anal. Chem.* 74:6114–6120.
39. Sapsford, K.E., Rasooly, A., Taitt, C.R., Ligler, F.S. (2004). Detection of *Campylobacter* and *Shigella* species in food samples using an array biosensor. *Anal. Chem.* 76:433–440.
40. Delehanty, J.B., Ligler, F.S. (2002). A microarray immunoassay for simultaneous detection of proteins and bacteria. *Anal. Chem.* 74:5681–5687.
41. King, K.D., Anderson, G.P., Bullock, K.E., Regina, M.J., Saaski, E.W., Ligler, F.S. (1999). Detecting staphylococcal enterotoxin B using an automated fiber optic biosensor. *Biosens. Bioelectron.* 14:163–170.
42. Anderson, G.P., King, K.D., Gaffney, K.L., Johnson, L.H. (2000). Multi-analyte interrogation using the fiber optic biosensor. *Biosens. Bioelectron.* 14:771–777.
43. Straitis-Cullum, D.N., Griffin, G.D., Mobley, J., Vass, A.A., Vo-Dinh, T. (2003). A miniature biochip system for detection of aerosolized *Bacillus globigii* spores. *Anal. Chem.* 75:275–280.
44. Vo-Dinh, T., Alarie, J.P., Isola, N., Landis, D., Wintenberg, A.L., Ericson, M.N. (1999). DNA biochip using a phototransistor integrated circuit. *Anal. Chem.* 71:358–363.
45. McBride, M.T., Gammon, S., Pitesky, M., O'Brien, T.W., Smith, T., Aldrich, J., Langlois, R.J., Colston, B., Venkateswaran, K.S. (2003). Multiplexed liquid arrays for simultaneous detection of simulants of biological warfare agents. *Anal. Chem.* 75:1924–1930.
46. Estes, C., Duncan, A., Wade, B., Lloyd, C., Ellis, Jr., W., Powers, L. (2003). Reagentless detection of microorganisms by intrinsic fluorescence. *Biosens. Bioelectron.* 18:511–519.
47. Mason, H.Y., Lloyd, C., Dice, M., Sinclair, R., Ellis, Jr., W., Powers, L. (2003). Taxonomic identification of microorganisms by capture and intrinsic fluorescence detection. *Biosens. Bioelectron.* 18:521–527.
48. Turnbough, Jr., C.L., (2003). Discovery of phage display ligands for species specific detection of *Bacillus* spores. *J. Microbiol. Meth.* 53:263–271.
49. Sato, A., Ida, N., Fukuyama, M., Miwa, K., Kazami, J., Nakamura, H. (1996). Identification from a phage display library of peptides that bind to toxic shock syndrome toxin-1 and that inhibit its binding to major histocompatibility complex (MHC) class 2 molecules. *Biochemistry* 35:10441–10447.
50. Bakker, E., Telting-Diaz, M. (2005). Electrochemical sensors. *Anal. Chem.* 74:2781–2880.
51. Radke, S.M., Alocilja, E.C. (2005). A high density microelectrode array biosensor for detection of *E. coli* 0157:H7. *Biosens. Bioelectron.* 20:1662–1667.
52. Dill, K., Montgomery, D.D., Ghindilis, A.L., Schwarzkopf, K.R., Ragsdale, S.R., Oleinikov, A.V. (2004). Immunoassays based on electrochemical detection using microelectrode arrays. *Biosens. Bioelectron.* 20:736–742.
53. Ertl, P., Wagner, M., Corton, E., Mikkelsen, S.R. (2003). Rapid identification of viable *Escherichia coli* subspecies with an electrochemical screen-printed biosensor array. *Biosens. Bioelectron.* 18:907–916.
54. Söderström, C., Winquist, F., Krantz-Rülcker, C. (2003). Recognition of six microbial species with an electronic tongue. *Sens. Actuat.* 89:248–255.
55. Gibson, T.D., Prosser, O., Hulbert, J.N., Marshall, R.W., Corcoran, P., Lowery, P., Ruck-Keene, E.A., Heron, S. (1997). Detection and simultaneous identification of microorganisms from headspace samples using an electronic nose. *Sens. Actuat.* 44:413–422.
56. Canhoto, O., Magan, N. (2004). Electronic nose technology for the detection of microbial and chemical contamination of potable water. *Sens. Actuat.* 44: 413–422.
57. Ebersole, R.C., Ward, M.D., (1998). Amplified mass immunosorbent assay with a quartz crystal microbalance. *J. Am. Chem. Soc.* 110:8623–8628.
58. Ruan, C., Zeng, K., Varghese, O.K., Grimes, C.A. (2003). Magnetoelastic immunosensors: amplified immunosorbent assay for the detection of *Escherichia coli* 0157:H7. *Anal. Chem.* 75:6494–6498.

20

Approaches to Allergy Detection Using Aptasensors

Emile Brys, Sara Tombelli, Maria Minunni, Marco Mascini, and Anthony P. F. Turner

CONTENTS

20.1 Immunoglobulins and Immunoglobulin E

Antibodies are glycoprotein molecules which are produced by plasma cells in response to an immunogen (1). They belong to a class of spherical proteins called globulins and are hence known as immunoglobulins (Igs). Their primary function is to mediate the host immune response by binding to antigens.

Immunoglobulins have been divided, on the basis of physical and functional properties, into the five following classes: IgG, IgM, IgA, IgD, and IgE. They have a Y-shaped structure and are composed of two identical light chains and two identical heavy chains, which are held together by disulfide bonds (2), as illustrated in Figure 20.1. Each heavy chain and each light chain has a variable (V) domain and a constant (C) domain. The V domain is responsible for binding to the antigen, and the C domain mediates the antibody's function (2).

20.1.1 Immunoglobulin E

Human IgE is a monomeric immunoglobulin of 190,000 Da (3). Initially termed IgND after its discovery by Johansson and Benich (4), it was soon confirmed to be identical to the γE discovered by Ishizaka and Ishizaka (5), and determined to be central in the immediate hypersensitivity reaction pathway (6). Of similar structure to the other immunoglobulins (IgG, IgM, IgD, and IgA), it is composed of two light chains and two heavy chains, as shown in Figure 20.2, and constitutes about 0.0005% of total serum immunoglobulins in adults (3).

20.1.2 The Hypersensitivity Reaction

The term "allergy" was originally introduced by von Pirquet in 1906, meaning "changed reactivity" of the host after the second or subsequent contact with an allergenic agent (7). The first evidence of a transferable/soluble factor as the mediator of an allergic reaction

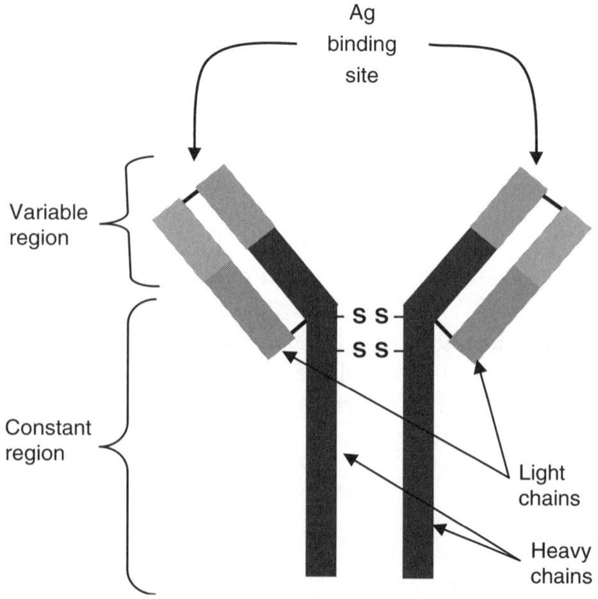

FIGURE 20.1
General structure of an immunoglobulin (Ig).

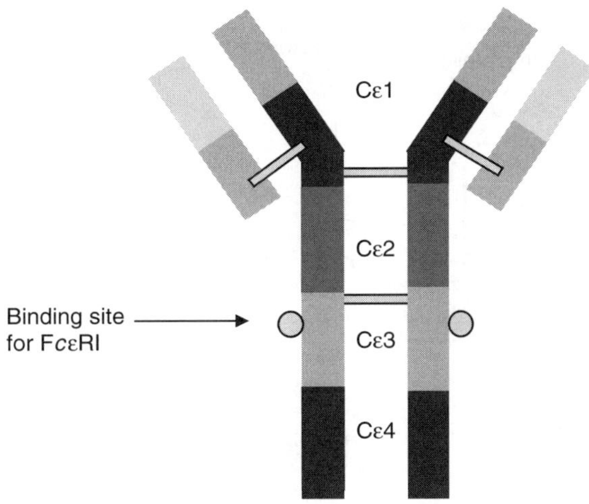

FIGURE 20.2
Structure of IgE. (Modified from Hamelmann E, Rolinck-Werninghaus C and Wahn U 2002. From IgE to Anti-IgE: Where do we stand? *Allergy* 57: 983–994.)

was published in 1919 by Ramirez (8), where the author reported on an incident during which a man entering a horse-drawn carriage experienced an acute asthmatic episode two weeks after having received a blood transfusion from a man with a known horse allergy (7). It was not until 1921, however, that Prausnitz and Küster gave the first scientific description of the mechanism of the allergic reaction, showing that hypersensitivity could be transferred from an allergic patient to a nonallergic patient through a serum factor (7). Type I hypersensitivity reactions, in which IgE plays a central role, include wheal and flare eruptions of the skin, sneezing, rhinorrhea, and conjunctival irritation (9). More serious conditions include asthma and anaphylaxis, which are believed to share a similar pathogenesis (9).

The production of IgE is mainly under control of T cells and T cell cytokines. T cells can undergo two separate differentiation pathways, namely Th1 and Th2, in response to distinct stimuli, and are characterized by the type of cytokines predominantly produced (7). Th1 cells secrete, amongst others, interleukin (IL)-2, interferon-γ (IFN-γ), tumor necrosis factor-α (TNF α), and lymphotoxin (LT). These cytokines mobilize the cellular and humoral defense mechanisms against intracellular pathogens and antagonise IgE responses (9). The cytokines secreted by Th2 cells include IL-4, IL-5, IL-6, IL-9, and IL-13. These cytokines coordinate the host defense against large, extracellular pathogens (9).

The hypersensitivity cascade pathway, depicted in Figure 20.3, begins with the presentation of an antigen/allergen to a T helper cell by an antigen-presenting cell (APC) and, in this case, native T cells undergo differentiation along the Th2 pathway, defined by the production of Th2 cytokines (7), as described above. IL-4 and IL-13 then direct the differentiation of B cells to IgE-producing plasma cells (10). This differentiation also requires interaction between the CD40 antigen, a cell-surface receptor expressed on B cells, and its ligand (CD154) on the T-cell surface (9). The biological activities of the free circulating IgE are then mediated through specific receptors (7): the high-affinity receptor (FcεRI) (see Figure 20.2) is mainly expressed on mast cells and basophiles, whereas the low-affinity receptor (FcεRII) is expressed on B cells. Following binding of IgE to the high-affinity receptors, mast cells remain sensitized for up to 12 weeks, thereby greatly increasing the half-life of free-circulating serum IgE, which is of 2½ days (7). Reexposure of sensitized patients to the allergen leads to specific binding of the allergen to the IgE–FcεRI complexes on the mast cells. Crosslinking of the receptors then triggers the release of inflammatory cytokines, chemokines, and mediators, such as histamine and heparin (7). It is these molecules that cause the symptoms observed in the allergic response. IgE is thus the "gatekeeper" of immediate type I hypersensitivity.

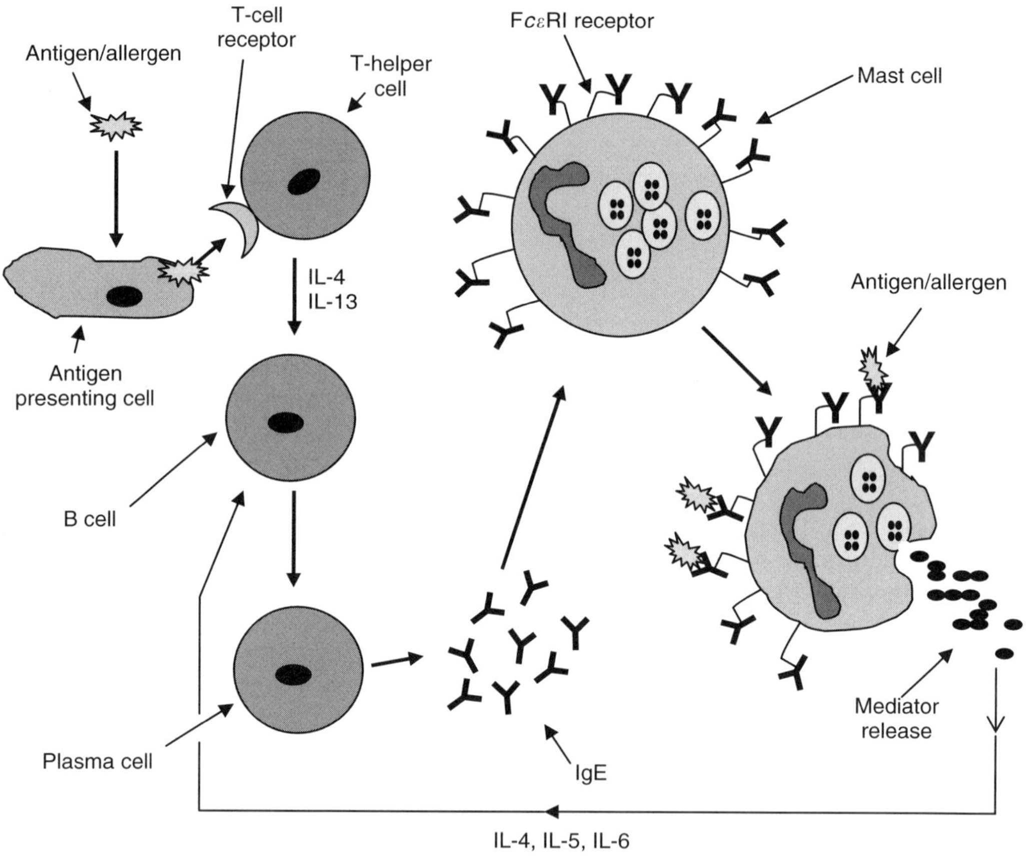

FIGURE 20.3
The allergic cascade. (Modified from Hamelmann E, Rolinck-Werninghaus C and Wahn U. 2002. From IgE to Anti-IgE: Where do we stand? *Allergy* 57: 983–994.)

20.1.3 Importance of Immunoglobulin E

Allergies represent a very serious health problem, with the prevalence of asthma, hay fever, and other IgE-mediated diseases dramatically increasing over the past two to three decades in industrialized countries (7; 11). In the United States, for example, asthma prevalence increased by 75% between 1980 and 1994 (12), with an estimated 15 million individuals affected and more than 5000 asthmatics dying annually (11). In Britain, 23% of children aged 6–7 years and 21% of children aged 12–14 years suffer from asthma (13). Furthermore, in Sweden, the number of children with allergic rhinitis, asthma, or eczema doubled during the years 1979–1991 (14). In addition, the high degree of morbidity associated with allergic disorders requires large quantities of costly medication to relieve symptoms. In the United States for example, the total annual costs for treating asthma were estimated at $6 billion in 1997 (15).

A number of theories attempting to explain the observed increase in allergic disease have been put forward, many of them centring on the role of the western lifestyle. Reasons put forth include the increase in air pollution as well as a lack of exposure of the immune system to infections due to increased sanitization and vaccinations, resulting in the failure of the immune system to "learn" optimally (10). Although there is much evidence to support these hypotheses, the causes remain complex and many unknowns still remain.

The above examples expose the severity and scope of allergic diseases and consequently highlight the importance of IgE as a medical analyte.

20.1.4 Current Diagnosis Methods

The primary aim of allergy testing is to determine whether an individual presenting with symptoms of an allergic hypersensitivity has demonstrable allergen-specific IgE. The first step in the diagnostic procedure is an assessment of symptom history and a physical examination. Once the clinician has concluded that there is a high degree of suspicion of an allergic disorder, in vivo and in vitro analyses for the presence of IgE can be performed to strengthen the likelihood of the diagnosis being correct (3). In vivo testing refers to skin testing, which has become the standard to which other detection methods are compared, and is the primary method for detection of specific IgE (6). In vitro tests, examined in more detail below, refer to IgE immunoassays which are used as an alternative to skin tests when these are not practical or possible, or when equivocal results are obtained (6). This could be the case, for example, when an allergenic substance is not available as a licensed extract for skin testing (e.g., latex and industrial chemicals), or when a patient is taking medication that would preclude a skin test (6). Generally, results of skin testing performed under optimal conditions agree with those obtained using specific IgE immunoassays, but experts agree that these two methods are not interchangeable (6). When discrepancies in results exist, the skin test is usually positive and the immunoassay negative, probably due to the limitations of the immunoassays (6).In contrast, in the case of a positive immunoassay test and a negative skin test, one should question whether the skin test was performed correctly (6). Most studies however show a sensitivity of ~70–90% when immunoassays are compared to skin testing (16; 17).

The first commercial in vitro assay designed for the detection of IgE was the Phadebas Radioallergosorbent test (RAST, Pharmacia, Uppsala, Sweden). In this format, an allergosorbent was prepared by covalently coupling an allergen of a particular specificity onto cyanogen bromide-activated cellulose paper disks. Allergen-specific antibody of all isotypes in the serum could then bind after addition of the sample. After washing, radio-iodinated antihuman IgE was used to detect bound IgE. After a second wash, radioactivity relating to the amount of IgE initially present in the sample was measured using a gamma-counter (3). Although the term "RAST" is still commonly used to refer to an immunoassay for allergen-specific IgE, it is in fact a trademark name and refers to an assay which is now very rarely used (6). So-called "second-generation" allergen-specific IgE antibody assays have almost all been based on the RAST chemistry, but use a larger number and higher quality of allergen extract for the preparation of allergosorbents. Furthermore, the shelf-life, ease of use, and safety of these tests have improved due to the use of nonradioactive, enzyme-labeled antibodies. Automation has also improved precision and reproducibility of results (3). Consequently, these assays have become a lot more competitive with the in vivo skin tests.

There are, however, a number of problems associated with immunoassays, the greatest of which is the lack of standardization. This is in most part due to the varying sources of raw allergenic materials, different methods for binding allergen to the detection matrix, and different detection systems (6). Although no official test standard exists, the Pharmacia CAP system is in worldwide use and is a de facto standard to which other tests are compared (6).

It is important to bear in mind that although these assays are often promoted as allergy diagnosis tests, they should best be regarded as tests indicating the presence or absence of detectable specific IgE (6). Furthermore, specific IgE can be found in patients with allergic diseases as well as in about 15% of normal asymptomatic individuals (6). Immunoassays or skin tests, therefore, cannot be used to determine whether the patient will exhibit symptoms of IgE-mediated hypersensitivity upon exposure to an allergen. Assays for total serum IgE, based on the same format as specific IgE immunoassays, are also available. High serum levels of IgE may correlate to an atopic (i.e., allergic) state in the individual, although the limitations of using total serum IgE as a diagnostic analyte include its age-related concentration and the wide overlap in concentrations between atopic and nonatopic individuals (6). The detection of total IgE, therefore, cannot result in a definite clinical diagnosis, but rather would serve as a first-line test to guide clinicians in their decision-making process.

20.1.5 Affinity Biosensors

Affinity-based biosensors are devices incorporating a molecular recognition element (MRE) such as an antibody, receptor protein, nucleic acid, molecular imprinted polymer (MIP), or aptamer. The use of affinity-based biosensors has increased significantly over the past decade (18). This is due to the progress made in signal transduction technologies, as well as to the fact that MREs, especially antibodies, have become easier to isolate and purify and hence have become more widely available. When an affinity biosensor includes an antibody or antibody-related substance as the MRE, it is known as an immunosensor (19). Affinity biosensors can be further classified based on the MRE incorporated into the device and on the type of transducer used. The following is a brief overview of affinity biosensors and their applications, in particular as applied to the detection of IgE.

20.1.5.1 Types of Affinity Biosensor

20.1.5.1.1 Immunosensors

Although immunosensors have been designed in a variety of different ways, incorporating different transduction technologies, they generally fit into one of three basic formats, namely direct noncompetitive, competitive (direct or indirect), or sandwich (20) (see Figure 20.4). In the direct noncompetitive format, the binding of the antigen to the antibody is directly detected based on the electrochemical or optical properties of the analyte (Figure 20.4A). Surface plasmon resonance (SPR), discussed in Section 20.2, is a pertinent example of direct optical detection of the binding event. Competitive assay formats are generally used when the analyte is of a low molecular weight and the binding event cannot be detected directly. In these formats, the signal is inversely related to the analyte concentration (18) (Figure 20.4B). The sandwich format is based on the enzyme-linked immunosorbent assay (ELISA) principle, where the analyte tracer is conjugated to an enzyme, such as horseradish peroxidase, and is detected after addition of a substrate (Figure 20.4C). Most electrochemical transducers are based on this format, where the secondary antibody is usually conjugated to an enzyme capable of generating an electroactive product after addition of a substrate (21).

Although electrochemical transducers are the most well characterized of the transducer technologies (18), they do not lend themselves well to affinity-based sensors. This is because most affinity-based reactions are not electrochemically active and, therefore, cannot participate directly in redox reactions (21). In the case of amperometric immunosensors, coupling of an immunoreaction to the transducer usually requires a labeled immunoagent to facilitate the production of electroactive species, which in turn can be detected electrochemically (21). The first report of an amperometric immunosensor was from Aizawa et al. (22), who used catalase as a label to detect human chorionic gonadotrophin. Major advantages of this type of electrochemical immunosensor include

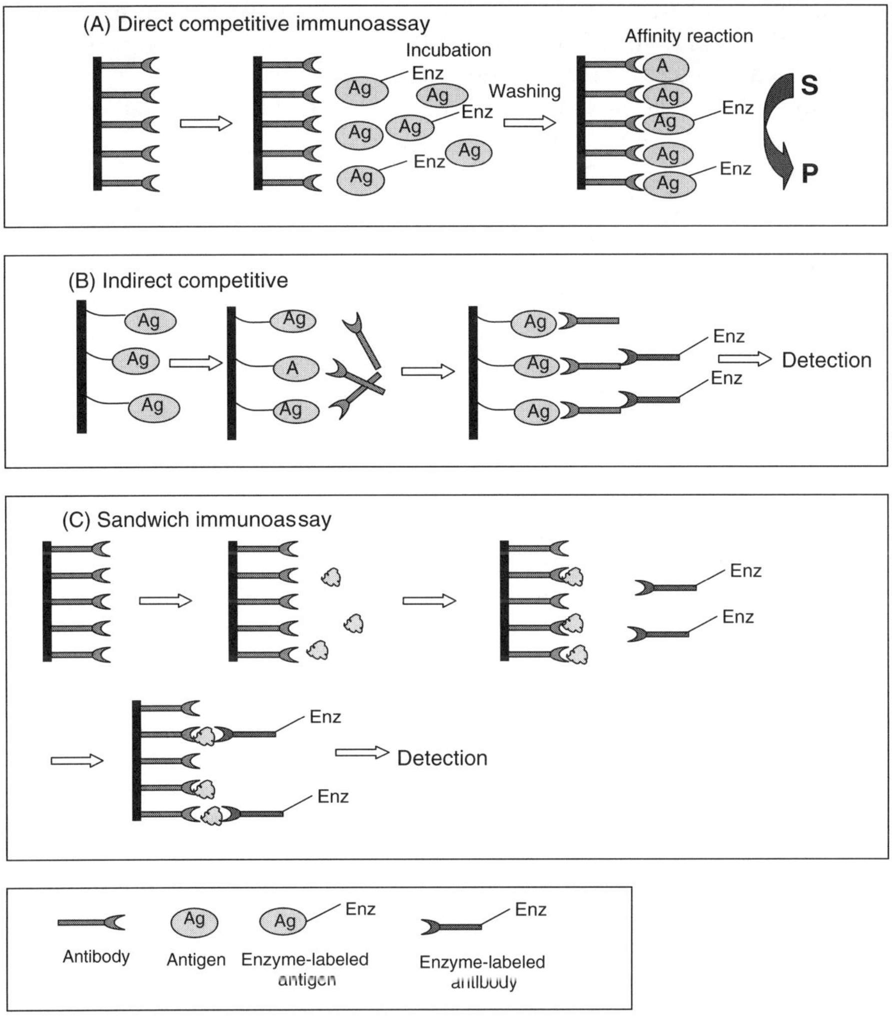

FIGURE 20.4
Immunosensor formats. (A) Direct competitive immunoassay. (B) Indirect competitive immunoassay. (C) Sandwich immunoassay. (Modified from Turner APF, Laschi S and Mascini M. 2002. *Biosensors*: *Kirk-Othmer Encyclopaedia of Chemical Technology*. Wiley.)

the ability to detect the label in turbid media, which is not possible with optical sensors, and the ability to increase the surface area to increase sensitivity (21). Potentiometric immunosensors, based on charge differences between antibody–antigen complexes and the antibody or antigen alone, ion-selective, or gas-sensing electrodes, have also been reported (21). They have, however, been less successful because of the lack of sensitivity caused by nonspecific binding and other background interferences with the transducer (21). They also require additional time due to the indirect nature of the configurations.

Piezoelectric immunosensors, the most common of which is the quartz crystal microbalance, have also been widely applied to detect antibody binding to an immobilized antigen (18). This technique is able to measure small changes in surface properties, such as bound surface mass and surface viscosity, which can then be related to the concentration of the analyte in the sample (18; 20). The major advantages of these devices include their small size, high sensitivity and stability, simplicity of construction and operation, and low-power requirement (20).

A recent development in the field of immunosensors is the use of antibodies labeled with paramagnetic particles (23; 24). With the antigen immobilized, the binding event can be detected by measuring the magnetic field induced by the magnetic particles. A major advantage of this method is the ability to permanently record the signal, which could, in principle, be remeasured at any time.

The last category is optical immunosensors, for which direct and indirect detection methods exist. Direct methods, requiring no labeling of the antibody or antigen, include internal reflectance spectroscopy, evanescent wave sensing, and SPR, which will be examined in detail in Section 20.2.

The above-mentioned methods have also been applied to indirect immunosensing, in which a label, usually fluorescent, is used to monitor the binding event (21). This approach improves the sensitivity and selectivity of the assay, given that the quantitative signal arises only from the presence of the label, thereby eliminating nonspecific binding interference.

20.1.5.1.2 Other Types of Affinity Biosensors

Other types of affinity-based biosensors include deoxyribonucleic acid (DNA) biosensors, also known as gene chips. These biosensors are used for recognition and quantitation of target DNA sequences, based on hybridization of complementary strands to immobilized single-stranded (ss) DNA (20; 21). Electrochemical and optical transduction methods have been applied, using electroactive or fluorescent labels specific for binding to hybridized sequences. Labelfree electrochemical detection is also possible through the direct oxidation of guanine bases (25; 26). Fluorescent tags are still the labeling method of choice for DNA biosensors and were first introduced by Affymetrix with the GeneChip, in 1996 (21). Piezoelectric DNA biosensors, capable of detecting the mass change associated with hybridization, have also been reported (27), as have electrochemical sensors, such as those from Nanogen.

Molecularly imprinted polymers are artificial ligands which possess steric and chemical memory for a template, which they can bind with specificity similar to that of an antibody–antigen interaction (28). They have been employed as nonbiological alternatives to antibodies in competitive binding assays, and applications for analytes such as cortisol, theophylline, and morphine have been investigated, suggesting that MIPs are a promising technique for use in affinity sensors (19).

Lastly, aptamers are also a very promising technology for the use in immunosensors, and will be discussed in detail in Section 20.3.

20.1.6 Recent Work on Immunoglobulin E Biosensors

Early work performed by Su et al. (29) made use of antihuman IgE antibodies immobilized on a piezoelectric sensor for the detection of total IgE in serum samples. The immunosensor proved to have a sensitivity of 12 μg/L and could be regenerated five times without appreciable loss of activity. Further work was performed by Kreuzer et al. (30), who described an amperometric immunosensor for the detection of IgE. The sensor used a disposable screen-printed carbon electrode and had a detection limit of 90 ng/L of IgE in whole blood, with an analysis time of 30 min.

More recently, Liss et al. (31) reported on work performed using a quartz crystal biosensor and compared the performance of a specifically designed aptamer and an antibody for the detection of IgE. The authors compared a well-established, published aptamer, which they subsequently improved on during experimentation, with a commercially available monoclonal anti-IgE antibody. Their work demonstrated an equal specificity and sensitivity

(100 μg/L) of the aptamer with respect to the antibody but with an extended linear detection range to tenfold higher concentrations of IgE. In addition to this, the study demonstrated for the first time that an aptamer-based biosensor could specifically detect an analyte in a complex protein matrix. Lastly, regeneration of the aptamer receptor layer was shown to be possible.

The work used to illustrate this chapter is based on SPR aptasensors, and as such the theory of SPR and aptamers is expanded on in the Sections 20.2 and 20.3, respectively.

20.2 Surface Plasmon Resonance

SPR, as applied to biosensors, is a relatively new technology which has been used in a number of fields to determine a desired quantity by measuring the change in refractive index occurring due to the interaction between, for example, a receptor molecule and its ligand (32). The development of SPR-sensing configurations and applications has been described for the measurement of physical, chemical, and biological quantities. It is, however, in the field of affinity biosensors that SPR has shown the greatest potential, allowing real-time (i.e., under continuous flow conditions) analysis of biospecific interactions without the use of labeled molecules (33).

Much of the success of SPR is because of the fact that it is a fast, reliable, and sensitive method that can be used to answer a number of fundamental questions about the interactions of the molecules being investigated. The information that can be gained includes the following (34):

- Concentration determination of the interacting molecules.
- Determination of the association and dissociation rate constants and affinity.
- Determination of active binding regions and relative binding patterns (e.g., epitope mapping).
- Specificity of the interaction (for a particular molecule or a class of molecules).

The potential of SPR for the characterization of thin molecular films and monitoring processes at metal interfaces was recognized in the late seventies (35; 36). Currently, several companies have commercialized SPR sensor technology, which has become a leading method for the direct real-time observation of biomolecular interactions (37).

20.2.1 Theory

A surface plasmon (SP) is a charge–density oscillation (wave) that may exist at the interface of two media with dielectric constants of opposite signs, such as a metal and a dielectric (e.g., water) (32). The wave is strongly localized and propagates along the interface between the metal and the ambient medium (34).

The basic principle behind using SPR as a biospecific-interaction analysis (BIA) technique is as follows. Incident light is used to excite an SP. At the SPR angle, where the energy and momentum of the incident light coincide with that of the charge density wave, the photon energy is transferred to the charge density wave (38). This phenomenon is observed by a sharp dip in intensity of the reflected incident light. The resulting wave propagates along the interface between the metal and the ambient medium and is extremely sensitive to changes in refractive index near the metal surface, for example due to receptor–ligand interactions (38). Changes in the SPR angle, for small angular shifts, are proportional to changes in the refractive index and consequently to the mass concentration of the biomolecules at the surface of the metal, which is the basis of the use of SPR for biosensing purposes (34).

The most common setup for SPR applications is the Kretschmann configuration, shown in Figure 20.5. In this configuration, SPs are excited at the surface of a gold film deposited on a prism. The thickness (*d*) of the gold film is crucial, as the efficiency of conversion of bulk waves into SPs decreases with increasing thickness and decreasing transparency of the gold film (39). In contrast, if the film is too thin, SPs are rapidly reconverted into bulk waves, significantly changing the width of the resonance curve and reducing the sensitivity of the SPR measurements (39). The optimal thickness for a gold film using, for example, a wavelength of 790 nm, is 45 nm (39).

As mentioned above, SPs occur at the interface between two media of dielectric permeabilities of opposite sign, such as between free electronlike metals including silver, aluminum and gold, and water. Given that the dielectric permeability of any material is a function of the wavelength of the incident light, it follows that for any given interface, the excitation of SPs will not be possible at all wavelengths (39). For metals, the dielectric constant (ε_m) is negative in the infrared–visible range of the spectrum, whereas it is positive for water (ε_b) within the same range. This therefore allows the existence of SPs at metal–water interfaces (39). Owing to its optical and chemical properties, gold is the most commonly used metal for SPR applications, a thin film of which is typically deposited onto a glass substrate (39).

SP waves are distinguished by the fact that they have maximal intensity at the interface, decaying exponentially in both the metal and the ambient medium, with distance from the surface. The SP wave is thus specifically associated with the metal–dielectric interface and is therefore different to a bulk wave. The propagation of electromagnetic waves is characterized by a wave, or propagation, vector. The propagation vectors for the SP wave (k_{sp}) and the bulk wave (k_b) are defined below (39) (see Figure 20.5):

$$k_{sp} = k_0[\varepsilon_m \varepsilon_b/(\varepsilon_m + \varepsilon_b)]^{1/2} \tag{20.1}$$

$$k_b = k_0 \varepsilon_b^{1/2} \tag{20.2}$$

where k_0 is the propagation vector in a vacuum.

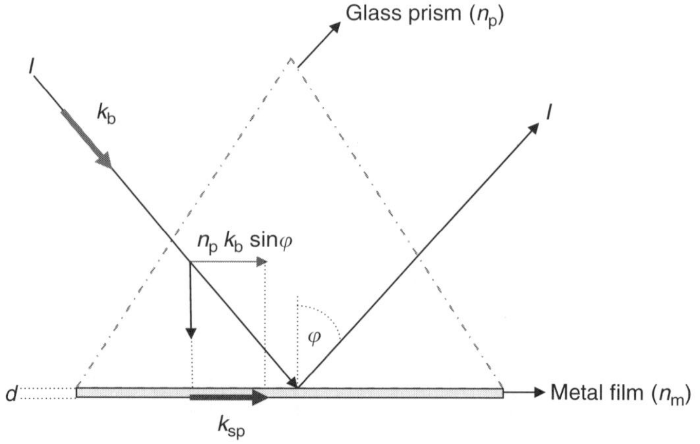

FIGURE 20.5
Kretschmann configuration for surface plasmon resonance (SPR). See text for details.

The condition required for resonance, that is to say the condition required for the bulk wave to be transformed to the SP wave, is that the bulk propagation vector be equal to the SP propagation vector (39):

$$k_{sp} = k_b \tag{20.3}$$

Furthermore, resonance depends on the angle of incidence, φ, in the following way (39):

$$k_{sp} = n_p k_b \sin \varphi \tag{20.4}$$

where n_p is the refractive index of the prism.

At a particular angle of incidence, k_{sp} will equal k_b and the conditions for resonance will be fulfilled (see Equation 20.3). At this angle, there is maximal coupling of incident light into SPs, which results in maximal absorption of the incident light by the gold film and a resultant sharp decrease (minimum) in the reflection coefficient of the incident light is observed (see Figure 20.9A). The angle at which the SPR minimum occurs is greatly dependent on all the refractive indices of all the boundary media, including the gold film, the bulk solution in contact with the gold-coated prism, and additional layers such as molecules deposited on the gold surface (39). Generally, k_{sp} increases proportionately with increasing biolayer thickness, resulting in increases in the resonance angle (see Equation 20.4) (39). Therefore, the thicker the biolayer at the surface, the greater the SPR-minimum shifts to higher angles will be. These shifts can be measured in real time during the adsorption of biomolecules at the gold film surface.

The SPR technique is extremely sensitive to optical thickness changes, and it has been shown that a change in protein surface concentration of 1 ng/mm^2 will generate a change of SPR coupling angle of 0.1° (40).

20.2.2 Concept

Generally, an SPR device consists of an optical system, a transducing medium, and an electronic system supporting the optoelectronic components of the sensor and allowing data processing (32). The optical component of the SPR device contains a source of optical radiation and an optical structure in which the SP wave is excited and interrogated. The transducing medium interrelates the optical and bio(chemical) domains, relating changes in the refractive index to changes in the quantity of interest. Its properties also determine the selectivity and response time of the sensor. The sensitivity, stability, and resolution of the sensor, in contrast, depend on the properties of both the optical system and the transducing medium (32). Most modern SPR devices use light-emitting diodes as light sources and linear arrays of charge-coupled devices to detect reflected light from the surface. The use of array detectors allows reflected light to be measured at a wide range of angles, thereby avoiding the need to mechanically control the angular position of the detector (39).

20.2.3 Applications

SPR has been applied to a number of fields, including physics, chemistry, and molecular biology. In the field of physics, SPR sensing devices have been developed to exploit certain physical phenomena that occur in various optical transducing materials, such as humidity-induced refractive index changes in porous thin layers and polymers; or temperature sensors based on the thermooptic effect in hydrogenated amorphous silicon (32).

Most chemical SPR sensors are based on the measurement of SPR variations due to adsorption or a chemical reaction of an analyte with a transducing medium, which results in changes in its optical properties (32). Applications of these sensors include, amongst others, monitoring of the concentration of vapors of hydrocarbons, aldehydes, and alcohols by adsorption in polyethylene glycol films, and the detection of vapors of aromatic hydrocarbons by their adsorption in Teflon films (32).

It is, however, in the field of molecular biology that SPR has been most successful, with applications including immunological analysis, studies of protein–protein interactions, molecular-biological studies on the mechanisms of gene expression, signal transduction and cell–cell interactions, screening of new ligands, quantification of protein adsorption and immobilization, the evaluation of surfaces for biocompatibility, epitope mapping, determining affinity constants, and the examination of binding kinetics (39). The first application of SPR to biosensing was demonstrated in 1983 (41). Since then, the technology and consequent breadth of applications have continued to develop. This is due in major part to the fact that it is possible to measure the kinetics of biomolecular interactions in real time with a high degree of sensitivity; and that no labeling of the biomolecules is necessary for their detection (39). Furthermore, analysis of receptor–ligand interactions with a wide range of molecular weights, affinities, binding rates, and in numerous different chemical environments is possible (42). Indeed, analyses of analytes with masses ranging from hundreds of daltons to whole cell binding have been reported (42).

Detection methods are most often direct, where analyte quantification is carried out by direct detection of the binding of the analyte to the immobilized receptor. In cases where the analyte is small and its binding to the receptor does not produce a measurable increase in the refractive index, indirect sandwich or competitive assay methods may be used (32).

Table 20.1 lists the most common molecular biological applications of SPR biosensors. Examples of the *italicized* applications are given below.

As mentioned previously, the use of SPR for the determination of kinetic parameters is a major contributor to its success as an analytical technology. Early work using SPR for the determination of reaction kinetics was performed to determine the effect of single base pair mismatches on DNA hybridization kinetics. In work performed by Gotoh et al. (44), it was shown that the association kinetics for DNA hybridization were inversely proportional to the number of mismatched base pairs, and that the dissociation kinetics were even more strongly influenced by the number of mismatches.

In more recent work, the determination of kinetic parameters has proven to be extremely valuable. Indeed, numerous examples exist in which interaction rates are more descriptive of a given biological process than the equilibrium binding affinities (42). For example, Leferink et al. (45) described growth factor interactions with ErbB-1 in which the

TABLE 20.1

SPR Biosensor Applications

Qualitative	Quantitative
Following molecular purification	Active concentration
Specificity	*Kinetics*
Epitope mapping	Equilibrium constants
Molecular assembly	*Thermodynamics*
Ligand fishing	Stoichiometry
Small molecule screening	*Mechanism*

Note: SPR, surface plasmon resonance.

Source: Modified from Rich R and Myszka D. 2000. Advances in surface plasmon resonance biosensor analysis. *Current Opinion in Chemical Biology* 11: 54–61.

dynamic rate constants correlated better with mitogenic activity than did the equilibrium constants (42). A more pertinent example, however, is that of McDonnell et al. (46), who identified a role for the Cε2 domain for IgE (see Figure 20.2) in allergic responses. Removal of this domain from IgE has little effect on the overall affinity of IgE for its receptor FcεRI, but has a clear effect on the off rate. It was thus demonstrated that Cε2 contributes to the slow off rate of IgE-mediated mast cell sensitization during the allergic response.

Insights into binding-reaction mechanisms can be gained by analyzing binding kinetics and thermodynamics over a range of conditions, such as temperature, ionic strength, and pH. Measurements over a range of ionic strengths, for example, can help describe the role of electrostatic interactions for a particular binding reaction. Thus, one can, for example, discriminate between specific and nonspecific protein–DNA binding, due to their difference in ionic strength dependence. Indeed, nonspecific protein–DNA binding interactions are highly dependent on ionic strength due to the significant electrostatic contribution of the negatively charged phosphates of the DNA backbone (42).

SPR has also been used for investigating biological membrane events. This has been made possible by the recent introduction of sensor surfaces specifically designed for this purpose, such as the HPA (hydrophobic) or L1 (lipophilic) surfaces from Biacore (Uppsala, Sweden) (43). These surfaces offer the ability to specifically orient immobilized ligands and have been shown to be a successful membrane mimic (42). Danelian et al. (47), for example, used a liposome-covered sensor surface for an assay for lipid absorption for a panel of 27 drugs and showed a strong correlation with passive intestinal absorption. Future advances in this field might include inserting ion channels, transmembrane receptors, and cell-signaling molecules within the immobilized lipid surface in an attempt to closer mimic in vivo systems (43).

Lastly, the general versatility, ease of automation, lack of labeling requirements and low sample consumption of SPR analysis make it a promising means for large-scale screening for binding events, both for small molecules in drug discovery and for macromolecules in large-scale ligand fishing experiments (42). It is able to perform functional characterization of "hits" from primary screens, providing information not obtainable from traditional screening methods and making it possible to rank ligands based on affinity as well as association and dissociation kinetics. Additionally, some of the more advanced instruments (BIAcore 2000 and 3000) have a throughput of 100–300 samples per day, depending on assay conditions (43), thus making this method capable of high-throughput analysis. SPR has in fact been used to monitor binding of thyroxin analogues to an immobilized antibody (48), as well as by the La Jolla Pharmaceutical Company for the development of a new drug for the treatment of systemic lupus erythromatosis (42).

20.2.4 Future Developments

Despite the many advantages offered by SPR, there are a number of advances that will need to be made in order for this analytical technique to become as widespread as, for example, immunoassays, which are the major competitor of SPR within the field of analysis and detection of biochemical substances, and which offer low cost and high sensitivity and specificity testing. These advances are likely to include (32):

- *Improvement of detection limits.* Current detection limits stand at about 1 pg/mm^2, which is not sufficient for detecting low concentrations of low-molecular weight analytes. Although optimization of SPR optical instruments and refining of data-processing methods may lower the current detection limits, no approach currently exists that will lower this limit by the necessary orders of magnitude. Progress will therefore need to be made in this area.

- *Multichannel performance.* This would be required for high-throughput screening and detection for new pharmaceuticals. Advances have already been made in this area, with the introduction of a four-channel chip that can be rotated by 90°, effectively providing 16 channels. Myszka and Rich (49) have more recently described a prototype micro array chip with 64 individual immobilization sites in a single flow cell.
- *Development of advanced recognition elements for applications involving complex realistic samples (e.g., blood).* Stable receptor matrices that allow sensor responses and nonspecific background effects to be resolved will also have to be developed.

All these developments may eventually lead to miniaturized, integrated, compact, and rugged sensing elements which would fulfill a number of the requirements of an ideal biosensor.

20.3 Aptamers

Aptamers have been defined by James (50) in the *Encyclopaedia of Analytical Chemistry* as: "...artificial nucleic acid ligands that can be generated against amino acids, drugs, proteins and other molecules. They are isolated from complex libraries of synthetic nucleic acid by an iterative process of adsorption, recovery and reamplification. They have potential applications in analytical devices, including biosensors, and as therapeutic agents." In essence, they are synthetic oligonucleotide sequences which are able to bind a wide array of molecules (ligands) with high affinity and specificity (51; 52). Their name is derived from the Latin word "*aptus*", meaning "to fit" (52).

Importantly, the ligand-binding capacity of aptamers is based on their three-dimensional conformation and not on nucleotide base pair complementarity (53). Indeed, upon association with their molecular targets, aptamers fold into molecular structures in which the target becomes an intrinsic part of the nucleic acid structure (54). They are therefore able to bind proteins and other molecules that would not normally interact with DNA or ribonucleic acid (RNA), making them extremely powerful as screening tools (53). They were initially reported, in 1990, quasi simultaneously by three independent laboratories: those of Robertson and Joyce (55), Ellington and Szostak (56), and Tuerk and Gold (57), and are obtained using an in vitro selection and purification technique, now known as systematic evolution of ligands by exponential enrichment (SELEX), as described in Section 20.3.2 (51).

20.3.1 Aptamer Libraries

Before the SELEX process is undertaken, an aptamer library has first to be created. Part of the enormous potential of aptamers lies precisely in the fact that libraries with vast numbers of potential ligands can be created and screened within a few days. Typically, aptamer libraries consist of 10^{13}–10^{18} random oligonucleotide sequences (51) and can be screened within a number of days. This is all the more impressive when compared to conventional libraries of potential drugs which consist generally of no more than 10^6 different molecules and may take months to screen (53).

When creating a library however, the following factors need to be taken into account (53):

- *The aptamer library complexity*: Given random manufacture of aptamers, the library complexity can be calculated relatively easily. For example, a library of oligonucleotides of N nucleotides in length generated from y different nucleotides (not solely limited to the four naturally occurring nucleotides) will have a complexity of y^N. However, for practical purposes, and as mentioned above, the maximum number of unique sequences that can be screened in a library is limited to 10^{13}–10^{18}. It is interesting to note, however, that further diversity may be introduced during the SELEX process itself (see Section 20.3.2), in particular due to the infidelity of the DNA polymerase used in the polymerase chain reaction (PCR) step, which may not be 100% accurate.

- *The nucleotide chemistry*: This is important because, firstly, it defines the range of possible three-dimensional structures into which the aptamer can fold, and secondly, the nucleotide chemistry plays an extremely important role in relation to the aptamer's stability to degradation. Indeed, the susceptibility of single-stranded oligonucleotides to enzymatic or chemical cleavage is a severe practical handicap to the use of aptamers that needs to be overcome for the molecules to be stable in biological fluids. Three different approaches have been adopted to overcome this problem:
  - Modification of nucleotide bases, which has been the most commonly used method. The modification of pyrimidines at the 5′ position with I, Br, Cl, NH_3, and N_3 and at the 2′ position with NH_2, F, and OCH, for example, has been described by Pieken et al. (58). These modifications can increase the RNA oligonucleotide half-life up to 15 h (54). It should be noted, however, that any modified nucleotides should still be compatible with the enzymes used in the SELEX protocol (e.g., DNA polymerase). Interestingly, some problems associated with modified nucleotides have been overcome by a modification of the SELEX protocol known as "Transcription Free SELEX", in which random RNA fragments bind random DNA templates, after which the fragments are ligated either enzymatically or using a standard chemical condensation reaction. The RNA aptamer can then be recovered by melting the duplex.
  - Modification of the phosphodiester backbone, for example through the use of α-thio substituted deoxynucleotide triphosphates, although this technique has been more successful with DNA as opposed to RNA aptamers.
- The use of enantiomeric aptamers, known as spiegelmers (from the German word meaning "mirror"). This technique consists in creating a mirror image of the target and selecting an aptamer for this mirror image. A stereo isomer of the selected aptamer is then created (i.e., the spiegelmer), which will be specific for the target but will not be susceptible to normal enzymatic degradation due to the substitution of the natural D-ribose with L-ribose (54).
- *The constant region primer design*: The random aptamer sequence has to be flanked by 5′ and 3′ constant sequences, usually 20–25 base pairs in length, which provide hybridization sites during a number of steps of the SELEX process (51). The 3′ flanking sequence generally acts as an attachment site for the reverse-transcriptase primer; and the 5′ flanking sequence acts as the attachment site for the PCR primers during the amplification step of the SELEX protocol. The design of the constant region for the SELEX protocol is even more important than for normal PCR given that a complete SELEX protocol may include up to 200 cycles of PCR. Any artifacts would thus be drastically amplified in the final aptamer population.

20.3.2 SELEX

After a suitable aptamer library has been prepared, it can undergo the SELEX protocol, as shown in Figure 20.6. This technique essentially consists of the repeated binding, selection, and amplification of aptamers from the initial library until one (or more) aptamer(s) displaying the desired characteristic(s) has (have) been isolated (53).

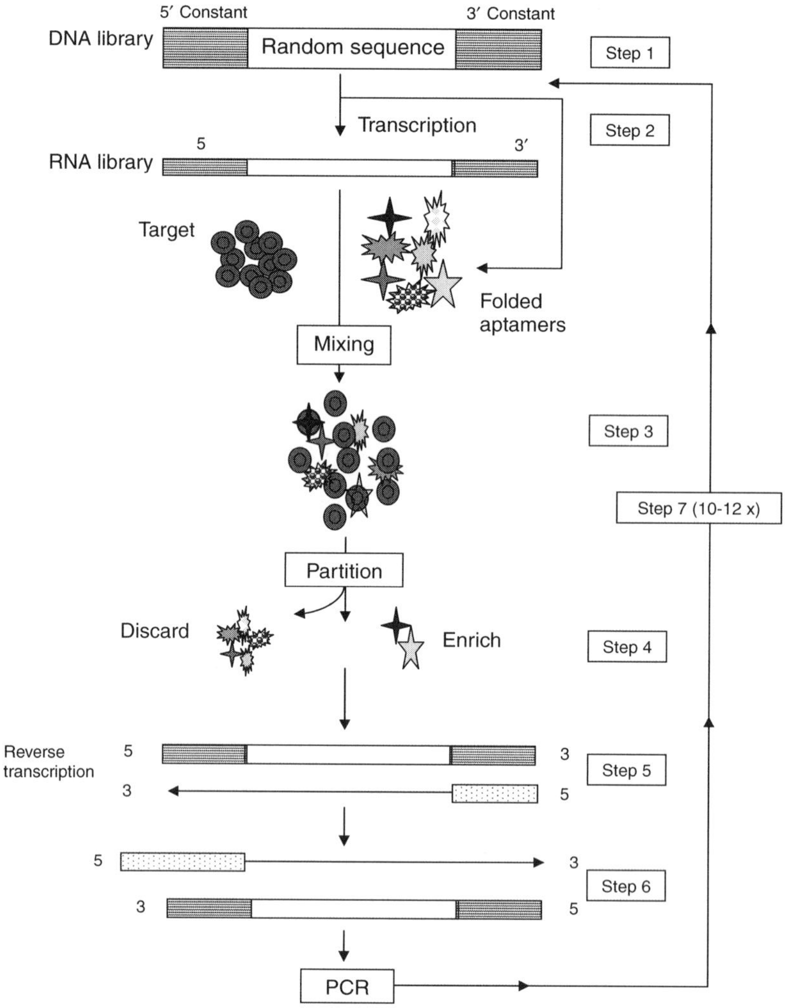

FIGURE 20.6
The SELEX Protocol. (Modified from Sampson T. 2003. Aptamers and SELEX: The technology. *World Patent Information* 25: 123–129) (see text for details).

Referring to Figure 20.6, Steps 1 and 2 involve the preparation of the aptamer library/candidate mixture, as detailed above. Note that the library would generally consist of double-stranded DNA, which needs to be either transcribed (for RNA selection) or strand-separated (for ss DNA selection), to be in a suitable form for selection (54).

In Step 3, the target and the aptamers are brought together under favorable binding conditions, where the aptamers with the highest affinity will bind the target. These aptamers are then partitioned from the aptamers with lower affinity in Step 4. This step can be performed by attaching the aptamers to a solid phase support, such as Sepharose, and specifically eluting the desired aptamers after binding has taken place. Immobilizing the aptamers, however, is likely to affect the binding interactions between the aptamer and the target. The same would be true if the target were immobilized instead of the aptamer (53). Alternatively, the aptamer and target could be allowed to interact freely in solution, after which the target–aptamer complex could be recovered by filtration through nitrocellulose (54). This method is commonly used, although it is important to note that it is only applicable when the target molecule is a protein. A negative selection step is also frequently used at this stage, in which the aptamers are passed over a cellulose filter in the absence of the target. This is to eliminate aptamers that bind the filter in a target-independent manner (54). Counter selection is also sometimes used, where aptamers that bind structures similar to that of the target are removed (51). Affinity chromatography is generally used to isolate aptamers for small molecular targets (54).

The high-affinity aptamers are then amplified by reverse transcription-PCR (RT-PCR) (for RNA aptamers) (Step 5) or by PCR for DNA aptamers (Step 6), to create a new aptamer library enriched with the aptamers of high affinity. The entire process is then repeated (Step 7), resulting in fewer and fewer unique sequences, with higher and higher affinity to the target, being retained. Note that during each round of selection, the binding conditions for the aptamer and the target are generally made more stringent to increase the selective pressure on the remaining aptamers. Generally, a complete SELEX process (between 8 and 15 cycles) will yield a final mixture of no more than ten aptamers (53; 54). The aptamers can then be cloned and sequenced, allowing further identical aptamers to be generated by chemical synthesis.

The SELEX procedure is clearly involved and can take weeks to months to produce a suitable aptamer when performed manually. Cox et al. (59) first reported on the automation of the procedure. A more recent report (60) however details the automation of the procedure where approximately 12 rounds of selection can be carried out in 2 days. Based on these figures, it has been estimated that one robot could produce aptamers against 120 targets in 1 month, exceeding manual throughput by 10–100 fold. It has also been suggested that further development of the automated procedure could lead to the production of aptamers to upwards of 1000 targets per month. Several robots working in parallel could therefore generate aptamers to an entire proteome within a relatively short period of time, further highlighting the immense potential of aptamers. An additional advantage of an automated procedure is the consistency of the repetitive tasks that it offers, creating a bench mark to allow comparisons between different laboratories and eliminating variations due to manual selection techniques (59).

20.3.3 Aptamers and Antibodies

Given the characteristics of aptamers described above, it is clear that they could be of great use for therapeutic, analytical, and diagnostic procedures. Currently, antibodies are most frequently used in procedures where high affinity and specificity for a particular target are required, such as ELISAs. The following is a discussion of the advantages and disadvantages of using aptamers or antibodies for these techniques.

The use of antibodies to detect analytes became widespread in the 1970s, when polyclonal sera from immunized animals were the most popular choice (52). It was not long thereafter till the discovery of monoclonal antibody technology which allowed the production of a unique antibody in large quantities (52). This technology allowed affinity-based assays to be further refined and optimized and was embraced throughout the scientific community. The many advantages of antibodies include their high affinity and specificity for their particular antigen, typically with very low dissociation constants. Furthermore, selected clones producing the antibody of choice can be cultured continuously, hence providing a limitless supply of a particular antibody (theoretically). Lastly, the immunogen used for the identification of a monoclonal antibody does not have to be pure (52).

There are, however, a number of disadvantages and limitations associated with the use of antibodies. These are listed below (52):

- The production of antibodies requires animals. The generation of antibodies against molecules that are not well tolerated by the animal (e.g., toxins) or against molecules that are not inherently immunogenic can therefore be problematic.

- The production and identification of monoclonal antibodies are laborious procedures and are likely to be costly in searches for rare antibodies requiring the screening of large numbers of colonies.
- High yields of antibodies may be problematic to obtain due to difficulties associated with growing certain hybridomas in vivo, the manner in which high yields are typically achieved.
- The performance of the same antibody may vary between batches, requiring the immunoassay to be reoptimized with each new batch of antibody.
- It is not feasible to identify antibodies that could detect targets under nonphysiological conditions.
- There is no control over the selection of the target protein to which the antibody binds (this is "determined" by the immune system of the animal) (54).
- The kinetic parameters of antibody–target interactions cannot be manipulated on demand.
- Antibodies are heat-labile, undergoing irreversible denaturation.
- They have a limited shelf-life.

It is important to note that various approaches addressing the shortcomings listed above are being investigated, including humanization of antibodies, antibody engineering, and in vitro immunization (52).

Listed below are the various advantages aptamers offer over antibodies for the use in analytical and diagnostic devices (52):

- Aptamers do not require the use of animal systems for their production.
- The properties of aptamers can be changed on demand due to the fact that they are produced through an in vitro process that does not depend on animals or on in vivo conditions.
- Selection conditions can be manipulated to obtain aptamers that bind the target under nonphysiological conditions, for example. Similarly, the kinetic parameters of the aptamer can also be manipulated on demand.
- Aptamers can be generated against targets that are toxic or that are not inherently immunogenic.
- Batch to batch variation is eliminated in aptamer production, given the fact that they are chemically synthesized.
- Aptamers can be easily engineered to include reporter molecules at a precise location, specified by the user, without affecting their binding characteristics.
- Small aptamers can be used to generate dense receptor layers, thus allowing increased sensitivity for a given affinity (31).
- The heat denaturation process in aptamers is reversible, allowing them to be easily regenerated.
- Aptamers are stable in long-term storage.

The main advantages of aptamers over antibodies are summarized in Table 20.2.

The main drawbacks of aptamers include the time required for selection using a manual SELEX process, although this is slowly being overcome by the introduction of automated SELEX processes. Additionally, their affinity constants are generally lower than those of antibodies and their structural stability is still questionable (19). Furthermore, the fact that the science of aptamers is relatively young and much of the work and knowledge base in the field is still highly experimental is an additional concern hindering their common use.

TABLE 20.2
Advantages of Aptamers Over Antibodies

Antibodies	Aptamers
Requires the use of animals	No animals required
Limitations against nonimmunogenic and toxic substances	Can be generated against nonimmunogenic and toxic substances
Kinetic parameters cannot be modified on demand	Kinetic parameters can be modified on demand
Labeling can cause loss of affinity	Variety of reporter molecules can be attached without affecting binding properties
Not feasible to identify antibodies that interact with targets under nonphysiological conditions	Selection conditions can be modified to select aptamers with particular properties
Problems with batch to batch variation	Little or no batch to batch variation
Limited shelf-life, temperature sensitive and susceptible to denaturation	Stable in long-term storage, can be transported at ambient temperature and can be regenerated after denaturation

Source: Modified from O'Sullivan CK. 2002. Aptasensors—the future of biosensing? *Analytical and Bioanalytical Chemistry* 372: 44–48.

20.3.4 Applications

20.3.4.1 Two Site Binding Assays

ELISA sandwich assays, using antibodies, are a very common diagnostic assay. The term enzyme-linked oligonucleotide assay (ELONA) is applied to the same assay format using aptamers (51). Various ELONA formats can be adopted, using for example (51):

- Aptamers as capture molecules and antibodies as reporter molecules.
- Antibodies as capture molecules and labeled aptamers against the antibody–antigen complex as reporter molecules.
- Aptamers as capture molecules and labeled aptamers against the aptamer–analyte complex as reporter molecules.

ELONAs have been demonstrated to be successful for the detection of VEGF (61) and $CD4^+$ cells (62).

20.3.4.2 Flow Cytometry

Flow cytometry is a powerful tool that is used both analytically and diagnostically for the multiparameter analysis of cells in suspension. Generally, this method is performed using labeled monoclonal antibodies. Recently however, aptamers have been used to perform the role of monoclonal antibodies in this process, with promising results. Indeed, it has been shown that aptamers can be relatively easily conjugated to small fluorophores, such as fluorescein, or even to larger proteins such as phycoerythrin, and still retain their binding characteristics, demonstrating their ability to be used with a wide variety of reporter molecules that are generally used in diagnostics (52). Generally, aptamers have been shown to perform as well as antibodies or can even be used in combination with them for flow cytometry analyses (52).

20.3.4.3 Biosensors

Affinity sensors should ideally fulfill at least three basic criteria (52), namely:

- The ability to transduce the binding event without adding extra reagent.
- The ability to detect and quantify the target within the desired concentration range and time period.
- The ability to make repeated measurements on the same transducer.

The use of aptamer-based biosensors would offer a number of advantages in this respect. Indeed, the function of immobilized aptamers could be regenerated using extremes of heat, salt concentration, or chelating agents, without loss of activity, resulting in a reusable transducer (52). Aptamers can also be easily modified for immobilization without affecting their function (52). Furthermore, they can be selected for under conditions resembling those of the real matrix, which is particularly useful for environmental and food applications (51). Lastly, aptamers can be easily labeled with a wide range of reporter molecules, allowing the design of a variety of detection formats (52).

Recent work on aptamers in affinity sensors includes the use of an RNA aptamer specific for L-adenosine (63) and a DNA aptamer for human thrombin (64). Results in both cases were promising, with the aptamers showing high specificity and the ability to be regenerated (52).

20.3.4.4 Molecular Beacons

Molecular beacons are a class of fluorogenic probe in which a fluorophore is attached to one terminus and a quencher is attached to the other. In the absence of the target molecule, the probe forms a hairpin loop, bringing the fluorophore and the quencher into close proximity, resulting in the absence of a fluorescent signal. In the presence of a target molecule, however, the hairpin loop structure is opened (the target is complementary to the loop region), the fluorophore and quencher are distanced from each other, and a fluorescent signal is produced, as illustrated in Figure 20.7. Until recently, these probes have only been available for the detection of nucleic acid targets, due to the inability of these probes to interact with other classes of targets, such as proteins (52).

Aptamers, however, have the ability to interact with a very wide range of targets and have, therefore, been exploited to extend this technique to include targets other than nucleic acids. For this purpose, molecular beacons, termed ligand beacons when used in this way, are designed to be complementary to a nucleotide region in the aptamer. In the absence of the target (Figure 20.8A), the ligand beacon binds the aptamer, resulting in the separation of the fluorophore and the quencher and the production of a fluorescent signal. In the presence of the target (Figure 20.8B), however, the binding site on the aptamer is no longer available to the ligand beacon because it is bound by the target. This results in the ligand beacon adopting an internal hairpin loop structure and in the absence of a fluorescent signal (52).

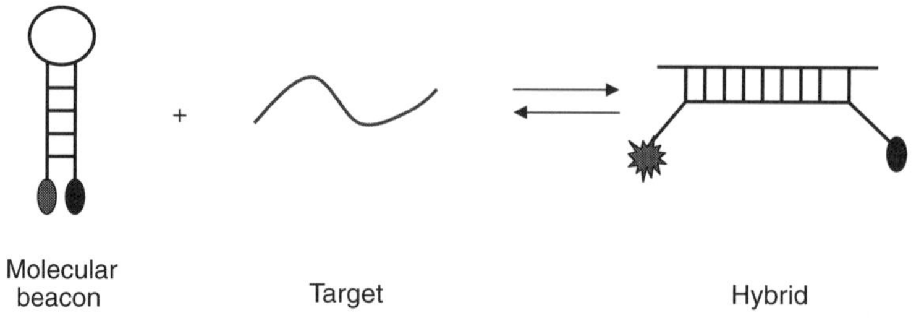

FIGURE 20.7
Functioning of a molecular beacon.

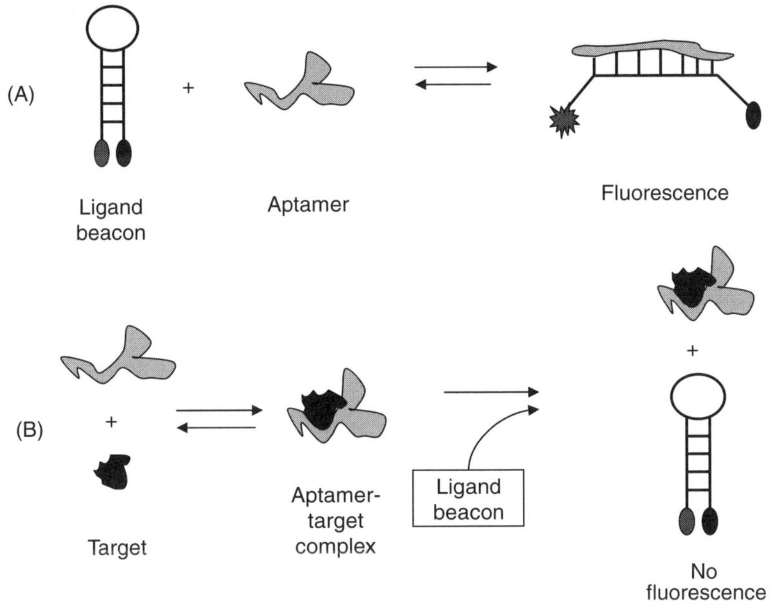

FIGURE 20.8
Functioning of a ligand beacon. (A) In the absence of the target analyte, a fluorescent signal is produced. (B) In the presence of the target analyte, no fluorescent signal is produced.

This format is also applicable to a multiplex system, in which a number of different analytes are detected simultaneously with the use of ligand beacons conjugated to fluorophores emitting at different wavelengths (52).

This type of assay once again clearly shows the potential of aptamers for analytical and diagnostic procedures, which cannot be fulfilled by antibodies.

20.3.4.5 Capillary Electrophoresis

Capillary electrophoresis (CE) is a separation technique based on the same principle as conventional electrophoresis. It offers the advantages of speed, small sample volumes, suitability for automation, sensitivity, and the possibility of performing multiplex assays (52). CE is now being adapted for immunoassays, in which antibody–antigen complexes are separated on a fluidic stream under an applied electric field (52). However, there are many practical difficulties associated with this method, such as poor separation between antibody–target complexes and free, labeled, antibodies when the targets are small and uncharged (52).

Recently, German et al. (65) adapted the technique to use a fluorescently labeled DNA aptamer for the detection and quantification of human IgE. The technique was able to separate free aptamer from the aptamer–IgE complex in the presence and in the absence of serum, indicating that the aptamer–target interaction was not affected by the complex medium.

20.3.4.6 Molecular Switches

A very relevant example, as described by Jayasena (52), of where aptamers can be used as molecular switches is in PCR. To avoid nonspecific amplification, which is often due to nonspecific binding of primers at suboptimum temperatures and to the ambient temperature activity of the DNA polymerase enzyme, an essential amplification ingredient can be withheld until the temperature is sufficient to ensure specific binding of the primers. This has been achieved, for example, by using a monoclonal antibody that neutralizes the activity of the DNA polymerase. When the temperature of the reaction mixture is increased to begin the amplification reaction, the antibody is denatured and the DNA polymerase regains its activity. One of the shortcomings of this method is the high temperature at which the antibody is denatured (>75°C). Although suitable for DNA amplification, these high temperatures would denature RNA targets, making this method of preventing nonspecific amplification unsuitable for RT-PCR.

For this purpose, aptamers able to inhibit the activity polymerase enzymes and which are "switched off" above 40°C have been selected using high-temperature selection conditions (52), demonstrating how aptamers can be used as temperature-sensitive molecular switches.

It is clear, however, that aptamers could be selected as molecular switches sensitive to other criteria such as pH, salt, or the presence of metal-chelating agents (52).

20.3.4.7 Aptamer Arrays

Although antibody-based microarrays for the use in proteomics are currently being developed, aptamer-based arrays would be attractive for the following reasons (52):

- Aptamer identification can be performed on an automated platform.
- Aptamers can be immobilized at defined densities and at precise locations using existing deposition technology.
- Homogenous preparations of aptamers are readily available from chemical synthesis.
- Aptamer-based arrays would be robust and have a long shelf-life.
- Proteins could be bound irreversibly using specifically modified aptamers.

These arrays will certainly play an increasingly important role in the study of proteomics as the wealth of information at the DNA level continues to increase with the multitude of genome screening projects.

20.3.4.8 Aptazymes

Aptazymes are aptamers linked to ribozymes (catalytic RNA molecules) possessing catalytic activity. They are able to directly transduce molecular recognition into a quantifiable catalytic event and can be generated using two different strategies (54):

- Preexisting aptamers and preexisting catalytic RNAs, with known specificity and catalytic properties, can be integrated. In this way, the resulting molecule will possess the desired affinity for the target as well as the expected catalytic effect.
- The aptazyme with the desired binding specificity is selected from a pool of aptazymes containing a random sequence receptor site and a catalytic region with established properties.

Aptazymes have been applied to the monitoring of posttranslational modification (66) and have also been adapted to a chip array able to distinguish between various metabolites (67).

20.4 Implementation and Illustration of Instrumentation

The Biacore X (Biacore) SPR analytical instrument is a typical device with which this type of work can be performed. The following is a brief description of the methods involved in this work.

20.4.1 Immobilization

The IgE-specific aptamer was immobilized to a dextran sensor chip (CM5) obtained from Biacore (Uppsala, Sweden) via a streptavidin–biotin reaction. The dextran chip allows for a more "biocompatible" environment, increased receptor activity, increased receptor density, as well as reduction in nonspecific sample interactions. Once immobilization was complete, binding experiments using IgE could then be performed by running appropriate sample concentrations and volumes over the sensor surface. After binding of the target molecule, the sensor surface could be regenerated using a solution of EtOH/NaOH. A typical result obtained from an SPR experiment, known as a sensogram, is shown in Figure 20.9.

20.4.2 Calibration

Using a range of concentrations, it is possible to obtain a calibration curve, which can be used to extrapolate the concentration of target analyte in solutions of unknown concentrations. Such a calibration curve is shown in Figure 20.10.

Although this work has been carried out successfully, there remain a number of areas where vast improvements can still be made, particularly with respect to the sensitivity of the measurements. Indeed, in this particular case, the sensitivity of the aptasensor has not

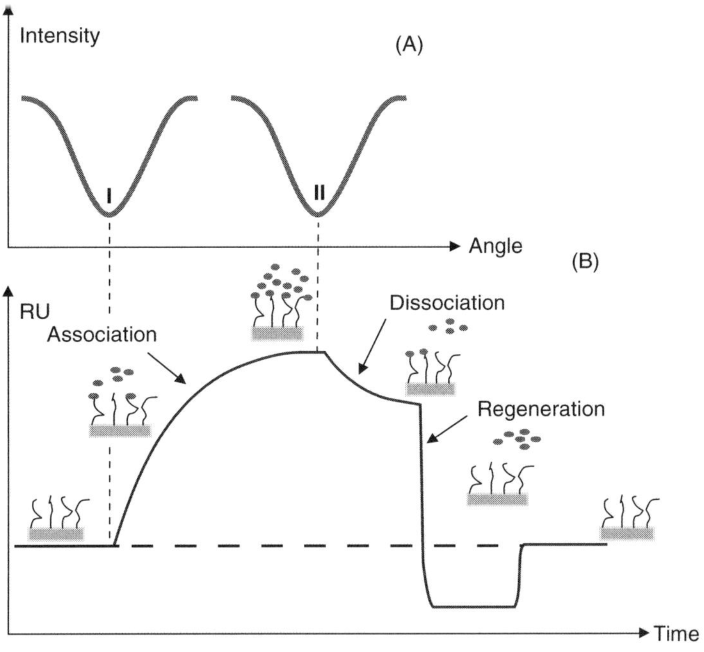

FIGURE 20.9
Sensogram from a typical SPR experiment. (A) Decrease in intensity of reflected light at resonance angles. (B) The shift in resonance angle (from position I to position II) is due to the binding of the analyte at the sensor surface. (Modified from Katsamba PS, Park S and Laird-Offringa IA. 2002. Kinetic studies of RNA–protein interactions using surface plasmon resonance. *Methods* 26: 95–104.)

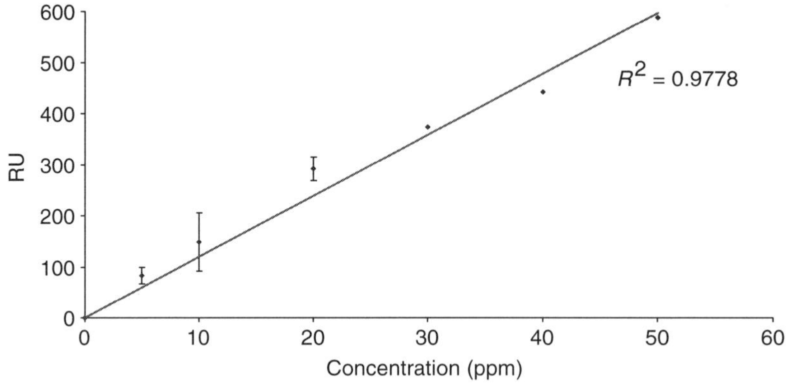

FIGURE 20.10
Calibration curve in the association phase (Biacore). This curve was constructed for IgE binding in the association phase at concentrations from 5 to 50 ppm. Standard deviations were calculated based on three samples at 5 ppm, three at 10 ppm, and four at 20 ppm, giving a CV_{ave} of 22%. Binding experiments at higher concentrations were only carried out once to conserve reagents.

yet been optimized and is not sufficient for the detection of IgE in real samples. For measurements of very low concentrations of target analyte to be feasible, a high association constant of the aptamer and IgE is necessary. This constant is very much dependent upon the three-dimensional stability of the aptamer and further developments in the design and selection of theses aptamers needs to be made.

20.5 Future Prospects

If an aptasensor for the detection of IgE were to be developed for routine clinical use, the aptamer selected would need to form a very stable complex with the target analyte to withstand the numerous washing steps, which would be necessary for its detection in a complex solution such as blood. Selection of an aptamer known to form a stable complex with IgE when immobilized to a solid support would therefore appear to be necessary.

Further work would also be needed to test the aptamer in complex protein solutions which would undoubtedly be necessary to develop an aptasensor capable of clinical application.

The ultimate aim would be to develop allergen-specific IgE aptamers and incorporate them into a commercially viable aptasensor that could be used by an individual to diagnose precisely to which allergen they are allergic.

References

(1) Mayer G. 2004. *The structure and function of immunoglobulins.* Available at http:// www.med.sc.edu:85/mayer/IgTypes2000.htm

(2) Stefan R-I, van Staden J and Aboul-Enein H. 2000. Immunosensors in clinical analysis. Fresenius *Journal of Analytical Chemistry* 366: 659–668.

(3) Hamilton R and Adkinson N. 2003. Clinical laboratory assessment of IgE-dependent hypersensitivity. *Journal of Allergy and Clinical Immunology* 111: S687–S701.
(4) Johansson SG and Benich H. 1967. Immunological studies of an atypical (myeloma) immunoglobulin. *Immunology* 13: 381–394.
(5) Ishizaka K and Ishizaka T. 1967. Identification of gamma-E-antibodies as a carrier of reaginic activity. *Journal of Immunology* 99(6): 1187–1198.
(6) Dolen WK. 2003. IgE antibody in the serum—detection and diagnostic significance. *Allergy* 58: 717–723.
(7) Hamelmann E, Rolinck-Werninghaus C and Wahn U. 2002. From IgE to Anti-IgE: Where do we stand? *Allergy* 57: 983–994.
(8) Ramirez M. 1919. Horse asthma following blood transfusion: Report of a case. *JAMA* 73: 984.
(9) Corry DB and Kheradmand F. 1999. Induction and regulation of the IgE response. *Nature* 402: B18–B23.
(10) Miescher S and Vogel M. 2002. Molecular aspects of allergy. *Molecular Aspects of Medicine* 23: 413–462.
(11) Leung DY. 1998. Molecular basis of allergic diseases. *Molecular Genetics and Metabolism* 63: 157–167.
(12) Doyle R. 2000. Asthma worldwide. *Scientific American* 282: 30.
(13) Mitchell EA. 1999. Asthma epidemiology: Clues and puzzles. *Pediatric Pulmonology Supplement* 18: 31–33.
(14) Aberg N, Hesselmar B, Aberg B and Eriksson B. 1995. Increase of asthma, allergic rhinitis and eczema in Swedish school children between 1979 and 1991. *Clinical and Experimental Allergy* 25: 815–819.
(15) Smith DH, Malone DC, Lawson KA, Okamoto LJ, Battista C and Saunders WB. 1997. A national estimate of the economic costs of asthma. *American Journal of Respiratory and Critical Care Medicine* 156: 787–793.
(16) Williams PB, Dolen WK, Koepke JW and Selner JC. 1992. Comparison of skin testing and three in vitro assays for specific IgE in the clinical evaluation of immediate hypersensitivity. *Annals of Allergy* 68: 35–45.
(17) Ownby DR, Magera B and Williams PB. 2000. A blinded, multi-centre evaluation of two commercial in vitro tests for latex specific IgE antibodies. *Annals of Allergy, Asthma and Immunology* 84: 193–196.
(18) Rogers K and Mulchandani A. 1998. *Affinity Biosensors: Techniques and Protocols*. Humana Press Inc., New Jersey, USA.
(19) Luppa PB, Sokoll LJ and Chan DW. 2001. Immunosensors—principles and applications to clinical chemistry. *Clinica Chimica Acta* 314(1–2): 1–26.
(20) Turner APF, Laschi S and Mascini M. 2002. *Biosensors: Kirk-Othmer Encyclopaedia of Chemical Technology*. Wiley.
(21) D'Orazio P. 2003. Biosensors in clinical chemistry. *Clinica Chimica Acta* 334: 41–69.
(22) Aizawa M, Morioka A, Suzuki S, Nagamura Y. 1979. Enzyme immunosensor. III. Amperometric determination of human chorionic gonadotropin by membrane-bound antibody. *Analytical Biochemistry* 94(1): 22–8.
(23) Richardson J, Hawkins P, Luxton R. 2001. The use of coated paramagnetic particles as a physical label in a magneto-immunoassay. *Biosensors and Bioelectronics* 16(9–12): 989–993.
(24) Gabig-Ciminska M, Holmgren A, Andresen H, Bundvig Barken K, Wumpelmann M, Albers J, Hintsche R, Breitenstein A, Neubauer P, Los M, Czyz A, Wegrzyn G, Silfversparre G, Jurgen B, Schweder T, Enfors SO. 2004. Electric chips for rapid detection and quantification of nucleic acids. *Biosensors and Bioelectronics* 19(6): 537–546.
(25) Mascini M, Palchetti I, Marrazza G. 2001. DNA electrochemical biosensors. *Fresenius Journal of Analytical Chemistry* 369(1): 15–22.
(26) Lucarelli F, Kicela A, Palchetti I, Marrazza G, Mascini M. 2002. Electrochemical DNA biosensor for analysis of wastewater samples. *Bioelectrochemistry* 58(1): 113–118.
(27) Mannelli I, Minunni M, Tombelli S, Mascini M. 2003. Quartz crystal microbalance (QCM) affinity biosensor for genetically modified organisms (GMOs) detection. *Biosensors and Bioelectronics* 18(2–3): 129–140.

(28) Piletsky S and Turner APF. 2004. *Molecular Imprinting*. Landes Bioscience, Georgetown, TX, USA.
(29) Su X, Chew FT and Li SFY. 1999. Self-assembled monolayer-based piezoelectric crystal immunosensor for the quantification of total human immunoglobulin E. *Analytical Biochemistry* 273: 66–72.
(30) Kreuzer MP, O'Sullivan CK, Pravda M and Guilbault GG. 2001. Development of an immunosensor for the determination of allergy antibody (IgE) in blood samples *Analytica Chimica Acta* 442: 45–53.
(31) Liss M, Petersen B, Wolf H and Prohaska E. 2002. An aptamer-based quartz crystal protein biosensor. *Analytical Chemistry* 74: 4488–4495.
(32) Homola J, Yee S and Gauglitz G. 1999. Surface plasmon resonance sensors: Review. *Sensors and Actuators* B 54: 3–15.
(33) Mullett WM, Lai EPC and Yeung JM. 2000. Surface plasmon resonance-based immunoassays. *Methods* 22: 77–91.
(34) Liedberg B and Johansen K. 1998. Affinity biosensing based on surface plasmon resonance detection. In: Rogers K and Mulchandani A. 1998. *Affinity Biosensors: Techniques and Protocols*. Humana Press Inc., New Jersey, USA.
(35) Brockman JM, Nelson BP and Corn RM. 2000. Surface plasmon resonance imaging measurements of ultrathin organic films. *Annual Review of Physical Chemistry* 51: 41–63.
(36) Hickel W, Kamp D and Knoll W. 1989. Surface plasmon microscopy. *Nature* 339: 186.
(37) Elkind JL, Stimpson DI, Strong AA, Bartholomew DU and Melendez JL. 1999. Integrated analytical sensors: The use of the TISPR-1 as a biosensor. *Sensors and Actuators B* 54: 182–190.
(38) Liedberg B, Nylander C and Lundström I. 1995. Biosensing with surface plasmon resonance—how it all started. *Biosensors and Bioelectronics* 10(8): i–ix.
(39) Silin V and Plant A. 1997. Biotechnological applications of surface plasmon resonance. *Trends in Biotechnology* 15: 353–359.
(40) Earp R and Dessy R. 1996. Surface plasmon resonance. In: *Commercial Biosensors: Applications to Clinical, Bioprocess, and Environmental Samples*. Graham Ramsay. 1998. Wiley, New York, USA.
(41) Liedberg B, Nylander C and Lundström I. 1983. Surface plasmon resonance for gas detection and biosensing. *Sensors and Actuators* 4: 299–304.
(42) McDonnell JM. 2001. Surface plasmon resonance: Towards an understanding of the mechanisms of biological molecular recognition. *Current Opinion in Chemical Biology* 5: 572–577.
(43) Rich R and Myszka D. 2000. Advances in surface plasmon resonance biosensor analysis. *Current Opinion in Chemical Biology* 11: 54–61.
(44) Gotoh M, Hasegawa Y, Shinohara Y, Shimizu M and Tosu M. 1995. A new approach to determine the effect of mismatches on kinetic parameters in DNA hybridization using an optical biosensor. *DNA Research* 2(6): 285–293.
(45) Leferink AE, van Zoelen EJ, van Vugt MJ, Grothe S, van Rotterdam W, van de Poll ML and O'Connor-McCourt MD. 2000. Superantagonistic activation of ErbB-1 by EGF-related growth factors with enhanced association and dissociation rate constants. *Journal of Biological Chemistry* 275: 26748–26753.
(46) McDonnell JM, Calvert R, Beavil RL, Beavil AJ, Sutton BJ, Gould HJ and Cowburn D. 2001. The structure of the IgE Ce2 domain and its role in stabilizing the complex with its high-affinity receptor FceRI. *Nature Structural Biology* 8: 437–441.
(47) Danelian E, Karlen A, Karlsson R, Winiwarter S, Hansson A, Lofas S, Lennernas H and Hamalainen MD. 2000. SPR biosensor studies of the direct interaction between 27 drugs and a liposome surface: correlation with fraction absorbed in humans. *Journal of Medicinal Chemistry* 43: 2083–2086.
(48) Adamczyk M, Moore JA and Yu Z. 2000. Application of surface plasmon resonance towards studies of low-molecular-weight antigen-antibody binding interactions. *Methods* 20: 319–328.
(49) Myszka D and Rich R. 2000. Implementing surface plasmon resonance biosensors in drug discovery. *Pharmaceutical Science and Technology Today* 3: 310–317.
(50) James W. 2000. Aptamers. In: *Encyclopaedia of Analytical Chemistry*. Mayers RA (ed.), pp. 4848–4871.
(51) O'Sullivan CK. 2002. Aptasensors—the future of biosensing? *Analytical and Bioanalytical Chemistry* 372: 44–48.

(52) Jayasena SD. 1999. Aptamers: An emerging class of molecules that rival antibodies in diagnostics. *Clinical Chemistry* 45(9): 1628–1650.

(53) Sampson T. 2003. Aptamers and SELEX: The technology. *World Patent Information* 25: 123–129.

(54) Luzi E, Minunni M, Tombelli S and Mascini M. 2003. New trends in affinity sensing: Aptamers for ligand binding. *Trends in Analytical Chemistry* 22(11): 810–818.

(55) Robertson DL and Joyce GF. 1990. Selection in vitro of an RNA enzyme that specifically cleaves single-stranded DNA. *Nature* 344 (6265): 467–468.

(56) Ellington AD and Szostak JW. 1990. In vitro selection of RNA molecules that bind specific ligands. *Nature* 346: 818–822.

(57) Tuerk C and Gold L. 1990. Systematic evolution of ligands by exponential enrichment: RNA ligands to bacteriophage T4 DNA polymerase. *Science* 249: 505–510.

(58) Pieken W, Tasset D, Janjic N, Gold L and Kirschenheuter G. 1997. High affinity nucleic acid ligand containing modified nucleotides. US Patent No 5,660,985.

(59) Cox JC, Rudolph P and Ellington AD. 1998. Automated RNA selection. *Biotechnology Progress* 14: 845–850.

(60) Cox JC and Ellington AD. 2001. Automated selection of anti-protein aptamers. *Bioorganic and Medical Chemistry* 9: 2525–2531.

(61) Drolet DW, Moon-McDermott L and Romig TS. 1996. An enzyme-linked oligonucleotide assay. *Nature Biotechnology* 14 (8): 1021–1025.

(62) Davis KA, Lin Y, Abrams B and Jayasena SD. 1998. Staining of cell surface human CD4 with 2¢-F-pyrimidine-containing RNA aptamers for flow cytometry. *Nucleic Acids Research* 26(17): 3915–3924.

(63) Kleinjung F, Klussmann S, Erdmann VA, Scheller FW, Fürste JP and Bier FF. 1998. High-affinity RNA as a recognition element in a biosensor. *Anaytical Chemistry* 70(2): 328–331.

(64) Potyrailo RA, Conrad RC, Ellington AD and Hieftje GM. 1998. Adapting selected nucleic acid ligands (aptamers) to biosensors. *Analytical Chemistry* 70: 3419–3425.

(65) German I, Buchanan DD and Kennedy RT. 1998. Aptamers as ligands in affinity probe capillary electrophoresis. *Analytical Chemistry* 70(21): 4540–4545.

(66) Vaish NK, Dong F, Andrews L, Schweppe RE, Ahn NG, Blatt L and Seiwert SD. 2002. Monitoring post-translational modification of proteins with allosteric ribozymes. *Nature Biotechnology* 20(8): 810–815.

(67) Seetharaman S, Zivarts M, Sudarsan N and Breaker R. 2001. Immobilized RNA switches for the analysis of complex chemical and biological mixtures. *Nature Biotechnology* 19(4): 336–341.

(68) Katsamba PS, Park S and Laird-Offringa IA. 2002. Kinetic studies of RNA–protein interactions using surface plasmon resonance. *Methods* 26: 95–104.

21

Biosensors for Virus Detection

Ebtisam S. Wilkins and Ravil A. Sitdikov

CONTENTS

21.1 Introduction

According to epidemiologic studies, viral diseases are the primary cause of serious disorders that do not require hospitalization among people who reside in developed countries. Among infants and children, they exact a heavy toll in mortality and permanent disability. Emerging viral diseases, such as those brought about by HIV, Ebola virus, and Hantavirus, appear regularly. In addition, while antibiotics effectively combat most bacterial-based infections, viral infections are not so readily controlled; by comparison, they pose a greater threat to people's health. Additionally, according to some data, the broad range of established viral diseases known today appears to be expanding into other serious human ailments including tumors, juvenile diabetes, rheumatoid arthritis, and a variety of neurological and immunological disorders [1].

Virus may even have had a role in the natural selection of animal species because like other microorganisms they have the ability to infect all forms of life including bacteria, plants, protozoa, fungi, insects, fish, reptiles, birds, and mammals. In a manner similar to the selective role smallpox virus played in humans, the natural selection of rabbits that were resistant to virulent myxoma virus occurred, and was documented during several epidemics deliberately induced to control the Australian rabbit population.

Another unproven but possible way in which virus may affect evolution [2] is through the introduction of viral genetic material into animal cells by mechanisms similar to those that govern gene transfer by bacteriophages. For instance, when genes from a virulent retrovirus are integrated into genomes of chickens or mice, they produce resistance to reinfection by related, virulent retroviruses. Reports of human leukemia-causing retroviruses indicate that the same relationship may exist for human retroviruses.

As small, subcellular agents, virus are unable to multiply outside a host cell (intracellular, obligate parasitism). Only one type of nucleic acid (RNA or DNA) is present in the assembled virus (virion), in addition to a protective protein coat in simple virus. The nucleic acid carries the genetic data needed to program the synthetic machinery of the host cell for viral replication while the protein coat serves two functions: it protects the nucleic acid from extracellular environmental insults such as nucleases, and permits attachment of the virion to the membrane of the host cell, which otherwise exhibits a negative charge that repels a naked nucleic acid. Once the host cell is infected through viral genome penetration, virus replication becomes dependent on that host cell for its energy and synthetic needs.

The basic structure of virus [3,4] seems to permit them to be simultaneously adaptable and selective, since the various virion components are synthesized separately within the cell, after which they are assembled to form progeny particles. Distinguishing them from all other small, obligate, intracellular parasites, this assembly type of replication is unique to the virus. Under experimental conditions, viral genomes are so adaptable that once they have penetrated the cell membrane, viral replication can occur in almost any cell. Alternatively, intact virus are so selective that most of them are able to infect only a limited range of cell types, which selectivity exists largely because penetration of the nucleic acid usually requires a specific reaction that enables the coat to attach to specific intracellular components and the host cell membrane.

Virion multiplication normally causes host cell damage or death, which is why virus tend to establish milder infections in which cell death is more of an aberration than the norm. Notable exceptions include HIV, Ebola virus, Hantavirus, and rabies virus; indeed, some virus are capable of establishing forms of silent infection [5].

Their extreme dependence on the host cell makes virus distinct among other microorganisms. A virus grows within a host cell and is viewed together with its host in case of any consideration of pathogenesis, epidemiology, host defenses, or therapy. Specific conditions for pathogenesis are imposed by the bilateral association between the virus and its host. Rhinoviruses, for instance, require temperatures not exceeding 34°C. This environment restricts their growth to cells in the cool outer layer of the nasal mucosa, thereby preventing their spread to areas of higher temperature in deeper cells [6].

Even though the virus is protected from some of the host's immune mechanisms by its intracellular location, it is vulnerable because of its dependence on the host cell's synthetic machinery. This state, however, may be altered by subtle physical and chemical changes produced by the viral infection itself, including inflammation, fever, circulatory alterations, and interferon [7].

The virus–host association is greatly influenced by the virus's epidemiologic properties [8]. As an example, certain arthropod-borne virus in insects multiply only within a narrow temperature range; as a result, it is only under specific seasonal and geographical conditions that the virus are found. Transmissibility of virus in aerosols and food is determined by other environmental conditions [9].

Virus replicate only within host cells, primarily utilizing many of the host cell's biosynthetic processes, which makes them difficult targets for chemotherapy. The similarity of host-directed and virus-directed processes makes it difficult to identify antiviral agents with sufficient specificity to exert their effect on viral replication in infected cells rather than on functions in uninfected host cells. Through experimentation, however, we are learning that each virus may have specific steps of replication that can be used as targets for selective, carefully aimed chemotherapeutic agents. Appropriate use of such drugs requires a thorough knowledge of suitable targets, based on an in-depth understanding of the virion's replication mechanisms and a correct diagnosis [10].

Successful vaccines are based on the knowledge of pathogenesis and immune defenses, while comparable considerations govern treatment with interferon. Correct diagnosis, prevention of the virion's spread in the environment, and effective treatment of the disease are complex issues. Knowledge of the pathogenetic mechanisms by which a virus enters, spreads, and exits the body is critical for antibody-containing immunoglobulin treatment and requires knowing when a virus is susceptible to an antibody (for instance, during viremic spread), and when a virus reaches target organs where the antibody may be less effective [11]. Viral infections are among the most difficult and demanding situations that a physician must address. While tremendous progress has been made over the past few decades, unfortunately, some of these issues still lack satisfactory solutions. More aspects of medical virology are understood now than previously, while others are being clarified gradually, and yet more are still obscure. Successful investigation and management of their pathologic processes is based on the knowledge of the properties of virus and the relationships they establish with their hosts [12].

21.1.1 Virion and Virus Structure

Virus are noncellular infectious agents consisting of a single type of nucleic acid (either RNA or DNA) surrounded by a protein coat (see Figure 21.1). Lacking any independent metabolism of their own and reproducing wholly within living host cells, simple virus are defined as noncellular infectious agents consisting of a single type of nucleic acid (either RNA or DNA) surrounded by a coat of protein. Because they are incapable of independent reproduction, they are labeled as nonliving or noncellular. All organisms, including other microbes, are susceptible to viral attack, although virus are generally specific for a given host [13].

A virion is therefore defined as nothing more than nucleic acid surrounded by a protein shell, the components of which taken together represent one infectious particle [14]. The outer protein shell is referred to as the capsid, which consists of repeating protein subunits called capsomeres. The architectural arrangement of the capsid, or its structure, determines a virus's general form, classified as polyhedral, helical, or binal (see Figure 21.2). Slightly more complex arrangements are present in virus that have animal hosts. A notable addition is the presence of a membranous envelope surrounding the capsid; the virus having capsid are termed "enveloped," whereas those that lack the feature are called "naked." Connecting the envelope to the capsid is a layer of matrix proteins. These proteins also serve to strengthen the envelope that itself comprises glycoproteins (a protein–carbohydrate complex) embedded within lipid from the host's cell membrane. Glycoproteins may project from the envelope's surface in the form of spikes of varying lengths and shapes, which serve an important role in attachment and infection [15].

The origin of each component determines the virus's resistance and survival, which means it is important to distinguish between components that are coded by the virus's genes (genes are segment of RNA or DNA) vs. those that are coded by the host's genes. Only the virus codes the proteins of the viral envelope. The host cell codes the carbohydrate and

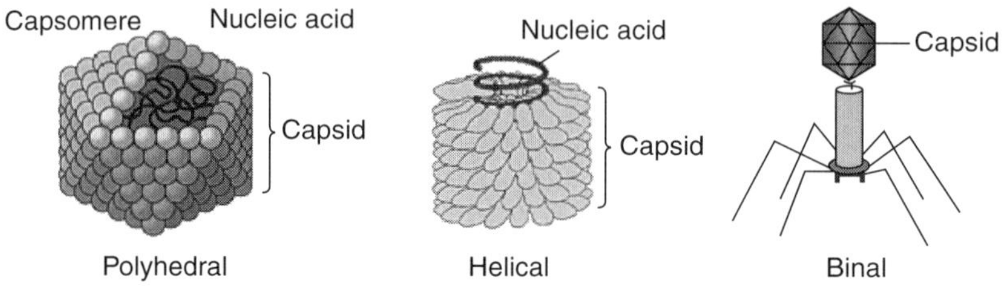

FIGURE 21.1
The types of virus structures. (From Indoor Environmental Quality Corporation. IEQ Fact Sheets. 2004. Virus. http://www.germology.com/fact_sheets.htm. Accessed December 20, 2004.)

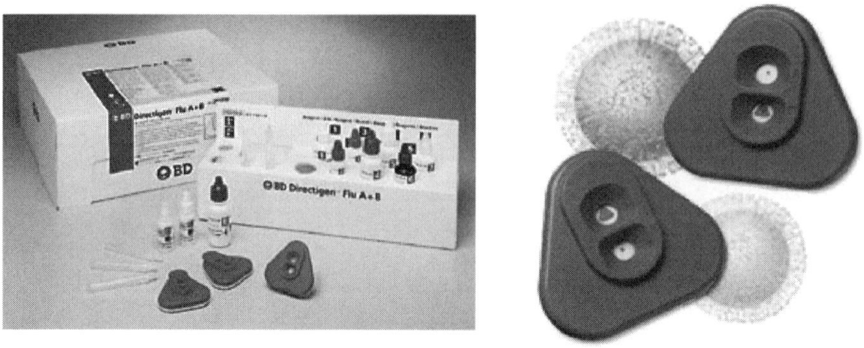

FIGURE 21.2
Directigen^{TM} Flu A+B Rapid Test Kit. (From www.bd.com/clinical/products/direct/fluaw.asp. Accessed December 18, 2004.)

TABLE 21.1
Characteristics and Comparison between Virus and Living Microorganisms

Characteristic	Virus	Living Microorganisms
Size	Generally less than 200 nm	Generally more than 200 nm
Nucleic acid	DNA or RNA	DNA and RNA
Outer covering	Protein coat; some with envelope	Complex membrane, wall, or both
Reproduction	Requires host	Generally self-replicating
Metabolism	Utilizes host metabolic machinery	Utilizes own metabolic machinery
Cultivation	Must be grown on media containing cells	Generally grown on cell-free media

Source: From Indoor Environmental Quality Corporation. IEQ Fact Sheets. 2004. Virus. http://www.germology.com/fact_sheets.htm. Accessed December 20, 2004.

lipids. Therefore, the type of lipid and carbohydrate within a given viral envelope is dependent on the particular host [16]. Distinguishing virus from other living microorganisms is the fact that they lack their own means of replication. The general comparisons between virus and living microorganisms along with their characteristics are given in Table 21.1.

21.1.2 Viral Taxonomy (Classification)

Animal virus (virus that require animal hosts) are classified based on

- size;
- the nature of the nucleic acid: DNA or RNA, single-stranded or double-stranded, segmented or single molecule;
- the type of viral replication: type of nucleic acid, type of participating enzymes, location of replication within the host cell;
- capsid structure: either polyhedral, helical, or binal; and
- presence or absence of viral envelope.

21.1.3 How Virus Cause Disease

A broad spectrum of disease in all living organisms (including humans) is caused by virus. To understand how virus cause these diseases, one must understand the basics of viral replication (i.e., reproduction involving the host's cellular mechanisms). There are essentially six steps to this process, details of which are varied and complex for a given type of virus. Each stage of the replication cycle influences the type of disease and its timing. The six basic steps are:

1. Attachment (adsorption): the virion attaches to the host cell.
2. Penetration: the virion breaks through the cell wall or membrane.
3. Uncoating: the virion releases its nucleic acid from the protein coat.
4. Replication: to produce more viral nucleic acid and viral proteins, the virion exploits the host's cellular mechanisms for reproduction.
5. Assembly (maturation): viral nucleic acids and viral proteins are reassembled.
6. Release: virions are released from the host cell; this generally results in cell and tissue damage and further activation of the immune system; the manifestations of this process (as well as its location within the host) determine the type and extent of disease.

Acute infections result when the virions remain localized, typically with short-lived symptoms. Cell and tissue damage results when new virions are released and the infected host cell dies. The host's defense mechanisms usually eliminate the virus over a period of days or months, bringing about the host's immunity to future infection. Acute infections include measles, mumps, and influenza (the flu) [13,17].

Persistent infections result when virions persist within the host organism, yet there are no disease symptoms. Several categories of persistent infections exist, manifestations of which are influenced by the process of replication as outlined above. In reality, persistent infections are usually the result of an acute infection. For instance, the infectious and disease phases of the measles are relatively short lived, yet in one-in-300,000 people late complications result in a fatal brain disorder known as subacute sclerosing panecephalitis (SSPE). SSPE, which is the result of a defective viral replication with brain cells, usually occurs within 10 years of having measles [9].

Latent persistent infections exhibit an extended, noninfectious stage between the original and subsequent disease. Chicken pox and shingles are the best examples of latent infection, both of which are caused by varicellazoster virus (VZV). Shingles or "zoster" appears after the virus is reactivated by unknown circumstances. Another example of latent virus is the herpes virus, both herpes simplex type 1 and type 2 [10].

Persistent infections may result in continuous or "chronic" infectious stages following a relatively brief period of disease. Examples of this type of infection are hepatitis B and C. Initial symptoms include nausea, fever, and jaundice; while the patient typically recovers from these early symptoms, he or she remains infectious. Over a period of time, the disease gets manifested as hepatitis, cirrhosis of the liver, or cancer. The opposite counterpart of a chronic infection is a "slow" infection in which no disease symptoms are initially observed, yet the infected person becomes more infectious over a period of time, and ultimately, a disease becomes apparent. An instance of the slow-infection type is AIDS, which is caused by the HIV virus [13,10].

21.1.4 Detection Methods

In general, diagnostic tests can be grouped into three categories: (1) direct detection, (2) indirect examination (virus isolation), and (3) serology [18]. In direct examination, the clinical specimen is examined directly for the presence of virus particles, virus antigen, or viral nucleic acids. Indirect examination is carried out in cell culture, eggs, or animals in an attempt to grow the virus; this is called virus isolation. Serology actually constitutes by far the bulk of the work of any virology laboratory. A serological diagnosis can be made by the detection of rising titers of antibody between acute and convalescent stages of infection, or the detection of IgM. In general, the majority of common viral infections can be diagnosed by serology. The specimen used for direction detection and virus isolation is very important. A positive result from the site of disease would be of much greater diagnostic significance than those from other sites. For example, in the case of herpes simplex encephalitis, a positive result from the CSF or the brain would be of much greater significance than a positive result from an oral ulcer, since reactivation of oral herpes is common during times of stress [18,19].

1. **Direct Examination of Specimen**

   - Electron microscopy morphology or immune electron microscopy
   - Light microscopy histological appearance, e.g., inclusion bodies
   - Antigen detection immunofluorescence, ELISA, etc.
   - Molecular techniques for the direct detection of viral genomes

2. **Indirect Examination**
    - Cell culture: cytopathic effect, hemadsorption, confirmation by neutralization, interference, immunofluorescence, etc.
    - Eggs pocks on chorioallantoic membrane (CAM): hemagglutination, inclusion bodies
    - Animal disease or death confirmation by neutralization
3. **Serology**

    Serology is about detection of rising titers of antibody between acute and convalescent stages of infection, or the detection of IgM in primary infection.

Diagnostic Methods for Detection of Viral Infection

Classical Techniques	Newer Techniques
1. Complement fixation tests (CFT)	1. Radioimmunoassay (RIA)
2. Hemagglutination inhibition tests	2. Enzyme-linked immunosorbent assay (EIA)
3. Immunofluorescence techniques (IF)	3. Particle agglutination
4. Neutralization tests	4. Western blot (WB)
5. Single radial hemolysis	5. Recombinant immunoblot assay (RIBA), line immunoassay (Liatek), etc.
	6. PCR methods.

21.1.5 Stains for Diagnosis of Viral Infections

Diagnosis of a viral infection based on microscopic examination of stained smears or tissue sections is limited to virus that induce characteristic morphologic changes during their replication in the cells they infect or for which specific antibodies to quantitatively significant and stable antigens are commercially available. Morphologic features indicative of a viral infection include the formation of inclusion bodies (masses of material consisting of viral particles or excess accumulation of products of viral synthesis) in the host cell nucleus, cytoplasm, or both, and for some virus, in the multinucleate giant cells [20].

21.1.6 Direct Detection in Smears

Microscopic examination of cells in stained smears as a mechanism for diagnosis of viral infection was popularized in the late 1940s by Tzanck, who used this technique to study cells scraped from the base of vesicular skin lesions as an aid to dermatologic diagnosis. In particular, the Tzanck preparation allowed differentiation of herpes simplex virus (HSV) or VZV infection, based on specific cytopathic changes, from other dermatologic diseases with a similar clinical presentation [21–23]. Cytologic preparations of material collected from a variety of sites have also proven useful for the detection of cytopathic changes of several different virus [24–26].

To perform the Tzanck test, the lesion selected for study is cleaned, the vesicle or pustule is opened and the crust is removed with a scalpel blade, the base is scraped vigorously with the edge of the blade, and the cellular material collected is evenly spread on a glass microscope slide. As originally described by Tzanck, the preparation is allowed to air-dry and then stained with Giemsa or a Romanowsky polychrome dye, such as Wright stain; Toluidine Blue O is an acceptable alternative. These stains are simple to use, but air-drying results in poor definition of nuclear detail, which could make recognition of cytopathic changes difficult. Modifications of the technique that allow better preservation of cellular detail and thus may be more sensitive than air-dried preparations include immediate fixation in 95% ethanol or 70% isopropyl alcohol and the use of alternative stains, such as Papanicolaou, modified Papanicolaou, Diff-Quik (another Romanowsky dye preparation), or Paragon multiple stain [27–29]. Smears may also be stained with monoclonal antibodies, which increase diagnostic sensitivity and specificity and allow the diagnosis of infection with virus that do not induce visible cytopathic changes.

Although the Tzanck preparation is useful for rapid diagnosis of cutaneous or mucocutaneous lesions caused by HSV or VZV, the test has limitations. It does not distinguish between HSV and VZV, and the sensitivity is less than 100%. Viral cytopathic changes are most likely to be detected in smears prepared from vesicles, followed by pustules, and then crusted lesions. When clinical impression is used as the standard, the overall sensitivity of Tzanck smear is about 50–60% for diagnosis of HSV infection and 64–100% for VZV, whereas the sensitivity of cell culture is approximately 80% for HSV and only 26–60% for VZV [29–32]. By staining with monoclonal antibodies, using immunofluorescence or immunoperoxidase techniques, the sensitivity of the direct smear increases but remains lower than that of the cell culture for HSV, and identification of the specific virus (HSV or VZV) is possible [29,33,34].

In certain situations, examination of cytologic preparations stained with nonspecific stains is useful for diagnosis of several viral infections other than those caused by HSV and VZV. For example, examination of a Papanicolaou-stained smear of cells collected from the uterine cervix is a valuable screening tool for diagnosis of infection with genital human papillomaviruses [35–39]. Diagnosis of molluscum contagiosum, although usually based on the clinical appearance of the lesion, may be confirmed by the examination of a smear of the cellular material collected from the central cavity and stained with lugol's iodine [40]. The carbohydrate-containing matrix takes up the stain, and the inclusions appear as dark-brown masses.

Evaluation of cytologic preparations of bronchoalveolar lavage fluid stained by the Papanicolaou technique or Giemsa stain is a rapid method, used predominantly with specimens from immunocompromised hosts, for diagnosis of pneumonias caused by some virus, including cytomegalovirus (CMV), adenovirus, respiratory syncytial virus (RSV), HSV or VZV, or measles virus [41,42]. For RSV, the changes are often subtle and are found only in a small proportion of epithelial cells, so that their recognition requires careful search. Prolonged RSV infection in immunocompromised hosts, however, may result in production of multinucleate giant cells containing eosinophilic cytoplasmic inclusions. To confirm infection with any of these virus, a smear may be stained with specific antibodies [43]. Examination of a smear of Wright- or Giemsa-stained bone marrow aspirate for characteristic erythroblast inclusions is often necessary to confirm parvovirus B19 as the cause of chronic anemia in an immunocompromised patient, especially a person with AIDS, because serologic tests generally are not helpful in such cases. Infection with BK virus or CMV may be diagnosed by detection of cells showing cytopathic changes characteristic of the respective virus in cytologic preparations of urine sediment stained by the Papanicolaou technique [22,44,45]. For CMV, however, shell-vial centrifugation-enhanced cell culture with detection of CMV nuclear antigens is a more sensitive and more frequently used diagnostic test. Microscopic examination of imprints of brain tissue stained with Seller stain (a mixture of basic fuchsin and methylene blue) for Negri bodies is a useful method for rapid diagnosis of rabies, especially in dogs [46]. However, rabies in wildlife, particularly skunks and bats, may be caused by strains of the virus that do not produce Negri bodies, and Negri-like cytoplasmic inclusion bodies may occur in neural tissues in the absence of rabies [47–49]. For these reasons, use of the Seller stain for diagnosis of rabies has for the most part been replaced by immunofluorescence staining with specific antibodies [46].

For some viral infections, diagnosis based on visualization of characteristic morphologic changes in cells stained with a nonspecific stain is not possible or is very difficult, but in many of these diseases a diagnosis can be made by immunofluorescence staining of smears with virus-specific antibodies. This approach to diagnosis is used most widely for detection of respiratory virus (RSV, Influenza A and B virus, and parainfluenza virus) in smears of nasopharyngeal cells collected by aspiration or swab. Advantages of immunofluorescence staining are that they have rapid turnaround time, high specificity, and the ability to assess specimen adequacy. For RSV, the sensitivity of immunofluorescence staining is equal to or greater than that of other available diagnostic tests (enzyme immunoassay and cell culture) [50–55]. In contrast, cell culture is more sensitive than immunofluorescence staining for detection if the immunofluorescence test is negative [50,52,56,57]. Other uses of virus-specific antibodies include rapid diagnosis of measles by detection of viral antigen in smears of nasopharyngeal cells [58], diagnosis of CMV pneumonia by detection of virus-infected cells in smears of bronchoalveolar lavage fluid [59], and differentiation of asymptomatic shedding of CMV during reactivation from active disease. The last is based on detection and quantitation of antigenemia by staining smears of peripheral leukocytes (buffy coat) with a monoclonal antibody against CMV pp65 structural antigen [60,61].

21.1.7 Detection in Tissue Sections

The first step in the routine diagnosis of any infectious disease from tissue specimens is the examination of sections stained with hematoxylin and eosin (H&E). Although the individual virus cannot be seen, many induce easily recognizable cytopathic changes that in most cases are sufficient for diagnosis [62,63]. Other features of viral infection are cell necrosis, which occurs during release of viral particles after replication, and an inflammatory infiltrate predominantly composed of lymphocytes; macrophages generally appear later in the infection cycle. In certain viral infections, such as HSV bronchopneumonia or bronchitis, polymorphonuclear leukocytes (PMN) may be present, even in the absence of secondary bacterial infection. The composition and degree of the inflammatory response to any infectious agent, however, depends primarily on the host immune status.

Occasionally, the changes observed in H&E-stained sections are insufficient for diagnosis and additional stains or other tests are necessary. Only a few nonspecific "special stains" have been useful for diagnosis of viral infections. Lendrum's phloxine–tartrazine method improves detection of viral inclusion bodies, staining them red against a yellow background [64], and Parson's stain and Schleifstein's method have been used for detection of Negri bodies. Orcein, modified trichrome, or Victoria blue-nuclear fast red were once used for confirmation of chronic hepatitis B virus (HBV) infection, apparently on the basis of the presence of disulfide bonds in HBV surface antigen [65–67]. Today, however, immunohistochemical assays with commercial virus-specific antibodies are preferred.

21.1.8 Foodborne Illness

Several groups of virus may infect persons after ingestion and then shed via stool. Of these, the norovirus (NoV) and hepatitis A virus (HAV) are currently recognized as the most important human foodborne pathogens with regard to the number of outbreaks and people affected in the Western world. NoV and HAV are highly infectious and may lead to widespread outbreaks [68]. Secondary infection of populations in countries with high standards of hygiene can take place. Molecular-based methods can detect virus in shellfish but are not yet available for other foods. The applicability of the methods currently available for monitoring foods for viral contamination is unknown. No consistent correlation has been found between the presence of indicator microorganisms (i.e., bacteriophages, *Escherichia coli*) and virus.

21.2 Biosensors for Detection of Virus and Bacteria

21.2.1 Genosensors

21.2.1.1 DNA Chips and DNA Biosensors

For the visual detection of HBV, hepatitis C virus (HCV), and human immunodeficiency virus type-1 (HIV-1) simultaneously, a qualitative DNA chip method, combining multiplex and nested polymerase chain reaction (PCR) with arrayed anchored primer PCR and a biotin–avidin alkaline phosphatase (Av-AP) indicator system, was developed. After pretreatment of infected blood samples and reverse transcription of the RNA virus genome, PCR was performed in a single tube by using the outer primer pairs. In the second round, nested multiplex PCR was performed on the DNA chip, on which the primers array had already been prepared [69].

Genosensor technology relying on the use of carbon and gold electrodes is reviewed. The key steps of each analytical procedure, namely DNA-probe immobilization, hybridization, labeling, and electrochemical investigation of the surface, are discussed in detail with separate sections devoted to label-free and newly emerging magnetic assays. Special emphasis has been given to protocols that have been used with real DNA samples [70].

Current Strategies for Electrochemical Detection of DNA with Solid Electrodes

A review of current strategies aimed at detecting nucleic acids (NA) using NA-modified solid electrodes reveals the versatility and potential of electrochemical detection in this field. What emerged at the beginning of 1990s as a very promising detection system in DNA technology is now resulting in the first commercial device. Many aspects of the experimental design, for example, surface immobilization and detection schemes, are outlined and evaluated. Although most approaches use hybridization devices as the recognized detection scheme, those not based on hybridization are also included [71].

Recent trends and challenges in the electrochemical methods for the detection of DNA hybridization are reviewed. Electrochemistry has superior properties over the other existing measurement systems because electrochemical biosensors can provide rapid, simple, and low-cost on-field detection. Electrochemical measurement protocols are also suitable for mass fabrication of miniaturized devices. Electrochemical detection of hybridization is mainly based on the differences in the electrochemical behavior of the labels toward the hybridization reaction on the electrode surface or in the solution. Basic criteria for electrochemical DNA biosensor technology, and already commercialized products, are also introduced. [72].

Influenza A virus, which are further subtyped on the basis of antigenic differences in external hemagglutinin and neuraminidase glycoproteins, and Influenza B virus are prominent among the viral causes of respiratory diseases and can cause a wide spectrum of illness. Each year these virus are responsible for recurrent epidemics, frequently in association with genetic variation. There is a requirement for sensitive and rapid diagnostic techniques to improve both the diagnosis of infections and the quality of surveillance systems. A new three-dimensional biochip platform (Flow-Thru Chip; MetriGenix) was used to develop a rapid and reliable molecular method for the typing and subtyping of influenza [73].

An electrochemical biosensor for the voltammetric detection of DNA sequences related to HSVs and discrimination of HSV Type I and Type II virus from PCR-amplified real samples has been described in this study. The biosensor relies on the covalent immobilization of the 22-mer single-stranded oligonucleotides (probe) related to both HSV Type I and Type II sequences and hybridization of these oligonucleotides with their complementary and four-base mismatch containing (four-base MM) sequences at pencil graphite electrodes (PEGE). The extent of hybridization between probe and target sequences was determined by using differential pulse voltammetry (DPV), and Meldola Blue (MDB) was used as the hybridization indicator [74].

A novel DNA quantification method by using a redox-active molecule, Hoechst 33258, 2′-(4-hydroxyphenyl)-5-(4-methyl-l-piperazinyl)-2,5′-bi(lH-benzimidazole), is reported. Hoechst 33258 was made to interact with DNA in solution without immobilization on the electrode surface; thus, the time-consuming probe immobilization step was eliminated by using cyclic voltammetry (CV) at a bare GCE to determine the most effective molecule for DNA aggregation. Hoechst 33258 was found to form an aggregate in the presence of DNA, and this phenomenon was confirmed by using atomic force microscopy (AFM). Detection of DNA sequences related to *Salmonella enteritidis*, *Streptococcus sobrinus*, and HBV were demonstrated by using DNA-Hoechst 33258 aggregation system [75].

Pulsed amperometric detection (PAD) of target DNA with platinum electrodes modified by single-stranded DNA (ssDNA) entrapped within polypyrrole (ssDNA/Ppy) is reported for the first time. Single-stranded DNA 20-mers complementary to the target DNA were used to construct the DNA biosensors. PCR-amplified bovine leukemia virus (BLV) provirus DNA was used as target DNA. Electrochemical impedance spectroscopic (EIS) investigation of ssDNA/Ppy before and after incubation in target has been discussed in the work of Ramanaviciene and Ramanavicius [76].

A fully electrical array for voltammetric detection of redox molecules was produced by enzyme-labeled affinity binding complexes. The electronic detection is based on ultramicroelectrode arrays manufactured in silicon technology. The 200-μm circular array positions have 800-nm wide interdigitated gold ultramicroelectrodes embedded in silicon dioxide. Immobilization of oligonucleotide capture probes onto the gold-electrode surfaces is accomplished via thiol-gold self-assembling. Spatial separation of probes at different array positions is controlled by polymeric rings around each array position. The affinity bound complexes are labeled with alkaline phosphatase, which converts the electrochemically inactive substrate 4-aminophenyl phosphate into the active 4-hydroxyaniline (HA) [77].

21.2.1.2 Differential Pulse Voltammetry and Electrochemical Genosensing

A rigid carbon–polymer composite material as a transducer for the electrochemical determination of label-free DNA based on DPV is reported. Graphite–epoxy composites (GEC) have an uneven surface allowing DNA, oligonucleotides, and free DNA bases to be adsorbed using a simple and fast wet-adsorption procedure. In contrast with other transducers commonly used for electrochemical genosensing, the oxidation potentials are much lower when GEC is used. Free guanine base is oxidized at +0.35 V while adenine oxidation occurs at +0.63 V (vs. Ag orAgCl). Cytosine- and inosine-free bases show no peaks within the experimental potential range. The oxidation of DNA guanine moieties occurs at a potential of +0.55 V while DNA adenine bases are oxidized at +0.85 V. A novel label-free hybridization genosensor using GEC as an electrochemical transducer for the specific detection of a sequence related to *Salmonella* spp. is also reported [78].

21.2.2 Impedance Sensor

Application of impedance spectroscopy: This study used colloidal named An to enhance the amount of antibody immobilized on a gold electrode and ultimately monitored the interaction of antigen–antibody by impedance measurement. Self-assembly of 6-nm (diameter) colloidal An onto the self-assembled monolayers (SAMs) of 4-aminothiophenol-modified gold electrode resulted in an easier attachment of antibody. The redox reactions of

[Fe(CN)(6)](4-)/[Fe(CN)(6)](3-) on the gold surface were blocked due to the procedures of self-assembly of 4-aminothiophenol and antibody immobilization, which were investigated by cyclic voltammetry and impedance spectroscopy [79].

21.2.3 Immunomagnetic Technique

The coupling of an immunological separation (using immunomagnetic beads) with amperometric now injection analysis detection of viable bacteria is presented [80]. Using a solution containing *E. coli* 0157, the electrochemical response with two different mediators [potassium hexacyanoferrate(III) and 2,6-dichlorophenolindophenol] was evaluated in the FIA system. Antibody-derivatized Dynabeads were used to selectively separate *E. coli* 0157 from a matrix. The kinetics and the capacity parameters regarding the attachment of bacteria to the immunobeads were studied. The immunomagnetic separation was then used in conjunction with electrochemical detection to measure the concentration of viable bacteria. A calibration curve of colony-forming units (du) against electrochemical response was obtained. The detection limit for this rapid microbiological method was 10^5 cfu/ml, and the complete assay was performed in 2 h. Some advantages over ELISA methods are the direct detection of viable cells (and not total bacterial load) and the need for only one antibody (not enzyme-labeled), thus making the assay faster (only one washing step is necessary) and less expensive.

An immunomagnetic capture reverse transcription-PCR (IMC-RT-PCR) assay was evaluated to recover and detect enteric virus in sewage and to remove PCR inhibitors [81]. The procedure was applied along with a simple sample processing consisting of an initial separation of solids followed by polyethylene glycol precipitation and solvent extraction. This procedure reduced sample volumes by about 65-fold without eliminating RT-PCR inhibitors. Paramagnetic beads coupled to pooled human immunoglobulins were used to simultaneously capture poliovirus 1 (PV-1) and HAV from seeded-sewage concentrates. The IMC was efficient in removing PCR inhibitors and further reducing sample volumes by approximately 10-fold, allowing the analysis of 6–7 ml of sewage sample per RT-PCR reaction. The detection limits of IMC-RT-PCR from seeded concentrates were 0.1–1 PFU for PV-1 and 1 MPNCU for HAV. The described procedure could be applied successfully for the detection of enteroviruses, HAV, and rotaviruses in field-sewage samples [81].

21.2.4 Atomic Force Microscopy-Immunosensor Assay

Bionanotechnology can be viewed as the integration of tools and concepts in nanotechnology with the attributes of biomolecules. We shall discuss here an atomic force microscopy-immunosensor assay (AFMIA) that couples AFM with solid-phase affinity capture of biological entities for the rapid detection and identification of group B coxsackievirus particles. Virus identification is based on type-specific immunocapture and the morphological properties of the captured viruses as obtained by the AFM. Representatives of the six Group B coxsackieviruses have been specifically captured [82].

21.2.5 Flow Cytometry

This group has developed a rapid, duplexed microsphere-based immunoassay for the characterization of influenza-virus types, which has the potential to overcome many of the limitations of the current detection methods. The assay uses microspheres of two sizes, each coupled to an Influenza type A- or type B-specific monoclonal antibody (MAb), to capture influenza virus in the sample. A cocktail of fluorescently labeled, influenza-specific polyclonal antibodies then binds the captured virus. The sandwich complexes are measured using a multiparameter flow cytometer. The assay can distinguish between Influenza types A and B in a single reaction with good reproducibility and high sensitivity [83].

21.2.6 Optical Biosensors

A report of the specific detection of a few hundred molecules of genetic material using a fluorescent polythiophene biosensor. Such recognition is based on simple electrostatic interactions between a cationic polymeric optical transducer and the negatively charged nucleic acid target and can be done in less than 1 h, simply and affordably, and without any chemical reaction. This simple system is versatile enough to detect nucleic acids of various lengths, including a segment from the RNA genome of the influenza virus [84].

Light-Addressable Potentiometric Sensor: A genetically biotinylated single-chain fragment variable antibody (scFv) against Venezuelan equine encephalitis virus (VEE) was applied in a system consisting of an immunofiltration enzyme assay (IFA) with a LAPS for the rapid identification of VEE. The IFA involved formation of an immunocomplex sandwich consisting of VEE, biotinylated antibody, fluoresceinated antibody and streptavidin, capture of the sandwich by filtration on biotinylated membrane, and labeling of the sandwich by antifluorescein urease conjugate [85].

21.2.7 Piezoelectric-Based Acoustic Wave Devices

Faradaic Impedance Measurements and Quartz Crystal Microbalance Assay: A drug composition consisting of nucleoside reverse transcriptase inhibitors (NRTIs), nonnucleoside reverse transcriptase inhibitors (NNRTIs), and protease inhibitors (PIs) is commonly used in AIDS therapy. A major difficulty encountered with the therapeutic composite involves the emergence of drug-resistant virus, especially to the PIs, regarded as the most effective drugs in the composition. The study presents a novel bioelectronic means to detect the appearance of mutated HIV-1 exhibiting drug resistance to the PI saquinavir. The method is based on the translation of viral RNA, the association of cleaved or uncleaved Gag polyproteins at an electrode surface functionalized with the respective antibodies, and the bioelectronic detection of the Gag polyproteins associated with the surface [86].

Piezoelectric-quartz-crystal resonators modified with oligonucleotide probes were used for detection of HCV in serum. The gold electrodes on either rough or smooth surface crystals were modified with a self-assembled monolayer of cystamine. After activation with glutaraldehyde, either avidin or streptavidin was immobilized and used for attachment of biotinylated DNA probes (four different sequences). Piezoelectric biosensors were used in a flow-through setup for direct monitoring of DNA resulting from the reverse transcriptase-linked polymerase chain reaction (RT-PCR) amplification of the original viral RNA [87].

Single Virus Particle Mass Detection Using Microresonators with Nanoscale Thickness: This study involves the microfabrication and application of arrays of silicon cantilever beams as microresonator sensors with nanoscale thickness to detect the mass of individual virus particles. The dimensions of the fabricated cantilever beams were in the range of 4–5 μm in length, 1–2 μm in width, and 20–30 nm in thickness. The virus particles we used in the study were vaccinia virus, which is a member of the Poxviridae family and forms the basis of the smallpox vaccine. The frequency spectra of the cantilever beams, due to thermal and ambient noise, were measured using a laser Doppler vibrometer under ambient conditions. The change in resonant frequency as a function of the virus-particle mass binding on the cantilever-beam surface forms the basis of the detection scheme [88].

Surface imprinting techniques on polymer-coated quartz-crystal microbalances (QCM) have been used to detect tobacco mosaic virus (TMV) in aqueous media. Molecularly imprinted polymers (MIP), tailor-made by self-organization of monomers around a template (TMV), were generated directly on the gold electrodes. Imprinted trenches on the polymer surface mimicking the shape and surface functionality of the virus serve as recognition sites for readsorption after washing out of the template [89].

Nucleic acid sequence-based amplification with electrochemiluminescent detection (NASBA/ECL) is an isothermal technique allowing rapid amplification and detection of specific regions of nucleic acid from a diverse range of sources. It is especially suitable for amplifying RNA. A NASBA/ECL technique has been developed allowing the detection of RNA from avian Influenza virus subtype H7 [90].

21.2.8 Capillary Electrophoresis (CE)

A review is presented on the CE analysis of colloidal or nano particles. Topics discussed include the CE separation of polymeric, inorganic, microbial (i.e., virus, bacteria, fungi, and whole cells), and subcellular particles (i.e., mitochondria and nuclei). Several of the encountered difficulties in analysis as well as the methods employed to overcome them are presented in the article by Rodriguez and Armstrong [91].

21.2.9 Biosensors for Environmental Applications

Biosensors can be excellent analytical tools for monitoring programs that implement legislation. In this chapter, biosensors for environmental analysis and monitoring are extensively reviewed. Examples of biosensors for the most important families of environmental pollutants, including some commercial devices, are presented in this chapter. Finally, future trends in biosensor development are discussed. In this context, bioelectronics, nanotechnology, miniaturization, and especially biotechnology seem to be the growing areas that will have a marked influence on the development of new biosensing strategies in the near future [92].

21.2.10 Electrochemical Biosensors

Potentiometric biosensors are usually based on ion-selective electrodes. These devices measure the change in ion concentration during a reaction. Generally, a simple sensor consists of an immobilized enzyme membrane surrounding the probe of a pH meter where the catalyzed reaction generates or absorbs hydrogen ions [93–95]. This leads to a change in pH that can be easily read. Three main types of ion-selective electrodes are often used in biosensors: normal-glass pH electrodes, glass-pH electrodes coated with a selective gas-permeable membrane, and solid-state electrodes consisting of a thin membrane of a specific ion conductor [96]. It is also possible to use metal-oxide semiconductors (MOS), which can measure charge on a surface that causes a current flow proportional to the charge. MOS devices are small and so they have fast response time due to reduced diffusion. However, the sensitivity of these can be affected by the ionic strength and concentrations of the solutions being analyzed. Potentiometric biosensors have been widely used for bacterial analyses. Examples include the detection of bacterial contamination in milk using an L-lactate biosensor, bacterial growth, and sequence-specific biosensing of DNA. Electrochemical detection of DNA hybridization involves the monitoring of a current under controlled potential conditions [97]. The hybridization is detected via increased current of a redox indicator or by changes in conductivity or capacitance. In this chapter, however, we would like to concentrate on the use of LAPS, which are proving to be popular as a platform for detecting microbes. These are semiconductor-based systems with an electrolyte-insulator-semiconductor (EIS) structure. When a current is applied across the EIS region, a depletion layer is formed at the insulator–semiconductor interface [98,99]. The capacitance of the depletion layer changes with the surface potential, which is a function of the ion concentration in the electrolyte. To determine the capacitance, the semiconductor is illuminated by modulated light and the current is measured. LAPS have several advantages when compared with other sensors: the surface is flat, there is no need for wires or passivation, and they can measure pH and concentration. Researchers at the USDA have used a LAPS system in combination with an immunoligand assay (ILA) to detect live *E. coli* O157:H7. They have reported that both live and dead bacteria can be detected in 30–45 min. In this system, bacteria are captured onto a filter membrane by using specific antibodies [100]. A silicone-based sensor is then placed adjacent to the membrane and upon illumination small changes in acidity are detected. The signal is proportional to the number of bacteria present, and it was possible to detect 710 dead or 25,000 live *E. coli* O157:H7 organisms/ml [101]. In a recent development, a LAPS approach was used to detect *E. coli* in drinking water [94].

An immunoassay was developed such that there was specificity to a particular capsular protein present in the bacterium. The transducer, based on the LAPS principle, was able to detect the production of ammonia by urease–*E. coli*-antibody conjugate. It was claimed that 10 cells/ml were detected in 1.5 h. Generally, amperometric biosensors work by enzymatically generating current between two electrodes. They have fast response time, dynamic ranges, and sensitivities similar to potentiometric biosensors. Many amperometric biosensors depend on dissolved oxygen concentration that can pose a major problem. To overcome this situation, mediators are employed. These mediators transfer electrons directly to the electrode thereby eliminating the need for the reduction of an oxygen cosubstrate [102]. The most commonly used mediators are the ferrocenes. Amperometric biosensors have been used for the detection of *E. coli* in water (screen-printed electrodes), bacterial vaginosis, studies of bacterial contamination, detection of agents of biological warfare (e.g., anthrax) and *E. coli* heat-labile enterotoxin, and other neurotoxins. Amperometric biosensors [103,104] have also been used to study bacterial luciferase reactions, nanoscale bacterial surface proteins and growth, and viability of bacterial populations.

Many biosensors are based on the *UIDA* gene, or the β-D-glucuronidase (GLUase) enzyme for which it encodes [105]. Although it is possible to target the *UIDA* gene directly [106], usually GLUase activity is used as an enzymatic marker for the identification of *Salmonella and Shigella*. The GLUase activity is detected by its ability to cleave specific chromogenic or fluorogenic artificial substrates added to the culture medium [107] or directly to filtered cells [108]. This approach has led to considerably faster and specific methods for the detection of *Salmonella and Shigella* contamination in water, although the analysis time is still measured in hours. Also, it still requires either extensive manipulation and incubation time [109–112] or sophisticated equipment [108].

According to the recent literatures cited, the method of immunoassay has become rapidly popular due to factors like high sensitivity, decrease in the time of analysis, and reproducible data that is reliable. The class of amperometric biosensors has become extremely successful as they can be used under field conditions without the need for skilled personnel. With these amperometric biosensors, the pathogens can be detected directly in the original samples without any preenrichment of the sample. Recently developed techniques for the detection of *Salmonella* and *Shigella* include methods based on the integration of several technologies like the microelectromechanical systems (MEMS), self-assembled monolayer (SAMS), DNA hybridization, and enzyme amplification [113]. The group of Jen-Jr Gau has tried to detect

Salmonella and *Shigella* by integrating several of these above-mentioned technologies like, MEMS [114–116], SAMS [113], DNA hybridization, and enzyme amplification. They have succeeded in detecting *Salmonella* and *Shigella* concentration of 1000 cells/ml. But the greatest disadvantage of this method is that the combination of several technologies makes it a multistep, rigorous analysis. The preparation of the MEMS detector array has to be done in specially equipped laboratory and by skilled personnel. Even though the analysis time is 40 min, the preparation time required for the MEMS and SAMS is time-consuming. The analyte cannot be used for direct detection. Only samples from the culture media of the analyte can be used, thus reducing the chances for the developed prototype to be used for field-testing. The developed prototype can only detect *Salmonella* and *Shigella* concentration of 1000 cells/ml. Lower concentrations of *Salmonella* and *Shigella* cannot be detected.

21.2.10.1 Potential Markets for Biosensors

Clinical or Medical: Metabolite analysis (glucose, cholesterol, enzymes, and triglyceride); drug analysis (salicylate, digoxin, paracetamol, and theophyline); hormone, bacterial, and viral analysis.

Industrial: Food and drug processing, quality control, cosmetics testing, and fermentation.

Security or Defense: Detection of harmful chemical and biological agents including explosives, nerve gases, pathogenic bacteria, and virus.

Bioterrorism: A review [117–119] considered the role of biosensors toward the detection of pathogenic bacteria. There is a recent heightened interest in developing rapid and reliable methods of detection especially where bioterrorism is concerned. In addition, biosensors can be used in food poisoning and clinical problems such as antibiotic resistance. Several different types of transduction modes can be used in high-frequency (surface acoustic wave) and optical detection. There are three systems that may make a great impact in the next few years: integrated (lab-on-a-chip) systems, molecular beacons, and aptamers [117–121].

Rapid detection could be achieved if the bacteria are purified before detection. When purified from the sample matrix, specific advantages of bacterial concentration may include facilitating the detection of different bacterial strains; removal of any inhibitor in the matrix; and insurance of adequately reduced sample size to represent food-sample sizes and small media volumes. Lower levels of pathogenic detection can be obtained through concentration and elimination of cultural enrichment prior to detection. Of all concentration methods that have been reported, including centrifugation, filtration, and immunomagnetic separation, none is ideal for one food system. Sample enrichment and separation of bacterial cells from food samples during sampling is a major problem in the advancement of molecular methods for the detection of foodborne pathogens.

A detailed understanding is provided in the review of the possibilities and limitations of separating and concentrating bacterial cells from the food matrix in an effort to advance the ability to harness molecular methods for the rapid detection of foodborne pathogens [117–121].

Agriculture and Veterinary: Diagnosis of plant and animal diseases and chemical monitoring.

Environmental: Detection of hazardous chemical in air, water, and soil and detection of personal contamination [122,123].

Obtaining results from traditional methods for detection and enumeration of bacteria in water samples (or any media) requires several days. Several new techniques that reduce the time of analysis have been developed. This chapter describes ways to test a rapid detection and enumeration of total viable bacteria using direct fluorescent labeling and detection by laser scanning. This method (TVC or total viable count) was compared to a culture method and the cyano-ditolyl-tetrazolium chloride (CTC) staining method for the analysis of samples before the final chlorination (after GAC filtration) and drinking water samples. The TVC method requires less time than others, is simple to use, and allows the analysis of large volumes (100 ml) of drinking water [124].

Other Uses: Research applications and understanding the relationship between the structure of a molecule and its biological function.

Device Development: At the University of New Mexico we have developed two techniques: flow-through using highly dispersed carbon particles and flow-through using porous carbon membranes. The system comprises a flow-through, immunofiltration technology and electrochemical-transduction principles. They are capable of detecting the bacteria in the original samples without any preenrichment. With an overall assay time of 30 min and 20 min, these analytical systems will be simple to operate under field conditions requiring minimal skills and can be easily adapted to detect other pathogenic bacteria. Previous work was done using the highly dispersed carbon particles for the detection of Hantavirus, which was very successful. The pore size of the disposable filter was determined on the basis of the size of the analyte. The process of selection of the filter-pore size was also based on the "hook effect" [93,94]. So, the previously developed procedure for the Hantavirus was adapted for the detection of *Salmonella* and *Shigella* after a few parameters were changed and optimized. The use of porous carbon membranes was decided upon following previous trials using nylon membranes. The nylon membranes were nonconductive, which led to a very small signal and correspondingly small detection range. The carbon membranes were then chosen for their ability to simultaneously function as the site of immunointeractionas well as the working electrode while maintaining the advantages that come with a disposable membrane. We will survey most of the commercially available equipment for monitoring and the detection of bacteria and virus, before introducing our recent work.

21.3 Commercial Equipment for Monitoring Virus

21.3.1 Commercial Kits

21.3.1.1 Directigen™ Flu A, Directigen™ Flu A+B (Becton and Dickinson, Inc.)

With BD Directigen™ Flu A, it is possible to have Influenza A test that is not only easy to perform and proven accurate, but also truly STAT (Figure 21.2) [125]. Most importantly, it will help your laboratory quickly and confidently support critical antiviral therapy recommendations. Truly STAT testing for Influenza type A offers the following features with Directigen Flu A:

- total hands-on time is about 5 min, thus reducing the technologist time;
- instant, clear-cut readings;
- helps rule out Influenza A from similar viral infections;
- room-temperature storage of the entire kit;
- good patient care with accurate results in minutes;
- 91% sensitivity and 95% specificity when compared to tissue culture; and
- clear, accurate results in about 6 min for positive specimens.

Directigen™ Flu A Rapid Test Kit

Various specimen types include nasopharyngeal washes, nasopharyngeal aspirates, and nasopharyngeal swabs. The Directigen Flu A rapid test is part of the line of rapid respiratory diagnostics that also includes Group A Strep, RSV and Flu A+B, which is the first differentiated test for Influenza A and Influenza B.

Now, there is a rapid, *in vitro* assay that detects and differentiates Influenza A and B viral antigens in less than 15 min. BD Directigen™ Flu A+B is the first rapid assay on the market that distinguishes Influenza A viral antigens from those of Influenza B. Differentiation of Influenza A and B can help physicians select appropriate, cost-effective treatment. Directigen™ Flu A+B offers a differentiated result, high accuracy, 15-min time-to-results, easy workflow, and ability to test a wide range of specimen types.

Directigen™ Flu A+B Rapid Test Kit

Various specimen types include nasopharyngeal wash, nasopharyngeal aspirate, nasopharyngeal swab, lower nasal swab, throat swab, and bronchoalveolar lavage. The Directigen™ Flu A+B rapid test is the latest addition to the BD line of rapid respiratory diagnostics that also includes RSV and Group A Streptococcus. Sensitivity and specificity are its chief characterstics.

- Influenza A: sensitivity of 95.7% and specificity of 91.4% for nasopharyngeal aspirates and
- Influenza B: sensitivity of 87.5% and specificity of 98.1% for nasopharyngeal aspirates.

21.3.1.2 The QuickVue Influenza A+B from QUIDEL Company

Influenza is a highly contagious viral infection. There are three types of Influenza virus: A, B, and C. Type A virus are most prevalent and are associated with the most serious epidemics. Type B virus produce a disease that is generally milder than that caused by A. Type C virus have never been associated with a large epidemic of human disease. Both type A and B virus can circulate simultaneously, but usually, one type is dominant during a given season. This test is specific to Influenza type A and B (Figure 21.3). The QuickVue Influenza A+B test allows for the rapid and qualitative detection of Influenza type A and type B antigens directly from nasal swab, nasal wash, and nasal aspirate specimens. The test is intended as an aid in the rapid differential diagnosis of acute Influenza type A and type B virus infections [126].

The QuickVue Influenza test detects Influenza type A and B viral antigens in three easy steps from a nasal swab or nasal-aspirate, or nasal-wash specimen. Test results are available in 10 min or less. The easy-to-use and easy-to-read kit can be stored at room temperature. No instrumentation is required. The 10-test kit is conveniently packaged for single-use testing and disposal. The 25-test kit offers bulk packaging that is ideal for batch testing. Both the 10-test and 25-test kits include everything to perform the test including positive and negative controls. Three types of results can be obtained.

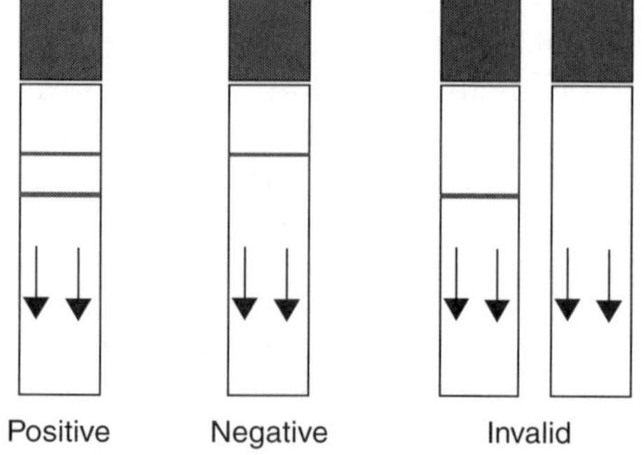

FIGURE 21.3
Results obtained with QuickVue Influenza A+B test. (From www.quidel.com/Products/productdisp.php?prod=101&section=pro. Accessed December 18, 2004.)

Positive: If two lines appear on the strip one red and the other blue then it is a positive result. The darker the red line, the stronger is the result, though any shade of a pink–red line is a positive result.

Negative: If there is only one visible blue line then it is a negative result.

Invalid: If no lines appear on the strip or there is only one visible red line then it is an invalid result. In this case, you should repeat the test with a new influenza test.

21.3.1.3 The BinaxNOW® Influenza A and B test (Binax, Inc.)

Binax, Inc. NOW® tests are rapid immunochromatographic tests (ICT) that detect specific analytes in clinical specimens using their patented book-shaped device [127]. These tests are easy to use and even easier to interpret. Binax has successfully developed NOW tests for bacteria, virus, and parasites. Use of NOW tests enable early and accurate identification of disease-causing pathogens, thus facilitating the implementation of appropriate and cost-effective treatment plans [127].

The BinaxNOW® Influenza A and B test simplifies rapid influenza testing. The easiest flu test on the market just got easier with accurate differentiation between Influenza A and Influenza B using one simple step in one test. Now, you can focus on the results and treating the patient, not running the test. The BinaxNOW Influenza A and B Test is an in vitro immunochromatographic assay for the qualitative detection of Influenza A and B nucleoprotein antigens in nasopharyngeal (NP) swab and nasal-wash or nasal-aspirate specimens. It is intended to aid in the rapid differential diagnosis of Influenza A and B viral infections. Negative test results should be confirmed by cell culture.

***Flu A and Flu B*:** Both the NOW Flu A and the NOW Flu B tests are the most frequently adopted rapid flu tests on the market today. Testing could not have been easier than these and the results are available in 15 min. Now, you can focus on the results and treating your patients, not running the test.

1. **Flu A**

   - Sensitivity (nasal wash or aspirate): 100%
   - Specificity (nasal wash or aspirate): 93%
   - Sensitivity (NP swab): 100%
   - Specificity (NP swab): 99%

2. **Flu B**

   - Sensitivity (nasal wash or aspirate): 92%
   - Specificity (nasal wash or aspirate): 94%
   - Sensitivity (NP swab): 100%
   - Specificity (NP swab): 99%
   - Sample type: Nasal wash or aspirate and NP swabs
   - Time for test results: 15 min

21.3.2 Commercial Equipment

21.3.2.1 Quartz Crystal Microbalance Analyzer: QCM100 (Stanford Research Systems, Inc.)

The QCM100 is a low-cost quartz-crystal microbalance system designed for real-time mass and viscosity measurements in processes occurring at or near surfaces, or within thin films

(Figure 21.4). It comes with controller, oscillator electronics, crystal holder, and three quartz crystals. An external frequency counter and precision voltmeter (not included) are needed to complete the QCM-measurement setup. The QCM series measures the resonant frequency and resistance of an AT-cut quartz crystal. The resonant frequency changes as a linear function of the mass of material deposited on the crystal surface. The resistance at resonance changes with the viscoelasticity of the material (film or liquid) in contact with the crystal's face [128].

As a gravimetric instrument, the QCM100 can measure mass ranging from micrograms to fractions of a nanogram, with detection limits corresponding to submonolayers of atoms. Measurement of resistance provides the opportunity to examine the viscosity and elasticity of films and liquids at or near the crystal surface. This makes it feasible to observe conformational changes, such as phase transitions, swelling, and cross-linking, in real time. The electronics are specifically designed to handle heavy loads (up to 5 kW) making these instruments ideal for studies involving glossy films and highly viscous liquids.

Applications of QCM100 instrument include the detection of virus capsids and bacteria, mammalian cells, immunosensors, sorption sensors, antigen–antibody reactions, and protein adsorption.

21.3.2.2 Packaged Interdigitated Microsensor Electrodes, (ABTECH Scientific, Inc.)

Central to ABTECH's suite of advanced products is its proprietary biochip technology. Based on impedemetric detection of chemically amplified biochemical signals, ABTECH's biochip technology gives rise to a suite of enzyme biochips (enzyme biosensors), immunodiagnostic biochips (immunosensors), and DNA biochips (genosensors) that are well suited to the physician's office, emergency room care, and the pathology lab markets (Figure 21.5) [129].

ABTECH is a biotechnology test and measurement company with interests in *point-of-concern* biomedical diagnostics. Engaged in research, product development and commercialization of proprietary biosensor-based diagnostics, and monitoring technologies, ABTECH is developing and marketing a suite of instrumented tests for the physician's office, emergency-room care, and the pathology laboratory. ABTECH's mission is to deliver chemical and biosensor-based test and measurement products that are cost-effective, reliable, and meet the sensitivity, detection limits, and convenience requirements of the markets that they serve.

The Laboratory Products Group manufactures, markets, and sells research and development products such as sensor devices, biochip substrates, transducers, electrodes,

FIGURE 21.4
QCM100. Quartz crystal microbalance analyzer. (From http://www.thinksrs.com/. Accessed December 20, 2004.)

FIGURE 21.5
Packaged interdigitated microsensor electrodes. (From www.abtechsci.com. Accessed December 20, 2004.)

research instruments, and associated software that are used by academic, industrial, and government researchers in developing and performing unique chemical and biosensor-based enzyme, immunodiagnostic, DNA, organelle, and whole-cell bioassays. A list for product selection is given as follows:

BioSenSys™, multianalyte diagnostics workstation: BioSenSys performs biosensor-based immunodiagnostic assays for the monitoring of blood metabolites, therapeutic drugs, and microbiological agents.

ToxSen™, chemical toxicity sensor: Based on a multielement array of differentiated whole cells, ToxSen comprises a quantitative screening test along with associated methods and apparatus for performing rapid, on-site screening of the chemical toxicity of water, wastewater, and other effluent streams.

EnVOCSys™, a VOC system for the monitoring of volatile organic constituents of the expired breath: The system comprises a 16-element array sensor (EnVOCSen™) that is responsive to gas-phase constituents at the sub-ppm level.

beChip™, the bioelectronic biochip is a 64-element impedimetric array for bioelectronic detection of DNA hybridization: This OncoChip™ is targeted at clinical cancer diagnostics and the rapid genomic staging of malignancies and as a complement to histopathology.

21.3.2.3 Biacore 2000 Analyzer, (Biacore Inc.)

Biacore 2000 is one of the high-sensitivity automated analysis system with surface-plasmon-resonance biosensor chips for rapid detection of several biological species (Figure 21.6). The sensor chip is a unique, integrated, but easily exchangeable component of Biacore systems. There is a range of sensor surfaces available to allow flexible immobilization of biomolecules through covalent coupling or capture. A wide range of biomolecules can be immobilized, including peptides, proteins, oligonucleotides, and membrane-bound receptors. In combination with optimized methods for efficient coupling of ligands, the immobilized surfaces are very stable, although this is ligand dependent. Biacore-sensor chips meet the highest demands for chemical stability, reproducibility, and low, nonspecific binding. Key benefits include high sensitivity (10^{-11} M), rapid analysis time per sample (2–10 min), and reproducible results [130].

The sensor chip surface consists of a glass support on which a gold film has been deposited. For most applications the gold layer is covered with a noncrosslinked-carboxylated-dextran hydrogel coupled through an alkyl-thiol-linker layer.

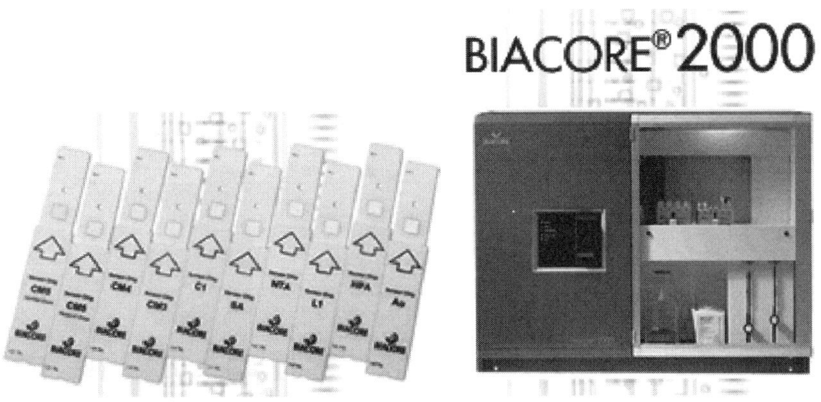

FIGURE 21.6
Biacore 2000 analyzer and sensor chips. (From http://www.biacore.com/products/systemsandsoftware/2000/. Accessed December 20, 2004.)

21.4 Future Commercially Available Immunosensors

Progress is being made in MESOSYSTEM Inc. for developing different types of sensors [94,131,132]. Micro fluidics cartridge and system has been shown in Figure 21.7.

A group has previously reported a sandwich-based strip immunoassay for rapid detection of *E. coli* O157:H7 [133]. In the present study, a microcapillary flow injection liposome immunoanalysis (mFILIA) system was developed for the detection of heat-killed *E. coli* O157:H7. A fused-silica microcapillary with anti-*E. coli* O157:H7 antibodies chemically immobilized on the internal surface via protein A served as an immunoreactor or immunoseparator for the mFILIA system. Liposomes tagged with anti-*E. coli* O157:H7 and encapsulating a fluorescent dye were used as the detectable label. In the presence of *E. coli* O157:H7, sandwich complexes were formed between the immobilized antibodies in the column, the sample of *E. coli* O157:H7 and the antibody-tagged sulforhodamine-dye-loaded liposomes. Signals generated by lysing the bound liposomes with 30-mM *n*-octyl-β-D-glucopyranoside were measured by a fluorometer. The detected signal was directly proportional to the amount of *E. coli* O157:H7 in the test sample. The mFILIA system successfully detected as low as 360 cells/mL (equivalent of 53 heat-killed bacteria in the 150 μL of the sample solution injected). MeOH (30%) was used for the regeneration of antibody-binding sites in the capillary after each measurement, which allowed the immunoreactor or immunoseparator to be used for at least 50 repeated assays. The calibration curve for heat-killed *E. coli* O157:H7 has a working range of 6×10^3 to 6×10^7 cells, and the total assay time was less than 45 min. A coefficient of variation for triplicate measurements was ≤8.9%, which indicates an acceptable level of reproducibility for this newly developed method [134–136].

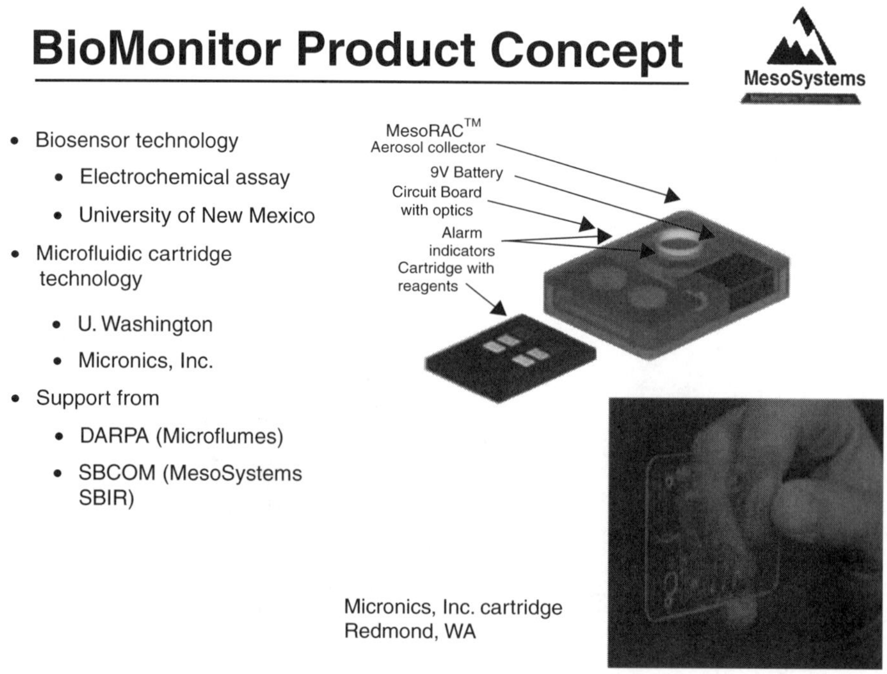

FIGURE 21.7
Micro fluidics cartridge and system.

In our laboratory, we have tested our sensor for Hantavirus detection of rodent's blood in the lab as well as in the field. We are now testing other virus such as Para Influenza and Influenza A.

21.5 Conclusions

Biosensor techniques are dynamic tools for determination of different analytes for environmental, clinical, agricultural, food, veterinary, or military applications. Finally, biosensor must be able to provide a lower and higher detection limit with a rapid analysis time at a relatively low cost.

Immunofiltration assay utilizing the carbon membrane can successfully determine concentrations of bacteria such as *Salmonella, Shigella, and E. coli* as low as 50 cells/ml. This system could be adapted for virus detection and can be easily automated and operated by a person with little or no training. The assay for virus detection takes about 22 min, which could be decreased with further optimization. Other benefits of this type of assay include a disposable sensing element, use of very little reagents, an increased signal relative to the immunosorbent assay, and the possibility for easy adaptation to detect other virus and bacteria [87,89,117–121].

An array of sensors should follow using microfluidics and nanotechnology. Most of the existing limitations could be directly related to operational and long-term stability of the sensor's components or the transducer itself. Other limitations could be attributed to reproducibility in complex matrices. For practical applications, the most important use of the sensors is encountered once the sensor is used for in situ real-sample monitoring. We have done that in the Hantavirus experiment referred to in the article of Vetcha, Wilkins, and Yates [137–159]. We have modified our immunoassay device, which contains a set of electronics that will drive an immunoelectrode and test a sample for Hantavirus. The photograph of the device is shown in Figure 21.8.

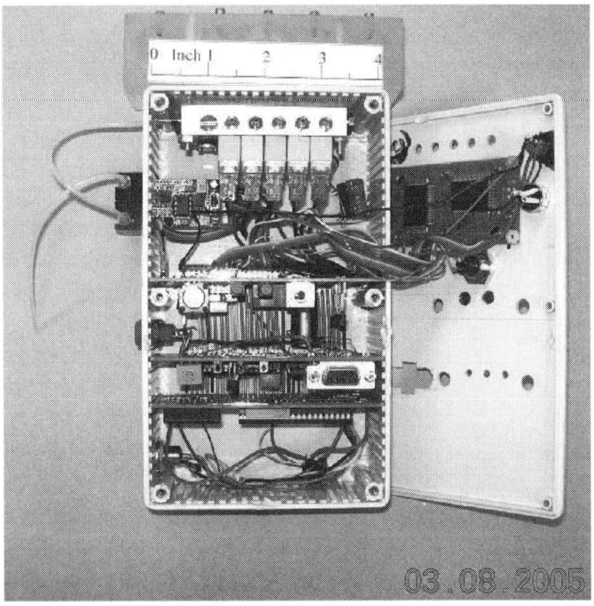

FIGURE 21.8
Photo of portable automated immunosensor device.

The power supply uses either batteries or 110 AC to generate all the voltages used in the device. The potentiostat is used to bias the immunoelectrode and convert its current into a voltage signal. The pumping system delivers the reagents used to test the sample. And the mircocontroller runs the program used to test the sample and display the results.

References

1. Dianzani F., Albrecht T. and Baron S. (1996). Introduction to Virology. In: Baron, S., ed. *Medical microbiology.* 4th ed. The University of Texas, Medical Branch at Galveston. http://www.ncbi.nlm.nih.gov/books/bv.fcgi?rid=mmed (accessed December 10, 2004).
2. Morse, S.S. (1994). *The Evolutionary Biology of Virus.* New York: Raven Press.
3. Mattern, C.F.T. (1977). Symmetry in virus architecture. In: Nayak, D.P., ed. *Molecular Biology of Animal Virus.* New York: Marcel Dekker.
4. Nermut, M.V. and Stevens, A.C. (1989). *Animal Virus Structure.* Amsterdam: Elsevier.
5. Palese, P. and Roizman, B. (1980). Genetic variation of virus. *Ann. NY Acad. Sci.* 354: 1.
6. Haywood, A.M. (1994). Virus receptors: binding, adhesion strengthening, and changes in virus structure. *J. Virol.* 68: 1–15.
7. Baron, S., Grossberg, S.E., Klimpel, G.R. and Brunell, P.A. (1984). Mechanisms of action and pharmacology: the immune and interferon systems. In: Galasso, G., ed. *Antiviral Agents and Viral Diseases of Man.* New York: Raven Press.
8. Evans AS: (1989). *Viral Infections of Humans: Epidemiology and Control.* 3rd ed. New York: Plenum Medical.
9. Nathanson, N. (1995). Epidemiology. In: Fields, B.N., Knipe, D.M., Howlett, P.M. ed. *Virology.* 3rd ed. New York: Raven Press.
10. Hayden, F.C. (1995). Antiviral Agents. In: Mandell, G.L., Bennett, J.E., Dolin, R. ed. *Principles and Practice of Infectious Diseases.* 4th ed. New York: Churchill Livingstone Inc. N.Y. p. 411.
11. Plotkin, S.A. and Mortimer, E.A. Jr. (1994) *Vaccines.* 2nd ed. Philadelphia: W.B. Saunders Co.
12. White, D.O. and Fenner, F. (1994) *Medical Virology.* 4th ed. San Diego: Academic Press.
13. Indoor Environmental Quality Corporation. IEQ Fact Sheets. 2004. Virus. http://www.germology.com/fact_sheets.htm (accessed December 20, 2004).
14. Gelderblom, H.R. (1996). *Structure and Classification of Virus.* In: Baron, S., ed. *Medical microbiology.* 4th ed. The University of Texas, Medical Branch at Galveston. http://www.ncbi.nlm.nih.gov/books/bv.fcgi?rid=mmed (accessed December 10, 2004).
15. Palmer, E.L. and Martin, M.L. (1988) *An Atlas of Mammalian Virus.* Boca Raton: CRC Press.
16. Baltimore, D. (1971). Expression of animal virus genomes. *Bacteriol. Rev.* 35: 234.
17. Murphy, F.A., Fauquet, C.M. and Bishop, D.H.L. (1995). *Virus Taxonomy: Sixth Report of the International Committee on Taxonomy of Virus.* New York: Springer-Verlag.
18. http://virology-online.com/ (accessed November 15, 2004).
19. http://virology-online.com/general/Tests.htm (accessed November 15, 2004).
20. Woods, G.L. and Walker, D.H. (1996). Detection of Infection or Infectious Agents by Use of Cytologic and Histologic Stains. *Clin. Microbiol. Rev.* 6:382–404.
21. Blank, H., Burgoon, C.F., Baldridge, G.D., McCarthy, P.L. and Urbach. F. (1951). Cytologic smears in diagnosis of herpes simplex, herpes zoster and varicella. *JAMA* 146:1410–1412.
22. Goldman, L., McCabe, R.M. and Sawyer, F. (1960). The importance of cytology technique for the dermatologist in office practice. *Arch. Dermatol.* 81:359–368.
23. Tzanck, A. and Melki, G.R. (1954). Cyto-diagnosis in dermatology. In: Mackenna, R.M.B., ed. *Modern trends in dermatology.* New York: Hoeber, pp.. 87– 102.
24. Ashton, P.R. (1983). Infectious organisms in cytologic material. *Lab. Med.* 14:227–233.
25. Bossen, E.H., Johnston, W.W., Amatulli, J. and Rowlands, D.T. Jr. (1969). Exfoliative cytopathologic studies in organ transplantation. *Am. J. Clin. Pathol.* 52:340–344.

26. Woods, G.L., Thompson, A.B., Rennard, S.L. and Linder. J. (1990). Detection of cytomegalovirus in bronchoalveolar lavage specimens. Spin amplification and staining with a monoclonal antibody to the early nuclear antigen for diagnosis of cytomegalovirus pneumonia. *Chest* 98:568–575.
27. Barr, R.J., Herten, J. and Graham, J. H. (1977). Rapid method for Tzanck preparations. *JAMA* 237:1119–1120.
28. Meisels, A., Fortin, R. and Roy, M. (1977). Condylomatous lesions of the cervix. II. Cytologic, colposcopic and histopathologic study. *Acta. Cytol.* 21:379–390.
29. Sadick, N.S., Swenson, P.D., Kaufman, R.L. and Kaplan, M.H. (1987). Comparison of detection of varicella-zoster virus by the Tzanck smear, direct immunofluorescence with a monoclonal antibody, and virus isolation. *J. Am. Acad. Dermatol.* 17:64–69.
30. Nahass, G.T., Goldstein, B.A., Zhu, W.Y., Serfling, U., Penneys, N.S. and Leonardi, C.L. (1992). Comparison of Tzanck smear, viral culture, and DNA diagnostic methods in detection of herpes simplex and varicella-zoster infection. *JAMA* 268:2541–2544.
31. Solomon, A.R., Rasmussen, J.E., Varani, J. and Pierson, C.L. (1984). The Tzanck smear in the diagnosis of cutaneous herpes simplex. *JAMA.* 251: 633–635.
32. Solomon, A.R., Rasmussen, J.E. and Weiss, J.S. (1986). A comparison of the Tzanck smear and viral isolation in varicella and herpes zoster. *Arch. Dermatol.* 122:282–285.
33. Gleaves, C.A., Lee, C.F., Bustamante, C.I. and Meyers, J.D. (1988). Use of murine monoclonal antibodies for laboratory diagnosis of varicella-zoster virus infection. *J. Clin. Microbiol.* 26:1623–1625.
34. Moseley, R.C., Corey, L., Benjamin, D., Winter, C. and Remington, M.L. (1981). Comparison of viral isolation, direct immunofluorescence, and indirect immunoperoxidase techniques for detection of genital herpes simplex virus infection. *J. Clin. Microbiol.* 13:913–918.
35. Bottone, E.J. (1980). *Cryptococcus neoformans*: pitfalls in diagnosis through evaluation of Gram-stained smears of purulent exudates. *J. Clin. Microbiol.* 12:790–791.
36. Casas-Cordero, M., Morin, C., Roy, M., Fortier, M. and Meisels, A. (1981). Origin of the koilocyte in condylomata of the human cervix. Ultrastructural study. *Acta. Cytol.* 25:383–392.
37. Koss, L.G. and Durfee, G.R. (1956). Unusual patterns of squamous epithelium of the uterine cervix: cytologic and pathologic study of koilocytotic atypia. *Ann. N. Y. Acad. Sci.* 63:1245–1261.
38. Meisels, A. and Fortin, R. (1976). Condylomatous lesions of the cervix and vagina. I. Cytologic patterns. *Acta. Cytol.* 20:505–509.
39. Meisels, A., Fortin, R. and Roy, M. (1977). Condylomatous lesions of the cervix. II. Cytologic, colposcopic and histopathologic study. *Acta. Cytol.* 21:379–390.
40. McCracken, A.W. (1980). Microscopic clues to viral and chlamydial diseases. *Lab. Med.* 11:152–158.
41. Parham, D.M., Bozeman, P., Killian, C., Murti, G., Brenner, M. and Hanif. I. (1993). Cytologic diagnosis of respiratory syncytial virus infection in a bronchoalveolar lavage specimen from a bone marrow transplant recipient. *Am. J. Clin. Pathol.* 99:588–592.
42. Woods, G.L., Thompson, A.B., Rennard, S.L. and Linder, J. (1990). Detection of cytomegalovirus in bronchoalveolar lavage specimens. Spin amplication and staining with a monoclonal antibody to the early nuclear antigen for diagnosis of cytomegalovirus pneumonia. *Chest* 98:568–575.
43. Stevens, D.A. 1983. Clinical and clinical laboratory aspects of nocardial infection. *J. Hyg.* 91:377–384.
44. Bossen, E.H., Johnston, W.W., Amatulli, J. and Rowlands, D.T. Jr. (1969). Exfoliative cytopathologic studies in organ transplantation. *Am. J. Clin. Pathol.* 52:340–344.
45. Coleman, D.V. (1975). The cytodiagnosis of human polyomavirus infection. *Acta. Cytol.* 19:93–96.
46. Lennette, E.H. and Schmidt, N.J. (1979). *Diagnostic procedures for viral, rickettsial and chlamydial infections.* 5th ed. Washington, D.C.: American Public Health Association.
47. Butts, J.D., Bouldin, T.W. and Walker, D.H. (1984). Morphological characteristics of a unique intracytoplasmic neuronal inclusion body. *Acta. Neuropathol.* 62:345–347.
48. Derakhshan, I. (1975). Is the Negri body specific for rabies? A light and electron microscopical study. *Arch. Neurol.* 32:75–79.

49. Derakhshan, I., Bahmanyar, M., Fayaz, A., Noorsalehi, S., Mohammad, M. and Ahouraii, P. (1978). Light-microscopical diagnosis of rabies: a reappraisal. *Lancet.* i:302–303.
50. Dominguez, E.A., Taber, L.H. and Couch, R.B. (1993). Comparison of rapid diagnostic techniques for respiratory syncytial and influenza A virus respiratory infections in young children. *J. Clin. Microbiol.* 31:2286–2290.
51. Johnston, S.L.G. and Siegel, C.S. (1990). Evaluation of direct immuno-fluorescence, enzyme immunoassay, centrifugation culture, and conventional culture for the detection of respiratory syncytial virus. *J. Clin. Microbiol.* 28:2394–2397.
52. Matthey, S., Nicholson, D., Ruhs, S., Alden, B., Knock, M., Schultz, K. and Schmuecker, A. (1992). Rapid detection of respiratory virus by shell vial culture and direct staining by using pooled and individual monoclonal antibodies. *J. Clin. Microbiol.* 30:540–544.
53. Minnich, L. and Ray, C.G. (1980). Comparison of direct immunofluorescent staining of clinical specimens for respiratory virus antigens with conventional isolation techniques. *J. Clin. Microbiol.* 12:391–394.
54. Takimoto, S., Grandien, M., Ishida, M.A., Pereira, M.S., Paiva, T.M., Ishimaru, T., Makita, E.M. and Martinez, C.H.O. (1991). Comparison of enzyme-linked immunosorbent assay, indirect immunofluorescence assay, and virus isolation for detection of respiratory virus in nasopharyngeal secretions. *J. Clin. Microbiol.* 29:470–474.
55. Thomas, E.E. and Book, L.E. (1991). Comparison of two rapid methods for detection of respiratory syncytial virus (RSV) (TestPack RSV and Ortho RSV ELISA) with direct immunofluorescence and virus isolation for the diagnosis of pediatric RSV infection. *J. Clin. Microbiol.* 29:632–635.
56. Leonardi, G.P., Leib, H., Birkhead, G.S., Smith, C., Costello, P. and Conron, W. (1994). Comparison of rapid detection methods for influenza A virus and their value in health care management of institutionalized geriatric patients. *J. Clin. Microbiol.* 32:70–74.
57. Rabalais, G., Stout, G. and Waldeyer, S. (1992). Rapid detection of influenza-B virus in respiratory secretions by immunofluorescence during an epidemic. *Diagn. Microbiol. Infect. Dis.* 15:35–37.
58. Minnich, L.L., Goodenough, F. and Ray, C.G. (1991). Use of immunofluorescence to identify measles virus infections. *J. Clin. Microbiol.* 29:1148–1150.
59. Stirk, P.R. and Griffiths, P.D. (1988). Comparative sensitivity of three methods for the diagnosis of cytomegalovirus lung infection. *J. Virol. Methods.* 20:133–142.
60. Boeckh, M., Woogerd, P.M., Stevens-Ayers, T., Ray, C.G. and Bowden, R.A. (1994). Factors influencing detection of quantitative cytomegalovirus antigenemia. *J. Clin. Microbiol.* 32:832–834.
61. Van der Bij, W., Torensma, R., vanSon, W.J., Anema, J., Schirm, J., Tegzess, A.M. and The, T.H. (1988). Rapid immunodiagnosis of active cytomegalovirus infection by monoclonal antibody staining of blood. *J. Med. Virol.* 25:179–188.
62. Cowdry, E.V. (1934). The problem of intranuclear inclusions in virus diseases. *Arch. Pathol.* 18:527–542.
63. Schwarz, T.F., Nerlich, A., Hottentrager, B., Jager, G., Wiest, I., Kantimm, S., Roggendorf, H., Schultz, M., Gloning, K.-P., Schramm, T., Holzgreve, W. and Roggendorf. M. (1991). Parvovirus B19 infection of the fetus. Histology and in situ hybridization. *Am. J. Clin. Pathol.* 96:121–126.
64. Lendrum, A.C. (1947). The phloxine-tartrazine method as a general histological stain for demonstration of inclusion bodies. *J. Pathol. Bacteriol.* 59:399–404.
65. Deodhar, K.P., Tapp, E. and Scheuer, P.J. (1975). Orcein staining of hepatitis B antigen in paraffin sections of liver biopsies. *J. Clin. Pathol.* 28:66–70.
66. Gubetta, L., Rizzetto, M., Crivelli, O., Verme, G. and Arico, S. (1977). A trichrome stain for the intrahepatic localization of the hepatitis B surface antigen (HBsAg). *Histopathology* 1:277–288.
67. Tanaka, K., Mori, W. and Suwa, K. (1981). Victoria blue-nuclear fast red stain for HBs antigen detection in paraffin section. *Acta. Pathol. Jpn.* 31:93–98.
68. Koopmans, M. and Duizer, E. (2004). Foodborne virus: an emerging problem. *Int. J. Food Microbiol.* 90:23–41.
69. Wen, J.K., Zhang, X.E., Cheng, Z., Liu, H., Zhou, Y.F., Zhang, Z.P., Yang, J.H. and Deng, J.Y. (2004). A visual DNA chip for simultaneous detection of hepatitis B virus, hepatitis C virus and human immunodeficiency virus type-1. *Biosens. Bioelectron.* 19:685–692.

70. Lucarelli, F., Marrazza, G., Turner, A.P.F. and Mascini, M. (2004). *Biosens. Bioelectron.* 19:515–530.
71. de-los-Santos-Alvarez, P., Lobo-Castanon, M.J., Miranda-Ordieres, A.J. and Tunon-Blanco, P. (2004). Current strategies for electrochemical detection of DNA with solid electrodes. *Anal. Bioanal. Chem.* 378:104–118.
72. Kerman, K., Kobayashi, M. and Tamiya, E. (2004). Recent trends in electrochemical DNA biosensor technology. *Meas. Sci. Technol.* 15:R1–R11.
73. Kessler, N., Ferraris, O., Palmer, K., Marsh, W. and Steel A. (2004). Use of the DNA Flow-Thru Chip, a Three-Dimensional Biochip, for Typing and Subtyping of Influenza Virus. *J. Clin. Microbiol.* 42:2173–2185.
74. Kara, P., Meric, B., Zeytinoglu, A. and Ozsoz, M. (2004). Electrochemical DNA biosensor for the detection and discrimination of herpes simplex Type I and Type II virus from PCR amplified real samples. *Anal. Chim. Acta.* 518:69–76.
75. Kobayashi, M., Takashi, K.B., Saito, M., Kaji, S., Oomura, M., Iwabuchi, S., Morita, Y., Hasan, Q. and Tamiya, E. (2004). Electrochemical DNA quantification based on aggregation induced by Hoechst 33258. *Electrochem. Commun.* 6:337–343.
76. Ramanaviciene, A. and Ramanavicius, A. (2004). Pulsed amperometric detection of DNA with an ssDNA/polypyrrole-modified electrode. *Anal. Bioanal. Chem.* 379:287–293.
77. Nebling, E., Grunwald, T., Albers, J., Schafer, P. and Hintsche, R. (2004). Electrical detection of viral DNA using ultramicroelectrode arrays. *Anal. Chem.* 76:689–696.
78. Erdem, A., Pividori, M.I., del Valle, M. and Alegret, S. (2004). Rigid carbon composites: a new transducing material for label-free electrochemical genosensing. *J. Electroanal. Chem.* 567:29–37.
79. Wang, M.J., Wang, L.Y., Wang, G., Ji, X.H., Bai, Y.B., Li, T.J., Gong, S.Y. and Li, J.H. (2004). Application of impedance spectroscopy for monitoring colloid Au-enhanced antibody immobilization and antibody-antigen reactions. *Biosens. Bioelectron.* 19:575–582.
80. Perez, F.G., Mascini, M., Tothill, I.E. and Turner, A.P.F. (1998). Immunomagnetic separation with mediated flow injection analysis amperometric detection of viable Escherichia coli O157. *Anal. Chem.* 70:2380–2386.
81. Casas, N. and Sunen, E. (2002). Detection of enteroviruses, hepatitis A virus and rotaviruses in sewage by means of an immunomagnetic capture reverse transcription-PCR assay. *Microbiol. Res.* 157:169–175.
82. Nettikadan, S.R., Johnson, J.C., Vengasandra, S.G., Muys, J. and Henderson, E. (2004). ViriChip: a solid phase assay for detection and identification of virus by atomic force microscopy. *Nanotechnology* 15:383–389.
83. Yan, X.M., Schielke, E.G., Grace, K.M., Hassell, C., Marrone, B.L. and Nolan, J.P. (2004). Microsphere-based duplexed immunoassay for influenza virus typing by flow cytometry. *J. Immunol. Methods.* 284:27–38.
84. Dore, K., Dubus, S., Ho, H.A., Levesque, I., Brunette, M., Corbeil, G., Boissinot, M., Boivin, G., Bergeron, M.G., Boudreau, D. and Leclerc, M. (2004). Fluorescent polymeric transducer for the rapid, simple, and specific detection of nucleic acids at the zeptomole level. *J. Am. Chem. Soc.* 126:4240–4244.
85. Hu, W.G., Thompson, H.G., Alvi, A.Z., Nagata, L.P., Suresh, M.R. and Fulton, R.E. (2004). Development of immunofiltration assay by light addressable potentiometric sensor with genetically biotinylated recombinant antibody for rapid identification of Venezuelan equine encephalitis virus. *J. Immunol. Methods.* 289:27–35.
86. Alfonta, L., Blumenzweig, I., Zayats, M., Baraz, L., Kotler, M. and Willner, I. (2004). Electronic transduction of HIV-1 drug resistance in AIDS patients. *Chembiochemistry* 5:949–957.
87. Skladal, P., Riccardi, C.D., Yamanaka, H. and da Costa, P.I. (2004). Piezoelectric biosensors for real-time monitoring of hybridization and detection of hepatitis C virus. *J. Virol. Methods.* 117:145–151.
88. Gupta, A., Akin, D. and Bashir, R. (2004). Single virus particle mass detection using microresonators with nanoscale thickness. *Appl. Phys. Lett.* 84:1976–1978.
89. Dickert, F.L., Hayden, O., Bindeus, R., Mann, K.J., Blaas, D. and Waigmann, E. (2004). Bioimprinted QCM sensors for virus detection - screening of plant sap. *Anal. Bioanal. Chem.* 378:1929–1934.
90. Collinsa, R.A., Koa L.-S., Funga K.-Y., Chana K.-Y., Xinga J., Laua L.-T. and Yu, A.C.H. (2004). Rapid and sensitive detection of avian influenza virus subtype H7 using NASBA. *Biochem. Biophys. Res. Commun.* 300:507–515.

91. Rodriguez, M.A. and Armstrong, D.W. (2004). Separation and analysis of colloidal/nanoparticles including microorganisms by capillary electrophoresis: a fundamental review. *J. Chromatogr. B.* 800:7–25.
92. Rodriguez-Mozaz, S., Marco, M.P., de Alda, M.J.L. and Barcelo, D. (2004). Biosensors for environmental monitoring of endocrine disruptors: a review article. *Anal. Bioanal. Chem.* 378:588–598.
93. Chen, Y.F., Yang, J.M., Gau, J.J., Ho, C.M. and Tai, Y.C. (2000). In: The 3rd International conference on the interaction of Art and Fluid Mechanics, Zurich, Switzerland.
94. Ercole, C., Del Gallo, M., Pantalone, M., Santucci, S., Mosiello, L., Laconi, C. and Lepidi, A. (2002). A biosensor for Escherichia coli based on a potentiometric alternating biosensing (PAB) transducer. *Sens. Actuat., B Chem.* 83:48–52.
95. Ivnitski, D., Wilkins, E., Tien, H.T. and Ottova, A. (2000). Electrochemical biosensor based on supported planar lipid bilayers for fast detection of pathogenic bacteria. *Electrochem. Commun.* 2: 457–460.
96. Ivnitski, D., Abdel-Hamid, I., Atanasov, P. and Wilkins, E. (1999). Biosensors for detection of pathogenic bacteria. *Biosens. Bioelectron.* 14:599–624.
97. Marrazza, G., Chianella, I. and Mascini, M. (1999). Disposable DNA electrochemical biosensors for environmental monitoring. *Analytica Chimica Acta.* 387: 297–307.
98. Wang, J. (1997). DNA electrochemical biosensors for environmental monitoring. A review. *Analytica Chimica Acta.* 347:1–8.
99. Ruzgas, T., Csöregi, E., Emnéus, J., Gorton, L. and Marko-Varga, G. (1996). Peroxidase-modified electrodes: Fundamentals and application. *Anal. Chimica Acta.* 330:123–138.
100. Darst, S.A., Ahlers, M., Meller, P.H., Kubalek, E.W., Blankenburg, R., Ribi, H.O., Ringsdorf, H. and Kornberg, R. (1991). 2-Dimensional crystals of streptavidin on biotinylated lipid layers and their interactions with biotinylated macromolecules. *Biophys. J.* 59:387–396.
101. Gehring, A.G., Patterson, D.L. and Tu, S.I. (1998). Use of a light-addressable potentiometric sensor for the detection of Escherichia coli 0157: H7. *Anal. Biochem.* 258:293–298.
102. Wagner, P., Hegner, M., Guntherodt, H-J. and Semenza, G. (1995). Formation and in-situ modification of monolayers chemisorbed on ultra flat template-stripped gold surfaces. *Langmuir.* 11:3867–3875.
103. Shah, J. and Wilkins, E. (2003). Electrochemical biosensors for detection of biological warfare agents. *Electroanalysis.* 15:157–167.
104. Carnes, E. and Wilkins, E. (2005). The development of a new, rapid, amperometric immunosensor for detection of low concentration of bacteria. Part I: Design of detection system and applications. *Amer. J. Appl. Sci.* 2:597–606.
105. Martins, M.T., Rivera, I.G., Clark, D.L., Stewart, M.H., Wolfe, R.L. and Olson, B.H. (1993). Distribution of uida gene-sequences in Escherichia-coli isolates in water sources and comparison with the expression of beta-glucuronidase activity. *Applied and Environmental Microbiology* 59:2271–276.
106. Farnleitner, A.H., Kreuzinger, A.H., Kavka, G.G., Grillenberger, S., Rath J. and Mach, R.L. (2000). Simultaneous detection and differentiation of Escherichia coli populations from environmental freshwaters by means of sequence variations in a fragment of the beta-D-glucuronidase gene. *Appl. Environ. Microbiol.* 66:1340–1346.
107. Manafi, M. (2000). New developments in chromogenic and fluorogenic culture media. *Int. J. Food Microbiol.* 60:205–218.
108. Van Poucke, S.O. and Nelis, H.J. (2000). Rapid detection of fluorescent and chemiluminescent total coliforms and Escherichia coli on membrane filters. *J. Microbiol. Methods.* 42:233–244.
109. Farrell, M.J. and Finkel, S.E. (2003). The growth advantage in stationary-phase phenotype conferred by rpoS mutations is dependent on the pH and nutrient environment. *J. Bacteriol.* 185:7044–7052.
110. Albrecht, A. and Kampfer, P. (2003). Microbial endangering of workers during cleaning places contaminated by faeces of pigeons—Part 2: Measurements of airborne microorganisms. Gefahrstoffe Reinhaltung Der Luft. 63:15–23.
111. Ogunbanwo, S.T., Sanni, A.I. and Onilude, A.A. (2004). Effect of bacteriocinogenic Lactobacillus spp. on the shelf life of fufu, a traditional fermented cassava product. *World J. Microbiol. Biotechnol.* 20:57–63.

112. Esiobu, N., Mohammed, R., Echeverry, A., Green, M., Bonilla, T., Hartz, A., McCorquodale, D. and Rogerson, A. (2004). The application of peptide nucleic acid probes for rapid detection and enumeration of eubacteria, Staphylococcus aureus and Pseudomonas aeruginosa in recreational beaches of S. Florida. *J. Microbiol. Methods.* 57: 157–162.
113. Alvarez, M., Carrascosa, L.G., Moreno, M., Calle, A., Zaballos, A., Lechuga, L.M., Martinez-A, C. and Tamayo, J. (2004). Nanomechanics of the formation of DNA self-assembled monolayers and hybridization on microcantilevers. *Langmuir.* 20:9663–9668.
114. Gau, J.J., Lan, E.H., Dunn, B., Ho, C.M. and Woo, C.S. (2001). A MEMS based amperometric detector for E-Coli bacteria using self-assembled monolayers. *Biosens. Bioelectro.* 16: 745–755.
115. Finsterwalder, F. and Hambitzer, G. (2001). Proton conductive thin films prepared by plasma polymerization. *J. Membr. Sci.* 15:105–124.
116. Alberti, G., Casciola, M., Massinelli, L. and Bauer, B. (2001). Polymeric proton conducting membranes for medium temperature fuel cells (110–160 degrees C). *J. Membr. Sci.* 185:73–81.
117. Levin, R.E. (2004). The Application of Real-Time PCR to Food and Agricultural Systems. A Review. *Food Biotechnol.* 18:97–133.
118. McKillip, J.L. and Drake M. (2004). Real-time nucleic acid-based detection methods for pathogenic bacteria in food. *J. Food Prot.* 67:823–832.
119. Ercolini, D. (2004). PCR-DGGE fingerprinting: novel strategies for detection of microbes in food. *J. Microbiol. Methods.* 56:297–314.
120. Deisingh, A.K. and Thompson, M. (2004). Biosensors for the detection of bacteria. *Can. J. Microbiol.* 50:69–77.
121. Stevens, K.A. and Jaykus, L.A. (2004). Bacterial separation and concentration from complex sample matrices: A review. *Crit. Rev. Microbiol.* 30:7–24.
122. Petrenko, V.A. and Sorokulova, I.B. (2004). Detection of biological threats. A challenge for directed molecular evolution. *J. Microbiol. Methods.* 58:147–168.
123. Cage, B.R., Schreiber, K., Barnes, C. and Portnoy, J. (1996). Evaluation of four bioaerosol samplers in the outdoor environment. *Ann. Allergy Asthma Immunol.* 77:401–406.
124. Lepeuple, A.S., Giloppe, S., Pierlot, E. and de Roubin, M.R. (2004). Rapid and automated detection of fluorescent total bacteria in water samples. *Int. J. Food Microbiol.* 92:327–332.
125. www.bd.com/clinical/products/direct/fluaw.asp (accessed December 18, 2004).
126. www.quidel.com/Products/productdisp.php?prod=101&section=pro (accessed December 18, 2004).
127. http:// binax.com/NOWflua_b.shtml (accessed December 14, 2004).
128. http:// www.thinkoro.com/ (accessed December 20, 2004).
129. www.abtechsci.com (accessed December 20, 2004).
130. http:// www.biacore.com/products/systemsandsoftware/2000/ (accessed December 20, 2004).
131. Yager, P., Afromowitz, M.A., Bell, D., Forster, F.K., Brody, J.P., Qin, D., Cabrera, C., Holl, M., Kamholz, A. and Weigl, B. (1998). *Proc.SPIE - Int. Soc. Optical Eng.* 3515:252.
132. Yang, J.M., Bell, J., Huang, Y., Tirado, M., Thomas, D., Forster, A.H., Haigis, R.W., Swanson, P.D., Wallace, R.B. and Martinsons, B. (2002). An integrated, stacked microlaboratory for biological agent detection with DNA and immunoassays. *Biosens. Bioelectron.* 17:605–618.
133. Ho, J.A. and Hsu, H.-W. (2003). Procedures for preparing Escherichia coli O157:H7 immunoliposome and its application in liposome immunoassay. *Anal. Chem.* 75: 4330–4334.
134. Ho, J.A., Hsu, H.-W. and Huang, M.R. (2004). Liposome-based microcapillary immunosensor for detection of Escherichia coli O157:H7. *Anal. Biochem.* 330:342–349.
135. Zhang, J. and Ji, H.F. (2004). An anti E. coli O157:H7 antibody-immobilized microcantilever for the detection of Escherichia Coli (E. coli). *Analytical Sciences.* 20:585–586.
136. Yang, L.J., Li, Y.B., Griffis, C.L. and Johnson, M.G. (2004). Interdigitated microelectrode (IME) impedance sensor for the detection of viable Salmonella typhimurium. *Biosensors and Bioelectronics.* 19:1139–1147.
137. CDC (1998). Update of emerging infection from the centers for disease control and prevention. *Ann. Emerg. Med.* 31:121.
138. Butler, J.C. and Peters, C.J. (1996). Lessons learned from the hantaviruses and other hemorrhagic fever virus. *Amer. J. Med. Sci.* 311:55–59.

139. Dietrich, N., Pruden, S., Ksiazek, T.G., Morzunov, S.P. and Camp, J.W. (1997). A small-scale survey of hantavirus in mammals from Indiana. *J. Wildlife Dis.* 33:818–822.
140. Center for Health Protection (CHP) 2004. http:// www.info.gov.hk/info/flu/eng/ (accessed December 2, 2004).
141. WHO, 2004. http://www.info.gov.hk/info/flu/eng/global.htm (accessed December 2, 2004).
142. Cordingly, P. 2004. http:// news.bbc.co.uk/1/hi/health/medical_notes/1337640.stm (accessed December 5, 2004).
143. Li, K.S., Wang, J., 2004. http:// www.osha.gov/dsg/guidance/avian-flu.html (accessed December 4, 2004).
144. PHA, http:// www.hc-sc.gc.ca/english/diseases/flu/avian.html (accessed December 15, 2004).
145. Ghindilis, A.L., Krishnan, R., Atanasov, P. and Wilkins, E. (1997). Flow-through amperometric immunosensor: Fast 'sandwich' scheme immunoassay. *Biosens. Bioelectron.* 12:415–423.
146. Vetcha, S., Abdel-Hamid, I., Atanasov, P., Ivnitski, D., Wilkins, E. and Hjelle, B. (2000). Portable immunosensor for the fast amperometric detection of anti-Hantavirus antibodies. *Electroanalysis.*12:1034–1038.
147. Hjelle, B., Jenison, S., TorrezMartinez, N., Herring, B., Quan, S., Polito, A., Pichuantes, S., Yamada, T., Morris, C. and Elgh, F. (1997). Rapid and specific detection of Sin Nombre virus antibodies in patients with hantavirus pulmonary syndrome by a strip immunoblot assay suitable for field diagnosis. *J. Clin. Microbiol.* 35:600–608.
148. Jenison, S., Yamada, T., Morrison, C., Anderson, B., Torrezmartinez, N., Keller, N. and Hjelle, B. (1994). Characterization of human-antibody responses to 4 corners Hanta virus infection among patients with Hanta virus pulmonary syndrome. *J. Virol.* 68:3000–3006.
149. Abdel-Hamid, I., Atanasov, P., Ghindilis, A.L. and Wilkins, E. (1998). Development of a flow-through immunoassay system. *Sens. Actuat, B Chem.* 49:202–210.
150. Krishnan, R., Ghindilis, A.L., Atanasov, P. and Wilkins, E. (1995). Fast amperometric immunoassay utilizing highly dispersed electrode material. *Anal. Lett.* 28:2459–2474.
151. Krishnan, R., Ghindilis, A.L., Atanasov, P. and Wilkins, E. (1996). Development of an amperometric immunoassay based on an highly dispersed immunoelectrode. *Anal. Lett.* 29:2615–2631.
152. Krishnan, R., Ghindilis, A.L., Atanasov, P., Wilkins, E., Montoya, J. and Koster, F.T. (1996). Fast amperometric immunoassay for Hantavirus infection. *Electroanalysis* 8:1131–1134.
153. Ma, Z.M., Monk, T.G., Goodnough, L.T., McClellan, A., Gawryl, M., Clark, T., Moreira, P., Keipert, P.E. and Scott, M.G. (1997). Effect of hemoglobin- and perflubron-based oxygen carriers on common clinical laboratory tests. *Clin. Chem.* 43:1732–1737.
154. Frank, J.J., Bermes, E.W., Bickel, M.J. and Watkins, B.F. (1978). Effect of Invitro hemolysis on chemical values for serum. *Clin. Chem.* 24:1966–1970.
155. Sonntag, O. (1986). Hemolysis as interference factor in clinical chemistry. *J. Clin. Chem. Clin. Biochem.* 24:127–140.
156. Mather, A. and Mackie, N.R. (1960). Effects of hemolysis on serum electrolyte value. *Clin. Chem.* 6:223–227.
157. Dogan, Y. (1992). Effect of in vitro hemolysis on 25 common biochemical tests. *Clin. Chem.* 38:575–577.
158. Frew, J.E. (1989). Measurement of alkaline phosphatase activity by electrochemical detection of phosphate esters, application to enzyme immunoassay. *J. Electroanal. Chem.* 266:317–336.
159. Vetcha, S., Wilkins, E. and Yates, T. (2002). Detection of hantavirus infection in hemolyzed mouse blood using alkaline phosphatase conjugate. *Biosens. Bioelectron.* 17:901–909.

22

Detection of Hanta, Influenza A, and Parainfluenza Virus

Ebtisam S. Wilkins and Ravil A. Sitdikov

CONTENTS

22.1 Introduction

22.1.1 Hanta Virus

Hanta virus cardiopulmonary syndrome (HCPS) is a severe life-threatening illness that was first recognized in the spring of 1993. More than 202 cases have been confirmed in 30 states, since the 1993 outbreak, 45% of the cases being fatal. Most of the incidences of the infection have occurred in the western United States. HCPS is caused by several closely related yet genetically distinct Hanta virus species, each of which is maintained in a single rodent host [1]. In the United States, the causative agent for the vast majority of cases is the Sin Nombre virus (SNV), which is transmitted to human beings from the deer mouse (*Peromyscus maniculatus*), its primary rodent reservoir. Inhalation of aerosolized virus particles excreted in rodent urine, saliva, or feces is believed to be the major route of transmission, although infection can also occur by means of direct inoculation of the virus into wounds or by bite.

SNV (genus of the family Bunyaviridae) was identified during a 1993 outbreak of an acute respiratory disease with high mortality in the Four Corners region of the southwestern United States (New Mexico, Utah, Colorado, and Arizona) [2]. HCPS occurs in humans presumably after accidental/incidental inhalation of virus-contaminated excretion from the natural host, the deer mouse (*P. maniculatus*). Although the deer mouse has been determined to be the primary reservoir of the Hanta virus in many areas of North America, serologic evidence of infection has also been found in other mammals. The illness is characterized by flulike symptoms, fever, rapid-progressing pulmonary edema, significant increase in hematocrit, and a decrease in platelets. The infection is spread by aerosolized excreta and transmission to humans occurs under the conditions of close contact with the rodent host [3]. While a few cases of HCPS have been reported in the urban East, the predominant case distribution is in the rural West. Exposure and infection most often occurs in rural areas remote from medical centers with sophisticated diagnostic capabilities.

22.1.2 Avian Influenza

Avian influenza (AI) is an infectious disease of birds caused by type A strains of the influenza virus. The disease, which was first identified in Italy more than 100 years ago, occurs worldwide. From a biotechnology standpoint, the AI is an interesting topic due to the mutating capabilities of the influenza virus. By gaining an intuitive understanding on how the avian virus works, we can confront the virus head on and prevent the occurrence of a pandemic. Vaccines may be produced, which in effect will kill the invading virus in our body without any harmful effects to our own immune system [4].

It is necessary to determine the rate at which the avian virus is capable of mutating and spreading. By deriving an appropriate model for the virus, it is possible to determine the governing factors on how and why the virus spreads. From this data, it will then be possible to slow down the rate at which the virus spreads, which in turn prevents the virus from getting out of control.

All birds are thought to be susceptible to infection with AI, although some species are more resistant to infection than others. Infection causes a wide spectrum of symptoms in birds, ranging from mild illness to a highly contagious and rapidly fatal disease resulting in severe epidemics. The latter is known as "highly pathogenic AI". This form is characterized by sudden onset, severe illness, and rapid death, with a mortality that can approach 100%. According to Centre for Health Protection (CHP) [4], 15 subtypes of influenza virus are known to infect birds, thus providing an extensive reservoir of influenza viruses potentially circulating in bird populations. To date, all outbreaks of the highly pathogenic form have been caused by influenza A viruses (IAVs) of subtypes H5 and H7. Migratory waterfowl—most notably wild ducks—are the natural reservoir of AI viruses, and these birds are also the most resistant to infection. Domestic poultry, including chickens and turkeys, are particularly susceptible to epidemics of rapidly fatal influenza. Direct or indirect contact of domestic flocks with wild migratory waterfowl has been implicated as a frequent cause of epidemics. Live bird markets have also played an important role in the spread of epidemics.

Recent research has shown that viruses of low pathogenicity can, after circulation for sometimes short periods in a poultry population, mutate into highly pathogenic viruses. During a 1983−1984 epidemic in the United States of America, the H5N2 virus initially caused low mortality, but within 6 months became highly pathogenic, with a mortality approaching 90%. Control of the outbreak required destruction of more than 17 million birds at a cost of nearly US$ 65 million according to WHO [5]. During a 1999−2001 epidemic in Italy, the H7N1 virus, initially of low pathogenicity, mutated within 9 months to a highly pathogenic form. More than 13 million birds died or were destroyed. The quarantining of infected farms and destruction of infected or potentially exposed flocks are standard control measures aimed at preventing spread to other farms and eventual establishment of the virus in a country's poultry population. Apart from being highly contagious, AI viruses are readily transmitted from farm to farm by mechanical means, such as by contaminated equipment, vehicles, feed, cages, or clothing. Highly pathogenic viruses can survive for long periods in the environment, especially when temperatures are low. Stringent sanitary measures on farms can, however, confer some degree of protection.

In the absence of prompt control measures backed by good surveillance, epidemics can last for years. For example, an epidemic of H5N2 AI, which began in Mexico in 1992, started with low pathogenicity, evolved to the highly fatal form, and was not controlled until 1995.

In 1997, the outbreak of a new AI-like virus, known as Bird Flu, affected 18 humans. The cases, many severe or fatal, highlighted the challenges of unique influenza viruses. Researchers struggled to track a new potentially deadly strain of influenza, which before was only thought to affect chickens. Lessons from this episode help to improve international and national planning for influenza pandemics. This proposal highlights the action taken to keep the Bird Flu from becoming a global pandemic and a way of detection as a preventive measure.

Fowl plague (FP) was described in 1878 as a serious disease of chickens in Italy. It was determined in 1955 that FP virus is actually one of the influenza viruses. The AI viruses, along with the other influenza viruses, make up the virus family Orthomyxoviridae. The virus particle has an envelope with glycoprotein projections with hemagglutinating and neuraminidase (NA) activity. These two surface antigens, hemagglutinin (HA) and NA, are the basis of describing the serologic identity of the influenza viruses using the letters H and N with the appropriate numbers in the virus designation, for example, H7N2. There are now 15 HA and 9 NA antigens described among the type A influenza viruses. The type designation (A, B, or C) is based upon the antigenic character of the M protein of the virus envelope and the nucleoprotein within the virus particle. All influenza viruses affecting domestic animals (equine, swine, avian) belong to type A, and type A influenza virus is the most common type producing serious epidemics in humans. Types B and C do not affect domestic animals. All type A influenza viruses, including those that regularly cause seasonal epidemics of influenza in humans, are genetically labile and well adapted to elude host defenses. According to Cordingly [6] Influenza viruses lack mechanisms for the "proofreading" and repair of errors that occur during replication. As a result of these uncorrected errors, the genetic composition of the viruses changes as they replicate in humans and animals, and the existing strain is replaced with a new antigenic variant. These constant, permanent and usually small changes in the antigenic composition of IAVs are known as antigenic "drift."

The tendency of influenza viruses to undergo frequent and permanent antigenic changes necessitates constant monitoring of the global influenza situation and annual adjustments in the composition of influenza vaccines.

Influenza viruses have a second characteristic of great public health concern: IAVs, including subtypes from different species, can swap or "reassort" genetic materials and merge. This reassortment process, known as antigenic "shift," results in a novel subtype different from both parent viruses. As populations will have no immunity to the new subtype, and as no existing vaccines can confer protection, antigenic shift has historically resulted in highly lethal pandemics. For this to happen, the novel subtype needs to have genes from human influenza viruses that make it readily transmissible from person to person for a sustainable period.

Conditions favorable for the emergence of antigenic shift have long been thought to involve humans living in close proximity to domestic poultry and pigs. Because pigs are susceptible to infection with both avian and mammalian viruses, including human strains, they can serve as a "mixing vessel" for the scrambling of genetic material from human and avian viruses, resulting in the emergence of a novel subtype. Recent events, however, have identified a second possible mechanism. Evidence is mounting that, for at least some of the 15 AI virus subtypes circulating in bird populations, humans themselves can serve as the "mixing vessel" [7,8].

22.2 Sensor Design and Immunoassay System

22.2.1 Detection of Hanta Virus

The immunosensor design concept, previously described [9], has been modified and adapted into the immunoassay prototype system. The immunocolumn, which is the disposable sensing element, consists of a plastic column with a filter membrane at the bottom. The immunosorbent is deposited on the filter membrane (by centrifugation) resulting in highly dispersed immunosorbent forming the measuring (working) electrode or immunoelectrode. The immunocolumn was prepared by adding 100 μL of 0.5 mg/mL immunosorbent to the microcentrifuge filters. The immunosorbent consist of carbon particles that are coated with a predetermined optimum quantity of recombinant nucleocapsid protein (RNP), usually 10 μg/0.5 mg of carbon particles. The immobilization of RNP of SN virus on carbon microdispersed powder was performed using Woodward's reagent [10]. A plastic column holder comprised of a supporting electrode assembly: a platinum counter electrode and Ag/AgCl reference electrode that are exposed into a channel, which serve as an outlet for the flow-through immunosensor. The flow-through immunoassay system is described in detail [9,11]. The immunocolumn adaptor contains a capillary tube that is inserted and fixed into a hollow carbon rod. The capillary tube serves as the inlet to the immunosensor while the carbon rod serves as a current collector for the working electrode (immunoelectrode). When assembled, the current collector rests on top of the immunoelectrode to collect the amperometric output. This flow cell serves as an immunoreactor as well as an electrochemical cell. The optimization of disposable immunosensing elements of the amperometric transducer is based on the principle of close attachment of the components of the ligand–receptor system to the electrode surface of the transducer. Critical parameters such as the size of the immunosorbent, amount of immobilized antigen molecules (Hanta virus RNP) and their surface orientation, the effect of pH, and the stability of the immunosensing elements have been investigated and optimized previously [9,12–14].

22.2.2 Detection of Parainfluenza and Influenza A Viruses

The immunosensor design concept, previously described [9,13], has been modified and optimized using the immunoassay prototype system for detection of parainfluenza virus (PIV) and IAV.

However, the immunocolumn, which is the disposable biosensing element, is prepared differently than the Hanta virus. This column consists of a plastic tube with a 0.22 μm pore size filter membrane at the bottom. The immunosorbent is deposited on the filter membrane (by vacuum) resulting in dispersed graphite immunosorbent forming the measuring (working) immunoelectrode. Another difference is the immunocolumns prepared by adding 100–250 μL of 20 mg/mL immunosorbent to the microcentrifuge filters.

22.2.3 Immunoassay Scheme and Amperometric Detection

A previously described technique, the "sandwich" scheme of immunoassay, was employed [13,15]. It is illustrated in Figure 22.1.

22.2.4 Optimization of the Assay Parameters

The parameters such as the flow rate of the reagents, working potential, pH of conjugate, and substrate solutions and their concentrations, and incubation times are investigated and optimized earlier [9,10,12,14]. Using these same conditions promising results were obtained during the application of the device in detecting antibodies against Hanta virus in human plasma samples. However, several parameters have been changed due to the hemolysis of blood in the case of mice blood samples including the labeled conjugate to prove the feasibility of the device under operating-field conditions. The parameters affecting both immunological reaction stage and amperometric transduction stage, type of conjugate, namely working potential, pH of substrate solutions, and substrate concentrations, were investigated. Optimization of the electrochemical detection was based on the following criteria: maximum enzymatic activity of the horse-radish peroxidase (HRP) enzyme, maximum substrate concentrations, and maximum spontaneous oxidation of substrates.

22.2.5 Optimization of the Amperometric Measurement Stage

22.2.5.1 The Cyclic Voltammetry of Horse-Radish Peroxidase and Alkaline Phosphatase Enzymatic Products

The cyclic voltammetry apparatus consisted of a BAS CV-1 Voltammograph (Bioanalytical Systems, Inc.), a BK Precision model 2832 digital multimeter (Dynascan Corporation), a Huston Instrument model 100XY recorder, and an electrochemical cell. The electrochemical cell with 10 mL of electrolyte volume included a graphite disc working electrode (5 mm diameter), a carbon counter electrode, and an Ag/AgCl reference electrode. For cyclic voltammetry experiments of HRP and alkaline phosphatase (AP) products the 0.1 M acetic-acid buffer solution pH 4.5 with 0.15 M NaCl (AcBS) and 0.05 M bicarbonate buffer solution pH 9.5 with 0.15 M NaCl (BcBS) was used accordingly. The scan rate potential

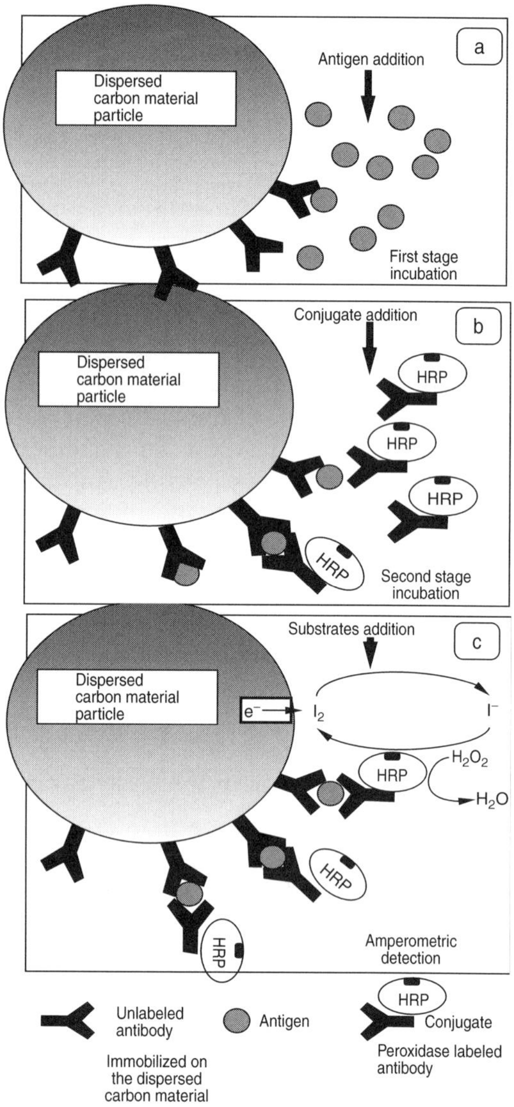

FIGURE 22.1
Sandwich scheme of immunochemical for carbon particle immobilization.

was 20 mV/s. The results of these experiments allowed choosing the potentials of working electrode for amperometric detection of products for HRP (iodine) and AP (α -naphthol).

22.2.5.2 Selection of a Working Potentials

The working potential is the potential at which the working electrode is biased with respect to the reference electrode. A suitable working potential has to be selected to achieve efficient product formation during the enzymatic reaction of enzyme labels (AP for Hanta virus and HRP for influenza viruses detection). Selection of a suitable working potential was performed using cyclic voltammetry. For amperometric detection of AP activity, the background and working voltammograms were obtained in the presence of 1 mM α-naphthyl phosphate (substrate of AP) in a bicarbonate buffer solution pH 9.6 (Figure 22.2.)

The maximum oxidation current of α-naphthol as AP enzymatic reaction product is observed at a potential of +100 mV (shown in Figure 22.2 as red arrow for peaks curves 1 and 2). Hence, this potential was selected as the working potential for amperometric registration of AP activity [11,13].

The same cyclic voltammograms for iodine–iodide as redox substrate–product system for amperometric detection of HRP-label activity were investigated. For amperometric detection of HRP activity, the background and working voltammograms were obtained in the presence of 1 mM potassium iodine (substrate-mediator for HRP) and different concentration of iodine in a sodium acetic-acid buffer solution pH 4.5 (Figure 22.3).

The range of reduction current of iodine, as HRP enzymatic reaction product, is observed at potentials from +200 to −200 mV (shown in Figure 22.3 as black arrows for cathodic curves 1 and 2). The −125 mV was selected as the working potential for amperometric registration of HRP activity.

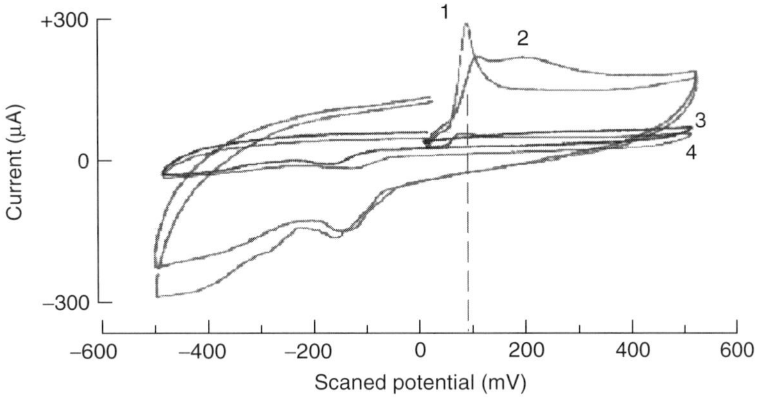

FIGURE 22.2
Cyclic volt–ampere diagrams obtained with two concentrations of α-naphthol (alkaline phosphatase (AP) product): 1, 4 mM; 2, 2 mM; 3, 0.0 in 0.05 M sodium bicarbonate buffer solution (BcBS) with 1.0 mM α -naphthyl (AP substrate) and 0.15 M NaCl; 4, 0.05 M BcBS with 0.15 M NaCl. Working electrode is graphite rod (5 mm diameter), reference electrode is Ag/AgCl wire (1 mm diameter, 10 mm length), and counter electrode is carbon rod (2 mm diameter, 10 mm length).

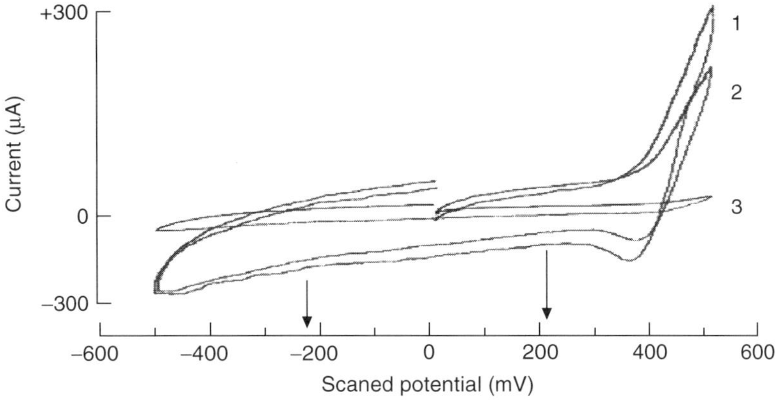

FIGURE 22.3
Cyclic volt–ampere diagrams obtained with different concentrations of iodine (HRP product): 1, 0.3 mM; 2, 0.5 mM; 3, 0.0 in 0.1 M AcBS with 1.0 mM KI and 0.15 M NaCl. Working electrode is graphite rod (5 mm diameter), reference electrode is Ag/AgCl wire (1 mm diameter, 10 mm length), and counter electrode is carbon rod (2 mm diameter, 10 mm length).

22.3 Experimental Design

22.3.1 Reagents and Materials

22.3.1.1 For Hanta Virus

22.3.1.1.1 Experimental Procedure

A "sandwich" scheme immunoprinciple of detection is employed [13,15]. The protocol for the experimental procedure using HRP-labeled enzyme marker is described previously [11]. The amperometric measurements were performed at a fixed electrode potential of 127 mV vs. Ag/AgCl to detect iodine produced by NaI and H_2O_2 in the presence of HRP. The modified procedure using AP-labeled conjugate involves the following stages.

- *Prewashing stage*: Flow of a washing solution 0.02 M Na–phosphate buffer (pH 7.8) containing 0.15 M NaCl and 0.1% Tween 20 (PBST) through the immunoelectrode for 1 min.
- *First stage of incubation*: Flow of PBST-containing human blood plasma to be tested for Hanta virus IgG antibodies (target analyte) through the immunocolumn for 3 min, thereby allowing the target analyte to be bound to the surface of the immunoelectrode.
- *First washing procedure*: Flow of the washing solution (PBST) for 3 min.
- *Second stage of incubation*: Flow of PBST containing AP-labeled antihuman IgG (immunoconjugate AP-labeled, concentration: 0.5 μg/mL) against the target analyte through the immunocolumn for 8.5 min, allowing immunoconjugate to form an immunosandwich with the antigen–antibody complex on the surface of the immunoelectrode.
- *Second washing stage*: Flow of a washing solution 0.05 M carbonate buffer (pH 9.6) containing 0.15 M NaCl through the immunoelectrode for 5 min. This washing allows the removal of conjugate molecules attached elsewhere except to the target analyte. An electrode working potential of +100 mV is applied to the immunoelectrode before starting the second wash causing polarization of the electrode.
- *Amperometric measurement stage*: Flow of 0.05 M carbonate buffer (pH 9.6) containing 0.15 M NaCl and 1 mM naphthyl phosphate (substrate of AP label), or 0.1 M acetate buffer solution (pH 4.5) containing 0.15 M NaCl, 1 mM KI, and H_2O_2 (substrate for HRP label), through the immunocolumn for 3 min.

The output is an amperometric signal determined by the extent of naphthol or iodine formed as a result of the enzymatic oxidation of naphthyl phosphate or reduction of iodine. The amperometric output was obtained as a steady state current due to electrooxidation/reduction of naphthol or iodine. The background current signal was measured using blood sample without antibodies to Hanta virus. The flow rate of the reagents through the sensor during all stages is 125 μL/min. The procedure for coating the Ag/AgCl electrode with Nafion layer contained two stages. (1) The Ag/AgCl electrode is prepared by anodizing silver (99.9%) electrode at +5.0 V for 5 min in a 0.15 M NaCl solution. (2) Then the Ag/AgCl electrode is coated with several layers of Nafion film by dipping in a series of Nafion solutions. The electrode is dipped in 0.5% Nafion solution, removed, and air-dried for 30 min. This step is repeated once with a 3% Nafion solution and then four times with 5% Nafion solution. Nafion solutions of 0.5 and 3% are prepared by dilution with 1:1 2-propanol and water. The final Nafion coating was cured in a drier at 40°C overnight.

The recombinant nucleocapsid protein of the Hanta virus envelope is formed from expression of plasmid constructs with the complete nucleocapsid (*Escherichia coli*) as previously described in [16]. The product is a fusion protein with phase T7 gene 10 protein and is purified by passage through a metal chelation column [16]. Woodward's reagent K (*N*-ethyl-5—phenylisoxazolium-3′-sulfonate), trypsin inhibitor, AP, and HRP-labeled antiperomyscus IgG (developed in goat), antihuman IgG antibodies, naphthyl phosphatase, and Tween-20 were obtained from Sigma Chemical Company (St. Louis, MO). Hanta virus positive and negative (control) human blood samples were obtained from clinical diagnostic specimens at the University of New Mexico Health Sciences Center's Hanta virus Diagnostic Unit. Positive samples were those that exhibited IgM antibodies to SN virus N antigen and IgG antibodies to SN virus N and G1 antigens, by Western blot or strip immunoblot tests [16,17]. Microcentrifuge filters were obtained from Sigma Chemical Company. Ultralow-temperature isotropic (ULTI) carbon microdispersed powder was provided as a gift from Carbomedics, Inc., (Austin, TX).

22.3.2.1 For Influenza A Virus and Parainfluenza Virus

Graphite powder was obtained from Fisher Scientific Company (Hampton, NH) Ultrafree-MC Centrifugal filter units pore size 0.22 μm, used as a basis for the measuring cell (immunocolumn) construction were obtained from Millipore Corp. (Bedford, MA). Carbon rods used as counter electrode were a courtesy of DFI Pultruded Composites, Inc. (Erlanger, KY). Woodward's reagent K (*N*-ethyl-5-phenyliso-xazolium-3′-sulfonate), trypsin inhibitor, sodium iodide, and Tween-20 were obtained from Sigma Chemical Company. 2-Propanol (US Industrial Chemicals Co., Division of National Distillers and Chemical Corporation, New York, NY.) was used without further purification. All chemicals used were of analytical grade. Goat antirabbit IgG, conjugate of antirabbit IgG with HRP and rabbit IgG, was obtained from Sigma Chemical Company. Inactivated PIV (Strain Sendai) and IAV (A/Panama/2007/99, H3N2), affinity-purified antibodies to PIV and IAV, and peroxidase-labeled affinity purified antibodies to viruses (conjugates) were obtained from Biodesign International (Saco, Maine). Other chemicals of analytical grade were obtained from the standard sources. For preparation of the aqueous solutions deionized water was used.

A previously described technique for immobilizing analyte related antibody on graphite powder has been adapted for immunosorbent preparation on staphylococcal protein A.

All previously stages described in Hanta virus detection of prewashing, first and second stages of incubation, and amperometric measurement stage are similar to flu viruses detection.

22.4 Results and Discussion for Hanta Virus Test

A flow-through amperometric immunosorbent assay system was developed based on disposable sensing elements for rapid detection of Hanta virus in mice blood. The system utilizes flow-through, immunosorbent and enzyme immunoassay techniques in conjunction with an amperometric sensor. The optimal amount of RNP on the surface of the immunosorbent was chosen on the basis of calculated kinetic constants [10,14].

The system has four main components: a pump, five valves, amperometric immunosensor assembly, and an electrochemical/data recorder interface. The immunosensor consists of a disposable centrifugal filter, which acts as the base for the antibody modified carbon particles. This disposable sensing element rests at the base of a hollow carbon rod that acts as the working electrode. This carbon rod is drilled longitudinally so that the fluid passes through the rod onto the carbon particles, which have been distributed evenly onto the filter by centrifugation. Another hollow carbon rod acts as the counter electrode and a hollow Ag/AgCl rod acts as the reference electrode as shown in Figure 22.4. The design is engineered so that the fluid flows from the inlet of the immunosensor through the hollow carbon rod onto the carbon particles on the filter and through the filter, and passes through the hole in the hollow Ag/AgCl to the hole in the hollow carbon rod and then through the outlet of the immunosensor. The design of the simple system is shown in Figure 22.5a. The cross-sectional view is given in Figure 22.5b.

22.4.1 Field Test of Hemolysis Blood Samples

Since the introduction of quantitative determinations in clinical chemistry, hemolysis has been recognized as an interference factor leading to increased levels of substances in serum (plasma). Hemolysis is a well-known interferent in many assays [18–22]. Hemolysis is common in blood specimens, which occurs when blood contacts foreign surfaces [22].

Hemolysis of samples is often encountered in clinical biochemistry. It results in a rosy to bright red appearance of the plasma after centrifugation or sedimentation of red blood cells and platelets. Hemolysis reflects the destruction of erythrocytes with release of hemoglobin into the plasma, giving the red color. It is a source of error in electrolytic and enzymatic

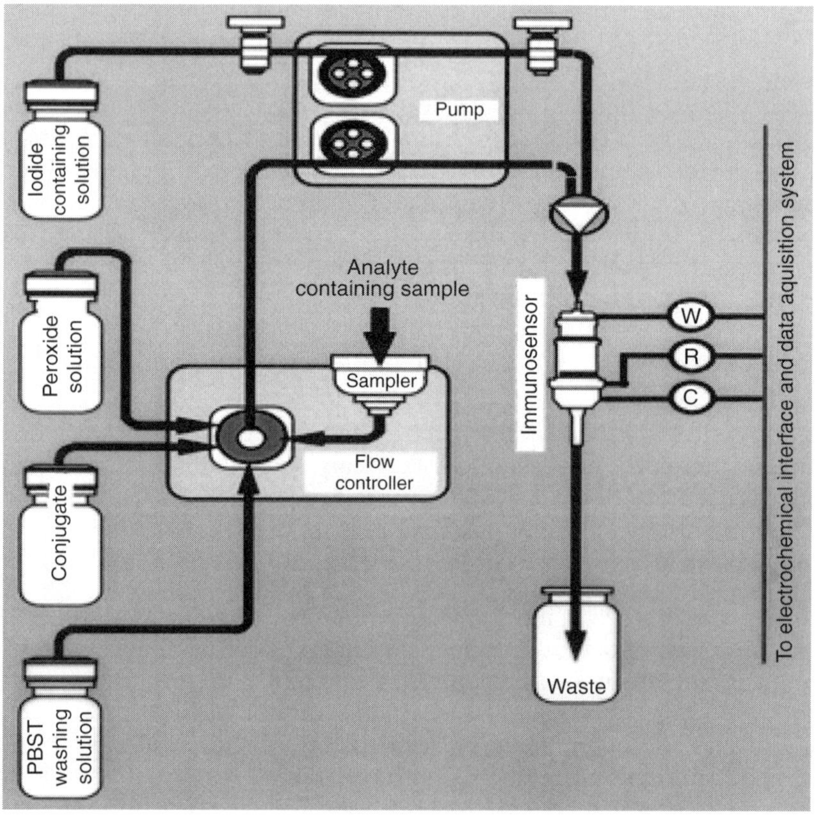

FIGURE 22.4
Schematic flow-through automated immunofiltration system.

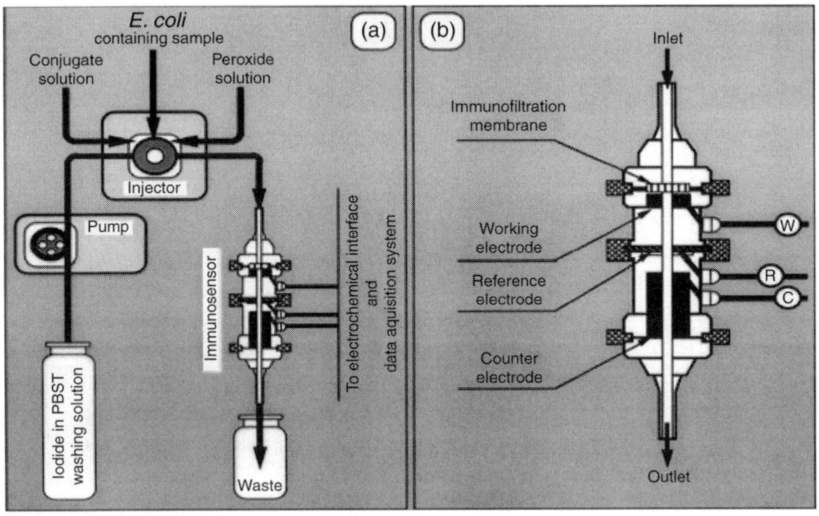

FIGURE 22.5
a) Flow-injection assay system; (b) immunofiltration sensor design.

determination in clinical biochemistry. This seems to be dependent on the material used for blood collection. When hemolysis occurs during blood sampling, it decreases the reliability of many results considerably. The use of partially or completely hemolyzed serum is unavoidable, however, especially during field testing. Hemolysis can result from improper drawing and handling of specimens and can also occur during centrifugation, separation procedures, and dry freezing.

Hemolysis leads to interference by intracellular constituents in reactions of the assay. The response of the device with hemolyzed samples can be seen Figure 22.6.

Experiments were performed under the same operating conditions as discussed earlier using HRP-labeled conjugate [13]. It is obvious that the difference between the signals obtained for negative and positive samples is insignificant, and they tend to overlap beyond the operating dilutions shown. All efforts of trying to record a response at dilutions other than that shown in Figure 22.6 have been failed due to the high degree of noise and instability of the device with the hemolyzed samples. One reason is that the broken erythrocytes have peroxidases that interfere with HRP. Since anodic oxidation of hydrogen peroxide in the substrate is utilized for the amperometric detection of anti-Hanta virus antibodies in blood samples, the presence of broken erythrocytes yield overlapping signals. Therefore, any chemical that can react with hydrogen peroxide will interfere with the accuracy of the test. Hemolysis of red blood cells implies the presence of debris (e.g., cell membranes), protein, peroxidases, and other potential interferents from cells. A comparison of the response obtained for hemolyzed and nonhemolyzed negative control samples is shown in [13]. It can be observed that in hemolyzed samples there is an increase in response with increasing concentration of the sample; this is due to the binding of contents of the blood on to the surface of carbon particles and also on to the surface of working, reference, and counter electrodes. However, we have not yet discovered which compounds present in the hemolyzed blood are responsible for the nonspecific signals. Also exposing an electrode poised at +100 mV vs. Ag/AgCl to biological fluids will produce a large background signal due to oxidation of endogenous sample components [23]. Hence, optimization of the analysis requires large dilution of serum to avoid electrode fouling and minimize the effect due to the presence of electro active species in hemolyzed blood samples.

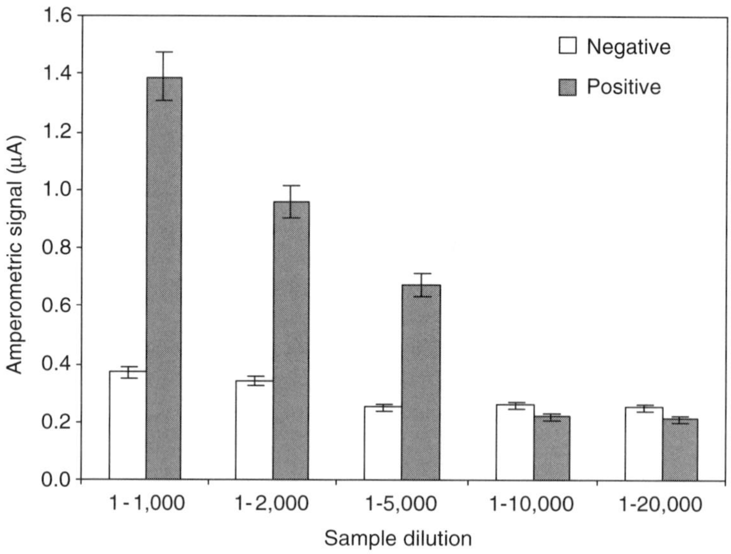

FIGURE 22.6
Amperometric signal obtained for control hemolyzed mouse blood sample using AP-labeled conjugate and nafion-coated electrodes at various sample dilutions. Error bars represent standard deviation from the mean of four runs, on a single control sample.

22.4.2 Effect of H_2O_2 Used in the Substrate

Approximately 10 μL of hemolyzed blood samples and nonhemolyzed samples were treated with 30 μL of 1 M H_2O_2 on a glass plate. Evolution of gas was observed suggesting decomposition of H_2O_2 in the case of hemolyzed samples while there is no reaction in the other case. This effect is observed in reactions involving hydrogen peroxide/peroxidase because of premature decomposition of hydrogen peroxide by hemoglobulin [22]. The plot of various background signals obtained can be seen in the Figure 22.6. The response of the device obtained for case 4 is very high compared to other background signals. This is due to the absorbance of contents of the hemolyzed blood on to the electrode surfaces interfering with the H_2O_2 in the substrate solution. Therefore, hydrogen peroxide is considered to be interfering with contents of the hemolyzed blood that can react or yield overlapping signals. Hence, we studied the possibility of using other enzyme markers such as AP instead of HRP. This was used to avoid the use of H_2O_2 thereby eliminating the interference of the contents of hemolyzed blood with the H_2O_2 used in the substrate.

22.4.3 Alkaline Phosphatase vs. Horse-Radish Peroxidase

Electrochemical enzyme analysis requires electrochemically active reaction products. α-Naphthyl phosphate was chosen as the suitable substrate for the electrochemical measurement of AP activity. It is electrochemically inactive while the reaction product α-naphthol (Equation 22.1) can be readily oxidized at a potential of +100 mV vs. Ag/AgCl.

$$\alpha\text{-Naphthyl Phosphate} + H_2O \xrightarrow{AP} \alpha\text{-Naphthol} + HPO_4^{-2} \quad (22.1)$$

The experimentl procedure performed with the use of AP-labeled conjugate is the same as described before, apart from the change of conjugate and substrate. The results attained with the use of AP-labeled conjugate are presented in Figure 22.7. It shows a summary of

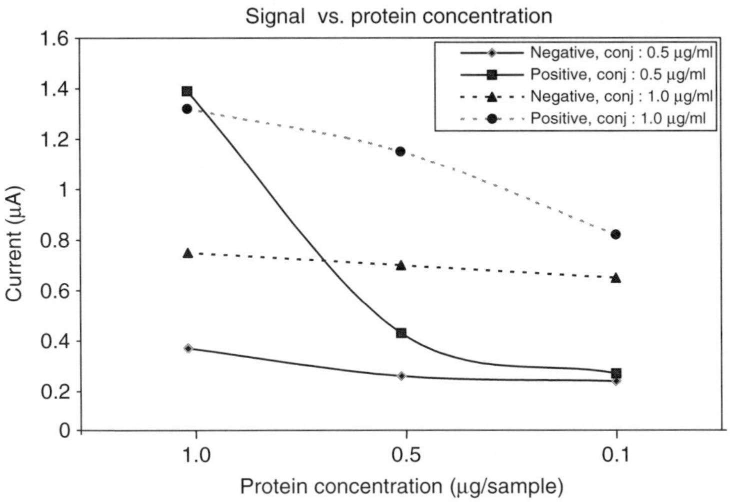

FIGURE 22.7
Optimization of antigen immobilized on the carbon particles: Amperometric response of the device at different concentrations of conjugate (AP) and recombinant protein of Sin Nombre virus (SNV) envelope at an analyte concentration of 1:1000.

the response of the device for negative and positive blood samples at various dilutions. It clearly shows the differentiation between the positive and negative samples, which is more than three times at a sample dilution of 1:1000.

The difference diminishes with increased dilution, due to the decrease in the antibodies in the analyte solution with a decrease in the analyte concentration. This modification provides an improved lower detection limit than the one previously described [11], which was confirmed by the working range of the sample dilutions. The present system was clearly able to distinguish the positive and the negative sample even at a dilution of 1:10,000. The substrate solution was prepared fresh everyday to minimize nonenzymatic hydrolysis.

22.4.4 Nafion Layer

The nonspecific sorption of the components in the hemolyzed blood samples on to the surface of the working, reference, and counter electrodes is further reduced by Nafion layer coating. We have studied the flow-through measurements with Ag/AgCl electrode covered with a Nafion layer (by dipping and drying a commercially available Nafion solution).

Covering the Ag/AgCl electrode with a Nafion layer improved the results even more. The Nafion layer protected the electrodes from contact with the interferents in the blood by acting as a barrier, reducing the binding of the interferents to the electrodes, and also minimizing fouling affect. A minor (twofold) decrease in the signal magnitude was observed but a significant improvement in reproducibility was obtained, as seen in Figure 22.6. Even though covering the reference electrode surface with a Nafion layer diminished the efficiency of the anodic oxidation process by twofold in comparison with the bare electrode, the difference in signal for the negative and positive samples is still satisfactory (more than threefold at 1:1000 sample dilution (Figure 22.6).

The decrease in signal is due to the presence of the Nafion layer, which restricts the substrate supply to the inner layer, where the enzymatic reaction is verified. The use of Nafion layer on the reference resulted in added stability to the device, also protecting the electrode against electrode fouling caused by the adsorption of macromolecules on the surface, which results in a gradual loss of response with time.

Optimization of the analysis required large dilution of serum to avoid electrode fouling and minimize the effect due to the presence of electro-active species at the working potential capable of producing faradaic interferences as well as macromolecules, like proteins or lipids, with an electrode fouling effect that may influence biosensor response. Also, it can be seen from the plot in Figure 22.6 that the amperometric signal for the negative samples remained almost constant at various dilutions, demonstrating the absence of the effect of hemolysis and nonspecific signals at higher dilutions of the analyte solution. Any attempts to increase the concentration of the analyte solution resulted in increasing response for the negative signal, thereby minimizing the difference between negative and positive analyte solutions. As the device is sensitive enough to work under such high dilutions, the operating dilution is chosen to be 1:1000. Optimization of the amount of recombinant SNV protein envelope used for one-sensing element was also examined. The performance of the device at two different concentrations of the conjugate solution and three different concentrations of the viral envelope immobilized on the highly dispersed carbon particles was described and is shown in Figure 22.7. The optimum results were obtained at a conjugate concentration of 0.5 μg/mL and a protein concentration of 1.0 μg/sensing element. Increased concentration of conjugate (1.0 μg/mL) had a more significant effect on the response of the negative sample than on the positive one. This is due to the increased nonspecific binding of the conjugate on the active electrode surface. Choosing the lower conjugate concentration for the working conditions minimizes this effect. Trial for use of higher concentrations of protein per sensing element is limited by the cost of the SNV protein envelope. The statistical performance of the device has been compared with the standard methods and described in detail [13].

22.4.5 Result and Discussion of Influenza Samples

The bird flu seems to be spreading, increasing the risk of a worldwide pandemic. With the bird flu taking the lives of nine Vietnamese since December 30, it looks like strain H5N1 is preparing to spread. Public officials say the disease has hit 18 Vietnamese provinces and cities in a rampage that has led to many deaths. The bird flu is spreading through Vietnam at a startling rate; with 232 communes in 23 cities reporting cases of influenza strain H5N1. Vietnam killed 450 ducks in Ho Chi Minh City in January as part of that nation's fight to stop the spread of the bird flu virus. Since the end of 2003, 26 Vietnamese people are on record as having died of influenza strain H5N1, more commonly known as the bird flu. Influenza strain H5N1, more commonly known as the bird flu, has been diagnosed in seven Vietnamese people since the start of the year [24].

It has been reported that five people died from Asian bird flu in the last 3 weeks; a Vietnam woman becomes the latest victim. A 3-week-old bird flu outbreak that has affected several Asian countries has claimed its fifth human victim. Thailand claims "80 percent ready" to handle major bird flu outbreak. There are three provinces in Thailand that are currently considered as bird flu stricken, with another 19 provinces on the danger list. When news that a dangerous bird flu virus had started to spread among humans reached top levels of Thailand's government, officials took it as immediate emergency [24].

It is now estimated that as many as 2000 people contracted bird flu during the 2000 epidemic in the Netherlands, which is far higher than the 69 cases. A Dutch investigation reveals that the 2003 Netherlands outbreak of bird flu was enormously more widespread than previously estimated. Japanese poultry farmers continue to suffer from the financial fallout created by last year's bird flu outbreak. Egg prices are 50% higher than before.

The Chinese government is stepping up inspection measures along its border amid concerns about a bird flu outbreak in Vietnam. The bird flu has most countries stockpiling vaccines in preparation for an outbreak. Britain however believes that it is impossible to determine what and who to vaccinate.

A Russian scientist has alarmingly announced that one billion people stand to die from the coming global flu pandemic. In the United States alone, as many seven million Americans may be killed by bird flu. Epidemiologists are racing to create a vaccine to protect humans from the bird flu as the disease continues to spread through Southeast Asia. New evidence shows that a deadly bird flu virus that has broken out across Asia may attack all of the human body, not just the lungs as previously known. Asian nations are stocking up on Tamiflu and making plans for civil-defense-type measures in the event of an outbreak of bird flu. The World Health Organization officials are concerned that February's Lunar New Year celebrations across Vietnam might lead to more bird flu outbreaks [24].

Bird flu is on the rise, and studies show that it may be able to be spread among humans. It is impossible to get the flu vaccine for everyone; however, the United Nations issued a strong warning recently to the growing number of Asian nations where a killer bird flu virus has broken out in recent weeks. Dozens of workers in Hong Kong's hospitals and clinics have come down with SARS or bird flu virus in recent months, but, until now, employees are yet to be compensated.

Therefore, detection and prevention are important. Biosensors for influenza viruses should meet or exceed certain requirements so as to make them comparable or even better than the traditional analytical systems. They must be simple to handle, small, cheap, and able to provide reliable information in real time. They also need to be sensitive and selective for the analyte of interest, and suitable for in situ monitoring [25]. The exceptional combination of a biological element in straight contact with a physical transducer makes it possible to fulfill all of these requirements. When designing biosensors, it is important to understand the multiple factors that influence the performance of the sensing system. It is also important to consider the limitations of the biosensor, especially when the final goal is for application in real-sample monitoring. Despite extensive research in biosensors, few biosensors are routinely used in real applications [26–28]

The major analytical parameters of amperometric immunoenzyme sensors, such as detection limit, analytical range, and signal to background ratio related with amount of bound conjugate and conditions of its detection. The results of optimization of graphite powder immunosorbent amount and composition of substrate solution for PIV detection are presented in Figure 22.8a,b.

It shows dependence of the amperometric immunosensor responses on concentration of H_2O_2 in substrate solution, amount of immunosorbent, and their influence on signal to background ratio. The best signal to background ratio was obtained with 5 mg of graphite anti-PIV immunosorbent, and substrate solution contains 0.5 mM H_2O_2 and 1 mM KI.

Figure 22.9 describes the amperometric sensor response to different concentrations of PIV in ng/mL. Lower detection limit 0.5 ng/mL with a coefficient of variation (COV) of 0.06. The upper detection limit is as high as 10,000 ng/mL. Once the sample is injected, it takes four min to reach a steady state, and the total time of the assay from the beginning to the end is about 22 min. The current is proportional to the concentration with the upper limit of our device of 15 μA.

Figure 22.10 of IAV response illustrates similar results to PIV and describes the amperometric sensor response to different concentrations of IAV in ng/mL (lower detection limit 0.4 ng/mL with a COV of 0.05).

The upper detection limit is high as 300 ng/mL. Once the sample is injected, it takes four min to reach a steady state (Figure 22.11), and the total time of the assay from the

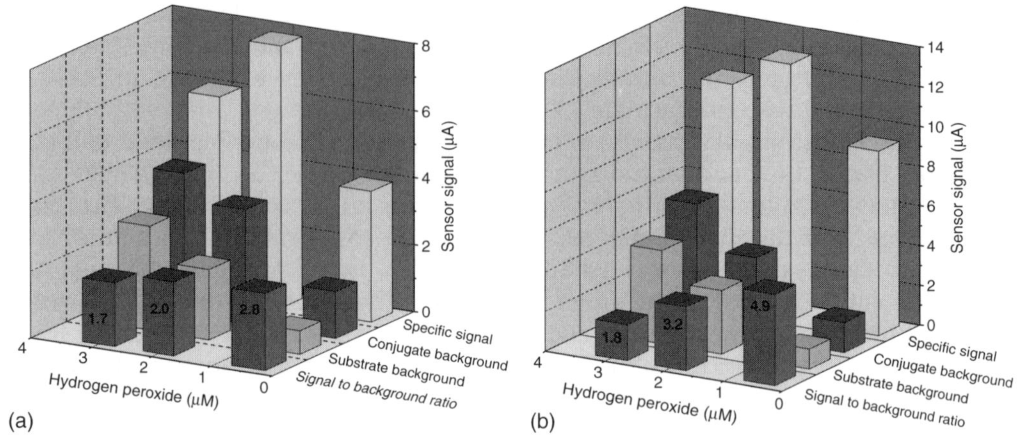

FIGURE 22.8
Diagrams showing the relationship between concentrations of hydrogen peroxide in substrate solution and specific signal to background ratio with different amounts of anti-PIV immunosorbent in each sensitive element: a for 2.5 mg and b for 5 mg of graphite powder immunosorbent. Concentration of PIV in the sample is 100 ng/mL.

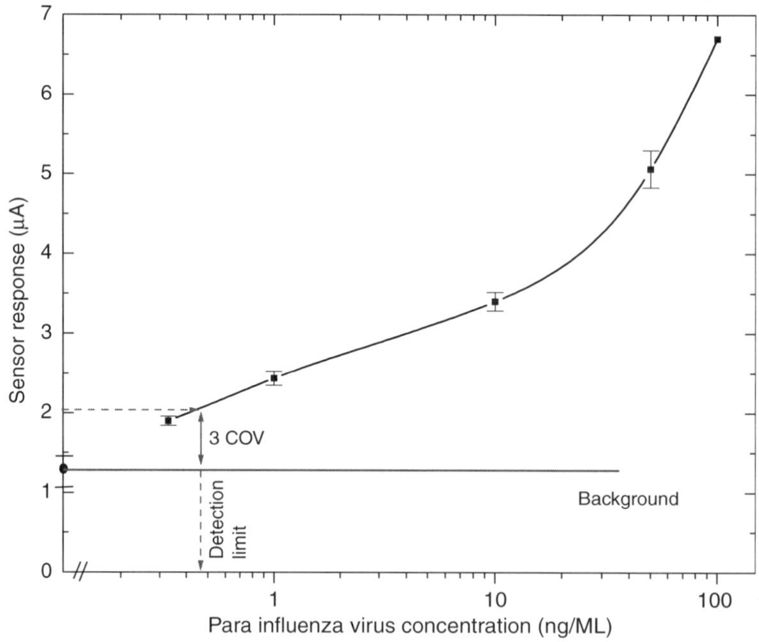

FIGURE 22.9
Determination of detection limit for parainfluenza virus (PIV) electrochemical immunoassay using three samples for each measurement with COV of 0.06.

beginning to the end is about 22 min. The current is proportional to the concentration with the upper limit of our device of 4 μA.

Therefore, the amperometric signal in IAV is smaller than in PIV, approximately half of the PIV values. This could be due to the outer protein shell, which is called capsid; this comprises repeating protein subunits. The architectural arrangement of the capsid, or its structure, as well as the presence of membranous envelope surrounding the capsid could explain these differences.

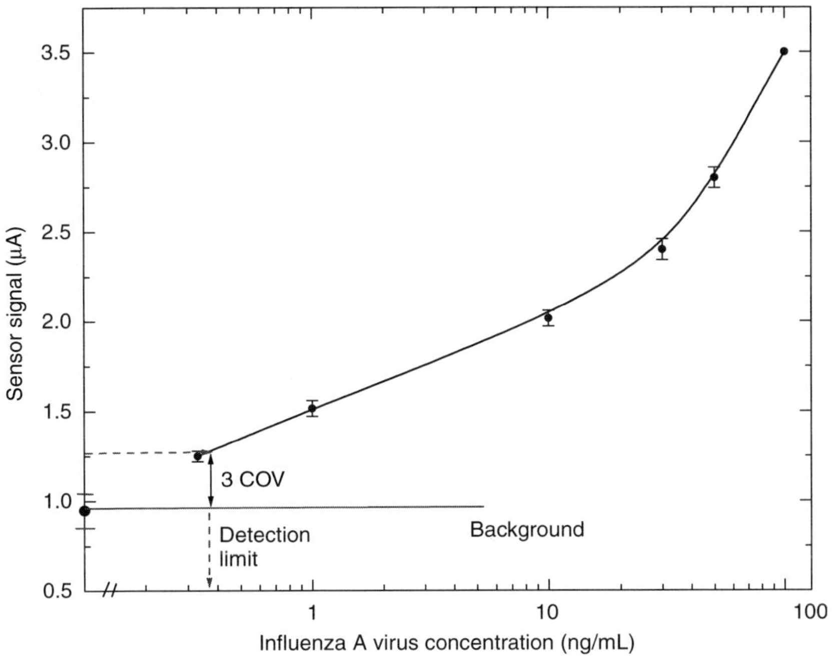

FIGURE 22.10
Calibration curve for sandwich IAV assay obtained from the flow-injection amperometric immunosensor. Conditions of detection: flow rate is 100 µL/min; the working electrodes (5 mg of graphite powder immunosorbent per 0.22 µm pores filter) were poised at −0.125 V vs. Ag/AgCl in 0.1 M acetate BS, pH 4.5 with 0.15 M NaCl, 1 mM KI, and 0.5 mM H_2O_2.

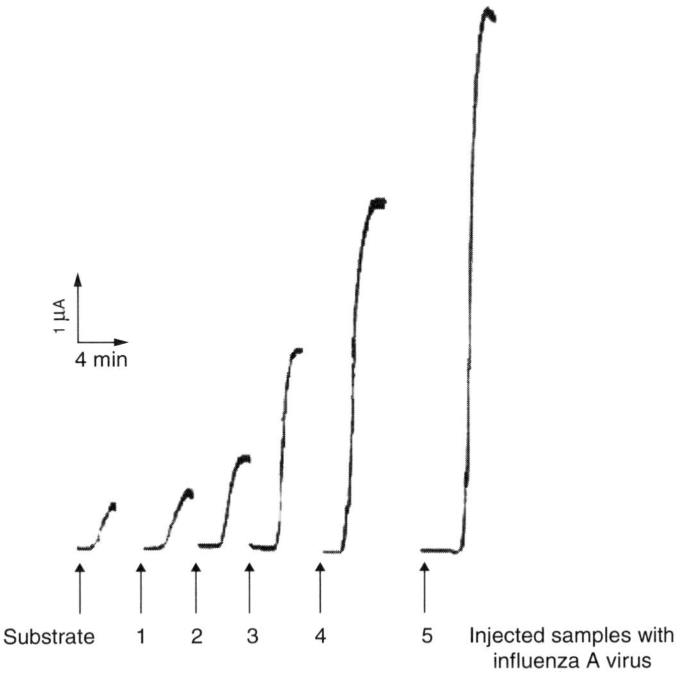

FIGURE 22.11
Sensor's response signals for different concentration of influenza A virus (IAV), ng/mL: 1, 0; 2, 2.0; 3, 30; 4, 100; 5, 300.

22.5 Conclusions

Electrochemical biosensors have emerged as a dynamic technique for qualitative and quantitative determination of different analytes for environmental, clinical, agricultural, food, or military applications. Despite the enormous potentials compared with laboratory-based analytical techniques, numerous problems still remained to be solved. Most biosensors have shown excellent characteristics for synthetic samples, pristine laboratory samples; they are not sufficiently robust in real samples. Most of the existing limitations could be directly related to operational and long-term stability of the biological receptor and the physical transducer. Other limitations could be attributed to poor reproducibility between sensors and selectivity in complex matrices. For practical applications, the most important obstacles are encountered once the sensor is used outside pristine laboratory conditions and is applied for in situ real-sample monitoring [25].

Elimination of various problems associated with hemolysis of sample was the subject of this study, in which a flow-through immunoassay system was developed on the basis of sandwich scheme of immunoassay using naphthol formed as a result of enzymatic hydrolysis of α-naphthyl phosphate in the presence of AP label, which has been quantified amperometrically. The main problems faced while analyzing antibodies against Hanta virus in hemolyzed mice blood samples were found to be due to various electroactive species present in the hemolyzed blood have been eliminated. The nonspecific signal due to the interaction of hydrogen peroxide used in the substrate solution with the hemolyzed blood samples was eliminated using AP-marked enzyme label. Our study showed improved results in minimizing the interferences by covering the Ag/AgCl electrode surface with a layer of Nafion. The analysis required large dilution of analyte sample to the order of 1:1000, to avoid electrode fouling and minimize the effect of interference, nonspecific binding due to the presence of electroactive species at the working potential. The device exhibited high degree of sensitivity capable of working up to an analyte dilution as high as 1:10,000. As a result extremely small blood samples can be analyzed increasing the utility of this device for working in the field. The biosensor has high selectivity and sensitivity for direct measurement of anti-Hanta virus antibodies in blood samples with minimum or no sample preparation. The total time of analysis is 25½ min.

It is evident also that our device can detect the PIV, IAV as well as Hanta virus. The amperometric signal for different types of flu is different, for each of the two types we tested. However, the signal is reliable and reproducible, even after 5-month immunosorbent was kept in refrigerator.

Each year influenza viruses are responsible for recurrent epidemics, frequently in association with genetic variation. There is a requirement for sensitive and rapid diagnostic techniques to improve both the diagnosis of infections and medication dispensed by the physician as soon as possible to enhance patient's recovery. The quality of surveillance systems will be trusted if the signal does not degrade fast [29].

Electrochemistry has superior properties over the other existing measurement systems because electrochemical biosensors can provide rapid, simple, and low-cost on-field detection. Electrochemical measurement protocols are also suitable for mass fabrication of miniaturized devices [30].

References

1. CDC. (1998). Update of emerging infection from the centers for disease control and prevention. *Annals of Emergency Medicine*. 31:121.

2. Butler, J.C., Peters, C.J. (1996). Lessons learned from the hantaviruses and other hemorrhagic fever viruses. *American Journal of the Medical Sciences.* 311:55–59.
3. Dietrich, N., Pruden, S., Ksiazek, T.G., Morzunov, S.P., Camp, J.W. (1997). A small-scale survey of hantavirus in mammals from Indiana. *Journal of Wildlife Diseases.* 33:818–822.
4. CHP (Center for Health Protection). (2004). http://www.info.gov.hk/info/flu/eng/(accessed December 2, 2004).
5. WHO. (2004). http://www.info.gov.hk/info/flu/eng/global.htm(accessed December 2, 2004).
6. Cordingly, P. (2004). http://news.bbc.co.uk/1/hi/health/medical_notes/1337640.stm (accessed December 5, 2004).
7. Li K.S., Wang J. (2004). http://www.osha.gov/dsg/guidance/avian-flu.html (accessed December 4, 2004).
8. PHA. (2004). http://www.hc-sc.gc.ca/english/diseases/flu/avian.html (accessed December 15, 2004).
9. Abdel-Hamid, I., Atanasov, P., Ghindilis, A.L., Wilkins, E. (1998). Development of a flow-through immunoassay system. *Sensors and Actuators B-Chemical.* 49:202–210.
10. Krishnan, R., Ghindilis, A.L., Atanasov, P., Wilkins, E. (1995). Fast amperometric immunoassay utilizing highly dispersed electrode material. *Analytical Letters.* 28:2459–2474.
11. Vetcha, S., Abdel-Hamid, I., Atanasov, P., Ivnitski, D., Wilkins, E., Hjelle, B. (2000). Portable immunosensor for the fast amperometric detection of anti-Hantavirus antibodies. *Electroanalysis.* 12:1034–1038.
12. Krishnan, R., Ghindilis, A.L., Atanasov, P., Wilkins, E. (1996). Development of an amperometric immunoassay based on an highly dispersed immunoelectrode. *Analytical Letters.* 29:2615–2631.
13. Vetcha, S., Wilkins, E., Yates, T. (2002). Detection of hantavirus infection in hemolyzed mouse blood using alkaline phosphatase conjugate. *Biosensors and Bioelectronics.* 17:901–909.
14. Krishnan, R., Ghindilis, A.L., Atanasov, P., Wilkins, E., Montoya, J., Koster, F.T. (1996). Fast amperometric immunoassay for Hantavirus infection. *Electroanalysis.* 8:1131–1134.
15. Ghindilis, A.L., Krishnan, R., Atanasov, P., Wilkins, E. (1997). Flow-through amperometric immunosensor: Fast 'sandwich' scheme immunoassay. *Biosensors and Bioelectronics.* 12:415–423.
16. Hjelle, B., Jenison, S., TorrezMartinez, N., Herring, B., Quan, S., Polito, A., Pichuantes, S., Yamada, T., Morris, C., Elgh, F. (1997). Rapid and specific detection of Sin Nombre virus antibodies in patients with hantavirus pulmonary syndrome by a strip immunoblot assay suitable for field diagnosis. *Journal of Clinical Microbiology.* 35:600–608.
17. Jenison, S., Yamada, T., Morrison, C., Anderson, B., Torrezmartinez, N., Keller, N., Hjelle, B. (1994). Characterization of human-antibody responses to 4 corners Hanta virus infection among patients with Hanta virus pulmonary syndrome. *Journal of Virology.* 68:3000–3006.
18. Ma, Z.M., Monk, T.G., Goodnough, L.T., McClellan, A., Gawryl, M., Clark, T., Moreira, P., Keipert, P.E., Scott, M.G. (1997). Effect of hemoglobin- and perflubron-based oxygen carriers on common clinical laboratory tests. *Clinical Chemistry.* 43:1732–1737.
19. Frank, J.J., Bermes, E.W., Bickel, M.J., Watkins, B.F. (1978). Effect of in vitro hemolysis on chemical values for serum. *Clinical Chemistry.* 24:1966–1970.
20. Sonntag, O. (1986). Hemolysis as interference factor in clinical chemistry. *Journal of Clinical Chemistry and Clinical Biochemistry.* 24:127–140.
21. Mather, A., Mackie, N.R. (1960). Effects of hemolysis on serum electrolyte value. *Clinical Chemistry.* 6:223–227.
22. Dogan, Y. (1992). Effect of in vitro hemolysis on 25 common biochemical tests. *Clinical Chemistry.* 38:575–577.
23. Frew, J.E. (1989). Measurement of alkaline phosphatase activity by electrochemical detection of phosphate esters, application to enzyme immunoassay. *Journal of Electroanalytical Chemistry.* 266:317–336.
24. http://www.newstarget.com/004314.html
25. Andreescu, S., Sadik, O.A. (2004). Trends and challenges in biochemical sensors for clinical and environmental monitoring. *Pure and Applied Chemistry.* 76(4):861–878.
26. Rogers, K.R., Lin, J.N. (1992). Biosensors for environmental monitoring. *Biosens. and Bioelectron.* 7:317–321.

27. (a) Rogers, K.R. (1995). Biosensors for environmental applications. *Biosens. and Bioelectron.* 10, 533–541; (b) Sadik O.A. and VanEmon J.M. (1996). Applications of electrochemical immunosensors to environmental monitoring. *Biosens. and Bioelectron.* 11:AR1–AR11.
28. Alocilja, E.C., Radke, S.M. (2003). Market analysis of biosensors for food safety. *Biosens. and Bioelectron.* 18:841–846.
29. Kessler, N., Ferraris, O., Palmer, K., Marsh, W., Steel, A. (2004). Use of the DNA flow-thru chip, a three-dimensional biochip, for typing and subtyping of influenza viruses. *Journal of Clinical Microbiology.* 42:2173−2185.
30. Kerman, K., Kobayashi, M., Tamiya, E. (2004). Recent trends in electrochemical DNA biosensor technology. *Measurement Science and Technology.* 15(2):R1−R11.

23

Perspectives on Biosensor Technology

Amarjeet S. Bassi and George K. Knopf

CONTENTS

23.1 Introduction

The field of smart biosensor technology is evolving rapidly, being driven by new materials, innovative fabrication processes, application of sophisticated information technology, and a multidisciplinary approach to product design and development. The contributors to this volume have identified and elaborated upon a number of new research directions. For example, single-molecule detection using intracellular nanosensors is an exciting and promising area of research. With these advancements, implantable nanosensors can provide unique insights into biological or cellular metabolism in the very near future. Also, innovative signal transducers fabricated using silicon nanowires have been investigated as a means to selectively detect individual virus particles in solution. Biochips containing high-density arrays are also being developed at an increasingly rapid pace. These technologies will make it easier to detect potential biological markers of many different diseases or illnesses with minimal sampling and cost.

New methods of layer-by-layer fabrication have enabled improved designs of biosensor platforms. These simple methods provide a mechanism for fine tuning the sensing layer and providing a greater degree of flexibility in biosensor design. Interfacing sensors with tissue, skin, and other biological structures is another important area of research. One related area in the design and development of porous silicon (PSi) interfaces that can carry out size-selective detection of analytes is how to better exploit the "bulk filtration" properties of the PSi matrix. However, developing viable biological interfaces requires additional fundamental studies in determining the stability of the system response.

The types and classes of materials used as the primary building blocks of biological-based sensors are also being expanded. A number of biosensor designs exploit neuronal tissue networks as the basic sensing element. These broadband biosensors provide electrophysiological and morphological data, and have been investigated for monitoring toxins. The networks can be applied in early warning systems for environmental threat detection. In addition, specialized materials such as bacteriorhodopsin (bR) are finding applications in many types of sensing platforms using light or voltage modulation. These include light-sensitive detectors, photocells, motion sensors, artificial retinas, and bioelectronic devices. Site-directed and semirandom mutagenesis can provide novel properties for custom building bR devices.

Biosensor technology is quickly filling an important niche in health and medicine by enabling rapid monitoring of pathogens and viruses. Advanced developments in biosensor arrays for toxicity monitoring and screening will provide the enhanced ability of recognizing and classifying specific chemical or biological toxins.

A window into future research directions and anticipated advances in biosensor development are presented in the final sections of this chapter. Through personal communication several contributors to the volume have offered their perspectives and insight to this dynamic discipline. The perspectives and comments are provided by leading researchers such as Ashok Mulchandani, Anthony P.F. Turner, Ulrich J. Krull, Claudio Nicolini, Won-Yong Lee, and Brian Cullum. The identified areas of key importance in the future are innovative technologies to enhance sensitivity and selectivity, achieve biomimetic sensor designs, develop optical nucleic acid sensors, utilize nanostructured organic biological matrices, exploit the functional capabilities of carbon nanotubes (CNTs), and rapidly advance medical research. Finally, Felix T. Hong's comments on reverse engineering biology challenges us to question our views on smart technology and whether embedded human-like intelligence is really possible in artificial devices and products. It is hoped that the following more speculative presentations will spur vigorous discussions and provide young researchers with valuable insight into future research directions. This presentation may also be of interest to individuals who wish to see a snap shot of views expressed by experts and leading-edge researchers in the early part of the twenty-first century.

23.2 New Directions of Research

23.2.1 Enhancing Sensitivity and Selectivity

Ashok Mulchandani (personal communication, 2005) and his colleagues associated with the University of California at Riverside remind us of the continual need to increase biosensor sensitivity and selectivity. Consider the area of clinical chemistry analysis that has received a great deal of attention over the past 30 years. Biosensors for the measurements of blood metabolites such as glucose, lactate, and urea, using both electrochemical and optical modes of transduction, are available commercially for point-of-care settings and in the case of glucose, for self-testing. However, the application of biosensors in other industrial sectors is still limited. In fact commercialization of biosensor technology has lagged far behind the output of research laboratories due to problems with stability and sensitivity, and quality-assurance issues.

More research is needed in the future to enhance the generated sensor signal by increasing the sensitivities of the transducers while reducing background noise. Some examples where enhanced sensitivity of the transducer has been demonstrated include the detection of a single virus particle (1). This has been possible due to the miniaturization of the transducers to one-dimensional (1D) nanostructured materials. Another advantage of 1D-nanostructured materials is the possibility of fabricating miniaturized sensor devices and the development of high-density arrays for the simultaneous analysis of a range of different analytes as demonstrated recently in the multiplexed detection of various cancer marker proteins up to 0.9 pg/mL levels (2). Also, 1D-nanostructured materials are extremely attractive for nanoelectronics because they can function both as devices and as the wires that access them.

Selectivity has also been a major issue in the development and application of biosensors. Selectivity basically depends on the biorecognition elements used by the biosensor. Hence, biotechnology and genetic engineering should provide improved molecular recognition elements that are tailored toward specific analytes of interest. Biotechnology will also be central in the future to improve the stability of the recognition elements, life usefulness, and for the development of aptamers. Aptamers are synthetic single-stranded DNA or RNA oligonucleotide sequences with the ability to recognize various target molecules with high affinity and specificity. Biosensor innovations could also benefit from the advancement of intelligent instrumentation, electronics, and multivariate signal-processing methods. In conclusion, we will definitely see an increasing role played by these biological-oriented devices in providing powerful analytical tools to the medical, environmental, military, agricultural, and food-safety sectors where rapid, affordable, and high sensitivity and specificity measurements are required.

23.2.2 Biomimetic Sensor Designs

Biosensors inherently comprise a biological or biologically derived sensing element intimately associated with, or integrated within, a physicochemical transducer. Over 1,500 articles are published each year describing the various permutations of sensing element and transducer (3,4). Professor Anthony Turner (personal communication, 2005) from Cranfield University at Silsoe informs us that these are applied mainly in medical diagnostics (5), in environmental diagnostics (6), in the food industry (7), and for crime prevention and security (8). The most significant impact of biosensors till date has been in the field of diabetes, where mediated amperometric biosensors account for over $4 billion in sales.

Until now, however, no device has come close to delivering a truly one-step procedure. In the immediate future we can expect to see new technology to deliver this objective. The development of suitably robust biosensors for many situations outside glucose monitoring has been hindered by several problems associated with the properties of biological material. The search for possible solutions to these problems has led to the development of biomimetic systems such as the electronic nose, which shows excellent practical potential for the detection of disease and infections (9).

An alternative approach has been to seek synthetic analogs of natural receptors and antibodies using supramolecular systems. If nature can produce nanomaterials with recognition and functional properties by evolution, molecular engineers should be able to accomplish comparable, but broader capabilities by design, guided by examples from living systems. One of the most promising areas of biomimetics is molecularly imprinted polymers (MIPs) (10). A key element here is the need for rational design (11). The ability to construct highly stable sensing structures, which are small enough to be implanted in the body, distributed widely in the environment, configured as high-density arrays, or coupled to modern information systems, offers exciting new horizons in the information revolution by furnishing our information and telecommunication systems with sophisticated "senses."

23.2.3 Optical Nucleic Acid Sensors

Optical nucleic acid sensors are another emerging technology that researchers with the University of Toronto at Mississauga, under the direction of Ulrich Krull (personal communication, 2005), believe will advance the role that biosensors play in our daily lives. The detection of nucleic acids is now used in many applications such as screening of genomes for mutations, detection and identification of bacteria and viruses, and efficient searching of compound libraries for potential therapeutic agents. Microarrays and multiplexed bead-based assays represent technologies that are suitable for large-scale screening, but are not ideally suited for practical applications involving a limited set of genomic targets. Enter the world of biosensors, where quantitative analysis can be achieved and the devices can concurrently address five practical priorities, namely reproducibility, reusability, speed of response, selectivity, and sensitivity. Novel configurations and applications in the development of optical nucleic acid sensors will continue. One can anticipate interest in a variety of areas including quantitative high-throughput (HT) nucleic-acid biosensors systems, development of spatially resolved biorecognition surfaces, and further development of generic nanosensors.

A quantitative HT nucleic-acid-biosensor diagnostic system will permit rapid detection of well-defined target nucleic acids in cell suspensions and solid tissues. New microbiosensors placed within microfluidic cartridges shall, with further refinement of surface chemistries and development of accelerated-target-delivery mechanisms, rapidly (in seconds) detect and determine low-copy number targets (10^3). Such systems will need to be complemented by new concepts in automated sampling, sample processing, and automation where throughput of ~1000 determinations per hour will be achieved.

Furthermore, new quantitative approaches to the development of spatially resolved biorecognition surfaces that would be suitable for screening and for SNP recognition will be developed. This requires improved methods of immobilization to improve confidence in the reliability of signals, as well as the use of smaller arrays to realize the full potential of quantitative microfluidics. Finally, the development of generic nanosensors for the direct, real-time determination of select nucleic acid levels in microenvironments with the ultimate aim of time-resolved intracellular determination of RNA levels is another direction. A significant leap in technology would be to measure and multiplex signals from inTrace.

23.2.4 Nanostructured Organic Biological Matrices

Recently, developed sensor systems that are based on nanostructured organic-biological matrices appear to have made a significant impact on the development of intelligent biosensors for monitoring gas and liquid with the potential for greatly effecting health care and environment control. According to Claudio Nicolini (personal communication, 2005), and his fellow researchers with the Nanoworld Institute at the University of Genoa, there appears to be four promising directions that are closely tied to industrial application of nanostructured organic matrices.

A unique opportunity exists for functionalized POAS and the PDMA-MWNT nanocomposites, which may lead in the near future to a spontaneous reversible sensor for acid vapors (12). These sensing devices utilize a simple comparative potentiometric circuit and can be easily extended to the sensing of most industrial relevant gases (13) by the ad hoc engineering of the proper organic sensitive element directly onto the circuit board.

A second promising direction of research is the nanostructure of rhodopsin matrices for liquid sensing. This approach has been recently extended to octopus rhodopsin from the early bacteriorhodopsin experimentation (14,15), pointing out the possibility of using self-assembled films as sensitive layers for optical biosensors for anesthetics and various hydrocarbon types as chloroform. An important point about this approach is that the optical properties of these layers appear reversible making it a good candidate for continuous monitoring.

Neural-network-based potentiometric stripping analysis (PSA) appears to be very sensitive for the simultaneous detection of several metal ions in aqueous samples (16) and is, therefore, a third major direction of research. The fourth and final major research opportunity with nanostructured organic-biological matrices involves the heme-enzymes P450s cytochrome isoforms immobilized br LS (16), gel-matrix or layer-by-layer (17), and solution casting with (18) and without (19,20) gold nanoparticles. This technology has proven potential as a sensing element for a wide range of organic substances important in the medical and ecological fields. Their pleiotropic properties and the LB/LS technology appear to yield the best stable working conditions for the sensor technology reaching the highest sensitivity (down to 0.4 mg/dl for cholesterol sensing).

23.2.5 Advances in Carbon Nanotubes

Won-Yong Lee (personal communication, 2005), of Yonsei University in South Korea, observes that electrochemical biosensors based on CNTs are rapidly developing because of the attractive physical properties exhibited by CNTs. Recent research has demonstrated the unique advantages of CNTs for the construction of a wide range of electrochemical enzyme-based biosensors, immunosensors, and DNA sensors. The remarkable electrocatalytic activity toward hydrogen peroxide and NADH permits the fabrication of low-potential amperometric biosensors. The electrical contacting of redox enzymes and electrode through CNT electrical wires offers great promise for the development of mediator-free biosensors. The use of CNT offers enhanced electrocatalytic activity toward target-DNA guanine bases as well as products of enzyme labels and thus provides great benefit for the development of PCR-free electrochemical DNA sensors.

Based on the recent advances in this field, the future of CNT-based electrochemical biosensors appears to be very bright. However, to expand the application domain of CNT-based biosensors it will be necessary to create better methods to control the chemical and physical properties of CNT, and develop a deeper theoretical understanding of their phenomena.

23.2.6 Future Impact on Medicine and Health Care

The role of biosensors in the area of medical research will dramatically expand according to Brian Cullum (personal communication, 2005) of the University of Maryland in Baltimore County. Since the development of biosensors in the 1960s, a large number of variations have been made in their design and applicability for the monitoring of many different environments. However, now, in the twenty-first century, biosensors are poised to change from being used primarily for simple quantitative measurements for chemical or biochemical monitoring applications, to providing insight into the brave new world of nanotechnology and nanoscience. With the increasingly small size of nanosensors today, and even smaller in the future, a whole new understanding of basic biological, chemical, and physical phenomena will become possible. With these small sensors and their ability to approach the molecular level, we are beginning to regain our greatest senses: sight. As science and technology have pushed to smaller and smaller scales, humans are no longer able to rely on their most trusted sense (i.e., sight). However, with the advent of nanoscale biosensors, and near-field optical sensing and imaging we are beginning to be able to visualize phenomena previously unobserved. This is leading to a better understanding of why particular events happen. One of the largest areas that will be impacted by the future of these nanobiosensors is that of medicine.

Medical research and treatment has long been studied through a top-down approach, in which physicians begin with the visual symptoms and trace their origin back to an organ(s) and eventually a more detailed biological reason for the symptoms. Once the symptoms are known, the standard treatment is prescribed. However, variability exists in how effective these treatments are for the patient, due to the interrelated nature of biological systems and the slight differences in these pathways from one individual to another. With nanobiosensors and arrayed biochips, we will be able to obtain a much greater understanding of systems biology and the interrelated nature of these pathways, by being able to monitor them all simultaneously. This will lead to a true understanding of the best means in which to treat a particular individual, fulfilling the ideal of personalized treatment for everyone.

23.3 Mimicking Biology: Is This a Realistic Goal?

Felix T. Hong (personal communication, 2006) of Wayne State University has been an active researcher in biosensing and bioelectronics for several decades and shares with us his views on reverse-engineering biology and whether human-like intelligence can be realized by artificial devices and products. Although the essay is inherently personal, he does ask us to challenge our perceptions on what we mean by "smart" technology and the goal of emulating biology.

The title of this volume "Smart Biosensor Technology" implies that the research and development of a new generation of biosensors emphasizes "smart" features. Before we plunge into a detailed discussion, we need to address the question of what constitutes smartness. Contemporary sensors and digital computers are smart based on the nineteenth- or twentieth-century standards and our contemporary expectations, but we expect them to be smarter in the future. Thus, the pertinent issue here boils down to *in what ways biosensors can become smarter than they have already been*. Perhaps a "smart sensor" ought to have some capability of "comprehension" or those important features that distinguish the human brain from the most advanced digital computers. One way to achieve this goal is to turn to biology for inspiration.

An earlier generation of information-technology professionals has already done just that in a multidisciplinary effort known as molecular electronics (21–25) and biocomputing (26–29). A frequently mentioned impetus for such endeavors was the continuing drive for device miniaturization. The well-known Moore's law predicts an exponential increase of the degree of miniaturization, in terms of chip densities and other related criteria. Miniaturization could make the sensors smarter because the increasing computing speed and the increasing number of computing elements would make it possible to perform tasks, which had previously been considered unrealistic for digital computers. For example, though still clumsy, character recognition, which requires a certain degree of analog-pattern recognition, is no longer a pipe dream. However, "discrete" microelectronic devices which operate on the basis of physical interactions will soon reach their physical limit at the nanometer scale (30,31). Yet, naturally occurring molecular devices found in living organisms seem to thrive on the nanometer scale, where "cross-talks" of devices are the normal mode of device operations.

Perhaps the nanometer-scale interactions are essential for the manifestation of intelligence. Size and densities matter but perhaps miniaturization alone is insufficient for true intelligence to arise. Perhaps there is something more in biology than that which meets the eyes. The nanometer-scale interactions in biology are nothing but the familiar biochemical interactions. Since biochemistry is ubiquitous in the living world, biochemical reactions may be responsible in part for intelligence. In addition, the almost exclusive use of carbon-based, organic materials as construction materials in the living world suggests that organic materials are more suitable than silicon-based materials for building intelligent devices, presumably because of the inherently diverse variety of organic compounds and the rich repertoire of biochemical reactions. The prominent role played by proteins in living systems suggests that proteins are more amenable to subtle and gradual changes of functions (26).

Comparison of various living organisms on the evolutionary scale further suggests that the emergence of intelligence might have something to do with the increasing complexity of "device architectures" (26,32). As compared to digital computer architectures, it is apparent that the device architectures in living organisms allow for massively parallel, distributed information processing—a feature that digital computers strive to emulate with limited success. These features provide a road map; device intelligence lies in exploiting both physical and chemical interactions of component materials, in the use of organic materials as building blocks, and in the configuration of life-like architectures. However, this road map does not prescribe detailed implementations. Numerous obstacles need to be overcome. Thus, the design goal is clear but what ought to be the optimal direction needs to be determined. From the very outset, molecular electronics research has been controversial and the approaches have frequently been disputed. For this reason, we should ask whether reverse-engineering biology is a viable approach.

The wisdom of reverse-engineering Nature has frequently been questioned. The skepticism was not totally unfounded. One of the best-known attempts to imitate Nature was Leonardo da Vinci's idea of designing a flying machine by imitating how a bird flies (33). As the history of technology transpires, all designs based on flapping wings failed. However, the argument against this particular example had an inherent flaw. A closer look at the history of science and technology reveals a sufficient number of successful examples of imitating Nature to warrant a second thought. It is important to remind ourselves that even the case of flying machines could be reinterpreted to favor reverse engineering, if one examines the case from a different angle.

Indeed, an important aspect of aircraft designs was a consequence of imitating Nature. Most, if not all, aircrafts, which glide in mid-air, have a wing profile similar to that of a bird: the curvature of the top surface is greater than that of the bottom surface, thus giving rise to a longer path of air flow on the top than on the bottom surface. It is well known that such a feature gives rise to a force of floatation (lifting) in air (Bernoulli principle). The departure from a "verbatim" imitation of birds was reflected in completely different designs of the propulsion mechanism. Birds use wings for both floatation and propulsion whereas using the wings for propulsion is ill conceived because moving parts, made of conventional materials, are usually the first to break. But then who said that both the propulsion and the floatation mechanisms must be implemented with a single device? Human ingenuity allowed past inventers to see through this superficial and imagined restriction by seeking separate devices for floatation and propulsion. The obvious lesson to learn is that imitation need not be performed at a superficial or "verbatim" level but rather at a deeper, more fundamental level of principles. To be able to do so requires some basic understanding. This is why basic research is important for technological endeavors.

Early molecular electronics researchers focused on molecular implementations of common microelectronic devices, such as switches. Critics from the biocomputing camp cited the fragility of biomaterials and argued that the molecular equivalent of diodes, flip-flops, etc., could never compete with their microelectronic predecessors. Instead, they advocated the development of devices for nontraditional computing. While the point was well taken, we held a different opinion (34). We believe that valuable lessons could be learned while investigators are "imitating" microelectronic devices. Imitating microelectronics was tantamount to what a beginner needs to master while ascending the learning curve. Sometimes as art imitates life and life imitates art, humans get inspired by engineering products. One of the most successful quantitative theories in biology—Hodgkin and Huxley's theory of nerve conduction—was based on imitation of marine cables.

A recent report by Koçer et al. (35) presented an interesting intermediate case. These investigators modified a mechanosensitive channel protein from *Escherichia coli* by attachment of synthetic compounds that undergo light-induced charge separation and reversibly open and close a 3-nm pore. This project was the outcome of convergence of reverse-engineering Nature, synthetic chemistry (and perhaps chemical engineering), and photochemistry. It is hard to decide whether it was science inspiring engineering or engineering inspiring science. The implication to liposome-based drug delivery via external photochemical control is obvious. Speaking about controlled drug delivery, Chen studied the feasibility of using bacteriorhodopsin to mediate light-induced drug release from a polymeric device (36). He found that the amount of released protons was insufficient to cause a gel–sol transition of the polymer. Here, reverse engineering offers a hint for future improvements. In visual transduction, it takes three consecutive biochemical reactions and a host of molecular mechanism to generate the well-known 100-fold amplification of energy (37). The lesson on bird's wings offers essentially a similar suggestion. Bacteriorhodopsin vesicles need not supply all of the needed protons. Rather, the surface potential associated with bacteriorhodopsin activation (38) can be used to trigger the release of a depot of protons, much like the action of light-controlled nanovalve of Koçer et al., which can trigger the release of drug encapsulated in liposomes.

Segregation of scientists into camps was probably fueled by human instinct of territorial claims and reinforced by human penchant for ideology. However, such social grouping sometimes fostered dogmatism, which could be deterrent to creativity. If we exercise the practice of reverse-engineering Nature at this point, we are likely to find that Nature actually adhered to no ideology. Take phosphorylation of proteins as an example, Nature recruited whatever mechanisms it could put its hands on, and enlisted a number of different molecular mechanisms in combination to activate a protein; the mechanisms include electrostatic interactions, steric hindrance, and allosteric effect (39). Nature was extremely explorative, and certainly recognized no disciplinary boundaries. Perhaps it was no coincidence that the earlier models of human creativity followed the analogy of evolution closely (40).

In this context, we extend the notion of reverse-engineering biology to address the question of how to make a sensor exhibit the kind of intelligence unique to higher animals. Artificial intelligence essentially stems from an attempt to imitate the intelligence exhibited by human beings. The enigma of human creativity has defied repeated attempts by psychologists, psychiatrists, and cognitive scientists over the past century. A century's worth of research has generated a wealth of detailed information, from the behavioral level all the way down to the molecular level. Yet, it is now commonly known that detailed local information does not always lead to the elucidation and understanding of the whole in a system as complex as the human brain. Here, artificial intelligence research could actually inspire our understanding of human creativity. It was reverse engineering in the opposite direction. For example, it is well known in artificial intelligence and in operations research that trial and error is a bad approach to problems with modest-to-high complexity due to combinatorial explosion. In contrast, the opposite approach of highly directed and focused attack on a difficult problem often lead to a dead end or cul de sac. The optimal approach is a compromise known as heuristic searching in AI or operations research jargon.

In our initial attempt to reverse engineering Nature's intelligence, a survey of biological information processing revealed that heuristic searching can be found in various hierarchies, from the submolecular level of protein folding to the systems level of conscious decision-making (41). The agent technology in problem-solving computer programs also demonstrates that giving the program a limited freedom yields handsome dividends (42). It is again the manifestation of heuristic searching that the computer programmer provides a general strategy but does not micromanage the detailed task. The general strategy prevents the computational resource from spreading out too thinly, whereas the limited freedom empowered to the software agent allows it to explore a wider range of options than the programmer can conceive ahead of time.

Here, we suggest that intelligence of the human brain lies in its ability to perform analog-pattern recognition and holistic judgment in addition to what a digital computer can do best—rule-based, step-by-step decision making (43). The capability of analog-pattern recognition is what intuition is all about. As compared to digital-pattern recognition, analog-pattern recognition is more error-prone. It seems that the ability to make small mistakes is the source of creativity and intelligence. Perhaps the educational community also has something to learn from the software agent technology. Even now the educational community still succumbs to the spell of behaviorist psychology. Students are trained to learn by boring repetitions only to be reinforced by means of rewards and punishments, much like training a rat to run a maze. The value of ruled-based reasoning is grossly exaggerated and students' innate intuition suffers accordingly. As a result, the public at large tends to get the short end of technological advances. Information explosion led to compartmentalization of disciplines and fragmentation of knowledge. The outcome restricts the range of human exploration. The development of smart sensors also demands a continuing supply of smart users. Otherwise, the enslavement by computers envisioned in science fictions may eventually turn out to be a nightmare coming true. As a saving grace, smart-sensor technology is promoting cross-fertilization of ideas across disciplines, one of the most effective ways to escape from the straightjacket of specialization.

From the point of view of smart-sensor designs, biomimetic science, which relies upon imitating Nature in an intelligent way, constitutes a kind of heuristic searching in engineering. Neither one abandons reverse engineering all together nor does one adhere strictly to Nature's implementation. From a big picture of science and technology policies, applied research, if coupled with sound basic research, constitutes heuristic searching at a higher level of R&D hierarchies.

23.4 Concluding Remarks

Biosensors provide an unlimited opportunity for developing smart, sophisticated technologies that enable the continual monitoring of individual health and overall living spaces. There is an increasing need in our dynamic society for inexpensive point-of-care diagnostics, portable field microanalyzers, and highly sensitive substance-recognition systems for a wide variety of security applications. In general, these biosensors must be compact devices that provide real-time, onsite detection and exhibit smart functions that eliminate the need for sample collection, preparation, and laboratory analysis. Many of the new products that incorporate biosensors must be small and robust, should operate quickly, and have low power requirements to be able to deliver the analytical information in a simple, inexpensive manner. The development of futuristic biosensors and sensory systems requires careful attention to both the recognition layer and physical transducer, and the coupling of these recognition transduction elements through the control of surface chemistry. Recent advances in understanding biological materials, the development of new manufacturing processes, and the exploitation of information or computing technology as described in this volume will all work toward achieving this goal.

Finally, it is also important to understand the constraints and limitations of the design approach taken. As scientists and engineers, we should always question our motivations and ask ourselves why we choose a particular direction of study to solve a problem. Mimicking biological sensory systems may lead to innovative solutions but may also be a pathway riddled by pitfalls.

References

1. Patolsky, F., Zheng, G., Hayden, O., Lakamamyali, M., Zhuang, X., and Lieber, C.M. (2004). Electrical detection of single viruses. *PNAS* 101:14017–14022.
2. Zheng, G., Patolsky, F., Cui, Y., Wang, W.U., and Lieber, C.M. (2005). Multiplexed electrical detection of cancer markers with nanowire sensor arrays. *Nature Biotech.* 23(10):1294–1301.
3. Turner, A.P.F. (2000). Biosensors—Sense and Sensitivity. *Science* 290(5495):1315–1317.
4. Newman, J.D., Tigwell, L.T., Turner, A.P.F., and Warner, P.J. (2004). *Biosensors—A Clearer View.* Cranfield University, http://www.cranfield.ac.uk/ibst/commercial/biosensor_report2004.htm
5. Newman, J.D., and Turner, A.P.F. (2005). Home blood glucose biosensors: a commercial perspective. *Biosensors and Bioelectronics*, 20:2435–2453.
6. Bilitewski, U. and Turner, A.P.F. (2000). *Biosensors for Environmental Monitoring.* Harwood Academic Publishers, pp. 1–409.
7. Tothill, I and Turner, A.P.F (2003). Biosensors. In: *Encyclopaedia of Food Sciences and Nutrition* (2nd Ed.), Benjamin Caballero B., Trugo L., and Finglas, P. (Eds.), Academic Press, pp. 489–499.
8. Turner, A.P.F. and Piletsky, S. (2005). Biosensors and Biomimetic Sensors for the Detection of Drugs, Toxins and Biological Agents. In: *Defense against Bioterror* (NATO STS series), Morrison, D., Milanovich, F., Ivnitski, D., and Austin, T.R. (Eds.), Springer, The Netherlands, pp. 261–272. ISBN-10: 1-4020-33850.
9. Turner, A.P.F and Magan, N. (2004). Electronic noses and disease diagnostics. *Nature Microbiol. Rev.* 2:161–166.
10. Piletsky, S. and Turner, A.P.F. (2006). *Molecular Imprinting, Landes Bioscience,* Georgetown, TX, USA. ISBN: 1 58706 2194.
11. Chianella, I., Lotierzo M, Piletsky, S.A., Tothill, I., Chen, B., Karim, K., and Turner, A.P.F. (2002). Rational design of the polymer specific for microcystin-LR using a computational approach. *Anal. Chem.* 74:1288–1293.
12. Bavastrello, V., Stura E., Carrara, S., Erokhin, V., and Nicolini, C. (2004). Poly(2,5-dimethylaniline-MWNTs nanocomposite: a new material for conductometric acid vapors sensor. *Sens. Actuat., B* 98:247–253.
13. Valentini, L., Bavastrello, V., Stura, E., Armentano, I., Nicolini, C., and Kenny, J.M. (2004). Sensors for inorganic vapor detection based on carbon nanotubes and poly(o-anisidine) nanocomposite material. *Chem. Phys. Lett.* 383:617–622.
14. Maccioni, E., Radicchi, G., Erokhin, V., Paddeu, S., Facci, P., and Nicolini, C. (1996). Bacteriorhodopsin thin film as a sensitive layer for an anaesthetic sensor. *Thin Solid Films* 284–285:898–900.

15. Nicolini, C., Erokhin, V., Paddeu, S., Paternolli, C., and Ram, M.K. (1999). Toward bacteriorhodopsin based photocells. *Biosens. Bioelectron.* 14:427–433.
16. Adami, M., Sartore, M., and Nicolini, C. (2005). A Neural Network based potentiometric stripping analyzer for multianalyte screening. *Biosensors and Bioelectronics.* submitted.
17. Paternolli, C., Antonimi, M., Ghisellini, P., and Nicolini, C. (2004) Recombinant cytochrome P450 immobilization for biosensor applications. *Langmuir* 20:11706–11712.
18. Ram, M.K., Bertoncello, P., Ding, H., Paddeu, S., and Nicolini, C. (2001). Cholesterol biosensors prepared by layer by layer technique. *Biosens. Bioelectron.* 16:849–856.
19. Shumantseva, V. De Luca, G., Bulko, T., Carrara, S., Nicolini, C., Usano,. S.A., and Archakov, A. (2004). Cholesterol amperometric biosensor based on cytochrome P450scc. *Biosens. Bioelectron.* 19:971–976.
20. Shumyantseva, V.V., Carrara, S., Bavastrello, V., Riley, J.D., Bulko, T.V., Skryabin, K.G., Archakov, A.I., and Nicolini, C. (2005). Direct electron transfer between cytochrome P450scc and gold nanoparticles on screen-printed rhodium-graphite electrodes. *Biosens. Bioelectron.* 21:217–222.
21. Carter, F.L. (ed.) (1982). *Molecular Electronic Devices*. New York and Basel: Marcel Dekker.
22. Aviram, A. (ed.) (1992). *Molecular Electronics—Science and Technology*. New York: American Institute of Physics.
23. Sienicki, K. (ed.) (1993). *Molecular Electronics and Molecular Electronic Devices*. Boca Raton, FL, Ann Arbor, MI, London, Tokyo: CRC Press.
24. Birge, R.R. (ed.) (1994). *Molecular and Biomolecular Electronics* (Advances in Chemistry Series No. 240). Washington, DC: American Chemical Society.
25. Barsanti, L., Evangelista, V., Gualtieri, P., Passarelli, V., and Vestri, S. (eds.) (2003). *Molecular Electronics: Bio-sensors and Bio-computers* (NATO Science Series II, Vol. 96). Dordrecht, Boston and London: Kluwer Academic Publishers.
26. Conrad, M. (1990). Molecular computing, In: Yovits, M. C. (ed.) *Advances in Computers*. Vol. 31. Boston, San Diego, New York, London, Sydney, Tokyo and Toronto: Academic Press, pp. 235–324.
27. Rambidi, N.G. (1997). Biomolecular computer: roots and promises. *Biosystems* 44:1–15.
28. Rambidi N. (2002). Roots and promises of chemical-based computing. *Biosystems* 64:169–178.
29. Zauner, K.P. (2005). Molecular Information Technology. *Crit. Rev. Solid State Mat. Sci.* 30:33–69.
30. Keyes, R.W. (1975). Physical limits in digital electronics. *Proc. IEEE* 63:740–767.
31. Chiabrera, A., Di Zitti, E., Costa, F., and Bisio, G.M. (1989). Physical limits of integration and information processing in molecular systems. *J. Phys. D: Appl. Phys.* 22:1571–1579.
32. Marr, D. (1982). *VISION, A Computational Investigation into the Human Representation and Processing of Visual Information*. New York: Freeman.
33. Grant, R.G. (2002). *Flight: 100 Years of Aviation*. London: DK Publishing, pp. 8–19.
34. Hong, F.T. (1994). Molecular electronics: science and technology for the future. *IEEE Eng. Med. Biol. Magaz.* 13(1):25–32.
35. Koçer, A., Walko, M., Meijberg, W., and Feringa, B.L. (2005). A light-actuated nanovalve derived from a channel protein. *Science* 309:755–758.
36. Chen, P.-C. (2004). Polymeric Drug Delivery Systems: Lidocaine Microspheres for Prolonged and Localized in vivo Anesthetic Effects and Light-Induced Drug Release from Polymeric Device Mediated by Bacteriorhodopsin. Ph.D. dissertation, University of Michigan, Ann Arbor, MI, pp. 61–90.
37. Stryer, L. (1986). Cyclic GMP cascade of vision. *Annu. Rev. Neurosci.* 9:87–119.
38. Hong, F.T. (2004). Molecular electronics switches in photobiology. In: Horspool, W., and Lenci, F., (Eds.) CRC Handbook of Organic Photochemistry and Photobiology. Boca Raton, FL: CRC Press, pp. 134–1 to 134–26.
39. Johnson, L.N., and Barford, D. (1993). The effects of phosphorylation on the structure and function of proteins. *Annu. Rev. Biophys. Biomol. Struct.* 22:199–232.
40. Simonton, D.K. (2004). *Creativity in Science: Chance, Logic, Genius, and Zeitgeist.* Cambridge: Cambridge University Press.
41. Hong, F.T. (2005). A multi-disciplinary survey of biocomputing: 1. molecular and cellular levels. In: Bajic, V.B., and Tan, T.W., (Eds.) *Information Processing and Living Systems.* (Series on Advances in Bioinformatics and Computational Biology, Vol. 2). London: Imperial College Press, pp. 1–137.

42. Jennings, N.R., and Wooldridge, M.J. (eds.) (1998). *Agent Technology: Foundations, Applications and Markets.* Berlin, Heidelberg and New York: Springer-Verlag.
43. Hong, F.T. (2003). The enigma of creative problem solving: a biocomputing perspective. In: Barsanti, L., Evangelista, V., Gualtieri, P., Passarelli, V., and Vestri, S., (Eds). *Molecular Electronics: Bio-sensors and Bio-computers.* (NATO Science Series II, Vol. 96). Dordrecht, Boston and London: Kluwer Academic Publishers, pp. 457–542.

Index

Q

R